MECHANICAL POWER TRANSMISSION COMPONENTS

MECHANICAL ENGINEERING

A Series of Textbooks and Reference Books

Editor

L. L. Faulkner

*Columbus Division, Battelle Memorial Institute
and Department of Mechanical Engineering
The Ohio State University
Columbus, Ohio*

1. *Spring Designer's Handbook,* Harold Carlson
2. *Computer-Aided Graphics and Design,* Daniel L. Ryan
3. *Lubrication Fundamentals,* J. George Wills
4. *Solar Engineering for Domestic Buildings,* William A. Himmelman
5. *Applied Engineering Mechanics: Statics and Dynamics,* G. Boothroyd and C. Poli
6. *Centrifugal Pump Clinic,* Igor J. Karassik
7. *Computer-Aided Kinetics for Machine Design,* Daniel L. Ryan
8. *Plastics Products Design Handbook, Part A: Materials and Components; Part B: Processes and Design for Processes,* edited by Edward Miller
9. *Turbomachinery: Basic Theory and Applications,* Earl Logan, Jr.
10. *Vibrations of Shells and Plates,* Werner Soedel
11. *Flat and Corrugated Diaphragm Design Handbook,* Mario Di Giovanni
12. *Practical Stress Analysis in Engineering Design,* Alexander Blake
13. *An Introduction to the Design and Behavior of Bolted Joints,* John H. Bickford
14. *Optimal Engineering Design: Principles and Applications,* James N. Siddall
15. *Spring Manufacturing Handbook,* Harold Carlson
16. *Industrial Noise Control: Fundamentals and Applications,* edited by Lewis H. Bell
17. *Gears and Their Vibration: A Basic Approach to Understanding Gear Noise,* J. Derek Smith
18. *Chains for Power Transmission and Material Handling: Design and Applications Handbook,* American Chain Association
19. *Corrosion and Corrosion Protection Handbook,* edited by Philip A. Schweitzer
20. *Gear Drive Systems: Design and Application,* Peter Lynwander

Additional Volumes in Preparation

Handbook of Turbomachinery, edited by Earl Logan, Jr.

Mechanical Engineering Software

Spring Design with an IBM PC, Al Dietrich

Mechanical Design Failure Analysis: With Failure Analysis System Software for the IBM PC, David G. Ullman

MECHANICAL POWER TRANSMISSION COMPONENTS

EDITED BY
DAVID W. SOUTH
Consultant
Madison Heights, Michigan

JON R. MANCUSO
Kop-Flex Power Transmission Products
Baltimore, Maryland

Marcel Dekker, Inc.   New York • Basel • Hong Kong

Library of Congress Cataloging-in-Publication Data

Mechanical power transmission components / edited by David W. South.
  Jon R. Mancuso.
      p.   cm. — (Mechanical engineering ; 92)
      Includes bibliographical references and index.
      ISBN: 0-8247-9036-7 (alk. paper)
      1. Power transmission. I. South, David W. II. Mancuso, Jon R.,
      III. Series: Mechanical engineering (Marcel Dekker, Inc.)
  ; 92.
  TJ1045.M38 1994   94-25795
  621.8'5–dc20   CIP

The publisher offers discounts on this book when ordered in bulk quantities. For more information, write to Special Sales/Professional Marketing at the address below.

This book is printed on acid-free paper.

Marcel Dekker, Inc.
270 Madison Avenue, New York, New York 10016

Current printing (last digit):
10 9 8 7 6 5 4 3 2 1

PRINTED IN THE UNITED STATES OF AMERICA

To the memory of my father, William W. South, and to my mother, Mary N. South; to my son, David, my brother, Jim, and my sisters, Shirley and Judy.

DWS

To the memory of my daughter Lynn
JRM

PREFACE

Textbooks and handbooks that deal with mechanical engineering and machine design, elements, and components usually present the fundamentals of design in very broad terms. This book deals with one specific area of mechanical engineering and machine design, mechanical power transmission, focusing on the fundamental design of mechanical power transmission elements and mechanisms that help to transmit power from one piece of equipment to another. This book will supply students, designers, engineers, and also machinists, technicians, and shop foremen with a storehouse of information— information needed daily on mechanical power transmission and related components. In one place it gathers the knowledge needed to make basic design, selection, replacement, installation, maintenance, and applications decisions.

Chapter 1, "Shafting," deals with the most basic element in power transmission. This chapter addresses the design of shafting, and includes sections on both steady-state and dynamic calculations, design practices, shaft material properties, and shaft rigidity.

Chapter 2 deals with flexible couplings, which are used to connect rotating equipment, transmit torque, accommodate misalignment, and compensate for end movement. This chapter discusses the types of couplings, coupling characteristics, capacities, how to select couplings, and design equations and parameters. Balancing, installation, and maintenance procedures are also discussed.

Clutches and brakes connect and accelerate or decelerate mechanical components and maintain them at appropriate velocities. Chapter 3 discusses the types of clutches and brakes, including nonfriction and friction types. Selection, design equations, and parameters are covered. Installation and maintenance procedures are also presented.

In Chapter 4, "Plain Bearings," these bearings are put into three classes: Class I bearings are oil and/or grease lubricated from external sources; Class II bearings are self-oiling and/or filled; Class III bearings are self-lubricated. Data and procedures pertaining to the design of bearings of the journal and thrust types are given. Sleeve bearings are discussed, including materials, mounting, load-carrying capacity, selection, and so on.

Rolling (or antifriction) bearings consist of two rings with rolling elements running on tracks or raceways. Chapter 5 covers types of rolling element bearings and details of construction and design, including clearances, materials, and lubrication. Information on handling, storage, mounting, and dismounting is also included.

Chapter 6, "Seals," deals with static and dynamic seal types. Machines require seals to retain lubricants, liquids, solids, and gases. Sealing devices also prevent foreign particles from entering and damaging machinery components. This chapter covers materials used for seals, gaskets, packings, diaphragms, radial lip seals, and mechanical face seals. Additional types of seals, design, installation, and troubleshooting are also discussed.

A gear is a machine element, the purpose of which is to transmit motion and power from its shaft, through another gear to its shaft, by means of gear teeth. Chapter 7 covers the types of gears, materials used, gear nomenclature, and manufacturing of gears. Lubrication and failure modes of gears are also covered.

A chain is composed of a series of links pinned together. It is a form of flexible gear connecting two toothed sprocket wheels mounted on parallel shafts. Chapter 8, "Chain Drives," covers types of chains, sprockets, chain standards, materials, and heat treatments. Sprocket system analysis, including forces and impact loading, is discussed. Also covered in detail are selection, installation, maintenance, and lubrication of chain drives.

Belts are used to transmit power between shafts that are spaced widely apart with a wide selection of speed ratios. Chapter 9, "Belt Drives," presents different types of belts and sheaves. Discussed are belt standards, materials, belt tension, failure modes, belt dynamics, and belt vibration. Also discussed are types of mounting, design, and selection, use of idlers, installation, and maintenance.

The study of hydraulics deals with the uses and characteristics of liquids. Chapter 10 examines the use of fluids under pressure in the transmission of power or motion under precise control. This chapter covers several principles, hydraulic fluids, hydraulic plumbing, connections, oil reservoirs, filters and strainers, cylinders, and directional controls used in hydraulic systems.

Reference books of this magnitude and nature are usually the result of the combined efforts of many people. Therefore, we are very grateful to the six other contributors (and the companies that they represent), who were selected because of their expert knowledge and many years of experience in the various areas of mechanical engineering and technology. The unselfish contribution of these authors has made it possible to complete this large task of putting together a different type of reference book.

Marcel Dekker, Inc., gave us (and the contributors) a totally free hand in developing the outline, arranging the contents, choosing topics, selecting contributors, and finalizing the manuscript, for what seemed to us at times a never-ending project; the professional assistance throughout this project was extremely helpful and most needed. Special thanks to Graham Garratt, Executive Vice President and Publisher, Simon Yates and Ruth Dawe, sponsoring editors, and other patient and understanding associates of Marcel Dekker, Inc., who contributed in their areas of expertise to this book.

David W. South
Jon R. Mancuso

CONTENTS

CONTRIBUTORS

Richard H. Ewert* Consultant, Factory Cost Systems and Gearing, St. Paul, Minnesota

Arthur J. Gunst Engineering Manager (retired), Engineering Department, FAG Bearings Corporation, Stamford, Connecticut

Leslie A. Horve Vice President, Industrial and Aerospace, CR Industries, Elgin, Illinois

Jon R. Mancuso Engineering Manager, Kop-Flex Power Transmission Products, Baltimore, Maryland

James D. Shepherd Manager, Power Transmission Product Application, The Gates Rubber Company, Denver, Colorado

David W. South Consultant, Madison Heights, Michigan

*Past president, American Gear Manufacturers Association, Alexandria, Virginia; past president (retired), Sewall Gear Manufacturing Company, St. Paul, Minnesota

Richard C. St. John Chief Engineer, Base Line Engineering, Inc., Louisville, Ohio

Richard H. Weichsel Managing Director, AB Consultants International, Inc., Naples, Florida

1

SHAFTING

Jon R. Mancuso

Kop-Flex Power Transmission Products, Baltimore, Maryland

I. INTRODUCTION

The strength of a shaft is only one of the considerations for its design. It must also be rigid enough to minimize deflection and must have sufficient angular and torsional stiffness to avoid vibratory movement. How rigorously one analyzes a shaft depends on past experience and practices with similar shafting and how much is known about the loading. If the designer considers the major types of loads applied to the shaft and designs with a relatively large safety factor, his analysis does not need to be as rigorous as when weight must be optimized, as in a helicopter transmission drive; then one must do a more rigorous analysis. It is up to the designer to determine how rigorous and "how safe is safe." This chapter will discuss shaft design for ductile materials, including types of loads on a shaft, its strength, and types of strengths of materials used. Also discussed will be fatigue strength and factors that affect it, including surface finish and stress concentration.

Design equations and procedures will be given for

1. Torsional shear stress
2. Bending stress
3. Combined bending and torsional stress

Also included will be guidelines for the design of shaft rigidity, torsion, and bending.

1

There are several different types of shafting. Power transmission shafting is usually round, either solid or tubular. For some small equipment (U-joints on agriculture machines, auxiliary equipment in mills), the shafting may be square or rectangular. This type of shafting is used to facilitate connecting; a matting part with a square or rectangle bore is used. This type of shaft is also used where length compensation may be required. The analysis and equations presented are for round solid shafts. Section IV.B shows how to apply the strength equations to tubular shafts.

II. STEADY-STATE DESIGN CRITERIA

A. Torsional Shear Stress

The basic equation for torsional shear stress for torsional loading (torque) is given in Eq. (1). For a round solid shaft see Eq. (2).

$$\tau = \frac{Tc}{J} \tag{1}$$

$$\tau = 16\frac{T}{\pi d^3} \tag{2}$$

The Von Mises stress, which is based on distortion energy, is given by the yield strength divided by the safety factor [Eq. (3)]:

$$\sigma_{VM} = \sqrt{3}\tau - \frac{S_{yld}}{N} \tag{3}$$

If we substitute for the shear stress (Eq. (2)) and then solve for the shaft diameter, we get

$$d = \left[\frac{16\sqrt{3}NT}{\pi S_{yld}} \right]^{1/3} \tag{4}$$

B. Bending Stress

$$\sigma = \frac{Mc}{I} \tag{5}$$

$$\sigma_{VM} = \frac{S_{yld}}{N} \tag{6}$$

$$d = \left[\frac{32NM}{\pi S_{yld}} \right]^{1/3} \tag{7}$$

C. Combined Bending and Torsion

The Von Mises stress for combined bending and torsion is

$$\sigma_{VM} = (\sigma_b^2 + 3\tau^2)^{1/2} \tag{8}$$

If we substitute the torsional and bending stresses, Eq. (1) and Eq. (5), into the Von Mises equation and solve for diameter, we get

$$d = \left[\frac{32N}{\pi S_{yld}} \left[M^2 + \frac{3T^2}{4} \right] \right]^{1/3} \tag{9}$$

III. DYNAMIC DESIGN CRITERIA

The dynamic stresses are usually more important than the static stresses. Dynamic loading exists whenever the loads alternate or the shaft rotates through a steady-state load and produces an alternating load on the shaft. The fatigue strength is handled by using a Goodman diagram. The alternating stresses are then plotted on the x-axis, and the steady-state stresses are plotted on the y-axis. A line is then used to connect the endurance strength to the ultimate strength. Any design point below this line is acceptable. The Goodman criterion is expressed by Eq. (10).

$$\frac{\sigma_a}{S_{end}} + \frac{\sigma_{ss}}{S_{ult}} = \frac{1}{N} \tag{10}$$

A. Reversed Bending with Constant Torque

Many shafts are loaded with reversed bending and constant torque. This is one of the most common conditions that exist in shaft design. Equations (11), (12), or (13) are all used to calculate the diameter of a shaft. Equations (11) and (12) are based on the distortion energy equation. Equation (11) uses the Goodman criteria, and Eq. (12) uses an elliptical strength line approach. Equation (13) uses maximum shear stress theory and the Soderberg criteria. All these approaches can be used in the sizing of shafting with reversed bending and with constant torque.

$$d = \left[\frac{32N}{\pi} \left[\frac{M}{S_{end}} + \frac{\sqrt{3}}{2} \frac{T}{S_{ult}} \right] \right]^{1/3} \tag{11}$$

$$d = \left\{ \frac{32N}{\pi} \left[\left(\frac{M}{S_{end}} \right)^2 + \frac{3}{4} \left(\frac{T}{S_{yld}} \right)^2 \right]^{1/2} \right\}^{1/3} \tag{12}$$

$$d = \left\{ \frac{32N}{\pi} \left[\left(\frac{M}{S_{end}} \right)^2 + \left(\frac{T}{S_{yld}} \right)^2 \right]^{1/2} \right\}^{1/3} \tag{13}$$

B.   Steady-State plus Alternating Bending and Torque

For steady-state plus alternating bending and torque, Eq. (14) based on distortion energy theory should be used.

$$
d = \left[\frac{32N}{\pi} \left\{ \left[\left(\frac{M_a}{S_{\text{end}}} \right)^2 + \frac{3}{4} \left(\frac{T_a}{S_{\text{end}}} \right)^2 \right]^{1/2} \right. \right.
$$
$$
\left. \left. + \left[\left(\frac{M_{\text{ss}}}{S_{\text{ult}}} \right)^2 + \frac{3}{4} \left(\frac{T_{\text{ss}}}{S_{\text{ult}}} \right)^2 \right]^{1/2} \right\} \right]^{1/3}
\tag{14}
$$

IV.   OTHER CRITERIA

A.   ASME Standard

The original ASME *Code of Design of Transmission Shafting* (ASA-B17C-1927) was withdrawn in 1954 because it was found to be very conservative in many cases and incomplete in others. The most recent standard, *Design of Transmission Shafting* [1] published by the American Society of Mechanical Engineers in 1985, uses Eq. (15):

$$
\left[\frac{\sigma_a}{S_{\text{end}}/N} \right]^2 + \left[\frac{\tau_{\text{ss}}}{S_{\text{sy}}/N} \right]^2 = 1
\tag{15}
$$

where S_{sy} is the shear yield

B.   Tubular Shafting

If the shaft is tubular, the diameter can be estimated by numerical methods by replacing d in Eq. (14) with

$$
d \left[1 - \left[\frac{d_i}{d} \right]^4 \right]^{1/3}
\tag{16}
$$

V.   MATERIAL PROPERTIES

A.   Typical Materials for Shafting

Shaft material is usually made from medium-carbon steel, such as AISI 1045, or heat-treated alloy steel, such as AISI 4140.

B.   Strength Limits

1.   Actual Limits

Actual values of material properties should be used whenever available. These values may be determined by actual physical property testing.

2. Estimated Steel Properties

The following can be used for rolled or forged (wrought) steel shafting. For through-hardened steel shafts, the ultimate tensile strength is based on the Brinell hardness of the shaft at mid-radius (for example, for a 12-in. diameter bar, mid-radius would be 3 in. from the surface of the bar or at a 6-in. diameter). For steel shafts that are surface hardened (nitriding, carburizing, etc.) the hardness of the core should be used (unless experience and/or testing indicate otherwise). The following are the approximations; see Table 1:

Ultimate tensile strength:

$$S_{ult} = 500\,HB \tag{17}$$

Tensile yield strength:

$$S_{yld} = 0.55S_{ult} \quad \text{where BHN} < 200 \tag{18}$$

$$S_{yld} = 0.70S_{ult} \quad \text{where BHN} > 200 \tag{19}$$

Fatigue strength:

$$S_e = 0.5S_{ult} \tag{20}$$

VI. FATIGUE STRENGTH

The fatigue strength of an actual part is influenced by several material conditions:

$$S_{end} = kS_e \tag{21}$$

Table 1 Approximate Strength for Materials Typically Used for Shafts

Material type	Brinell hardness (mid-radius)	Ultimate strength (psi)	Yield strength (psi)	Fatigue strength (psi)
1010–1020	100	50,000	27,500	25,000
1030–1040	140	70,000	38,500	35,000
1040–1050	170	85,000	46,750	42,500
4130–4140	255	127,500	89,250	63,750
4140–4150	270	135,000	94,500	67,500
4340	300	150,000	105,000	75,000

Ultimate strength = 500 × BHN; yield strength: <200 BHN = 0.55 × S_{ult}; >210 BHN = 0.70 × S_{ult}.
Source: Refs. 2 and 3.

where

S_{end} = a part's real endurance limit, psi

k = material/part modification factors

The fatigue strength modification factor k is the product of the product of several factors. Only two of these will be discussed in detail.

$$k = k_s \times k_f \times k_m \tag{22}$$

A. Surface Finish Factor (k_s)

The surface finish factor accounts for difference between a polished test specimen and the finish of actual shaft (see Fig. 1).

B. Stress Concentration Factor (k_f)

Actual applications show that shafts usually fail at a hole, keyway, shoulder, notch, or other discontinuity where stresses have been concentrated. The effect of this stress concentration on the fatigue strength of a shaft is called the fatigue stress concentration factor k_f:

$$k_f = \frac{1}{1 - q(k_t - 1)} \tag{23}$$

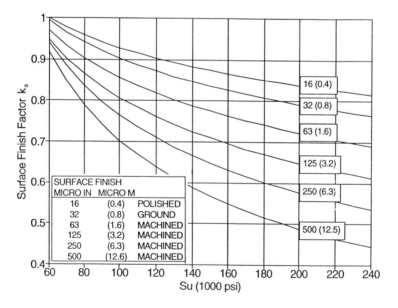

Figure 1 Surface finish factors. (Redrawn from Refs. 1 and 3.)

The notch sensitivity factor q accounts for the phenomenon that lower-strength steels are less sensitive to fatigue notches that are higher-strength steels. Figure 2 gives some values of q for through-hardened steels. If q is not known for a steel, a conservative approach is to assume $q = 1.0$. The following formula is used in Fig. 2.

$$q = \frac{1}{q + a/r} \qquad (24)$$

The following are approximations for a as a function of tensile strength:

S_{ult}, psi	a
50,000	0.015
100,000	0.007
150,000	0.0035
200,000	0.002

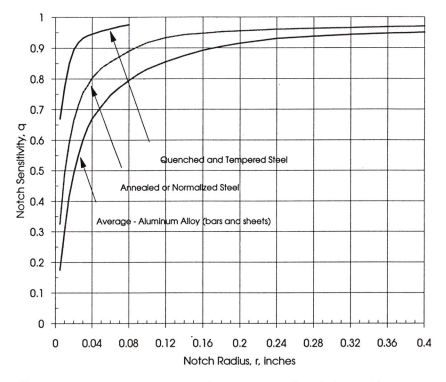

Figure 2 Average fatigue sensitivity factor. (Redrawn from Refs. 1 and 3.)

The theoretical stress concentration factors k_t for shafts with steps and grooves are shown (Figs. 3–6). Figures 7 and 8 show some typical stress concentration factors for keyways in solid shafts.

C. Miscellaneous Factors (k_m)

Several other factors affect shaft fatigue. If a shaft is designed rigorously because its weight needs to be optimized and it requires a low safety factor, the effect of the following factors on material fatigue properties may need to be considered:

1. Temperature
2. Residual stresses
3. Type of heat treatment
4. Corrosion and other environmental stresses

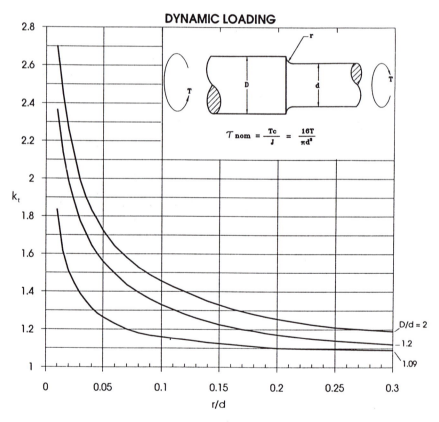

Figure 3 Shaft with fillet under torsional loading. (Redrawn from Refs. 1 and 3.)

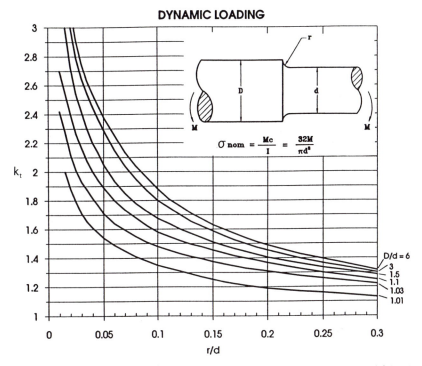

Figure 4 Shaft with fillet under bending load. (Redrawn from Refs. 1 and 3.)

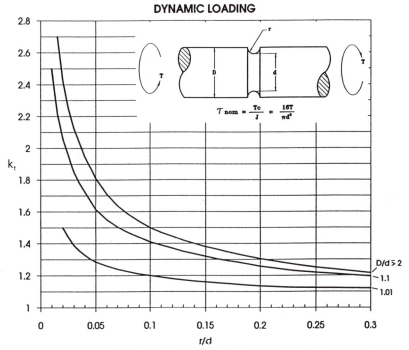

Figure 5 Grooved shaft under torsional loading. (Redrawn from Refs. 1 and 3.)

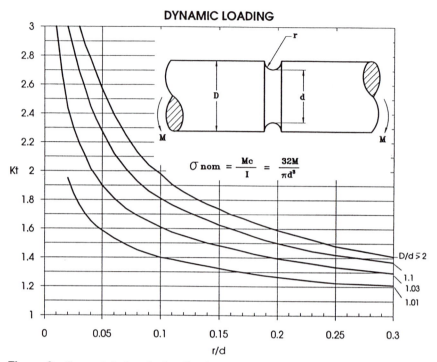

Figure 6 Grooved shaft under bending load. (Redrawn from Refs. 1 and 3.)

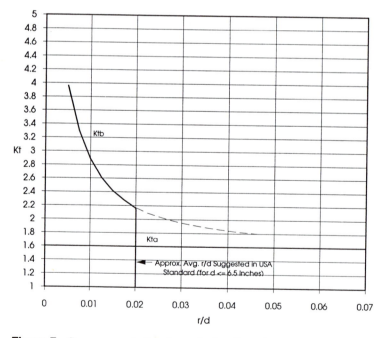

Figure 7 Stress concentration factor for bending of a shaft with a keyway. (Redrawn from Ref. 1.)

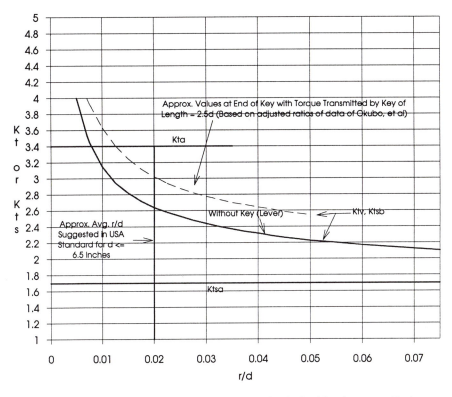

Figure 8 Stress concentration factor for torsion of a shaft with a keyway. (Redrawn from Ref. 1.)

5. Coating/plating on shaft
6. Shot peening

VII. SHAFT RIGIDITY

This section gives an overview of shaft rigidity. Deformation is the displacement of a shaft from its original shape. All shafts deform when they are loaded. The deformation can be small, large, restorable, or permanent. Shaft deformation can reduce the life of components mounted on the shaft (bearings, couplings, gears, and seals). Shafting needs to be designed rigidly enough to reduce these deflections. Also, a shaft's torsional rigidity can affect the system torsional loading characteristic and, if not properly designed, can cause components to resonate and thereby reduce their life.

The equations given show a simplified approach and may be used to determine the order of magnitude of the deformation of the shaft. There are more rigorous methods, such as finite element analysis (FEA), and in some cases testing may be required.

A. Torsional Rigidity (Windup)

Figure 9 shows a sketch of a shaft subjected to torsional loading (torque). The angle of twist for this shaft can be calculated by:

$$\phi_A = \frac{TL}{GJ} \tag{25}$$

where

ϕ_A = angle of twist, rad

T = torque, in.-lb.

L = length, in.

G = shear modulus (for steel $G = 11.5 \times 10^6$ lb/in.^{2})

J = polar moment in inertia, in.^{4}

For solid shafts,

$$J = \frac{\pi d_o^4}{32} \tag{26}$$

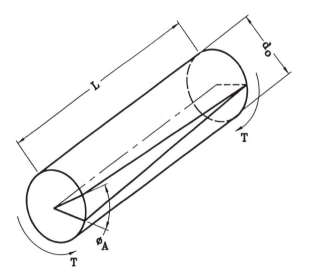

Figure 9 Torsional twist.

$$\phi_A = \frac{8.86 \times 10^{-7} TL}{d_o^4} \qquad (27)$$

The angle of twist between shaft sections can be calculated by adding the angle of twist of all the sections. For a shaft that consists of n different diameters and lengths we can calculate the angle of twist by

$$\phi_A = \frac{T}{G} \left[\frac{L_1}{J_1} + \frac{L_2}{J_2} + \cdots + \frac{L_n}{J_n} \right] \qquad (28)$$

where

L_n = length of nth section, in.

J_n = polar moment of nth section, in.^{4}

B. Bending Rigidity

Bending deflection is deformation perpendicular to the axis of the shaft. Slope is the rate of change of this deflection and is an important criteria. Two common simply supported cases are discussed: overhung load and intermediate concentrated load. These cases assume a uniform shaft section. Complex shafts can be evaluated by using a combination of these cases. Today complex shaft sections and loads can be analyzed by the use of a computer and finite element analysis.

1. Overhung Load

For a simply supported shaft loaded with a concentrated overhung load (Fig. 10), the equations for deflection and slope differ, depending upon the position of the section in relation to the supports.

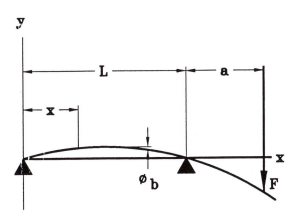

Figure 10 Overhung load.

If $x < L$,

$$y = \frac{Fax(L^2 - x^2)}{6EIL} \tag{29}$$

$$\phi_b = \frac{Fa(L^2 - 3x^2)}{6EIL} \tag{30}$$

If $x > L$

$$y = \frac{F(x - a)[(x - L)^2 - a(3x - L)]}{6EI} \tag{31}$$

$$\phi_b = \frac{F[3(x - L)^2 - 2a(3x - L)]}{6EI} \tag{32}$$

2. Intermediate Concentrated Load

For a simply supported shaft with a concentrated load applied between the supports (Fig. 11), the equations differ depending upon the position of the load in relation to the shaft.

For $x < a$,

$$y = \frac{Fx(L - a)(x^2 - 2aL + a^2)}{6EIL} \tag{33}$$

$$\phi_b = \frac{F(L - a)(3x^2 - 2aL + a^2)}{6EIL} \tag{34}$$

For $x > a$,

$$y = \frac{Fa(L - x)(x^2 + a^2 - 2Lx)}{6EIL} \tag{35}$$

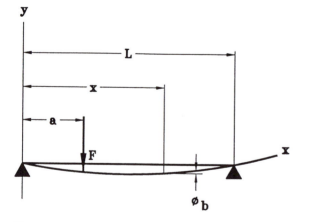

Figure 11   Intermediate concentrated load.

$$\phi_b = \frac{Fa(6Lx - 3x^2 - 2L^2 - a^2)}{6EIL} \tag{36}$$

where

y = deflection of shaft at n,

F = load, lb

L = length of shaft between supports, in.

a = distance from support to load, in.

x = distance from support to point of interest, in.

E = modulus of elasticity, lb/in.^{2}

I = area moment of interior, in.^{4}

ϕ_b = shaft slope at x, rad

SYMBOLS

c	Distance from the neutral axis, in.
d	Shaft diameter, in.
d_i	Shaft inner diameter, in.
E	Modulus of elasticity, psi
F_a	Alternating axial load, lb
F_s	Steady state axial load, lb
G	Shear modulus of elasticity, psi
HB	Brinell hardness
I	Moment of inertia, in.^{4}
J	Polar moment of inertia, in.^{4}
k	Fatigue strength modification factor
k_s	Surface finish fatigue modification factor
k_f	Fatigue modification factor for stress concentration
k_t	Theoretical stress concentration factor
L	Length, in.
M	Bending moment, in.-lb
M_{aa}	Reversing bending moment, in.-lb
M_{ss}	Steady state bending moment, in.-lb
N	Safety factor
n	Number of sections
q	Notch sensitivity factor
r	Radius of notch, in.
S_{end}	Actual fatigue limit, psi
S_e	Polished-specimen fatigue limit, psi
S_{sy}	Shear yield, psi
S_{yld}	Yield strength, psi
S_{ult}	Ultimate strength, psi
T	Torque, in.-lb
y	Deflection, in.

τ_{aa}	Alternating shear stress, psi
τ_{ss}	Steady state shear stress, psi
τ	Shear stress, psi
σ	Bending stress, psi
σ_a	Alternating stress, psi
σ_{ss}	Steady state stress, psi
σ_{VM}	Von Mises stress, psi
ϕ_A	Angle of twist, rad
ϕ_b	Shaft slope, rad

REFERENCES

1. American National Standard Institute/American Society of Mechanical Engineers, ANSI/ASME B106.1M-1985, *Design of Transmission Shafting.*
2. American Gear Manufacturers Association, *Design and Selection of Components for Enclosed Gear Drives,* AGMA 6001-C88.
3. Juvinall, Robert C., *Stress, Strain and Strength,* New York: McGraw-Hill, 1967.
4. Peterson, R.E., *Stress Concentration Factors,* New York: Wiley, 1974.
5. Shigley, J.E., and Mischke, C.R., *Power Transmission Elements,* New York: McGraw-Hill, 1986.

2

COUPLINGS

Jon R. Mancuso

Kop-Flex Power Transmission Products, Baltimore, Maryland

I. INTRODUCTION

Flexible coupling of connecting rotating equipment is a vital and necessary technique. Large shafts in loosely mounted bearings, bolted together by flanged rigid couplings, do not provide efficient and reliable mechanical power transmission. This is especially true in today's industrial environment, where equipment system designers are demanding higher speeds, higher torques, greater flexibility, additional misalignment, and lighter weights for flexible couplings. The need for flexible couplings is becoming more acute as is the need for technological improvements in them.

The basic function of a coupling is to transmit torque from the driver to the driven piece of rotating equipment. Flexible couplings expand the basic function by accommodating misalignment and end movement. During the initial assembly and installation of rotating equipment, precise alignment of the equipment shaft axes not only is difficult to achieve, but in most cases is economically prohibitive. In addition, misalignment during equipment operation is even more difficult to control. Misalignment in operation can be caused by flexure of structures under load, settling of foundations, thermal expansion of equipment and their supporting structures, piping expansion or contraction, and many other factors. A flexible coupling serves as a means to compensate for or minimize the effects of misalignment. Flexible couplings, however, have their own limitations. Therefore, calculations and predictions are required to know

17

what the maximum excursions can be. Only then can the correct coupling be selected.

A system designer or coupling user cannot just put any flexible coupling into a system and hope it will work. It is the responsibility of the designer or user to select a compatible coupling for the system. The designer must also be the coupling selector. Flexible coupling manufacturers are not system designers and should not be expected to assume the role of coupling selectors. Since they are not system experts, they can only take the information given them and size, design, and manufacture a coupling to fulfill the requirements specified. Some coupling manufacturers can offer some assistance, but they usually will not accept responsibility for a system's successful operation.

The purpose of this chapter is to aid the coupling selector (system designer or user) in understanding flexible couplings so that the best coupling can be selected for each system. To help coupling selectors do their jobs properly, the basics of couplings are included, which provide answers to often-asked questions, such as

Why use a flexible coupling?
What are the coupling's limitations and conditions?
How much unbalance will the coupling produce?
How can I compare couplings?
How do I install the coupling?
How is the coupling disassembled?
How is the coupling aligned?
How is the coupling lubricated?

There are two basic classes of couplings: rigid and flexible. Rigid couplings should only be used when the connecting structures and equipment are rigid enough so that very little misalignment can occur and the equipment is strong enough to accept the generated moments and forces.

There are hundreds of types of flexible couplings. The seven most common are

Gear couplings
Grid couplings
Chain couplings
Elastomeric shear couplings
Elastomeric compression couplings
Disk couplings
Diaphragm couplings

This chapter makes clear the importance of couplings in power transmission systems and, more specifically, provides the coupling selector with the basic tools required for the successful application of couplings to particular needs.

II. HISTORY OF COUPLINGS

A. Early History

The flexible coupling is an outgrowth of the wheel. In fact, without the wheel and its development there would have been no need for flexible couplings.

It has been reported that the first wheel was made by an unknown Sumerian more than 5000 years ago in the region of the Tigris and Euphrates rivers. The earliest record we have dates to 3500 B.C. History records that the first flexible coupling was the universal joint (see Fig. 1), used by the Greeks in or around 300 B.C. History also indicates that the Chinese were using this concept sometime around A.D. 25. The father of the flexible coupling was Jerome Cardan, who invented what was described as a simple device consisting of two yokes, a cross, and four bearings. This joint, the common ancestor of all flexible couplings, is still in use today and is continually being upgraded with the latest technology. Cardan did not design the Cardan shaft for rotating shaft applications, only as a suspension member. The Cardan joint is also known as the Hooke joint. Around 1650, Robert Hooke made the first application of this joint to a rotating shaft in a clock drive. Hooke write the equation for fluctuations in angular velocity caused by a single Cardan joint.

From 1700 to 1800, history records very little in the way of further developments in flexible couplings. The advent of the industrial revolution and, especially, the automobile precipitated the development of many flexible couplings.

In 1886, F. Roots theorized that if he thinned the flange section of a rigid coupling it would flex and prevent the equipment and shaft from failing. This is the forerunner of today's diaphragm coupling (see Fig. 2A). The Davis compression coupling (Fig. 2B) was developed to eliminate keys by compressing hubs onto the shaft. It was thought to be safer than other coupling devices since no protruding screws were required. What is believed to be the first chain coupling (Fig. 2C) was described in the May 1914 issue of *Scientific American.*

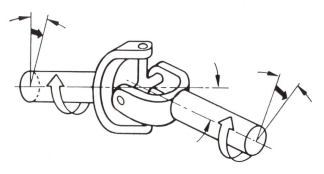

Figure 1 The ancestor of the flexible coupling—the universal joint.

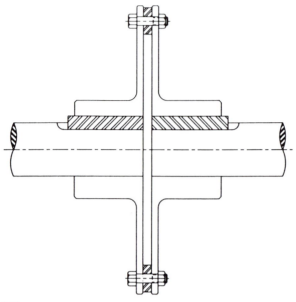

(A)

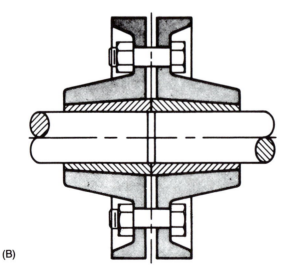

(B)

Figure 2 Early couplings: (A) diaphragm coupling patented in 1886; (B) Davis compression coupling; (C) early chain coupling.

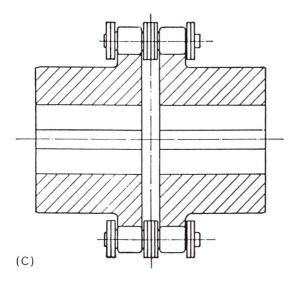

(C)

Figure 2 (Continued)

B. The Period 1900–1930

The coupling industry developed rapidly in the 1920s as a direct result of the invention of the automobile. Some of the coupling manufacturers established in this period are given in Table 1.

Table 1 Coupling Manufacturers

Original name	Date started	Today's name
Thomas Flexible Coupling Company	1916	Rexnord, Inc., Coupling Division
Fast Couplings, The Barlette Hayward Co.	1919	Kop-Flex, Inc., Power Transmission Products
Lovejoy Couplings	1927	Lovejoy, Inc.
Poole Foundry and Machine	1920	Deck Manufacturing
American Flexible Coupling	1928	Zurn Industries, Inc., Mechanical Drives Division
Ajax Flexible Coupling Company, Inc.	1920	Renold, Inc., Engineered Products
T. B. Wood's & Sons, Inc.	1920	T. B. Wood's & Sons Company
Bibby	1919	Bibby

C. The Period 1930–1945

From 1930 to 1945 many general-purpose flexible couplings were introduced into the industrial market. The most frequently used were chain, grid, jaw, gear, disk, and slider block couplings.

D. The Period 1945–1960

The late 1940s to the 1960s saw rapid technological advancement and use of rotating equipment. Larger and higher-horsepower equipment was used. This brought about the need for more "power dense" flexible coupling with greater misalignment.

Around this time the fully crowned gear spindle was developed and introduced to the steel industry. Also, the use of the gas turbine in industrial applications (generators, compressors) was becoming popular. With increased use of the gas turbine came the requirement for higher-speed couplings. Therefore, the gear and disk couplings were upgraded and improved to handle those higher-speed requirements.

One of the ever-increasing demands of rotating equipment is for higher and higher operating speeds. With increased speed of operation came system problems that required lighter-weight couplings, and the torsional characteristics of couplings became more important. This necessitated improvements in resilient couplings, as these couplings not only had to help tune a system but, in many cases, had to be able to absorb (dampen) anticipated peak loads caused by torsional excitations.

E. The Period 1960–1980

Higher horsepower and higher speed continue to be increasing requirements of rotating equipment. The 1960s saw the introduction of many new types of couplings. Many coupling manufacturers introduced a standard line of crowned tooth gear couplings. Today, the gear coupling is probably the most widely used coupling on the market. About six major manufacturers make a product line which is virtually interchangeable.

Grid and chain couplings are also very popular for general-purpose applications. The rubber tire coupling is widely used, with many companies offering one. More sophisticated resilient couplings were introduced in this period to help with the ever-increasing system problems, as were several flexible membrane couplings (disk or diaphragm). The application of nonlubricated couplings has grown rapidly in the past decade, and indications are that their use will continue to grow. The gear coupling has again been upgraded to meet the challenge of higher speeds.

The next sections discuss the types of couplings available today. Many of these flexible couplings will be around for years. As with the ancestor of flexible couplings, the universal joint, technological improvements in material, design, and manufacturing will help upgrade couplings so that they can handle the ever-increasing needs and demands of rotating equipment.

III. OVERVIEW OF COUPLINGS

A. Advantages of Using Flexible Couplings

Historically, rotating equipment was first connected by means of rigid flanges (Fig. 3). Experience indicated that this method did not accommodate the motions and excursions experienced by the equipment. As discussed in Section I, F. Roots was the first to thin these flanges and allow them to flex. Rigid couplings are used to connect equipment that experiences very small shaft excursions or with shafts made long and slender enough so that they can accept the forces and moments produced from the flexing flanges and shafts.

Basically a coupling

1. Transmits power (Fig. 4(A))
2. Accommodates misalignment (Fig. 4(B))
3. Compensates for end movement (Fig. 4(C))

1. Power Transmission

Flexible couplings must couple two pieces of rotating equipment: equipment with shafts, flanges, or both.

Flexible couplings must also transmit power efficiently. Usually, the power lost by a flexible coupling is small, although some couplings are more efficient than others. Power is lost in friction heat from the sliding and rolling of flexing

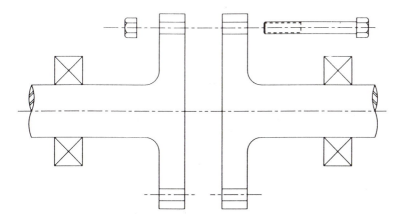

Figure 3   Rigid flanged connection.

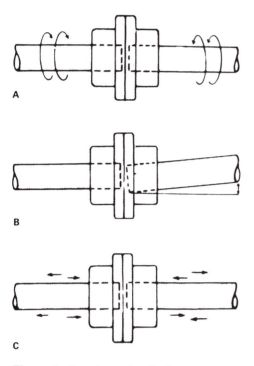

A

B

C

Figure 4 Functions of a flexible coupling: (A) transmit torque; (B) accommodate misalignment; (C) compensate for end movement.

parts and at high speed, windage and frictional losses indirectly cause lost efficiency. Most flexible couplings are better than 99% efficient.

2. Accommodating Misalignment and End Movement

Flexible couplings must accommodate three types of misalignment:

1. *Parallel offset*: Axes of connected shafts are parallel but not in the same straight line (see Fig. 5A).
2. *Angular*: Axes of shafts intersect at the center point of the coupling but not in the same straight line (see Fig. 5B).
3. *Combined angular and offset*: Axes of shafts do not intersect at the center point of the coupling and are not parallel (see Fig. 3C).

Most flexible couplings are designed to accommodate axial movement of equipment or shaft ends. In some cases (e.g., motors) couplings are required to limit axial float of the equipment shaft to prevent internal rubbing of a rotating part within its case.

Accommodation of misalignment and end movement must be done without inducing abnormal loads in the connecting equipment. Generally, machines are

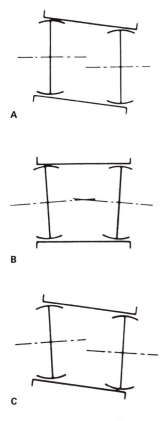

A

B

C

Figure 5 Types of misalignment: (A) parallel offset; (B) angular; (C) combined angular and offset.

set up at installation quite accurately. Many things force equipment to run out of alignment. The thermal effects of handling hot and cold fluids cause some movement in the vertical and axial directions, together with differentials of temperature in driver media such as gas and steam. The vertical motions could be a result of support structure expansions due to temperature differences, distortion due to solar heating, axial growth, or a combination of these. Horizontal motions are usually caused by piping forces or other structural movements, temperature differentials caused by poor installation practices, and expansions or contractions caused by changes in temperature or pressure differentials of the media in the system.

It is a fact of life that machinery appears to live and breathe, move, grow, and change form and position; this is the reason for using flexible couplings. A flexible coupling is not the solution to all movement problems that can or could exist in a sloppy system. Using a flexible coupling in the hope that it will compensate for any and all motion is naive. Flexible couplings have their limitations. The equipment or system designer must make calculations that will give a reasonable estimate of the outer boundaries of the anticipated gyrations. Unless those boundaries are defined, the equipment and system designer may just be transferring equipment failure into coupling failure (see Fig. 6).

One thing to remember is that when subjected to misalignment and torque, *all* couplings react on connected equipment components. Some produce greater reactionary forces than others and if overlooked, can cause shaft failures, bearing failures, and other failures of equipment components (see Fig. 7). Rigid couplings produce the greatest reactions. Mechanical element couplings such as gear, chain, and grid couplings produce high to moderate moments and forces on equipment that are a function of torque and misalignment. Elastomeric couplings produce moderate to low moments and forces that are slightly dependent on torque. Metallic element couplings produce relatively low moments and forces which are relatively independent of torque. The most commonly used flexible couplings today are those that produce the greatest flexibility (misalign-

Figure 6 Coupling failure.

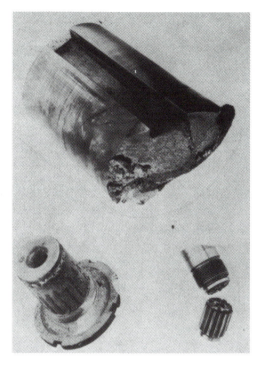

Figure 7 Equipment failure.

ment and axial capacity) while producing the lowest external loads on equipment.

B. Types of Couplings

There are many types of couplings. They can virtually all be put into two classes, two disciplines, and four categories (Fig. 8). The two classes of couplings are

1. Rigid (Fig. 9)
2. Flexible

The two disciplines for the application of flexible couplings are

1. Miniature, which covers couplings used for office machines, servomechanisms, instrumentation, light machinery, and so on (see Fig. 10).
2. Industrial, which covers couplings used in the steel industry, the petrochemical industry, utilities, off-road vehicles, heavy machinery, and so on.

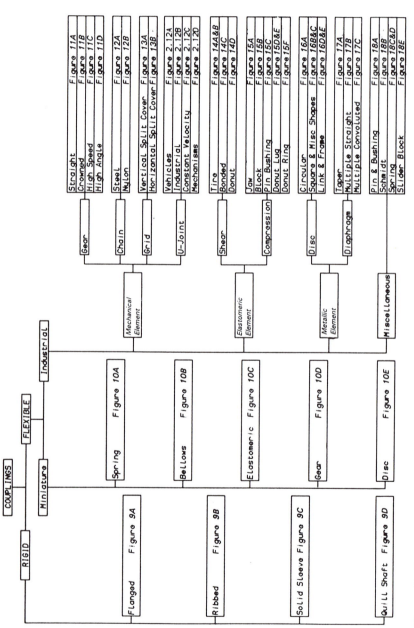

Figure 8 Types of couplings.

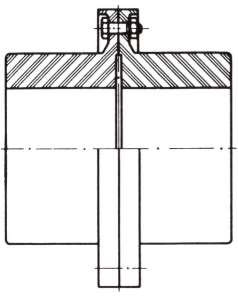

(A)

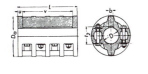

(B)

Figure 9 Types of rigid couplings: (A) rigid flanged coupling; (B) rigid ribbed coupling; (C) rigid sleeve coupling (courtesy of SKF Steel, Coupling Division); (D) quill shaft coupling.

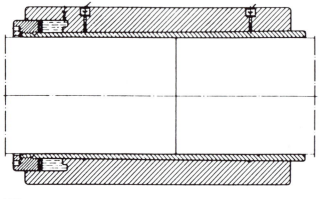

(C)

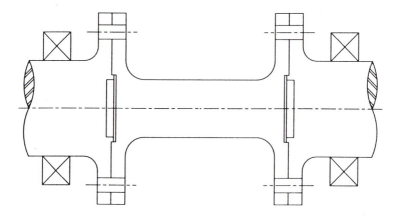

(D)

Figure 9   (Continued)

This chapter covers industrial couplings. The four categories of flexible industrial couplings are

1.   Mechanical element
2.   Elastomeric element
3.   Metallic element
4.   Miscellaneous

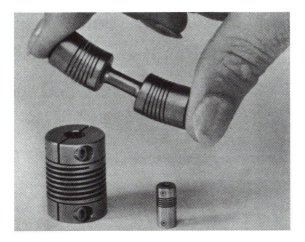

(A)

(B)

Figure 10   Types of miniature couplings: (A) miniature spring coupling (courtesy of Helical Products Company, Inc.); (B) miniature bellows coupling (courtesy of Metal Bellows Corporation); (C) miniature elastomeric coupling (courtesy of Acushnet Company); (D) miniature gear coupling (courtesy of Guardian Industries, Inc.); (E) miniature disk (courtesy of Coupling Division of Rexnord, Thomas Couplings).

(C)

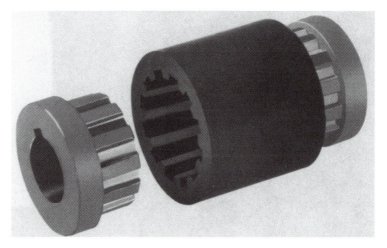

(D)

Figure 10   (Continued)

The general operating principles of the four basic categories of industrial couplings are as follows:

1. *Mechanical element couplings*: In general, these couplings obtain their flexibility from loose-fitting parts and/or rolling or sliding of mating parts (see Figs. 11 to 13). Therefore, they usually require lubrication

(E)

Figure 10 (Continued)

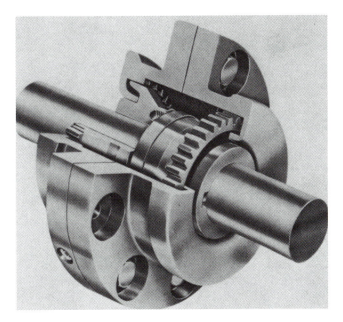

(A)

Figure 11 Types of gear couplings: (A) straight tooth gear coupling (courtesy of Kop-Flex, Inc.); (B) crowned tooth gear coupling (courtesy of Zurn Industries, Inc., Mechanical Drives Division); (C) high-speed gear coupling (courtesy of Kop-Flex, Inc.); (D) high-angle spindle gear coupling (courtesy of Zurn Industries, Inc., Mechanical Drives Division).

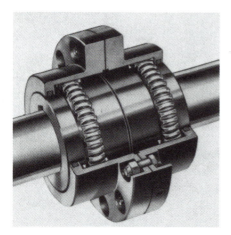

(B)

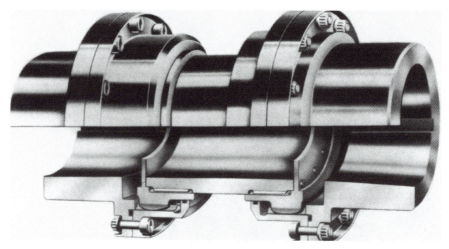

(C)

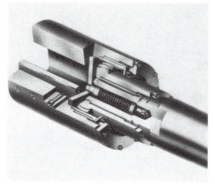

(D)

Figure 11   (Continued)

(A)

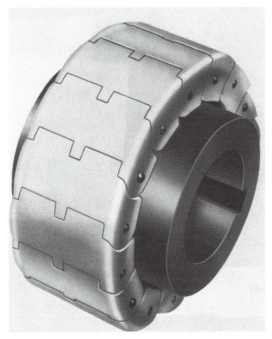

(B)

Figure 12 Types of chain couplings: (A) chain coupling (courtesy of Dodge Division of Reliance Electric); (b) nylon chain coupling (courtesy of Morse Industrial Corporation).

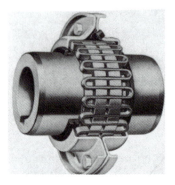

(A)

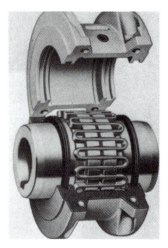

(B)

Figure 13 Types of grid couplings: (A) vertical split-cover grid coupling; (B) horizontal split-cover grid coupling. (Courtesy of Falk Corporation.)

unless one moving part is made of a material that supplies its own lubrication needs (e.g., a nylon gear coupling). Also included in this category are couplings that uses a combination of loose-fitting parts and/or rolling or sliding, with some flexure of material.

2. *Elastomeric element couplings*: In general, these couplings obtain their flexibility from stretching or compressing a resilient material (rubber, plastic, etc.) (see Figs. 14 and 15). Some sliding or rolling may take place, but it is usually minimal.

(A)

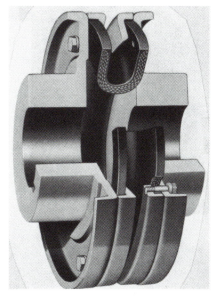

(B)

Figure 14 Types of elastomeric shear couplings: (A) rubber tire coupling (courtesy of Dayco Corporation); (B) rubber tire coupling (courtesy of Falk Corporation); (C) bonded shear coupling (courtesy of Lord Corporation; (D) donut elastomeric shear coupling (courtesy of T. B. Wood's Sons Company).

(C)

(D)

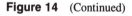

Figure 14   (Continued)

3. *Metallic element couplings*: In general, the flexibility of these couplings
    is obtained from the flexing of thin metallic disks or diaphragms (see
    Figs. 16 and 17).
4. *Miscellaneous couplings*: These couplings obtain their flexibility from a
    combination of the mechanisms described or through a unique mechan-
    ism (see Fig. 18).

(A)

(B)

Figure 15 Types of elastomeric compression couplings: (A) jaw elastomeric compression coupling (courtesy of Lovejoy, Inc.); (B) block elastomeric compression coupling (courtesy of Holset Engineering Co. Ltd.); (C) pin and bushing elastomeric compression coupling (courtesy of Morse Industrial Corporation); (D) donut lug elastomeric compression coupling (courtesy of Kop-Flex, Inc.); (E) donut lug elastomeric compression coupling (courtesy of Lovejoy, Inc.); (F) donut ring elastomeric compression coupling (courtesy of Dodge Division of Reliance Electric).

(C)

(D)

Figure 15   (Continued)

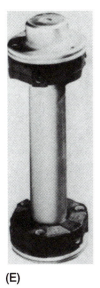

(E)

(E)

(F)

Figure 15   (Continued)

(A)

(B)

Figure 16 Types of disk couplings: (A) disk coupling (circular) (courtesy of Coupling Division of Rexnord, Thomas Couplings); (B) square disk coupling (courtesy of Formsprag Division of Dana Corporation); (C) hexagonal disk coupling (courtesy of Flexibox International, Inc., Metastream Couplings); (D) flexible frame coupling (courtesy of Kamatics Corporation, Kaflex Couplings); (E) link disk coupling (courtesy of TGW Thyssen Getriebe).

(C)

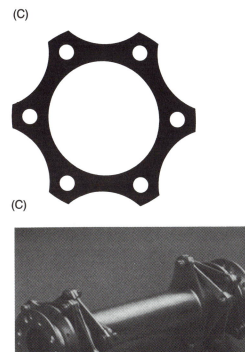

(C)

(D)

Figure 16 (Continued)

(E)

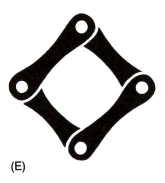

(E)

Figure 16 (Continued)

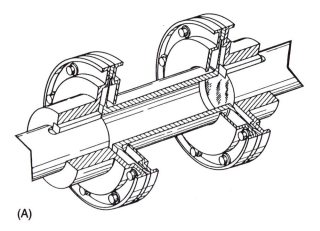

(A)

Figure 17 Types of diaphragm couplings: (A) tapered contoured diaphragm (courtesy of Lucas Aerospace Power Transmission Corporation); (B) multiple straight diaphragm coupling (courtesy of Flexibox International, Metastream Couplings); (C) multiple convoluted diaphragm coupling (courtesy of Zurn Industries, Inc., Mechanical Drives Division).

(B)

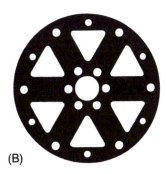

(B)

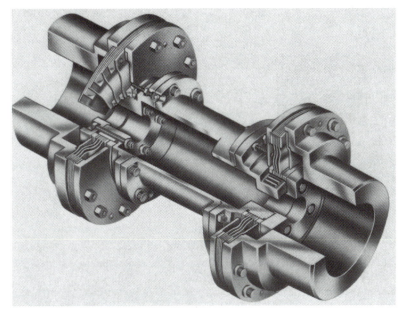

(C)

Figure 17 (Continued)

(A)

Figure 18 Miscellaneous types of couplings: (A) pin and bushing coupling (courtesy of Renolds, Inc., Engineered Products Division); (B) Schmidt coupling (courtesy of Schmidt Couplings, Inc.); (C) spring coupling (courtesy of Coupling Division of Rexnord, Thomas Couplings); (D) spring coupling (courtesy of Panamech Company); (E) slider block coupling (courtesy of Zurn Industries, Inc., Mechanical Drives Division).

(B)

(C)

Figure 18 (Continued)

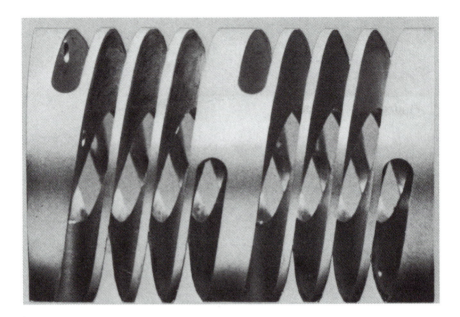

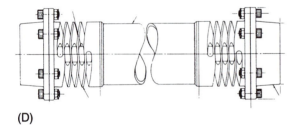

(D)

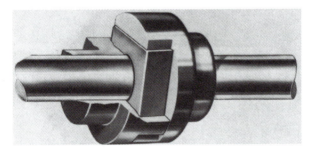

(E)

Figure 18   (Continued)

C. Coupling Functional Characteristics and Capacities

The *coupling selector* (equipment designer or system designer must decide what coupling is best for the system. The designer must review the possible candidates for a flexible coupling and make a selection. The person responsible for the selection of couplings should build a file of the most recent coupling catalogs. This file should be reviewed at regular intervals because designs, models, and materials are constantly being updated and improved.

Couplings are usually selected based on their characteristics and capacities. The two most important capabilities relate to torque and speed. Figure 19

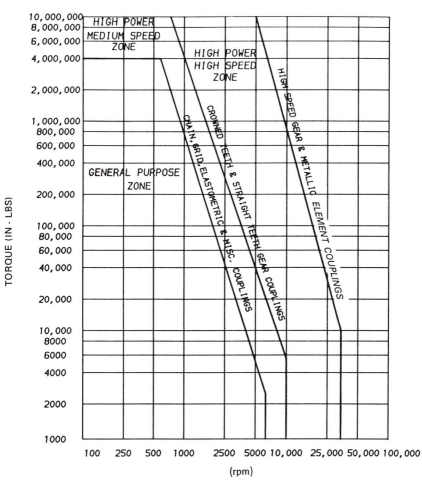

Figure 19 Torque versus speed (rpm) envelope of various types of couplings.

defines the torque and speed envelopes of the most common types of couplings. In addition, the functional capacities listed in Table 2 should be considered. Finally, the designer must take into account coupling characteristics (Table 3), which tell basically how the coupling will affect the system.

IV. SELECTION AND DESIGN

A. Selection Factors

Flexible couplings are a vital part of a mechanical power transmission system. Unfortunately, many system designers treat flexible couplings as if they were a piece of hardware.

The amount of time spent in selecting a coupling and determining how it interacts with a system should be a function not only of the cost of the equipment but also of how much downtime it will take to replace the coupling or repair a failure. In some cases the process may take only a little time and be based on past experience; however, on sophisticated systems this may take complex calculations, computer modeling, and possibly testing. A system designer or coupling user just cannot put any flexible coupling into a system with the hope that it will work. It is the system designer or the user's responsibility to select a coupling that will be compatible with the system. Flexible coupling manufacturers are not system designers; rather, they size, design, and manufacture couplings to fulfill the requirements supplied to them. Some coupling manufacturers can offer some assistance based on past exposure and experience. They are the experts in the design of flexible couplings, and they use their know-how in design, materials, and manufacturing to supply a standard coupling or a custom-built one if requirements so dictate. A flexible coupling is usually the least expensive major component of a rotating system. It usually costs less than 1%, and probably never exceeds 10%, of the total system cost.

Flexible couplings are called upon to accommodate the calculated loads and forces imposed on them by the equipment and their support structures. Some of these are motions due to thermal growth deflection of structures due to temperature, torque-load variations, and many other conditions. In operation, flexible couplings are sometimes called on to accommodate the unexpected. Sometimes these may be the forgotten conditions or the conditions unleashed when the coupling interacts with the system, and this may precipitate coupling or system problems. It is the coupling selector's responsibility to review the interaction of the coupling with that system. When the coupling's interaction with the system is forgotten in the selection process, coupling life could be shortened during operation or a costly failure of coupling or equipment may occur.

1. Selection Steps

There are usually four steps that a coupling selector should take to ensure proper selection of a flexible coupling for critical or vital equipment.

1. The coupling selector should review the initial requirements for a flexible coupling and select the type of coupling that best suits the system.
2. The coupling selector should supply the coupling manufacturer with all the pertinent information on the application so that the coupling may be properly sized, designed, and manufactured to fit the requirements.
3. The coupling selector should obtain the flexible coupling's characteristic information so that the interactions of the coupling with the system can be checked to ensure compatibility such that no determinal forces and moments are unleashed.
4. The coupling selector should review the interactions, and if the system conditions do change, the coupling manufacturer should be contacted so that a review of the new conditions and their effect on the coupling selected can be made. This process should continue until the system and the coupling are compatible.

The complexity and depth of the selection process for a flexible coupling depend on how critical and how costly downtime would be to the ultimate user. The coupling selector has to determine to what depth the analysis must go. As an example, a detailed, complex selection process would be abnormal for a 15-hp motor-pump application, but would not be for a multimillion dollar 20,000-hp motor, gear, and compressor train.

2. Types of Information Required

Coupling manufacturers will know only what the coupling selector tells them about an application. Missing information required to select and size couplings can lead to an improperly sized coupling and possible failure. At least three items are needed to size a coupling: horsepower, speed, and interface information. This is sometimes the extent of the information supplied or available. For coupling manufacturers to do their best job, the coupling selector should supply as much information as is felt to be important about the equipment and the system. See Table 4 for the types of information required. Tables 5 and 6 list specific types of information that should be supplied for the most common types of interface connections: the flange connection and the cylindrical bore.

3. Interaction of a Flexible Coupling with a System

A very important but simple fact often overlooked by the coupling selector is that couplings are connected to the system. Even if they are selected, sized, and designed properly, this does not ensure trouble-free operation. Flexible couplings generate their own forces and can also amplify system forces. This may

Table 2 Coupling Functional Capacities

Functional capacities	Mechanical element					
	Straight gear	Crowned gear	High-performance gear	Spindle	Chain	Grid
Max. continuous torque in.-lb. $\times 10^6$	40	44	13	25	1.5	4
Max. speed (rpm)	12,000	14,500	40,000	2,000	6,500	4,000
Max. bore (in.)	36	36	12	26	10	20
Angular misalignment (degrees)	1/2	1 1/2	1/4	7	2	1/3
Parallel offset (in./in.)	0.008	0.026	0.004	0.12	0.035	0.006
Axial travel (in.)	1/8-1[a] 12[c] ∞[c]	1/8-1[a] 12[c] ∞[c]	1/8-1[a]	1/2-3[a] 12[c] ∞[c]	±1/4	±3/16

Functional capacities	Elastomeric element		Metallic element			
	Shear	Compression	Laminated disk	Multiple diaphragm	Convolute diaphragm	Tapered diaphragm
Max. continuous torque in.-lb. × 10^6	0.5	15	4	4	6	6
Max. speed (rpm)	5,000	8,000	30,000	5,000	30,000	30,000
Max. bore (in.)	10	24	12	12	12	12
Angular misalignment (degrees)	3	3	1/4	1	1/2	1/2
Parallel offset (in./in.)	0.05	0.05	0.004	0.017	0.008	0.008
Axial travel (in.)	±5/16	±1/16	±3/8	±1/8	±1	±1

[a] Typical axial travels.
[b] Limit without very special designs.
[c] Limited by manufacturing today to 36 in.

Table 3 Coupling Characteristics

Coupling characteristics	Mechanical element						
	Straight gear	Crowned gear	High-performance gear	Spindle	Chain	Grid	
Lubrication	Yes	Yes	Yes	Yes	Yes	Yes	
Backlash	Med. high	Med.	Low	High	High	Med.	
Overhung moment	Med.	Med.	Low	High	Med.	Med.	
Unbalance	Med.	Med.	Low	High	Med. high	Med. high	
Bending moment	High	High	High	High	High	Med.	
Axial force	High	High	High	High	High	Med.	
Torsional stiffness	High	High	High	High	High	Med.	
Damping	Low	Low	Low	Low	Med.	Med. high	

Coupling characteristics	Elastomeric element		Metallic element			
	Shear	Compression	Laminated disk	Multiple diaphragm	Convolute diaphragm	Tapered diaphragm
Lubrication	No	No	No	No	No	No
Backlash	None	Low	None	None	None	None
Overhung moment	High	High	Low	Med.	Med. low	Med. low
Unbalance	Med. high	Med. high	Low med.	Low med.	Low	Low
Bending	Low	Med. low	Med. low	Med. low	Med. low	Med.
Axial force	Low	Med. low	Med. low	Med. low	Med. low	Med.
Torsional stiffness	Low	Med. low	High	High	High	High
Damping	High	Med. high	Low	Low	Low	Low

Table 4 Types of Information Required to Properly Select, Design, and Manufacture a Flexible Coupling

1. Horsepower
2. Operating speed
3. Interface connect information in Tables 4, 5, etc.
4. Torques
5. Angular misalignment
6. Offset misalignment
7. Axial travel
8. Ambient temperature
9. Potential excitation or critical frequencies
   a. Torsional
   b. Axial
   c. Laterial
10. Space limitations (drawing of system showing coupling envelope)
11. Limitation on coupling generate forces
    a. Axial
    b. Moments
    c. Unbalance
12. Any other unusual condition or requirements or coupling characteristics—weight, torsional stiffness, etc.

Note: Information supplied should include all operating or characteristic values of equipment for minimum, normal, steady state, momentary, maximum transient, and the frequency of their occurrence.

Table 5 Information Required for Cylindrical Bores

1. Size of bore, including tolerance or size of shaft and amount of clearance or interference required.
2. Lengths
3. Taper shafts
   a. Amount of taper
   b. Position and size of O-ring grooves if required.
   c. Size, type, and location of hydraulic fitting.
   d. Size and location of oil distribution grooves.
   e. Max. pressure available for mounting.
   f. Amount of hub draw-up required.
   g. Hub OD requirements.
   h. Torque capacity required (should also specify the coefficient of friction to be used).
4. Minimum strength of hub material or its hardness
5. If keyways in shift
   a. How many
   b. Size and tolerance
   c. Radius required in keyway (mininum and maximum).
   d. Location tolerance of keyway respective to bore and other keyways.

Table 6 Types of Interface Information Required for Bolted Joints

1. Diameter of bolt circle and true location
2. Number and size of bolt holes
3. Size, grade, and types of bolts required
4. Thickness of web and flange
5. Pilot dimensions
6. Others

change the system's original characteristics or operating conditions. The forces and moments generated by a coupling can produce loads on equipment that can change misalignment, decrease the life of a bearing, or unleash peak loads that can damage or cause failure of the coupling or connected equipment.

Listed are some of the coupling characteristics that may interact with the system. The coupling selector should obtain from the coupling manufacturer the values for these characteristics and analyze their affect on the system.

1. Torsional stiffness
2. Torsional damping
3. Amount of backlash
4. Weights
5. Coupling flywheel effect*
6. Center of gravity
7. Amount of unbalance
8. Axial force
9. Bending moment
10. Lateral stiffness
11. Coupling axial critical†
12. Coupling lateral critical†

4. Final Check

The coupling selector should use the coupling characteristics to analyze the system axially, laterally, thermally, and torsionally. Once the analysis has been completed, if the system's operating conditions change, the coupling selector should supply this information to the coupling manufacturer to ensure that the

*The flywheel effect of a coupling (WR^2) is the product of the coupling weight times the square of the radius of gyration. The radius of gyration is that radius at which the mass of the coupling can be considered to be concentrated.
†Coupling manufacturers will usually calculate coupling axial and lateral critical values with the assumption that the equipment is infinitely rigid. A coupling's axial and lateral critical values comprise a coupling's vibrational natural frequency in terms of rotational speed (rpm).

coupling selection has not been sacrificed and that a new coupling size or type is not required. This process of supplying information between the coupling manufacturer and the coupling selector should continue until the system and coupling are compatible. This process is the only way to ensure that a system will operate successfully.

B. Design Equations and Parameters

The intent of this section is to give the reader (coupling selector or user) some basic insights into coupling ratings and design. The equations and allowable values set forth should be used for comparison, not for design. They will help the reader compare "apples to apples"—specifically, in this case, couplings to couplings.

1. Coupling Ratings

Torque Ratings.   One of the most confusing subjects is that of coupling ratings. One reason for this is that *there is no common rating system.*

What Torque Ratings Mean.   Some coupling manufacturers rate couplings at limits based on the *yield strength* of the component's material and then require the application of *service factors.* Some couplings have ratings based on the endurance strength of their component parts and require the use of small or no service factors at all. The third basis for some coupling ratings is *life,* and this is the most confusing because it is based on experience and empirical data. We will not attempt to cover life ratings but leave their derivation and explanation to the specific coupling manufacturer.

Which Coupling Rating Is Correct.   All coupling ratings are correct—that is, as long as you use the recommended service factors and the selection procedure outlined by the coupling manufacturer.

How to Compare Couplings.   The following should be compared:

1.   The application factors related to the endurance strength (for cyclic-reversing loading) or yield strength (for steady-state loading) for normal operating torque and conditions
2.   The application factor related to yield strength for infrequent peak operating torques and conditions

Such a comparison will probably permit the coupling selector to select the safest coupling for the system. To complete the selection analysis, the coupling selector should ask the coupling manufacturer the following:

1.   What is the endurance torque (T_E) of the coupling?
2.   What is the weak link in endurance (examples, spacer, bolts, etc.)?
3.   What is the yield torque limit (T_Y) of the coupling?
4.   What is the weak link in yield (examples, gear teeth, bolts, etc.)?

Then a comparison of application factors should be made. Three application factors should be done.

1. $AF_E = T_E/T_C$, where T_C = maximum continuous operating torque.
2. $AF_Y = T_Y/T_C$.
3. $AF_L = T_Y/T_P$, where T_P = peak operating torque.

Speed Ratings. The true maximum speed limit at which a coupling can operate cannot be divorced from the connected equipment. This makes it very difficult to rate a coupling because the same coupling can be used on different types of equipment. Let us forget the system for a moment and consider only the coupling.

Maximum Speed Based on Centrifugal Stresses. The simplest method of establishing a coupling maximum speed rating is to base it on centrifugal stress (S_t):

$$S_t = \frac{\rho V^2}{g} \tag{1}$$

where
ρ = density of material (lb/in.^{3})
V = velocity (in./sec)
g = acceleration (386 in./sec)

$$V = \frac{D_o \times \pi \times \text{rpm}}{60} \tag{2}$$

$$\text{rpm} = \frac{375\sqrt{S_t/\rho}}{D_o} \tag{3}$$

where D_o is the outside diameter (in.) and

Material	ρ (lb/in.^{3})
Steel	0.283
Aluminum	0.100

$$S_t = \frac{S_{\text{yld}}}{\text{F.S.}} \tag{4}$$

where
F.S. = factor of safety (can be from 1 to 2, typically 1.5)
S_{yld} = yield strength of material (psi)

Maximum Speed Based on Lateral Critical Speed. For long couplings the maximum coupling speed is usually based on the lateral critical speed of the coupling. Couplings are part of a drivetrain system. If a system is very rigid and stiff, the maximum speeds might be greater than calculated, and if a system is soft and has a long shaft overhand, it may be much lower than the calculated value. Figure 20(A) is a schematic of a coupling as connected to a system. Figure 20(B) is a free-body diagram depicting the mass-spring system that models the coupling as connected to the system. The center section of a coupling can be represented as a simply supported beam.

$$N_c = 211.4\sqrt{\frac{1}{\Delta}} \tag{5}$$

$$\Delta = \frac{5WL^3}{384EI} \tag{6}$$

$$I = 0.049(D_o^4 - D_i^4) \tag{7}$$

$$K_c = \frac{W}{g}\left[\frac{\pi N_c}{30}\right]^2 \tag{8}$$

$$\frac{1}{K_e} = \frac{1}{2(K_c/2)} + \frac{1}{2K_L} \tag{9}$$

where

N_c = critical shaft whirl rotational frequency of the coupling shaft, cycles per minute (cpm)

Δ = shaft end deflection, in.

L = effective coupling length, in.

I = moment of inertia of the coupling spacer, in.[4]

W = center weight (supported) of the coupling, lb

$E = 30 \times 10^6$ for steel, psi

D_o = outside diameter of shaft, in.

D_i = inside diameter of shaft, in.

$g = 386$ in./sec^2

K_c = stiffness of coupling spacer, lb/in.

K_L = lateral stiffness of coupling, lb/in.

K_e = effective spring rate (lb/in.) for the system in Fig. 20(B) and calculated using Eq. (10)

These equations ignore the stiffness of the system which is usually how coupling manufacturers determine critical speed calculations. (Note: For some couplings (e.g., gear couplings), K_L can be ignored.) If we consider the stiffness of the connected shafts and bearings:

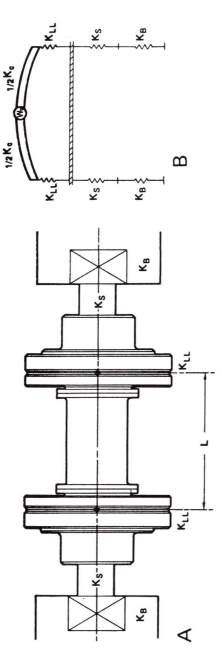

Figure 20 (A) Schematic of a coupling as connected to a system; (B) free body of a coupling as connected to a system.

$$\frac{1}{K_e} = \frac{1}{2(K_c/2)} + \frac{1}{2K_L} + \frac{1}{2K_s} + \frac{1}{2K_B} \tag{10}$$

where

K_s = equipment shaft stiffness, lb/in.

K_B = equipment bearing stiffness, lb.in.

Finally, the maximum speed based on the critical speed becomes

$$N_c = \frac{1}{2\pi\sqrt{K_e g/W}} \tag{11}$$

$$N_s = \frac{N_c}{F.S.} \tag{12}$$

F.S. = factor of safety (ranges from 1.5 to 2.0, usually 1.5)

Other Considerations. Many other considerations may affect a coupling's actual maximum speed capabilities, including

1. Heat generated from windage, flexing, and/or sliding and rolling of mating parts
2. Limitations on lubricants used (e.g., separation of grease)
3. Forces generated from unbalance
4. System characteristics

Service Factors. Service factors have evolved from experience based on past failures. That is, after a coupling failed, it was determined that by multiplying the normal operating torque by a factor and then sizing the coupling, the coupling would not fail. Since coupling manufacturers use different design criteria, many different service factor charts are in use for the same types of couplings. Therefore, it becomes important when sizing a coupling to follow and use the ratings and service factors recommended in each coupling manufacturer's catalog and not to intermix them with other manufacturers' procedures and factors. If a coupling manufacturer is told the load conditions—normal, peak, and their duration—a more detailed and probably more accurate sizing could be developed. Service factors become less significant when the load and duty cycle are known. It is only when a system is not analyzed in depth that service factors must be used. The more that is known about the operating conditions, the closer to unity the service factor can be.

Ratings in Summary. The important thing to remember is that the inverse of a service factor or safety factor is an ignorance factor. What this means is that when little is known about the operating spectrum, a large service factor should be applied. When the operating spectrum is known in detail, the service factor can be reduced. Similarly, if the properties of a coupling material are not exactly known or the method of calculating the stresses is not precise, a large safety factor should be used. However, when the material properties and the method of calculating the stresses are known precisely, a small safety factor can be used.

Sometimes the application of service and safety factors to applications where the operating spectrum, material properties, and stresses are known precisely can cause the use of unnecessary large and heavy couplings. Similarly, not knowing the operating spectrum, material properties, or stresses and not applying the appropriate service factor or safety factor can lead to undersized couplings. Ultimately, this can result in coupling failure or even equipment failure.

Sometimes service factors and safety factors are intermixed. The suggested service factors may have some consideration for the safety factor, and the safety factor of the coupling may have some service factor in it. Therefore when evaluating and comparing couplings it is best to look at what we will call an application factor. This shows the true margin of safety in a coupling.

2. Coupling Components Strength Limits

Allowable Limits.   Depending on type of loading, the basis for establishing allowable stress limits differ. Usually, loads can be classified into three groups:

1. Normal steady state
2. Cyclic reversing
3. Infrequent peak

Therefore, for comparison purposes we can conservatively establish three approaches to allowable stress limits.

Normal–Steady-State Loads:.

$$S_t = \frac{S_{\text{yld}}}{\text{F.S.}} \tag{13}$$

$$\tau = \frac{0.577 S_{\text{yld}}}{\text{F.S.}} \tag{14}$$

where
   S_t = allowable tensile stress, psi
   τ = allowable shear stress, psi
   S_{yld} = yield strength, psi
F.S. = safety factor

Safety factors are 1.25 to 1.75.

Cyclic Reversing, where loading produces primarily tensile or shear

$$S_{te} = \frac{S_{\text{end}}}{\text{F.S.} \times K} \tag{15}$$

$$\tau_e = \frac{0.577 S_{\text{end}}}{\text{F.S.} \times K} \tag{16}$$

where

S_{te} = allowable endurance limit tensile, psi

S_{end} = tensile endurance limit, psi = $.5 S_{\text{ult}}$

K = stress concentration factor (see Peterson, 1974)

τ_e = allowable endurance limit shear, psi

Safety factors are 1.25 to 1.75 and K usually ranges from 1.5 to 3.
Where combined tensile and shear stresses are present,

$$S_D \text{ or } S_M = \frac{S_t}{2} + \sqrt{\left(\frac{S_t}{2}\right)^2 + \tau^2} \tag{17}$$

$$\frac{1}{\text{F.S.}} = \frac{S_D}{S_{\text{end}}} + \frac{S_M}{S_{\text{ult}}} \tag{18}$$

where

S_D = combined dynamic stresses, psi

S_M = combined steady-state stresses, psi

S_{ult} = ultimate strength, psi

Safety factors are 1.25 to 1.75.

Infrequent Peak Loads (F.S. ranges from 1.0 to 1.25, typically 1.1):.

$$S_t = \frac{S_{\text{yld}}}{\text{F.S.}} \tag{19}$$

$$\tau = \frac{0.577 S_{\text{yld}}}{\text{F.S.}} \tag{20}$$

Typical Material Properties of Material for Couplings. Table 7 shows typical materials used for various components of couplings. The following are listed: material hardness, tensile ultimate, tensile yield, shear ultimate, and shear yield.

Compressive Stress Limits for Various Steels. Table 8 shows compressive stress limits for various hardnesses of steels. These limits are strongly dependent on whether relative motion takes place, type of lubricant used, and whether lubrication is present. The limits given in this table are typical limits for nonsliding components (e.g., keys, splines). For sliding parts the allowable

Table 7 Properties of Typical Coupling Components

	Hardness BHN	Tensile ultimate S_{ult} (psi)	Tensile yield S_{yld} (psi)	Shear ultimate τ_{ult} (psi)	Shear yield τ_{yld} (psi)
Aluminum die casting	50	25,000	12,000	18,500	6,900
	75	50,000	25,000	35,000	14,400
Aluminum bar—bar forging heat treated	100	60,000	35,000	45,000	20,200
	150	80,000	70,000	60,000	40,400
Carbon steels	110	50,000	30,000	37,500	17,300
	160	85,000	50,000	64,000	28,900
	250	110,000	70,000	82,000	40,400
Alloy steels	300	135,000	110,000	100,000	63,500
	330	150,000	120,000	110,000	69,200
	360	160,000	135,000	120,000	77,900
Surface hardened alloy steels	450	200,000	170,000		
	550	250,000	210,000		
	600	275,000	235,000		

$\tau_{ult} \cong .75 S_{ult}$; $\tau_{yld} \cong .577 S_{yld}$.

Table 8 Compressive Stress Limits of Steel

BHN	Normal psi	Maximum psi
110	32,500	65,000
160	45,000	90,000
250	65,000	130,000
300	72,500	145,000
330	80,000	160,000
360	90,000	180,000
450	125,000	250,000
550	137,500	275,000
600	150,000	300,000

value would be approximately one-fourth to one-third of the given values. It should also be noted that in some applications with good lubricants and highly wear resistant materials (nitrided or carburized) for parts that roll and/or slide (e.g., high-angle low-speed gear couplings), the values in the table can be approached.

3. General Equations

Torque

Torque (T) = horsepower (hp) $\times$ 63,000/rpm = in.-lb   (21)

Torque (T) = kilowatts (kW) $\times$ 84,420/rpm = in.-lb   (22)

Torque (T) = newton-meters (N $\cdot$ M) $\times$ 8.85 = in.-lb   (23)

Misalignment

Angular Misalignment. With the exception of some elastomeric couplings, a single-element flexible coupling can usually accommodate only angular misalignment (α) (see Fig. 21(A)). A double-element flexible coupling can handle both angular and offset misalignment (see Fig. 21(B)).

Angular $\alpha = \alpha_1 + \alpha_2$ (24)

Offset Misalignment (see Fig. 21(C)).

$S = L \tan \alpha$ (25)

where
  S = offset distance
  L = length between flex points

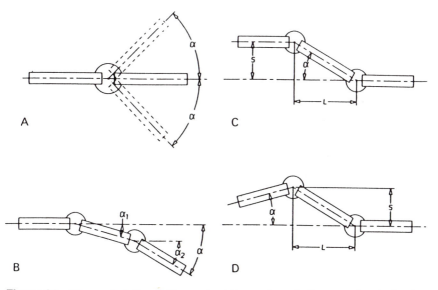

Figure 21   Types of shaft misalignments: (A) angular misalignment for single element; (B) angular misalignment for double element; (C) offset misalignment; (D) offset and angular misalignment combined.

α (deg)	Offset in.in. shaft separation	α (deg)	Offset in./in. shaft separation
1/4	0.004	2	0.035
1/2	0.008	3	0.052
3/4	0.012	4	0.070
1	0.018	5	0.087
1 1/2	0.026	6	0.100

Combined Offset and Angular Misalignment (see Fig. 21(D)).

$$\alpha_{\text{allowable}} = \alpha_{\text{rated}} - \tan^{-1}\left[\frac{S}{L}\right] \tag{26}$$

Weight.   Solid disk:

$$\text{Weight} = \text{lb} \tag{27}$$

$$\text{Steel } W = 0.223 LD_o^2 \tag{27}$$

$WR^2 = $ lb-in.^{2}

$$WR^2 = \frac{WD_o^2}{8} \tag{28}$$

where

$D_o = $ outside diameter, in.

$L = $ length, in.

Torsional stiffness $(K) = $ in.-lb/rad

Steel $K = 1.13 \times 10^6(D_o^4)$ \qquad\qquad (29)

Tubular (disk with a hole):

Weight $= $ lb

Steel $W = 0.223L(D_o^2 - D_i^2)$ \qquad\qquad (30)

$WR^2 = $ lb-in.^{2}

$$WR^2 = \frac{2}{8}(D_o^2 + D_i^2) \tag{31}$$

where

$D_i = $ inside diameter (in.).

Torsional stiffness $(K) = $ in.-lb/rad

$$\text{Steel } K = \frac{1.13 \times 10^6(D_o^4 - D_i^4)}{L} \tag{32}$$

Bending Moment.   Most couplings exhibit a bending stiffness. If you flex them, they produce a reaction. Bending stiffness (K_B) is usually expressed as

$K_B = $ in.-lb/deg \qquad\qquad (33)

Therefore, the moment reaction (M) per flex element is

$M = K_B\alpha = $ in.-lb \qquad\qquad (34)

(Note: Some couplings (e.g., gear couplings) exhibit bending moments that are functions of many factors (torque, misalignment, load distribution, and coefficient of friction).)

Axial Force.   Two types exist: force from the sliding of mating parts and force from flexing of material.

Sliding force (F) (e.g., gear and slider block couplings) is

$$F = \frac{T\mu}{R} = \text{lb} \tag{35}$$

where

T = torque, in.-lb

μ = coefficient of friction

R = radius at which sliding occurs, inc.

Flexing force (F) (e.g., rubber and metallic membrane couplings) is

$$F = K_A S = \text{lb} \tag{36}$$

where

K_A = axial stiffness, lb/in. per mesh

S = axial deflection, in.

Note: For two meshes, $K_A = K_A/2$.

4. Component Stress Equations

Shaft Stresses.

Circular Shaft. Solid (Figure 22(A)):

$$\tau_s = \frac{16T}{\pi D_o^3} \tag{37}$$

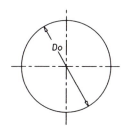

A

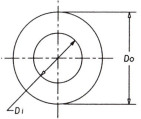

B

Figure 22 Types of shaft shapes: (A) circular shaft, solid; (B) circular shaft, tubular.

Tubular (Fig. 22(B)):

$$\tau_s = \frac{16TD_o}{\pi(D_o^4 - D_i^4)} \tag{38}$$

Spline Stresses (Fig. 23). Compressive stress:

$$S_c = \frac{2T}{n(\text{P.D.})Lh} \tag{39}$$

Shear stress:

$$S_s = \frac{4T}{\pi(\text{P.D.})^2 L} \tag{40}$$

Bending stress:

$$S_b = \frac{2T}{(\text{P.D.})^2 L} \tag{41}$$

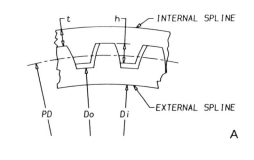

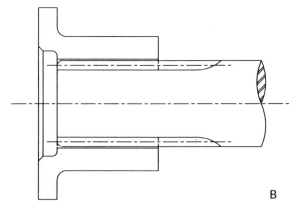

Figure 23 Spline dimensions: (A) spline teeth dimensions; (B) splined hub dimensions.

Bursting stress (hub):

$$S_t = \frac{T}{(\text{P.D.})L} \left[\frac{\tan \theta}{t} + \frac{2}{\text{P.D.}} \right] \qquad (42)$$

Shaft shear stress (shaft):

$$S_s = \frac{16 T D_o}{\pi (D_o^4 - D_i^4)} \qquad (43)$$

where

T = torque

n = number of teeth

P.D. = pitch diameter

L = effective face width (in.)

h = height of spline tooth

θ = pressure angle

t = hub wall thickness

D_o = diameter at root of shaft tooth

D_i = inside diameter (ID) of shaft (zero for solid shaft)

Key Stresses.   Square or rectangular keys (Fig. 24):

$$T = \frac{FDn}{2} \qquad (44)$$

$$F = \frac{2T}{Dn} \qquad (45)$$

$$\tau = \text{shear stress} = \frac{F}{wL} = \frac{2T}{WLDn} \qquad (46)$$

$$S_c = \text{compressive stress} = \frac{2F}{hL} = \frac{4T}{hLDn} \qquad (47)$$

where

L = length of key, in.

n = number of keys

D = shaft diameter, in.

w = key width, in.

h = key height, in.

T = torque, in.-lb

F = force, in.-lb

Allowable key stress limits (see Table 9).

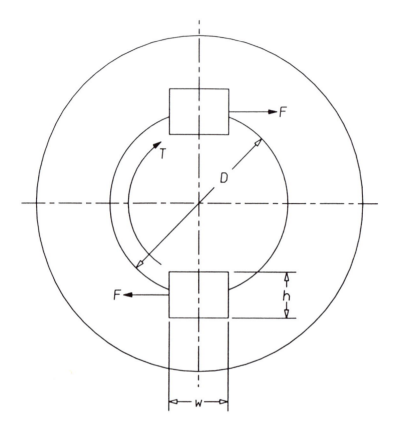

Figure 24   Square or rectangular key loading and dimensions.

Table 9   Key Limits

	Shear τ normal (psi)	Compressive S_c normal (psi)	Shear τ peak (psi)	Compressive S_c peak (psi)
Commercial key stocks	12,500	32,500	23,000	65,000
1045 (170 BHN)	15,000	45,000	27,500	90,000
4140–4130 HT (270 BHN)	30,000	70,000	55,000	140,000

5. Flange Connections (Fig. 25)

Friction Capacity.

$$T = F\mu R_f \qquad (48)$$

F = clamp load from bolts

$$F = \left[\frac{T_B}{\mu_B d} \right] n \qquad (49)$$

where

T_B = tightening torque of bolt, in.-lb

n = number of bolts

d = diameter of bolt, in.

μ_B = coefficient of friction at bolt threads (0.2 dry threads) (see Table 10)

μ = coefficient of friction at flange faces (0.1 to 0.2, usually 0.15)

R_f = friction radius, in.

$$R_f = \frac{2}{3} \left[\frac{R_o^3 - R_i^3}{R_o^2 - R_i^2} \right] \qquad (50)$$

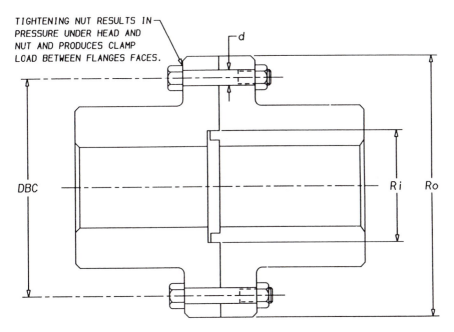

TIGHTENING NUT RESULTS IN PRESSURE UNDER HEAD AND NUT AND PRODUCES CLAMP LOAD BETWEEN FLANGES FACES.

d

DBC

Ri   Ro

Figure 25 Dimensions for bolted flange connection.

Table 10   Frictional Factors for Various Bolt Tread Conditions

Multiply torque (T_B by C)	Condition
1	Dry threads—unplated or black oxided bolts
$\mu = 0.2$	
0.75%	Lube threads—unplated or black oxided bolts
$\mu = 0.15$	
0.75%	Dry threads—phosphated or cadmium-plated bolts
$\mu = 0.15$	
0.56	Lube threads—phosphated or cadmium-plated bolts
$\mu = .11$	

where

R_o = outside radius of clamp, in.

R_i = inside radius of clamp, in.

Bolt Stresses for Flange Couplings (see Fig. 25).

Shear Stress.   Allowables:

$$\tau_A \text{ at normal operation} = \frac{0.30S_{\text{ult}}}{1.5} = 0.20S_{\text{ult}} \tag{51}$$

$$\tau_A \text{ at peak operation} = 0.55S_{\text{yld}} \tag{52}$$

where (see Fig. 26)

S_{ult} = ultimate strength of material, psi

S_{yld} = yield strength of material, psi

$$\tau_s = \frac{8T}{(\text{D.B.C.}) \, (\%n)\pi d} \tag{53}$$

where

T = operating torque, in.-lb

D.B.C. = diameter bolt circle, in.

d = bolt body diameter

n = number of bolts

$\%n$ = percentage of bolts loaded: fitted bolts 0.000 to 0.010 loose (100%); clearance bolts 0.010 to 0.032 loose (75%).

For very loose bolts with over 0.032 clearance or very long bolts of approximately 10d, such analysis may not be applicable. For very loose and long bolts, the bending stresses produced in a bolt should be considered.

S.A.E. GRADE	HEADSTYLE	BOLT SIZE DIA. IN.	PROOF LOAD	TENSILE STRENGTH	HARDNESS ROCKWELL	MATERIAL – HEAT TREATMENT
2	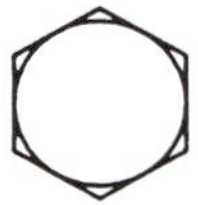	Up to 1/2 Incl. Over 1/2 to 3/4 Over 3/4 to 1-1/2	55,000 52,000 28,000	69,000 64,000 65,000	100 B Max. 100 B Max. 95 B Max.	Low or medium carbon steel; C 0.28 max., and S 0.05 max. For bolts over 6 in. Long, or over 3/4 in. dia., carbon may be as high as 0.55.
5	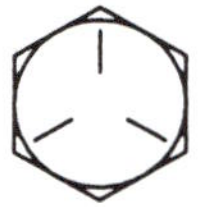	Up to 3/4 Incl. Over 3/4 to 1 Over 1 to 1-1/2	85,000 78,000 74,000	120,000 115,000 105,000	23-32 C 27-32 C 19-30 C	Medium carbon steel, C 0.28 to 0.55 P 0.04 max., and S 0.05 max., Quenched and tempered at a minimum temperature of 800°F.
8	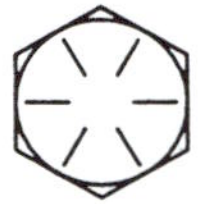	Up to 1-1/2 Incl.	120,000	150,000	32-38 C	Medium carbon fine grain alloy steel (b), C 0.28 to 0.55, P 0.04 max. and S 0.05 max., providing sufficient hardenability to have a minimum oil quenched hardness of 47 RC at the center of the threaded section one diameter from the end of the bolt. Oil quenched and tempered at a minimum temperature of 800°F.
12-point Aircraft Style		Up to 1-1/4	144,000	160,000	34-40 C	Alloy steel usually 4140, 4340, and others.

Figure 26 Properties of common grades of coupling bolts.

6. Hub Stresses and Capacities

Hub Bursting Stress for Hubs with Keyways (Fig. 27). The following
approach is applicable when the interference is less than 0.0005 in./in. of shaft
diameter and X is greater than $(D_o - D_i)/8$.

Stress:

$$S_t = \frac{T}{AYn} \tag{54}$$

where

    n = number of keys

    Y = radius of applied force, in.

    A = area = XL (L = length of hub, in.

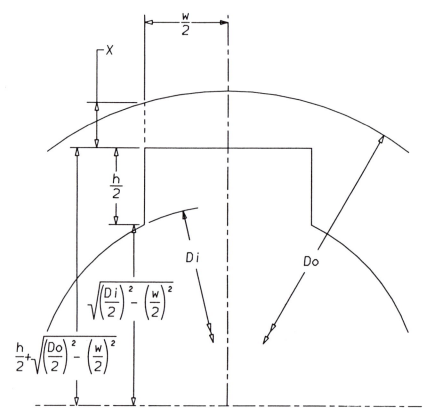

Figure 27   Dimensions for calculating hub bursting stress.

$$X = \sqrt{\left[\frac{D_o}{2}\right]^2 - \left[\frac{W}{2}\right]^2} - \frac{h}{2} - \sqrt{\left[\frac{D_i}{2}\right]^2 - \left[\frac{w}{2}\right]^2} \quad (55)$$

$$Y = \frac{2D_i + h}{4} \quad (56)$$

where

D_o = outside diameter

D_i = inside diameter

h = height of key

w = width of key

Hub Stress and Capacity for Keyless Hubs (Shrink Drives) (see Fig. 28). Shrink fit or interference fits are generally based on Lamé's equation for a thick-walled cylinder under internal pressure:

Stresses:

$$S_t = P \left[\frac{D_o^2 + D_i^2}{D_o^2 - D_i^2} \right] \quad (57)$$

$$P = \frac{Ei}{D_i^2} \left[1 - \left[\frac{D_i}{D_o} \right]^2 \right] \quad (58)$$

where

D_o = outside diameter, psi

D_i = inside diameter, psi

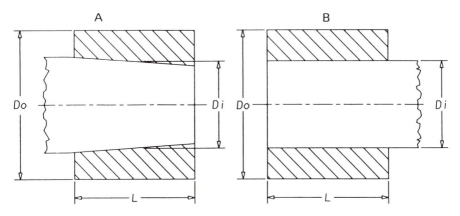

Figure 28 Dimensions for keyless hubs: (A) taper shaft; (B) straight shaft.

P = pressure, psi

E = modulus of elasticity for steel, 30×10^6 (psi)

i = diametral interference

Use maximum interference for stress calculations. Use minimum interference for torque capacity.

Torque capacity:

$$T = \frac{\pi D_i^2 L \mu P}{2} \tag{59}$$

where L is the length of the hub or sections of it (in.) and $\mu = 0.1$ to 0.2, usually 0.15.

Torque Capacity. For hubs with steps or flanges (Fig. 29), Section 1 requires maximum amount of pressure for mounting:

Section 1:

$$P_1 = \frac{Ei}{2D_3} \left[1 - \left(\frac{D_3}{D_o} \right)^2 \right] \tag{60}$$

$$S_{t_1} = P_1 \left(\frac{D_o^2 + D_3^2}{D_o^2 - D_3^2} \right) \tag{61}$$

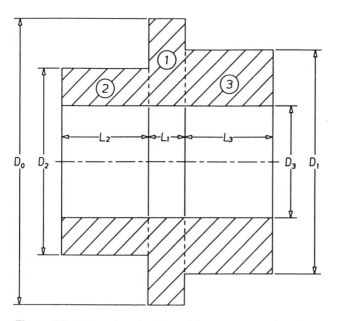

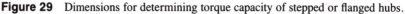

Figure 29 Dimensions for determining torque capacity of stepped or flanged hubs.

$$T_1 = \frac{\pi D_3^2 L_1 \mu P_1}{2} \qquad (62)$$

Section 2:

$$P_2 = \frac{Ei}{2D_3} \left[1 - \left[\frac{D_3}{D_2} \right]^2 \right] \qquad (63)$$

$$S_{t_2} = P_2 \left[\frac{D_2^2 + D_3^2}{D_2^2 - D_3^2} \right] \qquad (64)$$

$$T_2 = \frac{\pi D_3^2 L_2 \mu P_2}{2} \qquad (65)$$

Section 3:

$$P_3 = \frac{Ei}{2D_3} \left[1 - \left[\frac{D_3}{D_1} \right]^2 \right] \qquad (66)$$

$$S_{t_3} = P_3 \left[\frac{D_1^2 + D_3^2}{D_1^2 - D_3^2} \right] \qquad (67)$$

$$T_3 = \frac{\pi D_3^2 L_3 \mu P_3}{2} \qquad (68)$$

Total torque capacity:

$$T_1 = T_1 + T_2 + T_3 = \text{in.-lb} \qquad (69)$$

C. Balancing of Couplings

Like coupling ratings, when and to what level to balance a coupling appear to be almost as mystifying. In reality, however, they are rather simple once a basic understanding is achieved of what contributes to unbalance and how it affects a system. Once this understanding is obtained, the arbitrary balance limits can be put into perspective for a system's needs. The values given in this section are only guidelines. Much of the information in this section is from the American Gear Manufacturers Association (AGMA) 9000.

1. Basics of Balancing

Units for Balancing or Unbalance. The unbalance in a piece of rotating equipment is usually expressed in terms of unbalance weight (ounces) and its distance from the rotating centerline (inches). Thus we get unbalance (U) in oz-in. terms.

Example 1. Unbalance weight is 2 oz; the distance from the rotating centerline is 2 in. (see Fig. 30). Unbalance (U) equals 4 oz-in.

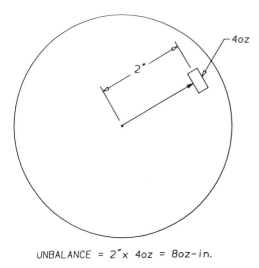

UNBALANCE = 2″x 4oz = 8oz-in.

Figure 30   Unbalance units.

For couplings, it is more convenient to express unbalance in terms of mass displacement. This usually helps bring balance limits into perspective with tolerances, fits, and the "real world." Mass displacement is usually specified in microinches, mils, or inches (1000 microinches (μin.) = 1 mil = 0.001 in.).

Example 2.   If a 100-lb part (1600 oz) is displaced 0.001 in. (1 mil or 1000 μin.) from the centerline (Fig. 31), a 1.6 oz-in. unbalance (1600 oz × 0.001 in.) would occur.

Balancing.   Balancing is a procedure by which the mass distribution of a rotating part is checked and, if required, adjusted (usually by metal removal, but on low-speed couplings metal may be added) so that the unbalance force or equipment vibration is reduced.

Types of Unbalance.   There are two types of unbalance: single-plane unbalance (sometimes called static balance) and two-plane (moment, coupling, or dynamic unbalance. Single-plane unbalance exists when the center of gravity of a rotating part does not lie on the axis of rotation (Fig. 32). This part will not be in static equilibrium when placed at random positions about its axis. Single-plane balancing can be done without rotation (much like static balancing of an automobile wheel), but most couplings are balanced on centrifugal balancing machines, which produce a much better balance. Single-plane balancing is usually limited to relatively short parts, usually with a length less than $1.5D_o$ (depending on part configuration), where D_o is the outside diameter of the part being balanced.

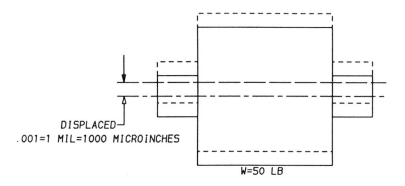

Figure 31 Displaced mass.

Two-plane unbalance is present when the unbalance existing in two planes is out of phase, as shown in Fig. 33, but not necessarily 180° out of phage. The principal inertial axis closest to the axis of rotation is displaced from the axis of rotation, and these two axes are askew with respect to each other. If not restrained by bearings, a rotating part with this type of unbalance will tend to rotate about its principal inertial axis closest to its geometric axis. If the moments are equal and opposite, they are referred to as a coupling unbalance. Figure 33 illustrates two-plane unbalance.

Rigid Rotor. A rotor is considered rigid when it can be corrected in any two planes and, after the part is balanced, its unbalance does not significantly exceed the unbalance tolerances limit at any other speed up to the maximum operating speed and when rubbing conditions closely approximate those of the final system. A flexible coupling is an assembly of several components having diametral clearances and eccentricities between pilot surfaces of those components. Therefore, it is not appropriate to apply standards and requirements that were written for rigid rotors in lieu of something more appropriate.

Axis of Rotation. The axis of rotation is a line about which a part rotates as determined by journals, fit, or other locating surfaces.

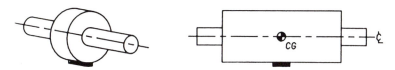

Figure 32 Single-plane unbalance.

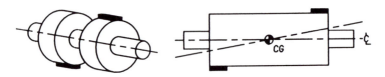

Figure 33   Two-plane unbalance.

Principal Inertial Axis Displacement.   The displacement of the principal inertial axis is the movement of the inertial axis that is closest to the axis of rotation with respect to the axis of rotation. In some special cases these two axes may be parallel. In most cases, they are not parallel and are therefore at different distances from each other in the two usual balancing planes.

Amount of Unbalance.   The amount of unbalance is the measure of unbalance in a part (in a specific plane) without relation to its angular position.

Residual Unbalance.   Residual unbalance is the amount of unbalance left after a part has been balanced. It is equal to or usually less than the balance limit tolerance for a part. (Note: A check to determine whether a part is balanced by removing it from the balancing machine and then replacing it will not necessarily produce the same residual unbalance as that originally measured. This is due to the potential differences in mounting and/or indicating surface runouts.)

Potential Unbalance.   The potential unbalance is the maximum amount of unbalance that might exist in a coupling assembly after balancing (if corrected) when it is disassembled and then reassembled.

Balance Limit Tolerance.   The balance limit tolerance specifies a maximum value below which the state of unbalance of a coupling is considered acceptable. There are balance limit tolerances for residual unbalance and for potential unbalance.

Mandrel.   A shaft on which the coupling components or assembly is mounted for balancing purposes is called a mandrel (see Fig. 34).

Bushing Adapter.   A bushing adapter (see Fig. 34) or adapter assembly is used to mount the coupling components or the coupling assembly on the mandrel.

Mandrel Assembly.   A mandrel assembly (see Fig. 34) consists of one or more bushings used to support the part mounted on a mandrel.

Mounting Surface.   A mounting surface is the surface of a mandrel, bushing, or mandrel assembly on which another part of the balancing tooling, a coupling

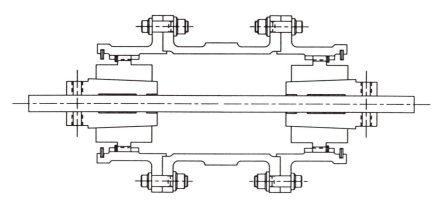

Figure 34   Mandrel assembly for balancing a gear coupling.

component, or the coupling assembly is mounted. This surface determines the rotational axis of the part being balanced.

Mounting Fixtures.   Mounting fixtures are the tooling that adapts to the balancing machine and provides a surface on which a component or coupling assembly is mounted.

Indicating or Aligning Surface.   The indicating surface is the axis about which a part is aligned for the purpose of balancing. The aligning surface is the axis from which a part is located for the purpose of balancing. (Note: Some coupling manufacturers do not component-balance. Not all couplings can be component- or assembly-balanced without mandrels or mounting fixtures.)

Pilot Surface.   A pilot surface is that supporting surface of a coupling component or assembly upon which another coupling component is mounted or located. Examples are shown in Table 11.

Table 11   Types of Coupling Pilot Surfaces

Components	Typical examples
Rigid hub	Bore, rabbet diameter, bolt circle
Flex hub (gear type)	Bore, hub body OD, tooth tip diameter
Flanged sleeve (gear type)	Tooth root diameter, end ring ID, rabbet diameter, bolt circle
Flanged adapter	Rabbet diameter or bolt circle
Flanged stub end adapter	Stub end (shaft) diameter, rabbet diameter, bolt circle
Spool spacer (gear type)	Tooth tip diameter
Flanged spacer	Rabbet diameter or bolt circle
Ring spacer	Rabbet diameter or bolt circle

2. Why a Coupling Is Balanced

One important reason for balancing a coupling is that the forces created by
unbalance could be detrimental to the equipment, bearings, and support structures. The amount of force generated by unbalance is

$$F = 1.77 \times \left[\frac{rpm}{1000} \right]^2 \times oz.\text{-in.} = lb \qquad (70)$$

$$F = \frac{microinches}{1,000,000} \times weight \times 16 \times \left[\frac{rpm}{1000} \right]^2 = lb \qquad (71)$$

According to these equations, the amount of force generated by unbalance is
proportional to the square of the speed.

Example 3.   Assume 2 oz-in. of unbalance at 2000 rpm and at 4000 rpm:

F at 2000 rpm = 14.1 lb
F at 4000 rpm = 56.6 lb

Therefore, if speed doubles, the same amount of unbalance produces four
times the force.

Another important reason for balancing is vibration. Sometimes this
unwanted vibration produces poor product quality. For example, spindle coupling vibration can produce chatter and rough finish on the rolled strip or sheet
being produced, making the product unacceptable. The next question is: If balance is all that important, why not forget all these complications and just balance all couplings? However, unless you have an unlimited budget, please read
on.

There are several reasons why manufacturers do not balance all couplings.
To balance a coupling in a balancing machine as an assembly, a shaft or a
fixture is needed. This means that the coupling must be assembled on this
special shaft or fixture and then balanced. (Note: Some couplings, including
disk and diaphragm couplings, can be rigidized and balanced without a shaft or
fixture.) This is expensive; also, the shaft or fixture does not exactly resemble
the equipment shaft, so inherent errors are introduced into the balancing.
Sometimes this is greater than the original "as manufactured" potential unbalance. Some coupling manufacturers do not assemble the entire coupling on a
shaft or fixture, but component-balance parts by aligning the indicating diameters of these individual parts, which again is costly and introduces some
inherent error from this eccentricity of the aligning diameters.

We must always remember that a coupling is an assembly of components
that are taken apart and put back together for assembly of equipment, maintenance of the coupling and the equipment, and so on. With the assembly and
disassembly of coupling components the relative position of mating parts can

change and therefore the coupling's state of unbalance will change. Both component and assembly balance can usually produce a coupling with equal potential unbalance, sometimes not what the coupling selector asked for but generally sufficient.

On very sensitive equipment the inherent errors introduced from the balancing method and assembly/disassembly unrepeatability may be far greater than the balance tolerance limits placed on the coupling and in some cases the actual requirements. When the real balance needs exceed practicality (potential unbalance capabilities), the coupling must be balanced on the actual equipment, which is very expensive and time consuming.

In summary, the basic reason why all couplings are not balanced is because a balanced coupling costs more, not only because of balance time and equipment but in most cases to assure some repeatability of the coupling potential unbalance, the tolerances and fits must be tightened, which can greatly increase the cost of the coupling. For example, to manufacture gear couplings with half the standard tolerances would cost 150% to 200% more.

3. *What Contributes to Unbalance in a Coupling*

Contributors to Potential Unbalance of Uncorrected Couplings.

   a. *Inherent unbalance*: If the coupling assembly or components are not balanced, an estimate of inherent unbalance caused by manufacturing tolerances may be based on one of the following:

   (1) Statistical analysis of balance data accumulated for couplings manufactured to the same tolerances, or

   (2) Calculations of the maximum possible unbalance that could theoretically be produced by the tolerances placed on the parts.

   b. *Coupling pilot surface eccentricity*: This is any eccentricity of a pilot surface that permits relative radial displacement of the mass axis of mating coupling parts or subassemblies.

   c. *Coupling pilot surface clearance*: This is the clearance that permits relative radial displacement of the mass axis of the coupling components or subassemblies. (Note: Some couplings must have clearances in order to attain their flexibility (e.g., gear couplings).)

   d. *Hardware unbalance*: Hardware unbalance is the unbalance caused by all coupling hardware, including fasteners, bolts, nuts, lockwashers, lube plugs, seal rings, gaskets, keys, snap rings, keeper plates, thrust plates, and retainer nuts.

   e. *Others*: Many other factors may contribute to coupling unbalance. The factors mentioned are those of primary importance.

Contributors to Potential Unbalance of Balanced Couplings.

   a. *Balance tolerance limits*: These limits are the largest amount of unbalance (residual) for which no further correction need be made.

b. *Balance fixtures of mandrel assembly unbalance*: This is the combined unbalance caused by all components used to balance a coupling, including mandrel, flanges, adapters, bushings, locking devices, keys, setscrews, nuts, and bolts.

c. *Balance machine error*: The major sources of balancing machine error are overall machine sensitivity and error of the driver itself. This unbalance is usually very minimal and can generally be ignored.

d. *Mandrel assembly, mounting surface eccentricity*: This is the eccentricity, with respect to the axis of rotation, of the surface of a mandrel assembly upon which the coupling assembly or component is mounted.

e. *Component or assembly indicating surface eccentricity*: This is the eccentricity with respect to the axis of rotation of a surface used to indicate or align a part on a balancing machine. On some parts this surface may be machined to the axis of rotation rather than indicated on balancing machines.

f. *Coupling pilot surface eccentricity*: This is an eccentricity of a pilot surface that permits relative radial displacement of the mass axis of another coupling part or subassembly, upon assembling subsequent to the balancing operation. This eccentricity is produced by manufacturing before balancing or by alternation of pilot surfaces after balancing. For example, most gear couplings that are assembled balanced are balanced with an interference fit between gear major diameters and after balancing are remachined to provide clearance. This may produce eccentricity.

g. *Coupling pilot surface clearance*: This is the clearance that permits relative radial displacement of the mass axes of the coupling component or subassemblies on disassembly/reassembly. The radial shift affecting potential unbalance is equal to half this clearance. (Note: If this clearance exists in a coupling when it is balanced as an assembly, the potential radial displacement affecting potential unbalance is equal to the full amount of the diametral clearance that exists in the assembly at the time of balance.)

h. *Hardware unbalance*: This is the unbalance caused by all coupling hardware, including fasteners, bolts, nuts, lockwashers, lube plugs, seal rings, gaskets, keys, snap rings, keeper plates, thrust plates, and retainer units.

i. *Others*: Many other factors may contribute to coupling unbalance.

4. How to Bring a Coupling into Balance

Couplings can be brought into balance by four basic methods:

1. Tighter manufacturing tolerances
2. Component balancing

3. Assembly balancing
4. Field balancing on the equipment

Tighter Manufacturing Tolerances. The majority of unbalance in most couplings comes from the tolerances and the clearance fits that are placed on components so that they can be competitively produced yet be interchangeable. Most couplings are not balanced. The amount of unbalance can be greatly improved by tightening fits and tolerances. For example, by cutting the manufacturing tolerances and fits approximately in half, only half the amount of potential unbalance remains. If a coupling is balanced without changing the tolerance, the unbalance is improved by only 5% to 20%. Remember that the tighter the tolerances and the fits, the more couplings are going to cost. A real-life practical limit is reached at some point. As tolerances are tightened, the price of the coupling not only increases, but interchangeability is lost, delivery is extended, and in some instances the choice of couplings and potential vendors is limited. Table 12 shows how tolerances affect cost.

Component Balancing. Component balance is usually best for couplings that have inherent clearances between mating parts or require clearances in balancing fixtures. Component balance offers interchangeability of parts usually without affecting the level of potential unbalance. In most cases component-balanced couplings can approach the potential unbalance limits of assembly-balanced couplings. In some cases it can produce potential unbalance levels lower; this is particularly true where large, heavy mandrels or fixtures must be used to assembly-balance a coupling.

Assembly Balancing of Couplings. Assembly balancing may provide the best coupling balance. This is true when no clearance exists between parts (e.g., disk or diaphragm couplings). The balancing fixture and mandrels are lightweight and are manufactured to extremely tight tolerances: 0.0003 to 0.0005 *total indicator runout* (TIR). This is 300 to 500 μin., or the coupling is somehow locked or rigidized without the use of a fixture or a mandrel. For

Table 12 Coupling Tolerances Versus Cost

	General tolerances[a]	Important tolerances[b]	DBC true location	Approximate cost
Low speed	± 1/64	±0.005	± 1/64	1
Intermediate speed	±0.005	±0.002	±0.005	1.5–2
High speed	±0.002	±0.001	±0.005	3–4

[a] General tolerance applied to noncritical diameters and lengths (example: coupling OD).

[b] Important tolerances apply to critical diameters and lengths (example: bores, pilots).

relatively large assemblies the mandrels and fixtures may introduce more error than if the coupling were not balanced at all. It is possible to balance into the coupling more unbalance than the original unbalanced coupling had. Assembly-balanced couplings are matchmarked, and components should not be interchanged or replaced without rebalancing.

Field Balancing. On very high speed and/or lightweight equipment, it may not be possible to provide a balanced coupling to meet the requirements due to the inherent errors introduced when balancing a coupling on a balancing machine. When this occurs, the coupling manufacturer can provide a means whereby weight can easily be added to coupling so that coupling can be balanced by trial and error on the equipment itself. Couplings can be provided with a balancing ring with radial setscrews and tapped holes so that bolts with washers can be added or subtracted. Other means can also be used.

5. When to Balance and to What Level

The amount of coupling unbalance that can be satisfactorily tolerated by any system is dictated by the characteristics of the specific connected machines and can best be determined by detailed analysis or experience. Systems insensitive to coupling unbalance might operate satisfactorily with values of coupling balance class lower than those shown. Conversely, systems or equipment that are usually sensitive to coupling unbalance might require a higher class than suggested. Factors that must be considered in determining the system's sensitivity to coupling unbalance include

1. *Shaft end deflection*: Machines having long and/or flexible shaft extensions are relatively sensitive to coupling unbalance.
2. *Bearing loads relative to coupling weight*: Machines having lightly loaded bearings or bearing loads primarily by the overhung weight of the coupling are relatively sensitive to coupling unbalance. Machines having overhung rotors or weights are often sensitive to coupling unbalance.
3. *Bearings, bearing supports, and foundation rigidity*: Machines or systems with flexible foundations or supports are relatively sensitive to the coupling unbalance.
4. *Machine separation*: Systems having a long distance between machines often exhibit coupling unbalance problems.
5. *Others*: Other factors may influence coupling unbalance sensitivity.

Table 13 is taken from AGMA 9000. In general, selection bands can be put into the following speed classifications:

Low speed: A and B
Intermediate speed: C, D, and E
High speed: F and G

Table 13 Coupling Balance Classes versus Selection Bands

| Selection band | Typical AGMA coupling balance class; system sensitivity to coupling unbalance | | |
	Low	Average	High
A	5	6	7
B	6	7	8
C	7	8	9
D	8	9	10
E	9	10	11
F	10	11	
G	11		—

Figure 35 is also taken from AGMA 9000. Superimposed on this graph are the three most common speed classifications. The graph has also been extended to 2000 lb.

Coupling Balance Limits. The balance limit placed on a coupling should be its *potential* unbalance and not its residual unbalance limit. The residual unbalance limit usually has little to do with the *true* coupling unbalance (potential). It can be seen in many cases that by cutting the residual unbalance limit in half, the coupling potential unbalance may change by only 5%. The best method of determining the potential unbalance of a coupling is by the square root of the sum of the squares of the maximum possible unbalance values. These unbalances are mostly from the eccentricities between the coupling parts, but include any other factors that produce unbalance. The coupling balance limit is defined by AGMA 9000 as a range of unbalance expressed in microinches. The potential unbalance limit classes for couplings are given in Table 14. They are given in maximum root-mean-square (rms) microinches of displacement of the inertial axis of rotation at the balance plane. Limits are given as per displacement plane. The residual unbalance limit of a part or an assembly-balanced coupling is shown in Table 15. Tolerances tighter than these usually do very little to improve the overall potential unbalance of a coupling assembly.

6. Other Coupling Balancing Criteria

What is meant by "arbitrary" balancing criteria? The limits (potential and residual) described in the preceding section give the most realistic values for unbalance limits. Other criteria are being used, generally referred to as *arbitrary limits*. They are used in several coupling specifications. The most common is to express unbalance (U) as

$$\text{oz-in. (unbalance)} = \frac{KW}{N} = U \text{ per balanced plane} \qquad (72)$$

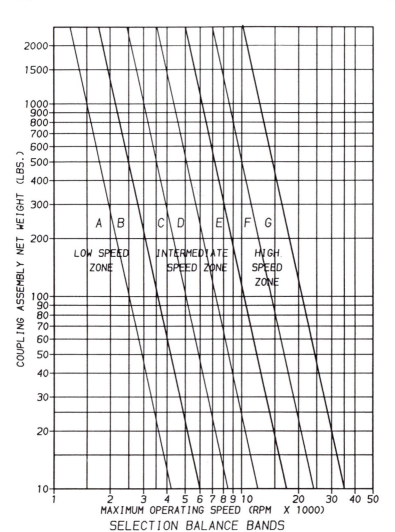

Figure 35 Balance selection weight versus speed.

where
   K = 40 to 120 for potential unbalance limits, 4 to 12 for residual unbalance
      limits
   W = weight of the part per balance plane (lb)
   N = operating speed of coupling (rpm)

Table 14 AGMA Coupling Balance Classes

AGMA coupling balance class	Maximum displacement of principal inertia axis at balancing planes (rms microinches)
4	Over 32,000
5	32,000
6	16,000
7	8,000
8	4,000
9	2,000
10	1,000
11	500

The two most common values for these arbitrary limits are given in American Petroleum Institute (API) Standard 671:

$$\text{oz-in.} = \frac{4W}{N} \tag{73}$$

Potential limit:

$$\text{oz-in.} = \frac{40W}{N} \tag{74}$$

In general, these limits are not too bad, but in some cases if users are not careful they may get a very expensively balanced coupling when they do not need one or pay extra and not gain anything. In other cases they will be specifying limits beyond the real world of practicality and will be faced with arguments, delays, and so on, while everyone involved regroups and tries to blame everyone else for not specifying correctly and/or not balancing correctly.

Table 15 Practical Residual Unbalance Limits

Speed class	Residual unbalance limits (microinches)
Low-speed couplings	500[a]
Intermediate-speed couplings	200
High-speed couplings	50

[a] Low-speed couplings are usually not balanced.

Arbitrary Potential Unbalance Limits. Applying the arbitrary limits to the three coupling speed classes results in the following values:

Speed class	Unbalance limit (oz-in.)
Low speed	$120W/N$
Intermediate speed	$80W/N$
High speed	$40W/N$

As stated before, it is best to put unbalance in terms of microinches. If this is done, another arbitrary criterion results, but now if a low limit tolerance is applied (in microinches) we can ensure that the balance limit has some relationship to the real world of manufacturing the coupling with the requisite tolerances and fits.

Speed class	Unbalance limit[a] (μin.)
Low speed	$7,500,000/N$ or 4000
Intermediate speed	$5,000,000/N$ or 2000
High speed	$2,500,000/N$ or 500

[a] The highest values become limits

Arbitrary Residual Unbalance Limits. The residual unbalance limits are approximately one-tenth of the potential unbalance limit.

Speed class	Unbalance limit (oz-in.)	Unbalance limit[a] (μin.)
Low speed	$12W/N$	$750,000/N$ or 400
Intermediate speed	$8W/N$	$500,000/N$ or 200
High speed	$4W/N$	$250,000/N$ or 50

[a] The highest values become limits.

7. Types of Coupling Balance

Most coupling manufacturers can and will supply couplings that are brought into balance. Some coupling manufacturers have what they call their "standard balancing procedure and practice" and if you request other than their standard practice they will charge extra. For example, some coupling manufacturers prefer to supply component-balanced couplings, whereas others prefer assembly-balanced couplings. What do these balance types mean?

1. *As manufactured*: Most couplings are supplied as manufactured with no balancing.
2. *Controlled tolerances and fits*: Usually provides the most significant improvement in the potential unbalance of coupling. This can also substantially increase the price of a coupling if carried too far.
3. *Component balance* (see Figs. 36 and 37): Can usually produce potential unbalanced values equal to assembly-balanced couplings. Offers the advantage of being able to replace components (as related to balance—some couplings are not interchangeable for other reasons) without rebalancing.
4. *Assembly balancing with mandrels or fixtures* (see Fig. 38): Usually offers the best balance, but is usually expensive because the mandrels and/or fixtures must be made extremely accurately. The coupling is basically rigidized with the mandrel or fixtures and then balanced. On assembly-balanced couplings, parts cannot be replaced without rebalancing the coupling.

Figure 36 Component balancing of a geared spacer (courtesy of Zurn Industries, Inc., Mechanical Drives Division).

Figure 37   Component balancing of a hub and spacer for a gear coupling (courtesy of Zurn Industries, Inc., Mechanical Drives Division).

5. *Component balancing with selective assembly*: Sometimes offers the best possible balance attainable without field balancing on the equipment. Parts are component-balanced and then runout (TIR) is checked. The highs of the TIR readings between controlling diameters for mating parts are marked. At final coupling assembly, the high spots are assembled 180° out of phase. This tends to negate eccentricities and reduces the potential unbalance of the coupling. The parts are still interchangeable as long as replacement parts are inspected and marked for their high TIRS.

6. *Assembly balancing without a mandrel* (see Fig. 39): This is usually limited to disk, diaphragm, and some types of gear couplings. The coupling is locked rigid with various locking devices, which are usually incorporated into the coupling design. The coupling is rolled on rolling surfaces that are aligned or machined to the coupling bores or alignment pilots. This type of balance can usually provide a better balance than with a mandrel. This is because there is no weight added to the

Figure 38 Assembly balancing of a gear coupling (courtesy of Sier-Bath Gear Co., Inc.).

Figure 39 Assembly balancing of a diaphragm coupling (without mandrel) (courtesy of Zurn Industries, Inc., Mechanical Drives Division).

assembly when it is balanced. On very large and long couplings a mandrel assembly can weigh almost one-half that of a coupling. This can introduce very significant balancing errors.

7. *Field or trim balancing on the equipment*: Offers the best balance but is usually the most costly, because of the trial and error and the time involved. The coupling cannot be disassembled/reassembled without rebalancing.

V. INSTALLATION AND MAINTENANCE

A. Introduction

In this section we discuss the installation and maintenance of flexible couplings in general terms and also the various coupling types available on the market. Most coupling manufacturers provide information and recommendations for installation, maintenance, and lubrication of their couplings. If a copy is not included with a coupling, the coupling manufacturer should be contacted before installation is begun. A good maintenance department has a file where this information is kept for reference. Why should you read this section? Because it presents general procedures as well as some of the reasons for these practices and provides a good starting place for the person who has not installed a coupling before.

B. Installation of Couplings

1. Preparation

Upon receiving a coupling, proceed as follows:

1. Obtain a drawing (if one exists) of the coupling and a copy of the coupling manufacturer's and/or equipment builder's installation and maintenance manual.
2. Read the information. If any questions arise, contact the coupling manufacturer or equipment builder.
3. Inspect the components to ensure that all parts ordered have been shipped.
4. If the coupling is going to be stored, make sure that the parts are properly protected. Steel parts are usually coated with oil or wax. Rubber parts are usually wrapped or packaged so that light and air exposure are minimized (light and air tend to harden rubber and some elastomers). If the storage time is to be more than one year, check with the coupling manufacturer for special instructions. Particular attention should be directed to seals, rubber elements, and any prelubricated parts when long-term storage is required.

5. Prior to assembling the coupling, clean and disassemble the components. *Note*: Some subassemblies should not be taken apart. Check specific coupling instructions as to what to disassemble. *This is very important.*
6. Check for burrs and nicks on the mating surfaces. If they exist, remove (stone, file, sand) them.
7. Measure and inspect
   a. Bore dimensions
   b. Keyways
   c. Coupling lengths
   d. Diameters of bolt circle (DBC), bolts, and holes of the mating flange
   e. Any other dimensions to ensure that the coupling will mate properly to the equipment
8. Obtain hardware, keys, tools, and anything else not supplied but needed to complete the assembly.

2. Hub Installation

Normally, hubs are mounted before the equipment is aligned. In some instances, hubs may be mounted by the equipment manufacturer. Next we describe the installation procedure for various fittings.

Straight Shafts

Clearance Fits. This type of installation is relatively simple.

1. Install the key(s) in the shaft keyway(s).
2. Make sure that any part that will not slide over the coupling hub is placed back on the shaft, such as seals, carriers, and covers, and on gear couplings, the sleeves.
3. Push the hub onto the shaft until the face of the hub is flush with the ends of the shaft. (Note: Some coupling hubs are not mounted flush. Check specific instructions.) If the hub does not slide onto the shaft, check the clearances between the bore and the shaft. In addition, verify that there is clearance between the keys at the sides of the coupling keyway and on top of the key.
4. Lock the hub in position (usually with setscrews). Make sure that setscrews have a locking feature such as a Nylok insert, or use locking compound. Some hubs use bolts, nuts, or other means to secure the hub in place. See the specific instructions.

Interference Fit. This type of installation is the same as that for the straight shaft, with the exception that the hubs must be heated before they slide onto the shaft. The coupling manufacturer usually supplies information as to how to heat the hub and to what temperature. For steel hubs, 160°F is required for every 0.001 in. of interference per inch of hub diameter (0.001 in./in.). For example, a steel hub with a 4-in. bore with an interference of 0.003 in. requires $0.00075/0.001 \times 160 = 120°F$. Therefore, if the shaft temperature is 80°F, the hub temperature must be 200°F. This does not account for human factors such as cooling due to handling time, errors in measurements, and so on. As a general rule, add 50 to 75°F to the calculated expansion temperatures to account for these factors. The hub should be heated in an oil bath or an oven; a torch or open flame should not be used. This could cause localized distortion or softening of the hub material. It could also cause an explosion in some atmospheres. Oil bath heating is usually limited to approximately 350°F, or under the flash point of the oil used. Special handling devices are required: tongs, threaded rods placed in taped holes in the hub, and so on. Oven heating offers some advantages over oil. Parts can be heated to higher temperatures (usually not exceeding 600°F) and handled with heat-resistant gloves. In any event, handling heated hubs requires extreme care to avoid injury to personnel.

It is also important when mounting interference hubs to make sure that clearance exists over the top of keys; otherwise, when the hub cools, it will rest on the key and produce high stresses in the hub that could cause it to fail.

Straight Shafts with Intermediate Bushings. Intermediate bushings come in two basic configurations: with and without flanges. To assemble, insert the bushing into the hub without tightening the screws or bolts; then slide the hub and bushing onto the shaft. Since the bushing is tapered, tighten the screws or the shaft. Once the hub is at the correct position, the screws should be tightened gradually in a crisscross pattern to the specific torque. Bolts are tightened on the wheel of an automobile. Refer to the specific instructions for further recommendations and the correct torquing value.

Taper Shafts. Tapered shafts have the advantage that the interference between the hub and the shaft can be accomplished by advancing the hub on the shaft. Depending on the amount of interference, the hub may be drawn up with nuts or heating. Removal of the hub is usually easier on tapered shafts than on straight shafts.

Applications using tapered bores require more attention than those using straight shafts because it is easier to machine two cylindrical surfaces that match than two tapered surfaces. The hub can be overstressed if it is advanced too far on the shaft. Dirt and surface imperfections can restrain the hub advance and give the false impression that the desired interference has been reached.

To determine the draw-up required to obtain the desired interference, use the following equation:

$$\text{Draw-up (in.)} = 12 \times \frac{i}{T}$$

where

  i = diametral interference (in.)

  T = taper (in./ft)

The area of contact between the bore of the hub and the shaft should be checked with machinist's bluing. Fifty to 80% contact is the range of acceptability; usually, 70% is the most desirable. If less than the required contact is achieved, the contact can be increased by lapping the bore and/or the shaft with a plug or ring made from a master plug and ring gage. It is not recommended that the master gages or shaft be used to lap the hub, as the gages could end up with ridges. Ridges in the hub or shaft will prevent proper hub installation and could cause the hub or shaft to fail because of stress concentrations.

1. *Light interference* (under 0.0005 in./in.): When the interference is less than 0.0005 in./in., the hub can usually be advanced without heating. Although heating the hub is the most common method, the hub can usually be advanced by tightening the retaining nut or plate on the shaft. It is also common practice when light interference is used with a combination of keys and a retaining nut or plate to use a light grease or antiseize compound between the hub, shaft, and threads on the shaft and nut. This should help facilitate installation and future removal and help prevent shaft and/or bore gauling.
2. *Medium interference* (usually 0.0005 to 0.0015 in./in.): When the interference is over 0.0005 in./in., the force required to advance the hub could become too large for manual assembly. When this occurs, the hub *must* be heat-mounted or hydraulically mounted. Heating hubs for mounting is the most common method. Regardless of the method used, the amount of draw-up must be measured.
3. *Heavy interference* (usually over 0.0015 in./in.): When the interference is over 0.0015 in./in., hubs are usually heat-mounted and removed hydraulically. Some users prefer to both mount and remove hydraulically.

The following is recommended as a general guide when installing hubs on an equipment shaft:

1. Install the hub on the shaft, ensuring that the parts mate properly and are burr-free and clean. Using a depth gage or dial indicator (see Fig. 40(A)), measure and record the initial reading.
2. Remove the hub and lubricate the bore or shaft if hydraulic assist is to be used; if not, heat in oil or an oven. When using a heating method for mounting hubs, it is best to provide a positive step, such as a clamp on the shaft, to ensure proper draw-up (see Fig. 40(B)). The reason for this is that a hub advanced too far may not be removable (too much force

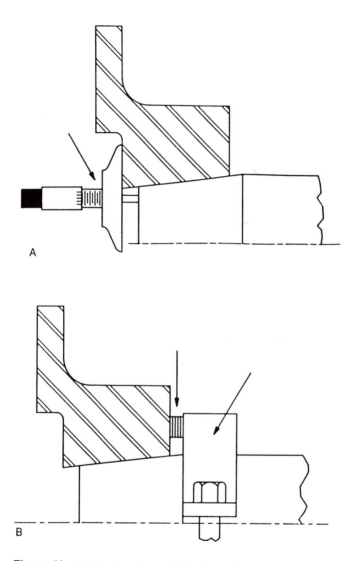

Figure 40   Positioning hubs: (A) hub position measurement with depth gage; (B) positive stop used when mounting a hub.

required or not enough hydraulic pressure available to remove the hub) and normally requires that the hubs be cut off.
3.   The hub is installed and advanced the required amount.
4.   The shaft nut is then properly tightened and locked in place. Locking is done with a tab washer or setscrew.

Rough-Bored Couplings. Most coupling manufacturers will supply couplings with rough bores. This is usually done with a spare coupling so that as its equipment shafts are remachined, the spare coupling can be properly fitted. This type of coupling also helps to reduce inventory requirements.

It is important that the user properly bore and key these couplings; otherwise, the interface torque-transmission capabilities can be reduced or the coupling balance (or unbalance) can be upset. Recommendations should be obtained from the specific coupling manufacturer. As a general guide, the hub must be placed in a lathe so that it is perpendicular and concentric to its controlling diameters. On rigid hubs the pilot and face are usually the controlling diameter and surface that should be used to bore (see Fig. 41A). On flex hubs (gear and chain) the gear major diameter (OD) and hub face act as the controlling diameter and surface. (Note: Some manufacturers use the hub barrel as the controlling diameter (see Fig. 41B).)

Some coupling manufacturers supply semifinished bore couplings. In this case, the finished bore should be machined using the semifinished bore as the controlling diameter. Indicate the bore-in, for concentricity and straightness.

Most couplings must have one or two keyways cut in the hub. Particular attention should be given to

1. Keyway offset (the centerline of the keyway must not intersect with the centerline of bore)
2. Keyway parallelism
3. Keyway lead
4. Keyway width and height

Keys. The fitting of keys is important to ensure the proper capacity of the interface. Refer to AGMA standard 9002 on keyways and keys. As a general rule, three fits must be checked:

1. The key should fit tightly in the shaft keyways.
2. The key should have a sliding fit (but not be too loose) in the hub keyway.
3. The key should have a clearance fit radial with the hub keyway at the top of the key (see Fig. 42A).

The key should be chamfered so that it fits the keyway without riding on the keyway radii (see Fig. 42B). A sloppily fitted key can cause the keys to roll or shear when loaded. The results of a sloppy key fit are shown in Fig. 42C. The forces generated by torque are at distance S, and this movement tends to roll the key and can cause very high loading at the key edges. On the other hand, too tight a fit will make assembly very difficult and increase the residual stresses, which could cause premature failure of the hub and/or shaft.

A key in the keyway that is too high could cause the hub to split (see Fig. 42D). When there is too much clearance at the top or sides of a key, a path is

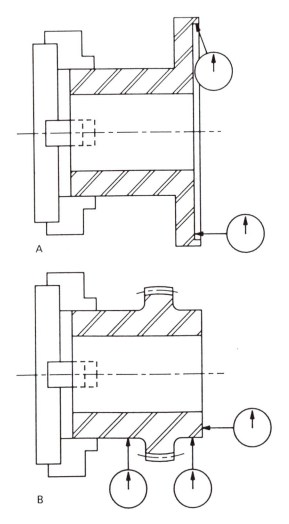

Figure 41 Setup for reboring hubs: (A) rigid hub setup; (B) flex hub setup.

provided for lubricant to squeeze out. For lubricated couplings, clearances between keys and keyways must be sealed to prevent loss of lubricant and thus starvation of the coupling.

C. Alignment of Equipment

Once the hubs have been assembled, the next step is usually to align the equipment, although some people prefer to assemble the coupling and then check

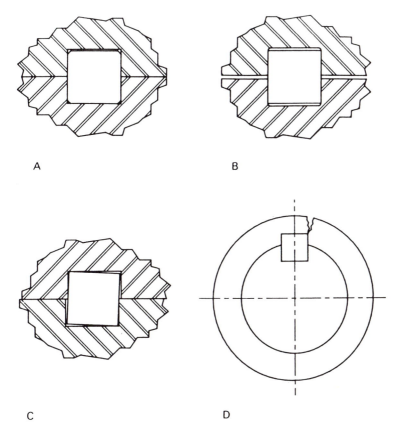

A   B

C   D

Figure 42   Improperly fitted keys: (A) proper key fit; (B) improper fit of key with no chamfers; (C) rolling of a sloppy key; (D) a split hub from an improperly fitted key.

and align the equipment with the coupling fully assembled. The danger here is that if the alignment is too far off, the static misalignment imposed on the coupling while assembling it may cause damage to the coupling, which could result in a premature failure. Therefore, it is suggested that at a minimum, the equipment should be rough-aligned before the coupling is installed.

For a full discussion of types of alignment and the steps required to align equipment, see John Piotrowski, *Shaft Alignment Handbook* (Marcel Dekker, 1986). Alignment tools vary from the simple straightedge to very sophisticated laser methods.

You may be wondering why we need to worry about alignment since flexible couplings are designed to allow some flexibility in alignment. One important reason is that the coupling's useful life depends on the amount of misalignment. If it were possible to align equipment perfectly and we could ensure that it would stay aligned, a flexible coupling would not be required. A second reason for accurate initial alignment is that alignment changes over time because of bearing wear, foundation settling, and thermal changes. When alignment is carefully done initially, generally to within one-eighth to one-third of the coupling's rated misalignment capacity, there should be enough capacity left in the coupling for it to handle foreseeable increases in misalignment with time.

Two rotating shafts can have the combinations of position shown in Fig. 43, although conditions A and C are unlikely to occur in real life. A flexible cou-

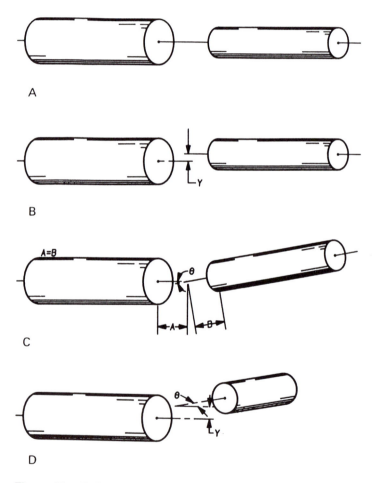

Figure 43   Shaft alignment conditions: (A) alignment; (B) parallel offset misalignment; (C) symmetrical angular misalignment; (D) combined angular and offset misalignment.

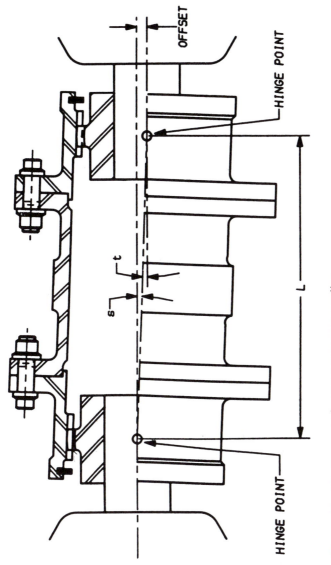

Figure 44 Misalignment of a spacer-type gear coupling.

pling is usually employed as a double flex (except for elastomeric couplings). A gear-type coupling incorporating a spacer is shown in Fig. 44. Each coupling end (flex element) will accommodate only angular misalignment α_1 at one end, and α_2 at the other; a double flex coupling can accommodate $\alpha = \alpha_1 + \alpha_2$ angular. It is important to remember that *there must be two flex elements* to accommodate any offset misalignment (this does not apply to elastomeric couplings). The amount of offset misalignment a double flex coupling can accommodate is determined as follows:

Offset $(S) = L \tan \alpha$

where

α = angular misalignment, deg

L = center distance between flex points

Example 1.   If a gear coupling is rated at $\alpha_1 = 1\ 1/2°$ per mesh and the distance between flex elements (flex points) is 6 in., the amount of offset capacity is

$6 \tan 1.5° = 0.157$ in.

A flexible coupling can accommodate more offset if it is used in conjunction with a spacer: the longer the spacer, the larger the distance between flex points and the greater the allowable offset misalignment allowed.

D.   Coupling Assembly

Once the hubs are installed on the shafts and the equipment is aligned, the coupling can be assembled. Each coupling type is assembled differently; however, many couplings have similar procedures. The following discussion covers some general procedures for several of the most common types available. For more specific instructions and guidelines, consult the manufacturer's instructions, which are usually shipped with the coupling.

1.   Lubricated Couplings

There are four types of lubricated couplings: (1) gear type (Fig. 11A, B), (2) chain type (Fig. 18A), and (3) steel grid type (Fig. 13A, B).

The gear, chain, and steel grid are similar and are covered together. The first step is the application of a coating of lubricant to all moving parts, such as gear teeth and chains or steel grids, as well as to the O-rings and their mating surfaces. Then the coupling can be closed by bringing the sleeves or the covers together. Most lubricated couplings use gaskets or O-rings between the flanges so that the lubricant is sealed in. The importance of properly sealing a coupling cannot be overemphasized, because the amount of lubricant contained in a coupling is small. Even the smallest leak would soon deplete the coupling of lubricant, resulting in rapid wear and ultimate failure.

To ensure proper sealing, a few rules should be followed: The flange surfaces must be clean, flat, and free of burrs or nicks. Any imperfections on the surfaces of the flange must be removed by using a honing stone or fine-grit sandpaper.

The gaskets are made of special material, as oil can seep through regular cardboard or paper. The gasket should be in one piece and be free of tears or folds. To facilitate assemble, some people glue the gasket to one of the flanges with grease. This procedure should be avoided, as it can prevent proper tightening of the bolts. It is suggested that a thin oil be applied to the gasket, which will help hold it in place.

Once the flanges are brought together, the bolts are inserted in the holes. The bolts must be assembled from the flange side, shown on the assembly drawing. The holes in the flanges must be oversized or countersunk to accept the bolt-head radii. Care should be taken in aligning the holes in the gaskets with those in the flanges. If this is not done, the gasket is torn by the bolts and pieces of material can stick between the flanges, preventing proper sealing.

Once all the bolts are in place, slide on the lockwashers and install the nuts. If self-locking nuts are used, lockwashers are not needed. If room permits, the nuts (not the bolts) should be tightened to specifications, which means that a torque wrench should be used. The nuts should be tightened in two or more steps in a crisscross fashion, similar to the way automobile lugs are tightened.

If the sleeves or covers incorporate lube plugs, the plugs on the two sleeves should be diametrically opposite. If the coupling was dynamically balanced, match marks may indicate the relative position not only between the two sleeves (or covers) but also between the sleeves and the hubs, unless the coupling was component-balanced; then alignment may not be required.

Care should be exercised not to mix parts of similar couplings unless the coupling manufacturer's instructions say that it is acceptable to do so. Particular attention should be paid to balanced couplings and special couplings, where parts mixing may affect not only balance but the functional operating characteristics of the coupling as well.

2. Metallic Element Couplings

Flexible disk couplings come in two basic configurations:

1. *Low speed* (Fig. 16A): The disks are usually taped or wired together. The disks are alternately fastened to the hub and spacer by bolts, nuts, and sometimes stand-off spacers. If stand-off washers are used, they must be assembled with the correct surface against the disk (usually curved). Sometimes, washers of different thicknesses are used. The washers must be placed in the correct position. See the specific instructions.
2. *High speed* (Fig. 16C): The disks are usually assembled as a pack, and the pack is bolted alternately to the hub and spacer.

Diaphragm couplings come in two basic configurations:

1. *Flex spacer section* (Fig. 17A): This coupling just drops out between the two installed rigids. The center section is one piece: a subassembly or welded assembly. The center section should not be disassembled unless so instructed by the coupling manufacturer. Some couplings have pilot rigids, which requires that the center section be compressed for assembly.

2. *Flex subassembly* (Fig. 17C): On this type of diaphragm coupling the center of the diaphragm coupling section consists of three pieces or assemblies. This construction is common where equipment is separated by a long distance, as it helps to reduce the cost of spares.

3. Elastomeric Element Couplings

The rubber donut coupling (Fig. 15D) is assembled with radial bolts, and, in the process of tightening the bolts, the rubber "legs" are precompressed. Precompression should be uniform around the coupling. The split insert should be the first to reach the final "seated" position in the hub's notch.

Rubber tire couplings (Fig. 14A) are similar to donut couplings. They have a split element which is wrapped around the hubs. However, the fasteners are arranged axially, and they clamp the side walls of the tire between two plates.

Pin-and-bushing couplings, as well as jaw-type couplings (Fig. 15A, F), have an elastomeric element trapped between the hubs. To install or remove the element, usually at least one of the hubs slides on the shaft; if not, the equipment must slide axially out of the way. The shaft separation must be larger than the thickness of the element to allow for installation; however, some elements are split so that close shafts can be used.

Most elastomer-type couplings can be obtained in a "spacer" configuration in which the two hubs and the flexing element can be "dropped out" as a unit. This unit is attached to the two shafts through two "rigid" hubs with axial bolts. To install the couplings between the rigid hubs, the element has to be squeezed so that it will fit between the rabbets of the rigid hubs.

4. Bolt Tightening (Flange Connections)

Most instruction sheets provided with couplings give information on bolt tightening. Unfortunately, these instructions are not always followed; many mechanics tighten the bolts by feel.

Couplings resist misalignment, and the resulting "forces and moment" put a strain on the equipment and connecting fasteners. If the fasteners are loose, they are subjected to alternating forces and may fail through fatigue. A bolt that is not properly tightened can become loose after a short period of coupling operation.

Few bolts work only in tension. Most coupling bolts also work in shear, which is caused by the torque transmission. Usually only some of the torque is transmitted through bolt shear; part of the load is transmitted through the friction between the flanges. Depending on the coupling design, as much as 100% of the torque can be transmitted through friction. If the bolts are not tightened properly, there is less clamping force, less friction, and more of the torque is transmitted through shear. Because of the combined shear and tensile stresses in bolts, recommendations for bolt tightening vary from coupling to coupling. Coupling manufacturers usually calculate bolt stresses, and their tightening recommendations should always be followed. If recommendations are not available, it is strongly suggested that a value be obtained from the coupling manufacturer rather than by guessing. *Find out what the specific coupling requires.*

Bolts should be tightened to the recommended specification in at least three steps. First, all bolts should be tightened to one-half to three-fourths of the final value in a crisscross fashion. Next, they should be tightened to specifications. Finally, the first bolt tightened to the final value should be checked again after all the bolts are tightened. If more tightening is required, all the bolts should be rechecked. Also, the higher the strength of the bolt, the more steps that should be taken:

Grade 2: two or three steps
Grade 5: three or four steps
Grade 8: four to six or more steps

If an original bolt is lost, a commercial bolt that looks similar to the other coupling bolts should not be substituted. It is best to call the coupling manufacturer for another bolt, or they may suggest an alternative.

If room permits, always tighten the nut, not the bolt. This is because part of the tightening torque is needed to overcome friction. The longer the bolt, the more important it is to tighten the nut rather than the bolt. As there is additional friction when turning the bolt, more of the effort goes into friction than in stretching the bolt.

Some couplings use lockwashers; others use locknuts. Whereas a nut-lockwasher combination can usually be used many times, a locknut loses some of its locking properties every time it is removed from the bolt. If not instructed otherwise, it is best to replace nuts after five or six installations. Some couplings use aircraft-quality bolts and nuts, which can generally be used 10 to 12 times before they lose their locking features.

E. Lubrication of Couplings

Since a large number of flexible couplings require lubrication, a separate section is devoted to lubrication. Couplings classified as mechanical element, namely gear, chain, and grid couplings, require lubrication. Although different, these types of flexible couplings have a very similar mechanism of lubrication.

The gear coupling is generally used as an example in this section; however, most discussions apply to all three types of couplings.

The gear coupling has four to five major parts: two hubs, two sleeves, and in many applications a spacer. It also has bolts and nuts and must have some type of seal to keep the lubricant in.

Each gear mesh acts like a spline. Some gear teeth are straight, both internal and external, and some couplings have straight internal and external teeth with crowns. The mesh has clearance (backlash). Misalignment results from sliding or rolling these loosely fitted parts. Because of this relative motion, the need for lubrication is clear. In some flexible couplings, motion is only a few thousandths of an inch, but in some couplings it may be as much as 4 to 6 in. From this range of motion a coupling can experience sliding velocities from less than 1 in. per second (ips) to as much as 200 ips.

It is important to remember the effect on lubricants of centrifugal forces. In a gear coupling the lubricant is trapped in the sleeve, which at high speeds acts as a perfect centrifuge. It tends to separate the grease and, for continuously lubricated couplings, it will separate foreign particles out of the lubricant, which tend to build up in the coupling (sludge). Chain and grid couplings ar much like gear couplings, except that the "sleeve" in one case is the chain and cover and in the other case is the grid and cover.

1. Types of Seals

For lubricated couplings the seals are almost as important as the lubricant itself. Almost as many coupling failures can be attributed to seal failure as to lube failure.

There are two basic types of seals used to retain lubricants: the metallic labyrinth seal and the elastomeric seal. Again we will use the gear coupling to discuss specifics, but most of these sealing devices are incorporated in the chain and grid couplings.

Metallic Labyrinth Seals.   When gear couplings were introduced about 75 years ago, synthetic positive seals, which can withstand repeated flexing motion between the hub and the sleeve and the deterioration caused by the lubricant, were not available. Thus, as shown in Fig. 45(A), the all-metal labyrinth seal plate with its center of contact in line with the center of the gear flex point was used. Metallic seals usually require clearance and are not considered to be a "positive" seal. Lubricant can leak out of a coupling if it is stopped or is constantly being reversed. Also, this seal does not adequately protect the coupling from contaminants which could cause the coupling or its lubricant to deteriorate. This type of seal is still used very successfully in many applications.

Elastomeric Seals.   The four basic elastomeric seals used in couplings are the following:

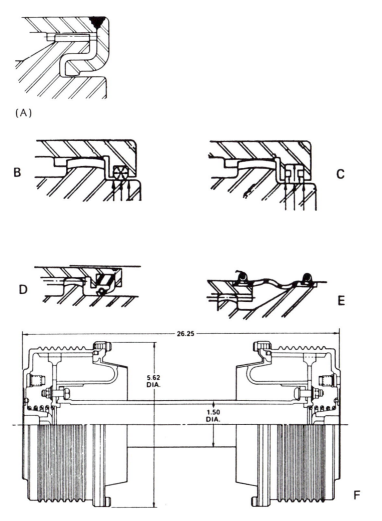

Figure 45 Types of seals: (A) metal labyrinth seal; (B) O-ring; (C) H or T cross-section seal; (D) lip seal; (E) boot seal; (F) high-speed boot seal.

1. O-ring
2. H or T cross-section seal
3. Lip seal
4. Boot seal

The O-ring seal shown in Fig. 45(B) was developed by the Air Force during World War II and is now very common in mechanical equipment. Most couplings requiring lubrication use O-rings, which are one of the most positive means of retaining lubricant. The major disadvantage with O-rings is that they usually allow relatively small amounts of motion between parts. To account for the oscillatory motion, seals can only be squeezed through approximately 10% of their cross section before they start to take a set and no longer function.

The H or T cross-section seal (Fig. 45(C)) functions and operates very similarly to an O-ring but has more surface area in contact with the sealing surface and for relatively small motion will usually do a slightly better job. For many applications that require moderate amounts of oscillatory motion together with axial movement, the H or T cross-section seal tends to roll and twist. When this occurs, a spiral path is created through which the lubrication can exit the coupling.

The lip seal (see Fig. 45(D)) is another very common seal used with couplings. This seal can be designed with "extended" lips that can accommodate high eccentricities and thus a large degree of misalignment. The lip seal can also be designed with spring retainers and steel inserts to prevent the lips from lifting off at high speed.

The boot seal (see Fig. 45(E)) is not in common use, but it is a very effective means of retaining lubricant. This type of seal is usually limited to operate at low speed, but with a special fabric reinforcement material (see Fig. 45(F)) it has been used on couplings that operate at $7\,1/2°$ of misalignment and 6000 rpm.

2. Methods and Practices of Lubrication

Grease Lubrication. There are two methods for lubricating flexible couplings with grease: lubrication before closing coupling (pack lubrication) or lubrication after closing coupling. To pack-lubricate a coupling, an appropriate quantity of grease is applied manually to each half coupling, making sure that the teeth and slots are coated. Then the coupling is bolted up or assembled. If couplings are supplied with lube plugs, before the coupling is assembled the working surfaces are wiped and a light coating of lubricant, then assembled. Usually, couplings are provided with two lube plugs in their periphery they should be removed before being filled. Only the amount specified by the coupling manufacturer should be used, as too much lubricant could cause seals to be damaged or the coupling to bind in operation. After filling, replace the lube plugs.

Periodic maintenance of grease-lubricated couplings is necessary if the coupling is to give satisfactory service. A coupling should be relubed at regular intervals, usually at least every 6 to 12 months. The coupling manufacturer's suggestions should always be followed. Most lubrication cycles are established by experience, so records should be kept of relubrication history.

It is also suggested that the couplings be disassembled and cleaned at regular intervals to get rid of foreign materials and to check for wear and deterioration.

Parts should be washed, dried, and inspected. If parts appear to be worn, they should be replaced, relubricated, and reassembled.

Note: When disassembling and relubrication a coupling, it should be inspected for the condition of the lubricant. If the grease is soapy or separated into oil and soap, it is evident that the lubricant previously used is not suitable for this application. It would be wise to contact the coupling manufacturer and obtain a better lubricant recommendation.

Oil Lubrication.

Continuous Flow. In this method oil is injected into the coupling by an oil jet(s) directed toward the collecting lips (fixed or removable) or rings. Oil may be injected at one end of the coupling and exit at the other, or each end is lubricated separately. Oil collectors may be provided on the hubs, sleeves, or spacer, depending on the coupling design. Oil jets are made from tubes that have been capped and drilled to provide the correct quantity of oil and exit velocity based on supply oil pressure and the size of the supply line. The oil flow for a given type of design depends on function coupling size, transmitted horsepower, speed, and possible coupling misalignment. Figure 46 shows a

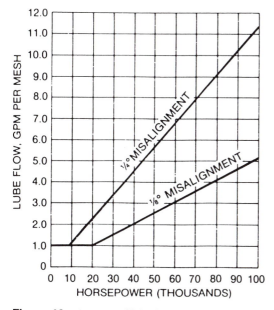

COUPLING HORSEPOWER
VS. LUBE FLOW

Figure 46 Amount of lube flow required for gear couplings.

curve for a high-speed gear coupling which gives the required lube flow as a function of horsepower and misalignment. Such curves differ for various types of couplings and from one manufacturer to another. Use whatever is recommended by the coupling manufacturer.

Several things influence required lubricant flow, such as material hardness or whether the coupling incorporates positive dams or no dams. Damless couplings usually require much more oil. Usually, continuously lubricated couplings use the same oil as does the connected machinery bearings. A sufficient quantity of clean oil (filtered and not contaminated with water or corrosive media) must be supplied to the coupling. We suggest that a separate filtration be used, usually 5 microns or smaller, and that the oil properties and contamination be monitored and a centrifuge used.

Sludge accumulation (Fig. 47) may be detrimental to equipment and the coupling. It can produce high moments and forces on the equipment and the coupling which could result in catastrophic failure. Sludge may affect a coupling in several ways: it can reduce axial movement between coupling parts, tends to accelerate wear, stop the flow of oil, causing the coupling to operate without sufficient lubricant, and in contaminated oil systems cause or

Figure 47   Sludge-contaminated gear coupling.

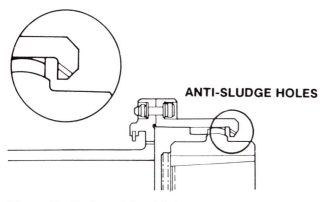

Figure 48 Design to help minimize sludge buildup: flow-through design.

encourage corrosion. Sludge can be controlled by filtration to remove foreign particles, and to prevent lube contamination (water and corrosive media) the use of separators and constant sampling of the lube is required, and/or the coupling housing system can be sealed from contaminated atmospheres.

Most couplings are designed to minimize sludge (see Fig. 48). Many people believe that the solution to sludge accumulation is to remove the coupling dams. Unless you can drastically increase the coupling lubricant flow and ensure a reliable lube source, the problems associated with damless couplings usually far exceed the possible benefits. Figure 49 shows what happens to the oil in a damless coupling. Unless the lubricant flow is substantially increased, most of the working portion of the teeth will be starved. If couplings are prop-

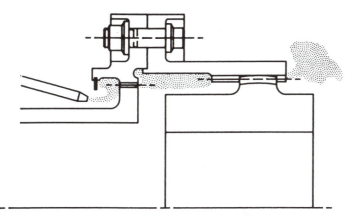

Figure 49 Oil flow in a damless gear coupling.

erly maintained, the sludging problem can be minimized. At regular intervals and every time the equipment is stopped the coupling should be opened and inspected for sludge accumulation. The sludge should be removed mechanically and/or chemically and the parts thoroughly cleaned and inspected for damage (tooth distress, corrosion, breakage, etc.).

Confined Lubrication. Whenever a coupling is confined in a housing together with other lubricated components, such as gears, it may use the oil from the housing for its lubrication. When the coupling operates above the oil surfaces, lubrication is provided either by an oil pump or through splashing. In either case the oil flow to the coupling is a function of operating speed. If the working surfaces of a coupling are partially or totally submerged in oil, the lubrication requirements are satisfied by the flow of oil from the housing into the coupling. To ensure adequate lubrication, submerged couplings are provided with holes in their covers or sleeves, or the coupling may incorporate pickup scoops or other arrangements to aid in ensuring adequate lubrication. Because the centrifugal action of these types of couplings tends to retain dirt from the system, it is advisable to clean the coupling when the equipment is down or during scheduled periodic cleaning intervals.

Self-Contained Lubrication. This type of coupling is supplied with seals and can be filled at assembly and operated for an extended period without further attention. The usual method of filling this type of coupling is to remove the lube plugs and pour the correct amount of recommended oil into the coupling. It is recommended that the working portions of the coupling be coated with lubricant before assembly. This will prevent bare metal contact between mating parts at assembly and at start-up before oil properly distributes. Maintenance of an oil-filled coupling consists basically of preventing the coupling from losing oil. It is therefore important to maintain flange surfaces, seals, and gaskets in good condition. Lube plugs should be properly seated (or sealed) so that no leakage occurs. A regular lubrication schedule should be established and maintained following all of the manufacturer's assembly, lubrication, maintenance, and cleaning instructions.

3. Types of Lubrication

Couplings are usually lubricated with oil or grease. Grease-lubricated couplings are usually lubricated by one of two methods. They are either self-lubricated or are lubricated from an external supply. For the self-contained method, oil or grease can be used and covers and/or seals are required. For the externally lubricated coupling, the lubricant is oil. The oil is supplied with a specific flow rate, or it may be dipped or intermittently lubricated.

Which type of lubrication is best? sealed lube or continuous? Gear, chain, and grid couplings may be filled with a specific quantity of lubricant and sealed. Many factors govern which method should be chosen, and each coupling application must be evaluated with respect to its requirements. Some of the factors that should be considered are listed next.

Sealed lubrication:

1. Affords the opportunity to choose the best lubrication available.
2. Outside contaminants are effectively prohibited from entering working surfaces.
3. Case design is simplified—not required to be an oiltight enclosure.
4. Couplings may operate for extended periods of time without servicing except for normal leakage checks.
5. Seals are required in connected equipment.
6. Seals tend to wear and age, and replacement is not always easy.
7. High ambient temperatures may adversely affect the lubricant.

Continuous lubrication:

1. Permits continuous operation but requires periodic sludge removal.
2. Removes generated heat effectively and increases coupling life significantly in applications subject to high ambient temperatures
3. May eliminate the need for costly seals in connected equipment.
4. Eliminates seals in the coupling itself.
5. Does not permit the choice of the best lubricating oils.
6. Requires oil supply filtration to 5 to 10 μm or less absolute particle size.
7. Requires an oiltight case.

4. Grease Lubrication

Because gear, chain, and grid couplings are similar as to how they accommodate misalignment and transmi torque, a grease that works satisfactorily in one should be good for the others. Various manufacturers have many recommendations for lubricants to use in their couplings, and it is very difficult to find a common denominator. Many tests and studies have been undertaken for gear couplings, and the conclusion should also apply to chain and grid couplings. The most important discovery of many of these studies is that the wear rate of a coupling is greatly influenced by the viscosity of the base oil of the grease: the higher the viscosity, the lower the wear rate.

One study shows that centrifugal force is not only important in helping to lubricate a coupling but can cause grease to deteriorate and create serious problems. The sliding parts of a lubricated coupling would run dry if it were not for the centrifugal forces created by the rotation of a coupling. Centrifugal force is generated by the coupling's rotation and is a function of the coupling's diameter and the square of the rotational speed. Centrifugal force generates pressure in the lubricant. This pressure helps the coupling in two ways. First, it forces the lubricant to assume an annular form, flooding the coupling's teeth (or chain or steel grid); and second, it forces the lubricant to fill all the voids rapidly.

Figure 50 can be used to calculate the magnitude of the centrifugal forces in a coupling. It can be seen that even in a coupling operating at motor speeds, the centrifugal force can exceed $500g$. Because the centrifugal force is a function of the square of the rotational speed and the frequency of the hub tooth oscillatory motion increases directly proportional to the speed, couplings get better lubrication when operating at very low speeds. The reverse is true when a coupling operates at very low speeds. The centrifugal force decreases rapidly and below a given level can no longer force a thick lubricant, such as grease, between the coupling teeth, with rapid wear as a result.

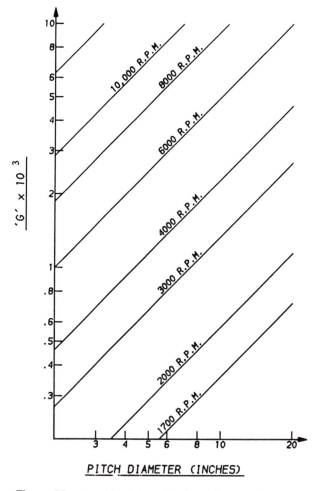

Figure 50   Centrifugal force developed in a coupling.

Lubricant selection depends on the type of coupling and particularly on the application. Even the best lubricant for one application can cause rapid wear if used in another application. This fact cannot be overemphasized. A coupling user who lubricates all the couplings in the plant with whatever grease is available cannot expect good performance from the couplings. In reality, a substantial reduction in maintenance costs can be realized through the use of proper lubricants.

The best approach is to follow the coupling manufacturer's recommendations. If lubricant recommendations are not available, the following can be used as a guide:

Steel grip coupling manufacturers generally recommended an NLGI no. 1 to 3 grease.

Gear and chain coupling manufacturers recommend an NLGI no. 0 to 3 grease.

Normal applications can be defined as those where the centrifugal force does not exceed 200*g*. In general, these couplings can operate at motor speeds up to 3600 rpm. Also, normal applications are those where the misalignment at each hub is less than 3/4° and where the peak torque is less than 2 1/2 times the continuous torque. For these applications, an NLGI no. 2 grease with a high-viscosity base oil (preferably higher than 198 centistokes at 40°C) should be used.

Low-speed applications can be defined as those where the centrifugal force is lower than 10*g*. If the pitch diameter (*d*) is not known, the following formula can be used:

$$\text{rpm} \leqslant \frac{200}{\sqrt{d}}$$

The same conditions for misalignment and shock torque as already mentioned are valid. For these applications, an NLGI no. 0 or no. 1 grease with a high-viscosity base oil (preferably higher than 198 centistokes at 40°C) should be used.

High-speed applications can be defined as those where the centrifugal force is higher than 200*g*, the misalignment at each hub is less than 1/2°, and the torque transmitted is fairly uniform. For these applications, the lubricant should have a very good resistance to centrifugal separation. This information is seldom published and can be obtained from either the lubricant or the coupling manufacturer.

High-torque, high-misalignment applications can be defined as those where the centrifugal forces are lower than 200*g*, the misalignment is larger than 3/4°, and the shock loads exceed 2.5 times the continuous torque. Many such applications also have high ambient temperatures (e.g., 100°C), at which only a few greases can perform satisfactorily. Besides the characteristics of a grease for 'normal applications," the grease should have antifriction and antiwear additives (such as polydisulfide), extreme pressure (EP) additives, a Timken load larger than 40 lb, and a minimum dropping point of 150°C.

5. Oil Lubrication

Oil is seldom used as a lubricant for chain and grid couplings. For gear couplings, oil lubrication is very common. The coupling can be oil filled or continuous, dipped or submerged. For oil-filled couplings the oil used should be of a high-viscosity grade, not less than 150 SSU at 100°C: the higher the viscosity, the better. For high-speed applications an oil viscosity of 2100 to 3600 SSU at 100°F can be used very successfully. Generally, an oil that conforms to MIL-L-2105 grade 140 is acceptable. The only problems with an oil of such high viscosity is that it is very difficult to pour and fill a coupling, but the time it takes is usually well worth the good operating characteristics obtained.

Continuously lubricated couplings use the oil from the system. This lubricant is seldom a high-viscosity oil. Because of this, it is best to cool the lubricant before it enters the coupling. This will help supply this lubricant to the coupling at its highest lubricity (viscosity). The viscosity should be a minimum of 50 SUS at 100°C. An oil that conforms to MIL-L-17331 is usually satisfactory.

F. Coupling Disassembly

1. Opening the Coupling

A visual inspection of couplings before opening them can sometimes help in planning the required maintenance procedure. For example, loose or missing bolts indicate poor installation practices; discolored (usually blue) coupling parts indicate that the coupling probably ran dry and lubrication procedures should be reviewed. If sleeves, covers, or seal carriers have to be slid off the hubs, the exposed hub surface should be cleaned with emery paper to remove rust or corrosion. Elastomer parts should be examined for wear, which indicates that they came in contact with stationary components, for localized melting or permanent set, and for signs of hardening or chemical attack. It is important to observe the state of a coupling even if it is damaged beyond repair. Understanding how or why a coupling failed can help to prevent similar occurrences in the future.

Before opening a coupling it is advisable to have on hand a container to store the parts. It is important to remember that most couplings use "special bolts" which should not be replaced with bolts of lesser quality, different grips, or different thread length. The coupling should be checked to see if there are match marks, particularly for some balanced couplings. If not and the coupling is to be reused, match marks should be scribed *before* the coupling is opened. Because many couplings wear in, it is best to ensure that match marks be made even if the coupling was not balanced, so that the coupling parts can be reassembled in the same relative position they were in when the coupling was running. To shorten downtime, it is also best to have on hand all the parts that might need replacing, such as gaskets, O-rings, lubricant, and possibly a spare flex element or an entire coupling.

After visual inspection, the bolts should be removed. Make sure that wrenches of the proper size are used; this is very important. Otherwise, the heads of the bolts or the nuts could be damaged. Do not use locking pliers or adjustable wrenches. If lockwashers are used, the bolt should be loosened rather than the nuts. This method should keep the lockwasher from digging into the flange and nut surface.

Once the bolts have been removed, one would expect the coupling to open up easily, but this is seldom the case. Parts usually stick together, particularly when gaskets are used. To separate two flanges it is best to use the jacking holes (if provided). Tapping the parts with a soft-face hammer can also help in separating parts. Prying the parts apart by hammering, chiseling, or by prying with a screwdriver is not recommended, as it will damage the surfaces of the flanges and the surfaces become difficult to seal when reassembled.

Be prepared for a mess, as some of the lubricant (if it is a lubricated coupling) will leak out of the coupling when it is opened. All the lubricant should be removed from the coupling and discarded, as lubrication properties may deteriorate when a coupling is opened.

When the connected machines are removed from their location or when couplings are left disconnected for long periods, it is best to bolt a piece of cardboard or plywood to the half coupling. There are two advantages to this procedure: the coupling is protected from dirt and contamination, and the coupling bolts that are used to hold the sleeves or cover will not get lost.

2. Removing the Hubs from the Shafts

Most installed hubs need some force to be removed. The hubs can be removed by application of a continuous force or through impact. Hammering will always damage the coupling, which would be all right only if the coupling is to be replaced. However, hammering is not recommended because it can damage the bearing in the equipment and even bend shafts.

The first thing to do before removing the hub is to remove the locking means, such as tapered keys, setscrews, or intermediate bushings, on the shaft or nut. Make sure that all retaining rings, bolts, setscrews, and so on, are loosened.

The most common way to remove the hubs is by using the puller holes (Fig. 51). If the hubs have no puller holes, two holes can be drilled and tapped before the hub is installed on the shaft. It is best to buy them equal to one-half the hub wall thickness to ensure proper sizing. To facilitate removal, use oil to lubricate the threads and use penetrating oil between the hub and the shaft. For

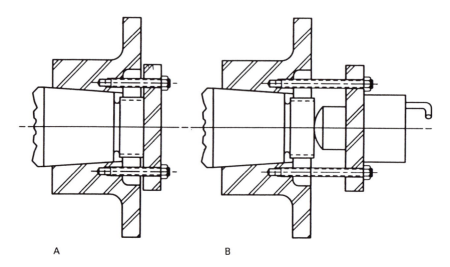

A   B

Figure 51   Methods of removing bulbs: (A) using puller holes; (B) using a hydraulic press.

straight shafts use a spacer between the shaft end and the flat bar, as force has to be applied for as long as the hub slides on the shaft. If there are no puller holes, a "wheel puller" can be used; however, care should be taken not to damage the coupling. Figure 52 illustrates some don'ts for this procedure.

If the hub does not move even when maximum force is applied, apply a blow with a soft-face hammer. If nothing happens, heat must be applied to the hub. It is important to use a low-temperature flame (not a welding torch) and to apply the heat uniformly around the hub (localized temperature should not exceed 600°F for most steel). Heat does not have the same effect in this case as it has at installation, when only the hub is heated. At removal the hub on the shaft and the heat will also expand the shaft. The secret is to heat only the out-side diameter of the hub, shielding the shaft from the heat. If possible, the shaft should be wrapped with wet rags. *The heat should be applied while the pulling force is being applied.*

Caution: If heat has been applied to remove the hub, hardness checks should be made at various sections before using the hub again, to ensure that softening of the material has not occurred. If there is any doubt, the coupling should be replaced. Rubber components should be removed during heating. If not removed, they will have to be replaced if they come into contact with excessive heat.

The force required to remove a hub is a function of its size (bore and length) and of the interference used at installation. For very large couplings, pressure can be used in two ways for hub removal. The first is to use a portable

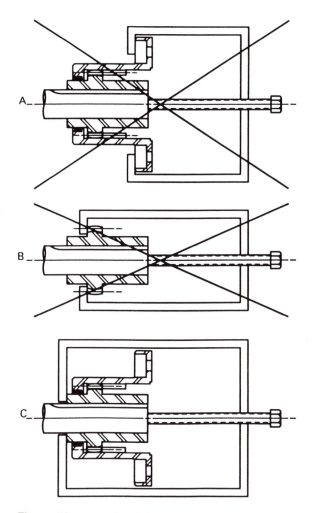

Figure 52 Removing hubs with a wheel puller: (A), (B) improper positioning of puller; (C) proper positioning of puller.

hydraulic ram such as the puller press shown in Fig. 51. The force that can be applied is limited only by the strength of all the threads and rods.

There are many cases where nothing helps to remove the hub; then the hub has to be cut. This is usually cheaper than replacing the shaft. The place to cut is above the key, as shown in Fig. 53. The cut should be made with a saw, but a skilled welder could also flame-cut the hub, taking precautions that the flame does not touch the shaft. After the cut is made, a chisel is hammered in the cut,

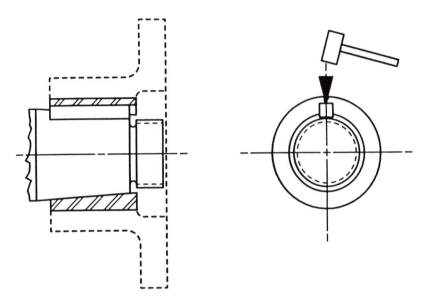

Figure 53   Splitting a hub for removal from a shaft.

spreading the hub and relieving its grip on the shaft. If the hub is too thick, it can be machined down to a thin ring, then split.

3. Intermediate Bushings

To loosen intermediate bushings, remove the bolts (usually more than two) and insert them in the alternate holes provided, making sure to lubricate the threads and the bolt (or setscrew) points. When tightened, the bolts will push the bushing out of the hub and relieve the grip on the shaft. The bushing does not have to be moved more than 1/4 in. If the bushing is still tight on the shaft, insert a wedge in the bushing's slot and spread it open. Squirting penetrating oil in the slot will also help in sliding the hub-bushing assembly off the shaft. Do not spread the bushing open without the hub on it, as this will result in the bushing breaking or yielding at or near the keyway.

4. Returning Parts to Service

Any part that is to be reused should be thoroughly cleaned and examined before using. The part should be dimensionally inspected to ensure that it is within the coupling manufacturer's limits. The part should be examined for dents, burrs, and corrosion. It should be nondestructively tested (magnaflux or dye penetrant tested) for surface cracks or other distress. If there is any question about the coupling integrity, *throw it away* and use a new one, or return it to the coupling manufacturer so they can determine its suitability for further service. *Using damaged or deteriorated parts could result in serious failure of the coupling or the equipment.*

3

CLUTCHES AND BRAKES

Richard C. St. John

Base Line Engineering, Inc., Louisville, Ohio

Mechanical clutches and brakes couple, decouple, and accelerate mechanical components and maintain them at appropriate velocities. The functions of the devices are often so similar that their roles are interchangeable; however, clutches generally accelerate and brakes generally decelerate the moving components. Regardless of whether the clutch/brake components are mechanically, electrically, or hydraulically actuated, they are mechanical clutches and brakes because they transmit mechanical power. There are a plethora of mechanisms with one discriminating feature—the drive is either direct or indirect as shown in Fig. 1.

A large variety of clutches and brakes of different designs exist which employ many principles of operations and are produced by a host of manufacturers. Some of these products are quite use specific, and others are more general purpose. This chapter cannot detail all of the variations in design, but addresses principles, some mathematical techniques and examples, and general descriptions that may use specific products for the purpose of illustration. Clutch and brake manufacturers tend to be quite generous in their descriptions of operation and operating principles, and users are well advised to collect vendor data in addition to technical publications to assist them in more detailed understanding of specific products and methods of selection. *Power Transmission Design* (Penton Publishing Co., Cleveland) publishes their "Product

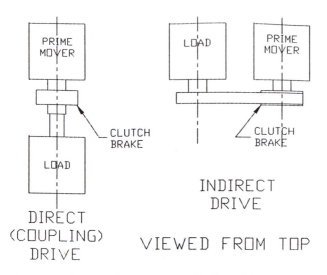

Figure 1   A comparison of direct and indirect drives.

Specification Guide" annually, and one of the features of the guide is a list of product categories and the names of manufacturers who produce the products.

Clutches and brakes are most generally used in rotary motion applications. Whatever the motion, the clutch or brake in any application must be designed and built to convert the mechanical energy absorbed during the relative motion or slip phase into heat energy and survive both the mechanical and thermal steady-state and shock loads imposed by the system during the remainder of the operational cycle without damage and stay within the physical and cost constraints required to make the system economically valuable to the end user.

I. POWER TRANSMISSION PRINCIPLES

As previously mentioned, properly operating clutches and brakes either accelerate or decelerate loads and connect loads to the prime mover to restrict relative motion between the load and the prime mover to an acceptable level. Motion is generally assumed to be uniformly accelerated and although sometimes rectilinear, clutches and brakes are usually associated with rotary motion. The parameters of rotary motion that are of most interest when dealing with clutches and brakes are

Torque (static and dynamic)
Angular velocity
Acceleration: angular and normal
Energy
Power
Moment of inertia

A. Torque

Torque twists, and its units are pound-inches, pound-feet, newton-meters, etc.

Torque = moment = force × radius

or when the term *moment* is interchanged with *torque,*

$$M_s = FR \qquad (1)$$

Torque is static when it twists, but there is no change in angular velocity, i.e., no angular acceleration. Consequently, a shaft that is delivering torque at a constant rpm is providing static torque even though the shaft is turning. If the shaft is providing torque and the rpm is changing, the component of torque that drives the load is static torque, and the component of torque that causes the system to change speed is the dynamic component. Either a couple or a torque causes rotation as shown by Fig. 2. Examination of the figure shows that a couple employs equal forces symmetrically displaced about the axis of rotation to cause pure rotation. Clutches and brakes that work on inertial loads and on symmetrical loads, such as turbines or generators, often actually form couples. A torque as shown by Fig. 2 consists of a couple that causes pure rotation plus a force that causes translation or bearing load. Mechanical arrangements with which torques are often encountered are devices such as belts, chains, films, webs, many wrenches, and gearing in which the force is located on one side of

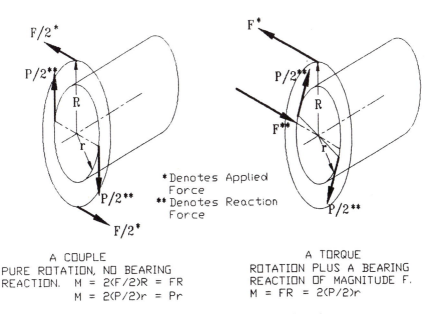

A COUPLE
PURE ROTATION, NO BEARING
REACTION. M = 2(F/2)R = FR
                    M = 2(P/2)r = Pr

A TORQUE
ROTATION PLUS A BEARING
REACTION OF MAGNITUDE F.
M = FR = 2(P/2)r

Figure 2   A demonstration of the difference between a couple and a torque.

the prime mover. Exactly the same equation (1) is used to quantify the moment in either case. These are generally thought of as static torque equations, i.e., equations that relate to a load that rotates at a constant (including zero) velocity. Dynamic torque is reviewed in an ensuing paragraph.

B. Angular Velocity

Angular velocity is the rate at which an object turns about some axis of rotation. It is generally referred to in revolutions per minute (rpm). The physics of the phenomena are such that a more natural way of expressing angular velocity is in radians/second or radians/per minute, but radians cannot be measured very well, so revolutions are used, and one revolution equals 2π radians. Equations of motion are usually derived with the units of radians/second, and the common term $2\pi/60$ is used to convert rpm's that can be easily measured into radians/second.

C. Angular Acceleration

Angular or tangential acceleration is the rate at which a rotating object changes velocity. For example, if an object is rotating at 1000 rpm and accelerates uniformly to 2000 rpm in 2 sec, the angular acceleration is $(2000 - 1000)/2 = 500$ rpm/sec. Were the situation reversed and it accelerated from 2000 to 1000 rpm in 2 sec, the acceleration would be $(1000 - 2000)/2 = -500$ rpm/sec. In the latter case, the negative acceleration is often, but not necessarily, called deceleration. Radians and seconds are also used with angular acceleration. The equations are derived with the units for angular acceleration as rad/$\sec^2$ = rad/sec/sec, and there are $60/2\pi$ rad/$\sec^2$ for each rpm/sec. The following is a useful formula for angular acceleration when the data are in rpm's and seconds:

$$\alpha = \frac{2\pi(U_2 - U_1)}{60t} \tag{2}$$

where

α = angular acceleration, rad/$\sec^2$

t = time for the object to accelerate from U_2 to U_1, sec

U_1 = initial angular velocity, rpm

U_2 = final angular velocity, rpm

D. Normal Acceleration

Normal acceleration is that which causes a rotating mass to experience centrifugal force, and is related only to angular velocity, not angular velocity change. The units of normal acceleration are linear, such as ft/$\min^2$, in/$\sec^2$, m/$\sec^2$. The value of normal acceleration as an entity unto itself is normally of little value; however, the product of it and the mass of the object upon which it is acting is the centrifugal force and is vital in many engineering applications. Discriminating between mass and weight and getting all of the conversion factors lined up can be messy, and the following formula has proven very useful in solving a host of centrifugal force problems:

$$C = W(2.84 \times 10^{-5})U^2 R \tag{3}$$

where

C = centrifugal force in the same units as W

W = weight of the rotating object in any weight (not mass) units

U = angular velocity of the rotating object, rpm

R = distance of the center of mass of the weight from the axis of rotation, in.

Note: The term $2.84 \times 10^{-5} \times U^2 R$ is the normal acceleration expressed in gravities or G's.

E. Inertia (Moment of)

The statement that a body at rest remains at rest and a body in motion remains in motion without the application of an external force is just as true here as in the physics lab. This tendency to resist a change in velocity is due to inertia, and in the case of rotating mechanisms the inertia is named and quantified as the moment of inertia. The key property of inertia is that it resists any and all changes in the velocity of a body; i.e., it resists acceleration. Consequently, when it is necessary to change the velocity of a rotating body, added torque beyond the static torque must be applied and is quantified as

$$M_d = I\alpha \tag{4}$$

where

M_d = dynamic torque (lb-in., lb-ft., etc.)

$I = kmr^2$ = moment of inertia (lb-in.-sec^2, etc.)

α = angular acceleration (rad/sec^2)

k = a constant that varies with the shape of the body, but for a thin hoop $k = 1$, for a solid cylinder $k = 1/2$

m = mass of the rotating body (not the weight)

r = outside radius of a cylindrical body or hoop

Equation (4) requires a lot of playing with conversion factors to be very useful, and the following simpler expression has been developed for dynamic torque to uniformly accelerate a rotating body from one velocity to another velocity

$$M_d = \frac{WK^2 \Delta U}{3686t} \tag{5}$$

where

M_d = moment, lb-in.

K = radius of gyration, in.

When mass is a thin hoop with OD − ID ≅ 0, $K^2 = D^2/4$.

When mass is a solid cylinder with an OD = D, $K^2 = D^2/8$.

When mass is a hollow cylinder with an outside diameter of D and an inside diameter of d, $K^2 = (D^2 + d^2)/8$.

The preceding shapes rotate about their longitudinal axis. When other shapes or conditions are encountered, consult engineering design texts or handbooks.

ΔU = change in angular velocity = $U_2 - U_1$, rpm

t = time interval in seconds for the body to accelerate from U_2 to U_1

W = weight of the rotating mass in pounds. For any cylinder

$$W = \pi(D^2 - d^2)Lw/4$$

and

L = length of cylinder, in.

w = weight density, lb/in.^{3}

$w \sim 0.265$ lb/in.^{3} for many cast irons

$w \sim 0.283$ lb/in.^{3} for steel

$w \sim 0.98$ lb/in.^{3} for many aluminum alloys

$w \sim 0.302$ lb/in.^{3} for many copper-bearing alloys

Figure 3 illustrates various common rotating configurations along with the formulas for K^2.

F. Energy

Work and energy can be considered synonymous in the context of most clutch and brake phenomena, and are the product of the amount of force expended over a distance that coincides with the line of action of the force and the distance, i.e.,

$$E = Fs \tag{6}$$

In (6) E = energy, F = force, and s = distance that is parallel to the line of action of F. When these linear terms are converted to angular relationships,

$$E = M\theta \tag{7}$$

where θ is the angle in radians. Since it is more likely that the angular velocity and the time interval are known than the total angle rotated,

$$E = \frac{2\pi MUt}{60} \tag{8}$$

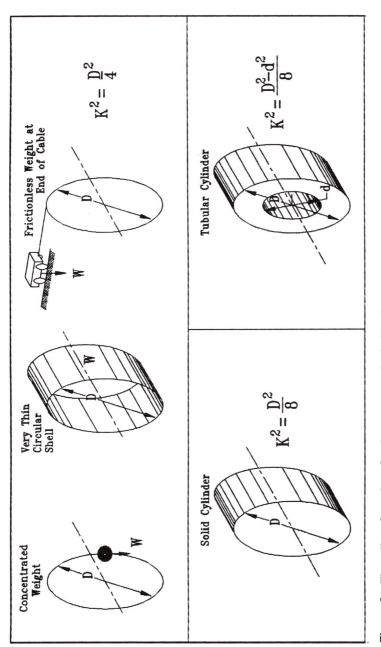

$$K^2 = \frac{D^2}{4}$$

Frictionless Weight at
End of Cable

$$K^2 = \frac{D^2 - d^2}{8}$$

Tubular Cylinder

Very Thin
Circular
Shell

$$K^2 = \frac{D^2}{8}$$

Solid Cylinder

Concentrated
Weight

Figure 3 The radius of gyration of various common physical bodies.

in inch-pounds. Many equations refer to energy in the context of mechanical power transmission, but they are all derived from (6). When dealing with clutches and brakes, the major concerns are being able to accelerate ($+$ or $-$) a mass from one velocity to another (mechanical work), and maintain that inertial mass at a target velocity. Since all of the mechanical energy that is expended by slippage of the clutch or brake is converted to heat energy, the amount of heat energy generated is also important. The amount of energy required to accelerate a rotating mass from one angular velocity to the other when K is in inches is

$$E = \frac{WK^2[U_2^2 - U_1^2]}{70,400} \quad \text{lb-in.} \tag{9}$$

$$E = \frac{WK^2[U_2^2 - U_1^2]}{90.5} \quad \text{btu} \tag{10}$$

In many applications when a clutch accelerates an inertial mass, it must also drive a load which must be added to the inertial load of Eqs. (7) and (8). The load may be constant torque, the torque may vary as the angular velocity, or the load may vary exponentially with the angular velocity. This energy can often be at least approximated by

$$E = E_1 + E_2 \tag{11}$$

where E is the total energy for the engagement, E_1 is the energy required to carry the static load from Eqs. (7) or (8), and E_2 is the energy to accelerate the inertial load from Eqs. (9) or (10).

G. Power

Power is the rate of change of energy with respect to time, or

$$P = \frac{E}{t} \tag{12}$$

where P is power in in.-lb/sec and t is the time in seconds. Horsepower is the common dimension for power, and 1 hp = 33,000 ft-lb/min = 550 ft-lb/sec = 660 in-lb/sec = 42.42 btu/min = 0.746 kW. The classic English formulas to determine the power required to drive a constant torque at a constant angular velocity are

$$\text{HP} = \frac{MU}{5252} \tag{13}$$

where
    M = torque, lb-ft
    U = angular velocity, rpm

$$HP = \frac{MU}{63000} \tag{14}$$

where M is in lb-in.

These equations are useful for determining how much torque a shaft should carry, or the torque capacity of the clutch or brake, but they do not tell much about the thermal capacity requirements of the clutch or brake. The thermal horsepower can, however, be determined from Eqs. (9), (11), and (14), and the conversion factors in the previous paragraph:

$$THP = \frac{WK^2(U_2^2 - U_1^2)/70{,}400t + MU/9.55}{6600} \tag{15}$$

where K is in inches, M is in lb-in., and THP is in horsepower.

H. Load Factors

Unfortunately, clutch and brake loads are not smooth and totally predictable. In some design applications, the designer can pretty well determine the loads analytically; however, in most commercial applications, experience has taught manufacturers how to specify their products so failures do not create an unacceptable financial burden. These products are generally overdesigned by usage, rather than intent, to some undetermined level. When they break, they are replaced with an identical or higher capacity unit. Manufacturers have, over the years, developed service factors that their experience causes them to apply to products to keep overload to an acceptable level. The procedure is to multiply the ideal *torque or power* values by the service factors shown in Tables 1 and 2 when selecting clutches or brakes and there is no better information

Table 1 Service Factors for Multiplying Ideal Torque or Power Values

Load class	Electric motors		I.C. engines No. cylinders		Load description
	AC	DC	Multi	Single	
A	1.5	1.5	1.5	2	Dead load, low inertia, nonpulsating
B	2	2	2	2.5	High inertia (flywheel) requiring prolonged acceleration period; heavy-duty dead load.
C	2.5	2.5	2.5	3–5	High breakaway torque; medium pulsating and unbalanced loads.
D	3–5	5–10	3–5	5–8	High breakaway torque; heavy pulsating loads.

Table 2 Classification of Various Applications[a]

Driven load	Class	Driven load	Class
Agitators		Generators	B
Vertical	A	Lawn mowers	A
Horizontal	B	Laundry machines	B
Bakery machines	B	Line shafts	B
Centrifugal blowers	B	Mangles	A
Calenders	B	Machine tools	
Clarifiers	A	Light	A
Classifiers	B	Heavy	B
Compressors		Rotary mills	B
Rotary	A	Motor scooters	A
Multicylinder	C	Presses	B
Single cylinder	D	Propeller shafts	A
Conveyors		Pulverizers	B
Belt	A	Pumps	
Oven	A	Centrifugal	A
Screw	A	Rotary gear	A
Reciprocating	C	Multicylinder	C
Shaker	C	Single cylinder	D
Crushers		Sanding machines	B
Coal	B	Screens	
Iron	B	Rotary	C
Stones	B	Vibrating	C
Hammer mill	B	Stokers	
Dryers	B	Screw feed	A
Elevators		Ram feed	A
Bucket	C	Washing machines	
Exciter	B	Centrifugal	B
Fans		Textile/paper	A
Centrifugal	A	Wire-winding machines	B
Propeller	B	Woodworking machines	B
Garden tractors	A		

[a] *Source*: R. St. John, *Power Transmission Design Handbook,* Penton, 1977–78.

available. This provides for a higher capacity product than would specifying to the ideal value.

Manufacturers also often rate their products by how much slip energy they can accommodate in a given period of time (thermal horsepower) because high temperatures that result from slipping are the second limit to brakes and clutches, torque being the first limiting factor. It can be seen from the preceding that the amount of heat energy is relatively easy to determine. Determining temperature is quite another problem, and some discussion of it is made in the section on frictional material. However, no serious attempt is made to treat temperature because of the complexity of the analysis and the multitude of factors that must be known, such as initial temperature, actual ambient temperature near the clutch or brake, heat flow rates, surface conditions, cooling fluid velocity, the nature of coolant flow (laminar, mixed, or turbulent), the kind of coolant, and the complexity of the calculations. If a new product is being developed, comparing the new product with a similar existing product, modern computer analysis, and an appropriate test program will help determine satisfactory energy/temperature limits.

The third limiting condition is the maximum speed at which rotating devices can be safely operated. A stress analysis followed by testing is the preferred course of action to establishing this limit. Different clutches and brakes present quite different requirements. Internal shoe-type centrifugal clutches present a much different problem than face-type clutches of the same capacity because in the drum-type clutch, centrifugal and self-energization forces cause the shoes to load the drum and very likely generate higher hoop and bending stresses than are generated by centrifugal force acting on the drum itself. Centrifugal stresses in the plates of the face clutches are much less of a problem; however, centrifugal stresses may present a problem in the friction material where they generally are not even considered in the drum-type clutch. Manufacturers also rate maximum operating speeds on their clutches and brakes because of thermal and centrifugal stresses.

II. NONFRICTION CONTACT-TYPE CLUTCHES AND BRAKES

A. Positive Contact Clutches

Positive contact clutches, of which the square jaw or dog clutch is the most fundamental version, are the simplest form of clutch. The primary difference between this type of clutch and similarly constructed couplings is that the clutch includes a mechanism such as an actuating lever and a throw-out bearing that permits the driving and driven member to be engaged and disengaged on demand rather than by an assembly/disassembly procedure. Figure 4 depicts the progression of the dog clutch to the multitoothed clutch. Figure 4a is a plan view of the dog clutch. One of the jaws slides along its shaft on a key or spline to permit engaging and disengaging the load from the prime mover. The square jaws restrict engagement of the prime mover to very low speeds, such as 0 to 10 rpm, and even then heavy shocks and difficult engagement may be encountered. However, the engagement is positive and there is no likelihood of slip as long as the torque transmitted through the clutch is within its physical limits.

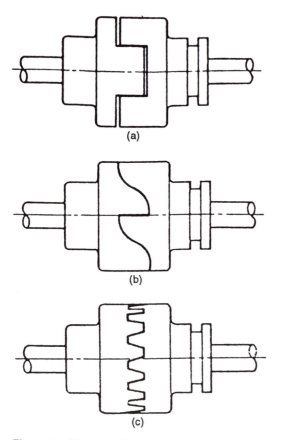

(a)

(b)

(c)

Figure 4   Three positive contact clutches: dog, spiral, and taper toothed. (From J. Shigley and C. Mischke, *Standard Handbook of Machine Design*, McGraw-Hill, New York, 1986.)

Figure 4b shows what happens when the driving side of the dog clutch is tapered to provide a spiral jaw clutch. This clutch will permit engagement with low-velocity relative rotation in the order of 150 rpm, but must be held into engagement by some form of locking mechanism because the driving torque will cause a disengagement force that is a function of the friction diameter, the magnitude of the torque transmitted, the spiral angle, and the coefficient of friction between the driving and driven faces. When the jaws are held into engagement by a spring, the spiral jaw clutch becomes an overload mechanism that will automatically disengage beyond some threshold torque. A simple helical compression spring that encircles the shaft is one of the easiest ways to perform this task; however, sufficient analysis of the system should be made to ensure that the spring load is high enough to allow no relative motion under normal operating conditions, and that the spring rate is low enough to ensure complete disengagement at the target overload condition. If the former condition is not met, fretting and abrasive wear will encourage short clutch life and catastrophic failure. If the latter condition is not met, the clutch will fail to provide the required or desired overload protection. The spiral jaw clutch provides the same reverse rotation operation as the dog clutch. The operation can also be reversed to provide positive lockup when driving and disengagement during the overrun condition, i.e., when the load drives the primer mover.

When both sides of the jaw are tapered, a toothed clutch as shown by Fig. 4c results. These clutches are best used to engage and disengage the load with no relative rotation; however, adequate surface hardness, tooth strength, and tooth configuration allow them to perform well under some conditions of relative rotation at speeds around 300 rpm, and they are often used as an overload clutch with various spring-loading techniques such as that just described. Magnetic attraction, such as that encountered with an electromagnetic clutch, or load retention with pneumatic and hydraulic devices are other means of maintaining a specific load. When both sides of the teeth are tapered, these clutches will engage and disengage in both directions. Figure 5 depicts an electrically actuated version of this clutch with a tooth pattern that is separated by smooth faces in an asymmetric pattern. The prime mover and the load can then be disengaged on command, or under overload conditions, but the engagement will be in a specific radial orientation when operating normally. Other patterns can provide two, three, or more specific orientations. The teeth can be replaced by rollers or balls to provide replaceable, economical, precise load carriers.

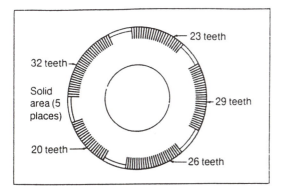

Figure 5   An electrically actuated asymmetric pattern tapered tooth face clutch. (From *Power Transmission Design,* February, 1991, Penton Publishing Co.)

B. Wrap Spring Clutches

Wrap spring clutches consist of a helical torsion spring that has one end attached to the prime mover or an external actuator and its inside diameter wrapped around and contacting both driving and driven shafts with the spring coils wound opposite to the direction of shaft rotation. Figure 6 shows such a clutch. In operation, the spring rotates with the prime mover. Torque is transmitted from driving to driven shaft when the load resists the rotation of the prime mover and friction between the inside diameter of the spring and the outside diameter of the shafts causes the spring to wind up on the shafts with ever-increasing friction forces that lock the shafts together preventing further relative rotation. All of this occurs with very little relative rotation of the prime mover and the load, often well less than a degree. The friction action is similar to that of the classic band brake, and is a function of $e^{\mu\theta}$, where μ is the coefficient of friction between the spring and the load shaft, and θ is the wrap angle in radians of the spring about the load shaft. Consequently, even if μ is

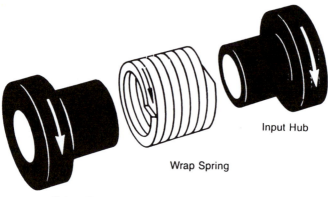

Input Hub

Wrap Spring

Output Hub

The basic wrap spring clutch consists of just three elements: an input hub, an output hub, and a spring whose inside diameter is just slightly smaller than the outside diameter of the two hubs. When the spring is forced over the two hubs, rotation in the direction of the arrow wraps it down tightly on the hubs, positively engaging them. The greater the force of rotation, the more tightly the spring grips the hubs.

Figure 6   A basic wrap spring clutch. (Courtesy of Warner Electric, South Beloit, IL.)

small ($<<0.1$), θ is generally large and the two shafts are locked together. The ability of the spring to carry tensile loads generally limits the torque transmitting ability of the clutch. Good practice dictates that there be as many coils of wire about the input shaft as about the output shaft to prevent slip in the maximum torque condition.

R. D. Lowery and A. W. Mehrbrodt describe several wrap spring clutch configurations and detail the mathematics to determine torque capacity and spring stresses, among other things, in the July 1976 *Machine Design* (Penton Publishing Co.) article "How to Do More with Wrap Spring Clutches." Three key equations from this reference are

$$M_D \cong (e^{2\pi\mu N} - 1)\frac{2EI\sigma}{D_m^2} \qquad \text{maximum slip-free driving torque} \qquad (16)$$

$$Mo \cong \frac{2EI\sigma}{Dm^2} \qquad \text{overrunning torque} \qquad (17)$$

$$\sigma_y \geqslant \frac{2TD/A + hE\delta/Df}{Dm} \qquad \text{minimum spring wire yield strength} \qquad (18)$$

where

μ = coefficient of friction between shafts and spring

N = number of turns of spring wire about load shaft

E = modulus of elasticity of spring wire

δ = diametral interference between hub and spring

I = moment of inertia (second moment of area) of spring wire cross section

Dm = mean diameter of spring coil when installed on shaft

Df = free mean diameter of spring coil

A = cross-sectional area of rectangular spring wire

h = depth of rectangular spring wire

When the direction is reversed, the clutch overruns with drag torque that is generally much less than the driving torque. Wrap spring clutches are useful in many applications and, as with most highly self-energized devices, engagement is very rapid. The rapid engagement and freedom from slip cause them to satisfy ratcheting-type applications nicely; however, the minute slip that does occur during engagement is cumulative, and a synchronous relationship between the driving and driven shafts will not be maintained through repeated stopping and starting.

The slip that occurs between the inside surface of the spring and the outside surface of the shafts during overrunning and engagement causes fretting and wear that reduces the cross section and degrades the strength of the spring and the performance of the clutch. Rectangular wire is most often used for the spring to reduce wear by providing the maximum bearing surface, and lubrication, viscous or dry, may be required to achieve satisfactory life. The inside diameter of the spring and the outside diameter of the shafts are generally controlled very closely to provide predictable performance and long life. In some cases, they are ground to size. The necessity for parts with tolerances that are generally much tighter than industrial standards makes good maintenance important and repair very difficult. Replacement rather than repair is likely to be the most economical solution if original equipment repair parts cannot be obtained.

These clutches can also be used in backstopping applications by fixing one shaft, and running the other in the overrunning direction. The spring will friction-lock on the fixed shaft immediately as reverse rotation is attempted.

C. Overrunning Clutches

Although wrap spring clutches can be used in overrunning applications, wear does not generally suit them for long-term operations. Ball, roller, cam, and sprag clutches often offer superior wear resistance. These clutches also operate in driving, backstopping, and intermittent operation modes and generate high Hertz stresses on the mating components under high-load conditions. Applications such as those involving internal combustion engines of few cylinders driving piston-type compressors may generate severe fretting that causes short life in such clutches. However, small roller clutches served well in the final freewheeling drive of millions of 2000-lb European automobiles with three-cylinder, two-stroke-cycle engines of up to 70 hp. Consequently, although history shows that these clutches can survive such severe service, the applications should be preceded by rigorous design, good engineering judgment, and meaningful testing. Clutch OEMs should be able to provide suitable assistance. The overrunning nature of these clutches generally causes them to engage automatically without external influence.

1. Roller and Ball Clutches

These clutches are identical in principle. The ball clutch has the advantage of lower cost in low-load applications, and the roller clutch has the advantage of much greater load capacity with the same dimensional envelope. Figure 7 depicts a typical clutch with four rolling elements, and shows a free-body diagram of the forces acting on the rolling element. In operation, the spring keeps the rolling element engaged with both the hub and the housing. There is very little lost motion between the overrun and drive states of these clutches, but they will not provide synchronous motion. Lost motion is minimized by high spring forces which force the rollers through lubricants and small misalignments and facilitate whatever component deformation might occur during engagement. However, high spring forces increase the amount of heating that occurs when

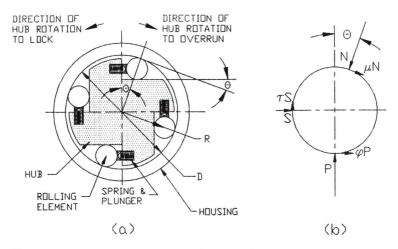

Figure 7 A typical roller clutch and free body diagram.

the clutch overruns and the rolling elements slide. When relative rotation occurs between the hub and the housing in a direction that causes further wedging, the clutch engages. When relative rotation in the opposite direction occurs, the clutch overruns. In backstopping applications, either the hub or the housing may be fixed. Rotation of the nonfixed component that causes the rolling element to move against the spring allows rotation to continue. Rotation in the other direction causes the rolling elements to immediately lock the hub and the housing together, and motion to cease.

There are various detail configurations of the hub/rolling element arrangement, but Fig. 7 is fairly typical. Although these devices are very simple on the surface, there are a number of hazards in their design that make a careful analysis worthwhile, and a net time saver. Figure 7 and the following analysis consider different coefficients of friction for each rubbing surface because designers often choose different materials as opposed to the old practice of making all of these clutches component from hardened steel. If all of the components are of the same material and the lubrication conditions are essentially the same at each rubbing surface, all three coefficients of friction may be the same value. However, if the designer chooses a hardened steel rolling element, a screw machine steel housing, a powdered metal hub, and a nylon plunger in a dry application, the coefficients of friction will be different at each rubbing surface and may be something like $\mu = 0.15$, $\phi = 0.25$, and $\tau = 0.31$ when the clutch is cold, and much different when it gets hot. On the other hand, if the clutch is run in oil, all three coefficients of friction may be around 0.02 during the sliding phase.

Lockup is key to good clutch operation, and an analysis of the free-body diagram of Fig. 7b shows that this will occur when

$$\theta \leqslant 2 \tan^{-1}(\mu) \tag{19}$$

The spring force (S) is ignored in this analysis because it is so small when compared to N and P and requires a long iterative solution without adding much to the final result. The same free-body analysis shows that the clutch can damage itself during normal operation, and possibly either not overrun in the application or release with difficulty when

$$\theta \leqslant \tan^{-1}(\mu) \tag{20}$$

The normal force that is required to drive the load is found by considering the torque (M) that the device must carry with n rolling elements, and the action between the housing and the rolling element as follows:

$$N = \frac{2M}{\mu D n} \tag{21}$$

With N now known from (21), P can be determined from

$$P = N(\cos\theta + \mu \sin\theta) \tag{22}$$

If θ is the value calculated from (20) as $\theta = \tan^{-1}(\mu)$, $P = N$; however, if another value has been chosen for whatever reason, P will be a different value.
If

$$\mu N < \phi P \tag{23}$$

the rolling element will rotate, and the device will not lock up. Consequently, this calculation must be made, and if $\mu N < \phi P$, the design and material selection must be revised until $\mu N > \phi P$ at all times.

Since the clutch has been designed to self-lock, knowledge of P and N may seem superfluous; however, these devices encounter very large compressive forces, and it is desirable to keep the compressive stresses in the rolling element, hub, and housing within acceptable levels. The following formulas from the third edition of J. Shigley's *Mechanical Engineering Design* (McGraw-Hill) provide an estimate of the compressive stress on these elements:

$$p_{\max} = \frac{2F}{\pi b L} \tag{24}$$

$$b = \left\{ \frac{2F[(1 - v_1^2)/E_1 + (1 - v_2^2)/E_2]}{\pi L(1/d - d/D)} \right\} \tag{25}$$

where
$F = N$ for the housing
$F = P$ for the hub

d = diameter of the rolling element

D = diameter of the housing for housing analysis

$D = \infty$ for the hub analysis → $1/D = 0$

E_1 = modulus of elasticity for the rolling element

E_2 = modulus of elasticity for the hub or housing, depending upon which is being analyzed

L = length of the rolling element

v_1 = Poisson's ratio for the housing element

v_2 = Poisson's ratio for the hub or housing, depending upon which is being analyzed.

It is desirable that when the device is overrunning, the rolling element actually rotates, because if it does not the housing will slide on and wear a flat on the rolling element and eventually cause the clutch to become useless. At this point, centrifugal force acting on the rolling element may come into play. The torque conditions on the rolling element during overrun must then be

$$\mu N > \phi P + \tau S \qquad (26)$$

where

$$N = \frac{S + C(\sin\theta + \phi\cos\theta)}{(\phi + \mu)\cos\theta + (1 - \mu)\sin\theta} \qquad (27)$$

$$P = N(\cos\theta - \mu\sin\theta) - C\cos\theta \qquad (28)$$

$$C = W(2.84 \times 10^{-5})U^2 R$$

W = weight of the rolling element

As long as there is sliding, heat will be generated, and although the analysis to determine temperature rise is beyond the scope of this book, the amount of heat energy that is generated during overrun can, and should, be estimated. The heat generated by the device will be the sum of the friction energy caused by sliding on the rolling element. If the device has been properly designed and is working correctly, there will be no sliding between the housing and the rolling element, but there will be sliding between the rolling element and the hub and the plunger. Given these assumptions, the rate of heat generation when the relative angular velocity between the hub and the housing is U can be estimated from

$$Q = 2\pi(\phi P + \tau S)UD \qquad (29)$$

If P and S are in lb, U is in rpm, and D is in inches, Q has the units of lb-in./min. This can be expressed in watts by dividing Q by 531.

2. Cam Clutches

There are a number of general cam clutch configurations, one of which is depicted by Fig. 8. This device is constructed so that a bearing between the housing and the hub permits them to rotate with respect to each other. A cam carried in the housing is configured to wedge between the housing and the hub by a spring-loaded plunger. When torque is applied to the hub, the wedged cam becomes self-locking and drives the housing in the same manner as a sprag clutch. In some applications, considerable overruning is required of the clutch, and excessive heating and wear of the cam and the hub will occur if the spring-loaded plunger continues to cause the cam to slide on the hub. Such sliding is prevented by configuring the cam to offset its center of mass from the point of contract with the housing so that the centrifugal force (C) acting on the cam will oppose the spring force (S) and at some angular velocity will cause the cam to rotate away from the hub. When the load comes back on, the housing will slow down and the diminished centrifugal force will permit the spring-loaded plunger to rotate the cam back into contact with the hub, and driving can take place again. When the cam is engaged and driving, the friction

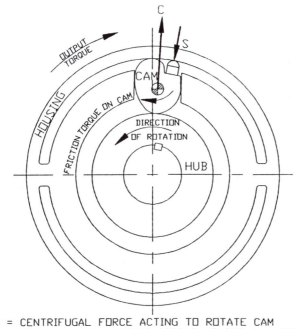

C = CENTRIFUGAL FORCE ACTING TO ROTATE CAM
S = FORCE ACTING AGAINST SPRING-LOADED PLUNGER

Figure 8   One form of a cam clutch.

torque on the cam is so large that the spring force (S) and the centrifugal force (C) are generally insignificant. More than one cam can be used, although only one is shown in Fig. 8. Multiple cams will reduce the load on the bearing between the hub and the housing.

3. Backstopping

Backstopping occurs when free rotation is allowed in one direction and totally inhibited in the other direction. It is accomplished by grounding—that is, attaching one element of the overrunning mechanisms just discussed to the machine base and permitting the other to rotate. When counterrotation occurs, the overrunning mechanism locks up instantly and acts as a brake to prevent equipment damage or other undesirable occurrences. As a rule, backstopping operations cause no problems for the overrunning device because there is little stored energy or other cause of dynamic load. However, the instantaneous braking action is so abrupt that even relatively small dynamic loads will cause some deflection in the braked system, and can easily cause damage. Therefore it is imperative to have knowledge of the worst-case condition, and use that knowledge in sizing and designing the backstopping. When the backstopping device is used to prevent the load from causing an inclined conveyor to run backwards in the event of overload or power failure, it is called a "holdback."

4. Sprag Clutches

In years past, wagons had poor brakes in the light of present standards, and a sprag (a pointed timber or iron bar with an effective radius somewhat larger than that of the wagon wheel) was pivotally attached to and extended rearward from the rear axle. Then if the wagon started to roll backwards the sprag would dig into the roadbed or ground and prevent the wagon from rolling backwards by wedging between the axis and the earth surface; however, the sprag presented no inhibition to motion in the forward direction. The sprags in a sprag clutch act similarly to prevent or permit relative rotation between an inner and outer race.

Figure 9a is a photograph of a sprag clutch rotor and Fig. 9b a complete sprag clutch with the rotor, inner race, and outer race. Figure 9c is a drawing of a typical sprag along with the spring, races, and some of the forces that act on the sprag. Some complete sprag clutch assemblies include a bearing to keep the two races oriented radially, and in others the bearing must be supplied by the machine on which the clutches are used. A multiplicity of sprags is placed in the annulus between the races and oriented circumferentially by a sprag retainer. The sprags are held in contact with the races by an energizing spring. Consequently, when the clutch is rotated in the engaging direction (the outer race rotates counterclockwise with respect to the inner race in Fig. 9b), friction causes the sprags to rotate in the same direction. However, since dimension a is larger than dimension b, the sprag rotation is nil and the sprags wedge

(a)

(b)

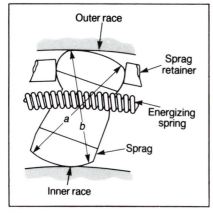

(c)

Figure 9 (a) A sprag clutch rotor. (b) A complete sprag clutch. (Courtesy of Georg Muller, Nurnberg AG, Germany.) (c) A single sprag and its geometry. (From *Power Transmission Design,* February, 1991, Penton Publishing Co.)

between the races with a torque that increases as the torque between the two races. When the relative rotation between the races is reversed, the sprags tend to rotate in the reverse direction and effectively disengage from the races to permit relative rotation, i.e., overrunning. Sprag clutches can be designed to be essentially unaffected, activated, or deactivated by centrifugal force as well, and may be designed to be disengaged upon command.

The spring configuration is generally either the expanding-type garter spring shown in Fig. 9c, a contracting garter spring that provides the same moment about the sprag, or leaf springs in some of the larger clutches. The retainer acts much as the separator in rolling element bearings, and the construction and manufacturing tolerances involved in industrial-quality sprag clutches are very similar to those of rolling element bearings.

Lubrication requirements for sprag clutches are different than for rolling element bearings because the lubricant film must be maintained in the presence of shock, vibration, and minute oscillatory motion without the opportunity for the sprags and races to move enough to replace lost lubricant. Consequently, the lubricants must be very pervasive, possess high film strength, and be capable of reacting appropriately with the base material to maintain protective inorganic surface films such as oxides and sulfides. Clutch and lubricant manufacturers will provide users with information regarding the best lubricant for a specific application.

Sprag clutches have three main areas of usage. The first is as an overrunning mechanism such as that used when two prime movers drive the same load. Electrically powered refrigeration units which must be capable of being driven by a standby internal combustion engine in the event of electrical power failure is an example. The sprag clutch permits automatic engagement and disengagement of either or both power sources as demanded by the system with no need for external control. The second usage is to convert rotary motion into intermittent motion such as an indexing mechanism in which the prime mover drives a crank and connecting rod that is attached to the load through a lever and a sprag clutch. The motion of the load will be related to the stroke of the crank divided by the radius of the lever, and is usually less than 180°. In this manner, rotation of the prime mover is converted to oscillation of the load. A third use is as a backstopping device as discussed previously. The multiplicity of sprags provides sprag clutches with large torque capacities in relatively small packages, and makes them particularly good candidates for holdback applications.

The problem of fretting corrosion and other surface failure problems encountered by other overrunning mechanisms are shared by sprag clutches. The very hard surfaces of, and high stresses encountered in, these clutches makes them peculiarly sensitive to motion that combines high loads and little or no motion, such as that encountered with some low-amplitude oscillations or vibrations. This is the type of motion that is encountered with internal combustion engines operating under high-constant-load conditions when the load is locked to the engine. Rolling element, particularly ball, bearings that operate in a nonrotating mode in other types of clutches encounter similar problems. By and large, if the lubricant does not break down, surface failure is not a problem. The likelihood of success is high, even in the light of these difficult applications, if the user consults with and receives the concurrence of the clutch manufacturer's engineering department.

D. Fluid Clutches

Clutches that transmit torque through both viscous fluids and particulate materials (dry fluids) are called fluid clutches. Although viscous fluids and particulates share the common property of generally assuming the shape of the interior of the container in which they are placed, their behavior in both static and dynamic situations is much different in that centrifugal force within practical industrial limits is not a major factor in the behavior of liquids. However, particulate materials do not flow well against acceleration because mechanical friction between particles and centrifugal force causes them to act ever more like a solid as rotational velocity increases. Consequently, viscous fluid clutches reject shock well, are usually considered to have zero torsional stiffness, and can generally be made to slip without damage to the mechanical parts, while the mechanical friction of the densely packed dry fluids at high angular velocities generally transmit more shock and effectively lock the driving and driven elements of the drive. Both clutch types are capable of soft starting loads, and the viscous fluid clutch will always operate at something well less than 100% efficiency because it cannot generate torque without slip, while the dry fluid clutch can operate without slip and so encounter no efficiency losses through slip.

1. Viscous Fluid Clutches

Figure 10a depicts a simplified version of a viscous fluid clutch which consists of an input shaft and pump (impeller), an output shaft and turbine (runner), a housing, and a viscous fluid that partially fills the cavity formed by the inner surface of the housing/impeller. It should be noted that the runner is housed within the impeller to retain the fluid. The impeller and the runner are similarly constructed and are often semitoroids made from stamped steel. The toroids have a multiplicity of radial vanes, often spot-welded in place, between their inner and outer surfaces. When the input shaft is rotated, centrifugal force slings the fluid outward from the axis of rotation. When the moving fluid contacts the vanes of the runner, which is rotating more slowly then the impeller, it imparts energy to the runner vanes, which has the effect of (a) accelerating the runner and (b) decelerating the fluid. Since the fluid contained within the semitoroid of the runner is moving more slowly than that within the impeller, it is being accelerated outward from the axis of rotation less than the fluid in the impeller semitoroid. Consequently, both flow continuity and force balance requirements cause the fluid to circulate back toward the center of rotation, and the circular flow pattern depicted by the arrows on the vanes of Fig. 10a is established.

The amount of fluid in the clutch is a matter of design and can affect clutch performance. The two basic designs are (a) constant filling, in which a constant

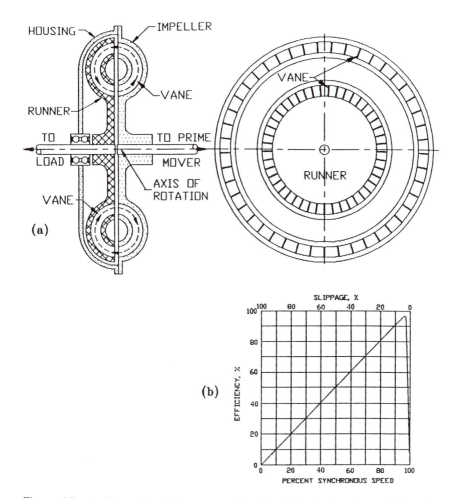

Figure 10 An illustration of the construction of a fluid clutch and a typical performance curve.

volume of fluid is in the clutch at all times, and (b) variable filling, in which external controls vary the fill volume as required by the application.

It is the nature of the viscous fluid clutches that they can transmit no more torque than the primer mover provides; however, the addition of a vane between the impeller and the runner can turn them into a torque converter in which torque can be multiplied over a broad range. Consequently, if the viscous fluid clutch is to be used in an application where an increase in torque is required, it must be coupled with a gear box or other means of torque multiplication.

Slippage is a requirement for the operation of a viscous fluid clutch and is usually in the range of 1% to 3%, which corresponds to an efficiency of 97% to 99%. Figure 10b provides a general picture of the relationship between slip and efficiency for a viscous fluid clutch operating at 2.5% minimum slip. From E. T. Baumeister and Marks, *Standard Handbook for Mechanical Engineers,* seventh edition (McGraw-Hill), the torque that would be provided by this clutch would be approximately

$$M = 36N^2D^5 \times 10^{-12} \text{ lb-in.} \tag{30}$$

where U is in rpm's and D is the outer wetted runner diameter in inches. These clutches are capable of handling anything from fractions to thousands of horsepower.

2. Dry Fluid Clutches

Figure 11 is a sketch of a dry fluid clutch which consists of a housing that is attached to the prime mover shaft, a rotor that is oriented with respect to the housing by a bearing that permits relative rotation between the rotor and the housing, and a flow charge, generally steel shot. When there is no rotation, the shot lies in a pile at the bottom of the housing; however, it becomes uniformly distributed around the internal periphery of the housing almost immediately as usable rotational velocities are encountered. As soon as the housing starts to move, the shot drags on the rotor, which often has vanes formed on it to increase drag at low speeds, and imparts torque to the rotor via friction between the housing and the shot, and the shot and the rotor. As the angular velocity of the housing increases, centrifugal force packs the shot ever tighter together, and the driving friction forces and torque increases accordingly. Consequently, when the clutch is operating within its rated capacity, the friction torques exceed the torque required by the drive and the input and output are locked together to provide a clutch that has 100% efficiency because there is no slip. In a properly sized application, the clutch should slip in the event of severe overload to reduce or prevent damage to the driving and driven machinery, so sizing of the drive is important, and there can be too much of a good thing. Sizing is also important to ensure that too much slip does not occur because the shot can be very abrasive and excess slip can cause premature failure of the rotor and possibly the housing. These clutches are very often a good solution to an economical soft-start, high-efficiency drive requirement, and come in many sizes from various vendors.

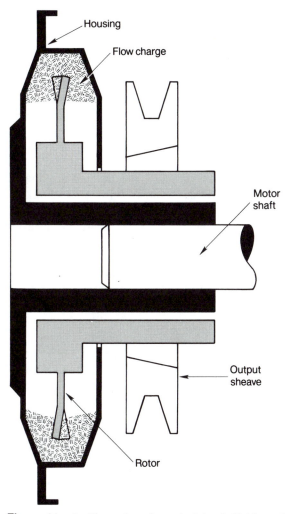

Figure 11 An illustration of a typical dry fluid (shot) clutch. (From *Power Transmission Design,* February, 1990, Penton Publishing Co.)

E. Magnetic Clutches and Brakes

Four general types of magnetic clutches are used in industry:

1. Friction clutch, brake, and clutch brake (often referred to simply as electric clutches and brakes) in which an energized electromagnetic causes or permits springs to cause clamping and unclamping of a friction disk(s) to an

opposing friction surface(s), and so generate friction forces that permit torque to be transmitted between the two components. Deenergizing the electromagnetic causes unclamping or clamping to take place to permit free relative rotation or driving as the detail design dictates. These clutches are generally not considered suitable for continuous slip operation.

2. Magnetic particle clutch that behaves similarly to the dry fluid clutch previously described except that a magnetic field instead of centrifugal force activates magnetic media to frictionally engage a rotor to a housing and thus transmit torque. The wear in these clutches is so low that they are generally treated as noncontact clutches that may be operated in either a slip or synchronous mode. Torque is nominally independent of slip speed.

3. Hysteresis clutch in which a noncontacting ferrous rotor is acted upon by a rotating magnetic field to transmit torque without slip below a threshold value and with slip above that value. These are noncontact clutches that may be operated in either slip or synchronous modes. Torque is generally independent of slip speed when operation is in the slip mode.

4. Eddy current clutch in which a noncontacting, nonferrous (generally copper) rotor is acted upon by a rotating magnetic field to produce rotation and torque that is a function of slip velocity. If there is no slip, there is no torque, so these noncontact clutches can operate in the slip mode only. Torque is nominally directly proportional to slip speed.

Any one of the preceding clutches can be used as a brake by causing coupling between a rotating and nonrotating member to retard motion. Eddy current brakes are of limited value because they provide no torque unless there is relative rotation between the stationary and rotating components, but they can be effective retarders. Wear and high temperature generally limit the amount of slip that can be sustained by the friction clutches and brakes, and high temperature alone is generally the limiting condition for the other clutches. Although the heat energy caused by slip that occurs both while accelerating the load and while the load is operating with a constant slip velocity can be readily approximated with the following formulas, the determination of important temperatures is quite another matter and its determination is far beyond the scope of this chapter. In cases where published data do not provide answers to thermal limits, OEMs should know the capabilities of their products in this important area and should be consulted.

$$Q' = \frac{TUt}{60} + \frac{kWD^2U\,\Delta U}{140,796} \tag{31}$$

where

Q' = slip energy per cycle, in.-lb

T = slip torque, lb-in.

t = duration of cycle, sec

k = shape constant for the mass being accelerated: $k = 1$ for a hollow cylinder; $k = 1/2$ for a solid cylinder

D = diameter of the mass being accelerated, in.

W = weight of the mass being accelerated, lb

U = slip speed, rpm

ΔU = speed change of the accelerated mass, rpm

1. Electromagnetic Friction Clutches and Brakes

Figure 12 depicts an electromagnetic friction clutch with a nonrotating magnetic coil. In this device, applied electrical power energizes the electromagnet to engage the clutch so that it can transmit torque. What occurred was that when electrical power was applied to the electrical coil (9) of Fig. 12, the resulting magnetic flux attracted the armature (6) and pulled it against the friction material (7) to frictionally couple the armature (6) to the drive carrier (8). Torque is transmitted to the torque plate (4), which is attached to the hub (2) and carries the drive pins and springs (5) to which the armature (6) is attached. This attachment and the splines or keys between the drive carrier (8) and the

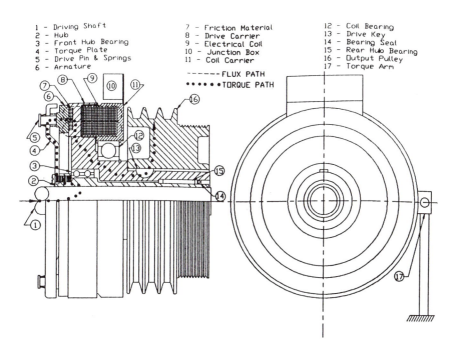

```
1 - Driving Shaft          7 - Friction Material     12 - Coil Bearing
2 - Hub                    8 - Drive Carrier         13 - Drive Key
3 - Front Hub Bearing      9 - Electrical Coil        14 - Bearing Seal
4 - Torque Plate          10 - Junction Box          15 - Rear Hub Bearing
5 - Drive Pin & Springs   11 - Coil Carrier          16 - Output Pulley
6 - Armature                                         17 - Torque Arm

                         ------ FLUX PATH
                         • • • • • TORQUE PATH
```

Figure 12   A nonrotating coil electric clutch construction with emphasized magnetic and torque paths.

output pulley (16) permit torque to be transmitted as shown when the armature (6) is held in engagement with the friction material (7). When the coil is deenergized, the springs force the armature out of engagement with the friction material and no torque is transmitted. The amount of torque transmitted is contingent upon the friction radius, the coefficient of friction, the spring forces, and the magnitude of the force that holds the armature in contact with the friction material. The engagement force is related to the number of ampere-turns produced by the electric coil and the magnetic linking between the coil and the armature. Manufacturers generally express the performance of these clutches by specifying them according to physical size and the static torque available at specific voltages.

There are myriad configuration details that cause one electric clutch to differ from the other. Rotating coils may be used with slip rings, and the friction material can be placed immediately adjacent to the electrical coil to provide minimal gap between the coil and the armature. The resulting shorter magnetic circuit generally increases torque, and the type of clutch selected becomes a trade-off between size, torque, cost, and maintenance. A spline between the hub and the armature may replace the drive pins. The friction material may be affixed to the armature or may even be a separate member altogether, and friction materials with lower coefficients of friction may be used to provide softer starts (lower angular acceleration), generally at the price of lower static torque. Springs may be used to engage the clutch, and electrical power may be used to disengage it in yet another configuration. Electric clutches are available in many configurations for direct, indirect, foot-mounted, shaft-mounted, and motor-mounted drives.

An electric clutch becomes an electric brake when the armature engages a nonrotating member so that when electric power is applied to the coil, the armature inhibits rotation of the driving shaft. For example, if the coil bearing (12) in Fig. 12 is eliminated, the coil carrier (11) and the drive carrier (8) become one so that when the armature (6) engages the friction material (7), the resulting friction torque causes the driving shaft (1) to stop rotating. Figure 13 illustrates an electric brake of this type. Again, the brake may be spring-applied and electrically released during events such as emergencies or power failures. Electric clutches and brakes are often combined in a single package to become a clutch-brake. Figure 14 is a photograph of such a clutch-brake that is configured for NEMA C-face compatibility. The clutch and the brake can be operated independently, or the same signal can cause the clutch to engage and the brake to disengage simultaneously and another signal can cause the clutch to disengage and the brake to engage. This type of action provides rapid start–stop that is limited only by the torque and thermal capacities of the clutch-brake. Electric clutch and brake design is generally both expensive and arcane, and considerable resources are required to provide a unit with predictable per-

Figure 13 An electric brake. (Courtesy of Warner Electric, South Beloit, IL.)

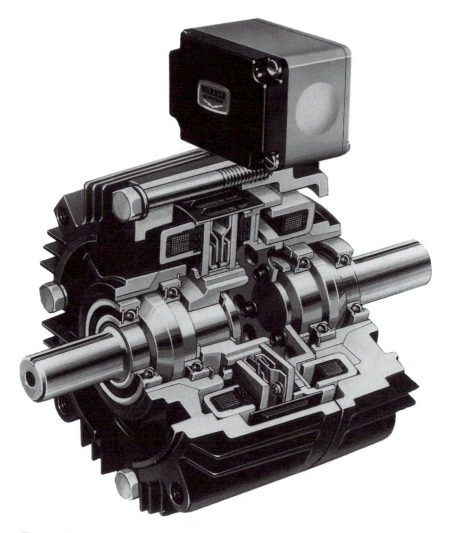

Figure 14 An electric clutch-brake. (Courtesy of Warner Electric, South Beloit, IL.)

formance and long life. However, so many variations of electric clutches are offered that manufacturers are generally able to provide almost any practical clutch off-the-shelf, or at least from standard components.

Electric clutches and brakes come in a broad range of sizes and configurations. They may be shaft, motor, or foot mounted, or they may be integral with another piece of equipment. They may drive belts, chains, gears, shafts, or gear boxes, and may also be used in two-speed and reversing-drive arrangements. Torque and power may range from a few ounce-inches and sub-fractional to over 100,000 in.-lb and hundreds of horsepower. High-volume production, multiplicity of manufacturers, broad range of sizes, and ease of use have caused them to become widely used.

2. Magnetic Particle Clutches

Figure 15 is a schematic of a magnetic particle brake, Fig. 16 a sketch of a magnetic particle clutch, and Fig. 17 is a photograph of an actual clutch. There are also clutch-brake combinations available off the shelf. The simplicity of the magnetic particle concept tends to belie the complexity of the two coaxial housings and the associated bearings that are necessary to implement that concept. This complexity may account for brakes constituting the major portion of the magnetic particle clutch-brake market. Actual clutches can be configured in many direct and indirect designs in much the same manner as other clutches. The typical magnetic particle clutch consists of a magnet or a magnetic coil surrounding a housing that in turn surrounds a rotor with a dry, fine, stainless steel powder in the cavity between the rotor and the housing. When the magnetic field is energized, the powder forms magnetic links along the flux lines

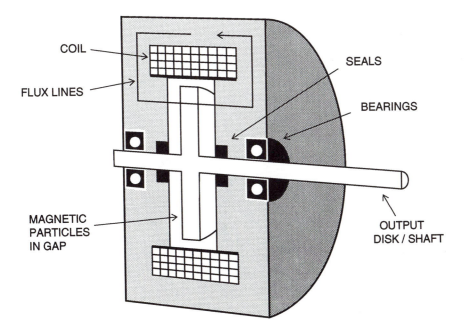

Figure 15 A magnetic particle brake schematic. (Courtesy of Placid Industries, Inc., Lake Placid, NY.)

1 - INPUT SHAFT   4 - ELECTROMAGNETIC COIL   7 - BEARING (4)
2 - OUTPUT SHAFT AND ROTOR   5 - MAGNETIC PARTICLES   8 - SLINGER
3 - HOUSING   6 - FLUX PATH   9 - SHAFT SEAL

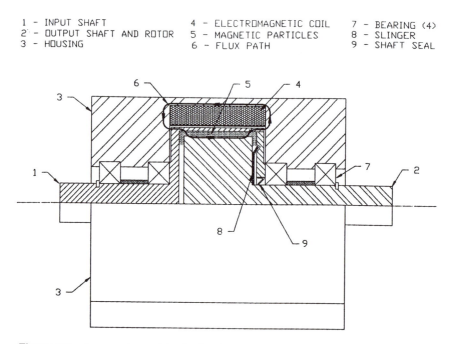

Figure 16   A magnetic particle clutch schematic.

and couples the housing to the rotor. As a result of the magnetic linking, torque is required to cause the housing to rotate with respect to the rotor. If there is no magnetic field, magnetic particle devices simply slip, transmit no torque, and generate little heat. Consequently, switching the field on and off permits magnetic particle devices to be used in intermittent motion and other command applications, and their rapid cycling time (as low as 3 msec on some smaller units) and low shaft inertias (45×10^{-7} lb-in.-sec^2 on the same small unit) are major reasons for using these clutches and brakes in intermittent applications. The breakaway torque at which the clutch starts to slip is very consistent and nearly linear with the applied current for most of the operating region of the clutch above about 30% of the rated current. Once the clutch has started to slip, the slip torque is essentially constant regardless of the slip velocity. Since there is no rubbing between the rotor and the housing, and literally no wear between the powder and either the rotor or the housing, the device is inherently wear free and suitable for intermittent motion applications and constant-slip operations such as film, foil, and wire winding and unwinding-type tensioning operations. There are two general limiting constraints to magnetic particle clutches:

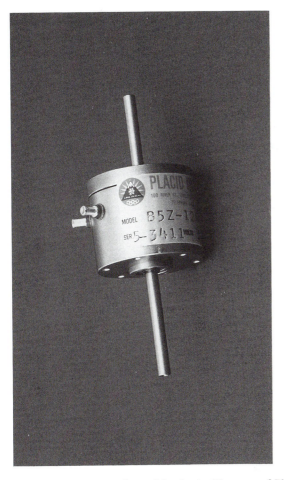

Figure 17 A magnetic particle clutch. (Courtesy of Placid Industries, Inc., Lake Placid, NY.)

1. The temperature rise that results from the heat generated by slip must not exceed a value that will damage the powder. The temperature/heat energy relationship is best determined by consulting with the clutch manufacturer, who should know both the temperature limitations and the thermal characteristics of his product, be it clutch or brake.

2. Too high rotational velocity generates centrifugal forces that causes the powder to clump and clutch torque to become erratic. Acceptable performance has been reported with normal acceleration in the general range of 160 to 180 g's, where

G = normal acceleration in g's = $1.42 \times 10^{-5} U^2 D$

U = rotational velocity, rpm

D = inner housing diameter, in.

The rotational velocity constraint may necessitate that the magnetic particle clutch rotational velocity be reduced by extra belting or gearing. Magnetic particle devices are used as brakes by restraining the output and leaving the coil deenergized during normal operation, then energizing the coil to retard or stop the drive. The same admonition about heat generation applies for the brake as for the clutch. Table 3 gives a general idea of what torques can be anticipated from commercially available magnetic particle clutches:

3. Eddy Current Clutches

Eddy currents are generated in conductors that are in a magnetic field and when there is relative motion between the field and the conductor. Eddy currents are generally pictured as magnetic whirlpools in the plane of the conductor as shown by Fig. 18. When the conductor is a thin disk, a multitude of these whirlpools is generated across the face of the disk, and flux links perpendicular to each of the whirlpools exert a magnetic drag between the disk and the magnet. The eddy currents cause $I^2 R$ heating in the metal disk that generally limits the usefulness of the clutch. These $I^2 R$ losses are the same losses that plague conventional electrical equipment, such as motors, generators, and transformers, and the major reason for laminations in those products is to minimize eddy current losses. In conventional electrical equipment, the eddy current $I^2 R$ losses are the result of the applied electrical current, while in the eddy current device the normal current is the result of the eddy currents. The disks of eddy current clutches are paragmatic (>1) and diamagnetic (<1) materials with permeabilities that approach unity from either direction. Paramagnetic copper and diamagnetic aluminum are two of the more common choices.

Table 3 Torques Available from Magnetic Particle Clutches

Housing		Rated torque			Max. heat
Dia. (in.)	Length (in.)	Nom.	Max. (lb-in.)	Drag	dissipation (watts)
1.44	1.63	2	4	1	9
3.48	3.31	50	80	16	45
5.25	4.01	120		20	140

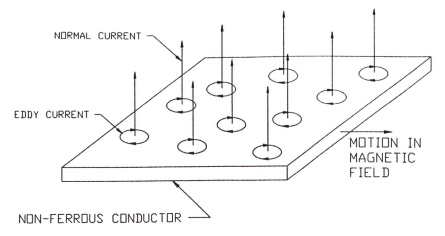

NORMAL CURRENT

EDDY CURRENT

MOTION IN
MAGNETIC
FIELD

NON-FERROUS CONDUCTOR

Figure 18   An illustration of eddy currents and the normal current that they generate.

Figure 19 is an adjustable torque eddy current clutch with two salient pole ceramic magnets, one on each side of a copper eddy current disk. The arrangement of these components and the housings is depicted schematically by Fig. 20. The torque adjustment is made by changing the relationship of the magnetic poles with respect to each other as shown by Fig. 21. Notice that the maximum torque occurs when opposite poles align, and minimum torque occurs when like poles align. Inasmuch as essentially all of the magnetic linking that occurs does so as the result of eddy currents, and the magnitude of eddy currents is proportional to the velocity of relative rotation (slip speed), eddy current clutches always slip, are never synchronous, and produce torques that are proportional to that slip speed. This need for relative rotation between rotor and stator generally makes eddy current devices a poor choice for a brake. Consequently, they often find drive applications in industry where overloads cause slip without harming the prime mover or coupling, the lack of synchronization between the input and the output is acceptable, and constant-torque output is desirable. Many tens of millions of these clutches have been used in automotive speedometers and tachometers with accuracies generally within 2% of full scale. In the speedometer a rotating magnet with a velocity indicator attached to it is driven by a flexible shaft that is linked to the final drive and rotates inside a nonferrous drag cup that is rotationally restrained by a small spiral torsion spring. The slip speed between the magnet and the drag cup is proportional to the automobile velocity so the torque on the spring is also proportional, and finally the velocity indicator displacement is proportional to automobile velocity.

Table 4 gives an idea of torque-carrying ability and the size of typical commercially available eddy current clutches.

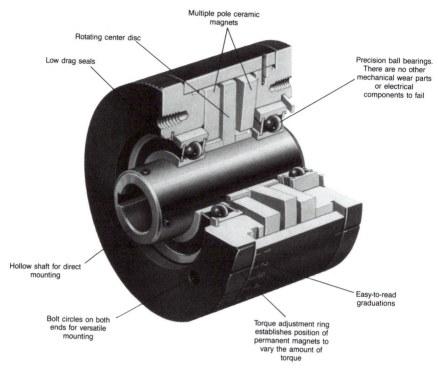

Figure 19 An adjustable torque eddy current clutch. (Courtesy of T. B. Wood's Sons Company, Mount Pleasant, MI.)

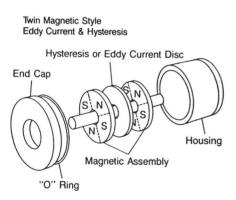

Figure 20 A display of the orientation of the magnets and rotor in an eddy current clutch. (Courtesy of T. B. Wood's Sons Company, Mount Pleasant, MI.)

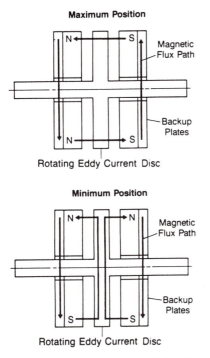

Figure 21 An illustration of the effect of relative orientation of the magnets on flux path and torque of an eddy current clutch. (Courtesy of T. B. Wood's Sons Company, Mount Pleasant, MI.)

Table 4 Torque-Carrying Ability of Eddy Current Clutches

O.D. (in.)	Length (in.)	Torque @ 1800 rpm (lb.-in.)	Thermal capacity (watts)	
1.88	1.61	0.072 min	10	
		0.38 max		
3.23	2.30	1.44 min	22	
		10.26 max		
4.65	2.89	5.13 min	72	
		48.60 max		
6.00	3.50	36.00 min	150	300 rpm max
		180.00 max		rated slip speed

4. Hysteresis Clutches

Hysteresis clutches are structurally almost identical to eddy current clutches, and Figs. 17 and 18 also depict hysteresis clutches and brakes. The differences between these two types of clutches are

1. Hardware: The hysteresis disk is of ferrous material.
2. Torque characteristic: Torque is independent of slip velocity, and the clutch is synchronous below a specific threshold torque value.
3. Magnetics: The magnetic field passes through the ferrous disk instead of being generated in the plane of a nonferrous disk.

Figure 22 depicts the orientation between the salient magnetic poles at maximum and minimum torque conditions on an adjustable hysteresis clutch or brake. Notice that the minimum torque occurs when opposite poles coincide and maximum torque is concurrent with the coincidence of like poles—again, the opposite of the eddy current clutch. These clutches are also typically used

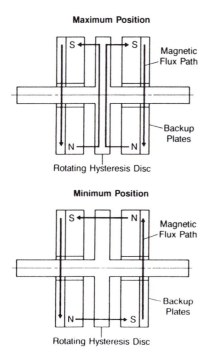

Figure 22   An illustration of the effect of relative orientation of the magnets on flux path and torque of a hysteresis clutch. (Courtesy of T. B. Wood's Sons Company, Mount Pleasant, MI.)

in overload and tensioning applications such as conveyor drives, coil winding, braiding, film and web tensioning, and rewinding.

Table 5 gives an idea of torque-carrying ability and the size of typical commercially available hysteresis clutches.

III. FRICTIONAL CONTACT-TYPE CLUTCHES AND BRAKES

A. Friction and Friction Material Fundamentals

Friction is a force that resists efforts to slide one surface across another and is generally viewed as the result of the force perpendicular to the plane of sliding and special properties of the surfaces of the sliding materials. Figure 23 is a schematic of two such sliding surfaces. The coefficient of friction (COF) is called μ and is simply the ratio of the force required to slide the movable object to the force that presses the object together, i.e.,

$$\mu = \frac{N}{F} \tag{32}$$

where

μ = coefficient of friction

N = normal force that presses the objects together

F = friction force, or force that resists sliding

There are a host of theories regarding the real mechanisms of friction. One theory still promulgated is that surface asperities, such as those shown in the enlarged view of Fig. 23, interfere with each other and both break and generate new asperities when motion occurs. Another theory is that minute welds between the sliding surfaces are continuously made and broken, and a third is the contention that friction is the result of repetitions of stick slip events. A more commonly accepted theory is that intermolecular activity actually occurs

Table 5 Torque-Carrying Ability of Hysteresis Clutches

O.D. (in.)	Length (in.)	Torque (lb-in.)	Thermal capacity (watts)
1.88	1.61	0.075–1.25	10
3.23	2.30	0.5–10.	22
4.65	2.89	1.0–25.	72
6.00	3.50	2.0–55.	150

$$F = \mu N$$

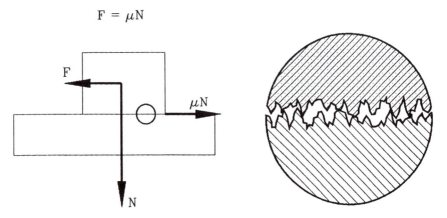

Figure 23 The relationship between friction force and normal force, and a view of the surface asperity concept of friction.

between the rubbing surfaces. Although $\mu = N/F$ is the expression that is normally accepted to describe the friction relationship because it is easy to handle mathematically and generally works quite well in real-world applications, the expression

$$F = \mu N + g \tag{33}$$

that was arrived at experimentally by Coulomb in about 1800 is more correct. In this expression g is called grippage and refers to a force that is required to slide the bodies with respect to each other even when there is no force pressing the bodies together—no self-weight, no anything. An article in the May 30, 1992, *Science News* reports interaction at the atomic level between the elements of the friction couple that supports Coulomb's contention and leads to the conclusion that friction is the cause of stick slip events rather than the result of them. In general practice, μ is considered completely independent of either pressure (p = normal force/contact area) or sliding velocity. Actually, μ is impacted by both pressure and velocity, but these are generally second-order effects and often small enough to be ignored, particularly given the tendency for variations of μ with temperature to be so dominant. The effect of velocity is often accounted for by considering μ_d for dynamic or sliding friction and μ_s for static friction, i.e., the condition when motion is just about to begin, but has not yet begun—the condition of impending motion. As long as μ_s is not exceeded, no motion will occur. When dealing with clutches and brakes where sliding goes from static to dynamic, and vice versa, the smoothest, most chatter-free engagement and disengagement occur when

$$\mu_s = \mu_d \tag{34}$$

Every friction material user, designer, and specifier must realize that the values given represent test data that is for one specific set of circumstances (materials, loads, velocities, and temperature), and that the test results are often very difficult to duplicate with precision. This is one of the major reasons that emphasis is placed on generating sensitivity curves for each clutch or brake application, particularly when self-energizing configurations are contemplated for use.

Figures 24, 25, and 26 demonstrate the effect of temperature, contact pressure, and velocity on wear and coefficient of friction for friction materials made by three different, commonly used processes. The test procedure and conditions are defined by specification SAE J661a, second fade test. All three samples were nonasbestos, organic materials. Examination of Figs. 24, 25, and 26 demonstrates that molded materials offer the most and woven materials the least stability. Tooling and manufacturing start-up costs follow the same order with molded materials requiring the largest initial investment. By and large, woven materials have the highest variable cost followed by molded, leaving

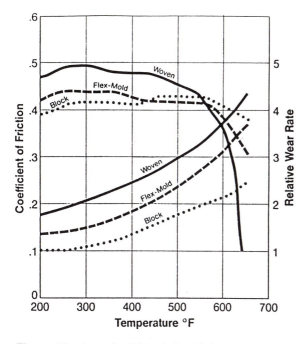

Figure 24 A graph of the relationship between temperature and coefficient of friction and wear rate for selected friction materials made by three major processes. (Reprinted with permission from SAE Technical Paper 851574 © 1985 Society of Automotive Engineers, Inc.)

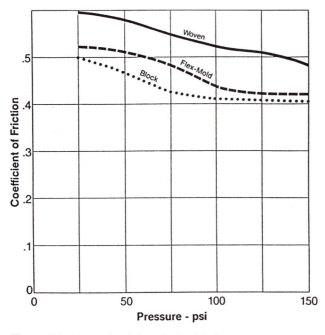

Figure 25 A graph of the relationship between contact pressure and coefficient of friction for friction materials made by three major processes. (Reprinted with permission from SAE Technical Paper 851574 ©1985 Society of Automotive Engineers, Inc.)

flex-mold products with the lowest variable cost. However, woven materials are the most flexible, readily available from stock, and the easiest to procure and apply (particularly to curved surfaces). Many low-volume applications are restricted to woven materials simply because the advantages of the other materials are outweighed by short-term availability, lower initial costs, and ease of application to curved surfaces in single-piece or small-volume lots.

Friction is not a purely surface phenomena, although it is often considered to be such, and a specific friction couple (friction material and the mating surface) tends to generate a typical surface roughness such that if the surfaces are smoother than the typical roughness, they will get rougher as wear progresses. Conversely, if the surfaces are rougher than the typical value, they will get smoother as wear progresses. An idea of such a value might be a typical surface roughness of $\sqrt{30}$ for an organic friction material against cast iron. Gouging, aligning, and pitting are not friction phenomena, which is generally considered as mildly abrasive wear, and characteristic wear rates are part of many friction material tests, including SAE J661a. Although wear rate, as is coefficient of friction, is generally affected by temperature, it can be expressed

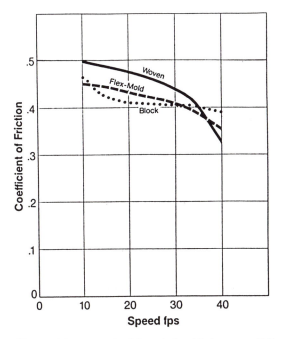

Figure 26 A graph of the relationship between sliding velocity and coefficient of friction for friction materials made by three major processes. (Reprinted with permission from SAE Technical Paper 851574 ©1985 Society of Automotive Engineers, Inc.)

in terms of friction material weight (or volume) loss and the energy expended in causing that weight (or volume) loss. A typical value of a wear rate of one organic friction material rubbing against type A, grade 30 cast iron at approximately 120°F is 10^8 ft-lb/in^3. Two of the equations that express this condition are

$$R_1 = \frac{dE}{dW} \tag{35}$$

where

R_1 = wear rate (ft-lb/lb)

dW = weight loss (lb)

dE = work expended (ft-lb)

$$R_2 = \frac{dE}{dV} \tag{36}$$

where

R_2 = wear rate (hp-hr/in.^{3})

dV = volume loss (in.^{3})

dE = work expended (hp-hr)

There is no attempt made here to rate friction materials by maximum working compressive stress or contact pressure, although many handbooks and design guides do so. They, in essence, tend to pressure rate organic friction materials between 75 and 200 psi with coefficients of friction between 0.2 and 0.6 (dry) and between 0.1 and 0.25 (wet). The author has successfully used material that would be pressure rated at 120 psi by most handbooks at over 600 psi and had end users report successful operation of the same material at even higher levels. These statements do not constitute a recommendation to ignore ratings, but rather to review application very carefully with the friction material manufacturer to actually determine what pressure should be used to arrive at the most effective design in the efficient use of friction material and to eliminate drum scoring.

Three general physical constraints that limit brakes and clutches are

1. Torque limits
2. Temperature limits
3. Physical strength

Staying with torque and physical strength limitations is pretty much a matter of mechanical design. The temperature limits are quite another matter and become very complex; however, friction surfaces are invariably limited by temperature. There are numerous finite element analysis programs that address thermal analysis and that are readily adaptable to clutch and brake heating that provide much more accurate and satisfactory results than the first order techniques that are presented in this chapter. Low-cost PC-based programs that can perform this type of analysis are available from many vendors such as Swanson Analysis Systems, Inc. (ANSYS) of Houston, PA. The rotating components, normally a drum or disk, are limited by their ability to resist the scoring and heat checking that may occur because of both high temperatures and high thermal gradients. The friction material must also be able to resist high temperatures without encountering excessive fade or physical degradation to the point where it is not effective in subsequent events. Temperatures around 600°F are often referred to in fade tests, but actual flash (instantaneous) temperatures of 1000 to 2000°F are not all that uncommon as witnessed by heat checking and the formation of spots of martensitic iron on the friction surface of fine, gray cast iron drums that have been subjected to high-energy stops.

Although relatively sophisticated analyses are required to predict heat dissipation and temperature rise, in many clutch and brake activities enough time lapses between events for the clutch or brake to cool to some base temperature, often the ambient temperature. In such cases, a rule of thumb is to calculate the energy of the stopping or starting event and convert that energy value into a bulk temperature rise of the rotating component that directly opposes the friction material without considering cooling. After all, many of these events take place in less than a second, and how much energy is going to be lost to an air environment in a second, or even 10 sec for that matter? Determining what is an acceptable temperature rise is always a problem, but product and application experience, the literature, and knowledgeable vendors can help provide workable answers. The following is an example of how this is done.

Rotating device: A solid steel flywheel, 24 in. in diameter and 24 in. thick (3073 lb) that must be brought to a stop from 1500 rpm.

Braking device: An internal expanding drum brake with a 2-in.-wide friction shoe that sweeps 240° of a 2.75-in.-wide cast iron drum that has a 12-in. I.D. and a 13-in. O.D. The brake has a rated torque capacity of 2400 lb-ft.

Maximum safe bulk temperature rise = 325°F.

Stopping energy

$$I = MR^2/2 = 0.5 \left[\frac{3073}{386} \right] \left[\frac{24}{2} \right]^2 = 573.2 \text{ in.-lb-sec}^2$$

$$E = I\Omega^2/2 = 0.5(573.2) \left[\frac{2\pi 1500}{60} \right]^2 = 7,071,572 \text{ lb-in.}$$

$$E = 589298 \text{ lb-ft} = 757.5 \text{ btu}$$

Bulk-temperature rise

Use the often accepted assumption that 25% of the energy is absorbed by the friction material and structure

$$E = (757.5)(0.75) = 568.1 \text{ btu}$$

$$W = \frac{0.26\pi(13^2 - 12^2)2.75}{4} = 14.04 \text{ lb}$$

$C = 0.13$ btu/lb-°F, the specific heat of cast iron

$$\Delta T = \frac{E}{WC} = \frac{568.1}{14.04 \times 0.13} = 311.25°F$$

This temperature rise is below the maximum safe bulk value, and the brake should be satisfactory.

The power consumed by the brake can be calculated from the preceding using the classic assumptions of uniformly accelerated motion, which experience shows are probably not very far off if the proper friction material has been chosen.

Stopping acceleration

$$\alpha = \frac{T}{I} = \frac{(2400)(12)}{18339} = 1.57 \text{ rad/sec}^2, \, or$$

$$\alpha = \left[\frac{30}{\pi} \right] 1.57 = 15 \text{ rpm/sec}$$

Stopping time

$$t = \frac{U}{\alpha} = \frac{1500}{15} = 100 \text{ sec}$$

100 sec is a relatively long time, and a significant amount of cooling should occur.

Stopping power

$$P = \frac{dE}{dt} = \frac{589298}{100} = 5893 \text{ lb-ft/sec} = 10.71 \text{ hp}$$

If this is expressed as P' (horsepower/square inch of lining area):

$$P' = \frac{\text{power}}{\text{lining area}} = \frac{P}{6[240\pi/180]2} \Rightarrow P' = 0.213 \text{ hp/in.}^2$$

The P' term is valuable because it gives the user some insights into whether or not a clutch or brake is being under- or overutilized. Table 6 lists P' values for organic friction materials that were generated by Mr. John Nordman, formerly of the Chrysler Corporation.

1. Self-Energization

Self-energization is a phenomena that occurs when the configuration of the brake is such that the friction force generated by the brake reinforces the exter-

Table 6 P' Values for Organic Friction Materials

Type of friction device	Extreme and critical duty, frequent use	Normal intermittent duty	Intermittent or infrequent full-duty use
Drum brakes	1/4	1/2	1½
Disk brakes[a] and plate clutches	1/10	1/5	1/2
Cone clutches	1/5	1/3	2/3

[a] These values of P' consider the entire braking surface to be swept by friction material.

nal actuation force. Consequently, a self-energized brake requires less actuation force than an identical non-self-energized brake. The downside to self-energization is that the more highly a brake is self-energized, the more sensitive it becomes to variations in coefficient of friction and mechanical characteristics that cause variations in brake torque. Figure 27 is a sensitivity curve of two configurations of a strap brake, one self-energized and the other a hypothetical identical non-self-energized brake. Both brakes have the same actuation force. Consider the brake with a friction material/drum that provides a nominal coefficient of friction (μ) 0.45. It is reasonable to expect the value of μ to vary from 0.40 to 0.50 (and possibly more) in actual use. In this case, the output torque of the non-self-energized brake would vary $\pm 11\%$. However, the percentile output torque of the self-energized brake would vary by about twice that much over the same variation in μ.

This example does not represent an extreme case, and the self-energized brake could easily be five times as sensitive to μ as the one shown. Highly self-energized brakes can be self-locking and provide torque that is harmful to the system. Most disk brakes are not self-energizing, but most strap and shoe (both internal and external) are. Good engineering judgment demands that a sensitivity curve such as Fig. 27 be constructed for each new brake design and

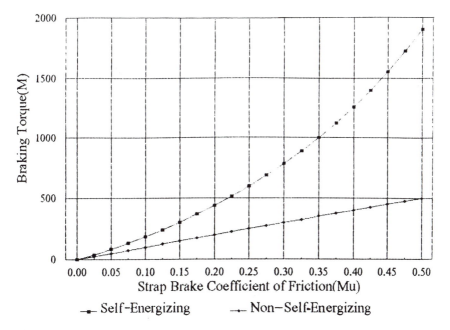

Figure 27 A sensitivity curve of a typical simple strap brake.

any application in which changes in friction material or brake configuration is contemplated. Any brake or clutch feature that affects brake torque should be investigated with a sensitivity curve, particularly when high torque from small packages is a goal. The following list identifies some of the logical parameters for investigation with shoe and strap-type brakes and clutches:

Shoe

1. Coefficient of friction and torque
2. Shoe pivot location and torque
3. Friction material arc of contact and angular location and torque

Strap

1. Coefficient of friction and torque
2. Wrap angle and torque
3. Lever ratios, particularly with differential brakes, and torque

2. Glazing and Burnishing

Driving and braking torque generally change with time after either or both of the friction couple components are new. After the initial wear-in period, the friction material becomes burnished and the coefficient of friction remains quite constant from that time until wear-out under the same set of friction conditions. Burnished friction material is generally smooth and shiny without appearing glassy. However, if excessive heat has been encountered, or contaminants such as oil, grease, or hydraulic fluid of any nature have gotten on the friction material, it will become glazed and appear glassy. The major user problem with glazing is that coefficient of friction becomes very low and stays low—often at an unusually low level. Glazing seems to be the result of changing the nature of the friction material by overcuring it in the case of simple overheating, adding a lubricant to the friction surfaces, or changing the chemical nature of the friction material by adding organic contaminants and curing them at high temperatures. There are times when the contaminant-affected zone can be rejuvenated with specific solvents that are commercially available. There are also times when the glaze-affected zone can be removed by sanding, grinding, or machining. However, once glazing has started it must be halted immediately because it not only causes improper friction action, but it also causes excessive heating through slip and is self-propagating. The most effective solution to glazing is probably to reface the metal half of the friction couple and replace the friction material.

When clean water is the contaminant, it is most often simply necessary to dry the friction material by letting it stand or by gentle slipping. Water can cause drastic reduction in the COF both by acting as a liquid lubricant and by generating steam which must escape through and across the friction surface. Further, it is generally best to leave the brake or clutch disengaged while drying because the combination of water, friction material, dust, and corrosion can cause the friction material to adhere to the disk or drum during drying and literally glue the two components together such that they may have to be pried apart. Such adhesion has been known to require machine disassembly.

Brake and clutch squeal and chatter problems are not as prevalent as they once were, but the problems still occur. Before in-depth investigations into squeal and chatter begin, the following factors should be investigated and rectified if necessary:

1. Static and dynamic coefficient of friction. μ_s and μ_d should be equal for the friction couple being used.
2. Ensure that provisions are made for evacuating wear product (dust) from the friction surface; kerfing the friction material may be adequate.
3. Ensure that the load carrying structure is sufficiently rigid and well enough damped to prevent ringing at objectionable frequencies.
4. Eliminate or damp mechanical clearance in the mechanism that permit unwanted oscillatory motion.
5. Chamfer the ends of the friction material that oppose rotation.

B. Axial Clutches and Brakes

Axial clutches and brakes generally encompass cone and disk devices. Disk clutches and brakes are special cases of the conical devices that occur when the cone angle goes to 180°. The primary difference between the clutches and brakes are that the driven member is restrained from rotating in brake applications, and permitted to rotate in clutch applications. In some applications, the exact same rotating members are used in clutch-brake mechanisms. In spite of the differences in appearance between these axial devices, the equations used to predict their torque and contact pressure are exactly the same, as will be discussed shortly.

1. Cone Clutches and Brakes

Figure 28 is a sketch of a cone clutch or brake that denotes the key geometric variables. This is the fundamental model of all of the axial clutches and brakes. The function of the cone is to

1. Create a large normal force with a modest axial force. The normal force is

$$F_n = \frac{F}{\sin \alpha} \tag{37}$$

2. Create the largest friction material contact area with the minimum outside diameter.

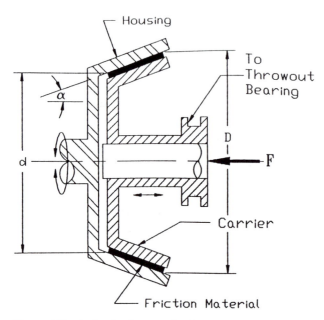

Figure 28 A sketch of a cone clutch.

Compressive stress on the friction material is one of the limiting conditions of any clutch and brake, and the two pressure cases that are generally considered are

1. Uniform pressure across the contact surface of the friction material
2. Actual operating conditions causing constant wear of the friction material

By and large, the second condition is the one actually encountered after the brake has passed through the constant-pressure phase of wear in. Consequently, the pressure distribution across its face cannot be uniform, and the highest pressure must occur at the inner circumference. Figure 29 is a sketch of the radial pressure distribution of a single-plate disk ($\alpha = 90°$) clutch (or brake) for both the constant-pressure and constant-wear assumptions. Figure 30 compares torque that is predicted by the uniform pressure theory with the uniform wear rate theory for three typical conical devices with semicone angles (α) of $10°$, $45°$, and $90°$. Items of particular interest about Fig. 30 are

1. Cone angles must be fairly small (about $\alpha = 10°$) if the cone is to be very effective. (Experience shows that cone clutches with small cone angles are hard to modulate and become too grabby for some applications.)

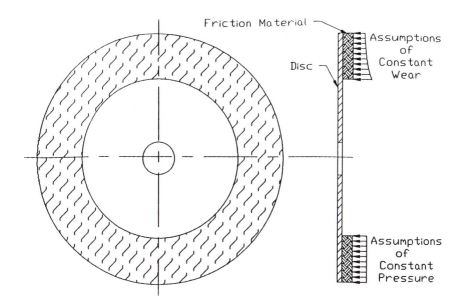

Figure 29 Pressure distributions of constant wear and constant pressure theories in face-type clutches.

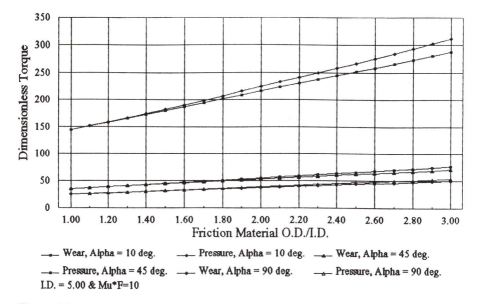

Figure 30 A comparison of the torque provided by constant pressure and constant wear clutches for various OD/ID ratios.

2. The difference in torque between the constant-pressure and constant-wear conditions is not extreme for the common D/d values of less than 2. Consequently, if the device is designed for the constant-wear condition, a little care during run-in will ensure that excessive slip does not occur.

J. Shigley's *Mechanical Engineering Design,* third edition (McGraw-Hill, 1977) presents the derivation of the following equations that predict torque capacity (M) and peak compressive stress (p) for all face-type clutches and brakes with annular friction material segments (refer also to Fig. 31):

For uniform pressure

$$F = \frac{3M(D^2 - d^2)\sin\alpha}{\mu n(D^3 - d^3)} \tag{38}$$

$$p = \frac{8F}{\theta(D^2 - d^2)} \tag{39}$$

For uniform wear

$$F = \frac{4M\sin\alpha}{\mu n(D + d)} \tag{40}$$

$$p = \frac{4F}{\theta d(D - d)}$$

where

d = inner diameter of friction material segment

D = outer diameter of friction material segment

F = axial force

n = number of friction surfaces

$\quad n = 1$ for cone clutch or brake

$\quad n = 2$ for double-sided disk clutch or brake

$\quad n = 2$ for single disk brake with one friction material segment on each side

$\quad n = 2 \times$ number of friction disks lined on both sides in multiple-disk clutches or brakes

α = semicone angle when $\alpha = 90°$, the clutch or brake has become a disk clutch or brake

μ = coefficient of friction

θ = included angle of annular friction material segment in radians

$\quad \theta = 2\pi$ for full 360° friction material segment

$\quad \theta = \dfrac{\pi}{2}$ for 90° friction material segment

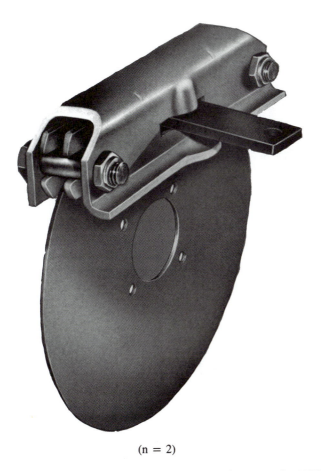

(n = 2)

Figure 31   A typical mechanical disk brake. (Courtesy of AUSCO Products, Inc., Benton Harbor, MI.)

Figures 31 through 33 depict disk clutches and brakes. Figure 31 is a mechanical floating-caliper disk brake that uses two friction segments, one on each side of the disk. When the lever is actuated, the caliper and friction segments (pucks) slide on the two transverse bolts to apply the force F essentially equally on both sides of the disk. Consequently, $n = 2$. Figure 32 is a hydraulic caliper brake that is similar to the mechanical brake of Fig. 31, the major difference being that an integral hydraulic cylinder performs the function of the mechanical lever. Fig. 33 is a hydraulically actuated multiple-disk clutch. In this case, hydraulic cylinders that are internal to the clutch clamp driven slidable friction disks that are on a common external slide between driving slidable metal disks that are mounted on a common internal slide. Torque is

(n = 2)

Figure 32   A typical hydraulic disk brake. (Courtesy of AUSCO Products, Inc., Benton Harbor, MI.)

transmitted through the multiple friction surfaces between the driving and driven shafts. Since there are 5 friction disks lined on both sides, there are a total of 10 friction surfaces, and $n = 10$.

Multiple-disk clutches and brakes are capable of transmitting large amounts of torque in small packages, but can become grabby and difficult to modulate when high coefficients of friction are used. Furthermore, the thin disks can neither absorb much heat energy nor dissipate it rapidly. Consequently, these clutches are best used in high-torque, low-inertia specifications. They are often immersed in oil to aid in cooling, increase wear life, and improve smoothness.

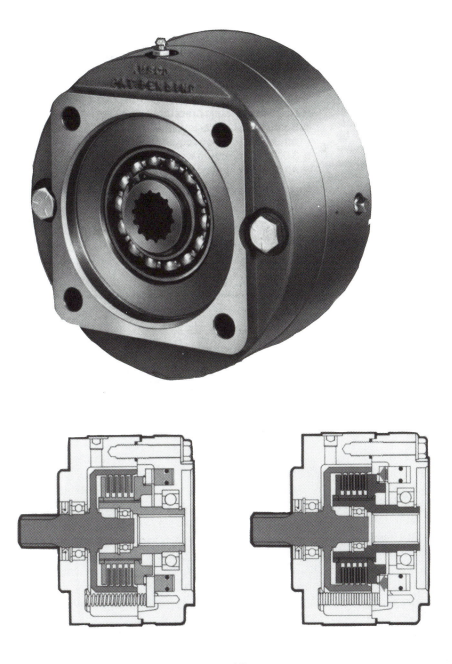

(n = 10)

Figure 33 A hydraulically actuated multiple disk clutch. (Courtesy of AUSCO Products, Inc., Benton Harbor, MI.)

Some wet clutches are further cooled by having oil circulated through the clutch and an external cooler to help dissipate heat. Oil shear clutches are a special case of the wet clutch.

The friction plates of oil shear clutches never touch, and so no wear of the friction material occurs during normal operation of the clutch even when slippage occurs. This feature of oil shear clutches permits relative motion between the input and output shaft without damage to the clutch, and so offers a workable means of infinitely variable speed control. Torque is transmitted between the plates by fluid friction, i.e., the viscous effect of a fluid acting upon surfaces that move in the same plane but at differing velocities. Viscosity is defined as the ratio of shear stress to the rate of shear strain. The torque transmitted in these devices is then proportional to the slip velocity, the active clutch-brake disk area, and the fluid viscosity. Consequently, if there are many disks that are kept close together, significant torque can be transmitted with no physical contact and little relative motion between the slipping disks. The fluid film between the disks can be maintained by keeping them constantly immersed in a fluid bath, pressure feeding, or feeding fluid through the clutch shafts and letting centrifugal force sling it onto the inside diameter of the disks. Just as in any other clutch-brake application, all of the slip energy is converted into heat energy and must be handled. The heat can be transmitted by conduction to and through the housings and then dissipated by radiation and convection to the environment, or hot fluid can be pumped out of the device, through a radiator, and then be returned to the device as cooled fluid.

The following examples demonstrate some of the techniques for using formulas (38) through (41).

Example 1. Your shop has an obsolete machine that is valuable to you, but that would be uneconomical to replace. The cone clutch (see Fig. 28) on the machine has not been serviced for years and has finally failed completely. You do not want to take the time or spend the money to recreate the cone clutch and prefer to replace it with a disk clutch that, for the sake of simplicity and ease of maintenance, has a single friction surface and can use the old actuating mechanism and mounting hardware with minor modifications. The housing of the old clutch had a 10 in. O.D. and just barely cleared the machine base. Other data concerning the old clutch: $D = 8$ in., $M = 400$ lb-ft., $w = 1.5$ in. (frictional material width), $d = 7$ in., $p = 100$ lb/in.^{2} max, $\mu = 0.4$, $n = 1$, $\theta = 2\pi$.

Equation (40) is used to determine the axial force necessary to arrive at the operating torque because we are primarily concerned with the uniform wear condition. First,

$$\alpha = \tan^{-1}\left[\frac{D-d}{2w}\right] = \tan^{-1}\left[\frac{8-7}{2 \times 1.5}\right] = 18.44°$$

Then

$$F = \frac{4M \sin \alpha}{\mu n(D + d)} = \frac{4 \times 4800 \sin(18.44)}{0.4 \times 1(8 + 7)}$$

$$= 1012 \text{ lb}$$

$$p = \frac{4F}{\theta d(D - d)} = \frac{4 \times 1012}{2\pi 7(8 - 7)} = 92 \text{ lb/in.}^2$$

So the existing mechanism provides the 400 lb-ft torque with an axial force of 1012 lb, and a 92 psi compressive stress on the friction surface that is capable of sustaining 100 psi. The new clutch will use a flex-molded friction material with a working compressive stress of 150 psi, and $\mu = 0.41$. The new clutch data: $D = 10$ in., $d = 8.8$ in., $p = 150$ psi, $\mu = 0.41$ and $\alpha = 90°$.

$$F = \frac{4M \sin \alpha}{\mu n(D + d)} = \frac{4 \times 4800 \sin(90)}{0.41 \times 1(10 + 8.8)}$$

$$= 2491 \text{ lb}$$

$$p = \frac{4F}{\theta d(D - d)} = \frac{4 \times 2491}{2\pi 8.8(10 - 8.8)} = 150 \text{ lb/in.}^2$$

This appears to be a workable solution for the single-plate clutch if the actuating mechanism and bearings can withstand the additional ($2491 - 1012 = 1479$ lb) actuating force. If such is not the case, a double-sided friction disk should bring F down to an acceptable level. The 8.8-in. value for d was determined by creating a table of F and p values with d as the variable and selecting d and F at $p = 150$ psi. A simultaneous solution of Eq. (40) and (41) to eliminate F given p and the other variables could have been used.

Example 2. A floating-caliper-type disk brake that is similar to the one shown in Fig. 32 is to have a 12-in.-diameter rotor and 2-in.-wide annular shoes, use a block friction material with a coefficient of friction of 0.45 and a working compressive stress of 200 psi, and must provide 10,000 lb-in. of torque. What should the angle of the annular segment be? A summary of the data follows: $D = 12$ in., $d = 8$ in., $p = 200$ psi, $n = 2$, $\alpha = 90°$, $\mu = 0.45$, $M = 10,000$ lb-in.

F is found by using Eq. (40):

$$F = \frac{4M \sin \alpha}{\mu n(D + d)} = \frac{4 \times 10000 \times 1}{0.45 \times 2(12 + 8)} = 2222 \text{ lb}$$

By rearranging Eq. (41),

$$\theta = \frac{4F}{pd(D - d)} = \frac{4 \times 2222}{200 \times 8(12 - 8)} = 1.389 \text{ rad} = 80°$$

Example 3. A wet multiple-disk clutch similar to that depicted in Fig. 33 is being considered for a new application that will require 60,000 lb-in. of torque. Both room constraints and angular velocity limit the friction material diameter to 12 in. The target compressive stress on the friction material is 120 psi, and the friction couple will provide a coefficient of friction of 0.2 when running in oil. The goal here is to find the minimum number of friction disks that will perform this clutching task assuming uniform wear conditions. A summary of the known data follows: $D = 12$ in., $M = 60,000$ lb-in., $\alpha = 90°$, $\mu = 0.2$, $\theta = 360°$, $p \leqslant 120$ psi.

There are two general ways to find d and n. The first is to eliminate F by solving Eqs. (40) and (41) simultaneously to get

$$f(p) = \frac{\mu n \theta}{16 \sin \alpha}(dD^2 - d^3) \Rightarrow \frac{d[f(p)]}{d(d)} = D^2 - 3d^2,$$

and when that first derivative is set to zero, $d = D/3^{1/2}$, or $d = 0.5774D$. The number of friction surfaces is then determined from the same equation as

$$n = \frac{16M \sin \alpha}{\mu p \theta (dD^2 - d^3)}$$

where $d = 0.5774 \times 12 = 6.929$ in. and

$$n = \frac{16 \times 60,000 \times 1}{0.2 \times 120 \times 2\pi[6.929(12^2 - 6.929^3)]} = 10$$

The number of double-sided friction disks is then $10/2 = 5$. Other methods of solving this problem are to iterate and create tables and graphs to seek the optimum values. Figure 34 is the result of the latter technique as executed with a spreadsheet program. The exact solution used is more precise and less time consuming, but the numerical method gives rapid insights into how sensitive the design is to variation and provides workable solutions.

C. Radial Acting Clutches and Brakes

Radial acting clutches and brakes generally consist of friction shoes or a strap acting against the inner or outer circumference of a brake drum or clutch housing. In some applications, double-faced shoes may act as a clutch when acting in one radial direction and as a brake when acting in the other direction. Similarly one of two separate similar or identical devices acting on the same axis may act as a clutch and the other a brake.

1. Air Tube Clutches and Brakes

Figures 35 and 36 depict devices that employ inflatable fabric-reinforced rubber tubes that have a carrier device on one circumference and friction shoes on the opposing circumference. Inflating the tube causes the tube to expand and force the friction shoes into the drum so that torque can be transmitted by the

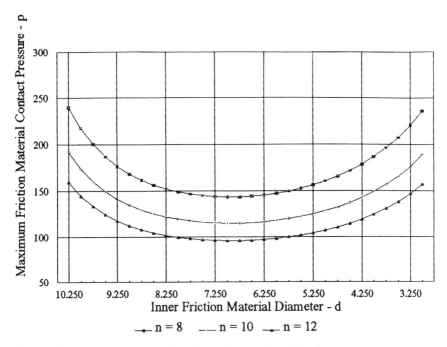

Figure 34 A demonstration of multiple-disk clutch optimization.

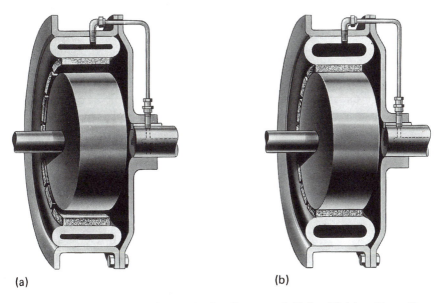

(a)

(b)

Figure 35 A constricting air tube clutch. (Courtesy of Airflex Division, Eaton Corporation, Cleveland, OH.)

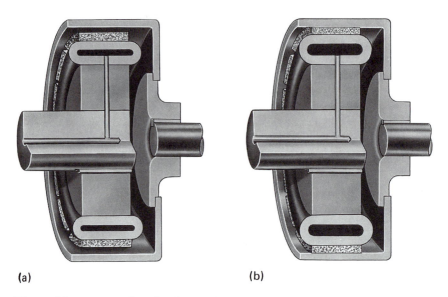

(a)   (b)

Figure 36   An expanding air tube clutch. (Courtesy of Airflex Division, Eaton Corporation, Cleveland, OH.)

resulting frictional forces. Figure 37 is a schematic cross section and force diagram of the constricting (Fig. 35a, b) and the expanding (Fig. 36a, b) devices with the friction shoes in the engaged position. These clutches engage when air enters the tube through plumbing that connects an air supply to the tube via a rotary union. When the air pressure is adequate to overcome spring (springs not shown) and centrifugal forces, the friction shoes frictionally engage the drum and transmit torque between the carrier and the drum to drive a load. The clutches use a number of friction shoes that either drive through the tube or cross members that transmit load between the shoes and the carrier. The multiplicity of friction shoes and the proximity of the tube and cross members to the friction surface results in little self-energization. The weight of the friction shoes, the unsupported portion of the tube, and other hardware that moves with the shoes is acted upon by normal acceleration to cause centrifugal forces as the carrier rotates. The centrifugal forces cause the friction shoes to disengage in the constrictive devices and to engage in the expanding devices. When this basic design is used as a clutch, the carrier must rotate; however, when it is used as a brake, either the carrier or the drum can be fixed and the other member rotated.

The details of manufacturing the air-tube-type air-actuated clutches and brakes along with the broad range of configurations that are offered as catalog items discourages independent and customized user designs. In the event that a first-order estimate of torque or more understanding of the operation of the devices is desired, the following formulas are keyed to Fig. 37 to quantify the approximate torque (M) that can be anticipated.

$$M = \mu N R_f \tag{42}$$

$$N = P - S - W(2.84 \times 10^{-5})U^2 R_c \quad \text{for constricting clutches} \tag{43}$$

$$N = P - S + W(2.84 \times 10^{-5})U^2 R_o \quad \text{for expanding clutches} \tag{44}$$

where

$P = 2\pi R_p w p$

p = air pressure in tube

R_c = radius of the center of mass of *a* friction shoe and the accompanying unsupported portion of the air tube about the axis of rotation

S = spring (not shown) force that opposes engagement

U = angular velocity, rpm

W = weight of the friction *shoes* and the unsupported portion of the air tube

μ = coefficient of friction between the friction shoe and the drum

Dimensional consistency is maintained in the equations.

2. Pivotal Shoe Brakes

Figures 38 and 39 depict internal expanding and external contracting pivotal shoe brakes respectively with the forces acting on the self-energized and the

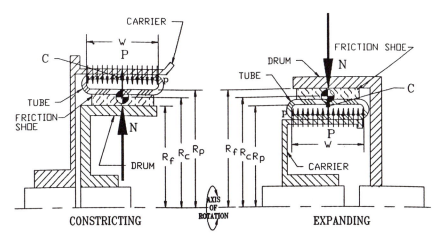

Figure 37 Air tube clutch radial force balances.

$$c = \cos(\alpha - \gamma) * \sqrt{m^2 + n^2} + a * \cos(\alpha + \beta)$$
$$\gamma = \tan^{-1}(n/m)$$

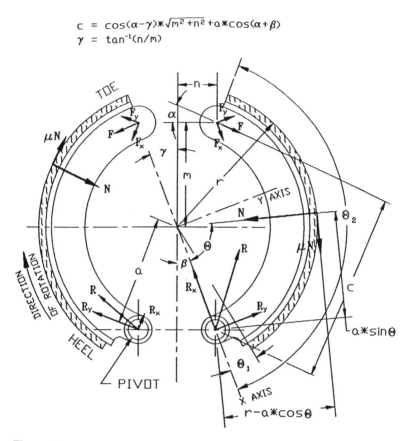

Figure 38   An external pivotal shoe brake free-body diagram.

self-deenergized shoe. Self-energization occurs when the direction of drum rotation causes the friction force to rotate the shoe about its pivot and into the drum so that the friction force becomes a part of the force that actuates the brake. Self-deenergization occurs when the friction force causes the shoe to rotate away from the drum and oppose actuation of the brake. Self-energization allows small brakes to provide large torques at the expense of smoothness and added sensitivity to changes in coefficient of friction. Self-deenergized brakes are less sensitive to changes in coefficient of friction and are generally smoother; however, they provide less torque with the same physical dimensions and actuating force. It is not uncommon for both shoes to be self-energized or self-deenergized as the application requires. It is possible to use this pivotal shoe design with externally applied actuation forces as a clutch, however the design difficulties make other designs, such as the air tube clutch,

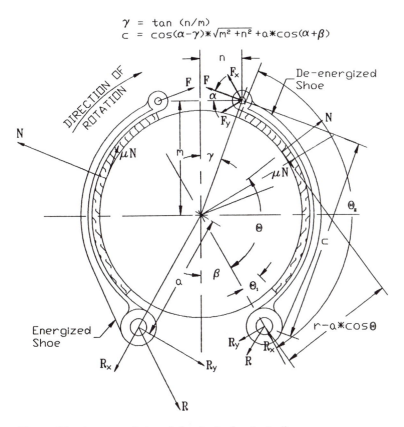

$$\gamma = \tan (n/m)$$
$$c = \cos(\alpha-\gamma)*\sqrt{m^2+n^2}+a*\cos(\alpha+\beta)$$

Figure 39   An external pivotal shoe brake free-body diagram.

more appealing. The pivotal shoe concept is readily applicable to the centrifugal clutches that are discussed in another section.

The following formulas are based upon the derivations and presentation in *Mechanical Engineering Design,* third edition, by J. Shigley (McGraw-Hill, 1977). That analysis is keyed to the maximum allowable compressive stress on the friction material (p_{max}) occurring at an angle θ_a that is the lesser of 90° or the angle of the toe end of the shoe from the pivot and integrating the incremental moments about the pivot of the shoe. This analytical technique also emphasizes that the normal moment (M_n) must be greater than the friction moment (M_f) about the pivot or the brake will become self-locking.

Care must be taken in the design and manufacture of these brakes to avoid single-point contact of any portion of the shoe with respect to the drum. Either heel contact or toe contact will probably cause excessive torque when the brake is new, and too much center contact has caused insufficient torque. Any of the single-point conditions promotes progressive localized glazing of the friction material that often results in inadequate brake performance, generally manifested as insufficient torque. Friction material at the heel does little to contribute to the braking effect, and eliminating it from the first few degrees helps eliminate the heel contact problem. The compressibility of the friction material and the compliance of the metal shoe parts permit well-controlled commercial machining and shoe grinding processes to provide adequately uniform contact between drum and shoe. If the process does not yield uniform contact on the new shoe, *slight* heel *and* toe contact have provided satisfactory new brakes. The construction of sensitivity curves of torque versus the pivot radius (a) and the coefficient of friction (μ) is most important to analyze these devices effectively and responsibly because either unexpectedly high or low torque values can result in improper performance and possible catastrophe.

The following equations are used to design and analyze the pivotal shoe brakes of Figs. 38 and 39.

$$A = p_{max} \frac{wr}{\sin \theta_a} \tag{46}$$

$$B = r(\cos \theta_1 - \cos \theta_2) + \frac{a}{2}(\sin^2 \theta_1 - \sin^2 \theta_2) \tag{47}$$

$$M_f = \mu AB \tag{48}$$

$$C = \frac{2(\theta_2 - \theta_1) - (\sin 2\theta_2 - \sin 2\theta_1)}{4} \tag{49}$$

$$M_n = AaC \tag{50}$$

$$T = \mu Ar(\cos \theta_1 - \cos \theta_2) \tag{51}$$

$$F = \frac{M_n - jM_f}{c} \tag{52}$$

$$F_x = F \sin(\alpha + \beta) \tag{53}$$

$$F_y = F \cos(\alpha + \beta) \tag{54}$$

$$R_x = \frac{A}{4} 2\{(\sin^2 \theta_2 - \sin^2 \theta_1) + j\mu[2(\theta_2 - \theta_1) - (\sin 2\theta_2 - \sin 2\theta_1)]\} - F_x \tag{55}$$

$$R_y = k\left\{\frac{A}{4}[2(\theta_2 - \theta_1) - (\sin 2\theta_2 - \sin 2\theta_1) - 2j\mu(\sin^2\theta_2 - \sin^2\theta_1) - F_y\right\} \tag{56}$$

$$R = (R_x^2 + R_y^2)^{1/2} \tag{57}$$

$$p_i - \frac{Fc \sin \theta_a}{rw(aC + j\mu_i B)} \tag{58}$$

where geometry and terms are shown by Figs. 35 and 36 and the x axis is defined by the center of drum rotation and the center of the pivot

$j = -1$ for the self-energized shoe and $j = 1$ for the self-deenergized shoe

$k = 1$ for internal expanding brakes and $k = -1$ for external contracting brakes

A, B, and C are often repeated constants

p_{max} = target maximum allowable compressive contact pressure for the friction material used in this application

p_i = a calculated instantaneous contact pressure

w = width of the friction material

θ_a = lesser of θ_2 or $90°$

M_f = friction force moment about the shoe pivot

M_n = normal force moment about the shoe pivot

T = torque that the shoe being analyzed exerts on the brake drum

F = actuating force applied to the shoes to arrive at the target contact pressure, p_{max}

F_x and F_y = components of F acting on the shoe along the x and y axes repeatedly

R = reaction at the shoe pivot of the vector sum of the other forces acting on the shoe

R_x and R_y = components of R acting on the shoe along the x and y axes respectively

μ = appropriate coefficient of friction

μ_i = arbitrary, assigned instantaneous coefficient of friction that coincides with p_i

There may seem to be little need to analyze existing brakes with formulas (46) through (58); however, it is generally desirable to do so when field changes are being made to the brake such as (a) considering a change in friction material, (b) considering a change in the cam or hydraulic cylinder that applies the force F, or (c) using a drum with a slightly larger or smaller diameter. Any of the preceding can cause dramatic differences in brake performance. The designer of a new brake faces similar problems and must determine if he can design a brake that will reliably provide adequate torque without being too sensitive to key variables and also have the physical strength to survive overloads and provide sufficient fatigue resistance.

The procedure for either evaluating or designing a brake with Eqs. (46) through (58) is similar. The design procedure is almost inherently iterative and best implemented by programming into a computer and executing. Figure 40 was programmed into a spreadsheet in less than an hour. The general procedure is as follows:

Two self-energized shoes or two deenergized shoes:

Step 1: Determine or assign values to all of the geometric and friction variables.

Step 2: Evaluate the constants A (46), B (47), and C (49).

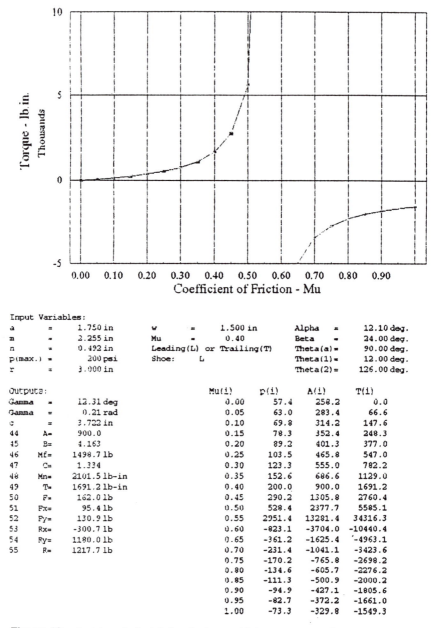

Input Variables:
a	=	1.750 in	w	=	1.500 in	Alpha	=	12.10 deg.
m	=	2.255 in	Mu	=	0.40	Beta	=	24.00 deg.
n	=	0.492 in	Leading(L) or Trailing(T)			Theta(a)=	90.00 deg.
p(max.)	=	200 psi	Shoe:	L		Theta(1)=	12.00 deg.
r	=	3.000 in				Theta(2)=	126.00 deg.

Outputs:
			Mu(i)	p(i)	A(i)	T(i)
Gamma	=	12.31 deg	0.00	57.4	258.2	0.0
Gamma	=	0.21 rad	0.05	63.0	283.4	66.6
c	=	3.722 in	0.10	69.8	314.2	147.6
44	A=	900.0	0.15	78.3	352.4	248.3
45	B=	4.163	0.20	89.2	401.3	377.0
46	Mf=	1498.7 lb	0.25	103.5	465.8	547.0
47	C=	1.334	0.30	123.3	555.0	782.2
48	Mn=	2101.5 lb-in	0.35	152.6	686.6	1129.0
49	T=	1691.2 lb-in	0.40	200.0	900.0	1691.2
50	F=	162.0 lb	0.45	290.2	1305.8	2760.4
51	Fx=	95.4 lb	0.50	528.4	2377.7	5585.1
52	Fy=	130.9 lb	0.55	2951.4	13281.4	34316.3
53	Rx=	-300.7 lb	0.60	-823.1	-3704.0	-10440.4
54	Ry=	1180.0 lb	0.65	-361.2	-1625.4	-4963.1
55	R=	1217.7 lb	0.70	-231.4	-1041.1	-3423.6
			0.75	-170.2	-765.8	-2698.2
			0.80	-134.6	-605.7	-2276.2
			0.85	-111.3	-500.9	-2000.2
			0.90	-94.9	-427.1	-1805.6
			0.95	-82.7	-372.2	-1661.0
			1.00	-73.3	-329.8	-1549.3

Figure 40 An internal pivotal shoe brake sensitivity-curve example.

Step 3: Determine if the brake defined in step 1 will provide the target torque by using Eq. (51). If so proceed, if not revise values in step 1.

Step 4: Determine M_f and M_n from Eqs. (48) and (50).

Step 5: Determine the magnitude of F from Eq. (52). If F is negative, M_f is greater than M_n, and the brake will be self-locking. The values of μ, a, and r in particular must be varied so that F is positive for all possible values with special attention being paid to μ. Varying the values of θ may also help.

Step 6: Construct sensitivity curves of torque (Eq. (51)) versus the appropriate variables such as μ, r, and a. To investigate the effect of a likely range of μ once F has been determined, assign values to μ_i and calculate the associated p_i using Eq. (58). Then calculate the shoe torque by substituting these values of μ_i and p_i into Eqs. (46) and (51) to determine A and ultimately T. A plot of T versus μ_i will demonstrate the sensitivity to T to μ. A satisfactory design is generally reached when all values of T over the anticipated range of μ are acceptable. The effect of varying r and a can be determined by calculating p_i with a range of the appropriate r or a values, then recalculating A (46) and T (51) and observing the behavior of T. Note that T will become very large as $\mu - aC$ approaches zero. Negative values of T have no useful significance and must not be used.

Step 7: Calculate the pivot reaction R by using Eqs. (55), (56), and (57). Use the R values (57) as the basis for the stress analysis that determines the adequacy of the structure. R should also be within acceptable limits at the extreme usable values found in the sensitivity studies of the previous step.

Step 8: Multiply the shoe torque value by 2 to arrive at the brake torque.

Note: If the brake is to be bidirectional, Steps 1–7 should be repeated for the other direction of rotation. Step 5 may not be necessary for deenergized shoes. Figure 40 is an example of this analysis performed on the energized shoe of a 6-in. brake and includes the recommended sensitivity curve. This is a fairly aggressive small brake with a nominal coefficient of friction of 0.4. It is reasonable to expect the coefficient of friction to vary from 0.3 to 0.5 at some time, and the designer must determine whether or not the excursion in torque from less than half of the target value to more than twice the target value is acceptable to the user. Notice that if a situation arises where μ approaches 0.55, the self-energized shoe will become self-locking.

One self-energized shoe and one self-deenergized shoe:

Step 1:   Repeat steps 1 through 7 in the preceding paragraph for the self-energized shoe.

Step 2:   Use the given data and the values calculated in step 1 to determine p_i for the self-deenergized shoe with Eq. (58).

Step 3:   Using the p_i value from the preceding step repeat step 1 for the self-deenergized shoe.

Step 4:   Add the torque from the self-deenergized shoe to that of the self-energized shoe to arrive at the brake torque.

Figure 41 depicts a pivotal shoe brake with floating anchors as opposed to the fixed pivots in Figs. 38 and 39. In such a configuration the anchor restrains the shoe in the horizontal direction, but offers no practical restraint in the vertical direction and requires all vertical restraint to be supplied by the brake drum. Consequently the normal force is always in the quadrant that permits the vertical component of the friction force to exactly oppose the vertical component of the normal force. The reaction force (R) is always greater than the

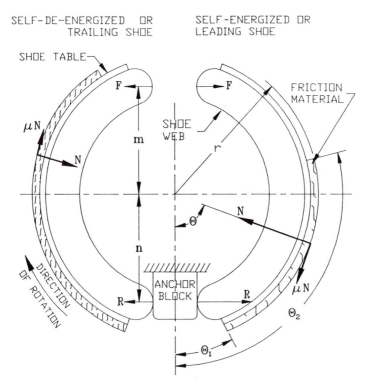

Figure 41   An external floating-anchor brake free-body diagram.

applied force (F) when the shoe is self-energized, and R is always less than F in the self-deenergized shoe. This set of circumstances causes uneven friction material wear when the shoe is fully lined, and a fully lined shoe that is completely worn out on the heel end and literally unworn on the toe end is not an uncommon sight when the brake has only been used in a single direction of rotation. Consequently, for purposes of calculation the friction material may be extended to the end of the shoe on the anchor end, and

$$\theta_2 = 3\theta_1 - \theta \tag{59}$$

or to the toe end of the shoe, *whichever is less*. The physical friction material may extend the full length of the shoe table, but the preceding upper limit on θ_2 is appropriate for the following design and evaluation calculations. Very high torques in small packages have been arrived at by replacing the anchor block with a pivotal ink or adjuster between the two shoes so the reaction force (R) from the first shoe becomes the actuation force (F) for the second shoe. The same design procedure used for the rigidly anchored brakes previously described may be used, but the following iterative procedure has been used successfully.

Step 1: Determine the torque required of the brake, and assign appropriate values of torque to each shoe. If one each of self-energized and self-deenergized shoes are used with a fixed anchor block, and no prior knowledge indicates to the contrary, assign two-thirds of the torque (T_1) to the self-energized shoe and one-third of the torque (T_2) to the self-deenergized shoe.

Step 2: Determine the general physical dimensions of the brake from a layout or prior knowledge.

Step 3: Calculate the normal force required for the first shoe with

$$N_1 = \frac{T_1}{\mu r} \tag{60}$$

where μ is the coefficient of friction for the friction material being used or considered.

Step 4: Calculate

$$\theta = \text{Arctan} \left[\frac{1}{\mu} \right] \tag{61}$$

Step 5: Determine the value of θ_1 from the layout, and calculate

$$\theta_2 = 3\theta_1 - \theta \tag{62}$$

or $\theta_2 = $ the angle to the toe end of the shoe, whichever is less.

Step 6: Calculate the maximum friction material contact stress with

$$p_{max} = \frac{N}{rb(\cos\theta_1 - \cos\theta_2)} \tag{63}$$

where b = the width of the friction material, and compare p_{max} with the specified value for the friction material.

Step 7: Adjust dimensions and repeat the previous steps until a satisfactory match of p_{max} is made.

Step 8: Calculate

$$F = \frac{N(n\sin\theta + \mu n\cos\theta + j\mu r)}{m + n} \tag{64}$$

where $j = -1$ for a self-energized shoe and $j = 1$ for a self-deenergized shoe.

Step 9: Determine

$$R = N(\sin\theta + \mu\cos\theta) - F \tag{65}$$

Step 10: If the first shoe was a self-energized shoe and the second shoe is a self-deenergized shoe, calculate the torque for the second shoe by

$$T_2 = \frac{\mu F(m + n)r}{n\sin\theta + \mu n\cos\theta + j\mu r} \tag{66}$$

Step 11: Sum the torques of the individual shoes and compare the sum to the target torque. If a satisfactory result has been arrived at, the analysis is complete. Otherwise revise the layout and other data and repeat steps 1 through 10 until a satisfactory match is made.

Good design practice requires sensitivity curves of T versus μ and T versus n in particular. As mentioned, these brakes can be highly self-energized to provide very high levels of torque in small packages, and so are quite susceptible to lock up with only small changes in key parameters.

3. Strap Brakes

Strap brakes employ one of the oldest concepts in power transmission and use the same theory as power transmission by belts and snubbing with chains or ropes. Figure 42 depicts the three general forms of strap brakes: the simple, self-energized differential, and self-deenergized differential. The simple brake is not only the easiest to use, but its performance is quite predictable and both small-volume and high-volume costs are low. Strap brakes can also often be incorporated with existing machine features. The torques of all three configurations vary with the direction of rotation with the single exception of the self-deenergized differential brake when the lever ratio $b/a = 1$. In this one case the torque is the same regardless of the direction of rotation, and the performance and torque can be very similar to a disk brake of equal size and actuating force. Although simple brakes are rarely self-locking unless the

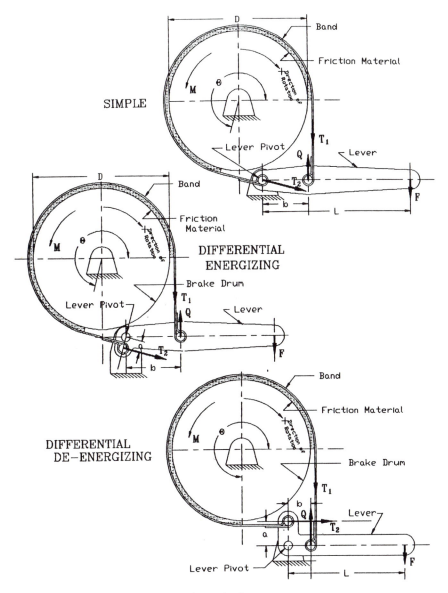

Figure 42 The three basic types of strap brake.

designer and/or manufacturer have erred badly, the self-energizing differential brake self-locks when

$$a \geqslant b/e^{\mu\theta} \tag{67}$$

and is sensitive to the coefficient of friction. Other factors to be considered when designing and applying strap brakes are to

1. Keep the line of action of the straps perpendicular to the line of action of the lever.
2. Minimize dead band in the linkage.
3. Minimize the amount of distortion of the band when it is installed in the application by procuring bands that are of the correct radius and by locating the pivots correctly.
4. Maintain alignment between the strap and the drum.
5. Maintain adequate rigidity in the actuation system to avoid vibration and squealing that is attributable to avoidable feedback and harmonics. Care must also be taken with the metal band design because it must be flexible enough to bend as required, but it also stretches longitudinally and so becomes a high-rate, low-amplitude spring during brake application.

One of the problems in applying strap brakes has been the proliferation of equations in handbooks for the three different types of brakes. Most strap brake design can be accomplished with the five following formulas from "Don't Forget Strap Brakes," *Power Transmission Design*, R. C. St. John (Penton Publishing Company, August 1991), and that are keyed to Fig. 42:

T_1: Lever side strap tension force

Forward rotation
$$\mathrm{T}_1 = \frac{\mathrm{FL}}{b + \mathrm{jae}^{\mu\theta}} \tag{68}$$

Reverse rotation
$$\mathrm{T}_1 = \frac{\mathrm{FLe}^{\mu\theta}}{\mathrm{be}^{\mu\theta} + \mathrm{ja}} \tag{69}$$

T_2: Anchor side strap tension force

$$T_2 = T_1 k e^{k\mu\theta} \tag{70}$$

M, brake torque

$$M = \frac{kD(T_2 - T_1)}{2} \tag{71}$$

p, maximum friction material to drum contact stress

$T_{\max}$ is the greater of T_1 or T_2

$$p = \frac{2T_{max}}{wD} \qquad (72)$$

where

   a = lever pivot to strap anchor end pivot distance

   b = lever pivot to band actuating end pivot distance

   D = brake drum diameter

   F = force applied to the brake lever

   L = distance between F and lever pivot

   w = friction material width where it contacts the drum

   μ = friction material to drum coefficient of friction

   θ = wrap angle of friction material in radians

   j:

      $j = 0$ simple brake

      $j = -1$ self-energing differential brake

      $j = 1$ self-deenergizing differential brake

   k:

      $k = 1$ forward rotation (in direction of T_1)

      $k = -1$ reverse rotation (in direction of T_2)

Figure 43 is a sensitivity curve for all three types of strap brakes with the same general physical dimensions and the same input force F. It is apparent from this curve that the general dimensions shown cause the self-energized brake when rotating in the forward direction to provide very high levels of torque with low coefficients of friction and to lock up when μ is approximately 0.25. The other two brakes are much less sensitive to μ but provide much less torque. None of the brakes exhibit locking tendencies in the reverse rotation mode, but there is vast disparity between forward and reverse torque, particularly with the self-energized differential brake. The self-deenergized differential brake shows the least sensitivity, and the torque ratio between forward and reverse when $\mu = 0.5$ is just a little over 2:1 in this particular configuration, while the torque ratio for the simple brake is close to 10:1 for the same conditions. The real purpose of Fig. 43 is to demonstrate the necessity for a study of torque variations in these applications.

Figure 44 depicts a unique bidirectional band brake that is constructed so that a simple brake (see Fig. 42) provides the same torque in either direction of rotation. The flexible, coaxial cable permits the brake band to rotate with the drum until the band is stopped by one of the two anchors that are mounted to the nonrotating machine base. This arrangement causes the cable force to be T_1 and of the same magnitude in either direction of rotation while simultaneously

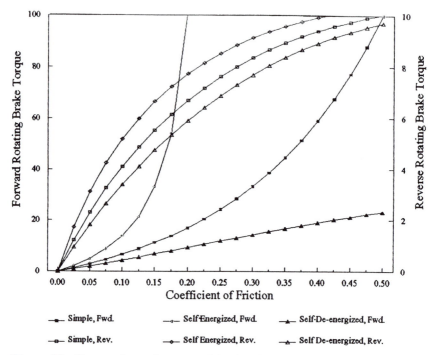

Figure 43   Comparative performance of the three basic strap brakes.

positioning the other end of the band against the fixed anchor. This design allows relatively high, symmetrical bidirectional torque to be generated from a small potentially modestly priced package.

4.   Centrifugal Clutches

There are a plethora of centrifugal clutch designs that employ centrifugal force as the means of actuation and often as a major source of the normal force. Most centrifugal clutches use springs to regard shoe motion so that the prime mover can begin to rotate before the clutch allows the driven system to place a load on it. For example, electric motors can accelerate beyond the high stall current starting condition before having to encounter system load, and internal combustion engines may use clutches that have engagement and release speeds that are above the idle speed of the engine to allow them to warm up and generate enough torque to drive a useful load before being accelerated to their operating speed. Some centrifugal clutches, primarily European, employ pivotal centrifugal weights that act upon the disks of face contact disks; however, most American centrifugal clutches are rim-type clutches that rely upon the weight of the clutch shoes and normal acceleration to force the friction material

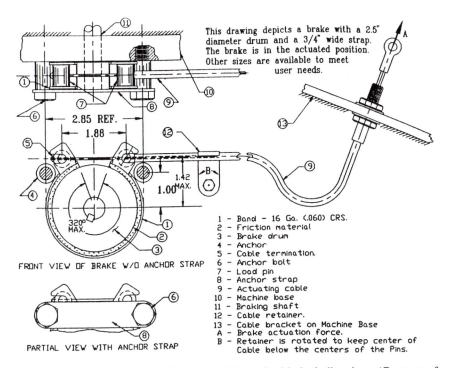

This drawing depicts a brake with a 2.5"
diameter drum and a 3/4" wide strap.
The brake is in the actuated position.
Other sizes are available to meet
user needs.

2.85 REF.

1.88

1.42

1.00 MAX.

320° MAX.

FRONT VIEW OF BRAKE W/O ANCHOR STRAP

PARTIAL VIEW WITH ANCHOR STRAP

1 – Band – 16 Ga. (.060) CRS.
2 – Friction material
3 – Brake drum
4 – Anchor
5 – Cable termination.
6 – Anchor bolt
7 – Load pin
8 – Anchor strap
9 – Actuating cable
10 – Machine base
11 – Braking shaft
12 – Cable retainer.
13 – Cable bracket on Machine Base
A – Brake actuation force.
B – Retainer is rotated to keep center of
Cable below the centers of the Pins.

Figure 44 A simple strap brake that is self-energized in both directions. (Courtesy of Component Product Sales Company, Fredricktown, OH.)

against the inner rim of a housing and so provide a torque transmitting mechanism. A rather notable exception to this technique was a mercury clutch in which centrifugal force caused liquid mercury to flow through small orifices and into a rubber tube gland which expanded under the force of the mercury and drove friction shoes into the inner rim of the clutch housing. These clutches provided respectable torque along with a soft slow start and experienced considerable success in the textile industry, on radial-engine-powered helicopters, and in other applications. Centrifugal clutch power ratings range from fractional to thousands of horsepower, although the majority of stock clutches are rated at less than 20 hp. Figure 45 shows a number of clutch shoe arrangements that are currently in use in American industry.

The clutch depicted by Fig. 45a employs a lined friction shoe that moves radially against the action of the spring to engage the housing. The mass of the shoe and the normal acceleration provide most of the force between the friction material and the housing, so there is little self-energization in most clutches of this configuration, and they generally provide quite soft starts. The short, articulated shoes of the clutch of Fig. 45b provide opportunities for self-energized

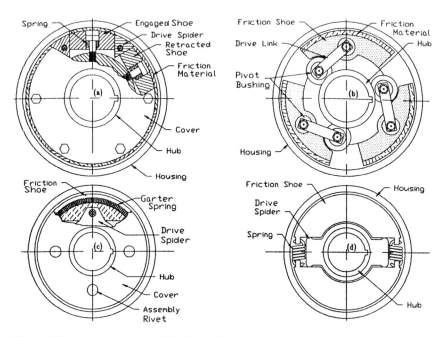

Figure 45   Some typical centrifugal clutch shoe arrangements.

or self-deenergized function and *can* provide very high torques in small packages when operating in an aggressive self-energized mode, or provide low torque and very soft starts when operating in the reverse direction of rotation, or the self-deenergized mode. The clutches of Fig. 45c and d are derivatives of these configurations, and generally used on relatively low power (fractional to maybe 10 hp) applications and are usually quite highly self-energized. The friction shoe is often made entirely of friction material and may be either organic or metallic. If metallic, it is generally powered metal. Self-energization in these two clutches is most heavily impacted by (a) the distance between the shoe pivot point and the center of clutch rotation and (b) the coefficient of friction. The clearance between the clutch shoe and the drive spider that comes about because the friction shoe displaces radially from the drive spider prior to and during operation can cause severe impacts between the friction shoe and the drive spider when fluctuations in load occur. In the case of I.C. engines, impacts will occur with each cylinder firing and possibly more often. In spite of such severe pounding, such clutches have been used successfully for many years at modest first cost.

$c = a*\cos(\alpha + \beta)$
$G = .284*\bar{U}_R^2*r_c$
Notice that α is negative as shown.

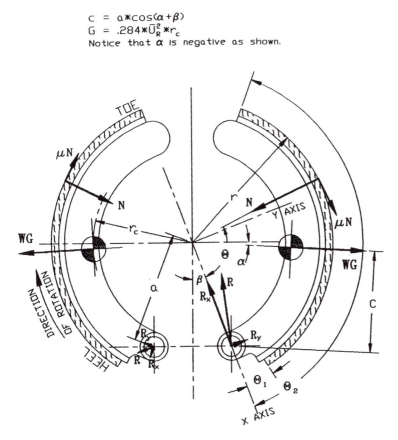

Figure 46 A pivotal shoe centrifugal clutch free-body diagram.

Centrifugal clutches that use the same general construction and principles of the brake of Fig. 41 as displayed by Fig. 46 have also been used with considerable success. The values of m and n from Fig. 41 are zero, and the formula for c is shown on Fig. 46. In these designs, centrifugal force causes the shoes to be rotated around their pivots and into contact with the housing. The higher the angular velocity, the greater is the centrifugal force and the friction force, and so the clutch torque. The brake mathematics using Eqs. (46) through (55) are perfectly suitable for the clutch application, and the force (F) to actuate the shoe becomes

$$F = WG \tag{73}$$

where

W = weight of the shoe

G = normal acceleration acting on the shoe in gravities $= 0.284U^2 r_{(c)}$.

$r_{(c)}$ = radial distance between the center of rotation and the center of gravity of the friction shoe assembly

Power is not an inherent characteristic of a clutch, but torque is. Torque can also be readily determined from power and angular velocity by using basic power transmission formulas. The frictional torque limitations of a centrifugal clutch are a function of the square of the angular velocity, and the following method has been used successfully to determine the torque that a centrifugal clutch will provide at any angular velocity:

$$T = T_b (U_O^2 - U_R^2) \tag{74}$$

where

T_b = basic torque of the clutch, i.e., the torque that the clutch will provide at 1000 rpm without springs or bias

U_O = angular velocity at which the clutch will be operating in thousands of rpm's

U_R = angular velocity at which the springs allow the clutch to disengage in thousands of rpm's

Solving Eq. (74) for values of T_b, U_O, and U_R and comparing the values with the needs of the application provides a rapid means of determining what is required of a centrifugal clutch for a specific application. The designer and specifier must take care to ensure that the clutch is structurally capable of handling the peak loads of the application, particularly when I.C. engines are being used. For example, four-stroke-cycle, single-cylinder diesel engines place heavy loads on centrifugal clutches. When highly self-energized centrifugal clutches are used with I.C. engines, an often startling phenomena occurs when a stroboscope is used to observe the clutch, or when engine speed and load speed are measured. Careful observation shows that the load may be running faster than the engine. This occurs because of within-cycle speed changes of the prime mover. When the prime mover accelerates, the clutch drives and accelerates the load, and then the prime mover decelerates. The inertia of the load may cause the load to decelerate less than the prime mover, so the clutch slips in the deenergized direction because for that short instant in time the load is turning faster than the prime mover. This phenomena cannot cause the load to run away because the maximum angular velocity it can reach is the maximum within-cycle angular velocity of the prime mover. For example, the engine when running at a nominal speed of 2500 rpm may actually have a within-cycle speed variation of ± 200 rpm. In this case, the ratcheting of the pivotal shoe centrifugal clutch will never cause the load to rotate faster than 2700 rpm.

ACKNOWLEDGMENT

I would like to thank Base Line Engineering Inc. for help in the production of many of the drawings.

4

PLAIN BEARINGS

Richard H. Weichsel

AB Consultants International, Inc., Naples, Florida

David W. South

Consultant, Madison Heights, Michigan

I. HISTORY OF THE PLAIN BEARING

The original plain bearing was cut out of a tree called lignum vitae. The earliest known use was among a seafaring people, the Phoenicians, who used the bearing material to make the whole shiv, the whole block, and the pintle and grudeon as well as the sphericals for the oars in the wall of the vessel. Instead of making the whole bearing out of something else and putting a bearing in place (as we do today), the entire structure was the oiled bearing material so as to better lubricate the bronze shafts, the only metallic shafts known at the time.

Over a thousand years later the ability to create high enough temperatures to melt iron gave the world cast iron, which, when properly made into round bars, became the shaft and bronze and sometimes wood became the bearing.

Another thousand years went by and the Bessemer converter gave us cheap steel, and rolling mills gave us rounds that became the shafts. Now we had a steel shaft and, to go on that shaft, we had cast iron bearings (grease lubricated), bronze bearings (oil lubricated), and wooden bearings. The state of the art pretty much stayed this way, which is what brought around the rolling element bearing in the industrial marketplace.

In the 1920s the world of antifriction devices basically consisted of rolling element bearings (balls, needles, rollers, spherical rollers, taper rollers, naked and mounted) and journal bearings (consisting of sleeve, flange, and thrust).

After World War II, with the brilliant breakthrough of the polytetrafluoroethylene (PTFE) bearing, the evolution got into high gear in favor of the sleeve bearing. The plain bearing marketplace climbed from 17.2% of all the dollars spent for bearings in 1974 to 22.0% from 1984 and, by 1988, 25.1% of all the money spent for bearings went into sleeve bearings. The driving force behind this was multifold.

1. In the area of nonferrous materials, Western Reserve developed Microcast.
2. In the area of PTFEs, filled-composition PTFEs were born.
3. In the area of polyimide/amide materials, Amoco and Phillips Products brought out Ryton and Torlon.
4. Next came DuPont Vespel ultrahigh temperature plastics and, most currently, the Celazole material from Hoechst Celanese, which can withstand 900°F.

There is no end in sight.

II. NOMENCLATURE

The bearing industry has created a language all of its own, so let us talk a little bit about nomenclature.

The cylindrical, tubular device called a bearing has been used for supporting a loaded shaft for over 4000 years. Some call the rotating member a journal. The supporting device is called many things, some of which are incorrect. For instance, it is not a friction bearing. Any unit put into a design to maintain rotation is obviously put in to remove friction. Therefore, all sleeve bearings can be considered antifriction devices.

There are actually two types of antifriction devices. First are rolling element bearings, which include ball bearings, needle bearings, roller bearings, spherical roller bearings, and taper roller bearings. These bearings come in various classes of fit, and they come naked or mounted in a housing (like pillowblocks).

These bearings are not bushings either. A bushing is used to maintain an aperture or entryway. For instance, contractors cut cardboard tubing of various diameters into 8-in. lengths and then place it between the inside and outside concrete mold for casting a cement wall so that, after the mold is removed, you have a nice round hole through which you can pull power lines, water lines, take sewer lines outward, or whatever. A bushing, however, does not imply rotation nor centerline maintenance.

They are, of course, journal bearings, but so is every bearing that supports a journal or shaft so that is not a good enough definition. A really good definition, European and dating back to the 1880s, was found in an engineering article by a German bearing designer, wherein the sleeve bearing was quite correctly called a "plane" bearing, meaning a single line of contact with the whole length of the bearing directly underneath the load. It is a geometry term if you will (i.e., a single plane of support for the oil film, the oil wedge, or the high-pressure system for supporting the journal during rotation.

If you have studied the art of the plain bearing, you know that it is not "plain" at all. Anybody who is familiar with a Morgan Engineering bearing called a Morgoil knows better. Anybody who is familiar with the products by KMC Division of Cookson America in the field of high-pressure deflecting pad bearings knows better. And, of course, there is the granddaddy of them all— the Kingsbury product line. Certainly, none of these are "plain" bearings.

So there are two classes of bearings, one of which is the rolling element series (which comes in five basic styles) and the other of which is the plain bearing (which sometimes is called a journal bearing and is generally in one of three silhouettes—a sleeve, a flange (combination of sleeve and thrust), and a thrust bearing). These plain bearings also may come naked or mounted in pillowblock designs.

The basic difference between the two, and what is hardest to grasp, is the materials in rolling element and plain bearings. It is not so difficult in rolling element because, thanks to the great engineering job done by the Timken bearing company of Canton, OH, SAE 52100 steel exists as an electric furnace product allowing phenomenal loads and, at the same time, super-hard-wearing surface conditions. In fact, probably 95% of all rolling element bearings are made up of this type of material. Another category of material in use is SAE 4140, which provides through-hardened condition. And, of course, there is probably 2% in stainless steel for those environmental characteristics that preclude the use of a pure ferrous base material.

Not so in the plain bearing field. There are about 84 plain bearing materials. In wood bearings you can find lignum vitae, straight-grain hard maple, and a few superhard Brazilian materials. In bronze there are over 20 different materials in the leaded bronzes and the hard bronze group in basic everyday uses. There are about 100 in total, counting all the specialty bronzes. With the advent of Western Reserve Manufacturing's new Microcast product, you just doubled every one of them.

This goes on and on. So it is best to simplify the various product lines by dividing them into classes. But before we do that, ponder this decisive difference between rolling element and plain bearings. If you have a situation where you must upspeed or overload the system, you change the L10 life chart rating on the rolling element bearing. It is then necessary to do something about the design. It is impossible to take a light-duty bearing out of a machine system and put a medium-duty bearing in its place without some major manufacturing alterations of the housing, which has quite an attendant expense.

However, in the plain bearing field, without changing the I.D., the O.D., or the length, you can pick one of about 84 different materials to resolve your problem without altering the housing or, generally, the shaft. Obviously, there is no attendant manufacturing cost impact if you are merely changing bearing for bearing, size for size.

The first category of plain bearings is known as Class I. Class I bearings have outside sourced lubrication systems to permit them to work. Any system requiring a liquid lubricant, a grease, or a plastic material that can work with any liquid is a Class I bearing system.

In Class II bearing systems materials contain within their own walls the lubricant necessary to get the job done. This category include porous powder metal structures that can be readily formed into sleeve, flange, or thrust or into raw material to be machined into any type of silhouette. There also is the first bearing that started it all—the lignum vitae bearing, later accompanied by the straight-grain hard maple. Also in this category are some wonderful new concepts in plastics, where plastic resin with a melting point lower than the burn point of the lubricant can be mixed with lubricant injection-molded or extruded through a hot zone melt system, giving a bearing that has microballoons of lubricant throughout the system. The plastic goes back to solidus, trapping minute droplets of oil throughout its system, and then waits for the shaft to wear a few millionths of an inch and crack open, allowing the oil with its fantastic clinging characteristics and its low-energy rupture characteristic to be exposed to the rotating member.

Last, and one of the fastest-growing fields, are Class III bearings, which include all PTFE bearings and all plastic bearings that have PTFE additives in sufficient quantities to make the apparent surface think that it is a full Teflon bearing to the shaft that has been presented to it. This area includes polyacetals, filled with Teflon, products like PEAK which are Teflon filled, as well as an entirely new development from Hoechst Celanese called Celazole.

III. CLASS I BEARINGS

Class I bearing systems are identifiable by the requirement for outside lubrication. Although these bearings have a high degree of apparent lubricity, they do not run dry. In order to have a long-wearing bearing combination (in plain bearings there are two components, the shaft and the bearing), it is necessary to have lubricant to serve as the interface.

Lubricants generally are in the petroleum class, and they are selected because of two inherent factors:

1. They have cling, which is the ability to coat the surface of the I.D. of the bearing and to thinly coat the surface of the O.D. of the journal.
2. They have the ability to rupture upon mechanical separation cleanly, smoothly, and with little effort.

The second factor is extremely important because of the need for low temperatures in the bearing system. Continuous rubbing characteristics, regardless of how light a force is in the equation, do contribute to thermal heat rise. Therefore, the coefficient of thermal expansion of all the materials in the system is affected (the bearing and the journal primarily but to some degree the housing).

The rating of the lubricant by whatever test method (ASTM, SAE, etc.) is very important to the design configuration: generally, the higher the velocity, the lower the SAE number, and vice versa. The higher the load and the lower the speed, the more important it is to have a heavier viscosity. When operating in the 5 sfm (surface feet per minute) and under characteristic, EP lubes with their high cling factor are necessary to maintain the oil wedge for lifting and separation of journal and bearing.

The characteristics of Class I bronze alloys are shown in Fig. 1. In this field there are many grades of material. In the leaded bronze grades, SAE 64 (which is 80% copper/10% tin/10% lead, all of which must be virgin-certified ingot material source) is the premier bearing material in this grade. Making use of its P and V capabilities and by using the proper grade of petroleum lubricant, the most shaft-kindly designs can be possible.

Copper has a high degree of radiation but a very low order of apparent lubricity and a very low order of load capacity, so it is necessary to alloy this product. The development of SAE 64 was the final step in designing a great line of standard bearings.

During World War II this line was abruptly broken because of the enormous worldwide need for tin for many applications and the impossibility of finding sufficient quantities fast enough to cover all of the commercial and wartime needs. As a wartime expedient, SAE 660 was developed. The three-digit number was to forever be a reminder that this was not one of the regular grades. SAE 660 (which is 83 copper/7 tin/7 lead/3 zinc) was derived by altering the British gunmetal grade of 85 copper/5 tin/5 lead/5 zinc with sufficient additional tin and lead (two points) and a reduction in the zinc (four points) to give a fair substitute.

Today, because of breakthroughs in the manufacturer of continuous-cast bronze bar, we have under the Microcast patent (property of Western Reserve Manufacturing Company, Inc.) a material that, because of its superfine grain structure, has increased the apparent lubricity to such an extent that SAE 660 is a passable replacement for SAE 64 metal. However, not to be overlooked is that Microcasting SAE 64 metal makes it superior to anything ever seen on the face of the industrial earth. (See Fig. 2 for some typical physical characteristics of alloys produced by various methods.)

In the field of bronze there is, in addition to the leaded bronze group, the "hard" bronze group, which covers aluminum bronze and manganese bronze. However, in this group there is competition from a leaded bronze product

Western Bronze CDA Alloy	% Cu	Sn	Pb	Zn	Ni	Tensile Ksi	Yield Ksi	Elongation % in 2"	Brinell Hardness	Typical Applications
836	85	5	5	5		45	21.4	28	72	Plumbing goods, pumps, valves, small gears
836† (Special)	85	5	5	5		51	26	12	83	Impellers, gears
844	81	3	7	9		37	16.5	23	55	Plumbing goods, low pressure valves
903	88	8		4		49	23	25	77	Bearings, valves, rings, pumps, gears
905	88	10		2		51.5	29	18	92	Piston rings, gears
907	89	11				51.5	29	18	95	Gears
916	88	10.5			1.5	51.5	29	14	100	Gears
922	88	6	1.5	4.5		45.5	26	35	76	Steam valves, fittings, vessels
927	88	10	2			48	20	18	80	Piston rings, gears
929	84	10	2.5		3.5	53	31	15	100	Gears, wear plates, cams. good machinability
932	83	7	7	3		44	24	16	72	Bearings, bushings
934	84	8	8			35	17	18	60	General bearings and bushings
937	80	10	10			41	24	10	80	High speed bearings for heavy loads Corrosive environments
938	78	6	16			34.2	23	12	62	Moderate demand bearings, Mine water pumps
941	75	5	20			28.7	22.8	8	57	High speed bearings for moderate loads
943	70	5	25			27	13	15	48	High speed bearings for light loads
943† (Special)	68	2	30			16.3			30	Fuel pump bearings
947 (HTTRT)‡	88	5		2	5	90	66	9	180	Wear resistant parts requiring very high strength

† Modified   ‡ Heat treated condition

Figure 1 Nominal chemical composition and typical physical properties of the most common bronze alloys. (Courtesy of Western Reserve Manufacturing Company, Inc., Lorain, Ohio.)

Alloy Cu-Sn-Pb-Zn-Ni	Type Casting	Tensile Strength psi	Yield Strength psi	Elongation % in 2"	Impact Strength Ft. Lb.	Brinell Hardness
CDA 929	Microcast	59,000	38,500	14	28	117
84-10-2 1/2-0-3 1/2	Continuous cast	51,500	29,000	18	25	92
Gears	Permanent mold	51,000	26,500	14	20.2	83
Wear Plates	Sand cast	48,400	21,600	42	8.7	63
CDA 836	Microcast	44,500	23,900	32	20	75
85-5-5-5	Continuous cast	42,000	21,400	28	15	72
Plumbing goods	Permanent mold	41,200	19,500	24	13.5	65
Pumps	Sand cast	36,400	16,900	35	12	49
Valves						
Small gears						
CDA 932	Microcast	45,500	27,600	21	12.5	77
83-7-7-3	Continuous cast	44,000	24,000	16	11	72
General use	Permanent mold	41,600	21,900	18	11	69
Bearings	Sand cast	40,400	19,500	39	9.3	65
Wear plates						
CDA 941	Microcast	31,300	18,600	26	7	60
75-5-20-0	Continuous cast	28,700	18,300	8	6	57
High speed	Permanent mold	28,500	18,200	12	5.3	53
Bearings	Sand cast	26,900	15,800	17	5.5	45

Figure 2 Typical physical properties of a series of foundry alloys cast by four methods. (Courtesy of Western Reserve Manufacturing Company, Inc., Lorain, Ohio.)

developed approximately 20 years ago called CDA (Copper Development Association) 929, which is a material that has a 3.5 nickel/2.5 lead/10.0 tin and the remainder copper. (There is no SAE equivalent for this grade.) I note this in passing because CDA 929 will outperform the hard aluminum bronze on almost every occasion while providing the shaft with a highly conformable bearing material.

While aluminum and manganese give hardness, that is not the perfect substitute for strength (which is acquired by the nickel additive in CDA 929). For ultrahigh hardness and ultimate strength belong to the Spinodole bronze category, which is out of the area of good shaft compatibility requiring ultrahard and superfinished shafts for maximum life.

It cannot be stressed too strongly that in the plain bearing world, the finish and hardness and the method of acquiring the finish in a plain bearing system for both the I.D. of the journal and the O.D. of the shaft is just as important as the finish, the tolerance, and the geometry in producing a Class 9 ball bearing system in the rolling element world. (See Fig. 3 for engineering data relating to Class I bearings.)

The Category of Class I bearings (i.e., those requiring exterior sourced lubricants) includes cast iron bearings, steel sleeve bearings, and bearings

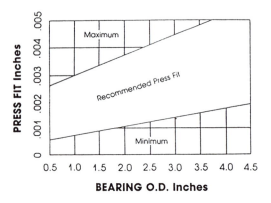

BEARING O.D. Inches

CLEARANCE
Greater than normal clearance must be allowed:
(1) if exceptionally high speed is involved, (2) if
higher than normal loading is encountered.

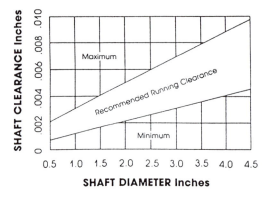

SHAFT DIAMETER Inches

Figure 3   Class I engineering data. (Courtesy of Western Reserve Manufacturing Company, Inc., Lorain, Ohio.)

designed by the B. F. Goodrich Company and sold under the Cutlass brand name, which is a very fancy rubber-lined bearing available in stainless steel jackets, hard bronze jackets, and fiberglass jackets, or can be naked and mounted in a housing. The lubrication is not a petroleum or a synthetic but a water-based lubricant which can range from the effluent in a sewer system to saltwater in the sea. The B. F. Goodrich Cutlass-type bearing can also be used in foot pump bearing supports and other wet applications.

Also in this category are butt-joint, steel-backed, rolled-up bearings. Some are steel and bronze, some are steel and babbitt, and a third kind is called trimetal, which is steel backed with a bronze lining with an overlayment of babbitt. There are also steel-backed bearings with plastic linings bonded thereto on the I.D., such as the Garlock DX bearing (which is a polyacetal-faced backer) to be used with grease lubricants.

There is, of course, a full line of these materials in mounted fashion with cast iron pillowblocks, stamped steel pillowblocks, machined steel pillowblocks, and molded plastic pillowblocks. These can be made available with carbon linings, graphite linings, bronze with graphite plugs, bronze with graphite grooves, babbitt linings, and various plastic bearing linings.

The Class I bearing system constitutes the largest percentage of sales at the marketplace but is not growing rapidly. That area is mostly growing only by the population relationship increases and the need for equipment. The real breakthroughs are being found in the Class III bearings.

IV. CLASS II BEARINGS

The plain bearing industry started in Class II. The original bearing was the wooden bearing made from the lignum vitae tree, and it is still in use. The plain bearing market in the United States is somewhere in the neighborhood of $12 million annually and is now serviced by two kinds of wood: the original lignum vitae for heavily loaded applications and straight-grain maple, such as that used in a bowling alley floor. This has to be kiln-dried to less than 3% moisture content and then heat-impregnated in a submerged vat of lubricating oil and petrolatum grease. Through a succession of baths at subsequently elevated temperatures, the wooden bearing slowly receives a charge over a four-day cycle period of heating and cooling in various tanks.

The frontrunner in this bearing field is the porous powdered metal bearing. As originally conceived by Chrysler Oilite under the efforts of Andrew J. Langhammer (who eventually became the president of the Amplex Manufacturing Company, a wholly owned subsidiary of Chrysler Corporation), the idea was to develop a line of powdered metal to go directly at the competition, which he considered not to be other powdered metal companies but the foundries of the world.

So today, thanks to this farseeing man, there are the following (see Fig. 4).

1. There are powdered metal bearings that are lead, tin, and copper that can be used against such high-lead bronzes as 30% and 40% cast or wrought materials.
2. There are standard copper tin materials that can be used against, for example, SAE 660.
3. By cranking up the pressure on the presses, you can compact the powder more heavily. So instead of 18–23% porosity for your oil reservoir (which is normal), you can have a 10–12% porosity, which is obtained by squeezing a little harder. You wind up with a 90 copper/10 tin bearing with a lesser density, but it can work against the high-strength bronzes.

ID CLOSE-IN

AS RELATED TO WALL THICKNESS (APPROXIMATE VALUES) FOR NORMAL PRESS FIT.

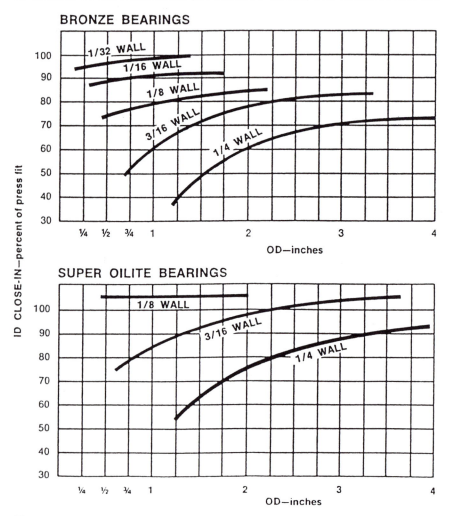

Figure 4 Class II bearing design information. (Courtesy of Amplex Division, Chrysler Corporation, Detroit, Michigan.)

4. By mixing a little iron with the copper you can make a bearing equal to the aluminum bronzes.

5. If you change the ratio by reversing percentages and quantities so that it is mostly iron with the remainder being copper, then you have a material that will go against manganese bronzes.

6. By adding 1.5 points free carbon and heat-treating after sizing, you get a bearing that is equivalent to a heavy-duty steel bearing.
7. Of course, by making your bearing pure iron, you can also wind up with something that carries oil and works against cast iron bearings.
8. Last, you can even make aluminum-powdered metal bearings that go up against tin babbitt aluminum castings.

Of course, throughout the whole thing the thread of the design is that if you want to make a 1-lb bearing, you put a pound of powder in the die, squeeze it, sinter it, size it, and you get a 1-lb bearing finished to close tolerances with no waste, which has to be a big time-cost reduction. From the lubrication engineer's standpoint, this bearing is 18–23% porous charged with the right kind of oil, protected from oxidation by being trapped in the labyrinth of the material, and ready, due to the vacuum directly opposite the load in any rotational application, to do the job.

Figure 5 shows nominal tolerances for this category of bearings.

When it comes to material strength modulus engineering, porous powdered metals employ the ability of the oil to withstand huge shocks hydraulically. Since oil is incompressible, a full-charged powdered metal bearing, when hit with an impact, resists because, in effect, the entire oil system through the chambers of the bearing is resistant to the shock. This will not work in high cycle shocks, but if you leave a full second or two between impacts the recovery gets the bearing ready to take the next shock. This is truly a wonderful material.

For Class II bearings known as self-oiling, there are, thanks to the wonderful efforts of people like Dr. Kawasaki of Oiles Bearing Company in Japan and James Cairns, chief engineer of Arguto in Philadelphia, patents where the porous powdered metal concept is taken into the plastic world by mixing plastic resin with synthetic oils or petroleum-based oils. After injection molding, the plastic cools and the oil, which has a burn rate higher than the melt rate of the plastic, causes the oil to be trapped in minute droplets throughout. These little balloons of oil are available for lubrication to the rotating member when it wears a few millionths off the bore of the bearing.

I am sure this is just the beginning. There will be other plastics, other synthetic lubricants, and higher-temperature capabilities. The self-oiling type of bearing will continue to grow.

Class II bearings are outracing Class I bearings in their growth percentage. The total amount of poundage has not yet caught up because bronze has been in existence for over 3000 years, but it is coming along.

Now to the class of bearings where the greatest gains are shown.

PLAIN AND FLANGE BEARINGS

Inside and Outside Diameters

OVER	UP TO & INCL.		Oilite Bronze, Bronze, BA Oilite,	Iron Oilite, Super Oilite,	Super Oilite 16
—	½	+.000	−.001	−.001	−.003
½	1	+.000	−.001	−.001	−.003
1	1½	+.000	−.001	−.0015	−.003
1½	2½	+.000	−.0015	−.002	−.004
2½	3½	+.000	−.002	−.003	−.005
3½	4½	+.000	−.0025	−.003	−.005
4½	5½	+.000	−.0035	−.0045	−.006
5½	6½	+.000	−.004	−.006	−.007

Length					
—	1½		±.005	±.010	±.010
1½	3		±.0075	±.015	±.015
3	4½		±.010	±.020	±.020
4½	6		±.015	±.030	±.030

Flange Diameter
Based on Flange OD:

—	1¼		±.005	±.010	±.010
1¼	2½		±.010	±.015	±.015
2½	4		±.015	±.020	±.020
4	6		±.025	±.025	±.025

Flange Thickness
Based on Flange OD:

—	1¼		±.0025	±.005	±.005
1¼	2½		±.005	±.0075	±.0075

Flange Fillets, Radii
Based on Body OD:

—	1		$\frac{1}{32} \pm .010$	$\frac{1}{32} \pm .010$	$\frac{1}{32} \pm .010$
1	2		$\frac{3}{64} \pm .010$	$\frac{3}{64} \pm .010$	$\frac{3}{64} \pm .010$
2	2½		$\frac{1}{16} \pm .010$	$\frac{1}{16} \pm .010$	$\frac{1}{16} \pm .010$
2½	4		$\frac{3}{32} \pm \frac{1}{64}$	$\frac{3}{32} \pm \frac{1}{64}$	$\frac{3}{32} \pm \frac{1}{64}$
4	6		$\frac{1}{8} \pm \frac{1}{64}$	$\frac{1}{8} \pm \frac{1}{64}$	$\frac{1}{8} \pm \frac{1}{64}$

Concentricity, ID with respect to OD
(Maximum Total Dial Indicator Reading)
Based on ID:

—	1		.003	.003	.003
1	1½		.003	.004	.004
1½	3		.004	.005	.005
3	4½		.005	.006	.006
4½	6		.006	.007	.007

Figure 5 Nominal tolerances (all figures are in inches). (Courtesy of Amplex Division, Chrysler Corporation, Detroit, Michigan.)

THRUST BEARINGS			
Inside Diameter			
—	1¼	±.005	±.005	±.005
1¼	2½	±.010	±.010	±.010
2½	4	±.015	±.015	±.015
4	6	±.020	±.020	±.020
Outside Diameter			
—	1½	±.010	±.010	±.010
1½	3	±.015	±.015	±.015
3	4½	±.020	±.020	±.020
4½	6	±.025	±.025	±.025
Thickness			
		±.0025	±.005	±.005
Parallelism of Faces Based on OD:			
—	1½	.002	.003	Not
1½	3½	.003	.004	Standard-
3½	6	.004	.005	ized

Figure 5 Continued

V. CLASS III BEARINGS

Once again, we become aware that the plain bearing has stood the test of time because, like the wooden bearing in Class II, the graphite bearing in Class III has been with us for a long, long time.

Pure natural-mined graphite is a flake in form, very big in one dimension, very thin in another, and extremely slippery. Unlike carbon (which is usually created by burning petroleum distillates), the graphite bearing is a remnant of when the world began 600 million years ago, the burned-out leftover debris from the beginning of time.

Graphite is totally inert. It will stand 1100° above and 200° below Fahrenheit. It can be molded similarly to powdered metal and makes a bearing symbolic of all dry-running, self-lubricating bearings. Graphite provides lubricant to the interface between the shaft and the I.D. of the bearing by the shaft rotating and the surface finish of the shaft carving off small amounts of particulate. This then transfers onto the surface of the journal and provides fill, if you will, of lubricant to separate the journal from the I.D. of the bearing. See Figs. 6–10 for the physical characteristics of graphite.

Graphite has great uses where the chemical environment is quite hostile. It has great uses where the temperature goes from one extreme to another because the thermal coefficient of expansion of the material is practically nil. It can be mixed with things like diatomaceous earth, which causes growth in the

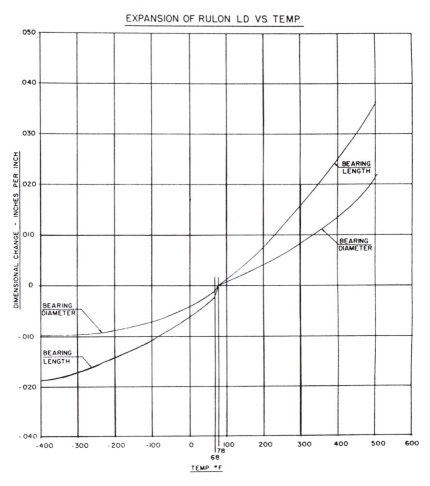

EXPANSION OF RULON LD VS. TEMP.

Figure 6 Shrinkage and expansion of Rulon versus temperature (inches of change per inch of diameter or length). (Courtesy of Dixon Industries Corporation, Bristol, RI.)

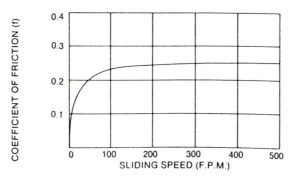

Figure 7 Coefficient of friction chart to be applied to rotary motion designs only. This chart was calculated at 100 psi, 10 fpm (applies to RULON LD). (Courtesy of Dixon Industries Corporation, Bristol, RI.)

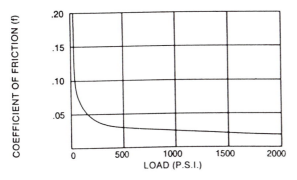

Figure 8 Coefficient of friction is for use in designing sleeve-type motion bearings, not to be applied to rotary motion applications (applies to RULON LD). (Courtesy of Dixon Industries Corporation, Bristol, RI.)

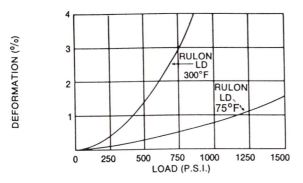

Figure 9 Deformation versus load. (Courtesy of Dixon Industries Corporation, Bristol, RI.)

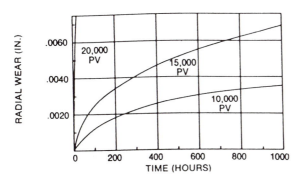

Figure 10 Radial wear versus time for RULON LD. (Courtesy of Dixon Industries Corporation, Bristol, RI.)

groove (filled with graphite) or the plug in the drilled hole (filled with graphite in bronze bearings), simply by accepting humidity or moisture within a submerged application. It has uses in textiles, ovens, and provides man with a very old, very universal bearing material.

Pure carbon and pure graphite are extremely friable. By mixing powdered bronze into the graphite mix prior to compaction into the bearing, you develop a metallized carbon bearing. This creates a great amount of strength and prevents the bearing from breaking or cracking at the edge when there are applications that have, as a characteristic of their design, an overhung load.

Often a higher strength unit is needed; then, a bronze (64 bronze, 660 bronze, aluminum bronze, manganese bronze) or a steel or a cast iron can be selected to give the load capacity. Then you groove the bore of the bearing with patterns like double figure of 8 or quad figure of 8, or you drill holes, giving a hole pattern of 25% up to 50% of the surface area covered with an open space represented by the hole. Then the groove or hole is filled with a graphite mix with two carriers. You can use a hardener and a resin (such as fiberglass construction would use) and mix 75% by volume of carbon or graphite into the mix. The thixotropic paint is applied to the holes or grooves prior to its hardening. After hardening, a single machining pass in the I.D. and the O.D. clean up the bearing, making it ready for use.

An alternative and one of the classic mix examples, and in my opinion a much preferred system, is to take the graphite and make a mixture of 80% graphite, 15% diatomaceous earth, and 5% molybdenum disulfide and compact this in a powdered-metal-type press in the form of a plug with a slight taper. This plug is then driven with a soft hammer into a hole and a final machine pass on the I.D. and the O.D. readies the bearing. However, the bearing must be wrapped in wax paper (not plastic) so that no moisture gets to the graphite plug (which has the diatomaceous earth fill), as the absorption will cause the plug to grow. The same thing is true in a graphite design.

This class of bearings was gifted with a tremendous technological breakthrough by the DuPont Corporation about 1936 when Teflon first saw the light of day. The advent of Teflon made dramatic differences in dry-running bearings. It has taken 30 years of design and evolution to reach what is currently considered the ultimate bearing (a phosphor bronze wire screen 50 mesh welded at the intersections with the 50 mesh plugged with a mix of PTFE molybdenum disulfide and amorphous graphite. This wire screen is then used naked (as in European car door hinges) or mounted by the use of FEP or PFA to a steel sheet and then cut, clipped, and rolled up in standard Finnster presses. Or, the wire screen is wrapped on a mandrel and overlay-wrapped with a double wrap of fiberglass and resin similar to the CJ-type bearing construction.

Between the 1936 invention of Teflon and its relatively low-load-carrying ability to the super-load-carrying capability of the wire screen design, many contributors to the engineering went into making the bearings what they are today.

Virgin Teflon gave way to filled composition Teflon. The original idea was to pick a fill that had a capability of matching the environmental characteristic of PTFE, which is basically chemically inert.

Obviously, one of the first fills had to be graphite. Graphite had a real plus for the bearing because normally Teflon allows liquid to bead up on it due to its low coefficient of surface tension, whereas graphite-filled Teflon thinks that it is dirty and the wetness lays as a sheet or film rather than rising up, which makes it much better for developing the hydrostatic wedge during operation.

The next selection was the use of fiberglass shards. These small fiberglass sticklets (0.001 in. long × 0.0001 in. in diameter and tubelike in construction to the Teflon resin) gave great strength and a great deal of chemical resistance except for substances that bother glass such as hydrofluoric acid, again, keeping pretty close to the inertness of Teflon.

Later it was thought that phosphor bronze powder might make an excellent fill. However, phosphor bronze powder has all the chemical problems that go with regular phosphor bronze, so it took away or detracted from the inertness of Teflon.

There were other filled applications. However, these are the major ones.

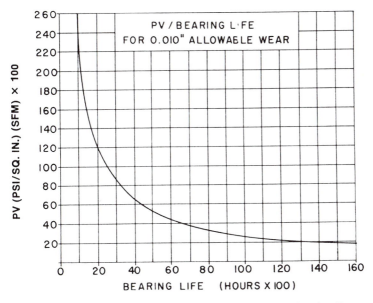

Figure 11 Class III: bearing life. (Courtesy of Dixon Industries Corporation, Bristol, RI.)

Not until about 20 years ago did a senior engineer with Dixon Corporation and a senior engineering with Pampus Plastics in Dusseldorft hit upon what has turned out to be the ultimate bearing—the phosphor bronze wire screen. However, while searching for load capacity, many people discovered that you could take other plastics and put the other plastic into a Teflon-like mode by mixing into the plastic resin a small amount of PTFE ranging from 15% to 25%. This then made the plastic bearing (Delrin or Celcon or the polyimide- or polyamide-based materials) think that they had Teflon-like characteristics, which are transferability and the lowered coefficient of friction because of the transfer. These bearings became very popular because of the inherent strength of the resin and the buffering characteristic of the PTFE as a friction reducer. (See Fig. 11.)

I am sure there will be an endless number of new materials and many changes in temperature applications and load applications at the high end from what we know today.

VI.   DESIGN CONSIDERATIONS*

In this section are given data and procedure for designing full-film or hydro-dynamically lubricated bearings of the journal and thrust types. However, before proceeding to these design methods, it may prove useful to review first those bearing aspects concerning the types of bearings available; lubricants and lubrication methods; hardness and surface finish; machining methods; seals; retainers; and typical length-to-diameter ratios for various applications.

The following paragraphs preceding the design sections provide guidance in these matters and suggest modifications in allowable loads when other than full-film operating conditions exist in a bearing.

A.   Classes of Plain Bearings

Bearings that provide sliding contact between mating surfaces fall into three general classes: *radial bearings* that support rotating shafts or journals; *thrust bearings* that support axial loads on rotating members; and *guide* or *slipper bearings* that guide moving parts in a straight line. Radial sliding bearings,

*Adapted from the book, *Machinery's Handbook,* 24th edition, by Erik Oberg, Franklin D. Jones, Holbrook L. Horton, and Henry H. Ryffel, edited by Robert Green, Copyright 1992 by Industrial Press, Inc., New York. The editors would like to express their sincere thanks to Mr. Robert Green and Industrial Press for their gracious permission in letting us incorporate their material in this chapter. The editors suggest that a copy of *Machinery's Handbook* be obtained by readers who are interested in further discussions on design and selection of mechanical components.

more commonly called sleeve bearings, may be of several types; the most usual being the plain full journal bearing which has 360° contact with its mating journal, and the partial journal bearing, which has less than 180° contact. This latter type is used when the load direction is constant and has the advantages of simplicity, ease of lubrication, and reduced frictional loss.

The relative motions between the parts of plain bearings may take place (1) as pure sliding without the benefit of a liquid or gaseous lubricating medium between the moving surfaces such as with the dry operation of nylon or Teflon; (2) with hydrodynamic lubrication in which a wedge or film buildup of lubricating medium is produced, with either whole or partial separation of the bearing surfaces; (3) with hydrostatic lubrication in which a lubricating medium is introduced under pressure between the mating surfaces causing a force opposite to the applied load and a lifting or separation of these surfaces; and (4) with a hybrid form or combination of hydrodynamic and hydrostatic lubrication.

Listed below are some of the advantages and disadvantages of sliding contact (plain) bearings as compared with rolling contact (antifriction) bearings.

Advantages

1. Require less space
2. Quiet in operation
3. Cost less, particularly in high volume production
4. Greater rigidity
5. Life generally not limited by fatigue.

Disadvantages

1. Higher frictional properties resulting in higher power consumption
2. More susceptible to damage from foreign material in lubrication system
3. More stringent lubrication requirements
4. More susceptible to damage from interrupted lubrication supply.

B. Types of Journal Bearings

Many types of journal bearing configurations have been developed; some of these are shown in Fig. 12.

Circumferential-groove bearings (a) have an oil groove extending circumferentially around the bearing. The oil is maintained under pressure in the groove. The groove divides the bearing into two shorter bearings which tend to run at a slightly greater eccentricity. However, the advantage in terms of stability is slight, and this design is most commonly used in reciprocating-load main and connecting-rod bearings because of the uniformity of oil distribution.

Short cylindrical bearings are a better solution than the circumferential-groove bearing for high-speed, low-load service. In many cases, the bearing

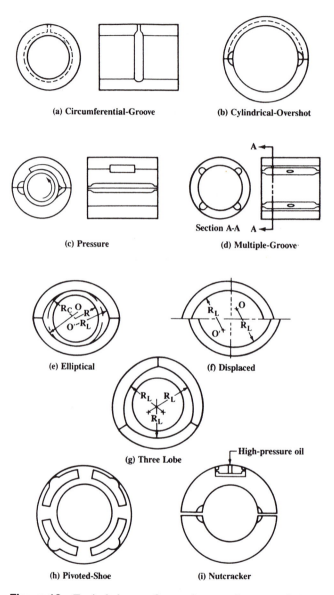

(a) Circumferential-Groove   (b) Cylindrical-Overshot

(c) Pressure   (d) Multiple-Groove

Section A-A

(e) Elliptical   (f) Displaced

(g) Three Lobe

(h) Pivoted-Shoe   (i) Nutcracker

Figure 12   Typical shapes of several types of pressure-fed bearings. (From Oberg et al., 1992.)

can be shortened enough to increase the unit loading to a substantial value, causing the shaft to ride at a position of substantial eccentricity in the bearing. Experience has shown that instability rarely results when the shaft eccentricity is greater than 0.6. Very short bearings are not often used for this type of application, because they do not provide a high temporary rotating-load capacity in the event some unbalance should be created in the rotor during service.

Cylindrical-overshot bearings (b) are used where surface speeds of 10,000 fpm or more exist, and where additional oil flow is desired to maintain a reasonable bearing temperature. This bearing has a wide circumferential groove extending from one axial oil groove to the other over the upper half of the bearing. Oil is usually admitted to the trailing-edge oil groove. An inlet orifice is used to control the oil flow. Cooler operation results from the elimination of shearing action over a large section of the upper half of the bearing and, to a great extent, from the additional flow of cool oil over the top half of the bearing.

Pressure bearings (c) employ a groove over the top half of the bearing. The groove terminates at a sharp dam about 45° beyond the vertical in the direction of shaft rotation. Oil is pumped into this groove by shear action from the rotation of the shaft and is then stopped by the dam. In high-speed operation, this situation creates a high oil pressure over the upper half of the bearing. The pressure created in the oil groove and surrounding upper half of the bearing increases the load on the lower half of the bearing. This self-generated load increases the shaft eccentricity. If the eccentricity is increased to 0.6 or greater, stable operation under high-speed, low-load conditions can result. The central oil groove can be extended around the lower half of the bearing, further increasing the effective loading. This design has one primary disadvantage: Dirt in the oil will tend to abrade the sharp edge of the dam and impair ability to create high pressures.

Multiple-groove bearings (d) are sometimes used to provide increased oil flow. The interruptions in the oil film also appear to give this bearing some merit as a stable design.

Elliptical bearings (e) are not truly elliptical, but are formed from two sections of a cylinder. This two-piece bearing has a large clearance in the direction of the split and a smaller clearance in the load direction at right angles to the split. At light loads, the shaft runs eccentric to both halves of the bearing and, hence, the elliptical bearing has a higher oil flow than the corresponding cylindrical bearing. Thus, the elliptical bearing will run cooler and will be more stable than a cylindrical bearing.

Elliptical-overshot bearings (not shown), are elliptical bearings in which the upper half is relieved by a wide oil groove connecting the axial oil grooves. They are analogous to cylindrical-overshot bearings.

Displaced elliptical bearings (f) shift the centers of the two bearing arcs in the horizontal and vertical directions. This design has greater stiffness than a cylindrical bearing, in both horizontal and vertical directions, with substantially higher oil flow. It has not been extensively used, but offers the prospect of high stability and cool operation.

Three-load bearings (g) are made up in cross section of three circular arcs. They are most effective as antioil whip bearings when the centers of curvature of each of the three lobes lie well outside the clearance circle, which the shaft center can describe within the bearing. Three axial oil-feed grooves are used. It is a more difficult design to manufacture, since it is almost necessary to make it in three parts instead of two. The bore is machined with shims between each of the three parts. The shims are removed after manufacture.

Pivoted-shoe bearings (h) are one of the most stable bearings. The bearing surface is divided into three or more segments, each of which is pivoted at the center. In operation, each shoe tilts to form a wedge-shaped oil film, thus creating a force tending to push the shaft toward the center of the bearing. For single-direction rotation, the shoes are sometimes pivoted near one end and forced toward the shaft by springs.

Nutcracker bearings (i) consist of two cylindrical half-bearings. The upper half-bearing is free to move in a vertical direction and is forced toward the shaft by a hydraulic cylinder. External oil pressure may be used to create load on the upper half of the bearing through the hydraulic cylinder. Or, the high-pressure oil may be obtained from the lower half of the bearing by tapping a hole into the high-pressure oil film, thus creating a self-loading bearing. Either type can increase bearing eccentricity to the point where stable operation can be achieved.

C. Hydrostatic Bearings

Hydrostatic bearings are used when operating conditions require full film lubrication that cannot be developed hydrodynamically. The hydrostatically lubricated bearing, either thrust or radial, is supplied with lubricant under pressure from an external source. Some advantages of the hydrostatic bearing over bearings of other types are low friction, high load capacity, high reliability, high stiffness, and long life.

Hydrostatic bearings are used successfully in many applications, among which are machine tools, rolling mills, and other heavily loaded slow-moving machinery. However, specialized techniques, including a thorough understanding of hydraulic components external to the bearing package is required. The designer is cautioned against use of this type of bearing without a full knowledge of all aspects of the problem. Determination of the operating performance of hydrostatic bearings is a specialized area of the lubrication field and is described in specialized reference books.

1. Design

The design of a sliding bearing is generally accomplished in one of two ways: (1) a bearing operating under similar conditions is used as a model or basis from which the new bearing is designed; (2) in the absence of any previous experience with similar bearings in similar environments, certain assumptions concerning operating conditions and requirements are made and a tentative design is prepared based on general design parameters or rules of thumb. Detailed lubrication analysis is then performed to establish design and operating details and requirements.

D. Modes of Bearing Operation

The load-carrying ability of a sliding bearing depends upon the kind of fluid film which is formed between its moving surfaces. The formation of this film is dependent, in part, on the design of the bearing and, in part, on the speed of rotation. It results in three modes or regions of operation designated as *full film*, *mixed film*, and *boundary* lubrication with effects on bearing friction as shown in Fig. 13.

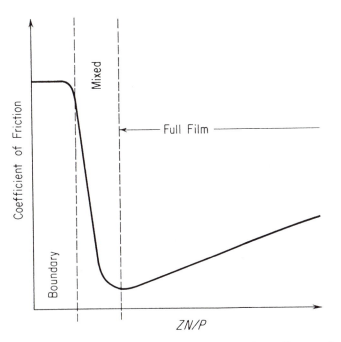

Figure 13   Three modes of bearing operation. (From Oberg et al., 1992.)

In terms of physical bearing operation these three modes may be further described as follows:

1.   Full film, or hydrodynamic lubrication produces a complete physical separation of the sliding surfaces. This results in low friction and long wear-free service life.

To promote full film lubrication in hydrodynamic operation the following parameters should be satisfied: (1) Lubricant selected has the correct viscosity for the proposed operation. (2) Proper lubricant flow rates are maintained. (3) Proper design methods and considerations have been utilized. (4) Surface velocity in excess of 25 fpm is maintained.

When fill film lubrication is achieved, a coefficient of friction between 0.001 and 0.005 can be expected.

2.   Boundary lubrication takes place when the sliding surfaces are rubbing together with only an extremely thin film of lubricant present. This type of operation is acceptable only in applications with oscillating or slow rotary motion. In complete boundary lubrication the oscillatory or rotary motion is usually less than 10 fpm with resulting coefficients of friction of 0.08 to 0.14. These bearings are usually grease-lubricated or periodically oil-lubricated.

3.   Mixed-film lubrication is a mode of operation between the full-film and boundary modes. With this mode there is a partial separation of the sliding surfaces by the lubricant film; however, as in boundary lubrication, limitations on surface speed and wear will result. With this type of lubrication a surface velocity in excess of 10 fpm is required with resulting coefficients of friction of 0.02 to 0.08.

A journal bearing in starting up and accelerating to its operating point passes through all three modes of operation. At rest, the journal and bearing are in contact and thus, when starting, the operation is in the boundary lubrication region. As the shaft begins to rotate more rapidly and the hydrodynamic film starts to build up, bearing operation enters the region of mixed-film lubrication. When design speeds and loads are reached, the hydrodynamic action in a properly designed bearing will now promote full film lubrication.

E.   Methods of Retaining Bearings

A number of methods are available to insure that a bearing remain in place within a housing. Which method to use depends upon the particular application but requires first that the unit lend itself to convenient assembly and disassembly; additionally, the bearing wall should be of uniform thickness to avoid introduction of weak points in the construction which may lead to elastic or thermal distortion.

1.   Press or Shrink Fit

One common and satisfactory technique for retaining the bearing is to press or shrink the bearing in the housing with an interference fit. This method permits the use of bearings having uniform wall thickness over the entire bearing length.

Standard bushings with finished inside and outside diameters are available in sizes up to approximately 5 in. inside diameter. Stock bushings are commonly provided 0.002 to 0.003 in. over nominal on outside diameter sizes of 3 in. or less. For diameters greater than 3 in., actual outside diameters are 0.003 to 0.005 in. over nominal. Since these tolerances are built into standard bushings, the amount of press fit is controlled by the housing bore size.

As a result of a press or shrink fit, the bore of the bearing material "closes in" by some amount. In general, this diameter decrease is approximately 70–100% of the amount of the interference fit. Any attempt to accurately predict the amount of close-in, in an effort to avoid final clearance machining, should be avoided.

Shrink fits may be accomplished by chilling the bearing in a mixture of dry ice and alcohol, or in liquid air. These methods are easier than heating the housing and are preferred. Dry ice in alcohol has a temperature of $-110°$ F and liquid air boils at $-310°$ F.

When a bearing is pressed into the housing, the driving force should be uniformly applied to the end of the bearing to avoid upsetting or peening of the bearing. Of equal importance, the mating surfaces must be clean, smoothly finished, and free of machining imperfections.

2. Keying Methods

A variety of methods can be used to fix the position of the bearing with respect to its housing by "keying" the two together. Possible keying methods are shown in Fig. 14, including (a) set screws, (b) Woodruff keys, (c) bolted bearing flanges, (d) threaded bearings, (e) dowel pins, and (f) housing caps.

Factors to be considered when selecting one of these methods are

1. Maintaining uniform wall thickness of the bearing material, if possible, especially in the load-carrying region of the bearing.
2. Providing as much contact area as possible between bearing and housing. Mating surfaces should be clean, smooth, and free from imperfections to facilitate heat transfer.
3. Preventing any local deformation of the bearing that might result from the keying method. Machining after keying is recommended.
4. Considering the possibility of bearing distortion resulting from the effect of temperature changes on the particular keying method.

F. Methods of Sealing

In applications where lubricants or process fluids are utilized in operation, provision must be made normally to prevent leakage to other areas. This is accomplished by the use of static- and dynamic-type sealing devices. In general three terms are used to describe the devices used for sealing:

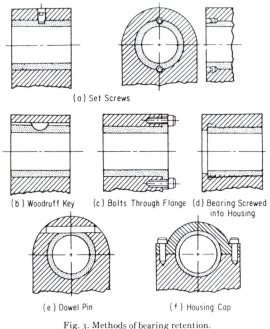

(a) Set Screws

(b) Woodruff Key   (c) Bolts Through Flange   (d) Bearing Screwed
                                                    into Housing

(e) Dowel Pin   (f) Housing Cap

Fig. 3. Methods of bearing retention.

Figure 14   Methods of bearing retention. (From Oberg et al., 1992.)

Seal: A means of preventing migration of fluids, gases, or particles across a
joint or opening in a container

Packing: A dynamic seal, used where some form of relative motion occurs
between rigid members of an assembly

Gaskets: A static seal, used where there is no relative motion between
joined parts

Two major functions must be achieved by all sealing applications: prevent
escape of fluid; and prevent migration of foreign matter from the outside.

The first determination in selecting the proper seal is whether the application is static or dynamic. To meet the requirements of a static application, there
must be no relative motion between the joining parts or between the seal and
the mating part. If there is any relative motion, the application must be considered dynamic, and the seal selected accordingly.

Dynamic sealing requires control of fluids leading between parts with relative motion. Two primary methods are used to this end: positive contact or
rubbing seals; and controlled clearance noncontact seals.

1. Positive Contact or Rubbing Seals

These are utilized where positive containment of liquids or gases is required, or where the seal area is continuously flooded. If properly selected and applied, contact seals can provide zero leakage for most fluids. However, because they are sensitive to temperature, pressure, and speed, improper application can result in early failure. These seals are applicable to rotating and reciprocating shafts. In many cases, the positive-contact seals re available as off-the-shelf items. In other instances, they are custom-designed to the special demands of a particular application. Custom design is offered by many seal manufacturers and, for extreme cases, probably offers the best solution to the sealing problem.

2. Controlled Clearance Noncontact Seals

Representative of the controlled-clearance seals, which includes all seals in which there is no rubbing contact between the rotating and stationary members, are throttling bushings and labyrinths. Both of these types operate by fluid-throttling action in narrow annular or radial passages.

Clearance seals are frictionless and very insensitive to temperature and speed. They are chiefly effective as devices for limiting leakage rather than stopping it completely. Although they are employed as primary seals in many applications, the clearance seal also finds use as auxiliary protection in contact-seal applications. These seals are usually designed into the equipment by the designer himself, and they can take on many different forms.

Advantages of this seal are that friction is kept to an absolute minimum and that there is no wear or distortion during the life of the equipment. However, there are two significant disadvantages: The seal has limited use when leakage rates are critical; and it becomes quite costly as the configuration becomes elaborate.

3. Static Seals

Static seals such as gaskets, "O" rings, and molded packings cover very broad ranges of both design and materials. Some typical types are (1) molded packings (lip type, squeeze-molded), (2) simple compression packings, (3) diaphragm seals, (4) nonmetallic gaskets, (5) "O" rings, and (6) metallic gaskets and "O" rings.

Detailed design information for specific products should be obtained directly from manufacturers.

G. Hardness and Surface Finish

Even in well-lubricated full-film sleeve bearings, momentary contact between journal and bearing may occur under such conditions as starting, stopping, or overloading. In mixed-film and boundary-film lubricated sleeve bearings, continuous metal-to-metal contact occurs. Hence, to allow for any necessary wearing-in, the journal is usually made harder than the bearing material. This allows the effects of scoring or wearing to take place on the bearing, which is more easily replaced, rather than on the more expensive shaft. As a general rule, recommended Brinnell hardness of the journal is at least 100 points harder than the bearing material.

The softer cast bronzes used for bearings are those with high lead content and very little tin. Such bronzes give adequate service in boundary and mixed-film applications where full advantage is taken of their excellent "bearing" characteristics.

High-tin, low-lead content cast bronzes are the harder bronzes and these have high ultimate load-carrying capacity: higher journal hardnesses are required with these bearing bronzes. Aluminum bronze, for example, requires a journal hardness in the range of 550 to 600 BHN.

In general, harder bearing materials require better alignment and more reliable lubrication to minimize local heat generation if and when the journal touches the shaft. Also, abrasives which find their way into the bearing are a problem for the harder bearing and greater care should be taken to exclude them.

1. Surface Finish

Whether bearing operation is complete boundary, mixed film, or fluid film, surface finish of journal and bearing must receive careful attention. In applications where operation is hydrodynamic or full film, peak surface variations should be less than the expected minimum film thickness; otherwise, peaks on the journal surface will contact peaks on the bearing surface, with resulting high friction and temperature rise. Ranges of surface roughness obtained by various finishing methods are: boring, broaching, and reaming, 32 to 64 μin., rms; grinding, 16 to 64 μin., rms; and fine grinding, 4 to 16 μin. rms.

In general, the better surface finishes are required for full-film bearings operating at high eccentricity ratios since full-film lubrication must be maintained with small clearances, and metal-to-metal contact must be avoided. Also, the harder the material the better the surface finish required. For boundary and mixed-film applications surface finish requirements may be somewhat relaxed since bearing wear-in will in time smooth the surfaces.

Figure 15 is a general guide to the ranges required for bearing and journal surface finishes. Selecting a particular surface finish in each range can be simplified by observing the general rule that smoother finishes are required for the harder materials, for high loads, and for high speeds.

H. Machining

The methods most commonly used in finishing journal bearing bores are boring, broaching, reaming, and burnishing.

2084 PLAIN BEARINGS

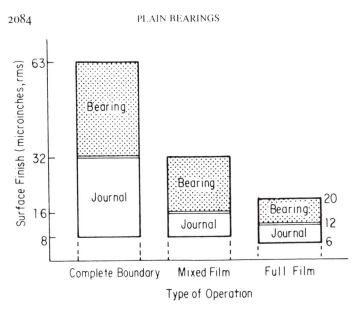

Figure 15 Recommended ranges of surface finish for the three types of sleeve bearing operations. (From Oberg et al., 1992.)

Boring of journal bearings provides the best concentricity, alignment, and size control; this being the finished method of choice when close tolerances and clearances are desirable. Broaching is a rapid finishing method providing good size and alignment control when adequate piloting is possible. Soft babbitt materials are particularly compatible with the broaching method. A third finishing method, reaming, facilitates good size and alignment control when piloting is utilized. Reaming can be accomplished both manually or by machine, the machine method being preferred. Burnishing is a fast sizing operation which gives good alignment control, but does not give as good size control as the cutting methods. It is not recommended for soft materials such as babbitt. Burnishing has an ironing effect which gives added seating of the bushing outside diameter in the housing bore; consequently, it is often used for this purpose, especially on a 1/32-in. wall bushing, even if a further sizing operation is to be used subsequently.

I. Methods of Lubrication

There are numerous ways to supply lubricant to bearings. The more common of these are described as follows:

Pressure lubrication, in which an abundance of oil is fed to the bearing from a central groove, single or multiple holes, or axial grooves, is effective and efficient. The moving oil assists in flushing dirt from the bearing and helps keep the bearing cool. In fact it removes heat faster than other lubricating methods and, therefore, permits thinner oil films and unimpaired actual load capacities. The oil supply pressure needed for bushings carrying the basic load is directly proportional to the shaft speed, but for most installations 50 psi will be adequate.

Splash fed is a term applied to a variety of intermittently lubricated bushings. It includes everything from bearings spattered with oil from the action of other moving parts to bearings regularly dipped in oil. Like oil bath lubrication, splash feeding is practical when the housing can be made oiltight and when the moving parts do not churn the oil. The fluctuating nature of the load and the intermittent oil supply in splash-fed applications requires the designer to use experience and judgment when determining the probable load capacity of bearings lubricated in this way.

Oil bath lubrication, in which the bushing is submerged in oil, is the most reliable of all methods except pressure lubrication. It is practical if the housing can be made oiltight, and if the shaft speed is not so great as to cause excessive churning of the oil.

Oil ring lubrication, in which oil is supplied to the bearing by a ring in contact with the shaft, will, within reasonable limits, bring enough oil to the bearing to maintain hydrodynamic lubrication. If the shaft speed is too low, little oil will follow the ring to the bearing, and, if the speed is too high, the ring speed will not keep pace with the shaft. Also, a ring revolving at high speed will lose oil by centrifugal force. For best results, the peripheral speed of the shaft should be between 200 and 2000 fpm. Safe load to achieve hydrodynamic lubrication should be one-half of that for pressure-fed bearings. Unless the load is light, hydrodynamic lubrication is doubtful. The safe load, then, to achieve hydrodynamic lubrication, should be one-quarter of that of pressure-fed bearings.

Wick or waste pack lubrication delivers oil to a bushing by the capillary action of a wick or waste pack; the amount delivered being proportional to the size of the wick or pack.

Grease packed in a cavity surrounding the bushing is less adequate than an oil system, but it has the advantage of being more or less permanent. Although hydrodynamic lubrication is possible under certain very favorable circumstances, boundary lubrication is the usual state.

J. Lubricants

The value of an oil as a lubricant depends mainly upon its film-forming capacity; that is, its capability of maintaining a film of oil between the bearing surfaces. The film-forming capacity depends to a large extent on the viscosity of the oil, but this should not be understood to mean that oil of the highest viscosity is in every case the most suitable lubricant. For practical reasons, an oil of the lowest viscosity which will retain an unbroken oil film between the bearing surfaces is the most suitable for purposes of lubrication. This is because a higher viscosity than that necessary to maintain the oil film results in a waste of power due to the expenditure of energy necessary to overcome the internal friction of the oil itself.

Figure 16 provides representative values of viscosity in centipoises for SAE mineral oils. Table 1 is provided as a means of converting viscosities of other units to centipoises.

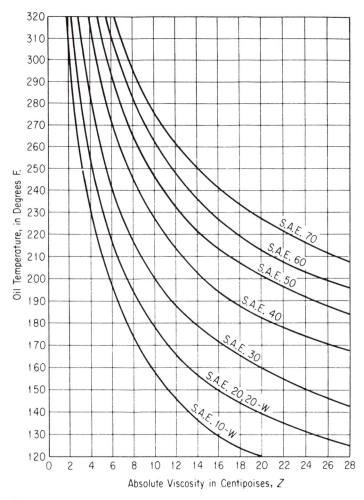

Figure 16 Viscosity vs. temperture—SAE oils. (From Oberg et al., 1992.)

Table 1 Oil Viscosity Unit Conversion

Convert from	Convert to				
	Poise (P)	Centipoise (Z)	Reyn (μ)	Stoke (S)	Centistoke (ν)
			Multiplying factors		
Poise (P)					
$\dfrac{\text{dyne-sec}}{\text{cm}^2}$ or $\dfrac{\text{gram mass}}{\text{sec-cm}}$	1	100	1.45×10^{-5}	$\dfrac{1}{\rho}$	$\dfrac{100}{\rho}$
Centipoise (Z)					
$\dfrac{\text{dyne-sec}}{100 \text{ cm}^2}$ or $\dfrac{\text{gram mass}}{100 \text{ sec-cm}}$	0.01	1	1.45×10^{-7}	$\dfrac{0.01}{\rho}$	$\dfrac{1}{\rho}$
Reyn (μ)					
$\dfrac{\text{lb force-sec}}{\text{in.}^2}$	6.9×10^4	6.9×10^6	1	$\dfrac{6.9 \times 10^4}{\rho}$	$\dfrac{6.9 \times 10^6}{\rho}$
Stoke (S)					
$\dfrac{\text{cm}^2}{\text{sec}}$	ρ	100ρ	$1.45 \times 10^{-5}\rho$	1	100
Centistoke (ν)					
$\dfrac{\text{cm}^2}{100 \text{ sec}}$	0.01ρ	ρ	$1.45 \times 10^{-7}\rho$	0.01	1

ρ = Specific gravity of the oil.
To convert from a value in the "Convert form" column to a value in a "Convert to" column, multiply the
"Convert from" column by the figure in the intersecting block, e.g. to change from Centipoise to Reyn, multi-
ply Centipoise value by 1.45×10^{-7}.
Source: Oberg et al., 1992.

K. Lubricant Selection

In selecting lubricants for journal bearing operation several factors must be
considered: (1) type of operation (full-, mixed-, or boundary film) anticipated,
(2) surface speed, and (3) bearing loading.

Figure 17 combines these factors and facilitates general selection of the
proper lubricant viscosity range.

As an example of using these curves, consider a lightly loaded bearing
operating at 2000 rpm. At the bottom of the figure locate 2000 rpm and move
vertically to intersect the light-load full-film lubrication curve which would
indicate an SAE 5 oil.

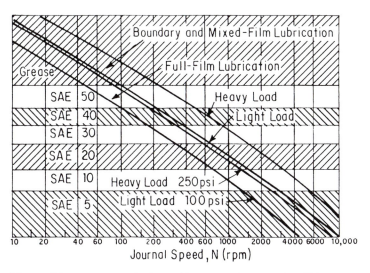

Figure 17 Lubricant selection guide. (From Oberg et al., 1992.)

As a general rule-of-thumb heavier oils are recommended for high loads and lighter oils for high speeds.

In addition, other than using conventional lubricating oils, journal bearings may be lubricated with either greases or solid lubricants. Some of the reasons for use of these lubricants are to

1. Lengthen the period between relubrication
2. Avoid contaminating surrounding equipment or material with "leaking" lubricating oil
3. Provide effective lubrication under extreme temperature ranges
4. Provide effective lubrication in the presence of contaminating atmospheres
5. Prevent intimate metal-to-metal contact under conditions of high unit pressure which might destroy boundary lubricating films.

1. Greases

Where full-film lubrication is not possible or is impractical for slow-speed fairly high-load applications, greases are widely used as bearing lubricants. Although full-film lubrication with grease is possible, it is not normally considered since an elaborate pumping system is required to continuously supply a prescribed amount of grease to the bearing. Bearings supplied with grease are usually lubricated periodically. Grease lubrication, therefore, implies that the bearing will operate under conditions of complete boundary lubrication and should be designed accordingly.

Lubricating greases are essentially a combination of a mineral lubricating oil and a thickening agent, which is usually a metallic soap. When suitably mixed, they make excellent bearing lubricants. There are many different types of greases which, in general, may be classified according to the soap base used. Information on commonly used greases is shown in Table 2.

Synthetic greases are composed of normal types of soaps but use synthetic hydrocarbons instead of normal mineral oils. They are available in a range of consistencies in both water-soluble and insoluble types. Synthetic greases can accommodate a wide range of variation in operating temperature; however, recommendations on special-purpose greases should be obtained from the lubricant manufacturer.

Application of grease is accomplished by one of several techniques depending upon grease consistency. These classifications are shown in Table 3 along with typical methods of application. Grooves for grease are generally greater in width, up to 1.5 times, than for oil.

Coefficients of friction for grease-lubricated bearings range from 0.08 to 0.16, depending upon consistency of the grease, frequency of lubrication, and type of grease. An average value of 0.12 may be used for design purposes.

2. Solid Lubricants

The need for effective high-temperature lubricants led to the development of several solid lubricants. Essentially, solid lubricants may be described as low-shear-strength solid materials. Their function within a bronze bearing is to act as an intermediary material between sliding surfaces. Since these solids have very low shear strength, they shear more readily than the bearing material and thereby allow relative motion. So long as solid lubricant remains between the moving surfaces, effective lubrication is provided and friction and wear are reduced to acceptable levels.

Table 2 Commonly Used Greases and Solid Lubricants

Type	Operating temperature (°F)	Load	Comments
	Greases		
Calcium or lime soap	160	Moderate	. . .
Sodium soap	300	Wide	For wide speed range
Aluminum soap	180	Moderate	. . .
Lithium soap	300	Moderate	Good low temperature
Barium soap	350	Wide	. . .
	Solid lubricants		
Graphite	1000	Wide	. . .
Molybdenum disulfide	−100 to 750	Wide	. . .

Source: Oberg et al., 1992.

Table 3 NLGI[a] Consistency Numbers

NLGI consistency no.	Consistency of grease	Typical method of application
0	Semifluid	Brush or gun
1	Very soft	Pin-type cup or gun
2	Soft	Pressure gun or centralized pressure system
3	Light cup grease	Pressure gun or centralized pressure system
4	Medium cup grease	Pressure gun or centralized pressure system
5	Heavy cup grease	Pressure gun or hand
6	Block grease	Hand, cut to fit

[a] National Lubricating Grease Institute
Source: Oberg et al., 1992.

Solid lubricants provide the most effective boundary films in terms of reduced friction, wear, and transfer of metal from one sliding component to the other. However, there is a significant deterioration in these desirable properties as the operating temperature of the boundary film approaches the melting point of the solid film. At this temperature the friction may increase by a factor of 5 to 10 and the rate of metal transfer may increase by as much as 1000. What occurs is that the molecules of the lubricant lose their orientation to the surface that exists when the lubricant is solid. As the temperature further increases, additional deterioration sets in with the friction increasing by some additional small amount but the transfer of metal accelerates by an additional factor of 20 or more. The final effect of too high temperature is the same as metal-to-metal contact without benefit of lubricant. These changes, which are due to the physical state of the lubricant, are reversed when cooling takes place.

The effects just described also partially explain why fatty acid lubricants are superior to paraffin-base lubricants. The fatty acid lubricants react chemically with the metallic surfaces to form a metallic soap that has a higher melting point than the lubricant itself, the result being that the breakdown temperature of the film, now in the form of a metallic soap is raised so that it acts more like a solid film lubricant than a fluid film lubricant.

L. Journal or Sleeve Bearings

Although this type of bearing may take many shapes and forms, there are always three basic components: journal or shaft, bushing or bearing, and lubricant. Figure 18 shows these components with the nomenclature generally used to describe a journal bearing: W = applied load, N = revolution, e = eccentricity of journal center to bearing center, θ = attitude angle, which is the angle between the applied load and the point of minimum film thickness, d =

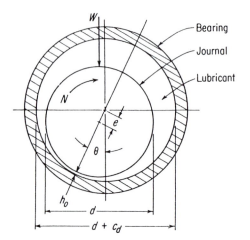

Figure 18   Basic components of a journal bearing. (From Oberg et al., 1992.)

diameter of the shaft, c_d = bearing clearance, $d + c_d$ = diameter of the bearing, and h_0 = minimum film thickness.

1. Grooving and Oil Feeding

Grooving in a journal bearing has two purposes: (1) to establish and maintain an efficient film of lubricant between the bearing moving surfaces and (2) to provide adequate bearing cooling. The obvious and only practical location for introducing lubricant to the bearing is in a region of low pressure. A typical pressure profile of a bearing is shown by Fig. 19. The arrow W shows the applied load. Typical grooving configurations used for journal bearings are shown in Fig. 20.

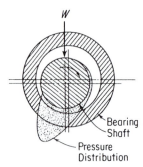

Figure 19   Typical pressures of journal bearing. (From Oberg et al., 1992.)

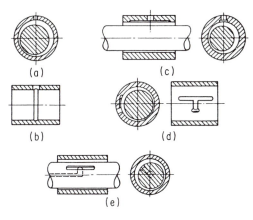

Figure 20 Types of journal bearing oil grooving: (a) single inlet hole, (b) circular groove, (c) straight axial groove, (d) straight axial groove with feeder groove, (e) straight axial groove in shaft. (From Oberg et al., 1992.)

2. Heat-Radiating Capacity

In a self-contained lubrication system for a journal bearing, the heat generated by bearing friction must be removed to prevent continued temperature rise to an unsatisfactory level. The heat-radiating capacity H_R of the bearing in foot-pounds per minute may be calculated from the formula $H_R = LdCt_R$ in which C is a constant determined by O. Lasche, and T_R is temperature rise in degrees Fahrenheit. Values for the product Ct_R may be found from the curves in Fig. 21 for various values of bearing temperature rise t_R and for three operating conditions. In this equation L = the total length of the bearing in inches and d = the bearing diameter in inches.

3. Journal Bearing Design Notation

The symbols used in the following step-by-step procedure for lubrication analysis and design of a plain sleeve or journal bearing are listed below:

c = specific heat of lubricant, Btu/lb/°F
c_d = diametral clearance, in.
C_n = bearing capacity number
d = journal diameter, in.
e = eccentricity, in.
h_0 = minimum film thickness, inc.
K = constants
l = bearing length as defined in Fig. 22, in.
L = actual length of bearing, in.
m = clearance modulus

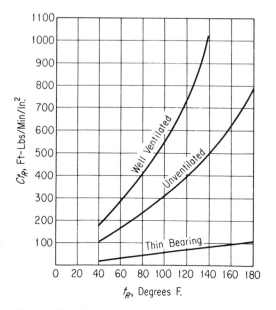

Figure 21 Heat radiating capacity factor, Ct_R, vs. bearing temperature rise, tg— journal bearings. (From Oberg et al., 1992.)

N = rpm
p_b = unit load, psi
p_s = oil supply pressure, psi
P_f = friction horsepower
P' = bearing pressure parameter
q = flow factor
Q_1 = hydrodynamic flow, gpm
Q_2 = pressure flow, gpm
Q = total flow, gpm
Q_{new} = new total flow, gpm
Q_R = total flow required, gpm
r = journal radius, in.
Δt = actual temperature rise in oil in bearing, °F
Δt_a = assumed temperature rise in oil in bearing, °F
Δt_{new} = new assumed temperature rise of oil in bearing, °F
t_b = bearing operating temperature, °F
t_{in} = oil inlet temperature, °F
T_f = friction torque, in.-lb/in.
T' = torque parameter
W = load, lb

X = factor
Z = viscosity, centipoises
ϵ = eccentricity ratio—ratio of eccentricity to radial clearance
α = oil density, lb/in.^{3}

M. Journal Bearing Lubrication Analysis

The following procedure leads to a complete lubrication analysis which forms the basis for the bearing design.

1. *Diameter of bearing d.* This is usually determined by considering strength and/or deflection requirements for the shaft using principles of strength of materials.

2. *Length of bearing L.* This is determined by an assumed l/d ratio in which l may or may not be equal to the overall length, L (see Step 6). Bearing pressure and the possibility of edge loading due to shaft deflection and misalignment are factors to be considered. In general, shaft misalignment resulting from location tolerances and/or shaft deflections should be maintained below 0.0003 in. per inch of length.

3. *Bearing pressure p_b.* The unit load in pound per square inch is calculated from the formula

$$p_b = \frac{W}{Kld}$$

where
$K = 1$ for single oil hole
$K = 2$ for central groove
W = load, lb
l = bearing length as defined in Fig. 22, in.
d = journal diameter, in.

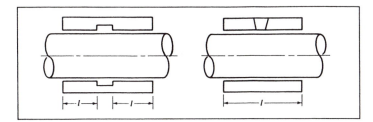

Figure 22 Length, l, of bearing for circular groove type (left) and single inlet hole type (right). (From Oberg et al., 1992.)

Typical unit loads in service are shown in Table 4. These pressures can be used as a safe guide in selection. However, if space limitations impose a higher limit of loading, the complete lubrication analysis and evaluation of material properties will determine acceptability.

4. *Diametral clearance c_d.* This is selection on a trial basis from Fig. 23 which shows suggested diametral clearance ranges for various shaft sizes and for two speed ranges. These are *hot* or *operating* clearances so that thermal expansion of journal and bearing to these temperatures must be taken into consideration in establishing machining dimensions. The optimum operating clearance should be determined on the basis of a complete lubrication analysis (see paragraph following Step 23).

5. *Clearance modulus m.* This is calculated from the formula

$$m = \frac{c_d}{d}$$

6. *Length-to-diameter ratio l/d.* This is usually between 1 and 2; however, with the modern trend toward higher speeds and more compact units, lower ratios down to 0.3 are used. In shorter bearings there is a consequent reduction in load carrying capacity due to excessive end or side leakage of lubricant. In longer bearings there may be a tendency towards edge loading. Length l for a single oil feed hole is taken as the total length of the bearing as shown in Fig. 22. For a central oil groove length, l is taken as one-half the total length.

Typical l/d ratio's use for various types of applications are given in Table 5.

7. *Assumed operating temperature t_b.* A temperature rise of the lubricant as it passes through the bearing is assumed and the consequent operating temperature in degrees Fahrenheit is calculated from the formula

Table 4 Allowable Sleeve Bearing Pressures for Various Classes of Bearings[a]

Types of bearing kind of service	Pressure (lb/in.^{2})	Types of bearing or kind of service	Pressure (lb/in.^{2})
Electric motor and generator		Diesel Engine	
bearings (general)	100–200	rod	1000–2000
Turbine and reduction		wrist pins	1800–2000
gears	100–250	Automotive,	
Heavy line shafting	100–150	main bearings	500–700
Locomotive axles	300–350	rod bearings	1500–2500
Light line shafting	15–35	Centrifugal pumps	80–100
Diesel engine, main	800–1500	Aircraft rod bearings	700–3000

[a] These pressures in pounds per square inch of area equal to length times diameter are intended as a general guide only. The allowable unit pressure depends upon operating conditions, especially in regard to lubrication, design of bearings, workmanship, velocity, and nature of load.
Source: Oberg et al., 1992.

$$t_g = t_{in} + \Delta t_a$$

where

t_{in} = inlet temperature of oil, °F

Δt_a = assumed temperature rise of oil in bearing, °F

An initial assumption of 20°F is usually made.

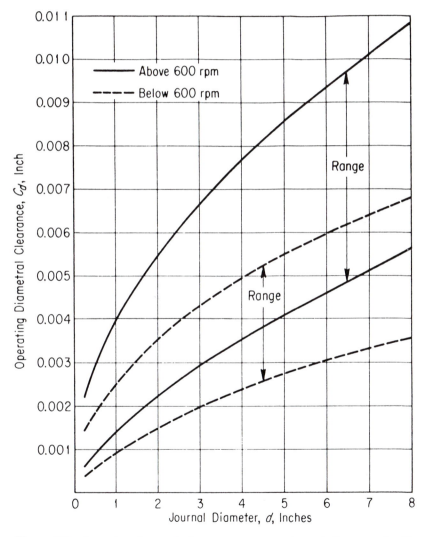

Figure 23 Operating diametral clearance, c_d, versus journal diameter, d. (From Oberg et al., 1992.)

Table 5 Representative *l*/*d* Ratios

Type of service	*l*/*d*	Type of service	*l*/*d*
Gasoline and diesel engine		Light shafting	2.5–3.5
Main bearings and crankpins	0.3–1.0	Heavy shafting	2.0–3.0
Generators and motors	1.2–2.5	Steam engine	
Turbogenerators	0.8–1.5	Main bearings	1.5–2.5
Machine tools	2.0–3.0	Crank and wrist pins	1.0–1.3

Source: Oberg et al., 1992.

8. *Viscosity of lubricant Z.* The viscosity in centipoises at the assumed bearing operating temperature is found from the curve in Fig. 16 which shows the viscosity of SAE grade oils versus temperature.

9. *Bearing pressure parameter P'.* This value is required to find the eccentricity ratio and is calculated from the formula

$$P' = \frac{6.9(1000m)^2 p_h}{ZN}$$

where N = rpm.

10. *Eccentricity ratio ϵ.* Using P' and *l*/*d*, the value of $1/(1 - \epsilon)$ is determined from Fig. 24 and from this ϵ can be determined.

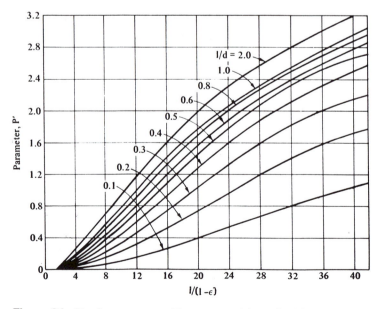

Figure 24 Bearing parameter, P', vs. eccentricity ratio, $1/(1 - \epsilon)$—journal bearings. (From Oberg et al., 1992.)

11. *Torque parameter T'.* This value is obtained from Fig. 25 or Fig. 26 using $1/(1 - \epsilon)$ and l/d.

12. *Friction torque T.* This value is calculated from the formula

$$T = \frac{T'r^2ZN}{6900(1000m)}$$

where r = journal radius, in.

13. *Friction horsepower P_f.* This is calculated from the formula

$$P_f = \frac{KTNl}{63,000}$$

where $K = 1$ for single oil hole, 2 for central groove.

14. *Factor X.* This factor is used in the calculation of the lubricant flow and can either be obtained from Table 6 or calculated from the formula

$$X = \frac{0.1837}{\alpha c}$$

where

α = oil density, lb/in.^{3}

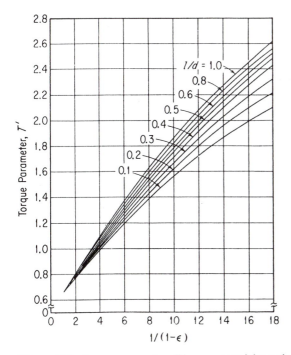

Figure 25   Torque parameter, T', vs. eccentricity ratio, $1/(1 - \epsilon)$—journal bearings. (From Oberg et al., 1992.)

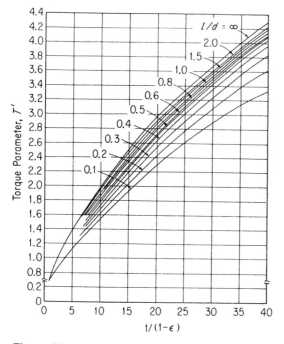

Figure 26   Torque parameter, T', vs. eccentricity ratio, $1/(1 - \epsilon)$—journal bearings. (From Oberg et al., 1992.)

c = specific heat of lubricant in BTU/lb/°F

15.   *Total flow of lubricant required* Q_R. This is calculated from the formula

$$Q_R = \frac{X(P_f)}{\Delta t_a}$$

16.   *Bearing capacity number* C_n. This is needed to obtain the flow factor and is calculated from the formula

Table 6 X Factor Versus Temperature of Mineral Oils

Temperature	X factor
100	12.9
150	12.4
200	12.1
250	11.8
300	11.5

Source: Oberg et al., 1992.

$$C_n = \frac{(1/d)^2}{60P'}$$

17. *Flow factor q.* This is obtained from the curve in Fig. 27.

18. *Actual hydrodynamic flow of lubricant Q_1.* This flow in gallons per minute is calculated from the formula

$$Q_1 = \frac{Nlc_d qd}{294}$$

19. *Actual pressure flow of lubricant Q_2.* This flow in gallons per minute is calculated from the formula

$$Q_2 = \frac{Kp_s c_d^3 d(1 + 1.5\epsilon^2)}{Zl}$$

where

$K = 1.64 \times 10^5$ for single oil hole

$K = 2.35 \times 10^5$ for central groove

p_s = oil supply pressure

20. *Actual total flow of lubricant Q.* This is obtained by adding the hydrodynamic flow and the pressure flow:

$$Q = Q_1 + Q_2$$

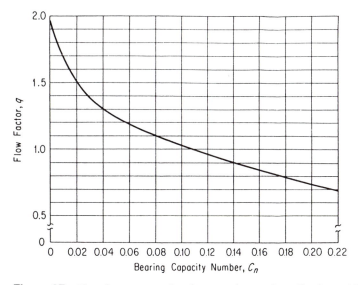

Figure 27 Flow factor, q, vs. bearing capacity number, C_n—journal bearings. (From Oberg et al., 1992.)

21. *Actual bearing temperature rise* Δt. This temperature rise in degrees Fahrenheit is obtained from the formula

$$\Delta t = \frac{X(P_f)}{Q}$$

22. *Comparison of actual and assumed temperature rises.* At this point if Δt_a and Δt differ by more than 5°F, Steps 7 through 22 are repeated using a Δt_{new} halfway between the former Δt_a and Δt.

23. *Minimum film thickness* h_0. When Step 22 has been satisfied, the minimum film thickness in inches is calculated from the formula $h_0 = c_d(1 - \epsilon)/2$.

A new diametral clearance c_d is now assumed and Steps 5 through 23 are repeated. When this repetition has been done for a sufficient number of values for c_d, the full lubrication study is plotted as shown in Fig. 28. From this chart a working range of diametral clearance can be determined that optimizes film thickness, differential temperature, friction horsepower, and oil flow.

N. Use of Lubrication Analysis

Once the lubrication analysis has been completed and plotted as shown in Fig. 28, the following steps lead to the optimum bearing design, taking into consideration both basic operating requirements and requirements peculiar to the application.

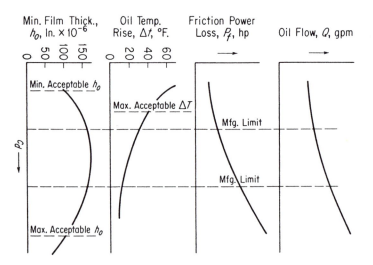

Figure 28   Example of lubrication analysis curves for journal bearing. (From Oberg et al., 1992.)

1. Examine the curve (Fig. 28) for minimum film thickness and determine the acceptable range of diametral clearance, c_d, based on (a) a minimum of 200×10^{-6} in. for small bearings under 1 in. diameter; (b) a minimum of 500×10^{-6} in. for bearings from 1 to 4 in. diameter; (c) a minimum of 750×10^{-6} in. for larger bearings. More conservative designs would increase these requirements.
2. Determine the minimum acceptable c_d based on a maximum Δt of $40°F$ from the oil temperature rise curve (Fig. 28).
3. If there are no requirements for maintaining low friction horsepower and oil flow, the possible limits of diametral clearance are now defined.
4. The required manufacturing tolerances can now be placed within this band to optimize h_0 as shown by Fig. 28.
5. If oil flow and power loss are a consideration, the manufacturing tolerances may then be shifted, within the range permitted by the requirements for h_0 and Δt.

Example. A full journal bearing, Fig. 29. 2.3 in. in diameter and 1.9 in long is to carry a load of 6000 lb at 4800 rpm, using SAE 30 oil supplied at $200°F$ through a single oil hole at 30 psi. Determine the operating characteristics of this bearing as a function of diametral clearance.

1. *Diameter of bearing.* Given as 2.3 in.
2. *Length of bearing.* Given as 1.9 in.
3. *Bearing pressure.*

$$P_b = \frac{6000}{1 \times 1.9 \times 2.3} = 1372 \text{ lb/in.}^2$$

4. *Diametral clearance.* Assume c_d is equal to 0.003 in. from Fig. 23 for first calculation.

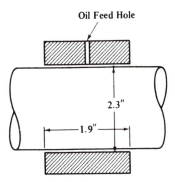

Figure 29   Full journal bearing example design. (From Oberg et al., 1992.)

5. *Clearance modulus.*

$$m = \frac{0.003}{2.3} = 0.0013 \text{ in.}$$

6. *Length-to-diameter ratio.*

$$\frac{l}{d} = \frac{1.9}{2.3} = 0.83$$

7. *Assumed operating temperature.* If the temperature rise Δt_a is assumed to be 20°F,

$$t_b = 200 + 20 = 200\,°F$$

8. *Viscosity of lubricant.* From Fig. 16, $Z = 7.7$ centipoises.
9. *Bearing pressure parameter.*

$$P' = \frac{6.9 \times 1.3^2 \times 1372}{7.7 \times 4800} = 0.43$$

10. *Eccentricity ratio.* From Fig. 24, $1/(1 - \epsilon) = 6.8$ and $\epsilon = 0.85$.
11. *Torque parameter.* From Fig. 25, $T' = 1.46$.
12. *Friction torque.*

$$T_f = \frac{1.46 \times 1.15^2 \times 7.7 \times 4800}{6900 \times 1.3} = 7.96 \text{ in.-lb/in.}$$

13. *Friction horsepower.*

$$P_f = \frac{1 \times 7.96 \times 4800 \times 1.9}{63,000} = 1.15 \text{ hp}$$

14. *Factor X.* From Table 6, $X = 12$, approximately.
15. *Total flow of lubricant required.*

$$Q_R = \frac{12 \times 1.15}{20} = 0.69 \text{ gpm}$$

16. *Bearing capacity number.*

$$C_n = \frac{0.83^2}{60 \times 0.43} = 0.0027$$

17. *Flow factor.* From Fig. 27, $q = 1.43$.
18. *Actual hydrodynamic flow of lubricant.*

$$Q_1 = \frac{4800 \times 1.9 \times 0.003 \times 1.43 \times 2.3}{294} = 0.306 \text{ gpm}$$

19. *Actual pressure flow of lubricant*

$$Q_2 = \frac{1.64 \times 10^5 \times 30 \times 0.003^3 \times 2.3 \times (1 + 1.5 \times 0.85^2)}{7.7 \times 1.9} = 0.044 \text{ gpm}$$

20. *Actual total flow of lubricant.*

$$Q = 0.306 + 0.044 = 0.350 \text{ gpm}$$

21. *Actual bearing temperature rise.*

$$\Delta t = \frac{12 \times 1.15}{0.350} = 39.4\,°F$$

22. *Comparison of actual and assumed temperature rises.* Since Δt_a and Δt differ by more than $5\,°F$, a new Δt_a, midway between these two, of $30\,°F$, is assumed and Steps 7 through 22 are repeated.

7a. *Assumed operating temperature.*

$$t_b = 200 + 30 = 230\,°F$$

8a. *Viscosity of lubricant.* From Fig. 16, $Z = 6.8$ centipoises.

9a. *Bearing pressure parameter.*

$$P' = \frac{6.9 \times 1.3^2 \times 1372}{6.8 \times 4800} = 0.49$$

10a. *Eccentricity ratio.* From Fig. 24, $1/(1 - \epsilon) = 7.4$ and $\epsilon = 0.86$.

11a. *Torque parameter.* From Fig. 25, $T' = 1.53$.

12a. *Friction torque.*

$$T_f = \frac{1.53 \times 1.15^2 \times 6.8 \times 4800}{6900 \times 1.3} = 7.36 \text{ in.-lb/in.}$$

13a. *Friction horsepower.*

$$P_f = \frac{1 \times 7.36 \times 4800 \times 1.9}{63,000} = 1.07 \text{ hp}$$

14a. *Factor X.* From Table 6, $X = 11.9$, approximately.

15a. *Total flow of lubricant required.*

$$Q_R = \frac{11.9 \times 1.07}{30} = 0.42 \text{ gpm}$$

16a. *Bearing capacity number.*

$$C_n = \frac{0.83^2}{60 \times 0.49} = 0.023$$

17a. *Flow factor.* From Fig. 27, $q = 1.48$.

18a. *Actual hydrodynamic flow of lubricant.*

$$Q_1 = \frac{4800 \times 1.9 \times 0.003 \times 1.48 \times 2.3}{294} = 0.317 \text{ gpm}$$

19a. *Pressure flow.*

$$Q_2 = \frac{1.64 \times 10^5 \times 30 \times 0.003^3 \times 2.3 \times (1 + 1.5 \times 0.86^2)}{6.8 \times 1.9} = 0.050 \text{ gpm}$$

20a. *Actual flow of lubricant.*

$Q_{new} = 0.317 + 0.050 = 0.367$ gpm

21a. *Actual bearing temperature rise.*

$$\Delta t = \frac{11.9 \times 1.06}{0.367} = 34.4\,°\mathrm{F}$$

22a. *Comparison of actual and assume temperature rises.* Now Δt and Δt_a are within 5°F.

23. *Minimum film thickness.*

$$h_0 = \frac{0.003}{2}(1 - 0.86) = 0.00021 \text{ in.}$$

This analysis may now be repeated for other values of c_d determined from Fig. 23 and a complete lubrication analysis performed and plotted as shown in Fig. 28. An operating range for c_d can then be determined to optimize minimum clearance, friction horsepower loss, lubricant flow, and temperature rise.

VII. UNDERSTANDING SLEEVE BEARINGS*

A. Materials and Load Capacity

Sleeve bearings are made in a vast array of shapes and sizes, and from wood, rubber, plastics, ceramics, and carbon as well as many ferrous and nonferrous metals and alloys. They have the advantages of quietness and freedom from fatigue. The development of wear (if it occurs) is gradual and can generally be detected before the equipment involved suffers damage. Often these bearings offer practical and unique solutions to severe operating conditions and hostile environments. For example, sleeve bearings operate well despite high temperatures or sprays of water or chemical solutions. They also work on slow shafts submerged in liquids, and on slow or oscillating shafts that operate only infrequently (Table 7).

   Like rolling element bearings, each application of sleeve bearings deserves a thorough analysis. The plant engineer should consider the operating environment and temperature, as well as the speed and load of the shaft and the life needed from the bearing. Operating conditions such as vibration, oscillating loads, and loads that vary in direction must also be taken into account. In particular, the direction and magnitude of the load will help to determine the

---

*Reprinted from Eugene R. Hafner, *Plant Engineering*, April 3, 1980, by permission of the author and publisher.

Table 7 Characteristics of Sleeve and Antifriction Bearings

Characteristic	Grease relubricable sleeve bearing	Antifriction ball or spherical roller bearings
Start-up torque	High	Low
Load-carrying capacity	Limited	Extremely high for roller bearings, moderate for ball bearings
Rate of failure	Generally gradual	Sometimes rapid or catastrophic
Sensitivity to contamination	Will absorb small amounts	Sensitive
Noise level	Low	Varies
Storage life	Long	Limited to effective life of grease or preservative
Immediate lubrication reservoir	Small	Variable, but greater than sleeve bearings
Frequency of lubrication	Frequent	Variable
Susceptibility to moisture and fretting corrosion	Insensitive	Sensitive
High-temperature operation	Depending on sleeve material, up to 1000 F	Depends on lubricant

Source: Hafner, 1980.

correct bearing material as well as the support housing design, the housing material, and the location of the lubricant entry.

1. Sleeve Bearing Materials

The reason sleeve bearings are manufactured from so many metals and nonmetals is that each material has certain advantages and weaknesses (Table 8).

Babbitt is a long-established material that provides dependable service with minimal maintenance despite the presence of moisture and chemicals. However, it is relatively soft and must be restricted to moderate loads at operating temperatures below 200°F. (It should be noted that it makes no difference to the bearing whether the heat is generated by the shaft's friction or picked up from an external source.)

The composition of the babbitts commercially available today has been developed through laboratory testing and reflects years of field experience. They provide the maximum strength compatible with resistance to seizure, good embeddability and formability, corrosion resistance, bondability, and low cost. Such babbitts are essentially alloys of antimony, copper, tin, and lead. The base material, which ranges from 65% to 85% of the whole, may be tin or lead.

Table 8 Material Selection Guide

Type of sleeve material	Load-carrying characteristics	Ambient temperature limit (°F)	Relubricable	Typical installations	Features
Babbitt	Moderate loads, light to moderate shock	200	Yes	Belt conveyors Bucket elevators Fan and blower applications Ball mills Mixer shafts Oscillating conveyors Mining and lumber handling machinery	Excellent service with minimum maintenance Low cost Low friction
Solid bronze	Moderate to heavy loads, moderate to heavy shock	300	Yes	Foundry shakeout equipment Hammer and ball mills Heavy-duty oscillating conveyors High-temperature fans and blowers Steel mill runout tables Large bucket elevators Mining and lumber handling machinery Crushers Hoists	Good resistance to shock loads and wear Replacement sleeves easily installed

Material	Load/Speed	PV	Self-lubricating	Applications	Remarks
Cast iron	Moderate loads at low speeds, light to moderate shock	300 (lubricated) 800 to 1000 (nonlubricated at light loads)	Yes	Oscillating shafts Flue damper support bearings Furnace conveyor rolls Adjustable deflection vanes	Can be used on light-duty service applications at high temperature with little or no lubrication (shaft hardness of 150 to 250 Brinell recommended for these sleeve)
Plastic, wood, rubber	Light loads at low speeds	200 for plastic; 160 for wood and most rubber	No	Equipment exposed to high concentrations of water and those handling some types of chemical solutions	Impervious to water and some chemical solutions
Sintered bronze	Moderate loads, light to mild shock	200	Yes	Hammer and ball mills Mixers Hoists Fans and blowers Gear support shafts	Contains own lubricant reservoir not subjected to rapid lubrication failure Can be relubricated as required Good resistance to shock loads and wear
Carbon	Light to moderate loads, light shock	300 to 700	No	High-temperature furnace rolls Applications subject to high concentration of water Applications that are inaccessible for lubrication Light duty oscillating or reciprocating shafts	Provides for continuous uniform, low friction film over entire surface of sleeve and shaft

Source: Hafner, 1980.

Babbitt bearings have the advantage of minimizing the danger of scoring and other damage to the shaft. Maintenance usually consists of rescraping the existing sleeve or pouring a new one at the jobsite.

Solid bronze performs well in applications in which the load and operating temperature exceed the recommended ranges for babbitt. Bronze sleeves can operate up to 300°F. They have good resistance to wear and shock loads.

Although bronze can be used at elevated temperatures, high-temperature operation requires that it be supplied with a lubricant that is also capable of withstanding the temperatures. Most multipurpose grease and petroleum oils are limited to 200 to 250°F. High-temperature lubricants that permit plant engineers to make use of the full capabilities of bronze are available, however.

Because bronze is considerably harder than babbitt, bronze sleeves can score or damage shafts. Such damage usually results from the use of the wrong lubricant or from a lack of lubrication.

Slightly damaged bronze sleeve bearings can often be repaired in the field by scraping the bore. Sleeves showing greater damage or wear should be replaced; replacement is seldom a different operation.

Sintered bronze sleeve bearings are made from a powdered composite material and impregnated with oil. The amount of oil used is usually equal to approximately 20% of the total sleeve volume. This oil is dispersed throughout the porous material and is drawn to the shaft area by capillary action when heat or pressure is applied by the shaft to the sleeve surface. The sleeve acts as a reservoir that feeds oils at a controlled rate to the rotating shaft.

Many sintered bronze sleeve bearings are designed to be relubricated. The oil used should be of the viscosity recommended by the bearing manufacturer. Sintered bronze bearings are satisfactory for applications involving moderate loads at operating temperatures up to 200°F.

Carbon bearings are primarily composed of carbon and graphite. (Graphite is a hexagonally crystallized allotrope of carbon.) They may be lubricated or operated as self-lubricated bearings at temperatures up to 700°F. Self-lubrication is provided by a very thin carbon-graphite coating that forms on the shaft, usually after the first few revolutions following installation. Unlubricated carbon bearings are classified as consumable bearings because the bearing itself is consumed to provide the lubricant. The life of such a bearing depends on the size of the load and on the shaft's surface finish and speed. As a bearing material, carbon is restricted to moderate loads and (because of its brittleness) to applications involving very light shock or none.

Cast iron can be used as a bearing material in the form of plain bore cast iron housings and sleeves. These are suitable for many slow oscillating and rotating shafts subject to light loads. Their major advantages is low cost, but in addition they require virtually no maintenance during the design service life. They may run dry or lubricated, depending on the service conditions. Their self-lubricating quality results from the free graphite flakes present in the cast iron.

Cast iron sleeve bearings can be used at temperatures as high as 1000°F. Additional shaft clearance must be specified with these high-temperature bearings to prevent seizure. For some applications, such as support bearings for flue dampers and reciprocating feed furnace conveyors, cast iron sleeve bearings provide the only practical approach.

Plastic bearings offer low coefficients of friction, low wear rates, resistance to impact and vibration, the elimination of lubrication, and cleanliness—an important consideration in food-processing machinery and other applications requiring low or no potential product contamination.

The most commonly used bearing plastics are polyamides, acetal, high-density polyethylene, polypropylene, reinforced phenolics, and fluorocarbons. The large variety of available plastic composites offers many different combinations of friction, strength, and heat, chemical, and wear resistance. Properly engineered, plastic bearings can be successfully used on a wide variety of applications, such as equipment handling corrosives, abrasives, and foods. Most plastic bearings require very smooth shafts to prevent damage to the relatively soft material.

Wood still furnishes sleeves that are used successfully on a limited number of applications. Light-duty machinery frequently employs small, oil-impregnated, hard maple bearings. Lignum vitae bearings are used for heavier loads and when the bearings must be operated submerged in water or some other liquid. Native only to the Caribbean, lignum vitae is the hardest and the most dense of all woods. Its specific gravity is 1.25; and its closely woven inner grain gives it good resistance to wear, compression, and splitting. Its resin content of 30% by volume provides remarkable self-lubrication. It is unaffected by saltwater, mild acids and alkalis, oils, bleaching compounds, and liquid phosphorus, and for this reason is often used in the chemical industry. The operating temperature of wooden sleeve bearings should not exceed 160°F.

Rubber sleeves are also used when a bearing must be submerged in water or chemical solutions. Rubber can also be used in applications in which abrasives are present.

Rubber sleeve bearings must never be permitted to operate dry, even at startup. For long life, they require smooth shafts. Their bore surfaces are often fluted to provide passages for the lubricants, for the extraction of contaminants, and to improve load distribution.

As a group, rubber, wooden, and plastic bearing materials are useful when water, high concentrations of moisture, or chemicals are present. They are generally restricted to light loads and moderate speeds. It is advisable to consult a bearing manufacturer for assistance in selecting these materials.

Exotic materials used for sleeve bearings include cobalt, nickel, and chrome. They demand exacting analysis of the application and are usually used in applications and are usually used in applications involving heavy loads, elevated temperatures, or highly abrasive or corrosive conditions, where their benefits are cost effective. The companies that offer bearings of exotic materials have a wide range of technical experience. The plant engineer should solicit their guidance to ascertain that the correct material is specified.

2. Mounted Sleeve Bearings

Although many sleeve bearings are pressed into holes bored in other parts, many more are supplied already mounted. The various styles of sleeve bearing housings provide flexibility in the mounting of the bearings and in equipment design. Rigid blocks are not self-aligning and require accurate alignment of the shaft. The shafts must also be sufficiently stiff to prevent excessive deflection.

Self-aligning units require less accurate shaft alignment and provide uniform load distribution for limited shaft deflection. The use of self-aligning sleeve bearings sometimes cuts costs by permitting a reduction in shaft diameter as a result of their greater tolerance for deflection.

3. Load-Carrying Capacity

The capacities of some types of materials, such as plastic and rubber, vary so widely that it is not productive to discuss them in a general article such as this one. For the most commonly used sleeve bearing materials, however, some general guidelines can be provided.

Babbitt and *cast bronze* sleeves are usually loaded in accordance with industry standards established by the Mechanical Power Transmission Association. These standards are expressed as unit pressure

$$P = \frac{F}{D \times L}$$

where

P = unit pressure, psi

F = load, lb

D = shaft diameter, in.

L = sleeve length, in.

For low-velocity shafts, the allowable unit loads are 250 psi for solid bronze and 150 psi for babbitt. For shafts of higher velocity, these unit loads should be decreased as shown in Fig. 30. Note that the shaft velocity is measured in surface feet per minute (sfm).

$$V = \frac{\pi \times D \times N}{12}$$

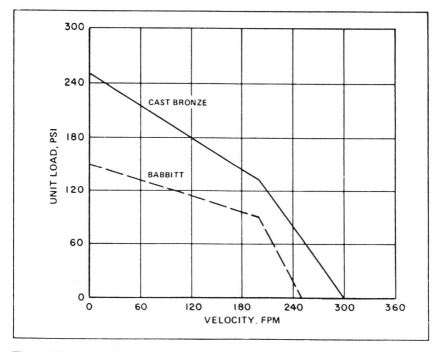

Figure 30 Lead ratings for babbitt and cast bronze fall off with increasing velocity. (Boundary lubrication is assumed.) (From Hafner, 1980.)

where

V = velocity, sfm

N = shaft speed, rpm

Thus, a large shaft that turns fairly slowly may have a higher surface velocity than a small shaft that makes a higher number of revolutions per minute. For these unit loads to apply, adequate lubrication must be provided without interruption. Proper alignment must be maintained, and starting and occasional peak loads should not exceed the unit load rating by more than 100%. The direction of the load should be no closer than 30° to the grease groove. (If the bearing has a cap that is not gibbed or doweled to the base, the load should be on the base and should not be closer than 30° to the joint between the cap and the base.) In addition, shaft finish should be in accordance with the bearing manufacturer's recommendations, and the bearing's operating temperature should not exceed the figures given in Table 8. A solid, vibration-free supporting structure should be provided, and the bearings should be protected from adverse operating conditions. In applications in which these conditions cannot be met, a service factor of up to 2.0 should be applied to the load.

Carbon and sintered bronze sleeve bearings are assigned load ratings according to the *PV* constant

$$PV = P \times V$$

Limiting values for these materials have been determined by field and laboratory tests. For carbon bearings, *PV* should be 10,000. For sintered bronze, it should be 20,000. When the *PV* constant for a carbon or sintered bronze bearing is known, its allowable radial load can be calculated from the equation

$$R = \frac{PVL}{0.262N}$$

where R = radial load, lb

This expression is valid as long as the value of pressure does not exceed 1000 psi for sintered bronze or 500 psi for carbon, and provided the velocity does not exceed 100 sfm for sintered bronze or 500 sfm for carbon.

Example. Determine the allowable load-carrying capacity for a sintered bronze sleeve bearing for a 2-in.-dia shaft operating at 150 rpm, if the sleeve has an effective length of 2.375 in.

Solution:

$$R = \frac{PVL}{0.262N}$$

$$= \frac{20,000 \times 2.375}{0.262 \times 150} = 1210 \text{ lb}$$

The pressure limit for sintered bronze is 1000 psi. Checking

$$P = \frac{E}{DL}$$

$$= \frac{1210}{2 \times 2.375} = 255 \text{ } \psi$$

The velocity limit for sintered bronze is 1000 sfm. Checking

$$V = \frac{\pi DN}{12}$$

$$= \frac{3.1416 \times 2 \times 150}{12} = 79 \text{ sfm}$$

Since both the pressure and the velocity values are within limits, the load-carrying capacity of the sleeve is 1210 lb.

4. Clearance

Establishing the correct clearance between the sleeve and the shaft is of prime importance. It must provide sufficient space for the lubricant to enter and form a film, and it must allow for radial shaft expansion, if the shaft's temperature will increase. In general, small, closely controlled clearances are more favorable to the formation of a strong hydrodynamic wedge and a separating oil film. Thinner oils can be used when clearances are small.

Larger clearances require lubricants of heavier viscosity to achieve fluid-film lubrication. Excessive clearances can result in slapping or knocking the shaft against the sleeve, which may disrupt the lubricant film. The bearing manufacturer's recommendations should be followed regarding tolerances. Table 9 furnishes guidelines that can be used if these are unavailable.

5. Selection

Reference to the sections on material and loads should enable plant engineers to choose the correct sleeve composition and load-bearing area. However, the housing, seals, shafting, fasteners, and so on must also be given consideration.

Housings must be selected as to the material (ductile iron, cast iron, or cast steel) and style required. For example, when the applied force will be located outside the housing-base area, as in horizontal or lifting applications with pillowblocks, ductile iron or cast steel housings, high-strength fasteners, extra-rigid mounting supports, and stop plates will be required. In many cases, however, cast iron can be safely used by relocating the mounting so that the load will pass through or near the center of the base of the housing.

Split housings will not be satisfactory for horizontal loads unless the housings are gibbed.

The point at which grease or oil is introduced influences the distribution of the lubricant in the clearance space. In a sleeve bearing, the lubricant should be introduced in the low-pressure area, unless the shaft's speed is very slow.

Thrust washers backed by thrust collars or shaft shoulders permit sleeve bearings to handle some thrust loading (Fig. 31). These washers should be of a suitable bearing material (usually bronze) and sufficiently large for the thrust involved.

Table 9 Shaft Clearances for Sleeve Bearings[a]

Shaft diameter (in.)	Shaft tolerance (in.)	Clearance (in.)
Up to 1	Nominal to -0.002	0.002 to 0.004
1 1/16 to 2	Nominal to -0.003	0.004 to 0.007
2 1/16 to 4	Nominal to -0.004	0.008 to 0.012
4 1/16 to 6	Nominal to -0.005	0.010 to 0.015
6 1/16 to 8	Nominal to -0.006	0.014 to 0.020

[a] For use with commercial cold-finished toleranced shafting.
Source: Hafner, 1980.

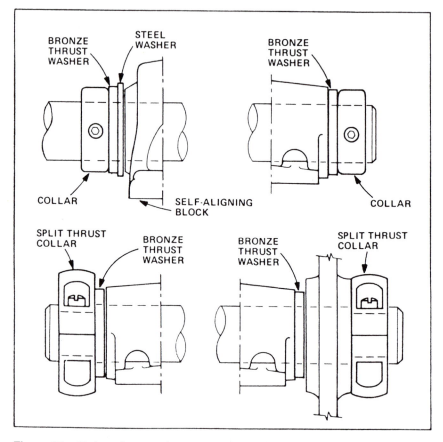

Figure 31   Various thrust-washer arrangements can be used to permit sleeve bearings to handle thrust. Collars and thrust washers are available from sleeve bearing sources. (From Hafner, 1980.)

Seals are often needed to protect sleeve bearings from abrasive contaminants. Felt seals help retain lubricant as well as exclude foreign matter.

Shafts must be of the correct size and finish. "Correct" finish will differ for various bearing materials; for example, carbon and sintered bronze sleeves demand a finish of 12AA μin., but shafts for babbitt and solid bronze sleeves may have finishes as rough as 32AA μin.

Important additional requirements are that the shaft be straight, and that it be free of seams, nicks, scratches, and burrs. These imperfections in the shaft's surface can disrupt the lubricant film and result in localized scoring of both the sleeve and the shaft. (Protruding shims on split bearing housings and sharp-edged grease grooves in the sleeve can act in the same fashion.) No keyways should extend into the bearing. Bent shafts create dynamic misalignment that can distort the normal pressure distribution in the bearing. This distortion typically causes rapid wear at the ends of the sleeve, followed by wear throughout the bore. Alignable housings compensate for static shaft deflection, but do not perform satisfactorily under conditions of dynamic misalignment. A rigid bearing mounting can sometimes be shimmed to provide alignment of the bearing with a deflected shaft, if the deflection does not vary significantly in operation.

Fasteners must prevent the housing from slipping under load, as well as be of sufficient strength to support loads not directed toward the base of the mounting. SAE Grade 5 mounting bolts can normally be used. High-strength bolts (above SAE 5) should not be used with cast-iron housings; the high torques recommended for these fasteners can induce undesirably high tensile stresses in the cast iron.

6. Installation

The following check lists may be of value in the initial assembly of sleeve bearings to shafts.

Solid Housings

- Position bearings. Slide the sleeves into position on the shaft and snug down the mounting bolts. Release the shaft, so that the bearings are carrying the applied load.
- Check for alignment and clearance. Bearing-to-shaft alignment has been achieved when uniform clearance exists at both ends of the bearing. The gap between the shaft and the sleeve bore (measured with shim shock) should be at least 0.002 in. per inch of shaft diameter. Large-area shims should be used to position the bearings, to prevent distortion of the housings. Flawed bearings that cannot be aligned should be removed and scraped.
- Complete assembly. Tighten the mounting bolts to the correct values for the grade used.
- Check for free rotation. If possible, turn the shaft by hand to make certain it does not bind in the bearing.

Split Housings

- Preparation. Remember that cap and base are machined as matched units—parts are not interchangeable. Matchmark the cap, base, and shims before disassembly.
- Improve lubrication. Chamfer the base and the cap at the split line, parallel to the shaft. Cut back the shims to suit. Making this alteration improves the lubrication of the bearing; however, it should be omitted if the joint will be in the load area.
- Position base. Put the base under the shaft and snug down the mounting bolts. Release the shaft's weight so that the bearing is carrying the load.

Test with Prussian blue to make sure the shaft rests on the entire surface of the load area. If necessary, shim the base, using large shims to prevent distortion of the housing. Some scraping of the bearing may be required for an accurate fit between the sleeve and the shaft.

Install cap and shims. Observing the matchmarks, replace the cap and shims. Make certain the shims do not touch the shaft. Bolt the cap securely to the base, torquing the cap bolts (or nuts) tightly enough to seat the base and cap firmly, but not tightly enough to distort the shims.

- Check for alignment and clearance. Test the gap between the sleeve bore and the shaft as described for solid housings. Remove the cap and scrape the bearing if necessary.
- Complete assembly. Tighten the mounting bolts in accordance with the grade of bolts used. If possible, provide stop bars or adjusting screws at the ends of the bearing units to assure proper location and prevent shifting under load; these devices are especially important when the direction of the load is not straight down through the base.
- Check for free rotation. If possible, bur the shaft by hand to make certain it does not bind.

7. Troubleshooting

Sleeve bearings, like other mechanical devices, sometimes suffer premature failure. When such failures occur, the plant engineer should determine their causes, so that recurrences can be prevented.

VIII. EXPLANATION OF PV

All the bearings discussed earlier in the chapter are designed basically around the PV formulation. Here in a simplistic form is the explanation for PV.

P means pressure. It is always shown in pounds per inch of projected area. (Projected area is the I.D. of the bearing in question times the length of the bearing, which gives you the support area for the journal.) All loads including the weight of the shaft and sudden surge or shock loads must be totaled and then divided by the number of inches of projected area to find P.

V is the maximum surface feet per minute that any material can handle in the way of top speed before, even when using the proper lubricants, you will reach a temperature rise induced by rubbing friction that will create, because of the coefficient of thermal expansion of the material in question, a "close-in" of the I.D. which will cause a seizure and early failure.

PV is the calculation which you yourself perform on the job at hand. It is derived by taking the actual P and V calculation of your particular situation. In other words, you gather up all of the load information, including surge and shock loads and the weight of the journal, spreading the load over the inches of projected area of your bearing, and then using that number for P, multiply it

Table 10 Plain Bearing Material *PV* Chart (Ranked by *PV*)

	Class I		
Max *PV* (*P* × *V*)	Max *P* (PPIPA)	Max *V* (sfm)	Material
150,000	8,000	100	Manganese bronze
125,000	4,500	225	Aluminum bronze
80,000	4,000	900	Microcast SAE 660
80,000	4,000	1,000	Microcast SAE 64
75,000	3,000	750	SAE 660
70,000	3,000	800	SAE 64
70,000	5,000	250	CDA 929
60,000	2,000	1,000	CDA 941
50,000	3,000	750	Concast Al/Sn/Bp
50,000	1,500	1,000	Babbitt high tin
30,000	1,500	1,200	Babbitt bimetal
17,000	1,500	60	Monocast nylon
12,000	700	800	High-lead Babbitt
3,000	380	475	Ext. nylon
1,000	50	1,000	Rubber deflection pad

Table 11 Plain Bearing Material *PV* Chart (Ranked by *PV*)

	Class II		
Max *PV* (*P* × *V*)	Max *P* (PPIPA)	Max *V* (sfm)	Material
80,000	3,000	1,200	Advanced-duty PM 8–10% porous
80,000	8,000	35	Super 16 iron, 18–23% porous
50,000	2,000	1,200	PM 80–10%, 10–23% porous
50,000	2,000	1,200	Aluminum PM
35,000	4,000	225	Copper iron PM
30,000	3,000	200	Iron copper PM
25,000	1,000	500	Oil-filled UHMWPE
25,000	600	500	Oil-filled UHMWPE, Nolu S
20,000	2,000	600	Oil-filled monocast 6/6
18,000	400	450	OIl-filled urethane
18,000	3,000	1,100	Oil-filled polyacetal
15,000	2,000	2,000	Wood, Lignum vitae
15,000	1,200	1,500	Wood, Maple
3,000	1,000	100	Nolu D UHMWPE

Table 12 Plain Bearing Material PV Chart (Ranked by PV)

Class II			
Max PV ($P \times V$)	Max P (PPIPA)	Max V (sfm)	Material
100,000	25,000	400	Steel bonded PTFE wire liner
100,000	25,000	400	Fiberglass bonded PTFE wire liner
100,000	20,000	50	Steel back woven PTFE bonded
100,000	20,000	50	Fiberglass back woven PTFE bonded
100,000	4,500	50	Graphite plug groove aluminum bronze
50,000	1,500	1,200	PTFE steel back PM
20,000	3,000	400	PTFE tape steel shell
20,000	1,500	400	PTFE tape no shell
15,000	600	2,500	Natural graphite
15,000	500	1,500	Man-made carbon
15,000	1,500	500	Metallized graphite
10,000	1,000	400	Rulon LR
10,000	200	225	PTFE 25% glass
10,000	200	275	PTFE 25% graphite
10,000	500	200	PTFE 25% bronze
7,500	750	400	Rulon J
3,000	1,000	100	Delrin/Celcon
3,000	1,000	1,000	Delrin AF
1,000	300	450	Virgin PTFE

by your velocity (surface feet per minute). Then you check to see if the answer derived is greater than the maximum PV allowed for the material in question.

As an example, a 90 Cu/10 Sn porous powdered metal bearing has a PV of 50,000. It has a maximum load capacity of 2000 pia and a maximum V of 1200 SFM, which, when multiplied together, would give 2,400,000 PV and obviously will not work. Hence, the limiting factor of 50,000 has been established by the manufacturers of this type of product and, as in all other materials where a maximum PV is shown, should be considered in any bearing calculations.

Tables 10 to 12 present PV charts for various materials.

GLOSSARY OF BEARING TERMS

arbor: A tool to insert bearings. Consisting of a shouldered stub that is 15/16 in. the length of the bearing, that is twice the radial wall of the bearing, and that has a driving head with a crowned surface made from a nonferrous material for plastic bearings and made from a ferrous material for nonferrous bearings, with the same diameter as the insertion stub and the same dimension as the maximum diameter of the shaft.

bearing, flange: A single-component system for containing rotary motion and light thrust loads.

bearing, plain: A single-component system for supporting rotating motion.

bearing, sleeve: A single component for supporting rotating motion.

bearing, thrust: A large-diameter thin-cross-section (short length) device for containing heavy-thrust loads.

boundary lubrication: The condition when there is no high-pressure system to force-feed the oil into the interface between the bearing and the shaft and lubrication occurs because the profile of the interface is irregular, allowing lubricant to be in the load zone, but not to be construed as a full-film system.

bushing: A sleeve for "armoring a hole."

CDA: Copper Development Association. A group that sets the standards for chemical analysis of bearing materials.

close-in: The amount of transfer of the interference fit to a reduction of the I.D. of the bearing, due to the compression of the radial wall of the bearing. Expressed in percentage of the interference fit, can be as little as 10% or as much as 110%, depending on the material of the bearing.

conformability: The tendency of the bearing material to assume, by wear, the profile configuration of the shaft of journal material, becoming "mirrorlike" in response to the rotation. Generally occurs in the first few revolutions.

embeddability: The tendency of the bearing material to absorb hard particulate within itself so as to not force a situation where the particulate will stand up on the I.D. of the bearing and create wear scars (usually rings) in the journal.

fretting: Wear caused by oscillation and/or vibration which removes minute particles from the interface of a bearing system.

hydrostatic wedge: Wedge or cushion of oil developed by the forceful rotation of the journal in the bearing in a horizontal mode caused by inadequate clearance in the load zone. An inability of the oil or grease or debris (in the case of self-lubricating bearings) to get by the load zone quickly enough.

P: P is the designation in bearing design for pressure and indicates that amount of load which may be applied to a specific material without galling occurring.

PBSA: Plane Bearing Standards Association. A group that sets the tolerances for housings, bearing materials, and shaft dimensions.

press fit: The range of tolerance between the O.D. of the bearing and the I.D. of the housing. Can be expressed in thousandths of an inch and is the sum of the distance between the I.D. of the housing and the O.D. of the bearing, considering the tolerance range of both.

PV: The maximum number of *P* times *V* for a particular material that can be achieved when multiplied together as a net answer without guaranteeing an early failure.

radial load: A load applied at right angles to a shaft (perpendicular to the journal's long axis).

running clearance: The sum of the tolerance of the air gap between the I.D. of the bearing and the O.D. of the journal.

SAE: Society of Automotive Engineers. A group that sets the standards for physical properties of materials and specifications.

static: Stationary, without rotation, without oscillation, and without lateral movement.

spall: The void or cavity remaining in a contact surface of a bearing system after a flake of metal has been removed. Generally caused by wear or load.

thrust load: A force applied parallel to the long axis of the journal or shaft.

V: Equating to velocity, it is the ultimate speed for a given bearing material before the rubbing friction will generate sufficient thermal impact on the bearing system so as to cause the bearing material, by coefficient of thermal expansion, to close-in and seize the shaft.

viscosity: A rating or measure of the flowability of a liquid at a specific temperature.

REFERENCES

Hafner, E.R., *Plant Engineering*, April 3, 1980.
Oberg, E., Jones, F.D., Horton, H.L., and Ryffel, H.H., *Machinery's Handbook*, 24th ed., edited by Robert Green. Industrial Press, New York, 1992.

5

ROLLING ELEMENT BEARINGS

Arthur J. Gunst

FAG Bearings Corporation, Stamford, Connecticut

I. HISTORICAL BACKGROUND

A. Roller to Wheel

Friction has always impeded man's effort to move things. Early efforts were most probably sledges dragged along the dry ground. When it was found that they moved easier over muddy, rather than dry, terrain, water was used as a lubricant to reduce tractive effort. A historical document nearly 4000 years old revealed that this method was used to move a large statue (Fig. 1(a)).

The next probable improvement was the use of loose rollers between the sledge and ground (Fig. 1(b)). Rolling had now replaced dragging. A refinement of this was to use a large square-sectioned stone to which were added wooden circular segments to form a continuous circle or wheel. Figure 2 shows a model with four circular "fins." No one knows just how or when the loose roller developed into the wheel. However, no doubt this was a most significant historical milestone. With the use of the wheel and axle came the use of the bearing.

B. Wheel Shafts/Axles Supported in Bearings

The origin of the wheel and cart is thought to come from central Asia, at least 6000 years ago. The material of the early wheels was wood or stone, which, of course, has changed but the basic concept has not.

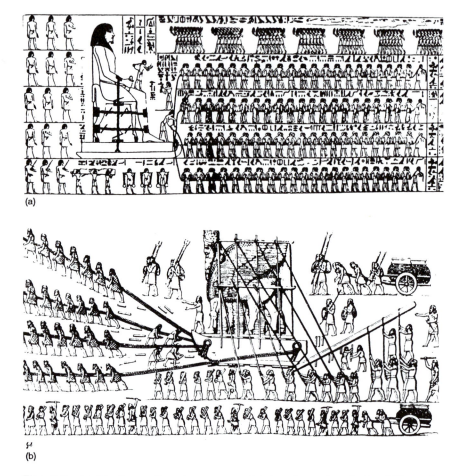

(a)

(b)

Figure 1   (a) Moving an Egyptian statue on lubricated wooden planks 1800 B.C. (b) Loose rollers move an Assyrian statue 700 B.C..

The roller became the wheel only after it could be positioned by a bearing assembled to a frame (Fig. 3). In the wheel center there is a bore into which the axle, axle journal, or bearing journal is inserted. A moving part now slides on a stationary part, and the relative motion between the two is sliding motion and is measured as sliding speed. Surfaces in sliding contact are subject to wear unless separated by a lubricating film.

In the case of the early wheel complete axles and wheels had to be replaced because bearing surfaces were worn. This undesirable condition led to the development of independent elements which could be inserted in the bore of the wheel or housing without replacing whole axles or wheels.

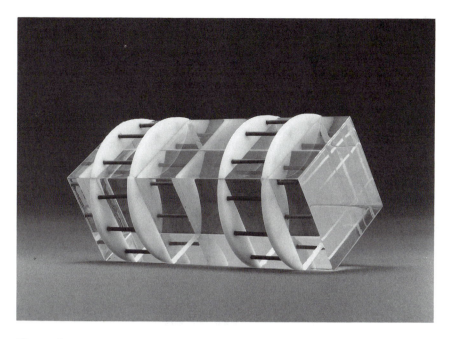

Figure 2 Model made with modern materials of circular wooden segments fastened to a square stone section 15th century B.C.

Figure 3 Wheelset and bearings.

C. Early Lubrication and Materials

The separation of bearing elements allowed for the selection of the most suitable material for the various components and for subsequent design improvements such as grooves/holes for introducing lubricants and for designing sealing arrangements against the environment.

Early advancement of bearings, bearing materials, and lubricants was made in small steps over hundreds of years. Early lubricants were usually water in stationary machines and animal fats in vehicles. Metal in times prior to the industrial era was used sparingly as it was considered a precious material. Because of the greater strength of metal, it allowed for smaller bearings and axles to be used for the same external loads (Fig. 4).

The development of the wheel and wheel bearing was the start of the modern antifriction bearing. Actually the term *antifriction bearing* is usually reserved for rolling element bearings, such as ball bearings, but this is a misnomer as all bearings are designed and used to reduce friction.

D. Rolling Bearings in the Industrial Era

About 200 years ago a better understanding of the rolling mechanism and the wider range of higher strength materials triggered a push in the development of rolling bearings. For example, Fig. 5 shows a post windmill, built around

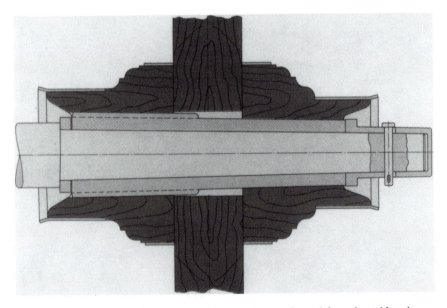

Figure 4   Section through wooden spoked wheel—use of metal for axle and bearings.

Figure 5   Slewing bearing from 1780 in a windmill from Great Britain.

1780, that had a guidance bearing (slewing ring bearing) mounted in a floor plate.

Figure 6 shows a carriage axle with two ball bearings in which the balls are loaded into the bearing through a filling hole which is then plugged. Figure 7 is a bearing mounting in a merry-go-round.

Different bearing designs were being developed for applications taking other than a simple radial load. Early in the 1900s the ball bearing was the most common type. Figure 8 shows various types of rolling elements.

E.   Developments in Europe and America

In Europe the impetus for developing the roller bearing industry was provided by the bicycle with ball bearings (Fig. 9), while in the United States the tapered roller bearing was replacing the plain bearing in the axles of horse-drawn wagons (Fig. 10). The first cylindrical roller bearing that proved superior in highly loaded rail cars was the flanged roller bearing in 1911 (Fig. 11). From these types of bearings sprang a need for bearings that were internally self-

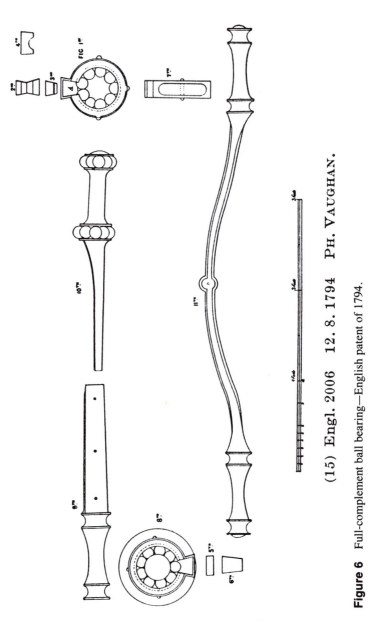

(15) Engl. 2006   12. 8. 1794   Ph. Vaughan.

Figure 6   Full-complement ball bearing—English patent of 1794.

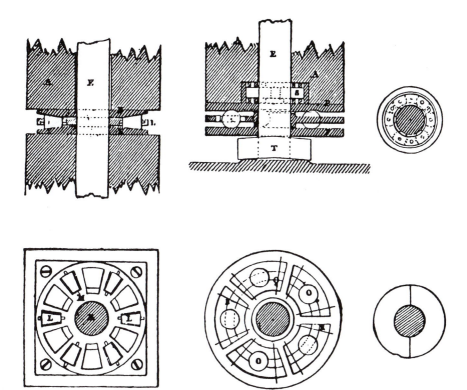

Figure 7 Caged thrust ball and roller bearings—merry-go-round, French patent 1802.

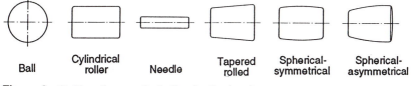

| Ball | Cylindrical roller | Needle | Tapered rolled | Spherical-symmetrical | Spherical-asymmetrical |

Figure 8 Rolling elements for ball and roller bearings.

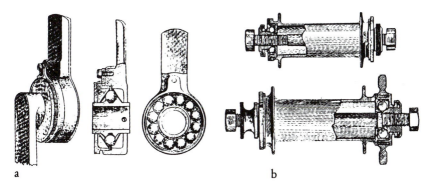

a   b

Figure 9 English-made bicycle bearings. (a) Pedal bearings, (b) front and rear wheel bearings.

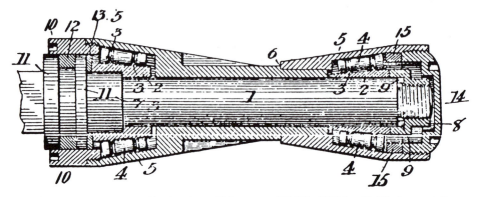

Figure 10   Tapered roller bearing—U.S. patent 606635, 1898 by H. Timken and R. Heinzelmann.

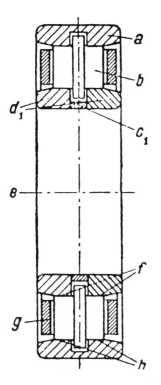

Figure 11   Flanged roller bearing made by G.H.J. Jaeger AG 1911, to take both radial and thrust loads.

aligning. Figures 12 and 13 show the self-aligning ball bearing and roller bearing. The development of the rolling bearing was closely allied to the development of machinery, and although it was prominent in the rapid evolution and invention of machines years ago, the refinements and understanding are still expanding today.

II. COMMON ROLLING BEARING TYPES

A. General Characteristics and Nomenclature

Usually a rolling bearing consists of two rings with rolling elements running on tracks or raceways. Commonly shaped rolling elements are ball, cylindrical, needle roller, tapered roller, symmetrical, and the asymmetrical barrel (spherical) roller (Fig. 8). Most often the rolling elements are in a cage or separator which spaces the elements and prevents contact between them. A secondary but important function is to hold the elements together with the ring or rings to facilitate bearing mounting.

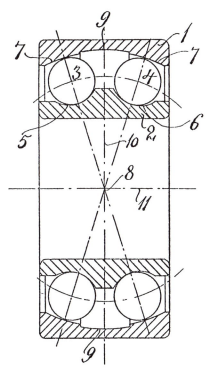

Figure 12   Self-aligning ball bearing Swedish patent of S.G. Wingquist of 1907.

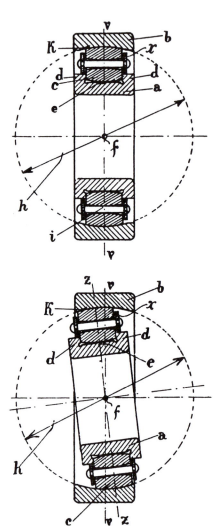

Figure 13   Self-aligning roller bearing, German patent No. 290038, February 16, 1912, by J. Modler.

The rings and rolling elements produced are predominantly made of through-hardened chromium steel, although many are also of case-hardened steel. Special bearings for hostile or extreme environments can be made of tool steel, stainless steel, plastic, or other materials.

The various types of bearings are classified by design features and intended usage. The direction of the predominant load creates two types, radial and

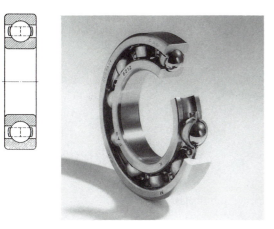

Figure 14   Deep-groove ball bearing without filling slot.

thrust bearings. Another distinction is by rolling element, either ball or roller. A further classification is whether the bearing has a design feature that permits angular movement or misalignment.

1.   Radial Ball Bearings

Deep-Groove Ball Bearings without Filling Slots.   The raceways consist of deep grooves which can support high radial as well as axial forces. This bearing can take axial or thrust loads in either direction. See Fig. 14.

Figure 15 shows how balls are installed. The inner ring is positioned to make space for the ball complement, after which the ring is positioned central to the outer ring. Then the balls are equally spaced to allow for cage installation. This type of assembly has also been called the Conrad assembly.

Bearings with seals or shields allow for compact designs. Ball bearings in Figs. 16 and 17 show double (two) shields and double (two) seals respectively.

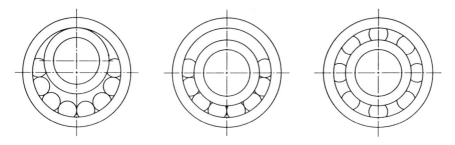

Figure 15   Filling balls into a deep-groove ball bearing without filling slots.

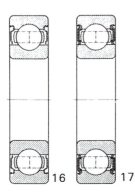

16   17

Figure 16   Deep-groove ball bearing with shields.

Figure 17   Deep-groove ball bearing with seals.

Bearings of either type are grease-lubricated by the bearing manufacturer. Note that single-shielded or single-sealed bearings are usually *not* supplied pre-greased. Shields are metal disks usually held in place by staking. Unlike seals, there is no contact with the inner ring shoulder but a small clearance is left.

Seals are made of rubber or plastic with an internal metal disk as stiffener. They have a lip that contacts and rubs on the inner ring shoulder. Seals are a more effective closure than shields. Some early seal designs had the sealing material sandwiched between metal disks. This arrangement was staked in place in the outer ring.

Externally Aligning Units.   The extended inner ring deep-groove bearings are made in a variety of configurations all of which have some form of locking device to fasten the inner ring to the shaft. The locking is usually done by clamping with an eccentric collar (Fig. 18) or by means of two set screws (Fig. 19).

The bearing outside diameter is spherical so that it can align itself in a like spherical housing. Another feature is that this bearing type is usually sealed.

Deep-Groove Ball Bearing with Filling Slots.   These bearings are also known as "Max" type because they carry a ball complement that is intended to give the maximum radial capacity. This is achieved by providing a filling slot in the inner and outer ring shoulders as shown in Figs. 20 and 21, through which the balls are loaded. These bearings are less suitable because under axial loads they tend to run over the filling slots with subsequent detrimental effects.

Figure 22 shows a double-row version of the Max-type bearing. Its use is limited, and it has been largely replaced by the double-row angular contact bearing.

Magneto Bearings.   These bearings are similar in construction to the Max type, except instead of a filling slot the outer ring has only one shoulder (Fig.

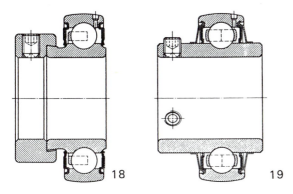

Figure 18 S-bearing unit with eccentric self-locking collar.

Figure 19 S-bearing unit with two set screws.

23(a),(b)), so the inner ring/ball assembly can be taken apart. This bearing can only take thrust in one direction, so it requires another to adjust against.

Single-Row Angular Contact Ball Bearings. These bearings are made to transmit forces from one raceway to the other through an angle, usually 40°. Various other angles (15°, 20°, 25°, and 30°) are also manufactured (Fig. 24(a),(b)). The higher contact angle of this type allows the bearing to sustain higher axial loads than deep-groove bearings. However, the axial load can be applied in one direction only. Radial load can be applied only with a simultaneously applied thrust load. Unless the axial load is constant or if the ratio of radial to thrust exceeds the limit, two opposed angular contact bearings must be installed for mutual axial support.

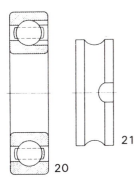

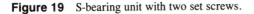

Figure 20 Deep-groove ball bearing with filling slot.

Figure 21 Inner ring of a deep-groove ball bearing with filling slot.

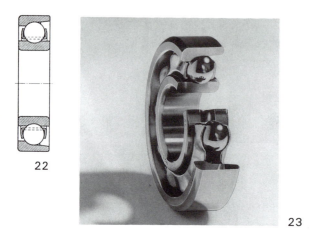

22

23

Figure 22   Double-row deep-groove ball bearing with filling slots.

Figure 23   Magneto bearing.

At one time, single-row angular contact ball bearings were matched in pairs for installation as , +, and tandem arrangements (Figs. 25, 26, 27). However, they are now made of a universal design which allows pairing in any configuration without matching or the use of shims.

Double-Row Angular Contact Ball Bearings.   There are various designs of this type. Figure 28 shows an O-arrangement with filling slots. Other versions

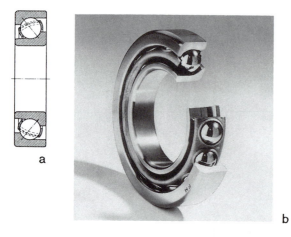

a

b

Figure 24   Single-row angular contact ball bearing.

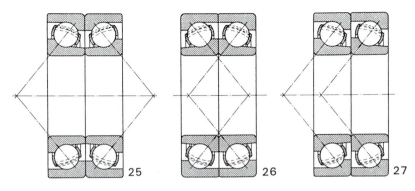

Figure 25   O-arrangement of two angular contact ball bearings of universal design.

Figure 26   X-arrangement of two angular contact ball bearings of universal design.

Figure 27   Tandem arrangement of two angular contact ball bearings of universal design.

are +-arrangements and designs without filling slots. The filling slot design requires mounting so that the row without slots takes the thrust. The nonfilling slot design has a lower capacity since it has a lesser ball complement.

Figure 29 shows a two-piece or split inner ring design for carrying higher loads and also reversing axial loads. This version allows a high contact angle and a maximum ball complement.

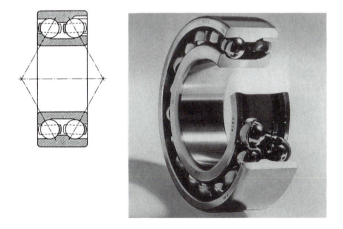

Figure 28   Double-row angular contact ball bearing with filling slots.

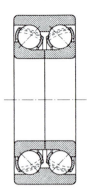

Figure 29 Double-row angular contact ball bearing with split inner ring.

Four-Point, Split Inner Ring, or Gothic Arch Bearings. This bearing type has arcs instead of radii for raceways so that if only radial load is applied, the ball makes contact at four points (Fig. 30). Therefore, this bearing is used for taking predominantly thrust loads. The major field of application is in power transmission to take thrust in either direction with a relatively narrow width bearing.

Self-Aligning Ball Bearings. The configuration has an outer raceway that is spherical with an inner ring that has two grooved raceways. The cage is usually common to the two ball sets. This allows the inner ring to align itself to the outer ring in the case of shaft misalignment (Fig. 31).

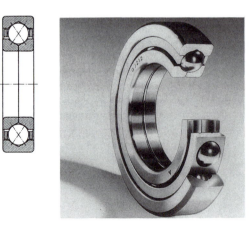

Figure 30 Four-point bearing with split inner ring.

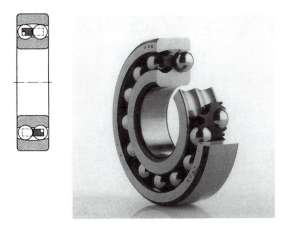

Figure 31 Self-aligning ball bearing.

2. Radial Roller Bearings

Cylindrical Roller Bearings. There are many types of these roller bearings with the rollers being guided by two flanges or lips on either the inner or outer ring. When this type of bearing is used to locate the shaft, there are either three or four flanges in the design. Most versions have one ring separable from the other. The cage and roller set are fixed to the doubled-flanged ring.

Full-complement (cageless) bearings are used in special applications. The various types of configurations are differentiated by the flange or lip arrangements as follows:

Type	Inner Lip	Outer Lip	Figure
NJ	—	Two	32
N	Two	—	33
NJ	One	Two	34
NUP	Two (one is loose)	Two	35
NH	Two (one is loose)	Two	36

In recent years the cylindrical roller bearing has had a capacity increase, which is indicated by the part number suffix E. The capacity increase was achieved by using rollers with larger diameters and lengths. This resulted in a

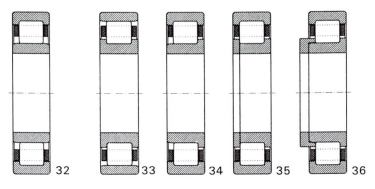

| 32 | 33 | 34 | 35 | 36 |

Figure 32   Cylindrical roller bearing, design NU.

Figure 33   Cylindrical roller bearing, design N.

Figure 34   Cylindrical roller bearing, design NJ.

Figure 35   Cylindrical roller bearing, design NUP.

Figure 36   Cylindrical roller bearing, design NH or NJ with angle ring.

bearing with the same external dimensions but with a substantial capacity increase (Fig. 37).

Double-row cylindrical roller bearings with increased running precision are used in machine tools (Figs. 38 and 39).

Needle Roller Bearings.   This type is a particular version of the cylindrical roller bearing made to take radial loads. The high diameter-to-length ratios vary from 1 to 2.5 to 1 to 10. The roller shape is the reason for the name needle roller. The low radial profile makes this bearing suitable for applications which are space limited.

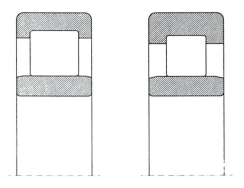

Figure 37   High-capacity design E and (right) normal design.

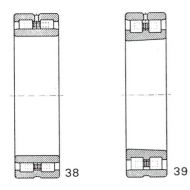

Figure 38 Double-row cylindrical roller bearing, type NNU.

Figure 39 Double-row cylindrical roller bearing, type NN.K.

Early-design needle roller bearings were without a cage, which restricted the bearing's speed capability. Today the preferred design is with a cage which retains and guides the rollers, allowing for higher speeds.

Needle roller cage assemblies can be used alone if provided with hardened and ground shaft and housings (Fig. 40). The drawn cup needle roller bearing has a hardened unground outer ring which is shaped by pressing (Figs. 41 and 42). The correct geometry is provided by the geometry of the housing when the bearing is mounted with a press fit.

Figure 43(a),(b) is a machined ring needle roller bearing, which usually has a double-lipped outer ring. Figure 44, however, shows a bearing with no lips.

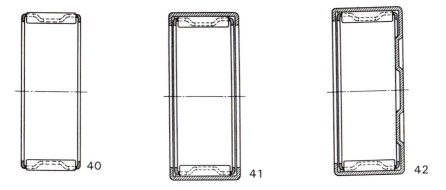

Figure 40 Needle roller and cage assembly.

Figure 41 Drawn cup needle roller bearing with open ends.

Figure 42 Drawn cup needle roller bearing with one closed end.

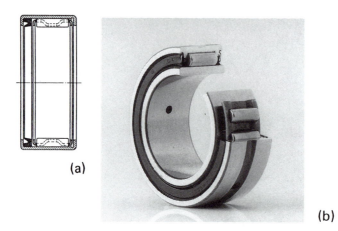

(a)

(b)

Figure 43   Needle roller bearing with integral lips.

Other special designs include the yoke-type track roller (Fig. 45) and stud-type track roller (Fig. 46). These have thick-walled outer rings which can transmit heavy or shock loads directly to the outer ring without a housing. They act as rollers in such applications as moving conveyors.

Tapered Roller Bearings.   Tapered roller bearings consist of an inner ring with two lips, cage, and rollers. This assembly is commonly called a cone. The outer ring is lipless and commonly known as a cup. The cone and cup are separable from each other. The contact lines between rollers and raceways intersect at a common point on the bearing axis (Fig. 47(a),(b)).

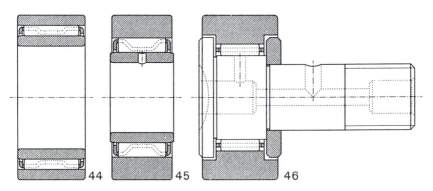

Figure 44   Needle roller bearing with lipless rings.

Figure 45   Yoke-type track roller.

Figure 46   Stud-type track roller.

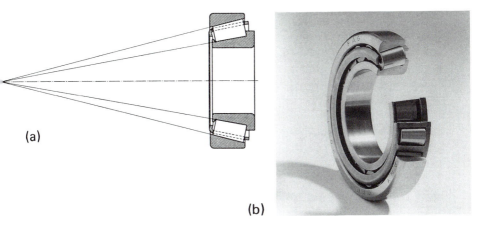

(a)

(b)

Figure 47   Tapered roller bearing.

The two lips or ribs on the cone serve different functions. The smaller rib retains the cage and rollers, while the large rib does the same and has a finely finished surface that guides the large end of the rollers.

Due to the inclined position of the raceways, a single, single-row tapered roller bearing can generally not accept pure radial load. It must have a seating axial load or be opposed by another bearing. In this way, they are similar to magneto bearings and single-row angular contact ball bearings.

Figure 48(a) shows a double-row tapered roller bearing with a solid cone spacer (no relubrication features), and Fig. 48(b) is a two-row bearing with a cup spacer. Both are designed for relubrication through the outer ring.

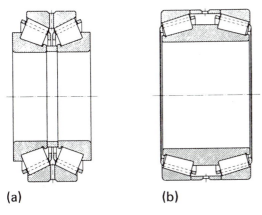

(a)   (b)

Figure 48   Double-row tapered roller bearing: (a) with split inner ring; (b) with split outer ring.

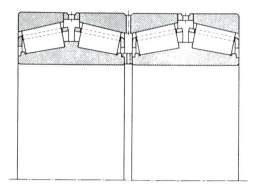

Figure 49   Four-row tapered roller bearing.

Figure 49 is a four-row bearing with cup and cone spacers. This type is used almost exclusively in rolling mills. The spacers used in either two or four bearings are ground at the manufacturing plant to give the correct bearing clearance when the bearing is mounted and running.

Barrel Roller Bearings.   These are one-row self-aligning roller bearings (Fig. 50(a),(b)). The rollers are barrel shaped and their profile is nearly the profile of the inner and outer raceways. The two lips on the inner ring guide the rollers. These bearings are designed mainly for radial load.

Spherical Roller Bearings.   These bearings have two rows of barrel-shaped rollers whose axes are inclined to the axis of rotation of the bearing. They are

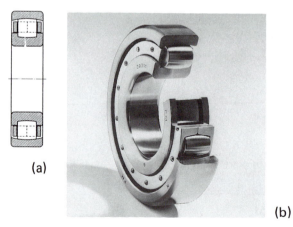

(a)

(b)

Figure 50   Barrel roller bearings.

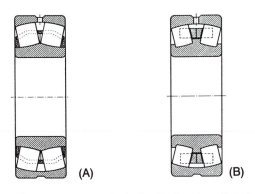

Figure 51 (A) Spherical roller bearings E design. (B) Spherical roller bearings conventional design.

designed to accept a certain degree of misalignment or shaft deflection by allowing the rollers to align on the outer raceway. Figure 51(A) shows the high-capacity E design, while Fig. 51(B) shows the conventional style with a center flange for roller guidance. The E design is characterized by a lipless inner ring which allows the use of longer rollers. Thinner rings allow for larger-diameter rollers.

Other spherical roller designs have rollers that are guided by the cage and a loose center guide ring installed between the roller ends. Another design feature found in practically all popular sizes is the circumferential groove with radial holes to allow the entrance of lubricant between the two rows of rollers.

3. Thrust Ball Bearings

Thrust Ball Bearings with Grooved Raceways. These are single acting or double acting, which refers to whether the bearing can accept the load in one direction or two. Figure 52(a),(b) shows a single-acting bearing, while Figure 53(a),(b) is a double-acting type. The smaller-bore inner ring is fastened between shoulders on a shaft, and the outer is clamped in a housing so that the assembly can function in two directions. The grooves limit the bearings to comparatively low speeds, because at low axial loads and high speed the centrifugal force tends to cause the balls to ride over the grooves.

Figure 54 shows a ball thrust bearing with a spherical housing washer used to accept initial mounting misalignment from out-of-square housing seats. Due to the high friction of the spherical washer, this type is usually not recommended for dynamic misalignment. The ball thrust bearing is used where axial loads are too high for radial bearings or where an axially stiff bearing is required.

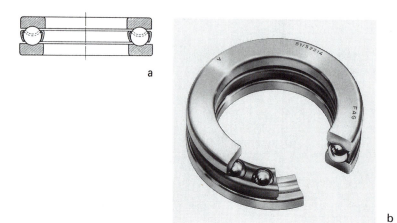

Figure 52   Single-acting thrust ball bearing.

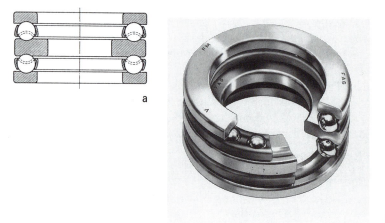

Figure 53   Double-acting thrust ball bearing.

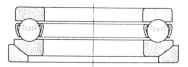

Figure 54   Single-acting thrust ball bearing, spherical housing washer, and seating ring.

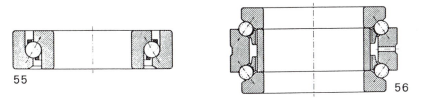

Figure 55   Single-acting angular contact thrust ball bearing.

Figure 56   Double-acting angular contact thrust ball bearing.

Angular Contact Thrust Bearings.   Figure 55 is a single-acting bearing, and Fig. 56 a double-acting bearing designed for machine tool applications. These bearings take higher speeds than the thrust bearings and are more accurate due to the shoulder guidance on the inner and outer rings.

Slewing Bearings (Fig. 57(a),(b).   These are large bearings used in cranes, excavators, dredges, and similar equipment to take radial, axial, and tilting moments. The bearing shown in Fig. 57(a) is a ball bearing, and Fig. 57(b) shows a roller bearing. Slewing bearings are mated closely to the application, as shown by the bolt holes for clamping to machine components and by the use of an integral gear.

4.   Thrust Roller Bearings

Spherical Roller Thrust Bearings.   The roller set in this type of bearing is made of asymmetrical barrel rollers inclined to the bearing axis so that the

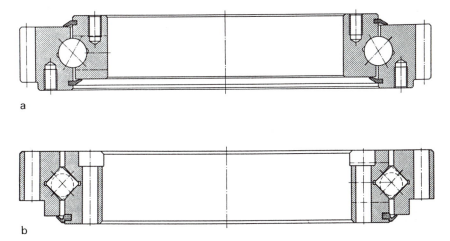

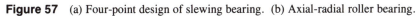

Figure 57   (a) Four-point design of slewing bearing. (b) Axial-radial roller bearing.

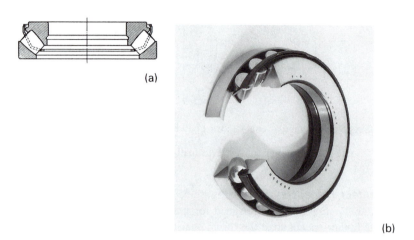

(a)

(b)

Figure 58 Spherical roller thrust bearing.

bearing can accommodate high axial loads as well as considerable radial loads (Fig. 58(a),(b)). The rollers are guided and supported by the inner ring flange. This bearing can take misalignment and shaft deflection.

Cylindrical Roller Thrust Bearings. These are mostly used as single-acting bearings and consist of two simple lipless washers or plates and a roller/cage assembly. Since there is nothing to locate the bearing in the radial direction, it usually requires a radial bearing for location.

Sliding occurs at the inner and outer ends of the rollers because there is not true rolling motion, unlike the tapered roller bearings (Fig. 47(a)). The long roller is therefore replaced by several short rollers to minimize the effects of sliding (Fig. 59). A shaft-guided cage is commonly used to retain and guide the rollers.

Needle Roller Thrust Bearings. There are used where space is limited and or low cost is a major consideration. They also can be used in conjunction with radial needle roller bearings. The washers are hardened but not ground and therefore are inexpensive. Figure 60 shows the bearing with a shaft-guided cage.

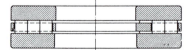

Figure 59 Single-acting cylindrical roller thrust bearing.

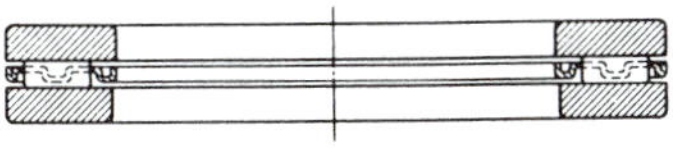

Figure 60   Needle roller thrust bearing.

Tapered Roller Thrust Bearing. Several designs are made. A common design is shown in Fig. 61(a), made with one tapered raceway and one flat washer. This design, unlike the cylindrical roller thrust bearing, does not have any sliding on the raceways. The rollers are guided by the flange. Figure 61(b) shows a cageless or full roller complement which is kept assembled by the pressed steel shroud.

III. BEARING CLEARANCE

A. Definitions

According to ANSI/AFBMA Standard 1-1990, radial internal clearance is in

> The arithmetical mean of the radial distances through which one of the rings or washers may be displaced relative to the other, from one eccentric extreme position to the diametrically opposite extreme position, in different angular directions and without being subjected to any external load. The mean value includes displacement with the rings or washers in different angular positions relative to each other and with the set of rolling elements in different angular positions in relation to the rings of washers.

> Shorter definition is this: "the clearance of a rolling bearing is the amount of possible displacement of one ring relative to the other in the radial or axial directions."

B. Clearance Tables

Radial internal clearance is found for the various types of metric rolling bearings in ANSI/AFBMA Standard 20 as follows:

Table 1 (Table 12, Section 20)   radial contact ball bearings

Figure 61   Tapered roller thrust bearings: (a) with conical shaft washer and flat housing washer; (b) with a full complement of rollers and a pressed steel shroud (bearings for steering knuckles).

Table 1   Radial Internal Clearance Values for Radial Contact Ball Bearings

PART I   Clearance values in micrometers

d mm over	incl.	SYMBOL 2 min.	max.	SYMBOL 0 (Normal) min.	max.	SYMBOL 3 min.	max.	SYMBOL 4 min.	max.	SYMBOL 5 min.	max.
2.5	6	0	7	2	13	8	23	-	-	-	-
6	10	0	7	2	13	8	23	14	29	20	37
10	18	0	9	3	18	11	25	18	33	25	45
18	24	0	10	5	20	13	28	20	36	28	48
24	30	1	11	5	20	13	28	23	41	30	53
30	40	1	11	6	20	15	33	28	46	40	64
40	50	1	11	6	23	18	36	30	51	45	73
50	65	1	15	8	28	23	43	38	61	55	90
65	80	1	15	10	30	25	51	46	71	65	105
80	100	1	18	12	36	30	58	53	84	75	120
100	120	2	20	15	41	36	66	61	97	90	140
120	140	2	23	18	48	41	81	71	114	105	160
140	160	2	23	18	53	46	91	81	130	120	180
160	180	2	25	20	61	53	102	91	147	135	200
180	200	2	30	25	71	63	117	107	163	150	230

PART II   Clearance values in micrometers

d mm over	incl.	SYMBOL 2 min.	max.	SYMBOL 0 (Normal) min.	max.	SYMBOL 3 min.	max.	SYMBOL 4 min.	max.	SYMBOL 5 min.	max.
2.5	6	0	3	1	5	3	9	-	-	-	-
6	10	0	3	1	5	3	9	6	11	8	15
10	18	0	3.5	1	7	4	10	7	13	10	18
18	24	0	4	2	8	5	11	8	14	11	19
24	30	0.5	4.5	2	8	5	11	9	16	12	21
30	40	0.5	4.5	2	8	6	13	11	18	16	25
40	50	0.5	4.5	2.5	9	7	14	12	20	18	29
50	65	0.5	6	3.5	11	9	17	15	24	22	35
65	80	0.5	6	4	12	10	20	18	28	26	41
80	100	0.5	7	4.5	14	12	23	21	33	30	47
100	120	1	8	6	16	14	26	24	38	35	55
120	140	1	9	7	19	16	32	28	45	41	63
140	160	1	9	7	21	18	36	32	51	47	71
160	180	1	10	8	24	21	40	36	58	53	79
180	200	1	12	10	28	25	46	42	64	59	91

Table 2 (Table 13, Section 20) cylindrical roller bearings
Table 3 & 4 (Tables 15 & 16, Section 20) self-aligning roller bearings
Table 5 & 6 (Tables 17 & 18, Section 20) double row, self-aligning ball
  bearings
Table 7 (per DIN 620) cylindrical roller bearings

Table 7 for cylindrical roller bearings is per DIN 620, which has eliminated
the need for "matched and unmatched" since the values shown are for interchangeable rings. Inner rings therefore can be mounted on shafts at a location
remote from the final assembly with the outer ring/roller set.

Radial internal clearance is usually marked on a face of the bearing, preceded with the letter C, for example C3 or C4. If the bearing is supplied with
normal or so-called standard clearance, then there is *no* marking. The radial
clearance groupings are

C1 Radial clearance smaller than C2
C2 Radial clearance smaller than normal
—   Radial clearance normal (no suffix)
C3 Radial clearance greater than normal
C4 Radial clearance greater than C3
C5 Radial clearance greater than C4

C. Selection of the Correct Radial Clearance

The initial radial clearance of a rolling bearing is reduced in the case of tightly
fitting ring/rings. The inner ring expansion can be set equal to j, and the outer
ring contraction is equal to A. The following relationships are used for mounting on steel shafts:

$$\Delta_j = \mu \frac{d}{h} \quad [\mu m] \tag{1}$$

Inner Ring on Hollow Steel Shaft:

$$\Delta_j = \mu \frac{d}{h} \frac{(d/\mu)^2 - 1}{(d/\mu)^2 - (d/h)^2} \quad [\mu m] \tag{2}$$

Outer Ring in Steel Housing:

$$\Delta_A = \mu \frac{H}{D} \quad [\mu m] \tag{3}$$

Outer Ring in Thin-Walled Steel Housing:

$$\Delta_A = \mu \frac{H}{D} \frac{(F/D)^2 - 1}{(F/D)^2 - (H/D)^2} \quad [\mu m] \tag{4}$$

See Fig. 61(c),(d).

Table 2 Radial Internal Clearance Values for Cylindrical Roller Bearings

PART I

Clearance values in micrometers

For each SYMBOL the four sub‑columns are, from left to right: Interchangeable Low, Matched Low, Matched High, Interchangeable High. ("Matched" = the inner pair; "Interchangeable" = the outer pair.)

d (mm) Over	d (mm) Incl.	S2 Int Low	S2 Mat Low	S2 Mat High	S2 Int High	S0 (Normal) Int Low	S0 Mat Low	S0 Mat High	S0 Int High	S3 Int Low	S3 Mat Low	S3 Mat High	S3 Int High	S4 Int Low	S4 Mat Low	S4 Mat High	S4 Int High
10	18	0	10	20	30	10	20	30	41	25	36	46	56	36	46	56	66
18	24	0	10	20	30	10	20	30	41	25	36	46	56	36	46	56	66
24	30	0	10	20	30	10	20	30	41	25	36	46	56	36	46	56	66
30	40	0	10	25	30	10	25	36	46	30	41	51	66	41	51	61	71
40	50	0	13	25	36	15	25	41	51	36	46	56	71	46	56	71	81
50	65	5	15	30	41	20	30	46	56	41	51	66	76	56	66	81	89
65	80	5	15	36	46	20	36	51	66	46	56	76	89	66	76	89	104
80	100	5	20	41	56	25	41	61	76	56	71	89	104	76	89	109	124
100	120	10	25	46	61	30	46	71	81	66	81	104	114	89	104	124	140
120	140	10	25	51	66	36	51	81	89	81	94	119	135	104	119	145	160
140	160	10	30	61	76	41	61	89	104	89	104	135	155	114	135	160	180
160	180	15	36	66	81	51	66	99	114	99	114	150	165	130	150	180	196
180	200	20	36	76	86	61	76	109	124	109	124	165	175	150	165	201	216
200	225	25	41	81	94	66	81	119	135	124	140	180	196	165	180	221	234
225	250	30	46	89	104	76	—	—	150	140	—	—	216	180	—	—	254
250	280	41	51	99	114	90	—	—	165	155	—	—	229	206	—	—	279
280	315	46	56	109	124	99	—	—	180	175	—	—	254	229	—	—	310
315	355	51	61	119	132	109	—	—	196	196	—	—	279	254	—	—	340
355	400	66	76	150	160	140	—	—	236	244	—	—	340	320	—	—	414
400	450	71	—	—	191	155	—	—	274	269	—	—	389	356	—	—	455
450	500	84	—	—	206	180	—	—	300	300	—	—	419	394	—	—	513

PART II

Clearance values in 0.0001 inch

d (mm) Over	d (mm) Incl.	SYMBOL 2 Interch. Low	SYMBOL 2 Matched Low	SYMBOL 2 Matched High	SYMBOL 2 Interch. High	SYMBOL 0 (Normal) Interch. Low	SYMBOL 0 Matched Low	SYMBOL 0 Matched High	SYMBOL 0 Interch. High	SYMBOL 3 Interch. Low	SYMBOL 3 Matched Low	SYMBOL 3 Matched High	SYMBOL 3 Interch. High	SYMBOL 4 Interch. Low	SYMBOL 4 Matched Low	SYMBOL 4 Matched High	SYMBOL 4 Interch. High
10	18	0	4	8	12	4	8	12	16	10	14	18	22	14	18	22	26
18	24	0	4	8	12	4	8	12	16	10	14	18	22	14	18	22	26
24	30	0	4	8	12	4	8	12	16	10	14	18	22	14	18	22	26
30	40	0	4	10	12	4	10	14	18	12	16	20	26	16	20	24	28
40	50	0	5	10	14	6	10	16	20	14	18	22	28	18	22	28	32
50	65	2	6	12	16	8	12	18	22	16	20	26	30	22	26	32	35
65	80	2	6	14	18	8	14	20	26	18	22	30	35	26	30	35	41
80	100	2	8	16	22	10	16	24	30	22	28	35	41	30	35	43	49
100	120	4	10	18	24	12	18	28	32	26	32	41	45	35	41	49	55
120	140	4	10	20	26	14	20	32	35	32	37	47	53	41	47	57	63
140	160	4	12	24	30	16	24	35	41	35	41	53	61	45	53	63	71
160	180	6	14	26	32	20	26	39	45	39	45	59	65	51	59	71	77
180	200	8	14	30	34	24	30	43	49	43	49	65	69	59	65	79	85
200	225	10	16	32	37	26	32	47	53	49	55	71	77	65	71	87	92
225	250	12	18	35	41	30	—	—	59	55	—	—	85	71	—	—	100
250	280	16	20	39	45	35	—	—	65	61	—	—	90	81	—	—	110
280	315	18	22	43	49	39	—	—	71	69	—	—	100	90	—	—	122
315	355	20	24	47	52	43	—	—	77	77	—	—	110	100	—	—	134
355	400	22	26	53	57	49	—	—	85	85	—	—	120	110	—	—	146
400	450	26	30	59	63	55	—	—	93	96	—	—	134	126	—	—	163
450	500	28	—	—	75	61	—	—	108	106	—	—	153	140	—	—	179

Table 3 Radial Internal Clearance Values for Self-Aligning Roller Bearings with Cylindrical Bore

PART I   Clearance values in micrometers

d mm		SYMBOL 2		SYMBOL 0 (Normal)		SYMBOL 3		SYMBOL 4		SYMBOL 5	
over	incl.	min.	max.	min.	max.	min.	max.	min.	max.	min.	max.
14	24	10	20	20	35	35	45	45	60	60	75
24	30	15	25	25	40	40	55	55	75	75	95
30	40	15	30	30	45	45	60	60	80	80	100
40	50	20	35	35	55	55	75	75	100	100	125
50	65	20	40	40	65	65	90	90	120	120	150
65	80	30	50	50	80	80	110	110	145	145	180
80	100	35	60	60	100	100	135	135	180	180	225
100	120	40	75	75	120	120	160	160	210	210	260
120	140	50	95	95	145	145	190	190	240	240	300
140	160	60	110	110	170	170	220	220	280	280	350
160	180	65	120	120	180	180	240	240	310	310	390
180	200	70	130	130	200	200	260	260	340	340	430
200	225	80	140	140	220	220	290	290	380	380	470
225	250	90	150	150	240	240	320	320	420	420	520
250	280	100	170	170	260	260	350	350	460	460	570
280	315	110	190	190	280	280	370	370	500	500	630
315	355	120	200	200	310	310	410	410	550	550	690
355	400	130	220	220	340	340	450	450	600	600	750
400	450	140	240	240	370	370	500	500	660	660	820
450	500	140	260	260	410	410	550	550	720	720	900
500	560	150	280	280	440	440	600	600	780	780	1000
560	630	170	310	310	480	480	650	650	850	850	1100
630	710	190	350	350	530	530	700	700	920	920	1190
710	800	210	390	390	580	580	770	770	1010	1010	1300
800	900	230	430	430	650	650	860	860	1120	1120	1440
900	1000	260	480	480	710	710	930	930	1220	1220	1570

Table 3 (Continued)

PART II   Clearance values in 0.0001 inch

d mm		SYMBOL 2		SYMBOL 0 (Normal)		SYMBOL 3		SYMBOL 4		SYMBOL 5	
Over	Incl.	Low	High	Low	High	Low	High	Low	High	Low	High
14	24	4	8	8	14	14	18	18	24	24	30
24	30	6	10	10	16	16	22	22	30	30	37
30	40	6	12	12	18	18	24	24	31	31	39
40	50	8	14	14	22	22	30	30	39	39	49
50	65	8	16	16	26	26	35	35	47	47	59
65	80	12	20	20	31	31	43	43	57	57	71
80	100	14	24	24	39	39	53	53	71	71	89
100	120	16	30	30	47	47	63	63	83	83	102
120	140	20	37	37	57	57	75	75	94	94	118
140	160	24	43	43	67	67	87	87	110	110	138
160	180	26	47	47	71	71	94	94	122	122	154
180	200	28	51	51	79	79	102	102	134	134	169
200	225	31	55	55	87	87	114	114	150	150	185
225	250	35	59	59	94	94	126	126	165	165	205
250	280	39	67	67	102	102	138	138	181	181	224
280	315	43	75	75	110	110	146	146	197	197	248
315	355	47	79	79	122	122	161	161	217	217	272
355	400	51	87	87	134	134	177	177	236	236	295
400	450	55	94	94	146	146	197	197	260	260	323
450	500	55	102	102	161	161	217	217	283	283	354
500	560	59	110	110	173	173	236	236	307	307	394
560	630	67	122	122	189	189	256	256	335	335	433
630	710	75	138	138	209	209	276	276	362	362	469
710	800	83	154	154	228	228	303	303	398	398	512
800	900	91	169	169	256	256	339	339	441	441	567
900	1000	102	189	189	280	280	366	366	480	480	618

Table 4 Radial Internal Clearance Values for Self-Aligning Roller Bearings with Tapered Bore

PART I   Clearance values in micrometers

d mm		SYMBOL 2		SYMBOL 0 (Normal)		SYMBOL 3		SYMBOL 4		SYMBOL 5	
over	incl.	min.	max.	min.	max.	min.	max.	min.	max.	min.	max.
18	24	15	25	25	35	35	45	45	60	60	75
24	30	20	30	30	40	40	55	55	75	75	95
30	40	25	35	35	50	50	65	65	85	85	105
40	50	30	45	45	60	60	80	80	100	100	130
50	65	40	55	55	75	75	95	95	120	120	160
65	80	50	70	70	95	95	120	120	150	150	200
80	100	55	80	80	110	110	140	140	180	180	230
100	120	65	100	100	135	135	170	170	220	220	280
120	140	80	120	120	160	160	200	200	260	260	330
140	160	90	130	130	180	180	230	230	300	300	380
160	180	100	140	140	200	200	260	260	340	340	430
180	200	110	160	160	220	220	290	290	370	370	470
200	225	120	180	180	250	250	320	320	410	410	520
225	250	140	200	200	270	270	350	350	450	450	570
250	280	150	220	220	300	300	390	390	490	490	620
280	315	170	240	240	330	330	430	430	540	540	680
315	355	190	270	270	360	360	470	470	590	590	740
355	400	210	300	300	400	400	520	520	650	650	820
400	450	230	330	330	440	440	570	570	720	720	910
450	500	260	370	370	490	490	630	630	790	790	1000
500	560	290	410	410	540	540	680	680	870	870	1100
560	630	320	460	460	600	600	760	760	980	980	1230
630	710	350	510	510	670	670	850	850	1090	1090	1360
710	800	390	570	570	750	750	960	960	1220	1220	1500
800	900	440	640	640	840	840	1070	1070	1370	1370	1690
900	1000	490	710	710	930	930	1190	1190	1520	1520	1860

Table 4 (Continued)

PART II											Clearance values in 0.0001 inch

d mm		SYMBOL 2		SYMBOL 0 (Normal)		SYMBOL 3		SYMBOL 4		SYMBOL 5	
Over	Incl.	min.	max.	min.	max.	min.	max.	min.	max.	min.	max.
18	24	6	10	10	14	14	18	18	24	24	30
24	30	8	12	12	16	16	22	22	30	30	37
30	40	10	14	14	20	20	26	26	33	33	41
40	50	12	18	18	24	24	31	31	39	39	51
50	65	16	22	22	30	30	37	37	47	47	63
65	80	20	28	28	37	37	47	47	59	59	79
80	100	22	30	30	43	43	55	55	71	71	91
100	120	26	39	39	53	53	67	67	87	87	110
120	140	31	47	47	63	63	79	79	102	102	130
140	160	35	51	51	71	71	91	91	118	118	150
160	180	39	55	55	79	79	102	102	134	134	169
180	200	43	63	63	87	87	114	114	146	146	185
200	225	47	71	71	98	98	126	126	161	161	205
225	250	55	79	79	106	106	138	138	177	177	224
250	280	59	87	87	118	118	154	154	193	193	244
280	315	67	94	94	130	130	169	169	213	213	268
315	355	75	106	106	142	142	185	185	232	232	291
355	400	83	118	118	157	157	205	205	256	256	323
400	450	91	130	130	173	173	224	224	283	283	358
450	500	102	146	146	193	193	248	248	311	311	394
500	560	114	161	161	213	213	268	268	343	343	433
560	630	126	181	181	236	236	299	299	386	386	484
630	710	138	201	201	264	264	335	335	429	429	535
710	800	154	224	224	295	295	378	378	480	480	591
800	900	173	252	252	331	331	421	421	539	539	665
900	1000	193	280	280	366	366	469	469	598	598	732

Table 5 Radial Internal Clearance Values for Double-Row Self-Aligning Ball Bearings with Cylindrical Bore

PART I

Clearance values in micrometers

d mm		SYMBOL 2		SYMBOL 0 (Normal)		SYMBOL 3		SYMBOL 4		SYMBOL 5	
over	incl.	min.	max.	min.	max.	min.	max.	min.	max.	min.	max.
2.5	6	1	8	5	15	10	20	15	25	21	33
6	10	2	9	6	17	12	25	19	33	27	42
10	14	2	10	6	19	13	26	21	35	30	48
14	18	3	12	8	21	15	28	23	37	32	50
18	24	4	14	10	23	17	30	25	39	34	52
24	30	5	16	11	24	19	35	29	46	40	58
30	40	6	18	13	29	23	40	34	53	46	66
40	50	6	19	14	31	25	44	37	57	50	71
50	65	7	21	16	36	30	50	45	69	62	88
65	80	8	24	18	40	35	60	54	83	76	108
80	100	9	27	22	48	42	70	64	96	89	124
100	120	10	31	25	56	50	83	75	114	105	145
120	140	10	38	30	68	60	100	90	135	125	175
140	160	15	44	35	80	70	120	110	161	150	210

PART II

Clearance values in 0.0001 inch

d mm Over	d mm Incl.	SYMBOL 2 min.	SYMBOL 2 max.	SYMBOL 0 (Normal) min.	SYMBOL 0 (Normal) max.	SYMBOL 3 min.	SYMBOL 3 max.	SYMBOL 4 min.	SYMBOL 4 max.	SYMBOL 5 min.	SYMBOL 5 max.
2.5	6	0.5	3	2	6	4	8	6	10	8.5	13
6	10	1	3.5	2.5	6.5	4.5	10	7.5	13	11	17
10	14	1	4	2.5	7.5	5	10	8.5	14	12	19
14	18	1	4.5	3	8.5	6	11	9	15	13	20
18	24	1.5	5.5	4	9	6.5	12	10	15	13	20
24	30	2	6.5	4.5	9.5	7.5	14	11	18	16	23
30	40	2.5	7	5	11	9	16	13	21	18	26
40	50	2.5	7.5	5.5	12	10	17	15	22	20	28
50	65	3	8.5	6.5	14	12	20	18	27	24	35
65	80	3	9.5	7	16	14	24	21	33	30	43
80	100	3.5	11	8.5	19	17	28	25	38	35	49
100	120	4	12	10	22	20	33	30	45	41	57
120	140	4	15	12	27	24	39	35	53	49	69
140	160	6	17	14	31	28	47	43	63	59	83

Table 6 Radial Internal Clearance Values for Double-Row Self-Aligning Ball Bearings with Tapered Bore

PART I   Clearance values in micrometers

d mm		SYMBOL 2		SYMBOL 0 (Normal)		SYMBOL 3		SYMBOL 4		SYMBOL 5	
over	incl.	min.	max.	min.	max.	min.	max.	min.	max.	min.	max.
18	24	7	17	13	26	20	33	28	42	37	55
24	30	9	20	15	28	23	39	33	50	44	62
30	40	12	24	19	35	29	46	40	59	52	72
40	50	14	27	22	39	33	52	45	65	58	79
50	65	18	32	27	47	41	61	56	80	73	99
65	80	23	39	35	57	50	75	69	98	91	123
80	100	29	47	42	68	62	90	84	116	109	144
100	120	35	56	50	81	75	108	100	139	130	170
120	140	40	68	60	98	90	130	120	165	155	205
140	160	45	74	65	110	100	150	140	191	180	240

PART II   Clearance values in 0.0001 inch

d mm		SYMBOL 2		SYMBOL 0 (Normal)		SYMBOL 3		SYMBOL 4		SYMBOL 5	
Over	Incl.	min.	max.	min.	max.	min.	max.	min.	max.	min.	max.
18	24	3	6.5	5	10	8	13	11	17	15	22
24	30	3.5	8	6	11	9	15	13	20	17	24
30	40	4.5	9.5	7.5	14	11	18	16	23	20	28
40	50	5.5	11	8.5	15	13	20	18	26	23	31
50	65	7	13	11	19	16	24	22	31	29	39
65	80	9	15	14	22	20	30	27	39	36	48
80	100	11	19	17	27	24	35	33	46	43	57
100	120	14	22	20	32	30	43	39	55	51	67
120	140	16	27	24	39	35	51	47	65	61	81
140	160	18	29	26	43	39	59	55	75	71	94

Table 7 Cylindrical Roller Bearings with Cylindrical Bore and Needle Roller Bearings

Nominal bore diameter, d		Radial internal clearance, in μm							
		C2		C0		C3		C4	
Over	Up to	min.	max.	min.	max.	min.	max.	min.	max.
-	24	0	25	20	45	35	60	50	75
24	30	0	25	20	45	35	60	50	75
30	40	5	30	25	50	45	70	60	85
40	50	5	35	30	60	50	80	70	100
50	65	10	40	40	70	60	90	80	110
65	80	10	45	40	75	65	100	90	125
80	100	15	50	50	85	75	110	105	140
100	120	15	55	50	90	85	125	125	165
120	140	15	60	60	105	100	145	145	190
140	160	20	70	70	120	115	165	165	215
160	180	25	75	75	125	120	170	170	220
180	200	35	90	90	145	140	195	195	250
200	225	45	105	105	165	160	220	220	280
225	250	45	110	110	175	170	235	235	300
250	280	55	125	125	195	190	260	260	330
280	315	55	130	130	205	200	275	275	350
315	355	65	145	145	225	225	305	305	385
355	400	100	190	190	280	280	370	370	460
400	450	110	210	210	310	310	410	410	510
450	500	110	220	220	330	330	440	440	550

The following can be used for rough estimates:

$$\frac{d}{h} \cong 0.75\text{--}0.85 \tag{5}$$

$$\frac{H}{D} \cong 0.85\text{--}0.90 \tag{6}$$

Lower values apply to bearings of heavier series and thus heavier section rings.

D. Axial Clearance

The conversion of radial clearance to axial can be done by means of graphs and/or conversion factors.

> *Example. Deep-Groove Ball Bearings*
> d = bearing bore (mm)
> e = radial clearance (μm)
> a = axial clearance (μm)
> Ball bearing 6008.C3, $d = 40$ mm
> Radial clearance $e = 10$ μm

$$a/e = 13 \tag{7}$$

$$a = 13 \times 10\mu m = 130\mu m$$

Table 8 shows the axial clearance a to radial clearance e for two row bearings. The values given in Table 8 for angular contact ball bearings are approximate; if exact results are required, the raceway curvatures must be taken into account. However, for those bearings having a nominal contact angle exceeding 30°, the table values are sufficiently accurate.

IV. MATERIALS

A. Ring and Rolling Element Materials

A rolling bearing is subject to very high local stresses, a general range being 1000–4000 N/mm^2 or about 145,000–580,150 psi. Since one, and sometimes both, bearing ring, is fitted with interference fits, additional stresses are induced. There are various amounts of sliding, in addition to rolling motion that can cause wear, so in addition to the bearing steel being strong, hard, and tough it must also have acceptable wear properties.

Table 8 Axial Clearance a to Radial Clearance c for Two Row Bearings

Bearing type	a/e	Remarks
Self-aligning ball bearing	$2.3Y_0$	Y_0 from catalog
Spherical roller bearing	$2.3Y_0$	Y_0 from catalog
Double-row tapered roller bearing	$2.3Y_0$	Y_0 from catalog
Paired single-row tapered roller bearing	$4.6Y_0$	Y_0 value of single-row bearing
Double-row angular contact ball bearing Series 32 and 33	1.4	35° contact angle
Paired single-row angular contact ball bearing Series 72B and 73B	1.2	40° contact angle
Four-point bearing	1.4	35° contact angle

Most bearings are made of through-hardening steel containing up to 2.5% chromium. Other bearing steels include case-hardening steels, induction hardening, stainless, and high-temperature steels. The most frequently used steel in the United States is AISI 52100, which is a through-hardening steel.

Commonly used case-carburizing steels are AISI 3310, 4320, 4620, 8620 and 9310. A typical induction hardening steel is AISI 1070. The stainless steel used is AISI 440C or 440C modified, and a typical high-temperature tool steel is AISI M-50. Hardness of rings and washers range from 58–64 Rockwell C scale and 60–65 for balls and rollers. Specifications for bearing quality steel are given in ASTM A295 and A485 for through-hardening steels and A534 for carburizing steels.

The preference of through hardening to case hardening, or vice versa, is open to disagreement. However, in cases where there is relative motion, such as in loose-fitting inner rings on rolling mill work roll necks, there is a danger of heat checking or cracking of the steel in which case the use of through-hardening steel could result in the crack propagating through the ring. This is much less likely to happen with case-hardening steel because of the softer core.

B. Separator Materials

The materials used in bearing separators or cages vary by function, volume, and size. Usually larger bearings are not high volume; for example, large spherical roller bearings use mostly machined brass cages, whereas bearing used in automobile front wheel hubs have cages of plastic material. High-volume bearings allow the use of expensive dies and/or molds because the payback on investment is possible in a relatively short time. Materials used in cages include glass-filled polyamide, pressed steel, machined steel, pressed brass, machined brass, resin-impregnated cloth (phenolic), aluminum- and silver-plated bronze, or steel.

V. LOAD-CARRYING/LIFE CALCULATIONS
A. Static Stressing

The term *static stressing* refers to bearings carrying a load when stationary or when subjected to operation with small oscillatory motions. The term *static* refers to the bearing's rotation not the type of load. Rolling bearings under static load can be stressed to a degree where minor plastic deformations occur in the rolling surfaces.

1. Admissible Static Load Rating—C_0
It is recognized that there is no clear boundary between elastic and plastic deformation. The definition of allowable static rolling element loading is that load which causes a permanent deformation of 0.0001 of the rolling element

diameter at the center of the most heavily loaded ball/raceway or roller/ raceway contact. Most bearing applications can operate with this load without impairment. This load is usually shown as C_0 in most bearing catalogs.

2. Static Equivalent Load

A force F acting on a bearing at an angle β (Fig. 62) is termed a *combined load*. This term conveys the fact that the bearing is stressed simultaneously by a radial and an axial component, namely $F_R = F \cos \beta$ and $F_A = F \sin \beta$. These two components combine to form the equivalent static load P_0, which is the force that produces, under the load rating condition stated above, the maximum plastic deformation as the actual combined load. The equivalent static load P_0 is

$$P_0 = X_0 F_R = Y_0 F_a \quad [\text{N}] \tag{8}$$

where

F_R = maximum radial load [N]

F_A = maximum axial load [N]

X_0 = radial factor (Table 8)

Y_0 = axial factor (Table 8)

B. Dynamic Stressing

Dynamic stressing refers to the loading on a rotating bearing. This load may be constant or varying.

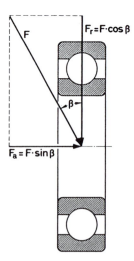

Figure 62 Combined load acting on a radial deep-groove ball bearing.

1. Material Fatigue and Failure Probability

The fatigue phenomena occurs on the operating surfaces of rotating rolling bearings depending on running time and load. Usually fatigue results from microcracks started below the surface. After further operation, the cracks enlarge, and material pitting and flaking develop on the operating surface, finally extending over large areas.

The first cracks emanate from weak points or inhomogeneities in the material. Inhomogeneities such as nonmetallic inclusions or nonuniform distributions of alloying elements are randomly distributed and of varying size. This plus the differing stress caused by variations in the bearing components due to manufacturing tolerances affect the time that fatigue appears on the operating surfaces.

Figure 63 shows the result of a 30-bearing test in which the load and speed were the same for each bearing. Life scatter is from 300 million revolutions for 5 bearings, with 11 bearings going only 13 million revolutions. These are typical test results.

2. Fatigue Theory

Fatigue is caused by the dynamic stressing of the material during cycling. The stress level dictates the time at which fatigue occurs. Presently the question of

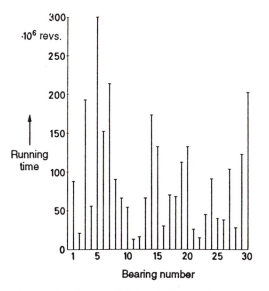

Figure 63   Scatter of fatigue life of 30 deep-groove ball bearings 6309.

which type of stress causes fatigue is not certain. However, three hypotheses have been put forward: maximum shear stress, distortion energy, and alternating shear stress.

The first theory makes the maximum shear stress the critical stress. The second theory states that material stressing is not ruled by the stress conditions at a single point in a body but by those in a certain zone, and the distortion process below the surface is the characteristic feature. Calculations showing the location of maximum distortion coincide more or less with that of the maximum shear stress. It is not agreed which of these gives that best explanation for the fatigue process. The third theory, that the maximum alternating shear stress or the orthogonal shear stress is responsible for material failure, is supported by the fact that most cracks in cycled components are found at a depth where this type of shear stress is at a maximum.

2. Fatigue Life

Fatigue life or rating is defined by ISO recommendation R281 and DIN 622 as

The rating life of a sufficiently large number of dimensionally identical bearings is expressed by the number of revolutions or number of hours at constant speed reached or exceeded by 90% of this bearing group before the first signs of material fatigue appear.

It is also defined in ANSI/AFBMA STD 9-1990 and STD 11-1990 as

Basic life, L10—for an individual rolling bearing or a group of apparently identical rolling bearings operating under the same conditions, the life associated with 90% reliability, with contemporary, commonly used material and manufacturing quality, and under conventional operating conditions.

The terms *rating life, L10 life,* and *B10 life* are used interchangeably.

C. Dynamic Load-Carrying Capacity

1. The Rating Life Equation

Testing has shown that the rating life of a group of bearings increases when the applied load decreases. Rating life is expressed by the equation

$$L10 = L = \left[\frac{C}{P} \right]^{p} \quad [10^6 \text{ revolutions}] \qquad (9)$$

where

$L10 =$ the nominal rating life expressed in millions of revolutions.

$C =$ Dynamic load rating [kN]. It is shown in most bearing catalogs. A load of this magnitude calculates an $L10$ of one million revolutions.

P = Equivalent dynamic load [kN]. This is a constant radial and/or axial load which is equivalent to the real loads as far as life is concerned.

p = Life exponent, differs for ball and roller bearings: $p = 3$ for ball bearings; $p = 10/3$ for roller bearings.

2. The Dynamic Load Rating C

The dynamic load rating is defined as that constant stationary radial load which a rolling bearing could theoretically endure for a basic rating life of one million revolutions. The dynamic load rating formulas use the following symbols:

d_w [mm]   Ball or roller diameter; in barrel rollers the largest diameter and in tapered rollers the mean

l_{eff} [mm]   Effective contact length of roller and raceway, use the shorter length

$T_{[mm]}$ Pitch circle diameter of balls or rollers

i Number of rolling element *rows* in a bearing

z Number of rolling elements in a row

α_0 Nominal contact angle

f_c Factor depending on the geometry, accuracy, and material of the bearing components.

The formulas for bearing capacity are per ISO recommendation R281, DIN 622, Sheet 2, ANSI/AFBMA STD 9-1990, and STD 11-1990.

Radial Ball Bearings.   For bearings with ball diameters equal to or less than 25.4 mm,

$$c = f_c(i \cos \alpha_0)^7 Z^{2/3} d_w^{1.8} \quad \text{[N]} \tag{10}$$

For bearings with balls greater than 25.4 mm,

$$c = f_c(i \cos \alpha_0) Z^{2/3} (3.647) d_w^{1.4} \quad \text{[N]} \tag{11}$$

The factor f_c depends on bearing type and the ratio $d_w \cos \alpha_0/T$ (see Table 9).

Radial Roller Bearings.   The dynamic load rating C is

$$C = f_c(il_{eff} \cos \alpha_0)^{7/9} Z^{3/4} d_w^{29/27} \quad \text{[N]} \tag{12}$$

The factor f_c depends on the ratio $d_w \cos \alpha_0/T$ [N] (see Table 10). The values in Table 10 are considered maximum values and should be applied only when there is a uniform stress distribution along the contact line.

Note: New research results have allowed an increase in dynamic load ratings, the factors in Tables 9–14 can be increased as follows:

30% ball bearing
15% spherical roller bearing
10% cylindrical and tapered roller bearing

Table 9 Factor f_c for Radial Ball Bearings

$d_w \cos \alpha_0 / T$	Single-row deep-groove ball bearings, single-row and double-row angular contact ball bearings f_c	Double-row deep-groove ball bearings f_c	Self-aligning ball bearings f_c	Magneto bearings f_c
0.05	46.7	44.2	17.3	16.2
0.06	49.1	46.5	18.6	17.4
0.07	51.1	48.4	19.9	18.5
0.08	52.8	50.0	21.1	19.5
0.09	54.3	51.4	22.3	20.6
0.10	55.5	52.6	23.4	21.5
0.12	57.5	54.5	25.6	23.4
0.14	58.8	55.7	27.7	25.3
0.16	59.6	56.5	29.7	27.1
0.18	59.9	56.8	31.7	28.8
0.20	59.9	56.8	33.5	30.5
0.22	59.6	56.5	35.2	32.1
0.24	59.0	55.9	36.8	33.7
0.26	58.2	55.1	38.2	35.2
0.28	57.1	54.1	39.4	36.6
0.30	56.0	53.0	40.3	37.8
0.32	54.6	51.8	40.9	38.9
0.34	53.2	50.4	41.2	39.8
0.36	51.7	48.9	41.3	40.4
0.38	50.0	47.4	41.0	40.8
0.40	48.4	45.8	40.4	40.9

Thrust Ball Bearings. The dynamic thrust rating C_a of single-acting and double-acting thrust ball bearings ($\alpha_0 = 90°$) where $d_w \leqslant 25.4$ mm:

$$C_a = f_c Z^{2/3} d_w^{1.8} \quad [\text{N}] \tag{13}$$

and for ball diameters over 25.4 mm

$$C_a = f_a Z^{2/3}(3.647)d_w^{1.4} \quad [\text{N}] \tag{14}$$

and for $d_w > 25.4$ mm

$$C_a = f_c (\cos \alpha_0)^7 (\tan \alpha_0) Z^{2/3} d_w^{1.4} \quad [\text{N}] \tag{15}$$

Roller Thrust Bearings—Single and Double Acting. For contact angle $\alpha_0 = 90°$,

Table 10   The Factor f_c for Radial Roller Bearings

$d_w \cos \alpha_0 / T$	f_c
0.01	52.1
0.02	60.8
0.03	66.5
0.04	70.7
0.05	74.1
0.06	76.9
0.07	79.2
0.08	81.2
0.09	82.8
0.10	84.2
0.12	86.4
0.14	87.7
0.16	88.5
0.18	88.8
0.20	88.7
0.22	88.2
0.24	87.5
0.26	86.4
0.28	85.2
0.30	83.8

$$C_a = f_c l_{\text{eff}}^{7/9} d_w^{29/27} \quad [\text{N}] \tag{16}$$

and for $\alpha_0 < 90°$,

$$C_a = f_c (l_{\text{eff}} \cos \alpha_0)^{7/9} (\tan \alpha_0) Z^{3/4} d_w^{29/77} \quad [\text{N}] \tag{17}$$

D. Equivalent Dynamic Load

The equation in Section V.3 assumes a pure radial load in radial bearings and a pure thrust load in thrust bearings. In many cases that applied load acts obliquely on the bearings or changes its magnitude. For these cases a constant radial or axial load must be determined, which represents with respect to the rating life an equivalent stress. This calculated force is termed *equivalent dynamic load P*.

1. Constant Combined Load

The term *combined load* indicates the substitution of a constant oblique load F by its horizontal and vertical components (Fig. 64). The equivalent dynamic *radial* load is P and the dynamic *thrust* load is P_A.

Table 11 The Factor f_c for Thrust Ball Bearings

d_w/T	$f_c(\alpha_0 = 90°)$
0.01	36.7
0.02	45.2
0.03	51.1
0.04	55.7
0.05	59.5
0.06	62.9
0.07	65.8
0.08	68.5
0.09	71.0
0.10	73.3
0.12	77.4
0.14	81.1
0.16	84.4
0.18	87.4
0.20	90.2
0.22	92.8
0.24	95.3
0.26	97.6
0.28	99.8
0.30	101.9
0.32	103.9
0.34	105.8

$$P = 0.407 Q_m Z \cos \alpha_0 \quad \text{for radial ball bearings} \tag{18}$$

$$P = 0.401 Q_m Z \cos \alpha_0 \quad \text{for radial roller bearings} \tag{19}$$

$$P_A = Q_m Z \sin \alpha_0 \quad \text{for ball and roller thrust bearings} \tag{20}$$

where α is the bearing contact angle.

Radial Factor X and Thrust Factor Y for Radial Bearings. The equivalent dynamic load X and Y factors have been determined to allow the calculation of P from the applied radial and thrust load. The term e is a constant depending on the contact angle of the bearing. Table 15 shows the values of X and Y for various bearing types.

$$P = XF_F + YF_a \tag{21}$$

where

P = equivalent dynamic radial load

Table 12 The Factor f_c for Angular Thrust Ball Bearings

$d_w \cos \alpha_0/T$	$\alpha_0 = 45°$	$\alpha_0 = 60°$	$\alpha_0 = 75°$
		f_c	
0.01	42.1	39.2	37.3
0.02	51.7	48.1	45.9
0.03	58.2	54.2	51.7
0.04	63.3	58.9	56.1
0.05	67.3	62.6	59.7
0.06	70.7	65.8	62.7
0.07	73.5	68.4	65.2
0.08	75.9	70.7	67.3
0.09	78.0	72.6	69.2
0.10	79.7	74.2	70.7
0.12	82.3	76.6	
0.14	84.1	78.3	
0.16	85.1	79.2	
0.18	85.5	79.6	
0.20	85.4	79.5	
0.22	84.9		
0.24	84.0		
0.26	82.8		
0.28	81.3		
0.30	79.6		

X = radial factor (Table 15)

F_R = radial load

Y = thrust factor (Table 15)

F_a = thrust load

Radial Factor X and Thrust Factor Y for Thrust Bearings. Thrust ball bearings, cylindrical roller thrust bearings, needle roller thrust bearings, and tapered roller thrust bearings with a nominal contact angle of 90° can sustain only purely axial forces. For centrally applied axial loads $P_a = F_a$. However, thrust bearings with a contact of less than 90° can carry both radial and axial loads. The X and Y factors are shown in Table 16 for the equation

$$P_a = XF_R + YF_a$$

Load Conditions in Radial Angular Contact Bearing. Figure 65 shows two single-row angular contact (also applies to tapered roller bearings). The

Table 13 The Factor f_c for Roller
Thrust Bearings with $\alpha_0 = 90°$

d_w/T	f_c ($\alpha_0 = 90°$)
0.01	105.4
0.02	122.9
0.03	134.5
0.04	143.4
0.05	150.7
0.06	156.9
0.07	162.4
0.08	167.2
0.09	171.7
0.10	175.7
0.12	183.0
0.14	189.4
0.16	195.1
0.18	200.3
0.20	205.0
0.22	209.4
0.24	213.5
0.26	217.3
0.28	220.9
0.30	224.3

radial load on bearing A induces an axial load on bearing B due to the contact angle. This internal axial force in the system must be taken into account. If only one angular contact ball bearing or tapered roller bearing is used with a floating bearing and the induced thrust is greater and opposite the applied thrust, the bearing could come apart. Table 17 shows the magnitude of the resultant axial force F_a, i.e., the sum of internal and external axial forces in the system. Figure 15 shows the thrust factor Y_A of bearing A and Y_B of bearing B to be used in Eq. (21).

2. Variable Load and Speed

For bearings with variable load and speed the curve pattern is approximated by a series of individual loads and speeds of duration $q\%$ (Fig. 66). In this case the equivalent dynamic load is

$$P = P_1^3 \frac{n_1}{N_m} \frac{q_1}{100} + P_2^3 \frac{n_2}{N_m} \frac{q_2}{100} + \cdots \quad \text{[kN]} \tag{22}$$

Table 14 The Factor f_c for Roller Bearings with $\alpha_0 < 90°$

	f_c		
$d_w \cos \alpha_0/T$	$\alpha_0 = 50°$	$\alpha_0 = 65°$	$\alpha_0 = 80°$
0.01	109.7	107.1	105.6
0.02	127.8	124.7	123.0
0.03	139.5	136.2	134.3
0.04	148.3	144.7	142.8
0.05	155.2	151.5	149.4
0.06	160.9	157.0	154.9
0.07	165.6	161.6	159.4
0.08	169.5	165.5	163.2
0.09	172.8	168.7	166.4
0.10	175.5	171.4	169.0
0.12	179.7	175.4	173.0
0.14	182.3	177.9	175.5
0.16	183.7	179.3	
0.18	184.1	179.7	
0.20	183.7	179.3	
0.22	182.6		
0.24	180.9		
0.26	178.7		

and the mean speed N_m is

$$N_m = N_1 \frac{q_1}{100} + N_2 \frac{q_2}{100} + \cdots \quad [\text{min}^{-1}] \text{ rpm} \tag{23}$$

If the load is variable and the speed constant:

$$P = P_1^3 \frac{q_1}{100} + P_2^3 \frac{q_2}{100} + \cdots \quad [\text{kN}] \tag{24}$$

The exponent 3 is used in the equations for ball and roller bearings.

If the loads grow linearly from a minimum value $P_{\min}$ to a maximum value $P_{\max}$ at constant speed, the following load P is obtained:

$$P = \frac{P_{\min} + P_{\max}}{3} \quad [\text{kN}] \tag{25}$$

3. Extended Rating Life Calculations

The rating life L or L_h applies to bearings made of standard rolling bearing steel and operating conditions usually prevailing in good practice (correct

Table 15 Factors X and Y for Radial Bearings

Bearing type	iF_a/C_0	e	Single-row bearings $F_a/F_r \le e$ X	$F_a/F_r \le e$ Y	$F_a/F_r > e$ X	$F_a/F_r > e$ Y	Double-row bearings[a] $F_e/F_r \le e$ X	$F_e/F_r \le e$ Y	$F_a/F_r > e$ X	$F_a/F_r > e$ Y
Deep-groove ball bearings with radial clearance normal[b]	0.025	0.022	1	0	0.56	2.0				
	0.04	0.24	1	0	0.56	1.8				
	0.07	0.27	1	0	0.56	1.6				
	0.13	0.31	1	0	0.56	1.4				
	0.25	0.37	1	0	0.56	1.2				
	0.50	0.44	1	0	0.56	1				
Deep-groove ball bearings with radial clearance C3[b] and angular contact ball bearings, $\alpha_0 = 15°$	0.025	0.31	1	0	0.46	1.75	1	2.0	0.75	2.8
	0.04	0.33	1	0	0.46	1.62	1	1.9	0.75	2.7
	0.07	0.36	1	0	0.46	1.46	1	1.7	0.75	2.4
	0.13	0.41	1	0	0.46	1.30	1	1.5	0.75	2.1
	0.25	0.46	1	0	0.46	1.14	1	1.3	0.75	1.9
	0.50	0.54	1	0	0.46	1	1	1.2	0.75	1.6
Deep-groove ball bearings with radial clearance C4[b] and angular contact ball bearings, $\alpha_0 = 15°$	0.025	0.40	1	0	0.44	1.42	1	1.6	0.72	2.3
	0.04	0.42	1	0	0.44	1.36	1	1.5	0.72	2.2
	0.07	0.44	1	0	0.44	1.27	1	1.4	0.72	2.1
	0.13	0.48	1	0	0.44	1.16	1	1.3	0.72	1.9
	0.25	0.53	1	0	0.44	1.05	1	1.2	0.72	1.7
	0.50	0.56	1	0	0.44	1	1	1.1	0.72	1.6

	2	3	4	5	6	7	8	9	10
Angular contact ball bearings $\alpha_0 = 20°$	0.57	1	0	0.43	1	1	1.09	0.70	1.63
$\alpha_0 = 25°$	0.68	1	0	0.41	0.87	1	0.92	0.67	1.41
$\alpha_0 = 30°$	0.80	1	0	0.39	0.76	1	0.78	0.63	1.24
$\alpha_0 = 35°$	0.95	1	0	0.37	0.66	1	0.66	0.60	1.07
$\alpha_0 = 40°$	1.14	1	0	0.35	0.57	1	0.55	0.57	0.93
$\alpha_0 = 45°$	1.33	1	0	0.33	0.50	1	0.47	0.54	0.81
Self-aligning	$1.5 \cdot \tan \alpha_0$						1	$0.42 \cdot \cot \alpha_0$	0.65
Magneto bearings	0.20	1	0	0.50	2.5				
Spherical roller bearings	$1.5 \cdot \tan \alpha_0$	1	0	0.40	$0.4 \cdot \cot \alpha_0$	1	$0.42 \cdot \cot \alpha_0$	0.67	$0.67 \cdot \cot \alpha_0$
Tapered roller	$1.5 \cdot \tan \alpha_0$	1	0	0.40	$0.4 \cdot \cot \alpha_0$	1	$0.42 \cdot \cot \alpha_0$	0.67	$0.67 \cdot \cot \alpha_0$
Column	2	3	4	5	6	7	8	9	10

[a] Applicable to bearings with the standard ring fits k5–j5 and J6. X and Y are the same for single-row and double-row deep-groove ball bearings.

[b] In X and O arrangements.

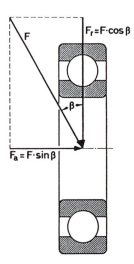

Figure 64 Combined load acting on a radial deep-groove ball bearing.

Table 16 Radial Factor X and Thrust Factor Y for Thrust Bearings

	Column						
	2	5	6	7	8	9	10
Bearing type	e	Single-acting $F_a/F_r > e$		Double-acting $F_a/F_r \leqslant e$		1 $F_a/F_r > e$	
	e	X	Y	X	Y	X	Y
Thrust $\alpha_0 = 45°$	1.25	0.66	1	1.18	0.59	0.66	1
ball $\alpha_0 = 60°$	2.17	0.92	1	1.90	0.55	0.92	1
bearings $\alpha_0 = 75°$	4.67	1.66	1	3.89	0.52	1.66	1
Spherical roller thrust bearings	$1.5 \tan \alpha_0$	$\tan \alpha_0$	1	$1.5 \tan \alpha_0$	0.67	$\tan \alpha_0$	1
Tapered roller bearings	$1.5 \tan \alpha^0$	$\tan \alpha_0$	1	$1.5 \tan \alpha_0$	0.67	$\tan \alpha_0$	1

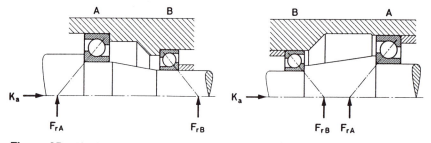

Figure 65 The load conditions in radial angular contact bearings in O arrangement (left) and X arrangement (right).

Table 17 Resultant Axial Force F_a for Bearing Arrangements as Shown in Fig. 65

		Resultant axial force	
Case	Load conditions	Bearing A	Bearing B
1	$\dfrac{F_{rA}}{Y_A} \leqslant \dfrac{F_{rB}}{Y_B}; \ K_a \geqslant 0$	$F_a = K_a + 0.5\dfrac{F_{rB}}{Y_B}$	$F_a = 0.5\dfrac{F_{rB}}{Y_B}$
2	$K_a \geqslant 0.5\dfrac{F_{rA}}{Y_A} - \dfrac{F_{rB}}{Y_B}$	$F_a = K_a + 0.5\dfrac{F_{rB}}{Y_B}$	$F_a = 0.5\dfrac{F_{rB}}{Y_B}$
	$\dfrac{F_{rA}}{Y_A} > \dfrac{F_{rB}}{Y_B}$		
3	$K_a \leqslant 0.5\dfrac{F_{rA}}{Y_A} - \dfrac{F_{rB}}{Y_B}$	$F_a = 0.5\dfrac{F_{rA}}{Y_a}$	$F_a = 0.5\dfrac{F_{rA}}{Y_A} - K_a$

mounting, adequate lubrication, reliable sealing, no extreme temperatures, high standards of cleanliness). Circumstances which deviate from these conditions usually lead to a decrease of the service life as compared with the rating life.

Bearings can even be fail-safe if loads are not too high and lubrication and cleanliness conditions are positive. The adjusted life calculation takes into account effects which change the life. With this calculation, the capacity of the

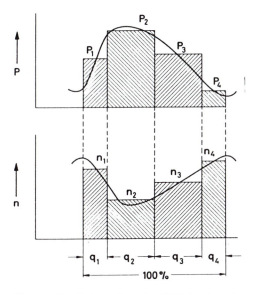

Figure 66 Substitution of variable loads and speeds by load and speed constants.

Table 18 Adjustment Factor a_1

Failure probability %	10	5	4	3	2	1
Life adjustment factor a_1	1	0.62	0.53	0.44	0.33	0.21

bearings can be assessed much more precisely than before. The adjusted life calculation is used predominantly where existing constructions must be optimized and not enough comparable applications exist or are known.

Adjusted Life Calculation to DIN ISO 281. With the life calculation to DIN ISO 281, refined rolling bearing steels, improved production methods, the effects of the operating conditions, and, in particular, more precise information , on the effect of lubrication on the fatigue process can all be taken into account. The attainable fatigue life L_{na} is calculated as follows:

$$L_{na} = a_1 a_2 a_3 L \quad [10^6 \text{ revolutions}] \tag{26}$$

or, expressed in hours,

$$L_{hna} = a_1 a_2 a_3 L_n \quad [\text{h}] \tag{27}$$

Factor a_1 for Failure Probability. Rolling bearing failures due to fatigue are subject to statistical laws. Therefore the fatigue life calculation must include the failure probability. The factor a_1 was introduced to enable failure probabilities other than 10% to be included in the calculation. Table 18 shows factors for failure probability values between 10% and 1%, L_{10} is the nominal life. Generally, the bearings reach a much longer life: The mean life L_{50} amounts to roughly five times the nominal life.

Factor a_2 for Material. Factor a_2 accounts for the properties of material and its heat treatment (Table 19).

Factor a_3 for Operating Conditions. Factor a_3 takes into account the suitability of the lubrication under operating speeds and operating temperatures, and

Table 19 Temperature Factor f_c

Operating temperature	Temperature factor f_c
150°C	1
200°C	0.73
250°C	0.42
300°C	0.22

conditions which can cause changes of material properties, e.g., high temperatures reducing the hardness.

Factor a_{23} and New Values. Due to the interdependence of life adjustment factors a_2 for material and a_3 for operating conditions, one uses only the values for the common factor a_{23}:

$$a_{23} = a_2 a_3$$

The factor a_{23} takes into account not only the effects of material and lubrication but also the effect of the load and bearing type, as well as the cleanliness in the lubricating gap.

The temperature factor f_t (Table 19) accounts for the effect of the operating temperature on the life. Consequently, the life calculation formula is

$$L_{na} = a_1 a_{23} f_t L [10^6 \text{ revolutions}] \tag{28}$$

and

$$L_{hna} = a_1 a_{23} f_t L_h [\text{h}] \tag{29}$$

The effect of the lubricating film formation can be indicated as a defined value for the selection of the a_{23} factor. The highest life values are reached under hydrodynamic lubricating conditions, i.e., if the contact areas between rolling elements and raceways are fully separated by the lubricating film and metal-to-metal contact does not occur. With these lubricating conditions, a high degree of cleanliness in the lubricating gaps, and higher load, the bearing life is terminated by the formation of pitting which has its origin in the area of maximum material stressing below the raceway surface. If the loads are moderate, which is frequently the case in practice, fatigue damage is not likely to occur under these ideal operating conditions; the bearings are fail-safe. Decreasing lubricating film thickness, increasing metal-to-metal contact between functional areas, and contaminants in the lubricant result in life reduction.

The effect of the lubricating film on the a_{23} factor is described in Fig. 67 by the viscosity ratio $\kappa = v/v_1$, where v stands for the lubricant viscosity at operating temperature and v_1 is the rated viscosity at definite speeds. As shown in Fig. 68, v_1 is obtained from the mean bearing diameter $(D + d)/2$ and the speed n. The operating viscosity v of a lubricating oil is obtained from the viscosity-temperature diagram with the help of the operating temperature and the nominal viscosity (ISO VG) of the oil at 40°C. In many cases, only the temperature of the stationary ring is known and not the real temperature of the stressed surfaces of the components in rolling contact.

For bearings with positive kinematics it is acceptable to assess the viscosity with the temperature of the stationary ring. If the bearings are heavily stressed $(f_s < 4)$, or larger sliding friction portions must be taken into account, higher temperatures must be assumed. If the individual $V - T$ diagram is not available

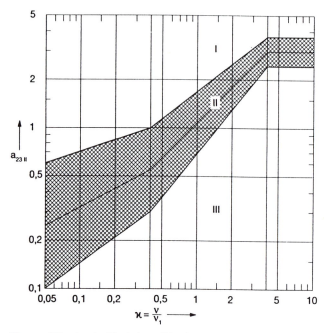

Figure 67 Attainable fatigue life.

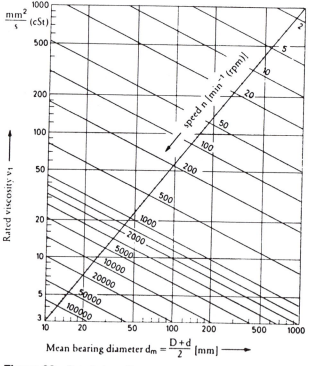

Mean bearing diameter $d_m = \dfrac{D+d}{2}$ [mm]

Figure 68 Rated viscosity v_1.

for an oil, approximate values can be taken from the diagram of Fig. 69. As a rule, the lubricating oil which is used for the gear wheels is also suitable for the transmission bearings. With lubricating greases, v is the operating viscosity of the base oil.

The diagram of Fig. 67 for the determination of factor a_{23} is divided into zones I, II, and III. The majority of applications in rolling bearing engineering can be allocated to zone II. This is the zone of good cleanliness standards. In zone II the factor a_{23} can be determined by means of the ratio χ and the values K_1 and K_2 in Table 20. Value K, the sum of K_1 and K_2, covers the influences of bearing type, lubricant, and additives, depending on the degree of surface separation, and evaluates the effect of the load on these influences by the index f_s:

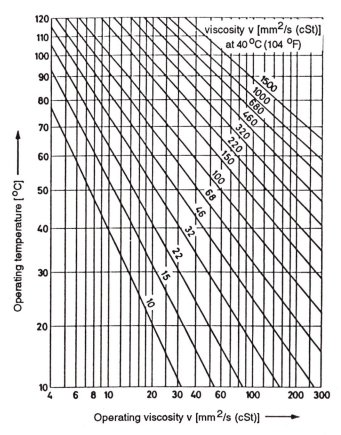

Figure 69 Average viscosity-temperature behavior of mineral oils.

Table 20 Values K_1, K_2, and K for Qualifying the a_{23} Factor

		Load			
		$F_s \geqslant 8$	$8 > f_s > 4$	$f_s < 4$	Values
	Ball bearings	0	0	0	
	Tapered roller bearings	0	1	2	K_1
	Cylindrical roller bearings				
Bearing type	Spherical roller bearings	1	2	3	
	Spherical roller thrust bearings				
	Cylindrical roller thrust bearings				
	Full-complement cylindrical roller bearings				
Lubricant	Doping without additives,	1	2	3	K_2
	$v/v_1 \geqslant 0{,}4$	2	4	6	
	without additives,				
	$v/v_1 < 0{,}4$	7	>7	>7	

K values $= K_1 + K_2$	Allocated a_{23} value
$\leqslant 2$	a_{23} value on the upper curve of zone II
3 to 4	a_{23} value in the middle of zone II
5 to 6	a_{23} value on the lower curve of zone II
$\geqslant 7$	zone III, i.e., try to improve operating conditions

$$f_s = \frac{C_0}{P_0}$$

Here C_0 is the static load rating of the bearing. The equivalent load P_0 is

$$P_0 = X_0 F_r + Y_0 F_a \quad [\text{N}] \tag{8}$$

The dynamic radial and dynamic axial loads must be entered in the formula for F_r and F_a.

If $K \geqslant 7$, the factor a_{23} will be found in zone III. The same applies to poor cleanliness standards independent of the K value. In this case it should be considered with which improvements the conditions of zone II can be reached.

With utmost cleanliness in the lubrication gap and a K value $\leqslant 6$, the a_{23} factor can become very large and belong to zone I of Fig. 67. According to the latest findings an S/N curve is applicable to the bearing life under these conditions. For such a case an a_{23} factor for "good cleanliness" is determined from Fig. 67. By multiplying this factor by f (Fig. 70), the factor a_{23}^* is obtained:

$$a_{23}^* = a_{23}f \tag{30}$$

Factor f is 1 for $x \leqslant 0.4$. For higher x values, f can be determined, depending on index f_s, from Fig. 70.

Factor a_{23}.
1. Transition range to the endurance strength.
   Preconditions: utmost cleanliness in the lubricating gap, loads not too high.
2. High degree of cleanliness in the lubricating gap, suitable additives in the lubricant.

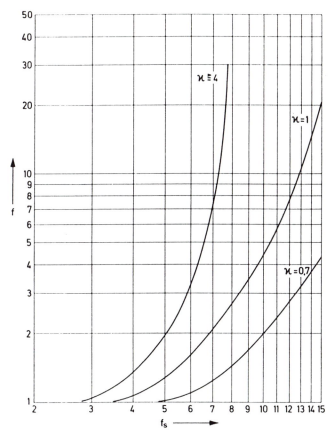

Figure 70   Factor f for determining the a_{23} at maximum cleanliness in the lubricating gap.

3. Unfavorable operating conditions, contaminations in the lubricant, unsuitable lubricants.

Attainable Fatigue Life L_{hna}.

$$L_{hna} = a_1 a_{23} f_t L_h \quad [\text{h}] \tag{29}$$

where

a_1 = life adjustment factor for failure probability;

a_1 = 1 for a 10% failure probability

a_{23} = life adjustment factor for material and operating

f_t = temperature factor

L_h = nominal life

v = operating viscosity of the lubricant

v_1 = rated viscosity

Factor f for Determining a_{23}^* at Maximum Cleanliness in the Lubricating Gap. This requires, above all, an exact as possible assessment of the equivalent dynamic load P. The equivalent dynamic load P is calculated with the known formula (DIN ISO 281)

$$P + XF_r + YF_a \tag{21}$$

If load P, speed n, and the other life-influencing parameters are not constant during the entire operating time, the adjusted rating life L_{hna} must be determined for all proportionate times q [%] under constant conditions. The total life is then obtained by the formula

$$L_{hna} = \frac{100}{q_1/L_{hna1} + q_2/L_{hna}2 + \cdots} \tag{31}$$

Limits of Fatigue Life Calculation. Since the fatigue life calculation modified by the adjustment factors a_1, a_2, and a_3 takes into account the material fatigue as the cause of failure, the assessed life corresponds to the service life of the bearing only if the following requirements are met:

1. The calculation is based on the lubricating conditions occurring in practical operation.
2. Loads and speeds correspond to the actual operating conditions.
3. The amount of contaminants in the lubricant is limited during the entire running time of the bearing.
4. Mounting instructions recommended by the rolling bearing manufacturer are followed.
5. The service life of the lubricant or the service life limited by wear is not shorter than the fatigue life.

VI. FRICTION, TEMPERATURE, AND LUBRICATION

Minimum friction and, as a rule, the small demands on lubrication are two important advantages of rolling bearings. The friction conditions are, however, different with individual rolling bearing types, as various kinds of sliding friction occur aside from rolling friction and lubricant friction. The heat which develops through friction affects the operating temperature of the bearing.

The lubrication of rolling bearings, and other bearings, prevents or minimizes metal-to-metal contact of the rolling and sliding surfaces and, therefore, keeps friction and wear to a minimum. In addition, the lubricant should also protect the bearing from corrosion and, in certain cases, dissipate heat.

A. Friction

The resistance of a rolling bearing to its rotation consists of rolling friction, sliding friction, and lubricant friction. Rolling friction occurs when rolling elements roll on the raceway. Sliding friction occurs at the guiding surfaces of the rolling elements in the cage, on the lip contact areas of the cage as well as, with rolling bearings, at the roller faces and at the lips. Lubricant friction is the result of the internal friction of the lubricant at the contact areas as well as of the churning and worked energy of the lubricant. This is expressed in the different friction values of the various bearing types.

The total running resistance of a rolling bearing is very small compared to the transferred forces, so that the friction energy loss of the rolling bearing can be neglected as a rule when designing a machine. The friction has significance due to the fact that it controls the heat in the bearing and, therefore, has an effect on the temperature of both the bearing parts and the lubricant.

1. Rolling Contact Friction

The process of rolling contact friction occurring between rolling elements and raceways in a loaded rolling bearing is complex. Rolling contact friction is partly due to the sliding resistance of various kinds and partly to elastic hysteresis. Figure 71 shows the contact relation between a rotating rolling element and the raceway. Both show deformations at the pressure area width $2b$; with this the roller is being upended and the raceway stretched. Sliding movements result from the different deformations, causing friction.

More extensive sliding movements occur at the pressure surface when the rolling element rolls in a groove and the pressure surface is curved laterally in the rolling direction (Fig. 72). As the distances of the individual pressure areas from the ball rotary axis are different, the circumferential speeds vary. Therefore, the middle section of the ball surface slides against the rolling direction, while the outer sections roll in the rolling direction. At points D and D' there is no sliding. The sliding and the resulting friction between rolling element and raceway profile increase with higher curvature ratio and load.

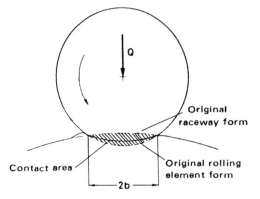

Figure 71 Deformation of a rolling element and the raceway in the direction of rolling.

An important part of the rolling contact friction can be attributed to the hysteresis of the material. Namely, during the rolling movement in the circumferential direction (Fig. 71) the parts lying ahead of the rolling elements are being deformed; during decrease of pressure the needed energy is only partially used for the rolling movement at the back part of the rolling element. The remaining energy is converted into heat.

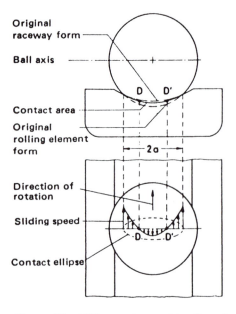

Figure 72 Sliding motion resulting from the curvature of the contact area.

In addition to the described friction effects there is also the spinning friction of ball bearings and axial roller bearings which are loaded axially, which is caused by the sliding resulting from the rotation of the rolling element around its axis at 90°.

Kinetic resistance is summarized within the concept of rolling contact friction. Many times experiments tried to calculate the share of the individual effects and to prove it through experiments.

Individual rolling contact friction forces and the moments of a rolling element have a constant friction value of u. The hysteresis and spinning friction losses at the individual balls of a ball bearing increase by $Q^{4/3}$, and the remaining rolling contact friction losses increases by $Q^{5/3}$, whereby Q represents the ball load. Based on this the friction of angular contact ball bearings and self-aligning bearings is determined to be dependent on the load angle.

Ball bearings with four or three contact points have greater friction than bearings with two contact points. Also, the effect of the surface curvature is easily recognized when comparing spherical roller bearings with cylindrical roller bearings. Furthermore, a ball bearing with a narrow curvature ratio between ball and groove shows greater friction than a bearing with a wider curvature ratio. Basically, rolling contact friction is proportional to the size of the contact surfaces between rolling elements and raceways. Also the size of the pressure angle affects the friction, as the sliding friction increases with an increasing pressure angle.

2. Sliding Friction

Besides the sliding motions discussed in the rolling contact friction section, there are also sliding movements at the guiding surfaces of the cage and, with bearings without a cage, at the common contact areas of the rolling elements. Rolling bearings with lips also have sliding motions where the rolling element end contacts the lips.

Guiding surfaces of the cage are the cage pockets and, with bearings with lip-guided cage, the bore or outside diameter of the cage. Sometimes the face areas of the cage are used for axial guidance (needle and roller cage assemblies). Forces affecting these guiding surfaces result from the weight of the cage, the shifting of the center of gravity, which depends on the cage pocket clearance, and the accelerations acting on the cage when the rolling elements enter the loaded area. In addition, there are the mass forces occurring at the start, at constant speed, and when variations in speed occur. Further forces occur with ball bearings when the contact or pressure angle varies at the bearing raceway, causing some balls to forge ahead and some to drag behind in relation to the cage.

Under normal operating conditions and with adequate lubrication the forces affecting the cage are small; for this reason the friction is also minimal. When there is poor lubrication, contamination, or increased speed, the friction can, however, be considerably higher. This applies especially when the running conditions are disturbed, such as the bearing rings of a ball bearing being misaligned with each other.

In rolling bearings without a cage, as in full-complement bearings (Fig. 73), sliding friction occurs at the contact areas of the rolling elements with each other instead of between rolling element and cage pocket. This friction is stronger than the friction in the cage pocket, because the sliding movement at the contact areas is in the opposite direction.

3. Lubricant Friction

Lubricant friction in a rolling bearing consists of the inner friction of the lubricant at the contact areas and of churning and worked energy occurring when there is too much lubricant and/or higher speed. Total lubricant friction depends most of all on the amount of lubricants and the viscosity of the lubricant. The geometry of the bearing (size of rolling element, size of the guiding gap between cage and bearing ring lip, space between bearings) influences the lubricant friction. At low speed it is generally minimal. However, it increases with higher speed, depending on the oil viscosity or oil consistency. Lubricant friction is higher if, for instance, larger amounts of oil are pumped through the bearing to dissipate heat. With grease lubrication one has to expect higher friction if the empty spaces of the bearing are filled to capacity and there is not enough free space at the sides where the surplus grease can be deposited. The enclosed grease amount is then being worked by cage and rolling elements, which leads to increased lubricant friction. How high the lubricant friction can become can be realized by the fact that there is danger of overheating when there is too much grease. If the grease which has been displaced by the rotating cage and rolling elements can settle outside on the sides, then low lubricant friction occurs, which is similar to the low friction of oil using a minimum amount of lubrication.

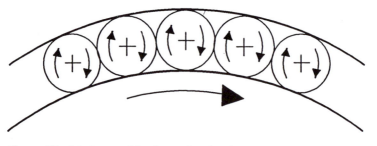

Figure 73 Motion condition in cageless bearings.

4. Friction Torque

The total friction of a bearing, which is the sum of rolling contact, sliding, and lubricant friction, is the resistance of the bearing against its movement. This resistance represents a torque which is designated as frictional moment M. Figure 74 shows the basic course of a friction value curve and the pertinent friction moment.

Estimation of the Friction Torque. For a rough calculation of the total friction torque of a rolling bearing the equation

$$M = \mu F \frac{d}{2} \tag{32}$$

where

M = total friction moment of bearing [N · mm]

μ = friction correction value table (Table 21).

F = resulting bearing load [N]

d = diameter of bearing bore [mm]

and the friction value μ from Table 21 can be used. They are valid for the following conditions:

Medium load ($P/C \cong 0.1$).
No additional stress due to tilting and detrimental axial load.
Load angles β (see Fig. 75), which are usual with various bearing types.
  Radial bearings (only cylindrical roller bearings and needle bearings) mainly loaded radially, axial bearings only loaded axially.

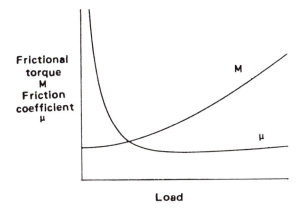

Figure 74   Frictional torque M and friction coefficient μ.

Table 21   Friction Correction Value μ

Bearing type	Friction correction value μ of bearings with $P/C \sim 0.1$
Deep-groove ball bearing	0.0015
Self-aligning bearing	0.0013
Angular contact ball brg. 1-row	0.0020
Angular contact ball brg. 2-row	0.0024
4-point bearing	0.0024
Cylindrical roller bearing	0.0013
Cylindrical roller bearing, full complement	0.0020
Needle bearing	0.0025
Spherical roller bearing	0.0020
Tapered roller bearing	0.0018
Thrust ball bearing	0.0013
Spherical roller thrust brg.	0.0020
Cylindrical roller thrust brg.	0.0040
Thrust needle bearing	0.0050

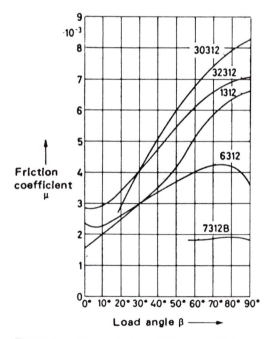

Figure 75   The coefficient of friction coefficient μ as a function of the load angle β.

Good lubricating conditions ($x = v/v_1 \cong 1$), medium speed, characteristic values nd_m.

Bearings without sliding seals.

The friction value μ depends on many factors. Important factors are

Bearing type and size
Lubrication condition
Load angle β
Load (little consequence for cylindrical roller bearings and needle bearings)
Speed

With large cylindrical roller thrust bearings the effect of sliding friction on the friction moment is much less than with small ones, as the rolling element in relation to the bearing diameter is much shorter. One can consider this when estimating the friction moment of cylindrical roller thrust bearings by replacing μ with the value μ^* in the equation:

$$\mu^* = \mu \frac{6L}{d} \tag{33}$$

where
μ^* = friction correction value for ratios $d/L > 6$
L = rolling element length $\cong 0.75(D - d)/2$
D = bearing outer diameter
d = bearing bore diameter

Example. For a cylindrical roller thrust bearing $\mu = 0.004$ (Table 21) with the cylindrical roller thrust bearing 812/500, $L = 56$, $d = 500$ and

$$\mu^* = \mu \frac{6L}{d} = \frac{(0.004)/(6)(56)}{500} = 0.003$$

When starting rolling bearings, friction values can be two- or threefold of the values shown in Table 21. If a more precise calculation of the friction moment is necessary, one proceeds according to Section VI.A.5.

5. Calculation of the Friction Moment

The friction moment M of a rolling bearing increases linearly or slightly progressively (Fig. 76) with increasing load according to bearing type. Therefore one can write

$$M = M_0 + M_1 \quad [\text{N} \cdot \text{mm}] \tag{34}$$

where
M = total friction moment of bearing
M_0 = portion not depending on load
M_1 = portion of total friction moment depending on load

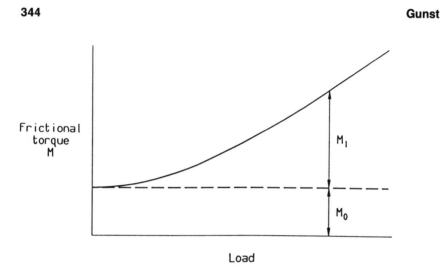

Load

Figure 76   Frictional torque and load.

The friction moment portion M_0 which is not depending on load relies mostly on the operating viscosity v of the lubricant. This is again affected by bearing friction and, consequently, by temperature. With decreasing viscosity the friction moment does not decrease proportionally because the thickness of the lubricating film decreases and the mixed friction increases. The demands on the cage lubrication and the bearing life require a certain minimum operating viscosity. In addition, speed, bearing size, the friction of the rolling elements in the cage pockets, and the friction of the cage on its guiding surface have a different effect on M_0, depending on the type of bearing (Table 22).

With the following formula for friction moment portion M_0 which is not dependent on load, a good relation was achieved with the results of experiments:

$$M_0 = f_0 \times 10^{-7}(Vn)^{2/3}d_m^3 \quad [\text{N} \cdot \text{mm}] \qquad \text{when } vn \geqslant 2000 \qquad (35)$$

$$M_0 = F_0 \times 10^{-7} \times 160d_m^3 \quad [\text{N} \cdot \text{mm}] \qquad \text{when } vn < 2000 \qquad (36)$$

where

   f_0 = correction value which takes into account the type of bearing and lubrication (Table 22)

$v\,[\text{mm}^2/\text{s}]$ = operating viscosity of oil or the grease-based oil

$n\,[\text{min}^{-1}]$ = speed of bearing

$d_m\,[\text{mm}]$ = $(D + d)/2$, pitch circle diameter

The correction value f_0 increases with the size of the rolling element and with the size of the bearing cross section. So the smaller values of the different

Table 22 Correction Factor f_0 for Calculating M_0, Depending on Bearing Type and Row

Bearing type row	Correction factor f_0 oil bath lubrication
Deep-groove ball bearing[a]	
160,62,63,64	1.5
619,618,60	1.75
628,638,639	2
Self-aligning bearing	
12	1.5
13	2
22	2.5
23	3
Angular contact ball bearing 1-row:	
72	2
2-row:	
32	3.5
33	6
Cylindrical roller bearings with cage	
2,3,4,10	2
22	3
23	4
Full complement:	
NCF18V	5
NCF22V	8
NCF29V	6
NCF30V	7
NNC48V	9
NNCL48V	9
NNC49V	11
NNCF49V	11
NNCL49V	11
NJ23VH	12
NN50V	13
Needle bearing	
NA48	5
NA49	5.5
NA69	10
Tapered roller bearing	
302,303,313	3
329,320,322,323	4.5
330,331,332	6
213	3.5
222	4
223,230,239	4.5
231	5.6
242	6
240	6.5
241	7

(*continued*)

Table 22 Continued

Bearing type row	Correction factor f_0 oil bath lubrication
Thrust ball bearing	
511,512,513,514	1.5
522,523,524	2
Cylindrical roller thrust bearing	
811	3
812	4
Axial needle bearing	5
Spherical roller thrust bearing	
292E	2.5
293E	3
294E	3.3

[a] When the ratio D_w/d_m is very small, smaller values have to be used, with thin section bearings 30–10% of these values.

ranges apply for bearings with light sections and the bigger values for bearings with heavy sections according to Fig. 76. If conditions are unfavorable one has to count on twofold values: for instance when the shaft is positioned vertically, or bearings have a heavy machined cage, or when there is too much lubricant.

When the lubricating oil for a certain application is chosen properly, the friction moment M_0 results mostly from the inner friction of the base oil. Therefore, the f_0 value of the grease lubrication is about the same as with the oil throwaway lubrication.

For bearings that have been newly lubricated or have rubbing seals, the starting friction moment M_0 is higher until the superfluous grease has been removed and an equilibrium has been reached. If necessary, the friction moment of the rubbing seals has to be adjusted by increasing M_0.

The friction moment M_1 is the result of the rolling contact friction, particularly the effects of hysteresis and sliding friction. It hardly changes with speed, but it does change with the size of the contact areas and, therefore, with the curvature ratio of the rolling element/raceway and with the load of the bearing. Further it is influenced by bearing type and size. The calculation is

$$M_1 = f_1 P_1^a d_m^b \quad [\text{N} \cdot \text{mm}] \tag{37}$$

where

M_1 = friction moment part dependent on load

f_1 = correction factor which considers the magnitude and direction of the load (Table 23)

Table 23 Factors for the Calculation of the Load-Dependent Friction Moment M_1

Bearing type, row	f_1	P_1[a]
Deep-groove ball bearing		
row 160, 62, 63, 64	0.0005	
619, 618, 60	$0.0007(P_0/C_0)^{0.5}$	F_r or $3.3F_a - 0.1F_r$[b]
628, 638, 639	0.0009	
Self-aligning bearing	$0.0003(P_0/C_0)^{0.4}$	$1.37 F_a/c - 0.1F_r$
Angular contact ball bearing		
1-row, $\alpha = 15°$	$0.0008(P_0/C_0)^{0.5}$	F_r oder. $3.3F_a - 0.1F_2$[b]
1-row, $\alpha = 25°$	$0.0009(P_0/C_0)^{0.5}$	F_r or $1.9F_a - 0.1F_r$[b]
1-row, $\alpha = 40°$	$0.001(P_0/C_0)^{0.33}$	F_r or $F_a - 0.1F_r$[b]
2-row or Paired 1-row	$0.001(P_0/C_0)^{0.33}$	$1.4F_a - 0.1F_r$
4-point bearing	$0.001(P_0/C_0)^{0.33}$	$1.5F_a + 3.6F_r$
Cylindrical roller brg. with cage		
Row 10	0.0002	
Row 2	0.0003	
Row 3	0.00035	F_r[c] (valid for all rows)
Row 4, 22, 23	0.0004	
Cyl. roller brg. full complement	0.00055	
Needle bearing	0.0015	F_r
Tapered roller bearing, 1-row	0.004	$2YF_a$
Tapered roller bearing, 2-row or		
two 1-row in X or O arrangement	0.0004	$1.21F_a/e$
Spherical roller bearing		
Row 213	0.00022	$1.36F_a/e$, if $F_r/F_a > c$
Row 222	0.00015	
Row 223	0.00065	$F_r[1 + 0.36(F_a/(cF_r)^3]$,
Row 230, 241	0.01	where $F_r/F_a \leqslant e$
Row 231	0.00035	(valid for all rows)
Row 232	0.00045	
Row 239	0.00025	
Row 240	0.0008	
Thrust ball bearing	$0.0008(F_a/C_0)^{0.33}$	F_a
Cylindrical roller thrust bearing		
Thrust needle bearing	0.0015	F_a
Spherical roller thrust bearing		
Row 292E	0.00023	F_a (where $F_r \leqslant 0.055F_a$)
Row 293E	0.0003	(valid for all rows)
Row 294E	0.00033	

Explanations: P_0 = equivalent static load (N)
  C_0 = static loading rate (N)
  F_a = axial components of the dynamic bearing load (N)
  F_r = radial components of the dynamic bearing load (N)
  Y,e = factors (see bearing tables)
    = pressure angle (see bearing tables)
[a] If $P_1 < F_r$, one has to calculate with $P_1 = F_r$.
[b] The larger value of each has to be used.
[c] Only radially loaded. For bearings that are also axially loaded see Section III.F.

P_1 [N] = load relevant for the frictional moment (Table 23)

d_m [mm] = $(D + d)/2 \cong$ pitch circle diameter of the bearing

   a,b = exponents dependent on bearing (Table 24)

With ball bearings the frictional correction factor f_1 is proportional to the expression $(P_0/C_0)^s$ because of the contact surface curvature; with rolling bearings f_1 remains constant with variable load. Thus, P_0 designates the equivalent static load and C_0 the static loading rate. The size of the exponent s depends on the amount of spinning friction; for ball bearings with low spin, such as deep-groove ball bearings, $s = 1/2$, and for ball bearings with high amounts of spin, as for instance for angular contact ball bearings with a contact angle of $\alpha_0 = 40°$, $s = 1/3$. The frictional correction factor f_1, as stated in Table 23, requires a separation of the rolling contacts ($x \geqslant 1$) by a lubricating film.

The larger the bearing, the smaller are the rolling element's relation to bearing pitch diameter d_m. The spinning friction between rolling element and raceway does, therefore, not increase proportionally to d_m. According to the above-mentioned ratios, friction moments M_1 can be too high with large bearings, especially when the bearing cross sections are small.

B. Temperature

The operational temperature of a bearing sets in as soon as the heat produced by bearing and sealing friction plus possible external heating remains in equilibrium with heat given off to the surroundings parts or air.

1. Operating Temperature

If, for the time being, one disregards external heating, the operational temperature which a rolling bearings assumes during medium speed and load is not high due to minimal bearing friction. Guiding factors for the average opera-

Table 24 Exponents for the Calculation of M_1

Bearing type	Exponent	
	a	b
All except spherical roller bearings	1	1
Spherical roller bearing		
Row 213	1.35	0.2
Row 222	1.35	0.3
Row 223	1.35	0.1
Row 230	1.5	−0.3
Row 231,232,239	1.5	−0.1
Row 240,241	1.5	−0.2

tional temperature of bearings which are differently subjected to stress can be found in Table 25.

The load P_1, which is relevant to the load-dependent frictional moment M_1, takes into account that M_1 changes with the load angle $\beta = \arctan(F_a/F_r)$. To simplify the calculation the axial factor Y was introduced as a reference value which also depends on F_a/F_r and the contact angle.

With the stated relations the frictional moment of a bearing can be sufficiently estimated. In practice, deviations are possible if the aimed-for hydrodynamic lubrication cannot be sustained and mixed friction occurs. The most favorable lubrication condition cannot always be reached.

The operating temperature a rolling bearing can have through external heating is sometimes considerably higher. As an example, Table 26 lists the operating temperatures of some bearings which are heated externally.

As the knowledge of the expected bearing operating temperature for the design of the bearing, lubrication, seals, etc., is important, it was tried repeatedly during equilibrium to calculate the bearing temperature t from the equation of the heat flow Q_R [W] produced by the bearing and the heat flow Q_L [W] carried off into the environment. The bearing temperature t depends heavily on the heat transmission conditions of the bearing and the surrounding parts as well as from environmental conditions. The equation is explained in the following. If the required data K_t and q_{LB} are known (possibly through experiments), the bearing temperature t can be determined from the heat balance.

The heat flow produced by the bearing is calculated from the frictional moment M [N · mm] and the speed n [min^{-1}] to

$$QR = 1.047 \times 10^{-4} nM \quad [\text{W}] \tag{38}$$

The heat flow Q_L given off to the environment is calculated from the difference of bearing temperature t and the environmental temperature t_u, from the size of the surfaces transferring heat—outer ring outside diameter $(D\pi B)$ and inner ring bore diameter $(d\pi B)$—and the customary heat surface power density Q_{LB} (Fig. 77) at normal operating conditions as well as the cooling factor K_t.

For the heat dissipation conditions of conventional pillowblock housings K_t is valid. In cases of better or worse heat dissipation see below.

$$Q_L = q_{LB} \left[\frac{t - t_u}{50} \right] K_t (2 \times 10^{-3}) d_m \pi B \quad [\text{W}] \tag{39}$$

where

$d_m = (D+d)/2$ [mm]

B = bearing width [mm]

K_t = cooling factor (0.5–2.5): 0.5 at poor heat dissipation; 1 at normal heat dissipation; 2.5 at very low heat dissipation; 2.5 at very good heat dissipation

q_{LB} = heat flow density [kW/m^2]

Table 25 Bearing Operating Temperature of Different Machines at 20° Environmental Temperature

Bearing arrangement	Operating temperature (approx. °C)	Bearing	Operating temperature (approx. °C)
Cutter block of a planing machine	40	Surface grinding machine	55
Bench drill	40	Jaw crusher	60
Horizontal boring machine	40	Axle bearing locomotive or railroad car	60
Circular saw shaft	40	Hammer Mill	60
Ingot slab structure	45	Roller bearing of wire rod mill	65
Lathe mandrel	50	Vibrating motor	70
Vertical boring machine	50	Stranding machine	70
Double shaft circular saw	40	Shaker screen	80
Wood shaper spindle	50	Beater mill	80
Calender roll of a paper machine	55	Marine screw propeller pressure bearing	80
Double roll arrangement of hot strip mills	55	Vibrating roller	90

Table 26 The Operating Temperature of Externally Heated Bearings

Bearing application	External heating	Operating temperature of bearing (approx. °C)
Electric traction motor	Electric heating from rotor, cooling of housing through airstream	80–90
Dry cylinder of paper machines	Steam from 140–150° through the bearing journal	120–130
Hot-air fans	Thermal conduction from the impeller wheel, admitted with hot gas, through the shaft to the bearing	90
Water pump in car engine	Heat from cooling water and motor	120
Turbo compressors	Compression heat carried off through shaft	120
Crankshaft of combustion engines	Combustion heat carried off through crankshaft; cooled housing	120
Calender for plastic masses	Delivery of 200–240°C heating liquid through bearing journal	180
Wheel bearing of furnace wagon	Heat radiation and conduction of furnace room	200–300

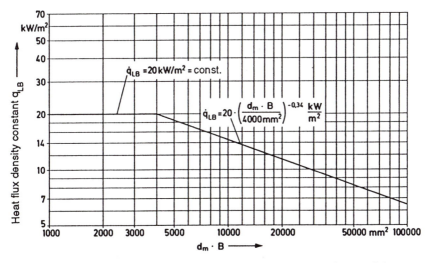

Figure 77 Heat surface power density unique to bearing at operating conditions.

If there is oil circulation, the oil carries off additional heat. The carried-off heat flow Q_0 is the result of the inflow temperature t_{in} and the outflow temperature t_{out}, the density p, and the specific heat capacity c of the oil as well as the oil amount m [cm³/min] flowing through within a time unit. The density is normally 0.86 to 0.93 kg/dm³, while the specific entropy c is between 1.7 and 2.4 kJ(kg · K) depending on oil type.

$$Q_0 = \frac{1}{60} mpc(t_{out} - t_{in}) \quad [W]$$

The bearing temperature t can be calculated if one equates

$$Q_R = Q_L + Q_0 \quad [W]$$

The result of such a temperature calculation is almost always incorrect as the quantities used in the calculation, especially the cooling factor K_t, are not known precisely. Little changes if a more refined calculating system is used. A useful basis can only be achieved if a steady-state temperature is established during a test run and, out of this, the heat dissipation factor can be determined. Thereby the steady-state temperature can be sufficiently estimated provided the mounting and operating conditions are comparable.

2. Temperature Differences between Inner and Outer Ring

The temperature of the inner ring of a bearing under normal operating conditions is always somewhat higher than the temperature of the outer ring, provided the shaft is not cooled or the housing is not heated. The reason for this is that the heat dissipation of the outer ring is higher than the inner ring because of its bigger surface. According to this the heat expansion of the inner ring is also greater than the heat expansion of the outer ring. Consequently, the radial clearance of the bearing heated by the operation is smaller than when it was mounted. This fact needs to be considered when the bearing design and the radial clearance group are being determined.

When the bearing and its surroundings are not cooled or heated, a temperature difference of 5–10 K can be generally expected. A temperature difference of approximately 15–20 K is only expected with bearings whose housings are being cooled (e.g., airstream cooling of vehicles). Accordingly, a greater temperature difference should occur in this case.

It there is external heating of the shaft or housing, it is difficult to estimate the temperature. Usually one has to be satisfied with determining which ring heats up more during the operation. For instance, in gearbox shafts heat flow travels from the gear through the shaft to the bearing inner ring. The bearings of the crankshaft of a combustion engine are affected by the output and radiant heat of the combustion chamber. With heated calender rollers, heat transfer fluid is led through the hollow roll neck on which the bearing inner ring is positioned. With these considerations the frictional heat of contact seals must not be overlooked. The temperature differences in these cases can only be estimated through values measured at equal or similar operating conditions of machines and instruments of the same design.

C. Basics of Lubrication

For lubricating rolling bearings, greases, oils, and, under special circumstances, solid lubricants are possible. Because grease lubrication does not require special aggregates and makes simple sealing possible, most bearings are lubricated with grease. But there are also cases where oil lubrication has advantages or is used for heat dissipation.

With rolling bearings the lubricating film represents a load-transferring component which separates the surfaces in motion. With moderate load and high cleanliness in the lubricating gap, complete separation of the contact areas and a nonfailing bearing is possible. Since elastic deformation of the surfaces plays a part in the forming of the lubricating film, this is called an elastohydrodynamic (EHD) lubrication.

The formulation of a carrying lubricating film is greatly influenced by the properties of the lubricant.

1. Properties of Lubricants

In technical applications the viscosity as well as the dependency of density and viscosity on pressure and temperature plays a part.

Viscosity. Viscosity is the basic physical property of lubrication oils; the carrying ability of the lubricating film depends most of all on this. Thickness and additives are also decisive for greases. Viscosity is the measure of "flowability." Fluid oils have low viscosity; sluggish oils have high viscosity.

In the lubrication theory one uses the dynamic viscosity n. It is expressed in milli-Pascal-second $[\text{mPa} \cdot \text{s}]$. Technical applications often use the kinematic viscosity v $[\text{mm}^2/\text{s}]$. The relation between both is

$$n = vf \quad [\text{mPa} \cdot \text{s}] \tag{40}$$

where the density f is expressed in kg/dm^3 and the kinematic viscosity v in mm^2/s.

The density of most lubricating oils lies between 0.86 and 0.93 kg/dm^3. It depends on the chemical composition of the oil. With increasing degree of refining of the oils, the density f decreases with the viscosity. With rising temperature f decreases slightly. The viscosity v decreases with rising temperature and increases with falling temperature. Therefore, the respective temperature has to be given with every viscosity value.

In the trade the kinetic viscosity of oils according to ISO 3448 (DIN 51519) is given in relation to 40°; for thin oils a temperature of 20° together with the viscosity is given, and for thick oils a temperature of 100° together with the viscosity. This value is designated as nominal viscosity.

Operating Viscosity. The viscosity of a lubricating oil at operating temperature is called operating viscosity. The operating viscosity can be established
from the viscosity-temperature diagram ($V-T$ diagram). If the $V-T$ diagram
of the oil in use is not known, the operating viscosity can be estimated from
Fig. 69. If, for instance, the oil has viscosity 6 mm^2/s at 100°C and 70 mm^2/s
at 20°, then the respective points on the $V-T$ diagram are connected with lines
and the operating viscosity of approximately 25 mm^2/s shows at an operating
temperature of 45° (see Fig. 69).

Dependency of Density and Viscosity on Pressure. In the contact areas of
rolling elements and raceways a load-carrying lubrication film forms even
when the Hertz pressure is several thousand N/mm^2. This fact, proven in
theory and with tests, is explained by the increase of density and viscosity with
pressure.

The density of lubricating oils increases about 20% when the pressure
increases to 1000 N/mm^2. This change of density has no strong effect on the
formation of the lubricating film. The dependence of the viscosity on pressure
can at first be described with the exponential equation

$$n = n_0 e^{\alpha p} \quad [\text{mPa} \cdot \text{s}] \tag{41}$$

where

n [mPa $\cdot$ s] = dynamic viscosity at pressure p

n_0 [nPa $\cdot$ s] = dynamic viscosity at atmospheric pressure

α [mm/N] = pressure-viscosity coefficient (generally between 0.01 and
0.02 mm^2/N)

p [N/mm^2] = pressure

Equation (41) does not take into account the temperature, so the viscosity
values are too high at high pressure.

2. Functions of the Lubricant

Lubricants have two primary functions:

To avoid, or at least reduce, wear and changes on the surfaces of the parts
in motion relative to one another
To reduce friction in the bearing

These functions are accomplished both in sliding bearings and in rolling bearings by feeding the lubricant between the contact areas where a load-
transmitting lubrication film builds up. The oil, adhering to the surface of the
parts in rolling contact, is fed between the contact areas. The oil film separates
the contact surfaces, preventing metal-to-metal contact ("physical lubrication"). In addition to rolling, sliding occurs in the contact areas. The amount of
sliding is, however, much less than in sliding bearings. Sliding is caused by

elastic deformation of the bearing components and by the curved form of the functional surfaces. Under pure sliding contact conditions, existing for instance between rolling elements and cage or roller faces and lip surfaces, the contact pressure is far lower than under rolling contact conditions.

Since sliding motions in rolling bearings play only a minor role, energy losses due to friction and wear are still acceptable, even if lubrication conditions are unfavorable. Therefore it is possible to lubricate rolling bearings with greases of different consistency and oils of different viscosity. In addition, wide speed and load ranges do not create any problems.

Sometimes the contact surfaces are not completely separated by the lubricating film. Even in these cases, low-wear operation is possible if the high local temperatures release chemical reactions between the additives in the lubricant and the metal surfaces of rolling elements or rings (tribochemical reaction layers) and generate reaction products which have a lubricating effect ("chemical lubrication"). Also solid lubricants added to the oil or grease and even the grease thickener have a lubricating effect. In special cases, it is possible to lubricate rolling bearings with solid lubricants only.

Additional functions of rolling bearing lubricants are protection against corrosion, dissipation of heat from the bearing (oil lubrication), discharge of wear particles and contaminations from the bearing (circulating oil lubrication; the oil is filtered), and sealing (grease collar, oil-air lubrication).

3. Different Lubricating Conditions

Friction and wear behavior and the attainable fatigue life of a rolling bearing are influenced by lubrication. In a rolling bearing, the following lubricating conditions exist.

Full Fluid Film Lubrication. The surfaces of the components in relative motion are completely or nearly completely separated by a lubricating film (Fig. 78(a)). Under these conditions almost pure elastohydrodynamic lubricating conditions exist. For continuous operation this type of lubrication should always be aimed at.

Mixed Lubrication. When the lubricating film becomes too thin, local metal-to-metal contact occurs, resulting in mixed friction (Fig. 78(b)).

Boundary Lubrication. With mixed lubrication, very high contact pressures and temperatures occur at the contact areas. If the lubricant contains suitable additives, reactions between the additives and the metal surfaces are released, generating reaction products which form a thin boundary layer with a lubricating effect (Fig. 78(c)).

Full fluid film lubrication, mixed lubrication, and boundary lubrication occur with grease and oil lubrication. The grease film thickness mainly depends on the viscosity of the base oil. The grease thickener also has a lubricating effect.

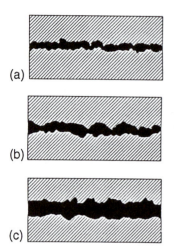

(a)

(b)

(c)

Figure 78   (a) Full fluid film lubrication. (b) Mixed lubrication. (c) Boundary lubrication.

Dry Lubrication. Solid lubricants, such as graphite and molybdenum disulfide, applied as a thin layer on the functional surfaces can prevent metal-to-metal contact. This layer can, however, be maintained over a long period only at moderate rotational speeds and low contact pressure. Solid lubricants added to the oil or grease also improve lubricating efficiency in cases of local metal-to-metal contact.

4. Lubricating Film with Oil Lubrication

The main criterion for the analysis of lubricating conditions is the lubricating film thickness between the load-transmitting rolling and sliding contact surfaces. The lubricating film between the rolling contact surfaces can be described by means of EHD lubrication. The lubrication under sliding contact conditions which exist, e.g., between the roller faces and lip surfaces of tapered roller bearings, is, however, adequately described by the hydrodynamic lubrication theory because the contact pressure in the sliding contact areas is smaller than in the rolling contact areas.

The calculation of the lubricating film thickness between surfaces in rolling contact according to the EHD theory is shown at the top of Fig. 79 for line contact. The equation at the bottom of Fig. 79 shows that the lubricant film thickness of a bearing with defined geometry primarily results from the rolling speed v, the dynamic viscosity n at operating temperature and atmospheric pressure, and by the pressure-viscosity coefficient α. The load Q has little influence because the viscosity rises with increasing loads and the contact surfaces are enlarged due to elastic deformation.

$$h_o = \frac{0.1 \cdot \alpha^{0.6} \cdot (\eta \cdot v)^{0.7}}{\left(\dfrac{1}{r_1} + \dfrac{1}{r_2}\right)^{0.43} \cdot \left(\dfrac{Q}{l}\right)^{0.13}} \cdot \left(\dfrac{E}{1 - \dfrac{1}{m^2}}\right)^{0.03} \quad [\mu m]$$

where

h_o	[microns]	minimum lubricating film thickness in the area of rolling contact
α	[mm²/N]	pressure-viscosity coefficient
η	[mPa·s]	dynamic viscosity
v	[m/s]	$= (v_1 + v_2)/2$ speeds
r_1	[mm]	roller radius
r_2	[mm]	inner ring raceway radius
Q	[N]	roller load
l	[mm]	roller length
E	[N/mm²]	modulus of elasticity = $2,08 \cdot 10^5$ for steel
$1/m$		Poisson's ratio = 0.3 for steel

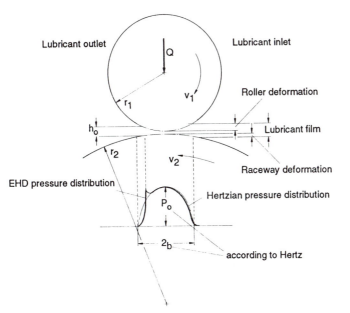

Figure 79 Elastohydrodynamic lubricating film. Example of roller–inner-ring contact.

In the case of "starved lubrication" the oil quantity is too small to form a lubricating film with a thickness h_0. The lubricating film is corresponding thinner.

In practice, a simplified method based on EHD theory is used for lubrication analysis. A rated viscosity v_1 can be assessed with Fig. 80 using the mean bearing diameter d_m and the rotational speed n. The operating viscosity value of the oil used, i.e., its kinematic viscosity at operating temperature, is supplied by the oil manufacturers. If only the viscosity at 40°C is known, Fig. 69 shows the viscosity at the operating temperature of mineral oils with average viscosity-temperature behavior. Figure 81 shows a comparison between viscosity grades ISO VG, AGMA grades, SAE crankcase oils, and SAE gear oils.

In many cases the temperature of the stationary ring is assumed as the operating temperature because the real temperature on the surface of the stressed elements in rolling contact is unknown. The temperature of the stationary ring can be accepted for determining the viscosity if bearing kinematics are advantageous. For heavily stressed bearings ($f_s < 4$) and bearings with a

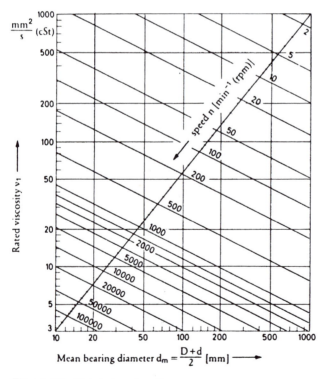

Figure 80 Rated viscosity of v_1.

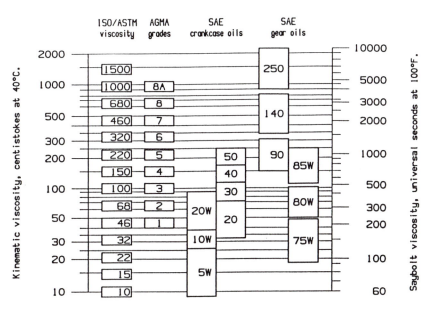

Figure 81 Viscosity grade comparisons.

higher amount of sliding (e.g., axially loaded roller bearings, full-complement cylindrical roller bearings, spherical roller bearings, and spherical roller thrust bearings), higher temperatures must be assumed.

If the bearing is exposed to external heat, the mean temperature of both rings must be assumed as the bearing temperature; with a high rate of sliding friction, the temperature for viscosity determination must be assessed in a similar way as for the operating temperature due to bearing friction where extraneous heat is excluded. The bearing temperature assessed by means of simple calculation methods is inaccurate because it depends largely on the heat transition conditions of the bearing, the mating components, and the environmental conditions. If these values are known (possible from tests), the bearing temperature can be derived from the heat balance.

D. Lubricant Selection

When selecting the lubricant, the operating conditions, such as load, speed, and operating temperature of the rolling bearing, are of prime importance. The environmental conditions must, however, also be taken into account. Added to these facts are the demands of the bearing running properties and the service life of bearing and lubricant. The lubrication system and the lubricant selected influence each other. When using highly doped mineral oils, such as hypoid gear oils, and with synthetic oils, their compatibility with seal and bearing materials (particularly the cage material) must be checked.

1. Greases

Table 27 lists principal grease types suitable for rolling bearing lubrication. The data in the table provide average values; manufacturers supply the precise data regarding the individual greases. Most of the greases listed are available in several consistency classes (worked penetration). More details on grease selection are given in the following text and in Table 28 and Fig. 82.

2. Running Properties

A *low, constant friction* is vital for bearings having to perform stick-slip-free motions, such as the bearings for telescopes. For these applications EP lithium greases with a base oil of high viscosity and MoS_2 additive are used. Low friction is also required from bearings applied to machines whose driving power is primarily determined by the bearing friction to be overcome, as is the case with frictional horsepower motors. These cases where the bearings start up rapidly from cold are best served by greases of consistency class 2 with a synthetic base oil of low viscosity. It must, however, be borne in mind that the various products differ widely in their frictional behavior (up to 1:3). The starting friction of rolling bearings at low temperatures is especially high if the environmental temperature is below the minimum operating temperature of the grease. The starting friction can be reduced by filling small amounts of grease in the bearings or by means of several starts, resulting in a favorable grease distribution. After some minutes of operation, the grease and bearing temperature due to bearing friction lead to a considerable reduction from the starting friction moment.

At normal temperatures, low friction can be obtained by selecting a stiffer grease of consistency class 3 to 4, the short period of grease distribution being excluded. These greases do not tend to circulate in the bearing if excess grease can settle in the housing cavities. If, in order to obtain low friction, bearings are filled with very small grease amounts—about 5% of the free bearing space in the case of thin-film lubrication, produced, for example, by a lubricant dispersed in a solvent—it must be remembered that small amounts of grease also mean a short grease service life.

Grease for *low-noise bearings* should not contain any solid particles. The greases should therefore be filtered and homogenized. A higher base oil viscosity reduces the running noise, especially in the upper frequency range. The bearing friction and the bearing temperature which rise with increasing viscosity, however, limit the increase in viscosity. In most cases, base oils of soft greases have a high viscosity. Soft greases have a shorter service life, and consequently the relubrication intervals are shorter. The standard grease for low-noise deep-groove ball bearings at normal temperatures is a filtered, lithium soap base grease of consistency class 2 with a base oil viscosity at approximately 60 mm^2/s at 40°C.

3. Special Operating Conditions

High temperatures occur if the bearings are exposed to high stressing and to extraneous heating. In this case, high-temperature greases should be selected. It must be taken into account that the grease service life is strongly affected by the temperature if the upper temperature limit of the grease is exceeded. Above this temperature every temperature rise of 15 K halves the service life of the grease.

The critical temperature limit is approximately 70°C for lithium soap base greases and 40–60°C for sodium and calcium soap base greases. High-temperature greases which contain mineral base oil and a thermally stable thickener can be used for short periods of up to 190°C; their limiting temperature varies between 80 and 110°C according to the grease type.

High-temperature greases with synthetic base oil can be used at higher temperatures than high-temperature greases with mineral base oil because synthetic oils evaporate less and do not deteriorate so quickly. A higher base oil viscosity increases the service life of the grease at high temperatures; the suitability of the grease for higher rotational speeds, however, is reduced. Greases with high-viscosity alkoxyfluoro oil are suitable for deep-groove ball bearings up to a speed index of $nd_m = 140,000$ min^{-1} even at temperatures of up to 250°C. If less heat-stable greases (i.e., cheap greases) are used for bearing lubrication at high temperatures, frequent relubrication is necessary. Greases must be chosen which do not solidify in the bearing and consequently impair the grease exchange.

At low temperatures a low starting friction can be obtained with low-temperature greases, such as gel-type greases or lithium soap base greases with thin mineral oil, polyalphaolefin, or ester as the base oil. Multipurpose greases are very stiff at temperatures below their lower temperature limit and cause an extremely high starting friction. If, at the same time, bearing loads are low, slippage can occur, resulting in wear on the rolling elements and rings. However, after a short running period, the friction decreases to normal values.

Condensate can form in the bearings and cause corrosion if the machine operates in a humid environment or in the open air, and the bearings cool down during longer idle times of the machine. In these cases, greases are recommended which emulsify with water, such as sodium and lithium soap base greases. Sodium grease absorbs larger amounts of water, but may soften to such an extent that it flows out of the bearing. Lithium grease only slightly emulsifies with water.

If the seals are exposed to splash water, a water-repellent grease of consistency class 3 should be used. The pressure of the water just not be so high that the grease is washed out from the sealing gap. Calcium soap base greases

Table 27 Grease Properties

Grease type			Properties							
Thickener			Temperature range °C	Drop point	Water resistance	Corrosion resistance	Load carrying capacity	Price relation[a]	Suitability for rolling bearings	Remarks
Type	Soap	Base oil								
Normal	Aluminum	Mineral oil	−20–70	120	c	d	d	2.5–3	d	Swells with water
	Calcium		−30–50	80–100	b	c	d	0.8	d	Good sealing action against water
	Lithium		−35–130	170–200	b	c	d	1	b	Multipurpose grease
	Sodium		−30–100	150–190	e	b	c	0.9	c	Emulsifies with water, may solidify
Complex	Aluminum	Mineral oil	−30–160	260	b	b	d	2.5–4	b	Multipurpose grease
	Barium		−30–140	220	c	b	c	4–5	b	Multipurpose grease, resistant to vapor
	Calcium		−30–140	240	c	c	c	0.9–1.2	b	Multipurpose grease may harden
	Lithium		−30–150	240	c	d	c	2	c	Multipurpose grease for higher temperatures
	Sodium		−30–130	220	d	b	d	3.5	b	
Normal	Lithium	Ester	−60–130	190	d	d	d	5–6	b	For low temperatures, high speeds
Complex	Barium	Ester	−60–130	200	c	b	c	7	b	For low temperature and high speeds at moderate loads

Thickener	Base oil	Temperature range							Remarks
Lithium	Ester	−40–180	240	c	d	d	10	b	For especially wide temperature range
Lithium	Silicone oil	−40–180	>240	c	d	d	20	c	For especially wide temperature range, $P/C < 0.03$
Bentonites	Mineral oil	−20–150	Without	b	e	d	2–6	c	For higher temperatures at low speeds
Polyurea		−25–160	250	b	d	c	3	b	For higher temperatures at medium speeds
Polyurea	Silicone oil	−40–200	250	b	d	e	35–40	c	For high and low temperatures, low loads
Polyurea	Fluorosilicone	−40–200	250	b	d	d	100	b	For high and low temperatures, moderate loads
PTFE or FEP	Alkoxyfluoro oil	−50–250	Without	b	d	c	150–400	b	Both greases for very high and low temperatures
	Fluorosilicone oil	−40–230	Without	b	d	c	120	b	Very good resistance to chemicals and solvents

[a] Reference grease: lithium soap base grease/mineral base oil = 1
[b] Very good
[c] Good
[d] Moderate
[e] Poor

Table 28 Criteria for Grease Selection

Criteria for grease selection	Properties of the grease to be selected
Operating conditions	
Speed index nd_m	Grease selection according to Fig. 82
Load ratio P/C	
Running properties	
Low friction, also during starting	Grease of consistency class 1 to 2 with synthetic base oil of low viscosity
Low constant friction at steady-state condition, higher starting friction admissible	Grease of consistency class 3 to 4, grease quantity <30% of the free bearing space or consistency class 2 to 3, grease quantity <20% of the free bearing space
Low noise level	Filtered grease (very clean) of consistency class 2; for especially high demands on quiet running filtered grease of consistency class 1 to 2 with high-viscosity base oil
Mounting conditions	
Inclined or vertical position of bearing axis	Grease with good adhesion properties of consistency class 2 to 3
Outer ring rotation, inner ring stationary, or centrifugal force on bearing	Grease of consistency class 3 to 4 with a large amount of thickener
Maintenance	
Frequent relubrication	Soft grease of consistency class 1 to 2
Infrequent relubrication, for-life lubrication	Grease retaining its consistency class 2 to 3 under stressing, resistant to temperatures which are higher than the operating temperature
Environmental conditions	
High temperature, for-life lubrication	Heat resistant grease with synthetic base oil and heat-resistant (e.g., synthetic) thickener
High temperature, relubrication	Grease which does not form any residues at high temperatures
Low temperature	Grease with thin synthetic base oil and suitable thickener, consistency class 1 to 2
Dusty environment	Stiff grease of consistency class 3
Condensate	Emulsifying grease, e.g., sodium or lithium soap base greases
Aggressive media (acids, bases, etc.)	Special grease, please consult bearing or lubricant manufacturer
Radiation	Up to absorbed dose rate 2×10^4 J/kg, rolling bearing greases to DIN 51825, up to absorbed dose rate 2×10^7 J/kg, consult lubricant manufacturer
Vibratory stressing	EP lithium soap base grease of consistency class 2, frequent relubrication. With moderate vibratory stresses, barium complex grease of consistency class 2 with solid lubricant additives or lithium soap base grease of consistency class 3.
Vacuum	Up to 10^{-5} mbar rolling bearing greases to DIN 51825, in the case of higher vacuum lubricant/bearing manufacturer should be consulted

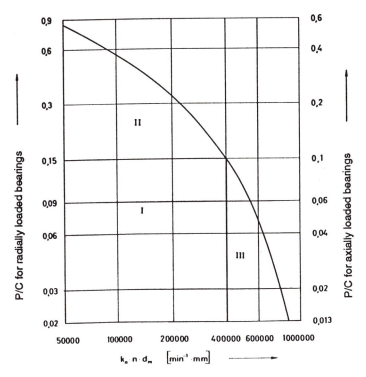

Figure 82 Grease selection from the load ratio P/C and the relevant bearing speed index $k_a nd_m$.

are water repellent. Since they do not absorb any water, a rust inhibitor is added to protect the bearings against corrosion.

Certain greases are resistant to *special media* (boiling water, vapor, acids, bases, aliphatic and chlorinated hydrocarbons). In these situations, FAG should be consulted.

The grease, acting as a *sealing agent,* prevents contaminations from penetrating into the bearing. Stiff greases (consistency class 3 or higher) form a protective grease collar at the shaft passage, remain in the sealing gap of labyrinths, and retain foreign particles. If the seals are the rubbing type, the grease must also lubricate the surfaces of the sealing lip and the shaft which are in sliding contact. The compatibility of the grease with the seal material must be checked.

Radiation can affect the bearings (e.g., in nuclear power plants). The total absorbed dose is the measure for radiation stressing of the grease—that is, either the radiation of low intensity over a long period of time or of a high intensity over a short period of time (absorbed dose rate). The absorbed dose rate must, however, not exceed a value of 10 J(kg · h). The consequences of stressing by radiation are change in grease consistency and drop point, evaporation losses, and the development of gas. The service life of a grease stressed by radiation is calculated from $t = S/R$, unless the service life is still shorter due to other stresses. Here t is the service life in hours, S the absorbed dose in J/kg permissible for the grease, and R the absorbed dose rate J/kg · h. Standard greases resist an absorbed dose of up to $S = 2 \times 10^4$ J/kg; special greases resist an absorbed dose of up to $S = 2 \times 10^7$ J/kg with gamma rays.

In the primary circle of nuclear power plants, certain substances such as molybdenum disulfide, sulfur, and halogens, are subjected to strong changes. It must therefore be ensured that greases used in the primary circle do not contain these substances.

In the case of *vibratory stresses,* the grease is moved and displaced in and around the bearing which has the effect of frequent irregular regreasing of the contact surfaces. This is true of many greases. On the other hand, the vibrations sooner or later result in a separation of the base oil and the thickener. Lithium soap base grease of consistency class 2 doped with EP additives has proven itself suitable for the heavy stressing in vibrating screens. It is good practice to relubricate vibrating screen bearings at short intervals, e.g., once a week. Barium complex soap greases of consistency class 2 containing dry lubricants, lithium soap base greases, and lithium complex soap base greases of consistency class 3 are appropriate, e.g., in vibratory motors.

The base oil of the grease gradually evaporates in *vacuum.* At room temperature, satisfactory service lives are still reached with standard greases in vacuums up to 10^{-5} mbar. Shields and seals improve grease retention in the bearing and reduce evaporation losses. If the bearings are exposed to higher vacuums, FAG should be consulted.

Inclined or *vertical shafts* can cause the grease to escape from the bearing due to gravity. Therefore, a grease with good adhesive properties should be used and be retained by means of baffle plates beside or below the bearing.

In the case of *for-life lubrication* or where there is a need for maximum grease service life, greases retaining their consistency class 2 or 3 are preferred. The upper temperature limit up to which the grease retains its lubricity should be clearly above the operating temperature, at least by 20 K.

If *frequent replenishment* is required and the lubricant pipelines are relatively long, a soft pumpable grease (consistency class 1 to 2) should be selected. If a stiffer grease is used, short lubricant pipelines with an adequate diameter must be provided.

4. Selection of Suitable Oils

Straight and additive mineral oils and synthetic oils are generally suitable for the lubrication of rolling bearings. The basic properties of these oils are specified in DIN standards and other national and international standards.

Apart from these oils, special lubricants are required for special branches of industry. The technological properties of lubricants required for these branches of industry are specified in standards and guidelines.

In modern lubrication engineering, lubricating oils based on mineral oils are the ones most commonly used. These mineral oils must at least meet the requirements indicated in DIN 51501. Special oils, often synthetic, are used for extremely punishing operating conditions or special demands on the stability of the oil under aggravating environmental conditions (temperature, radiation, etc.). The major chemicophysical properties of oils are listed in Table 29.

Recommended Oil Viscosity. Attainable fatigue life and safety against wear increase with better separation of contact surfaces by a lubricant film. Since the lubricating film thickness increases with rising viscosity of the oil, an oil with a high operating viscosity should be selected. A very long fatigue life can be reached if the operating viscosity v equals $3-4v_1$ (see Fig. 67). High-viscosity oils, however, have also disadvantages. A higher viscosity means higher lubricant friction; at low and normal temperatures, supply and drainage of the oil can cause problems (oil retention). Therefore, the oil viscosity should be selected so that a maximum fatigue life is attained and an adequate supply of oil to the bearings is ensured.

In isolated cases, the required operating viscosity cannot be attained

If the oil selection also depends on other machine components which require a thin-bodied oil

If, for circulating oil lubrication, the oil must be thin enough to dissipate heat and carry off contaminants from the bearing

If, in the case of temporarily higher temperatures or very low speeds, the required operating viscosity cannot be obtained even with an oil of the highest possible viscosity

In these cases an oil with a lower viscosity than recommended for the application can be used. A reduced fatigue life and wear on the functional surfaces must, however, be taken into account; see the adjusted life calculation (Fig. 67).

5. Oil Selection by Operating Conditions

Normal Operating Conditions. Under normal operating conditions (atmospheric pressure, max. temperature 120°C, load ratio $PC < 0.01$, speeds up to limiting speed), straight oils and preferably inhibited oils are used (corrosion and deterioration inhibitors, letter L in DIN 51502). If the recommended viscosity values cannot be maintained, oils with EP additives and antiwear additives should be selected.

High Rotational Speeds. For high speeds, an oil should be used which is stable to oxidation, has good defoaming properties, and a positive viscosity-

Table 29 Properties of Various Oils

Oil type	Mineral oil	Polyalpha-olefin	Polyglycol in(water soluble)	Ester	Silicon oil	Alkoxyfluoro oil
Viscosity at 40°C (mm²/s)	2–4500	15–1200	20–2000	7–4000	4–10000	20–650
Maximum temperature (°C) for oil sump lubrication	100	150	100–150	150	150–200	150–200
Maximum temperature (°C) for circ. oil lubrication	150	200	150–200	200	250	240
Pour point (°C)	-20^b	-40^b	-40	-60^b	-60^b	-30^b
Flash point (°C)	220	$230–260^b$	200–260	220–260	300^b	–
Evaporation losses	Moderate	Low	Moderate to high	Low	Low^b	Very low^b
Resistance to water	Good	Good	Good,^{b} hard to separate due to same density	Moderate to good^b	Good	Good
V-T behavior	Moderate	Moderate to good	Good	Good	Very good	Moderate to good
Pressure-viscosity coefficient (m²/N³)	$1.1–3.5 \times 10^8$	$1.5–2.2 \times 10^8$	$1.2–3.2 \times 10^8$	$1.5–4.5 \times 10^8$	$1.0–3.0 \times 10^8$	$2.5–4.4 \times 10^8$
Suitability for high temperatures (~150°C)	Moderate	Good	Moderate to good^b	Good^b	Very good	Very good
Suitability for high loads	Very good^a	Very good^a	Very good^a	Good	Poor^b	Good
Compatibility with elastomers	Good	Good^b	Moderate, to be checked when used with paint	Moderate to poor	Very good	Good
Price comparison	1	6	4–10	4–10	40–100	200–800

a With EP additives.
b Depending on the oil type.
c Measured up to 200 bar, depends on oil type and viscosity.

temperature behavior. On starting, when the temperature is generally low, a high friction and consequently high temperature due to a high viscosity are avoided; the viscosity at steady-state operating temperature is sufficient to ensure adequate lubrication.

Figure 81 shows the average V-T behavior of mineral oils. The viscosity of some oils decreases at a slower rate with rising temperature. Suitable synthetic oils with positive V-T behavior are esters and polyglycols.

High Loads. If the bearings are heavily loaded ($P/C \geqslant 0.1$) and if the operating viscosity v is smaller than the rated viscosity v_1, oils with antiwear additives should be used (EP oils, letter P in DIN 51502). EP additives reduce the harmful effects of metal-to-metal contact.

High Temperature. Selection of appropriate oils is guided by their properties which are described in the next section.

6. Oil Selection according to Oil Properties

Mineral oils are stable only up to temperatures of 150 to 180°C. According to the temperature and the period of time passed in the hot area, deterioration products form which impair the lubricating efficiency and settle as solid residual matter in or near the bearing. Mineral oils are suitable to a limited extent only, if contaminated with water, if they contain detergents to improve the compatibility with water. However, the water in the form of a stable emulsion leads to a reduction in rolling bearing life, depending on the amount of water. The permissible amount of water can vary between a few per mil and several percent, depending on the oil composition and the additives. Under radiation stressing (e.g., in nuclear power plants), normal mineral oils are stable up to an absorbed dose of $S = 5 \times 10^5$ J/kg, special mineral oils up to $S = 3 \times 10^6$ J/ kg. The service life of oil under radiation stressing is calculated in the same way as the service life of greases.

Esters (diesters and sterically hindered esters) are thermally stable, have a positive V-T behavior and low viscosity, and are therefore recommended for high speeds and temperatures. In most cases, esters are miscible with mineral oils and can be treated with additives. The various ester types react differently with water. Some esters saponify and split into their various constituents, especially if they contain alkaline additives.

Polyalkyleneglycols have a good V-T behavior and a low setting point. They are therefore suitable for high and low temperatures (-50–$+200$°C). Because of their high oxidation stability oil exchange intervals in high-temperature operation can be two to five times the usual interval for mineral oils. Depending on the type, polyalkyleneglycols are not water soluble, and their ability to separate water is poor. Their pressure-viscosity coefficient is lower than that of the other oils. The most commonly used polyalkyleneglycols are not miscible with mineral oils. They can attack seals and lacquered surfaces in the housing and aluminum cages.

Polyalphaolefins are synthetic hydrocarbons which can be used in a wide temperature range. Because of their good oxidation stability they attain a multiple of the life of mineral oils of similar viscosity under identical conditions. They are easily miscible with mineral oils. Polyalphaolefins have positive viscosity-temperature behavior.

Silicone oils (methyl phenyl siloxanes) can be used at extremely high and extremely low temperatures, because their *V-T* behavior is positive; they have a low volatility and a high thermal stability. Their load-carrying capacity, however, is low ($P/C < 0.03$), and their antiwear properties are poor.

Chlorofluorine compounds resist oxidation and water. Their pressure-viscosity coefficient and density are higher than those of mineral oils of the same viscosity.

Fire-resistant, aqueous *hydraulic fluids* play a special role. Their lubricating efficiency is moderate. Consequently, the service life of rolling bearings lubricated with these fluids is reduced considerably. Depending on the fluid, the life values vary between 5% and 30% of the value reached with mineral oils of the same viscosity, even if bearing kinematics are favorable. Hydraulic fluids are

Oil-in-water emulsions (HFA). The oil content is only a few percent. The attainable fatigue life is less than 10% of the normal fatigue life.

Water-in-oil emulsions (HFB). The oil content is about 60% and the attainable fatigue life less than 15%.

Aqueous glycol solutions (HFC). The attainable fatigue life is less than 15%.

Phosphate esters (HFD). The attainable fatigue life is up to 30%.

Chlorinated hydrocarbons. The attainable fatigue life is up to 30%.

7. Selection of Dry Lubricants

Dry lubricants are of interest only where oils and greases are unsuitable for rolling bearing lubrication, for example,

In vacuum where oil evaporates intensively

Under extremely high temperatures

If oil or grease is retained in the bearings only for a short period, e.g., blade bearings in controllable pitch blade fans which are exposed to centrifugal forces.

The most commonly used dry lubricants are graphite and molybdenum disulfide (MoS_2). They are applied as powder, bonded with oil as paste, or together with plastics material as sliding lacquer. Other solid lubricants are polytetrafluoroethylene (PTFE) and soft metal films (e.g., of copper and gold), which are, however, used rarely. In special cases, bearings can also be provided with self-lubricating cages, i.e., cages with embedded dry lubricants. The rolling elements transfer the lubricant to the raceways (transfer lubrication).

Bearings operating at extremely low speeds ($nd_m < 1500$ min^1 mm) can be lubricated with molybdenum disulfide or graphite pastes. The oil contained in the paste evaporates at a temperature of about $200°C$, leaving only a minute amount of residue. Rolling bearings running at a speed higher than $nd_m = 1500$ min^{-1} mm are in most cases lubricated with powder or sliding lacquer. A smooth powder film is formed by rubbing solid lubricant into the microscopically rough surface. The surfaces are usually bonderized in order to ensure better adhesion of the powder film. More stable films are obtained by applying sliding lacquer on bonderized surfaces. These sliding lacquer films can, however, be used only with small loads. Especially stable are metal films which are applied by electrolysis or by cathodic evaporation in an ultrahigh vacuum. It is advantageous to additionally treat the surface with molybdenum disulfide. The bearing clearance is reduced by four times the amount of the dry lubricant film thickness in the contact area. Therefore, bearings with larger than normal clearance should be used if dry lubrication is provided. Thermal and chemical stability are limited.

Graphite can be used for operating temperatures of up to $450°C$ because it is stable to oxidation over a wide temperature range. Graphite is not very resistant to radiation.

Molybdenum disulfide can be used up to $450°C$. It keeps its good sliding properties even at low temperatures. In the presence of water, it can cause electrolytic corrosion. It is only slightly resistant to acids and bases.

The compatibility of *sliding lacquers* with the environmental media must be checked. Organic binders of sliding lacquers soften at high temperatures, affecting the adhesive properties of the sliding lacquer. Inorganic lacquers contain inorganic salts as binder. These lacquers have a high thermal stability and do not evaporate in a high vacuum. Protection against corrosion, only moderate with all lacquers, is less with inorganic lacquers than with organic lacquers.

Pastes become doughy and solidify if dust penetrates the bearings. In a dusty environment, sliding lacquers are better.

8. Dry Lubricant Application

The most currently used dry lubricants are graphite and molybdenum disulfide. These lubricants are applied to the raceway surfaces in the form of loose powder, sliding lacquer, or paste. When applying a powder coating, a brush, leather, or cloth can be used. Sliding lacquers are sprayed on the functional surfaces (spray gun). The service life of many sliding lacquers can be increased by baking the lacquer on the surfaces. Pastes are applied with a paint brush.

Generally the bearings are bonderized (magnesium phosphate coating, phosphate coating) before the dry lubricants are applied. The phosphate coating allows for better adhesion of dry lubricants, protects against corrosion, and provides, to a certain extent, for emergency running properties. If high standards of protection against corrosion are required, the bearings are cadmium plated.

Powders and lacquers only partially adhere to greasy bearings, if at all. Perfect and uniform application is only possible during bearing manufacture before the individual components are assembled.

Pastes can be applied prior to bearing mounting. Paste layers can be touched up or renewed. Overlubrication with pastes should be avoided.

E. Lubricant Supply—Grease

To operate reliably all contact areas in the bearing must be continuously supplied with a sufficient amount of clean lubricant. For-life lubricated bearings with shields or seals have proven to be highly successful. Generally, adequate lubricant supply is ensured by

The lubrication system and the related equipment
The design of the bearing location
Selection of the right lubricant quantity
Taking into account the lubricant service life, lubricant replenishment, or exchange intervals

Only a few lubricating tools, if any, are required for adequate bearing lubrication with grease. Unless greased by the manufacturer, the bearings are greased on mounting, generally by hand. In some cases, grease syringes or guns are used. The grease is replenished by means of a grease gun connected via a high-pressure hose to the grease nipple in the housing. Frequent replenishment requires the use of grease pumps.

1. Grease Amounts
For bearing greasing, the following instructions should be carried out:

Pack bearings to capacity with grease to ensure that all functional surfaces are supplied with grease.
Fill the housing space on both sides of the bearing with grease to such an extent that it can still accommodate the grease expelled from the bearing. Excessive circulation of the grease is prevented.
Fill high-speed bearings, e.g., spindle bearings, only partially (30–40% of the free space) in order to facilitate and accelerate the grease distribution during bearing start-up.
Pack low-speed bearings ($nd_m < 50,000$ min$^{-1} \cdot$ mm) and the housing cavities to capacity with grease. The lubricant friction due to working is negligible.

If for-life lubrication is required, the grease must be retained in or near the bearing by means of seals or baffle plates. Grease deposited in the bearing vicinity can result in longer lubrication intervals because

Fresh grease is occasionally fed into the bearing due to machine vibrations (replenishment).

At higher temperatures, the external grease separates oil which contributes to bearing lubrication.

For high bearing temperatures, the largest possible grease volume should be deposited beside the bearing. This can be done, e.g., by an angular baffle plate (see Fig. 83). Higher rotational speeds usually produce higher bearing temperatures during the run-in period and even during several running hours occasionally. These effects are even more intensive the larger the grease quantity in and beside the bearing and the more the movement of the grease out of the bearing is obstructed.

Deep-groove ball bearings sealed on either side with seals design (.2RSR) or shields (.2ZR) are supplied pregreased. About 25–40% of the free bearing space should be filled with grease. This amount will last for the service life of the bearing because the seals and shields prevent grease from escaping.

If the pressure on one side of the bearing is different from the pressure on the other side, the grease is likely to be expelled. Openings and holes must be machined into the surrounding structure for pressure compensation.

2. Grease Replenishment

Grease replenishment or exchange is required if the service life of the grease is much shorter than the anticipated bearing life. This can be due to higher temperature, radiation, contaminants (water, aggressive media), or high mechano-dynamic stressing of the grease film on the functional surfaces. It is advisable to regrease the bearings at intervals which are clearly shorter than the service life of the grease.

The service life of greases is assessed in laboratory and field tests. Lubricant tests must be carried out on a statistical basis. Even under comparable test con-

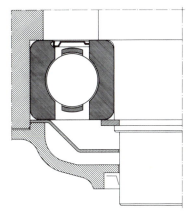

Figure 83 A grease deposit can form between the baffle and the bearing.

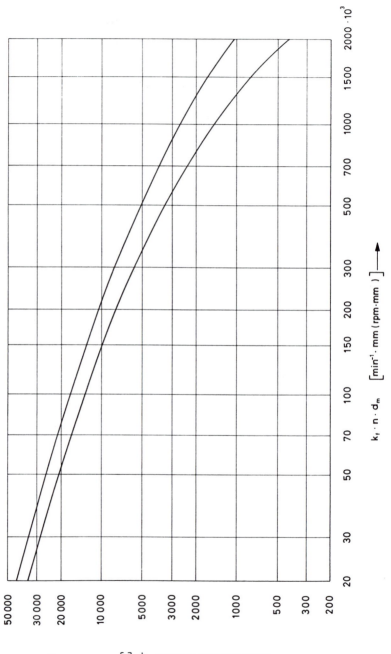

$k_f \cdot n \cdot d_m$ $\left[\text{min}^{-1} \cdot \text{mm (rpm} \cdot \text{mm)} \right]$

Relubrication Interval t_f [h]

BEARING TYPE		k_f
Deep groove ball bearing	single row	0.9...1.1
	double row	1.5
Angular contact ball brg	single row	1.6
	double row	2
Spindle bearing	α = 15°	0.75
	α = 25°	0.9
Four-point bearing		1.6
Self-aligning ball bearing		1.3...1.6
Thrust ball bearing		5...6
Angular contact thrust ball brg, double row		1.4

BEARING TYPE		k_f
Cylindrical roller bearing	single row	1.8...2.3
	double row	2
	full complement	25
Cylindrical roller thrust bearing		90
Needle roller bearing		3.5
Tapered roller bearing		4
Barrel roller bearing		10
Spherical roller brg without lips (E design)		7...9
Spherical roller brg with centre lip		9...12

Figure 84 Lubrication intervals (grease service life) for bearings lubricated with lithium soap base grease under favorable environmental conditions. Failure probability 10–20%. Reduced lubrication intervals t_{fq} must be taken into account for adverse operating and environmental conditions. The replenishment interval must be shorter than the lubrication interval (usual values: $0.5–0.7t_{cq}$).

ditions (identical operating conditions, bearings of the same quality), a scatter of the grease life values of up to 1:10 must be taken into account. Therefore, the calculation of the grease service life values and lubrication intervals is based on a certain failure probability, similar to the calculation of the bearing fatigue life.

Figure 84 shows the lubrication intervals as a function of the speed index of the bearings ($k_f n d_m$). They apply to a failure probability of 10% to 20%. A f_f value range is indicated for some bearing types. The higher values apply to the heavier series (higher load-carrying capacity) and the smaller values to the lighter series of a bearing type. The lubrication intervals in Fig. 84 apply to lubrication with a lithium soap base grease and temperatures up to 70°C, measured at the bearing outer ring, normal environmental conditions, and a mean bearing load corresponding to $P/C < 0.1$. In the case of higher bearing loads and temperatures, lubrication intervals are shorter. Every rise in temperature by 15 K over 70°C halves the lubrication interval of lithium soap base greases with mineral base oil. Also vibrations acting on the bearing reduce the lubrication intervals because they result in a separation of the grease into thickener and base oil.

Contaminants (including water) penetrating through the seals also affect the lubrication intervals. With gap-type seals, an air current passing through the bearing considerably reduces the lubrication interval. The air current deteriorates the lubricant, carries oil or grease from the bearing, and conveys contaminants into it.

For poor operating and environmental conditions, a reduced lubrication interval t_{fq} is obtained from the equation

$$t_{fq} = f_1 f_2 f_3 f_4 f_5 t_f \qquad (41)$$

The reduction factors f_1 to f_5 take into account the effect of contamination, shock loads, vibrations, increased temperatures, higher bearing loads, and air current. Figure 85 shows the corresponding reduction factors.

An overall reduction factor q which takes into account all poor operating and environmental conditions (Table 30) can be applied to certain bearing applications. The reduced lubrication interval f_{fq} is obtained from

$$t_{fq} = q t_f \qquad (42)$$

In the case of unusual operating and environmental conditions (high and low temperatures, high loads, and high speeds) the use of special greases appropriate for these operating conditions results in the lubrication intervals shown in Fig. 86. The effect of the operating and environmental conditions is balanced by the grease. Practical experience with these greases is described in brochures supplied by the lubricant manufacturers.

If the lubrication intervals determined according to Fig. 84 are exceeded considerably, a higher rate of bearing failures due to starved lubrication must

```
Effect of dust and moisture on the bearing
contact surfaces
moderate              f₁ = 0.7...0.9
strong                f₁ = 0.4...0.7
very strong           f₁ = 0.1...0.4

Effect of shock loads and vibrations
moderate              f₂ = 0.7...0.9
strong                f₂ = 0.4...0.7
very strong           f₂ = 0.1...0.4

Effect of high bearing temperature
moderate (up to 75°C)     f₃ = 0.7...0.9
strong (75 to 85°C)       f₃ = 0.4...0.7
very strong (85 to 120°C) f₃ = 0.1...0.4

Effect of high loads
P/C = 0.1 to 0.15     f₄ = 0.7...1.0
P/C = 0.15 to 0.25    f₄ = 0.4...0.7
P/C = 0.25 to 0.35    f₄ = 0.1...0.4

Effect of air current passing through the
bearing
slight current        f₅ = 0.5...0.7
strong current        f₅ = 0.1...0.5
```

Figure 85 Reduction factors f_1 to f_5 for poor operating and environmental conditions.

be taken into account. Therefore, grease renewal or replenishment must be scheduled in time. Grease renewal intervals should be shorter than the reduced lubrication intervals f_{fq}. In most cases it is difficult to remove the used grease entirely from the bearing when *relubricating* it. Consequently, the relubrication intervals must be shorter (usual relubrication intervals 0.5 to $0.7f_{fq}$).

Replenishment is required where the used grease cannot be removed during relubrication (no empty housing spaces or grease valve). Care should be taken in selecting the amount of grease required. Overgreasing must be prevented. The appropriate amounts for grease replenishment are shown in Fig. 84.

Large relubrication amounts are recommended with large free housing spaces, grease valves, or low rotational speeds. In these cases, the risk of overgreasing is reduced. Larger grease amounts have the advantage of improving the exchange of used grates for fresh grease and of contributing to the sealing against the ingress of dust and moisture. Relubrication with the bearing rotating at operating temperature is favorable. If lubrication intervals are longer, it is recommended removing the used grease completely and replacing with *fresh grease*. This is achieved by pressing in a greater amount of grease than that

Table 30 Overall Reduction Factor q for Certain Applications

	Dust moisture	Shocks vibrations	High running temperature	Heavy loads	Air current	q
Stationary electric motor	-	-	-	-	-	1
Tailstock spindle	-	-	-	-	-	1
Grinding spindle	-	-	-	-	-	1
Surface grinder	-	-	-	-	-	1
Circular saw shaft	o	-	-	-	-	0.8
Flywheel of a car body press	o	-	-	-	-	0.8
Hammer mill	o	-	-	-	-	0.8
Dynamometer	-	-	o	-	-	0.7
Axle box roller bearings for locomotives	o	o	-	-	-	0.7
Electric motor, ventilated	-	-	-	-	o	0.6
Rope return sheaves of aerial ropeway	oo	-	-	-	-	0.6
Car fore wheel	o	o	-	-	-	0.6
Textile spindle	-	ooo	-	-	-	0.3
Jaw crusher	oo	oo	-	o	-	0.2
Vibratory motor	o	ooo	o	-	-	0.2
Suction roll (paper making machine)	ooo	-	-	-	-	0.2
Press roll in the wet section (paper making machine)	ooo	-	-	-	-	0.2
Work roll (rolling mill)	ooo	-	o	-	-	0.2
Centrifuge	o	-	-	oo	-	0.2
Bucket wheel reclaimer	ooo	-	-	o	-	0.1
Saw frame	o	ooo	-	-	-	< 0.1
Vibrator roll	o	ooo	ooo	-	-	< 0.1
Vibrating screen	o	ooo	-	-	-	< 0.1
Slewing gear of an excavator	oo	-	-	ooo	-	< 0.1
Pelleting machine	o	-	o	ooo	-	< 0.1
Belt conveyor pulley	ooo	-	-	o	-	< 0.1

o   = moderate effect
oo = strong effect
ooo = very strong effect

which is in the bearing, thereby expelling the spent grease. A particularly large amount of grease is required, if, due to high temperatures, the old grease, or part of it, has solidified.

In order to drain as much used grease as possible by flushing, an amount of up to three times the grease quantity indicated in Fig. 84 is used for relubrication. FAG can provide information on the most suitable greases for flushing. The grease exchange is facilitated by a grease feed and flow which ensures uniform grease exchange over the bearing circumference. Design examples are shown in Figs. 87 to 91. During grease exchange, the grease must be able to escape easily to the outside or into a space of sufficient size for the accommodation of the spent grease.

Relubrication quantity m_1 for weekly to yearly lubrication

$$m_1 = D \cdot B \cdot x \ [g] = D \cdot B \cdot \frac{x}{28.35} \ [oz]$$

Relubrication	x
weekly	0.002
monthly	0.003
yearly	0.004

Quantity m_2 for extremely short relubrication interval

$m_2 = (0.5 \text{ to } 20) \cdot V \ [kg/h] = (17.6 \ldots 705) \cdot V \ [oz/h]$

Relubrication quantity m_3 prior to restarting after several years of standstill

$$m_3 = D \cdot B \cdot 0.01 \ [g] = D \cdot B \cdot \frac{0.01}{28.35} \ [oz]$$

V = free space in the bearing

$$\approx \frac{\pi}{4} \cdot B \cdot (D^2 - d^2) \cdot 10^{-9} - \frac{G}{7800} \ [m^3]$$

$$\approx \frac{\pi}{4} \cdot B \cdot (D^2 - d^2) \cdot 10^{-9} - \frac{G \cdot 0.4536}{7800} \ [m^3]$$

d = bearing bore diameter [mm]
D = bearing outside diameter [mm]
B = bearing width [mm]
G = bearing weight [kg]
G'= bearing weight [lb]

Figure 86 Amounts of relubrication grease.

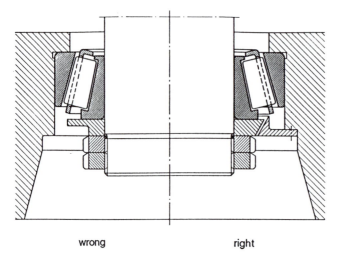

wrong   right

Figure 87   A baffle plate retains the grease inside and near the bearing.

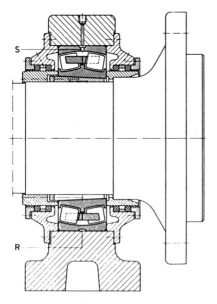

Figure 88   The grease is fed through the bearing outer ring.

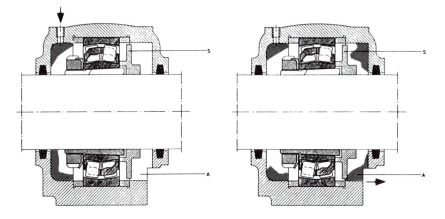

Figure 89 Relubrication from the side. Grease retention is prevented by grease valve S.

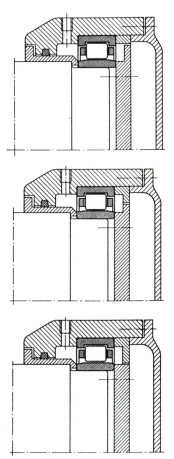

Figure 90 The pumping effect of the grease valve can be varied.

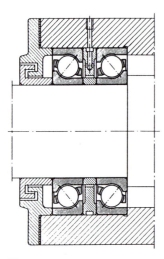

Figure 91   Grease supplied between a bearing pair.

Relubrication intervals must be *extremely short* if stressing is very high. Under these conditions, the use of a grease pump is justified. Care must be taken that the grease in the bearing, housing, and feed pipe maintains a pumpable consistency. At extremely high temperatures, the grease may solidify, obstructing the passage of fresh grease and resulting in seizure of the metering valves. The grease amount per unit time depends on the grease service life, the bearing size, the condition of the spent grease, and the required sealing action of the grease. Suitable relubrication amounts can be taken from Fig. 86. In a rolling bearing packed to capacity with grease, the right grease quantity is automatically adjusted, depending on the speed if the excess grease can be accommodated by the housing cavities on both sides of the bearing.

The escaping grease can act as a *sealing agent* if small quantities are continuously supplied at short intervals. A quantity of grease which the cavities of the bearing can hold or a multiple of it is used for hourly relubrication. By applying the quantities m_2 recommended in Fig. 86 for extremely short relubricating intervals, the grease escapes at a rate of 0.2 m/day or more depending on the sealing gap which.

If *temperatures* are *high,* either a cheap grease which is stable only for a short period or an expensive, thermally stable grease can be used. When using the former greases, relubrication quantities of 1–2% of the free bearing space per hour have proved appropriate. With thermally stable and very expensive special greases, amounts as small as 0.2% of the free bearing space will do. Direct grease supply into the bearing is absolutely essential for such small quantities. One advantage of small relubrication amounts is that the frictional moment and the temperature rise only slightly. Metering the grease quantity is advantageous especially for high rotational speeds; it requires more maintenance and complicated lubricating equipment. Very small amounts of grease can be directly fed to the bearing by spray lubrication (Fig. 92).

3. Grease Mixing

In the case of relubrication, a *mixture of the various grease types* cannot be avoided. Mixtures which are relatively safe are

Greases of the same soap base
Lithium greases/calcium greases
Calcium greases/bentonite greases

Mixtures to be avoided are

Sodium greases/lithium greases
Sodium greases/calcium greases
Sodium greases/aluminum greases
Sodium greases/bentonite greases
Aluminum greases/bentonite greases

When mixed, the structure of these greases can change drastically and the greases can soften. If a different grease type is selected, the old grease should be flushed out with a large amount of the new grease, provided this can be done with the existing design of the bearing location. To ensure that the old grease is entirely expelled, another supply of the new grease should be pressed

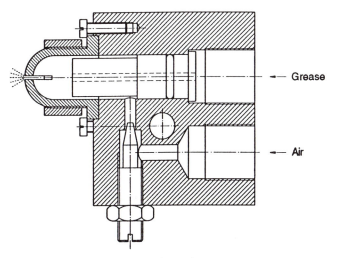

Figure 92 Lubricant–air mixing unit.

in after a fairly short period. Then normal relubrication intervals can be resumed.

4. Lubricant Supply—Oil

Unless oil sump lubrication is provided, the oil must be fed to the bearing locations by means of lubricating devices depending on the selected lubrication system. Larger and smaller oil volumes are fed to the bearings by means of pumps; small and very small oil volumes are supplied by oil-mist, oil-air, and central lubrication plants. The oil volume can be measured by means of metering elements, flow restrictors, and nozzles.

5. Oil Sump Lubrication

In an oil sump or, as it is also called, an oil bath, the bearing is partly immersed in oil. The oil level should not be too high. When the shaft is in the horizontal position, the bottom rolling element should be half or completely covered, when the bearing is stationary (Fig. 93). When the bearing rotates, the oil is conveyed by the rolling elements and the cage and distributed over the circumference. For bearings with asymmetrical cross section which, due to

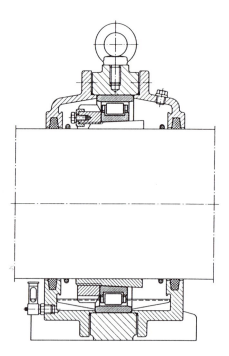

Figure 93   Oil level in an oil pump.

their geometry, have a pumping effect, oil return holes or ducts should be provided to ensure circulation of the oil.

If the *oil level* rises above the bottom roller and, especially, if speeds are high, the friction due to churning raises the bearing temperature and can cause foaming. At speed indices according to $nd_m < 150,000$ min^{-1} mm, the oil level can be higher. If complete immersion of a bearing in the oil sump cannot be avoided, as is the case with the shaft in the vertical position, the friction moment doubles or triples depending on the oil viscosity. With rising viscosity, the portion of the bearing friction which is due to the lubricant, increases proportionately by $v^{0.7}$.

As a rule, oil sump lubrication can be used up to a speed index of $nd_m = 300,000$ min^{-1} mm; when the oil is renewed frequently, a speed index of up to 500,000 min^{-1} mm is possible. At a speed index of $nd_m = 300,000$ min^{-1} mm and above, the bearing temperatures often exceed 70°C. The oil sump level should be checked regularly.

The *oil renewal schedule* depends on contamination and aging of the oil. Aging is accelerated by the presence of oxygen, metal abrasion (catalyst), and high oil temperatures. The alteration of the neutralization number NZ and of the saponification number VZ indicates to the oil manufacturer and the engineer to what degree the oil had deteriorated.

Under normal conditions, the oil renewal intervals indicated in Fig. 94 should be followed. It is important that the bearing temperature does not

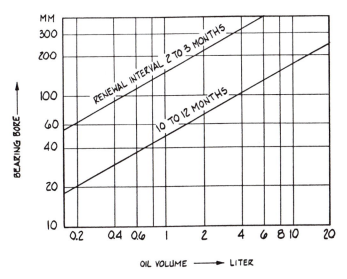

Figure 94   Oil volume and renewal interval versus bearing bore.

exceed 80°C and contamination due to foreign particles and water is low. As the figure shows, frequent oil changes are necessary if the oil volume is small. During the run-in period, an early oil change may be required due to the higher temperature and heavy contamination by wear particles. This applies particularly to rolling bearings lubricated together with gears.

Increasing content of solid and liquid foreign particles in the oil can require premature oil renewal. The permissible amount of solid foreign particles depends on the size and hardness of the particles. The permissible amount of water in the oil depends on the oil type, and the oil manufacturer should be consulted.

Water in oil leads to corrosion, accelerates oil deterioration by hydrolysis, and affects the formation of a load carrying lubricating film, no matter whether the water has emulsified in the oil. Water which has entered the bearing through the seals or condensate having formed in the bearing must be rapidly separated from the oil; an oil with positive water separation ability is advantageous. Water is separated by treating the oil in a separator. The separation of water and oil is, however, difficult with polyglycol oils because their density is approximately 1. Therefore, the water does not settle in the oil reservoir; at oil temperatures above 90°C the water evaporates.

For extreme applications it is advisable to determine the oil change intervals individually. It is good practice to analyze the oil after one to two months and, depending on the results of the first analysis, to determine, after a certain period, the neutralization number NZ, saponification number VZ, content of solid particles and water, and the viscosity of the oil. The oil change intervals are set according to the results of the second inspection. This complicated method of determining the oil change intervals is justifiable where large volumes of oil are in the housing. A simpler way of assessing the condition of the oil is by comparing a drop of fresh oil and one of used oil on a sheet of blotting paper. Major differences in color are indicative of oil deterioration or contamination.

6. Circulating Oil Lubrication

Having passed the bearings, the oil is collected in an oil reservoir and recirculated to the bearings. The oil volume required is determined by the operating conditions. Figure 95 shows the appropriate volumes of oil.

Only a very small amount of oil is required for lubricating the bearings, because the lubricating film thickness in the contact area is only 0.2 to 0.8 micron. In view of this small amount, the volumes of oil indicated in zone a (Fig. 95) seem to be rather large. The large oil volumes are recommended to ensure appropriate lubrication of all contact areas even if the oil supply to the bearings is inadequate; i.e., oil is not fed directly into the bearings. The minimum volume indicated is used for lubrication if low friction is required. The resulting temperatures are the same as with oil sump lubrication. If wear

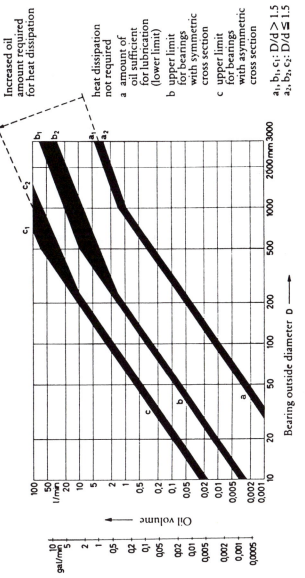

Figure 95 Oil volumes for circulating lubrication.

particles must be carried off or heat dissipated, larger oil volumes are provided. Since each bearing offers a certain resistance to the passage of oil, there are upper limits for the oil volume.

For bearings with asymmetrical cross section (angular contact ball bearings, tapered roller bearings, spherical roller thrust bearings), larger flow rates are permissible than for bearings with a symmetrical cross section, because their flow resistance is lower due to their pumping action. For the oil volumes indicated in Fig. 95, oil supply and retention at the feed side is supposed to take place without pressure up to an oil level of just below the shaft. Smaller oil volumes are required for larger bearings; i.e., the curves indicated in Fig. 95 rise less steeply with increasing bearing cross section; compared with their size the cross section of larger bearings is relatively smaller than that of small bearings and the oil cannot flow as freely through them.

The conditions of heat generation and dissipation in the individual cases dictate the volume of oil needed to keep the bearings at their required temperatures. The oil volume is determined by recording the bearing temperatures during machine start-up.

The flow resistance of bearings with symmetrical cross section increases with rising speed. If, in this case, larger oil volumes are required, the oil is injected directly into the gap between cage and bearing ring. *Oil jet lubrication* reduces the energy losses due to churning. Figure 96 shows the recommended oil volumes versus the speed index and the bearing size. The diameter and number of nozzles are indicated in Fig. 97. Oil entrapment in front of the bear-

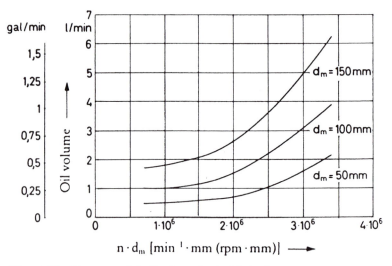

Figure 96   Recommended oil volume for jet lubrication.

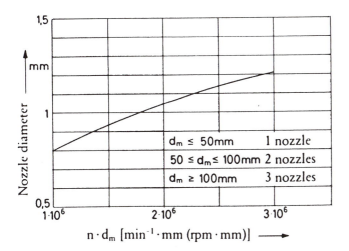

Figure 97   Diameter and number of nozzles for oil jet lubrication.

ing is prevented by injecting the oil into the bearings where free passage is assured. Discharge ducts with sufficient diameter allow the oil to drain freely.

For the high rotational speeds usual with oil jet lubrication, oils with an operating viscosity of 5 to 10 mm^2/s ($x = 1$–4) have proven efficiency. When designing the lubrication system, it must be taken into account that the drop in oil pressure in the nozzles can amount up to 10 bar.

Rolling bearings should be lubricated before going into operation. With circulating oil lubrication, this is achieved by starting the oil pump before the machine is put into operation. This is not necessary where provisions have been made to ensure that the oil is not entirely drained from the bearing and a certain amount of oil is present. A combination of an oil sump with a circulation system increases the operational reliability, because, in the case of pump failure, the bearing continues to be supplied with oil from the sump for some time. At low temperatures, the oil flow rate can be reduced to the quantity required for lubrication until the oil has heated in the reservoir (Fig. 95, curve a). This helps to simplify the circulating oil system (pump drive, oil return pipe).

If larger oil quantities are used for lubrication, retention of the oil must be avoided because this would lead to substantial energy losses due to churning and friction especially at high speeds. The diameter of the discharge pipe depends on the oil viscosity and the angle of inclination of the discharge pipes. For oils with an operating viscosity up to 500 mm^2/s, diameter d_a is calculated as follows: $d_a = (15$–$25)$ m [mm], where d_a [mm] is the inside diameter of the discharge pipe and m [l/min] is the oil flow rate.

The amount of oil M in the oil reservoir depends on the flow rate m. As a rule, the fill of the reservoir should be circulated $z = 3$ to 8 times per hour

$$M = \frac{60m}{z} \, [1] \tag{44}$$

If the z value is low, foreign matter settles in the reservoir, the oil can cool down and does not deteriorate so quickly.

7. Throwaway Oil Lubrication

Throwaway lubrication is also known as total loss lubrication. The oil volume fed to the bearing can be reduced below the lower limit indicated in Fig. 95, if a low bearing temperature is required with a high volume of oil. However, the energy losses due to friction and heat dissipation at the bearing location must allow bearing operation at moderate temperatures without additional cooling.

The state of minimum friction and minimum temperature—that is, when full fluid film lubrication sets in—is reached with an oil volume of 0.01 to 0.1 mm^3/min. The bearing temperature rises up to an oil volume of 10^4 mm^3/min. Beyond that volume heat is dissipated from the bearing.

The oil quantity required for an adequate oil supply largely depends on the bearing type. Bearings where the direction of the oil flow coincides with the pumping direction of the bearing, require a relatively large oil supply. Double-row bearings without conveying effect require an extremely small amount of oil if it is fed between the two rows of rolling elements. The rotating rolling elements prevent the oil from escaping. If very small amounts of oil are used, all sliding contact areas in the bearing (lip and cage guiding surfaces) must be adequately covered with oil. In the case of machine tools, it is advantageous to feed oil directly to ball bearings and cylindrical roller bearings and in the direction of conveyance of angular contact ball bearings. Figure 98 shows minimum oil quantities versus the bearing size and the contact angle (conveying effect) for these bearings. For bearings with lip-roller face contact (e.g., tapered roller bearings in rolling mills), direct oil supply to the roller faces, opposite to the conveying direction, is suitable.

The oil quantities permissible for total-loss lubrication also depend on the rotational speed of the bearings. The minimum oil volume required increases slightly with rising speeds. Continuous supply of too large an oil quantity or the intermittent supply even of small quantities at high speeds lead to a sharp rise in lubricant friction and a temperature difference between inner and outer rings. This can result in detrimental radial preloading and eventually in the failure of bearings which have a small radial clearance (e.g., machine tool bearings).

This is illustrated in Fig. 99 with the example of cylindrical roller bearing NNU4926 which, as a rule, is used in machine tools. Line a shows the minimum oil volume as a function of the speed index. Line b represents the

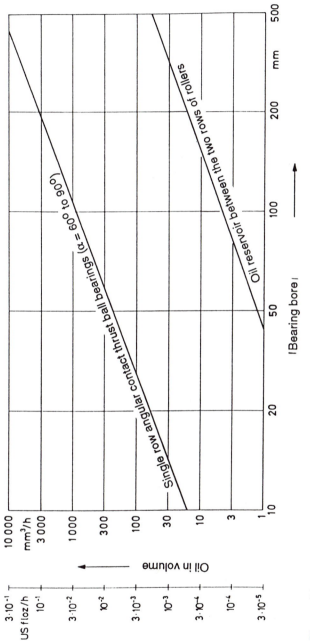

Figure 98 Oil volumes for throwaway lubrication. Applicable to $nd_m < 600,000$ min^{-1} mm.

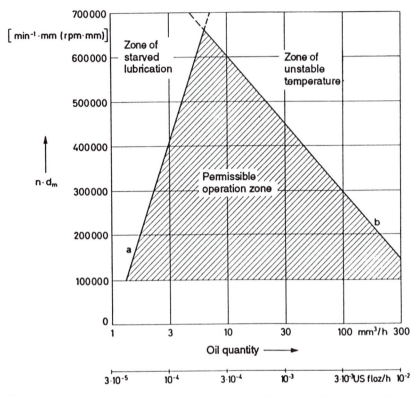

Figure 99   Selection of oil volume for throwaway lubrication [example: double row cylindrical roller bearing NNU4926 (d = 130 mm, small radial clearance) mainly used in machine tools]. Line a = minimum oil volume. Line b = permissible oil volume with uniform oil supply.

maximum oil volume; beyond this line excessive radial preloading can occur. The diagram is based on a continuous oil supply (oil-air lubrication) and average heat dissipation. The point of intersection of lines a and b represents the maximum speed index for throwaway lubrication. Since the minimum and maximum oil volumes permissible for throwaway lubrication depend not only on the bearing, but also on the oil, the oil supply, and heat dissipation, it is not possible to furnish a general rule for determination of the speed index and the optimum small oil quantities.

8.   Bearing Damage Due to Imperfect Lubrication

More than 40% of rolling bearing damage is due to imperfect lubrication. In numerous other cases which cannot be directly traced back to imperfect lubrication, it is one of the underlying causes of damage. Imperfect lubrication in the contact areas leads to wear, smearing, scoring, seizure, and fretting corrosion. In addition, fatigue damage (flaking) and heat cracks can occur. Sometimes bearing overheating occurs if, in the case of starved lubrication or overlubrication, the bearing rings are heated to different temperatures due to unfavorable heat dissipation, resulting in a reduction of radial clearance or leading eventually even to detrimental preload.

The main causes for damage due to imperfect lubrication are

Unsuitable lubricant (oil of too low a viscosity, lack of additives, unsuitable additives, corrosive action of additives)
Starved lubrication in the contact areas
Contaminants in the lubricant
Alteration of lubricant properties
Overlubrication

Starved lubrication and overlubrication can be remedied by selecting a lubricant supply system adapted to the relevant application. Damage due to unsuitable lubricant or changes of the lubricant properties can be avoided by taking into account all operating conditions in lubricant selection and by renewing lubricant in good time.

9. Contaminants

Hardly any lubrication systems are completely free from contaminants. Contaminants usually found in differing applications are taken into account in determining the fatigue life and service life because the methods of calculation are based on empirical values gained from experience in the field and test results. More contaminants in the lubricant than usual lead to reduced running times or premature fatigue. On the other hand, longer lives are obtained under particularly high standards of cleanliness (see Fig. 67).

All lubricants contain a certain amount of contaminants stemming from their manufacture. The minimum requirements for lubricants specified in DIN standards list, among others, have limits for the permissible contamination at the time of lubricant supply. In most cases, contaminants enter the bearing on mounting due to insufficient cleaning of the machine components, oil pipelines, etc., and during operation due to insufficient seals or openings in the lubrication unit (oil reservoir, pump). During maintenance, contaminants can also penetrate into the bearing, for example, through dirt on the grease nipple and on the mouthpiece of the grease gun, during manual greasing, etc.

For assessing the detrimental effect of contaminants it is essential to know the

Type and hardness of foreign particles
Concentration of the foreign particles in the lubricant
Size of foreign particles

10. Solid Foreign Particles

Solid foreign particles lead to wear and premature fatigue. Hard particles in rolling bearings cause abrasive wear, particularly at contact areas with a high rate of sliding friction, for example, between the contact surfaces of tapered roller bearings or between the contact surfaces of raceway edges and rollers in cylindrical roller thrust bearings. Wear increases with the particle hardness and more or less proportionately with the concentration of the particles in the lubricant and the particle size. Wear even occurs with extremely small particles. Abrasive wear in rolling bearings is acceptable to a certain extent, the permissible amount of water depending on the application.

Cycling of larger particles (approximate size 0.1 mm) causes indentations in the raceways. Plastically deformed material is rolled out at the edges and only partly removed during subsequent rolling. Each following load cycle causes higher stresses in the area of the indentation which result in a reduced fatigue life. The greater the hardness of the cycled particles (e.g., file dust, grinding chips, mold sand, corundum), and the smaller the bearings, the shorter the life.

The following steps are recommended for reducing the concentration of foreign particles in the bearing:

Thorough cleaning of the bearing mating parts
Cleanliness in mounting
With oil lubrication, filtering the oil with filters which have an adequate mesh
With grease lubrication, sufficiently short grease renewal intervals

Oil filters should have the following mesh widths:

60 μm with low demands on service life, oils containing MoS_2, and large-size bearings
25 microns with higher demands on service life (25 microns correspond to the requirements for industrial hydraulic systems)
< 10 microns with high demands on service life, preferably of small-size bearings (bore diameter < 50 mm) and with bearings which have a higher rate of sliding friction (e.g., tapered roller bearings, spherical roller bearings)

11. Liquid Contaminants

The main liquid contaminants in lubricants are water or aggressive fluids, such as acids, bases, or solvents. Water may be free, dispersed, or dissolved in oils. With free water in oil, visible by the oil discoloration (white-gray), there is the risk of corrosion. This risk is accelerated by hydrolysis of the sulfur bonded with the lubricant. Dispersed water in form of a water-in-oil emulsion affects the lubrication condition. Experience has shown that the fatigue life of bearings lubricated with these aqueous oils decrease considerably. It can be reduced to a very small percentage of the normal fatigue life. If water is dissolved in oils, the lubricity is not affected noticeably, provided the water quantity is small enough. Certain special oils are excluded.

Water in greases causes structural changes depending on the soap type as is the case with water-in-oil emulsions, the fatigue life is reduced. With contamination by water, the grease renewal intervals must be shortened, depending on the amount of water.

Aggressive media (acids, bases) solvents, etc., can alter drastically the chemicophysical characteristics and eventually deteriorate the lubricant. Information and recommendations on the compatibility of lubricants with these media, which are given by the lubricant manufacturers, must be followed. On areas in the bearings which are not protected by the lubricant, corrosion develops and finally destroys the surface.

12. Cleaning Contaminated Bearings

For cleaning rolling bearings, naphtha, petroleum, ethanol, dewatering fluids, chlorinated hydrocarbons, aqueous neutral, or alkaline cleansing agents can be used. Petroleum, naphtha, ethanol, and dewatering fluids are inflammable, chlorinated hydrocarbons develop vapors which are harmful to health, and alkaline agents are caustic.

When washing out bearings, lint-free cloth, paint brushes, or brushes should be used. Immediately after washing and evaporation of the solvent which should be as fresh as possible, the bearings must be preserved in order to avoid corrosion. If gummed oil and grease residues stick to a bearing, it should be soaked longer in chlorinated hydrocarbons or in an aqueous strong alkaline cleansing agent and then cleaned by hand.

VII. MOUNTING ROLLING BEARINGS

A. Care, Handling, and Storage

1. Store bearings in their original package in order to protect them against contamination and corrosion. Open package only at the assembly site immediately prior to mounting.
2. Larger bearings with relatively thin-walled rings should not be stored upright but flat and supported over their whole circumference; otherwise they could go out-of-round in shape.
3. Prior to packing, rolling bearings are dipped in anticorrosive oil. This oil does not gum and harden and is compatible with all commercial rolling bearing greases. In their original package rolling bearings are safely protected against external influences.
4. If the bearings are kept in their original package in a dry (max. relative air humidity 60%) and frost-protected room this protection is effective

for longer periods. In addition, no aggressive chemicals, such as acids, ammonia, or chlorinated lime, may be stored in the same room.

B. Preparation Before Mounting

Prior to mounting of rolling bearings, several preparatory steps should be taken. Study the shop drawing to be familiar with the design details of the application and the assembly sequence. Phase the individual operations and get reliable information on heating temperatures, mounting forces, and the amount of grease to be packed into the bearing.

Whenever rolling bearing mounting requires special measures, the bearing serviceman should be provided with comprehensive instructions on mounting details, including means of transport for the bearing, mounting equipment, measuring devices, heating facilities, type, and quantity of lubricant. In preparation, the bearing serviceman must make sure that the bearing number stamped on the package agrees with the designation given on the drawing and in the parts list. He should therefore be familiar with the bearing numbering and identification system. Standard bearings are identified by the bearing number listed in the pertinent standards and rolling bearing catalogs.

Rolling bearings are preserved, in their original package, with an anticorrosive oil. The oil need not be washed out when mounting the bearing. In service, the oil combines with the bearing lubricant and provides for sufficient lubrication in the run-in period. The seats and mating surfaces must be wiped clean of anticorrosive oil before mounting. Wash out anticorrosive oil with cold-cleaning agent from tapered bearing bores prior to mounting in order to ensure a safe and tight fit on the shaft or sleeve. Then thinly coat the bore with a machine oil of medium viscosity. Before mounting, wash used and contaminated bearings carefully with kerosene or cold-cleaning agent and oil or grease them immediately afterwards. Do not perform any rework on the bearing. Subsequent drilling of lubricating holes, machining of grooves, flats, and the like will disturb the stress distribution in the ring, resulting in premature bearing failure. There is also the risk of chips or grit entering the bearing.

Absolute cleanliness is essential! Dirt and humidity are dangerous offenders, since even the smallest particles penetrating into the bearing will damage the rolling surfaces. The work area must, therefore, be dust-free, dry, and well removed from machining operations. Avoid cleaning with compressed air. Ensure cleanliness of shaft, housing, and any other mating parts. Castings must be free from sand. Bearing seats on shaft and in housing should be carefully cleaned from antirust compounds and residual paint. Turned parts must be free from burrs and sharp edges. After cleaning, the housing bore should receive a protective coating.

All surrounding parts should be carefully checked for dimensional and form accuracy prior to assembly. Nonobservance of the tolerances for shaft and housing seat diameters, out-of-roundness of these parts, out-of-square of abutment shoulders, etc., impair bearing performance and may lead to premature failure. The responsibility of such faults for bearing failure is not always easy to establish and much time can be lost in looking for the cause of failure.

C. Mechanical Methods

1. Hammer and Sleeve

Small bearings with a cylindrical bore can be driven onto the shaft by means of a hammer and a mounting sleeve. The mounting sleeve is to be applied at the ring which has the tighter fit; its abutment surface should not be wider than the bearing ring. It is recommended to use soft steel as the sleeve material. Brass drifts will not damage the bearing when impacted, but the brass is liable to shave or chip and pieces may get into the bearing. The abutment surface must be flat in order to allow an even distribution of the mounting force over the entire ring. The crowned impact surface ensures a centric application of load. Mounting sleeves are also suitable for driving in small withdrawal sleeves.

2. Mechanical and Hydraulic Presses

Presses enable steady mounting and dismounting of small and medium-size bearings with a cylindrical base. They are particularly useful for series mounting of bearings without heating. Mechanical and hydraulic presses are available in a great variety of types. Sleeves used for hammer mounting can also be used here.

3. Nut and Hook (Spanner) Wrench) (see Fig. 100)

Locknuts or the nuts of adapter sleeves suffice for mounting small bearings (bore diameters of up to about 80 mm) on tapered seats. The nuts are tightened by means of hook wrenches. A locknut is also suitable for pressing in a small

Mounting with nuts

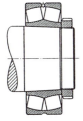 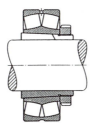 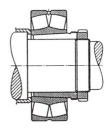

a: Fastening on a tapered shaft   b: Adapter sleeve fastening   c: Withdrawal sleeve fastening

Figure 100 (a–c) Mounting bearings by using nuts.

withdrawal sleeve if the shaft is threaded. Nuts are usually protected against loosening by means of lock washers and locking clamps.

4. Nuts and Thrust Bolts (see Fig. 101)

Nuts and thrust bolts are used for mounting small numbers of medium-size and large bearings with a tapered bore. For E-type spherical roller bearings no nuts with thrust bolts are available because the bolts would press onto the cage. The thrust bolts allow application of considerably greater forces than with a simple nut. The nut is equipped with eight tapered holes for the bolts. It is important that all bolts be uniformly tightened during mounting. In order to prevent the bearing or (if a withdrawal sleeve is mounted) the sleeve from being damaged by the bolts, a mounting plate or pieces of sheet metal are placed between the faces of the bolts and the bearing or the withdrawal sleeve. When the bearing

Mounting with thrust bolts

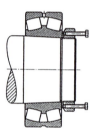 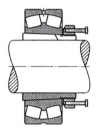 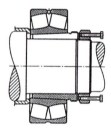

a: Fastening on a tapered shaft   b: Adapter sleeve fastening   c: Withdrawal sleeve fastening

Mounting with thrust bolts and hydraulic method

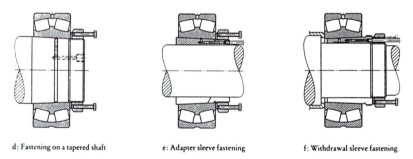

d: Fastening on a tapered shaft   e: Adapter sleeve fastening   f: Withdrawal sleeve fastening

Figure 101   (a–c) Mounting bearings by using thrust bolts. (d–f) Mounting bearings by using thrust bolts and hydraulic method.

has a tight seat on the shaft, the nut and the mounting plate are removed, and the bearing or the withdrawal sleeve is secured by means of the actual locknut.

If possible, the hydraulic method should be used. If a bearing is to be mounted on a withdrawal sleeve by means of the hydraulic method, the locknut must also have, besides the tapped holes for the pressure bolts, one or two through-holes for the oil supply (Fig. f). If the shaft end has no thread for a locknut, the mounting force can be applied by means of the bolts of an axle cap. If the bearing is mounted by means of the hydraulic method, the axle cup must have holes for the pressurized oil line. The axle caps are made by the users themselves.

5. *Mounting Mandrel for Drawn Cup Needle Roller Bearings (see Fig. 102)*

In order to prevent the thin-walled drawn cup needle roller bearings from being deformed during mounting and dismounting, the mounting force has to be applied as near to the outside diameter as possible. The recommended mounting mandrels center at the needle roller and cage assembly; their contact surface takes into account the thin cup wall. Usually the tool is applied at the stamped face of the bearing; in smaller designs it is hardened. But the occurrence of deformation or clamping of the needle roller and cage assembly when pressing it in at an unhardened tip can be prevented if the mounting mandrel is dimensioned for each size.

6. *Extractor or Bearing Puller*

Mechanical extractors are supplied by numerous manufacturers. These devices have two or more hook-shaped arms which are applied at the bearing. The

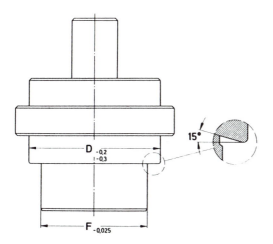

Figure 102 Mounting mandrel for drawn cup needle roller bearings.

extraction force is applied by means of a thread spindle or a hydraulic spindle. The extractor always has to be applied to the tight-fitting ring. If extraction over the rolling elements cannot be avoided, a ring of unhardened steel has to be placed around the outer ring (thicker than one-fourth of the height of the bearing cross section). This holds particularly for rolling bearings with a small cross section height and small contact angle (e.g., tapered roller bearings and spherical roller bearings).

D. Hydraulic Methods

1. Hydraulic Nuts (see Fig. 103)

Hydraulic nuts are screwed on like nuts and are hydraulically driven. Their use is recommended for medium-size and larger bearings with a tapered bore, and particularly for production mounting. Hydraulic nuts can be used if the bearing is directly seated on the shaft taper (Fig. a) and if adapter sleeves (Fig. b) or withdrawal sleeves (Fig. c) are used. If the shaft has no thread, the nut is sup-

Mounting with a hydraulic nut

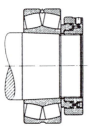
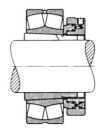
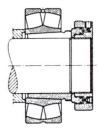

a: Fastening on a tapered shaft   b: Adapter sleeve fastening   c: Withdrawal sleeve fastening

Mounting with a hydraulic nut and hydraulic method

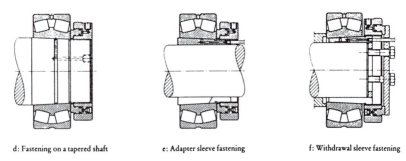

d: Fastening on a tapered shaft   e: Adapter sleeve fastening   f: Withdrawal sleeve fastening

Figure 103   (a–c) Mounting bearings using a hydraulic nut. (d–f) Mounting bearings using a hydraulic nut and hydraulic methods.

ported by an axle cap or a support plate (Fig. f). If the nut is to remain on the shaft after mounting, it has to be locked. Particularly high forces, such as are required for mounting ship's propellers, rudders, couplings, gearwheels, rims, etc., are generated by means of the reinforced type of the hydraulic nut.

2. Hydraulic Shaft Modifications

The hydraulic method is particularly suitable for mounting and dismounting bearings with a tapered bore, which are seated immediately on the shaft, or for bearings with bore diameters from approximately 160 mm which are fastened by means of adapter or withdrawal sleeves. If bearings with a cylindrical bore are used, the hydraulic method facilitates dismounting; for mounting, the bearings are usually heated. With only a few exceptions, the hydraulic method requires oil grooves and holes in the shaft or in the adapter sleeves and withdrawal sleeves, and connecting threads for the pressure generators (see Fig. 104).

The selection of the pressure generator is determined by the required oil quantity and the contact pressure in the fitting joint. The oil pressure has to be 1.5–2 times as high as the contact pressure for the hydraulic method. The oil groove NB is adapted to the shaft diameter. All the other dimensions, such as the groove depth NT, the diameter of the oil duct LD and the groove radius $K/2$ are determined by the groove width.

E. Thermal Methods

Bearings with a cylindrical bore for which tight fits on a shaft are specified and which cannot be pressed mechanically onto the shaft without great effort are heated before mounting. Tight bearing seats are obtained at a shaft tolerance of k6(k5)–p6 (p. 5). Figure 105 shows the heat-up temperature T_M required for trouble-free mounting versus the bearing bore diameter d. The lower limit of the range applies to the shaft tolerance k, the upper limit to tolerance p.

The recommendations for the heat-up temperature T_M (°C) take into account the maximum interference, a room temperature of 20°C, and to make sure, an overtemperature of 30 K. Table 31 gives the advantages and disadvantages of this method.

1. Oil Bath

Bearings of all sizes and types are heated in an oil bath (except sealed and greased bearings, spindle bearings, and other precision bearings). The heat-up temperature, which has to be controlled, is 80–100°C. Thermocouple control is required. In order to prevent a one-sided heating of the bearings, they must not touch the bottom of the oil bath container. Therefore they are placed on a screen or suspended in the bath. This measure also protects the bearing from contaminants settling on the tank bottom. Drawbacks are pollution of the environment by oil vapors and the inflammability of the hot oil. Prior to

Profile of the oil ducts

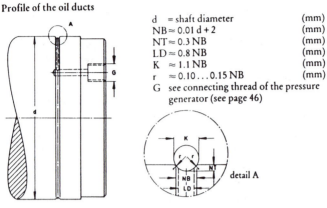

d = shaft diameter (mm)
NB ≈ 0.01 d + 2 (mm)
NT ≈ 0.3 NB (mm)
LD ≈ 0.8 NB (mm)
K ≈ 1.1 NB (mm)
r ≈ 0.10 ... 0.15 NB (mm)
G see connecting thread of the pressure
  generator (see page 46)

detail A

Oil groove for bearings with a tapered bore (a)
Oil groove for narrow bearings (B ≦ 80 mm) with a cylindrical bore (b)

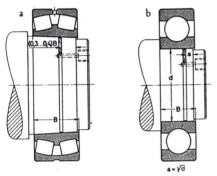

a ≈ √d

Oil grooves for wide bearings (B > 80 mm) with a cylindrical bore (a)
Oil grooves for several bearings with a cylindrical bore arranged in a row (b)

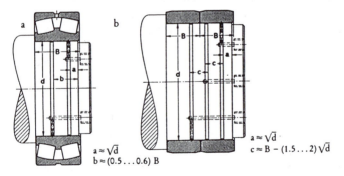

a ≈ √d
b ≈ (0.5 ... 0.6) B

a ≈ √d
c ≈ B − (1.5 ... 2) √d

Figure 104   Oil groove information for hydraulic method.

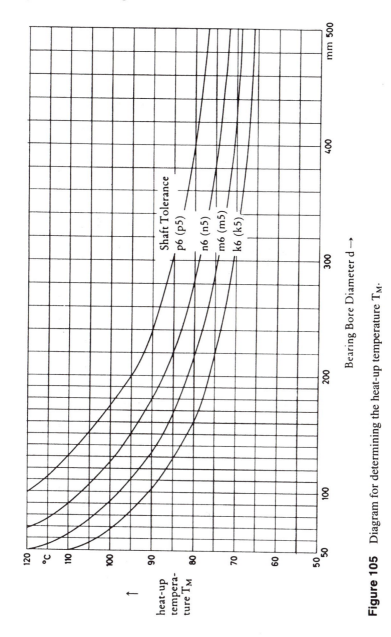

Figure 105 Diagram for determining the heat-up temperature T_M.

Table 31 Advantages and Drawbacks of the Thermal Methods

Heating device	Advantages	Drawbacks
Oil bath	Suitable for all sizes and bearing types (not for sealed or shielded bearings), temperature control	Air pollution by vapors, contamination of the bearings, not suitable for spindle bearings and other precision bearings, not for greased and sealed bearings
Heating plate	Clean, temperature control	Turn bearings, cover large bearings, use seating rings for protruding cages, be careful with bearings with polyamide cages
Hot air cabinet	Clean, temperature control	Long heating period, only to be recommended for small and medium-sized bearings
Induction heating device	Suitable for all bearing types, fast, clean, temperature control, also for sealed or shielded bearings (80°C), for series mounting	Limited by bearing size and weight
Heating ring	Uniform heating, suitable for dismounting medium-size cylindrical roller bearing and needle roller bearing inner rings	Heating ring must be adapted to the width and diameter of the ring
Induction coil	Fast, uniform heating, for inner rings of cylindrical roller bearings and needle roller bearings, suitable for mounting and dismounting	Purchase expensive, profitable for series mounting only, device must be adapted to the width and diameter of the ring

mounting, the fitting surfaces of the bearings must be wiped. Mineral oils with a viscosity of 60–75 mm^2/s at 40°C are recommended for the oil bath; flash point over 250°C. Oil baths can be purchased for small and medium-size bearings. Large bearings require special designs of the oil bath tanks.

2. Heating Plate

Individual bearings can be heated provisionally on an electric heating plate which should be thermostatically controlled. It is important to turn the bearings over several times in order to ensure a uniform heating. It is recommended to check the temperature of the bearings (80–100°C). In case the temperature of the heating plate exceeds + 120°C in an uncontrollable way during heating an E-type spherical roller bearing, the cage must not touch the plate. This can be prevented by placing a ring between heating plate and inner ring. Thermostatically controlled heating plates are available commercially.

3. Hot-Air Cabinet

A safe and clean method of heating rolling bearings is to use a thermostatically controlled hot-air cabinet; it is particularly advantageous to use a circulating air cabinet. The method is clean and is used for small and medium-size bearings. The heat-up times are, however, relatively long. Hot-air cabinets are available commercially.

4. Induction Heating Devices

Induction heating methods are fast and clean. Therefore they are particularly suitable for production mounting. They are used for heating complete bearings, rings of cylindrical roller bearings or needle roller bearings, and other rotational symmetric steel parts such as labyrinth rings, roll couplings, rims, etc. The smaller commercially available induction heating device for rolling bearings is suitable for bearings with a minimum bore diameter of 20 mm. The outside diameter can be, depending on the section height, up to 260 mm. The bearings can be sealed and greased. Other rotational symmetric steel components can also be heated.

In principle, the heating device consists of a live coil with an iron core (primarily coil) which generates a high induction current at a low voltage in a short-circuited secondary circuit (rolling bearings or other steel parts). The part to be mounted is heated quickly. The heating device has a sturdy, scratch-resistant polyurethane housing which is stable at temperatures of up to 140°C. The yoke with the part to be heated is placed on the two lateral supports.

The device operates with temperature control or time control. The heat-up temperature of temperature-controlled heating devices can be continuously adjusted up to 120°C. The temperature of the part to be heated is controlled by a sensor. The heat-up temperature of time-controlled heating devices can be continuously adjusted up to 300 s. Time control is recommended particularly if whole series of identical bearings or workpieces have to be heated. In such cases it is advisable to heat a part in the temperature control mode first. The time measured is then specified as a standard for the time control mode.

The induction heated parts become magnetic. After heating, or if the heating process is interrupted, the parts are automatically demagnetized. The retained magnetism of less than 2 A/cm is far below the permissible value.

5. Induction Coils

These are used primarily for heating inner rings of cylindrical roller bearings and needle roller bearings without lips (NU design) or with one lip only (designs NJ, NUP). For labyrinth rings, shrunk-on roll couplings, etc., special devices are used. Induction coils are suitable for heating the rings while they are shrunk on and extracted, and for demagnetization. The induction coils can be operated in two modes.

Mode of operation "main voltage".   For heating rings with bore diameters from 100 mm; ring weight may not exceed 250 kg.

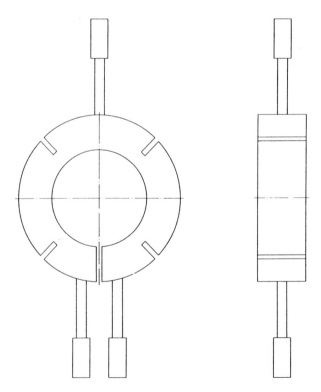

Figure 106   A heating ring.

Mode of operation "low voltage" (with an inserted transformer). For heating rings with bore diameters from 100 mm; ring weight may not exceed 1200 kg (600 kg for roll couplings).

Since induction heating devices are adapted to the prevailing application, precise operation data are needed because these units are usually custom designed.

VIII. DISMOUNTING—HEATING RINGS

Heating rings made of a special aluminum alloy are used for heating inner rings of cylindrical roller bearings or needle roller bearings so they can be extracted (see Fig. 106). The heating rings are heated to 200–300°C on an electric heating plate and slipped over the inner ring to be extracted. After the bearing ring is extracted it must be removed from the heating ring at once in order to prevent it from overheating. Heating rings are particularly disadvantageous for the occasional extraction of small or medium-size bearing rings. Each bearing size requires a hearing ring of its own. Therefore the following information is required:

1. Bearing number or dimension
2. Drawing of the bearing location with fit data

6

SEALS

Leslie A. Horve

CR Industries, Elgin, Illinois

I. INTRODUCTION

Modern society is highly dependent upon machinery to perform numerous essential tasks that range from transportation to household chores. These machines require seals to retain lubricants, liquids, solids, and gases. Special seal designs contain pressure and vacuum. Sealing devices also prevent foreign particles from entering the sealed cavity which can damage machinery components and lead to premature failure. The application conditions can vary considerably and many seal designs have evolved to satisfy these conditions (Table 1). Some applications will tolerate a small amount of leakage, others cannot allow any leakage. In general, seal complexity and cost will increase as the need for zero leakage increases.

Seals are typically classified as static or dynamic. Static seals are used where there is little or no relative motion between mating surfaces. Typical static seals include deformable gaskets, O rings and liquid sealants. Dynamic seals are used to seal shafts that rotate, oscillate, and reciprocate. A variety of seal types exist to satisfy the need of various dynamic applications (Table 2). This chapter is designed to assist engineers and maintenance personnel make the proper seal selection to satisfy their application needs.

Table 1 Application Variables to Consider for Seal Selection

Static or dynamic	System pressure, vacuum, and range
Sealed media properties	System temperature and range
Type of shaft speed and motion	Cyclic operation
Shaft eccentricity	Expected life
Shaft diameter	Ambient temperature and range
Shaft finish	Dust/mud
Shaft material	Ozone
Housing O.D./I.D.	Toxic materials
Housing material	Vibration
Housing eccentricity	Expected leakage rate
Housing finish	Economics
Component tolerance stackup	

Source: Chicago Rawhide.

Table 2 Seal Types

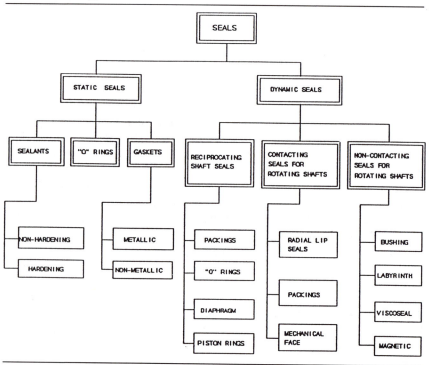

Source: Chicago Rawhide.

II. HISTORICAL OVERVIEW

Seal development has kept pace with the developments in the transportation, lubricant, appliance, and machinery industries. It is interesting to note that many early seal designs are used today in certain applications because they still provide adequate performance at a reasonable cost.

The earliest known seals were crude leather strips wrapped around the end of wagon wheel axles to hold grease made from animal fat in place. These seals and lubricants were not adequate when the industrial revolution began. Rotating and reciprocating shafts for machines of the eighteenth century were sealed with stuffing boxes using compression packings made with organic fibers such as flax, hemp, cotton, and wool (Fig. 1a). Animal fats were mixed with the fibers to act as fillers, binders and lubricants. These organic materials were not adequate when new machines evolved with higher speeds, loads, and temperatures. Asbestos packings were developed in the early 1900s that increased the chemical resistance and extended the temperature range beyond that obtainable with organic packings. Synthetic materials are now available to replace asbestos which has been recognized as a possible health hazard. Animal fats have been replaced with petroleum and synthetic lubricants, mica, graphite, polytetrafluoroethylene (PTFE), and molybdenum disulfide materials. New fabrication techniques have resulted in compression packings with extended capabilities made from materials such as metal foil, braided fibers, and braided PTFE filled cloth. Compression packings are still used in many modern applications even though the design concept dates back several hundred years.

Packings were not able to meet the high-speed requirements of the developing automotive industry. Radial lip seals made of leather clinched between metal cases were developed in the 1920s (Fig. 1b). These seals could follow the eccentricities of the rotating shaft without allowing leakage of the heavy lubricants used during this era.

During the 1920s the refrigeration industry began a rapid expansion using mechanical methods to cool rather than ice. Packings and lip seals allowed too much leakage of the potentially toxic and foul-smelling refrigerants (ammonia, sulfur dioxide, methyl chloride) from the compressor. The development of the mechanical face seal minimized leakage which helped sustain the rapid growth of the refrigeration industry (Fig. 1c).

Antifriction ball bearings were also developed during this period and rapid improvements were made in the design of machines with rotating members. Shaft speeds, load-bearing capabilities, and operating temperatures all increased. Lubricants become lighter, more sophisticated and, in many cases, more detrimental to existing seal materials. Synthetic oil-resistant elastomers were developed and used to replace leather and other organic materials used in packings and lip seals. The oil-resistant elastomers could be molded in a

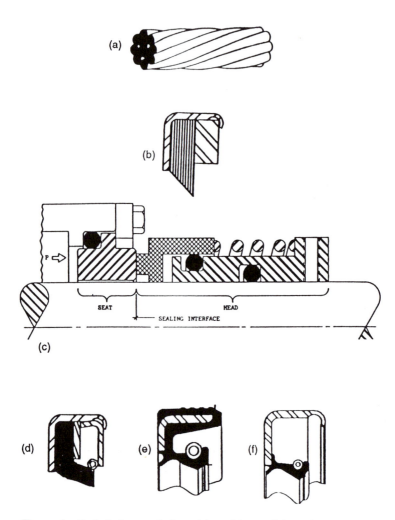

Figure 1   Seal design evolution: (a) packings, eighteenth century to present; (b) leather radial lip seals, 1920s; (c) mechanical face seal, 1920s to present; (d) assembled leather/synthetic elastomeric seals, 1930s to 1960s; (e) bonded synthetic elastomer seals, 1950s to present; (f) lightly loaded elastomeric seals with reduced bond area, 1980s to present. (Courtesy of Chicago Rawhide.)

variety of shapes and new designs such as O rings, V rings, U rings, and chevrons could now be made with lip configurations not previously obtainable. Many of these seal shapes helped extend the operating pressure range of hydraulic systems while reducing leakage. Elastomers replaced leather in radial lip seals used in automotive applications (Fig. 1d). Leakage was reduced even though automobile road speeds increased.

The 1950s saw the beginning and development of the jet and space age. New labyrinth and bushing seal designs and new exotic materials were required to seal jet engine applications. The assembled radial lip seal design used in a variety of applications was replaced by simpler and cheaper designs with the elastomer bonded directly to a single metal case instead of clinched between two metal components (Fig. 1e). The potential leakage path between the two components was eliminated.

The zero leakage requirements of the 1980s and 1990s resulted in elastomeric materials with improved oil resistance and broader temperature ranges. Seal designs have become more sophisticated and new concepts such as magnetic sealing systems have evolved. Bond areas to attach elastomers to metal were reduced as better cement systems evolved (Fig. 1f). This evolution must continue to develop new economic sealing solutions to protect the environment from the leakage of harmful chemicals and lubricants.

III. ELASTOMERIC MATERIALS FOR SEALS

A. Chemical Resistance of Polymer Types

Elastomeric materials are used in many of the seal types shown in Table 2. They are often used to make molded gaskets or as filler materials for composite gaskets. Elastomers are used to make lip seals, O rings, molded packings, and many other seal types. Base polymers are mixed with other ingredients to make proprietary elastomeric compounds that will meet the application requirements. The polymer selection process must be done with care and knowledge of the application conditions. High-temperature, oil-resistant polymers are selected for high-speed automotive engine requirements. Polymers and compounds that spring back near the original product dimensions after being compressed are selected for O ring and other static applications. This ability to spring back is known as resistance to compression set. Polymers with low gas permeability rates are selected for diaphragms or other applications that seal gases of various types. The polymer must also be compatible with the gas and fluid being sealed. It is very difficult to predict whether a polymer will be compatible with a given oil because chemicals or additives are added to improve the properties of the base petroleum oil. Some oils are synthetic and have no petroleum products. The chemical resistance of typical seal polymers to various fluids appears in Table 3.

Table 3 Characteristics of Sealing Polymers

	Polymer type														
	Nitrile (Buna-N)	Ethylene-propylene	(Chloroprene) neoprene	Fluorocarbon elastomer	Silicone	Fluoro silicone	Styrene butadiene (SBR)	Poly-acrylate	Poly-urethane	Butyl	Polysulfide	Chloro-sulfonate polyethylene (Hypalon)	Epi-chlorohydrin (Hydrin)	Phosphonitrile fluoroelastomer (PNF)	Polytetra fluoroethylene (PTFE)
Hardness range, A scale	40–90	50–90	40–80	70–90	40–80	60–80	40–80	70–90	60–90	50–70	50–80	50–90	50–90	50–90	
Relative polymer cost	Low	Low	Low/moderate	Moderate/high	Moderate	High	Low	Moderate	Moderate	Moderate	Moderate	Moderate	Moderate	High	High
Temperature range (°C) for static applications	−55 to 156	−55 to 160	−55 to 140	−40 to 225	−75 to 260	−65 to 175	55 to 100	−55 to 100	−55 to 100	−55 to 100	−55 to 100	−55 to 125	−55 to 125	−65 to 175	−95 to 200
Temperature range (°C) for dynamic applications	−40 to 110	−40 to 135	−40 to 125	−40 to 150	−75 to 150	−65 to 150	−55 to 100	−40 to 135	−40 to 100	−40 to 100	−40 to 100	−40 to 100	−40 to 100	−40 to 150	−40 to 200
Compression set resistance	Very good	Very good	Good	Very good	Excellent	Very good	Good	Fair	Fair	Fair/good	Fair	Fair/poor	Fair/good	Good	Good
Aging (oxygen, ozone)	Fair/poor	Very good	Good	Very good	Excellent	Excellent	Poor	Excellent	Excellent	Very good	Excellent	Very good	Very good	Excellent	Excellent
Impermeability to gases	Good	Good	Good	Very good	Poor	Poor	Fair/good	Very good	Fair	Excellent	Very good	Very good	Excellent	Fair	Excellent

Fluid														
Acid, inorganic	Fair	Good	Fair/good	Excellent	Good	Good	Fair/good	Poor	Poor	Poor	Excellent	Fair	Poor	Excellent
Acid, organic	Good	Very good	Good	Good	Excellent	Good	Good	Poor	Poor	Good	Good	Fair	Fair	Excellent
Alcohols	Very good	Excellent	Very good	Fair	Very good	Very good	Very good	Poor	Poor	Very good	Very good	Good	Fair	Excellent
Aldehydes	Fair/poor	Very good	Fair/poor	Poor	Good	Poor	Fair/poor	Poor	Poor	Fair/good	Fair/good	Fair/good	Poor	Excellent
Alkalis	Fair/good	Excellent	Good	Good	Good	Good	Fair/good	Poor	Poor	Poor	Excellent	Excellent	Good	Excellent
Animal oils	Excellent	Good	Good	Very good	Good	Excellent	Poor	Good	Good	Poor	Good	Good	Fair	Excellent
Esters, alkyl phosphate (skydrol)	Poor	Excellent	Poor	Poor	Fair/poor	Fair/poor	Poor	Excellent	Very good	Poor	Poor	Poor	Poor	Excellent
Esters, aryl phosphate	Fair/poor	Excellent	Fair/poor	Excellent	Good	Poor	Poor	Fair/poor	Excellent	Good	Fair	Poor	Excellent	Excellent
Esters, silicate	Good	Poor	Fair	Excellent	Poor	Very good	Poor	Fair/poor	Poor	Fair/poor	Good	Good	Fair	Excellent
Esters	Poor	Fair	Poor	Poor	Fair	Fair	Very good	Very good	Fair/poor	Good	Poor	Good	Poor	Excellent
Hydrocarbon fuels, aromatic	Excellent	Poor	Excellent	Excellent	Fair	Excellent	Poor	Very good	Poor	Excellent	Fair	Very good	Excellent	Excellent
Hydrocarbon fuels, aromatic	Good	Fair/poor	Poor	Excellent	Poor	Very good	Poor	Fair/good	Poor	Good	Fair/poor	Very good	Excellent	Excellent
Hydrocarbons, halogenated	Fair/poor	Poor	Excellent	Excellent	Poor	Very good	Fair/good	Excellent	Excellent	Good	Fair	Excellent	Fair	Excellent
Hydrocarbon oils, high aniline		Poor	Good	Excellent	Very good	Excellent	Excellent	Excellent	Excellent	Very good	Excellent	Excellent	Excellent	Excellent
Hydrocarbon oils, low aniline	Very good	Poor	Fair/poor	Excellent	Fair	Very good	Excellent	Very good	Excellent	Good	Very good	Excellent	Excellent	Excellent
Ketones	Poor	Poor	Poor	Poor	Poor	Fair/poor	Poor	Poor	Poor	Good	Fair	Fair	Poor	Excellent
Silicone oils	Excellent	Excellent	Excellent	Excellent	Good	Excellent	Excellent	Excellent	Excellent	Excellent	Excellent	Excellent	Excellent	Excellent
Vegetable oils	Excellent	Good	Excellent	Excellent	Excellent	Excellent	Good	Fair	Fair	Poor	Good	Excellent	Fair	Excellent
Water/steam	Good	Excellent	Fair	Fair	Fair	Fair	Poor	Poor	Excellent	Fair	Fair	Good	Fair	Excellent

B. Standard Classification and Qualification Tests for Elastomers

Elastomers must be resistant to the fluids and temperatures of the application to provide good sealing during the required life. Samples of compounded materials are typically tested by immersing the material in a specified fluid or air at the required temperature for a predetermined period of time. The physical properties of the tested compounds are measured and compared to the values obtained before test. The standard tests used are specified by the American Society for Testing and Materials (ASTM) and appear in Table 4.

Since application conditions can vary widely, not all of the tests shown in Table 4 may be required. A line call-out system has been developed to describe the test requirements and specifications that an elastomer must meet for a given application. This system is described in SAE J200 and ASTM D2000. The line call out begins with the letter M if the classification system is based on SI units. If English (inch/pound) units are used, then no letter will appear. The next position of the line callout indicates the grade of the material. A 1 indicates that basic minimum rest requirements are all that are required. If 2 through 8 appears, then additional requirements beyond the minimum will be specified.

The basic minimum test requirements are defined by the next two letters. The first letter ranges from A through J and specifies the test temperature in

Table 4 ASTM Test Methods for Vulcanized Elastomers

D395	Compression set method B
D412	Elongation
D412	Tensile modulus @ 100%, 200%, 300%
D412	Tensile strength die C (common)
D429	Adhesion bond strength method A
D430	Flex resistance
D471	Fluid resistance aqueous, fuels, oils, lubricants
D573	Heat age or heat resistance
D575	Compression deflection, method A
D624	Tear strength, die B or C
D813	Crack growth
D865	Deterioration by heating
D945/D2632	Resilience
D1053	Low-temperature torsional
D1171	Ozone or weather resistance
D1329	Low-temperature retraction
D1418	Elastomer classification
D2000/SAE J200	Common elastomer specification for auto and truck components
D2137	Low-temperature resistance A
D2240	Hardness durometer A

degrees Celsius. The material will pass the heat requirements if the change in tensile strength is within ±30% of the initial values after heat aging at the test temperature for 70 h. Other requirements are elongation decrease must be less than 50%, and the material hardness must be between ±15 points of the original values. The second letter (A to K) determines the resistance of the elastomer to swell after soaking in a standard test fluid (ASTM oil #3) for 70 h at the test temperature specified by the first letter (Table 5).

These letters are followed by a three-digit number to specify the initial hardness and tensile strength for the material. The first number indicates the material durometer hardness (6 for 60 ± 5, 7 for 70 ± 5, etc.) and the next two numbers indicate the minimum tensile strength (10 for 10 MPa, 20 for 20 MPa, etc.).

An example of the simplest line call out is: M1 CG 710. The M means the test results are reported in SI units. The 1 means only the basic test requirements need to be performed, the letter C indicates the test temperature is 125°C and the letter G means the maximum volume swell of the material in ASTM #3 oil after a 70-h soak at 125°C is 40%. The letter C also means the elastomer will be heat aged in air at 125°C for 70 h. After heat aging, the elongation loss will be less than 50%, the hardness change will lie within ±15 points from the original values and the tensile strength change will be within ±30%. The 7 indicates the material initial durometer specification is 70 ± 5, and the 10 means the tensile strength must be greater than 10 MPa.

Table 5 Basic Requirements for Heat Resistance and Volume Swell in ASTM Oil #3 for Vulcanized Elastomers

First letter of callout (establishes test temperature)		Second letter of callout (establishes max volume swell in ASTM #3 at test temperature)	
Letter	Test temperature (°C)	Letter	Maximum volume swell (%)
A	70	A	No requirement
B	100	B	140
C	125	C	120
D	150	D	100
E	175	E	80
F	200	F	60
G	225	G	40
H	250	H	30
J	275	K	10

Source: American Society for Testing and Materials (ASTM D2000).

If a 2 to 8 appeared on the first digit in the line callout, then additional tests are specified. The additional tests are defined by a suffix letter (S) which is followed by two numbers. The first number defines the ASTM test method (Table 6), and the second number defines the test temperature (Table 7). As an example, consider the line callout:

M 2 CG 710 A26 C12 E016 E036 F27

In addition to the basic requirements already discussed for the CG 710 callout, five additional test requirements have been added. The tests and specifications can be determined from the codes with the aid of Tables 6 and 7.

A26 is a heat-resistance test using method ASTM D865 for 70 h at a temperature of 150°C.

C12 is an ozone resistance test using ASTM test method D1171 (method A) at 38°C.

E016 is an oil-resistance test using ASTM test method D471 in ASTM oil #1 at 150°C.

E036 is an oil-resistance test using ASTM test method D471 in ASTM oil #3 at 150°C.

F27 is a low-temperature test using ASTM D1053 test method at −40°C.

IV. STATIC SEALS: GASKETS

A. Gasket Applications

A gasket is a material or combination of materials that is clamped between mating surfaces of mechanical assemblies to develop a barrier that will seal internal pressure, liquids, gases, and prevent external contaminants from entering the assembly. Gaskets can be made from a variety of elastomers (Table 3), metals, or formable sealants. Material selection and gasket design depend on the conditions of the application and the list of applications is almost endless. Application conditions must be considered when selecting gasket materials, designing gasket shape, and designing the joint configuration. Application conditions for the various systems for a typical automotive gasoline engine are shown in Table 8.

1. Media

Substances include oils of various types, gases, water, and a host of chemicals. Typically, the media will contact the gasket only at the edge. Reactions between the gasket material and the media can still damage the gasket and cause failure. The media and the gasket material must be compatible.

2. Operating Temperature Range

Gasket materials must be selected that will perform throughout the expected temperature range of the application. Temperature ranges for vulcanized elas-

Table 6 Suffix Letters and Test Methods for Vulcanized Elastomers

Suffix Letter	Test Description	First Suffix Number (ASTM Test Method)							
		1	2	3	4	5	6	7	8
A	Heat Resistance	D573 70 hr	D865 70 hr	D865 168 hr	D573 168 hr	-	-	-	-
B	Compression Set	D395 22 hr solid	D395 70 hr solid	D395 22 hr piled	D395 20 hr piled	-	-	-	-
C	Ozone/weather resistance	D1171 ozone "A"	D1171 weather	D1171 ozone "B"	-	-	-	-	-
D	Compression-deflection resistance	D575 "A"	D575 "B"	-	-	-	-	-	-
EO	Fluid resistance coils & lubricants	D471 oil #1 70 hr	D471 oil #2 70 hr	D471 oil #3 70 hr	D471 oil #1 168 hr	D471 oil #2 168 hr	D471 oil #3 168 hr	D471 fluid 101 70 hr	D471 designated fluid 70 hr
EF	Fluid resistance (fuels)	D471 Fuel A 70 hr	D471 Fuel B 70 hr	D471 Fuel C 70 hr	D471 Fuel D 70 hr	D471 Gasohol 70 hr	-	-	-
EA	Fluid resistance (aqueous)	D471 distilled water 70 hr	D471 water/glycol 70 hr	-	-	-	-	-	-
F	Low temperature resistance	D2137 "A"-9.3.2 3 min	D1053	D2137 A-9.3.2 22 hr	D1329	D1329	-	-	-
G	Tear resistance	D624 Die B	D624 Die C	-	-	-	-	-	-
H	Flex resistance	D430 A	D430 B	D430 C	-	-	-	-	-
J	Abrasion resistance		To be specified						
K	Adhesion	D429 A	D429 B	-	-	-	-	-	-
M	Flammability resistance			To be specified					
N	Impact resistance			To be specified					
P	Staining resistance	D429 A	D429 B	-	-	-	-	-	-
R	Resilience	D945	-	-	-	-	-	-	-
Z	Special; specified by user/supplier			To be specified					

Source: American Society for Testing and Materials (ASTM D2000).

Table 7 Numbers to Indicate Test Temperatures for Vulcanized Elastomers

Applicable suffix letters	Second suffix number	Test temperature (°C)
A,B,C,EA,EF,EO,G,K	11	275
	10	250
	9	225
	8	200
	7	175
	6	150
	5	125
	4	100
	3	70
	2	38
	1	23
	0	Ambient
F	1	23
	2	0
	3	-10
	4	-18
	5	-25
	6	-35
	7	-40
	8	-50
	9	-55
	10	-65
	11	-75
	12	-80

Source: American Society for Testing and Materials (ASTM D2000).

tomers appear in Table 3. Table 9 provides guidance for other nonmetallic materials and Table 10 gives upper temperature limits for metallic gasket materials.

3. Operating Pressure Range

The clamp load or flange pressure exerted on the gasket when the bolts holding the unit together are tightened develops a compressive stress on the gasket. This stress will normally decrease with time as the gasket material relaxes. The compressive load on the gasket must be greater than the maximum internal system pressure to insure a proper seal. The initial compressive stress on the gasket should be about double that of the maximum internal system pressure to allow for material relaxation. Materials with low relaxation are preferred since a lower initial flange pressure can be employed to provide an adequate margin

Table 8 Environment Conditions for some Gasket Systems for Automotive Gasoline Engine

System	Gasket Applications	Media	Expected Environment		Motion
			Temperature Range (°C)	Pressure kP$_A$	
Engine coolant	Water pump Crankcase	Water/Glycol	-40 to 140	Up to 550	Thermal expansion and contraction
Engine Lubrication / Oil sump	Crankcase Oil pan cavity	Engine oil	-40 to 160	Vacuum	Thermal expansion and contraction
Oil inlet suction	Oil pick up Mounting flange	Engine oil	-40 to 160	Vacuum	Joint motion during vehicle operation
High pressure oil discharge	Oil cooler Oil filter	Engine oil	-40 to 160	550	Thermal expansion and contraction
Engine Intake Air System / Carburetor/ Fuel Injection	Intake Manifold	Air/Fuel	-40 to 55	Vacuum	Thermal expansion and contraction
Turbo charged	Intake manifold	Air/Fuel	-40 to 55	10 - 100	Thermal expansion and contraction
Engine exhaust	Exhaust Manifold	Hot combustion exhaust	Up to 800	25	Thermal expansion and contraction
Cylinder head	Cylinder head gasket	Hot combustions gases	Up to 1000	3500 to 19000	Dynamic deflection
		High and low pressure oil	-40 to 160	0 to 1000	Thermal expansion and contraction
		Coolant	-40 to 140	Up to 550	

of safety. The internal system pressure tends to force the bolted units apart which will reduce the gasket compressive stress. The gasket material must have sufficient flexibility, resiliency, and compressive stress to maintain a seal as the internal pressure fluctuates. Pressure is also exerted on the gasket edges and tends to blow the gasket out when the flanges are glandless. A machined gland (Fig. 2) captures the gasket and locks it in place which prevents blowouts. This flange is more expensive to manufacture than the glandless type.

4. Mating Flange Characteristics

The mating flanges must be flat, parallel, and rigid to prevent distortion when the bolts are tightened. The surface finish of the flange will dictate the gasket thickness, the compressibility of the gasket material, and the type of material required to give a good seal. If the surface is rough, a thick, highly compres-

Table 9 Nonmetallic Gasket Materials

Gasket material	Properties	General usage
Untreated paper	Highly permeable, low cost, noncorrosive	To exclude dust and water splash
Vulcanized elastomer	Incompressible, impermeable, can vary in hardness, can be stretched, can be molded to special shapes	Used when stretching is required for installation. Seal alkalis, hot water and some acids. See Table 3 for temperature limitations
Rubber-asbestos/ asbestos substitute	Tough, durable, dimensionally stable, resistant to steam and hot water	Water and steam fittings to 260°C
Cork compositions	Lightweight, inert, high compression, will not deteriorate, low cost, excellent oil and solvent resistance, high friction, poor resistance to acids and alkalis	Mates irregular surfaces up to 70°C and 200 kPa. Used to join glass and ceramic to metal
Cork-rubber	Can compound material to give various levels of compression, high coefficient of friction	General-purpose gaskets. Temperature and pressure limits depend upon elastomer
PTFE	Chemically inert, wide temperature range, resistant to most solvents, expensive	Used in special applications where heat and fluid resistance is required up to 500°C

Table 10 Temperature Limitations for Metallic Gasket Materials

Material	Maximum temperature (°C)
Lead	100
Common brasses	260
Copper	315
Aluminum	425
Stainless steel	875
Soft iron, low-carbon steel	535
Titanium	535
Nickel	760
Monel	815
Inconel	1000
Hastelloy	1100

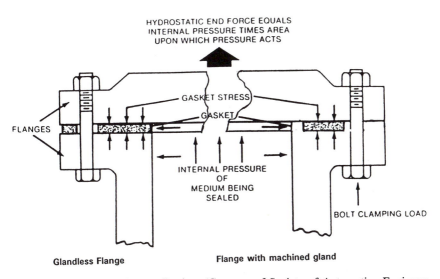

HYDROSTATIC END FORCE EQUALS
INTERNAL PRESSURE TIMES AREA
UPON WHICH PRESSURE ACTS

GASKET STRESS

GASKET

FLANGES

INTERNAL PRESSURE
OF
MEDIUM BEING
SEALED

BOLT CLAMPING LOAD

Glandless Flange   Flange with machined gland

Figure 2 Typical gasket application. (Courtesy of Society of Automotive Engineers, AE-13.)

sive material may be required to flow into flange surface. The material may extrude at high internal pressures. This can be solved by using a finer surface finish on the flange which will allow the selection of a thinner, harder gasket. Turned flanges have peaked grooves that result from the machining operation. This tends to reduce gasket area, increase local surface stresses, and increase gasket compression. The grooves tend to lock the gasket in place and prevent extrusion, so thinner gaskets can be employed. Very smooth surfaces can be undesirable since they will not grip the gasket and extrusion may occur. Minimum gasket thicknesses for various flange finishes and surface stresses at system pressure appear in Table 11.

Flange pressure is developed by tightening the bolts that hold the assembly together. The location, number and the sequence of tightening the bolts is very important in developing a uniform pressure without cocking, crushing, and distorting the gasket or flange. The bolt hole pattern can be checked by drawing a series of straight lines from one bolt to the adjacent bolt and continuing until the circuit is complete. If the sealing area is far outside the line connecting the bolts, then the flange pressure may be too low to provide an adequate seal. The usual solution is to add another bolt in this area or rearrange distribution of bolts to minimize the amount of area outside the projected lines (Fig. 3).

To insure an even distribution of pressure around the gasket circuit, the bolts should be tightened simultaneously. This is usually not practical, therefore, each bolt should be partially tightened using a repetitive sequence until all

Table 11   Minimum Gasket Thickness for Different Surface Finishes

Surface stress (kPa) at working pressure	Minimum gasket thickness (mm)				
	160 microns		40 microns		16 microns
	Ground	Lathe turned	Ground	Lathe turned	Ground
10,000	5.0	2.0	1.5	1.0	0.5
20,000	4.0	2.0	1.0	0.75	0.5
50,000	3.0	1.5	0.75	0.75	0.3
70,000	—	1.5	0.75	0.50	0.3
100,000	—	1.0	0.50	0.50	0.3

bolts reach the proper predetermined torque level. If the bolts are arranged in a circle, then the proper pattern is a crisscross pattern. Noncircular bolt patterns are usually tightened using a spiral pattern that begins in the middle (Fig. 4).

Approximately half of the initial tightening torque on a bolt is used to overcome collar friction, and another 30–40% is lost to friction in the threads. The remaining 10–20% creates the load on the gasket. Some of this load will be

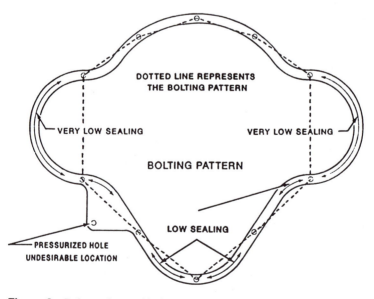

Figure 3   Bolt number and bolt pattern affects sealing. (Courtesy of Society of Automotive Engineers, AE-13.)

lost with time due to creep relaxation of the gasket. Increasing bolt length will reduce this loss.

The amount of torque required to compress the gasket to insure a good seal must be determined and the bolt loading should be calculated and checked to insure bolt failure will not occur. The process begins by determining what flange pressure is required to compress the gasket. This can be estimated from charts for various materials or by experimentation (Fig. 5). Once the flange pressure requirement is estimated, the required torque for each bolt can be calculated with Eq. (1). The bolt clamp load developed during the tightening process appears in Eq. (2). Proof load tables (SAE J49 for inch sizes and SAE J1199 for metric sizes) are then used to determine if the bolts are overloaded (Table 12). The bolt clamp load should be less than 80% of the proof load.

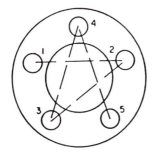

CRISS-CROSS FASTENING SEQUENCE

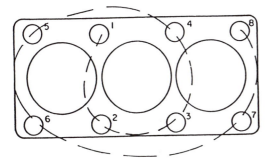

SPIRAL FASTENING SEQUENCE

Figure 4 Bolt fastening sequences. (Courtesy of Society of Automotive Engineers, AE-13.)

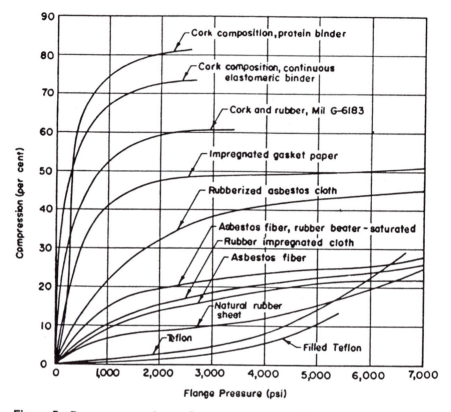

Figure 5   Percent compression vs. flange pressure. (Courtesy of *Machine Design.*)

$$T = \frac{A_g K D P_F}{N_B} \tag{1}$$

$$L_B = \frac{A_g P_F}{N_B} = \frac{T}{KD} \tag{2}$$

where

   A_g = gasket area, m²

   D = nominal bolt diameter, mm

   K = bolt friction factor: clean dry bolts, $K = 0.20$; lubricated steel bolts, $K = 0.11$ to 0.17

   L_B = bolt clamp load, kN

   N_B = number of bolts

   P_F = flange pressure, kPa

   T = initial bolt torque, Nm

Table 12 Portion of a Metric Proof Table for Bolts

Nominal thread dia and thread pitch	Class 4.6		Class 4.8	
	Proof load (kN)	Tensile strength min (kN)	Proof load (kN)	Tensile strength min (kN)
M8 × 1.25	8.24	14.6	11.3	15.4
M10 × 1.5	13.1	23.2	18.0	24.4
M12 × 1.75	19.0	33.7	26.1	35.4

Example. It is required to seal a circular cavity with five bolts and an internal system air pressure of 3000 kPa at 100°C. The flanges are lathe-trimmed flat to a surface finish of 160 microns. The inner diameter of the gasket is 100 mm (0.1 m), and the outer diameter is 150 mm (0.15 m). The bolt is metric (M10 × 1.5) and will be installed dry ($K = 0.20$). Determine material, gasket thickness, flange pressure, bolt class, bolt clamp load, and initial bolt torque for the application.

Cork rubber materials can generate about 6200 kPa of stress with a compression of 53% (Fig. 5). This pressure will provide a margin of safety which is twice that of the system pressure to compensate for creep. Nitrile or NBR can be chosen for the rubber material as it has a low cost, will resist temperatures of 125°C for static applications, and has good resistance to gas permeability (Table 3). The minimum gasket thickness for surface stress up to 10,000 kPa is 2 mm for lathe-turned flanges with a surface finish of 160 microns.

$$T = \frac{\frac{1}{4}\pi[0.15^2 - 0.10^2](0.2)(10)\,6200}{5} = 24.3 \text{ Nm} \tag{3}$$

$$L_B = \frac{24.3}{(0.2)(10)} = 12.14 \text{ kN} \tag{4}$$

From a portion of a typical proof load chart (Table 12), the proof load for a class 4.6 bolt is 13.1 kN and 18.0 kN for a class 4.8 bolt. The class 4.8 bolt must be chosen since the load of 12.14 kN is 93% of the proof load for a class 4.6 bolt.

Load charts (Fig. 6) can be used to analyze how gasket loads change during operation. When the bolts are tightened, the gasket is compressed and the bolts are elongated. The force generated by elongating the bolts is identical to the force required to compress the gasket material (point 1 of Fig. 6). This force must be greater than the minimum load required to ensure sealing. After the initial tightening, the stresses in the gasket may relax, which will allow the gasket to compress some more. The elongation of the bolt will be reduced and a load loss will occur (point 2 of Fig. 6). When the cavity is pressurized, the

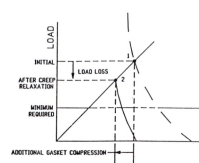

Effects of creep relaxation

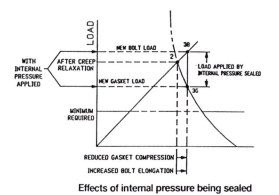

Effects of internal pressure being sealed

Figure 6 Bolt and gasket deflection under load. (Courtesy of Society of Automotive Engineers, AE-13.)

bolts will be elongated and gasket compression will be reduced. The bolt load will be increased (point 3B) and the gasket load will be reduced (point 3G). In the case shown, the gasket load has been reduced to a level very close to the minimum load required to ensure sealing. If the internal pressure of the application can not be reduced, a new gasket material or design may be required.

B. Nonmetallic Gaskets

Nonmetallic gaskets provide good sealing over a broad range of temperatures and pressures. Material choice is the primary factor to the success of the gasket. The temperature range, gas permeability, fluid resistance, and cost must be considered when selecting a material to meet application conditions (Table 3). Material compression is also important, and the amount of compression required depends upon the finish of the mating faces and the amount of internal pressure. Low-compression materials (such as PTFE) are used for high-pressure applications with smooth faces. High-compression materials (cork) are used for low pressures and rough face surface finishes.

The simplest gaskets are made of paperlike sheets which are formed of blends of organic materials and mineral fibers. Untreated paper gaskets are highly permeable and are not suited for containing fluids. They are low-cost methods to prevent dust and water from entering a cavity. The paper gaskets can be treated with synthetic rubber latex or glues to reduce permeability. Typical applications for treated paper gaskets include sealing oils or gasolines at temperatures of 70°C or less at pressures up to 140 kPa.

Granulated cork is combined with binders to produce sheets of material. Gaskets of all shapes and sizes are punched out of these sheets and are used to seal rough irregular surfaces at low temperatures (70°C or less) and pressures (200 kPa or less). Cork-based gaskets have a high compression under load and are used to seal oils, gasolines, and other petroleum-based products. They are attached by acids and bases. Cork is sometimes treated with oils to prevent drying which can cause shrinkage, hardening, load loss, and leakage.

Fluid-resistant elastomerics are used to produce molded gasket shapes which reduce material waste. Rubber is incompressible and will have very low compression under load. The gasket designer must allow room for the gasket to deform when the load is applied. The proper elastomer is selected from Table 3 for the fluid, temperature, and pressure of the application. PTFE and filled PTFE gaskets also exhibit low compression under load and are used in applications where the resistance to harsh fluids over a broad temperature range is required.

Different materials are often combined with one another and with fillers to provide an almost infinite range of gasket properties. Rubber and cork are vulcanized together in various ratios to provide gaskets that are almost as compressible as rubber. Asbestos fillers were used for many years to provide mechanical strength, chemical resistance, and temperature resistance at low cost. There are four types of asbestos (chrysotile, crocidolite, amosite, and anthophyllite), and they differ in fiber structure and physical properties. Health concerns have eliminated the use of all asbestos types except for chrysotile (white asbestos). Other more expensive materials such as graphite and aramid are being used to replace asbestos gaskets.

Gaskets can be treated to enhance properties. Oil-resistant rubber is often dipped or sprayed on cork or other materials to improve chemical resistance. Fungicides can be compounded directly into the gasket material or dusted onto the surfaces to prevent mold growth during storage in humid climates. A reflective coating (aluminum paint or lacquer) is often placed on the exterior of the gasket to reflect heat. Graphite powder is dusted on the exterior of the gasket to prevent adhesive of the gasket to metal surfaces. Glues are also placed on the gasket surfaces if the user wants to bond the gasket directly to the flanges.

The properties and general usage of gaskets made from various nonmetallic materials appear in Table 9.

C. Metallic and Composite Gaskets

Composites or combinations of materials are often used to combine the properties of a rigid, incompressible substrate such as steel with the properties of a compressible, soft material such as cork. The steel provides rigidity and strength, while the cork provides the compression that creates the seal. Gaskets made entirely from metal are used to seal high-pressure gases at high temperatures. They are required when the product of pressure and temperature exceeds 1,000,000 kPa · °C. A typical application would be an automotive head gasket.

Soft metals such as lead, brass, and copper are used when the mating surfaces of the gasket flange are rough. Other metals, such as stainless steel and aluminum, are used to minimize corrosion. The upper temperature limits for metallic gasket materials appear in Table 10.

The designs of all metal and composite gaskets vary widely. The simplest gasket is a flat, circular, all-metallic washer that is sometimes serrated on the faces to provide points of high load. The tips will deform and flow into the flange surface irregularities (Fig. 7a). The surface roughness of the flange should be 2 μm RA for the best results. Flat metal gaskets are normally limited to narrow faces and do not result in high unit loads. Metal gaskets with specialized cross sections will concentrate the load over a small area and thus allow higher internal pressures. Round, oval, octagonal, and lens ring cross sections (Fig. 7b–e) are used in high-pressure piping systems (7000 to 70,000 kPa) with moderate bolt loads. Metallic rings of various cross sections are designed to expand when the internal pressure increases. The U ring (Fig. 7f) is typical of these designs and is often coated with a soft elastomeric material that actually does the sealing. Hollow metallic O rings (Fig. 7g) made from tubing are also compressed in a flange to form a seal. Vent holes are often provided at the ID to allow the O ring to be pressure actuated.

Many composite designs of metal, rubber, and asbestos rope or fibers are used as gaskets. Flat sheets of thin metal (0.25 to 0.30 mm) are corrugated to give peaks that develop high local stresses when compressed. The corrugations are often filled with a soft rubber compound (Fig. 8a). Some gaskets have soft cores made of nonmetallic materials that are completely encased in thin metal sheets. The metal cladding provides protection for the soft cores and creates a rigid frame to provide mechanical strength (Fig. 8b). Other gaskets have a soft exterior with a rigid metallic core (Fig. 8c). Spiral wound gaskets are versatile since they combine strength with resilience. A V-shaped metal strip is combined with a nonmetallic filler and is wound into a spiral. The first few inner

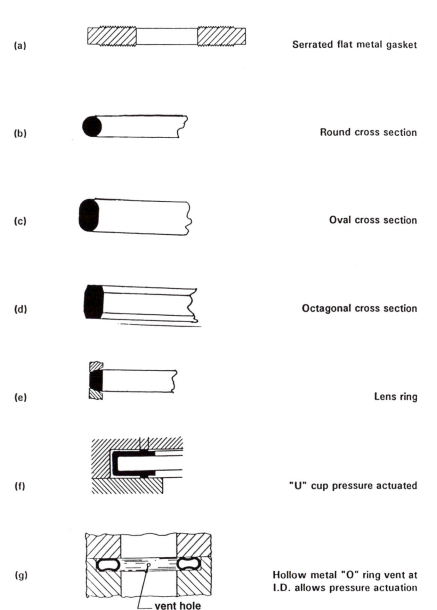

(a)   Serrated flat metal gasket

(b)   Round cross section

(c)   Oval cross section

(d)   Octagonal cross section

(e)   Lens ring

(f)   "U" cup pressure actuated

(g)   Hollow metal "O" ring vent at
                                                                        I.D. allows pressure actuation

vent hole

Figure 7   All-metal gasket types.

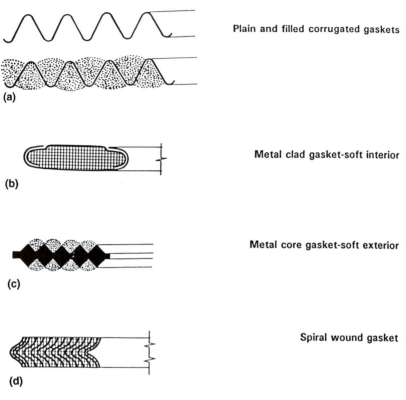

Plain and filled corrugated gaskets

(a)

Metal clad gasket-soft interior

(b)

Metal core gasket-soft exterior

(c)

Spiral wound gasket

(d)

Figure 8 Composite gaskets.

and outer turns do not have fillers and are sometimes welded together to provide rigidity (Fig. 8d). Additional solid metal rings can be used at the I.D., O.D., or both. The inner ring acts as a compression stop and helps prevent the accumulation of solid contaminants. The outer ring also acts as a compression stop, centers the gasket on the flange face, and prevents gasket blowouts. Engine head gaskets are typically complex compositions with a metallic core, a soft gasket material around the metal core, wire rings around the combustion area, and printed elastomeric sealing beads on the surface of the gasket. The bead prevents water and/or oil leakage. The wire rings provide heavy unit loads around the cylinders that prevent the escape of combustion gases.

D. Liquid Gaskets

Liquid gaskets fall into three broad categories: nonhardening, semihard but flexible, and rigid hardening. They are used to form gaskets for applications with relatively low pressure and temperature requirements. The nonhardening materials are typically used for caulking in the construction industry and as a general-purpose sealant. Semihardening materials cure like a synthetic rubber and still maintain flexibility. Hard materials undergo a chemical reaction and take a set which provides a rigid joint. Semihard materials such as silicone are often used as formed-in-place gaskets that replace traditional gasket types. The material can be applied by hand from a tube or the process can be automated. In some cases, the gasket material can be printed on a substrate using a silk screen process.

When the material is applied to a surface, the high-viscosity fluid will flow over the flange surfaces, filling voids and leveling the contact areas. Close metal-to-metal tolerances are not necessary, and, in some cases, costly machining operations can be eliminated. If the parts are joined before the material cures, the joint is essentially metal to metal with a thin layer of material between the surfaces. As the material cures, the two mating surfaces and the gasket material are essentially bonded to each other. If the material is allowed to cure before the joints come together, then the gasket behaves like any other material that is permanently fastened to one flange surface.

Some sealants cure in the absence of air (anaerobic). They retain as well as seal and are typically used as a pipe thread sealer. Anaerobic sealants are used for a gap size of 0.25 mm or less. In contrast, a flexible, cured sealant such as RTV (room temperature vulcanizing) silicone can seal gaps up to 2.5 mm. The bead size depends upon the joint width as well as the gap size. The data shown in Fig. 9 can be used to estimate the proper bead size. The amount of pressure that can be sealed depends upon the gap size, the cure time allowed before pressure is applied, the sealant type, and the cleanliness of the flange surfaces (Figs. 10, 11). Sealant types, properties, relative costs, and recommended uses appear in Table 13.

E. Troubleshooting Gasket Problems

Premature gasket leakage is caused by poor gasket design, assembly mistakes, improper flange design, improper flange face finish, and improper materials for the application. Nonmetallic gaskets should not be reused since they take a set and lose resilience. Metallic gaskets can be reused if they are annealed to reduce work hardening. A guide for solving gasket problems appears in Table 14.

F. Standard Classification System for Nonmetallic Gasket Materials

Nonmetallic gaskets can be made with a variety of materials such as asbestos, cork, cellulose, vulcanized rubber, and other organic or inorganic materials.

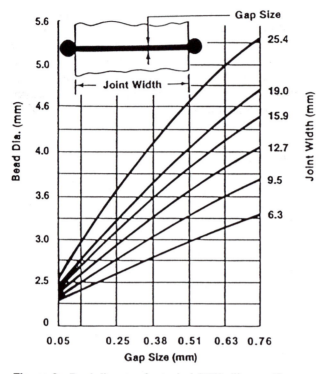

Figure 9 Bead diameter for typical RTV silicone. (Courtesy of Society of Automotive Engineers, J-1497.)

These materials are often combined with each other, binders, and impregnants to produce a finished gasket material. A system has been developed to describe and specify materials used to make nonmetallic gaskets. Rubber materials are specified by ASTM D2000. All other materials are specified by ASTM F104. Test methods used to evaluate materials for nonmetallic gaskets appear in Table 15.

The line callout is used to describe the specific physical and mechanical characteristics of materials used for nonmetallic gaskets. The first part of the callout is ASTM F104 to indicate which specification is referenced. This is followed by the letter F and a basic six-digit number. Each digit of the number represents a specific characteristic (Table 16). The first two digits of the basic six-digit number also define thickness tolerances (Table 17).

Additional specifications and descriptions are obtained by adding one or more suffix letter/numerical symbols (Table 18) to the basic six-digit callout. Consider the following callout as an example:

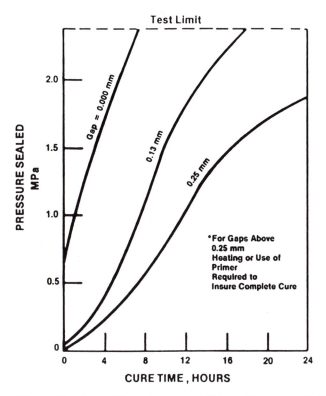

Figure 10 Anaerobic sealant capabilities. (Courtesy of Society of Automotive Engineers, J-1497.)

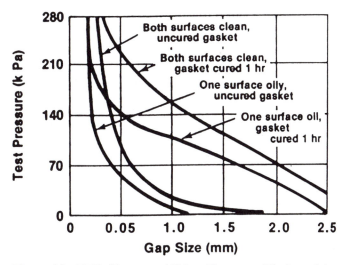

Figure 11 RTV silicone capabilities. (Courtesy of Society of Automotive Engineers, J-1497.)

435

Table 13 Sealant Types

Material		Relative cost	Uses
Nonhardening	Oleo-resin	2-6	Seal concrete joints; glazing; cable splices; glass to metal
	Butyl	5-10	Expansion/contraction joints; metal to metal; metal to glass; caulking
	Acrylic	8-10	Pipe joints, glazing, masonry, pipe dope
	Polybutene	3-7	General construction caulking
	Asphalt/bituminous	1-4	Expansion/contraction joints; caulking
Hardening/rigid	Epoxy (liquid/powder)	7-12	Potting electrical connectors, caulking, pipe sealant
	Polyester	6-12	Potting; encapsulating, gasket and pipe thread sealants
	Oleo-resin	1-6	Caulking, general-purpose sealant
Hardening/flexible	Neoprene	8-12	Seal dissimilar metals, caulking, general sealing
	Hypalon	8-12	As above
	Butyl	5-10	Glazing, caulking of metal glass and masonry joints
	Silicone (one part)	20-40	Construction sealant, caulking, and glazing
	Silicone (two part)	40-100	Formed-in-place gaskets, encapsulation, seal heat shields
	Polysulfide (one part)	17-25	General-purpose sealant, caulking, and glazing
	Polysulfide (two part)	10-22	Seal fuel tanks and pressure tanks, general construction, caulk plexiglass
	Polyurethane (one part)	12-18	General construction caulking/glazing
	Polyurethane (two part)	45-100	Seal heat shields, formed-in-place gaskets
	Modified epoxy	7-12	Caulking, potting, and encapsulating
	Fluorinated hydrocarbon	80-120	Gasket—high temperature and oil resistant

Table 14 Troubleshooting Gasket Problems

Problem	Cause	Suggested care
Gasket material deterioration	Improper material for application	Change materials to be compatible with fluid sealed, pressure, and temperature
Gasket extrusion	Seating stress too high. Internal pressure too high. Gasket too thick or width too small	Change to lower tensile material, reduce bolt load
Gasket can't seat properly	Incorrect flange surface texture	Use proper finish
	Dirty, corroded	Clean faces with wire brush, be sure old gasket is removed completely
	Metal faces uneven, deformed	Flanges too thin, increase flange thickness
Lack or loss of compression	Bolts not tightened properly	Use proper tightening sequence—tighten to proper torque specifications
	Excessive stress relaxation	Retighten gasket bolts after break-in period. Select stiffer gasket material
	Not enough bolt load	Increase number of bolts, increase diameter of bolts
	Improper bolt pattern	Change bolt pattern to eliminate low-stress areas
Gasket too thin or too wide	Change design, use thicker gasket; reduce gasket width	

Table 15   ASTM Test Methods for Gaskets

F36	Test method for compressibility and recovery of gasket materials
F37	Test method for sealability of gasket materials
F38	Test method for creep relaxation of a gasket material
F112	Test method for sealability of enveloped gaskets
F145	Recommended practice for evaluating flat-face gasketed joint assemblies
F146	Test method for fluid resistance of gasket materials
F147	Test method for flexibility of nonmetallic gasket materials containing asbestos or cork
F148	Test method for binder durability of cork composition gasket materials
F152	Method for tension testing of nonmetallic gasket materials
F363	Method for corrosion testing of enveloped gaskets
F433	Recommended practice for evaluating thermal conductivity of gasket materials
F434	Method for blowout testing of performed gaskets
F495	Test method for ignition loss of gasket materials containing inorganic substances
F586	Test method for leak rates versus y stresses and m factors for gaskets
F607	Test method for adhesion of gasket materials to metal surfaces
F806	Test method for compressibility of laminated composite gasket materials
F1087	Linear dimensional stability of a gasket material to moisture

ASTM F104	F225334	B2 E45
Classification system	Line callout	Suffixes

The designation ASTM F104 indicates the gasket is specified and described by the standard classification system for nonmetallic gasket materials. The first digit, 2, indicates the primary material is cork and the second digit, 2, indicates the cork is blended with an elastomer. The third digit, 5, signifies the gasket will have a compression loss from 20% to 30% when tested using ASTM F36. The fourth digit, 3, signifies the material thickness will increase 10% to 25% when immersed in ASTM #3 oil using ASTM F146 test procedures. The fifth digit, 3, means the material will have a maximum weight increase of 20% in ASTM #3 oil using ASTM F146. The final digit, 4, specifies the maximum weight increase will be 30% when immersed in water using ASTM F146 test procedures.

The suffix B2 (Table 18) indicates a maximum of 15% stress loss is expected under ASTM F38 test procedures. The suffix E45 specifies the max-

Table 16 Basic Physical and Mechanical Characteristics for Nonmetallic Gasket Materials

Basic Six Digit Number	0	1	2	3	4	5	6	7	8	9
					BASIC CHARACTERISTICS					
First number type of material	NS	Asbestos	Cork	Cellulose	Fluoro-carbon polymer	Flexible graphite	-	non asbestos	-	AS
Second number class of material when first number is 1 or 7	NS	Compressed sheeter process	Beater process	Paper and millboard	-	-	-	-	-	AS
Second number when first number is 2	NS	Cork composition	Cork and elastomer	Cork & cellular rubber	-	-	-	-	-	AS
Second number when first number is 3	NS	Untreated fiber	Protein treated fiber	Elastomeric treated fiber	Thermoset resin treated	-	-	-	-	AS
Second number when first number is 4	NS	Sheet PTFE	PTFE with expanded structure	PTFE filaments braided or woven	PTFE felts	Filled PTFE	-	-	-	AS
Second number when first number is 5	NS	Homogeneous sheet	Laminated sheet	-	-	-	-	-	-	AS
Third number compressibility ASTM F36 % compression loss	NS	0-10	5-15	10-20	15-25	20-30	25-40	30-50	40-60	AS
Fourth number % thickness increase when immersed in ASTM #3 oil ASTM F146	NS	0-15	5-20	10-25	15-30	20-40	30-50	40-60	50-70	AS
Fifth number maximum % weight increase when immersed in ASTM #3 oil ASTM 146	NS	10	15	20	30	40	60	80	100	AS
Sixth number maximum % weight increase when immersed in water ASTM F146	NS	10	15	20	30	40	60	80	100	AS

NS = not specified.
AS = as specified and agreed upon between supplier and user.
Source: Society of Automotive Engineers J90.

Table 17 Thickness Tolerances for Nonmetallic Gaskets

Material type and class (first two digits of basic six-digit number)	Specified thickness (mm)	Tolerance (mm)
11 and 12	Up to 0.41	−0.05 + 0.13
	Over 0.41 under 1.57	±0.13
	Over 1.57	±0.20
13	Up to 3.18	±0.13
	Over 3.18	±0.25
21	All thickness	±10% or ±0.25 (whichever is greatest)
22	Up to 1.57	±0.25
	Over 1.57	±0.38
23	Over 1.57	±0.38
00, 31, 32, 33, 39	Up to 0.41	±0.09
	Over 0.41 under 1.57	±0.13
	Over 1.57 under 2.39	±0.20
	Over 2.39	±0.41
		±0.51
51	Up to 1.6	±0.51
52	Up to 12.7	±10%

Source: Society of Automotive Engineers J90.

imum weight increase to be 30% with a thickness increase of 10% to 25% when immersed in ASTM fuel B using ASTM F146 test procedures.

The thickness tolerances for the gasket are also specified with the first two digits of the line callout (22) and Table 17. If the gasket thickness is less than 1.57 mm, the tolerance is ±0.25 mm. If the gasket thickness is greater than 1.57 mm, the tolerance is ±0.38 mm.

G. Standard Classification System for Laminated Composite Gasket Material

Flat, parallel layers of two or more different materials are often laminated together to form a composite gasket which will have properties not obtained if only one material is used. These materials may be organic, inorganic, and combinations of materials with various binders or impregnants. The layers can be bonded or mechanically attached to one another. A classification system (ASTM 868) is used to specify or describe the properties of laminated composite gasket materials.

The classification system consists of a line callout that describes the use, composition, and combining methods. The callout begins with ASTM F868

Table 18 Supplemental Physical and Mechanical Characteristics for Nonmetallic Gasket Materials

Suffix Symbol	Test Description	0	1	2	3	4	5	6	7	8	9
A	Sealability	-	-	-	-	-	-	-	-	-	AS
B	Creep Relaxation, ASTM F38, Maximum % Stress Loss	NS	10	15	20	25	30	40	50	60	AS
D	Corrosion & Adhesive effects of gasket materials on metal surfaces, ASTM F607	-	-	-	-	-	-	-	-	-	AS
E-First Number	ASTM Fuel B, ASTM F146 % Weight increase	NS	10	15	20	30	40	60	80	100	AS
E-Second Number	ASTM Fuel B, ASTM F146, % Thickness increase	NS	0-5	0-10	0-15	5-20	10-25	15-35	25-45	30-60	AS
H	Adhesion, ASTM F607	-	-	-	-	-	-	-	-	-	AS
K	Thermal conductivity, ASTM F433, 100°C (W/M.°K)	NS	0-0.09	0.07-0.17	0.14-0.24	0.22-0.31	0.29-0.38	0.36-0.45	0.43-0.53	0.50-0.60	AS
M	Tensile strength, ASTM F152 Minimum value (mPa)	NS	0.689	1.724	3.447	6.845	10.342	13.790	20.684	27.579	AS
R	Binder durability, ASTM F148	NS									AS
S	Volume change, ASTM #1 oil, ASTM #3 oil, ASTM Fuel A, ASTM F146	-	-	-	-	-	-	-	-		AS
T	Flexibility, ASTM F147	-	-	-	-	-	-	-	-	-	AS
Z	Other Engineering Specifications	-	-	-	-	-	-	-	-	-	AS

NS = Not specified.
AS = As specified and agreed upon between supplier and user.
Source: Society of Automotive Engineers J90.

and follows with numbers and letters that describe the gasket. The first number of the callout indicates the typical end use of the gasket. This number is followed by a letter group that indicates the type of material used by each layer in order. The first letter is the top material. The letter group is followed by a number that indicates how the layers are attached to one another. The codes used for the line callout appear in Table 19. The line callout is often followed by one or more suffix letter-numeral symbols that specify and describe the gasket more fully (Table 20). A typical line callout for an engine cylinder head gasket is

ASTMF868	4 FMF 3	E35G35H2K5; F = F112440, M = ASTM A109
Classification system	Line callout	Suffixes

The designation ASTM F868 indicates the gasket is specified and described by the standard classification system for laminated composite gasket materials. The first number, 4, tells the user the gasket is intended for engine cylinder head gaskets. The first layer (F) is an ASTM F104 material and is specified by F = F112440. The middle layer is a steel material that must meet ASTM A109 specifications. The third layer is also an ASTM F104 material which is identical to the top layer and must satisfy the same specifications. The metal portion of the gasket has tanged perforations and a chemical bond is also used to attach the upper and lower layers to the metallic core. This method of

Table 19 Classification System for Laminated Composite Gaskets

First digit describes gasket application	Letter group describes gasket composition by layer, first letter is top layer	Second digit describes method of attaching layers
0 - Not specified	N - Not specified	0 - Not specified
1 - Engine carburetor	B - Board	1 - Tanged perforation
2 - Engine intake manifold	M - Metal	2 - Chemical bond
3 - Engine exhaust manifold	F - ASTM F104 material	3 - Tanged perforation
4 - Engine cylinder head	R - ASTM D2000	plus chemical bond
5 - Engine transmission	rubber material	4 - Grommets
6 - Ducts and piping	P - Plastic	5 - Overlapped
7 - Compressors	T - Textiles	6 - Bonded
9 - As specified	S - As specified	9 - As specified

Source: American Society for Testing and Materials (ASTM 868).

Table 20 Suffix Letters and Test Methods for Laminated Composite Gasket Materials

Suffix Symbol	Test Description	0	1	2	3	4	5	6	7	8	9
A	Sealability	-	-	-	-	-	-	-	-	-	AS
C	Compressibility	-	-	-	-	-	-	-	-	-	AS
D	Release	-	-	-	-	-	-	-	-	-	AS
E-First Number	ASTM Fuel B, ASTM F146 % Weight increase	NS	10	15	20	30	40	50	60	100	AS
E-Second Number	ASTM Fuel B, ASTM F146 % Thickness increase	NS	0-5	0-10	0-15	5-20	10-25	15-35	25-45	30-60	AS
G-First Number	ASTM #3 oil, ASTM F146 Max % weight increase	NS	10	15	20	30	40	60	80	100	AS
G-Second Number	ASTM #3 oil, ASTM F146 % Thickness increase	NS	0-15	5-20	10-25	15-30	20-40	30-50	40-60	50-70	AS
H	ASTM F38 % Stress Loss	NS	10	15	20	25	30	40	50	-	AS
K	Thermal Conductivity, W/M °K ASTM F433	NS	0-0.09	0.07-0.17	0.14-0.24	0.22-0.31	0.29-0.38	0.36-0.45	0.43-0.53	0.50-0.60	AS
L	Laminated Bond Characteristics	-	-	-	-	-	-	-	-	-	AS
X	Crush-Extrusion Resistance	-	-	-	-	-	-	-	-	-	AS
Y	Coatings	-	-	-	-	-	-	-	-	-	AS
Z	Drawing Detailed Properties	-	-	-	-	-	-	-	-	-	

NS = not specified.
AS = as specified and agreed upon between supplier and user.
Source: American Society for Testing and Materials (ASTM 868).

attaching the layers together is specified by the final number, 3. These codes for the line callout appear in Table 19.

The suffixes describe additional specifications the gasket must meet (see Table 20). The first suffix, E35, indicates the gasket must be tested in ASTM fuel B using test method ASTM F146. The percent weight increase must not exceed 20% and the thickness increase must be 10% to 25%. The second suffix, G35, indicates the gasket must be soaked in ASTM #3 oil using ASTM test method F146. The weight increase must be less than 20% and the thickness increase is 20% to 40%. The third suffix, H2, indicates the loss of stress of the gasket must be less than 15% using ASTM F38 test procedures. The fourth suffix, K5, tells the user that the thermal conductivity of the gasket lies between 0.29 and 0.38 W/m · K.

V. STATIC SEALS: ELASTOMERIC

A broad variety of shapes can be molded or machined from elastomerics to provide effective static seals. The seals are fitted into notches or grooves machined into the housing components and are compressed when the surfaces are brought together to form a gland. The deformation of the elastomer produces forces on the walls of the housing that prevents leakage. Since elastomers are incompressible, the notches or grooves must be larger than the seal to allow the material room to deform when the units are assembled. The polymer used to make the elastomer is selected to meet the application temperature and pressure and be compatible with the sealed fluid (Table 3). The most common and widely used static seal is the O ring design (Fig. 12a) and is often used in dynamic applications as well. They are molded in sheets and the rubber flash is removed mechanically by cold tumbling with an abrasive media. O rings are identified by the inner diameter (I.D.), cross-section diameter, and material. Typical O ring gland shapes appear in Figs. 12b–d. The surface finish of these grooves should be 0.8 to 1.6 μm to prevent the O ring from twisting during pressure surges. Gland depth must be less than the diameter of the O ring to allow the ring to compress when the gland is tightened. The minimum amount of squeeze is 0.15 mm regardless of the O-ring diameter. The maximum squeeze is 35% of the O-ring cross-section diameter, and the typical squeeze is about 20%.

The rectangular- or square-cut ring is also popular. They are cut or ground to the desired size from vulcanized billets on a lathe. They are used for only static pressure applications up to 10,000 kPa. Square-cut rings are fit into a groove to serve as a typical squeeze gasket (Fig. 13). Many other shapes can be molded or machined. Some of these are pressure-actuated (Fig. 14).

Leakage of elastomeric static seals can occur because of installation damage to the seal, improper selection of material, sealing surfaces improperly

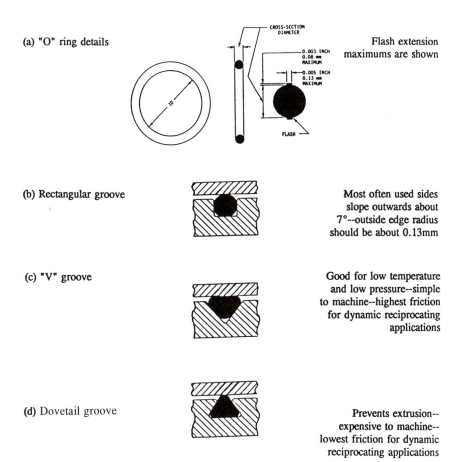

(a) "O" ring details — Flash extension maximums are shown

(b) Rectangular groove — Most often used sides slope outwards about 7°--outside edge radius should be about 0.13mm

(c) "V" groove — Good for low temperature and low pressure--simple to machine--highest friction for dynamic reciprocating applications

(d) Dovetail groove — Prevents extrusion--expensive to machine--lowest friction for dynamic reciprocating applications

Figure 12   O-ring static seals.

prepared, and improper fit to the groove dimensions (Table 21). Another source of failure is extrusion. Pulsating pressure forces portions of the seal into and out of the gap between components. This results in tears or "nibbling' which can lead to failure. Increasing elastomer hardness will increase the resistance to extrusion and seal life will increase (Fig. 15).

VI. DYNAMIC SEALS: PACKINGS

A. Compression Packings

Compression packings are used in high-speed rotating and reciprocating shaft applications as well as low-speed, infrequent actuations such as valves. The

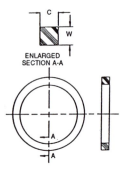

(a)   Square and
      rectangular cut
      ring details

ID and OD can be
carefully controlled
by machining

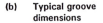

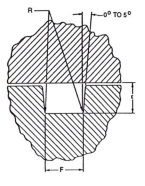

(b)   Typical groove
      dimensions

Groove surface finish
is 1.6 to 3.2 micrometer

Figure 13   Square and rectangular cut rings.

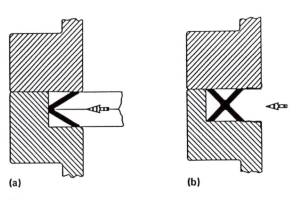

Figure 14   Pressure-actuated static seals: (a) Internal pressure expands V-ring; (b) Internal pressure expands quad ring.

Table 21 Troubleshooting Elastomeric Static Seals

Problem	Probable cause	Remedy
Seal leaks for no apparent reason	Seal may not be the right size for the groove	Check dimensions—install proper parts
Seal is hard or has taken an excessive permanent set	Temperature too high for material	Change material (Table 3)
	Material not compatible with sealed fluid	Change material (Table 3)
Seal damaged	Flange surfaces are cut, gouged, scratched, or have tool marks	Polish out defects or replace parts
	Pinched or cut during assembly	Use shims to protect seal when assembled over threads
Seal extrusion causes early leakage	Sealing surfaces not flat	Must be flat to 0.13 mm, replace parts if necessary
	Initial bolt torque too low	Check bolt torque and retorque if necessary
	Presure pulses too extreme	Change relief valve setting
	Pressure exceeds 10,000 kPa	Use backup rings
Seal surfaces abraded, worn	Sealing surface too rough	Polish to 0.4 μm
	Seal material too soft	Change to harder durometer material
	Excessive flange movement or "breathing"	Increase bolt torque

operating principle is quite simple. Material is installed around the shaft in a stuffing box and the gland follower is tightened. The compression packing is forced against the moving shaft and against the stuffing box I.D. Leakage does not usually occur or is expected in valve applications where shaft movement is slow and actuations infrequent. Compression packings will exhibit some leakage when used in rotating shaft applications. If the gland follower is over-tightened, leakage will stop, lubrication will be reduced, excessive friction will result, and overheating will occur. If the gland follower is not tightened sufficiently, excessive leakage will occur. Leakage is controlled by routine maintenance, which includes periodic tightening of the gland follower and eventual packing replacement.

A stuffing box arrangement for a rotating shaft application sealing a caustic fluid is shown in Fig. 16. A lantern ring is used to provide positive lubrication for the primary packings. Some of the process fluid is injected in the area to provide both lubrication and coolant. In some cases, a high-pressure buffer fluid is injected in the lantern ring area to flush contaminants or caustic process

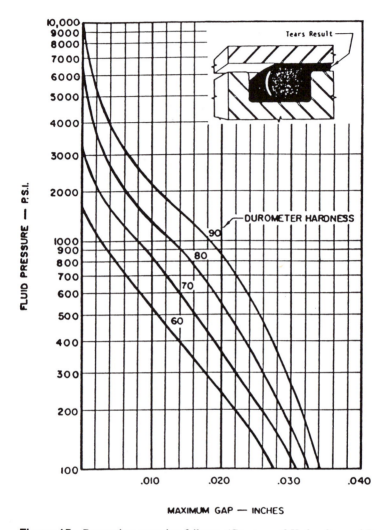

Figure 15   Preventing extrusion failures. (Courtesy of *Hydraulics and Pneumatics*.)

fluids away from the primary packings. A controlled amount of the buffer fluid will leak into the process fluid and into the atmosphere. In extreme cases, a secondary seal is used to create a cavity for the high-pressure buffer fluid.

Leakage is controlled by periodically tightening the nuts for the gland followers. These maintenance costs can be reduced by using bolts that are preloaded with Bellville spring packages that continue to compress the packing as it wears (Fig. 17).

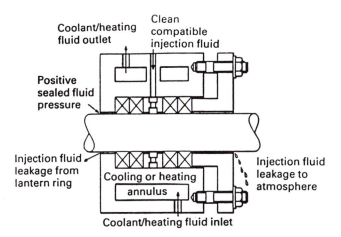

Figure 16 Arrangement to seal caustic fluid.

Packing cross sections may be round, square, or rectangular. They are made from fibers, cloth, yarn, and metal. These materials are braided, plaited, laminated, spiral-wrapped, or folded to form the desired packing shape (Fig. 18). The packings are often impregnated with mineral oil, grease, animal fat, graphite, or other lubricants to reduce wear, retain packing flexibility, and help effect a seal. Yarn, fabric, or cloth packings are made from plant fibers (flax, jute, cotton), mineral fibers (asbestos), or artificial fibers (graphite, PTFE). Metallic packings are used for high-temperature applications. Lead is soft and pliable and is used in applications where the temperature is less than 100°C. Soft pure copper is used in hot oil, tar, or asphalt pump applications. Aluminum and copper can withstand temperatures up to 425°C. Pure nickel is used

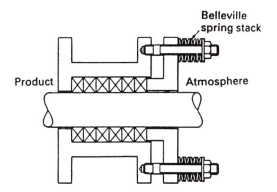

Figure 17 Preloaded bolts (live loaded packings).

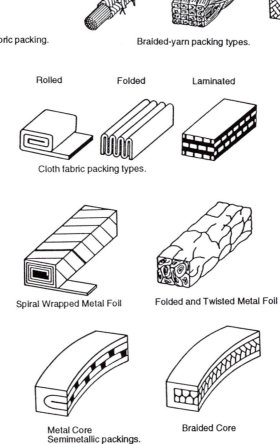

Figure 18   Packing design configurations.

in steam or caustic alkali fluids up to 760°C. A material selection chart is found in Table 22, design guidelines in Table 23, and a troubleshooting guide in Table 24.

B.   Molded or Automatic Packings

Packings can be molded from compounds that are made with elastomeric polymers selected from Table 3 to meet the needs of the application. The material is often reinforced with fabrics to extend the high-pressure capabilities to approximately 55,000 kPa. When temperatures are less than 120°C, then cotton duck is used. Asbestos cloth or other high-temperature synthetic materials are used when temperatures exceed 120°C. Nylon cloth is sometimes used to add strength to the seal while maintaining flexibility. The upper pressure limit for nonreinforced rubber packings is about 35,000 kPa. Automatic packings are used primarily to seal reciprocating applications for hydraulic and pneumatic cylinders. Rough surfaces can abrade the packings and cause premature failure. It is recommended that the sliding surfaces be finished to 0.2 to 0.4 μm RA for homogenous rubber packings and 0.4 to 0.8 μm RA for fabric-reinforced packings.

The U cup is one of the earliest designs used in pressure-energized lip seal applications and is effective at low to moderate pressures in hydraulic and pneumatic applications (Fig. 19). U cups can be installed in the cylinder housing or can be used to seal reciprocating pistons (Fig. 20). The upper pressure limit for a single U cup is about 17,500 kPa. The primary failure mode is extrusion of the heel section into the gap between the rod or piston and the housing. This can be avoided with a backup washer or an antiextrusion ring snapped directly into the U cup. The ring is usually made of a low-friction material such as nylon or PTFE.

The V ring or chevron design is typically installed in a gland and is usually found in sets to extend the upper pressure range to as much as 140,000 kPa. The number of rings required depends upon rod size, internal operating pressure, and ring material (Table 25). Many combinations and configurations of V rings can be utilized in a gland set. A typical set consists of a gland ring, a set of V rings, and a header ring which is sometimes vented to improve the actuation of the chevron packings (Fig. 21). The rings can be made of metal, glass-filled nylon, or PTFE-filled materials.

Exclusion seals or rod wipers are used to prevent contaminants from entering the seal area around a rod or piston. They are sometimes combined with U cups and V packings to provide a single seal. Special seal configurations combine an elastomer with a brass scraper that removes dried mud from the rod upon start-up. Boots are also placed over critical parts to shield them from outside contaminants (Fig. 22).

C. Floating Packings (Split-Ring Seals)

Expanding split-ring seals are installed in reciprocating pistons to seal gases in internal combustion engines, fluids in piston pumps and contain vacuum in vacuum pumps. The internal pressure forces the ring to mate on the side wall of a groove in the piston and on the inside wall of a cylinder (Fig. 23a).

Contacting split-ring seals are used to seal linear actuator rods when space limitation limits the use of packings. The internal pressure expands the ring

Table 22 Material Selection Chart for Packings (Maximum Performance Parameters)

Type	Construction	Rubbing speed (m/s)	Temp. (°C)	Pressure (kPa)	Application
Cotton	Plaited or braided with grease, graphite, or mica lubricant	7.5	90	—	Seals for water
Hemp	Plaited or braided with grease, graphite, or mica lubricant	—	—	—	Seals for water
Flax	Plaited or braided with grease, graphite, or mica lubricant	—	70–120	—	Rotary seals or pumps and slow reciprocating seals; ships' stem glands, etc.
Textile	Plaited or braided cotton, nylon, rayon, hemp, etc., with PTFE impregnation	—	—	—	Seals for water services, solvents, weak alkalis, oils, greases, foodstuffs, etc.
Asbestos (dry)	Round or square plaited or braided white asbestos	—	500	—	Autoclaves, boiler stop valves, etc.
Asbestos (lubricated)	Round or square plaited or braided white asbestos with graphite, mica, or mineral oil lubricant	—	−40–+300	3000+	Steam services, evaporators, cannisters, liquors, etc.
Acrylic	PTFE coated, lubricated, and plaited or braided	10	120–200	10,00	For low temperature, low general service slightly acidic fluids
PTFE	Yarns or tape	—	200–250		Soft packing, services involving corrosive media

Material	Description	Size	Temperature (°C)	Speed/Pressure	Application
PTFE and lubricant	PTFE braided and impregnated with graphite and/or molybdenum disulfide lubricant	20	-100-+250	3000+	Water, steam, acids, alkalis, solvents, oils, greases, hydrocarbons, etc.
	Braided pure PTFE fibers treated with PTFE dispersion	10	-200-+300	10,000	Sealing all media
PTFE (graphite)	Extruded fibrillated PTFE and graphite with mineral oil lubricant	10	-100-+250	10,000	Water, acids, alkalis, solvents, oils, degreasing fluids, oxidizing agents, foodstuffs, etc.
Aramid-PTFE	Braided PTFE coated kevlar yarn impregnated with lubricant and PTFE	10-20	-220-+300	20,000-100,000	High duty "superpacking"
Graphite yarn	Woven or braided graphite yarn treated with graphite powder	—	-200-+600	—	High-temperature services
Carbon fiber	Amorphous carbon yarns treated with graphite powder	—	-200-+600	—	High-temperature services
Expanded graphite	Flexible plait or tape form	35	-200-+600	30,000	High-temperature services, foodstuffs, etc.
Glass fiber	Braided glass fiber yarns with added lubricant	—	—	—	Suitable for use with acid and corrosive fluids (except caustic solutions)
Aluminum metal	Foil	5	280	3000+	High temperature
Copper wire	Braided	5	340	3000+	High-pressure steam
Lead	Foil	7.5	230	Depends on speed	High speeds
White metal	Foil	Slow rotary or reciprocating	240	3000+	Air, water steam oils

Table 23 Design Guidelines for Packings

Shaft	Must be well supported. Free of pits, grooves and shaft lead or machining spirals. Surface finish to be less than 0.5 μm RA. Dynamic runout less than 0.025 mm. Radial clearance between shaft and stuffing box—0.25 mm maximum. Shaft hardness to be 180 Brinell for soft packings and 500 Brinell for metal packings.
Stuffing box	Surface finish or stuffing box I.D. walls to be less than 1.6 μm RA. A tapered lead-in chamfer (15° with length ½ of packing cross section) to protect packings during installation.
Gland follower	Radial clearance between gland follower I.D. and shaft to be less than 0.4 mm. Radial clearance between gland follower O.D. and stuffing box to be less than 0.2 mm.

Packing size (dependent on shaft size)

Shaft diameter^{DS} (mm)	Packing section (mm)
DS $\leqslant$ 16	3
16 < DS $\leqslant$ 25	5
25 < DS $\leqslant$ 50	6.5
50 < DS $\leqslant$ 90	8
90 < DS $\leqslant$ 150	10
DS > 150	12.5

Number of packing rings (dependent on system pressure)

System pressure (Ps) kPa	Number of rings
PS $\leqslant$ 3500	4
3500 < Ps $\leqslant$ 7000	6
7000 < Ps $\leqslant$ 14000	8
Ps > 14000	10

Installation	Old packings should be removed carefully to minimize shaft damage; check shaft, stuffing box and gland follower to insure tolerances are within specifications. Remove corrosion, check for pittings, scoring. Replace damaged components; cut rings to proper length. Tamp hole to seat properly; stagger joints in rings by 120°.
Material selection	Must withstand temperature of the application. Must resist application fluid. Must be noncorrosive. Must conform to shaft and bore under moderate pressure.

Table 24 Troubleshooting Compression Packings for Dynamic Applications

Problem	Possible causes	Corrective action
Little or no leakage upon start-up (will create excessive heat)	Bolts overtorqued	Loosen bolts
Excessive leakage upon start-up	Bolts undertorqued. Wrong packing for application. Excessive shaft runout	Tighten bolts. Check packing size. Check bearings
Part or all of end ring missing	Excessive clearance between shaft and stuffing box allows extrusion	Replace worn parts. Check for excessive shaft to bore misalignment
Wear on outside of rings	Rings rotating with shaft, rubbing on stuffing box I.D.	Packing corroded and stuck to shaft-clean shaft. Packing O.D. too small and not generating enough force on stuffing box I.D.—change packing size. Packing I.D. too small and generating too much force on the shaft—change packing size
Ring I.D. charred and dry	No lubrication. Incorrect packing material	Loosen bolts. Upgrade packing to take higher speeds, temperature
Packing next to process fluid is deteriorated	Packing material not compatible with fluid	Change packing material
Rings next to gland follower are excessively worn	Gland follower is too tight	Follow correct installation procedures
Packing seizes on shaft	Process fluid solidifies in stuffing box after shutdown. Process fluid corrodes packing and shaft together	Heat box to prevent crystallization. Change shaft material, plate shaft. Change packing material

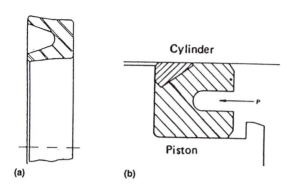

(a)   (b)

Figure 19   U-cup designs: (a) pneumatic application (thin lips); (b) hydraulic application (thick lips with antiextrusion ring.)

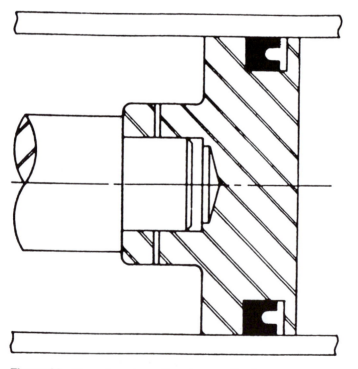

Figure 20   U-cup in reciprocating piston application.

Table 25   Maximum Pressure Ratings on V-Ring Sets

Number of V rings	Homogenous rubber (kPa)	Rubber/fabric composite (kPa)
2	3500	7000
3	7000	10000
4	14000	21000
5	21000	35000
6	35000	70000

and forces it against the outside diameter of a reciprocating rod and the side wall of a groove cut in the housing (Fig. 23b).

The joints of the rings can be made in several ways. The straight cut or butt cut is the cheapest and most widely used. Other methods of preparing the joint are shown and discussed in Fig. 24.

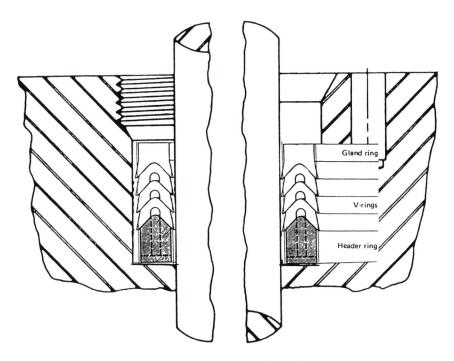

Figure 21   V or chevron packing set on reciprocating rod.

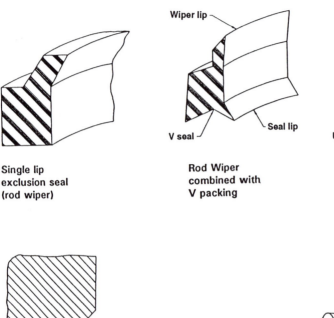

Single lip
exclusion seal
(rod wiper)

Rod Wiper
combined with
V packing

Rod wiper
combined with
U cup

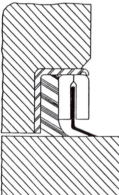

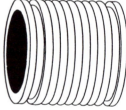

Wiper scraper

Boot

Figure 22 Exclusion seals. (Courtesy of Chicago Rawhide.)

When system pressure is high, small circumferential grooves are machined in the ring outer diameter. Fluid flows into the groove to form a thin dam and act as a pressure reducer (Fig. 25). A ring that has been machined in this manner is referred to as a balanced ring. Balanced rings have more leakage than unbalanced rings. Multiring systems are used to seal against high internal pressure (Table 26).

Metal rings are usually used in lubricated applications. PTFE has low friction and can be used in dry applications with a temperature range of -100 to $250°C$. Fillers such as glass, bronze, molybdenum disulfide, and graphite are

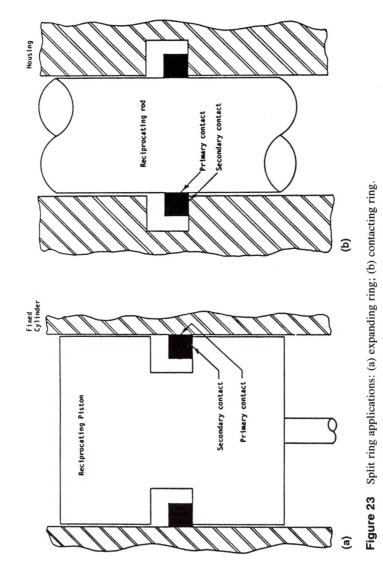

Figure 23 Split ring applications: (a) expanding ring; (b) contacting ring.

Gap Type	Comments
Uncut	Must be made of material that is flexible enough to be stretched over a piston or snaked into a groove in a cylinder wall (nylon or PTFE).
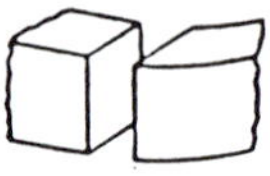 Straight or butt cut	Easy to manufacture, cheapest and most widely used. Used as piston seals, seldom used for rod or contracting ring applications.
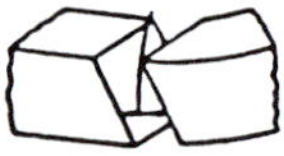 Beveled or scarf cut	Similar leakage as butt cut. Used as piston seals. Seldom used as rod seals.
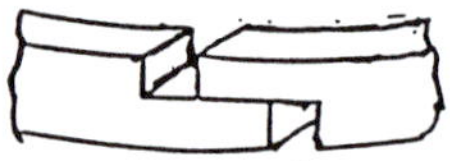 Stepped cut	Reduces leakage by eliminating straight leakage path. May be butt or scarf cut. Scarf cut has much lower leak rate, but is more expensive. Can be used as piston and rod seal.
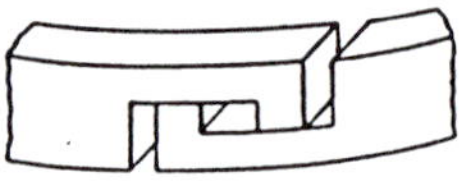 Interlocking joint	Expensive to manufacture. Used in blind assemblies. Used to limit amount of expansion.

Figure 24 Split ring seal joint design.

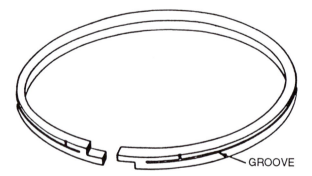

Figure 25 Balanced ring.

Table 26 Number of Rings Required to Seal System Pressure

Internal pressure (kPa)	Number of rings required
Up to 2000	2
2000 to 6000	3
6000 to 10,000	4
	5 balanced
10,000 to 20,000	5
	6 balanced
Over 20,000	6 balanced

used to increase the strength of the pure PTFE ring. Metals are sometimes plated to enhance properties. PTFE and graphite are used to reduce friction, phosphate reduces corrosion, and a variety of coatings will reduce corrosion and improve wear resistance (chromium, silver, tin, copper, and cadmium). Properties of ring materials are given in Table 27.

Table 27 Ring Material Properties

Material	Upper temperature limit (°C)	Ultimate tensile strength (kg/mm^2)	($\times 10^5$ kg/cm^2) modulus elasticity	Comments
Gray iron	340	25	10.9	Water resistant, self-lubricating
Malleable iron	370	46	14	Wear resistant
Ni-resist	425	25	10	Nonmagnetic
Ductile iron	370	55	20	Good heat resistance
Aluminum bronze	260	55	10.5	Good for marginal lubrication
Stellite	500	90	21	Resistant to heat and corrosion
Filled PTFE	250	Depends on type and amount of filler	Depends on type and amount of filler	Low friction, corrosion resistant

VII. DYNAMIC SEALS: DIAPHRAGMS

Diaphragms are used to exclude contaminants and contain pressure or vacuum in reciprocating applications. The seal is typically fastened to the housing at the outer edge and will move in the axial direction.

Flat diaphragms are the simplest and cheapest designs, but have a limited stroke that will stretch and stress the material. The stroke is often limited by a mechanical stop to insure the diaphragm is not stretched beyond material limitations. This is done to prevent premature failure due to fatigue and bursting. Springs return the diaphragm to the original position when the pressure is relieved. Convoluted or dished diaphragms are used to extend the stroke without overstressing the material. The acceptable travel is typically about twice the length of the diaphragm convolutes. Rolling diaphragms are used for long stroke applications with internal pressures up to 3500 kPa. The diaphragms are molded in the shape of a top hat with walls that range from 0.25 to 0.90 mm thick. The top hat is clamped at the outer edge and attached at the center to a piston. The top hat is inverted by a return spring when the pressure is relieved. When the cavity is pressurized, the piston is pushed down and the diaphragm is expanded to its original shape (Fig. 26). The rolling diaphragm cannot be used if the high- and low-pressure chambers can be reversed. Excessive wear can result. Rolling diaphragms can invert or balloon out in the area where the diaphragm is attached to the piston. This will result in

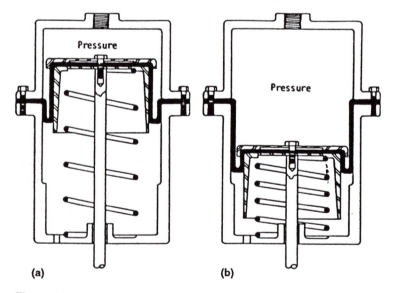

Figure 26   Rolling diaphragm: (a) pressure relieved; (b) pressure applied.

high wear and premature failure. This problem is eliminated by providing a curved lip on the retainer plate (Fig. 27).

Elastomeric materials are selected to meet the application requirements (Table 3). Fabrics are combined with the elastomers to provide a material with a high burst strength. Cotton and nylon materials are used at temperatures of 120°C or less. Other synthetic fibers such as Darvon can extend the temperature range up to 175°C. The elastomer used to coat the fabric must also have low permeability to gases so the gas will not leak through the material. Nitrile and epichlorohydrin have very low air permeability rates. Silicone has high air permeability rates which are as much as 1700 times greater than epichlorohydrin.

VIII. DYNAMIC SEALS: ELASTOMERIC RADIAL LIP

A. General Seal Design

Elastomeric radial lip seals are widely used to retain lubricants and exclude contaminants in rotating, reciprocating, and oscillating applications. These seals can be used in a variety of applications, will fit in small spaces, are easy to install and are low cost when compared to other seal types. The radial lip seal must maintain an interference between the elastomeric lip member and the moving shaft to function properly (Fig. 28). Some elastomeric materials will

Pressure Side

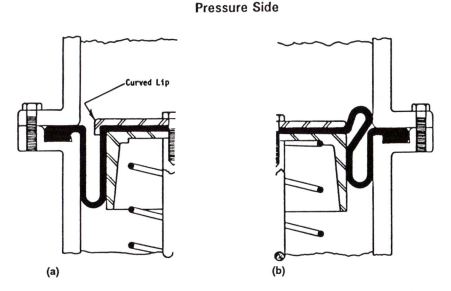

(a)   (b)

Figure 27   Preventing lip inversion: (a) curved lip prevents inversion; (b) inversion.

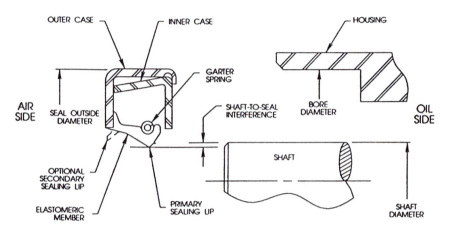

Figure 28 Typical radial lip seal. (Courtesy of Chicago Rawhide.)

harden and lose interference rapidly when exposed to excessive temperatures and harsh fluids. The proper material for the application can be selected from Table 3. A garter spring is sometimes used to help maintain lip interference and extend seal life. Leakage can also occur between the seal outer diameter and the housing bore. An interference fit prevents the seal from spinning in the bore, insures seal retention within the bore, and prevents leakage.

Some seal designs incorporate molded ribs or projections on the outside lip surface or the air side that will pump escaping oil from the air side back into the oil side (Fig. 29). The helix seal provides the largest pump rate, but can be used only if the shaft rotates in one direction. If the rotation is reversed, the helices will screw the oil out of the sump causing gross leakage. Seals with triangular pads, parabolic ribs, and sinusoidal ribs will pump less than helix ribs, but will function in both directions of shaft rotation. Some seals have a molded, wavy seal edge that pumps oil in both directions of shaft rotations without rib projections that create excessive heat and sweep contaminants into the sump (Fig. 30).

Changes in lip seal geometry can also effect performance (Fig. 31).

B. Seal Types

A multitude of seal types exist (Fig. 32), but specific designs are chosen to satisfy specific requirements. Springless seals are the least expensive solution for light-duty applications such as retaining grease at speeds less than 10 m/s. Springless seals are also used to exclude contaminants when the primary lip faces the atmosphere. Spring loaded seals are used to retain oils at surface speeds up to 25 m/s. Special seal designs are required for high pressure, high

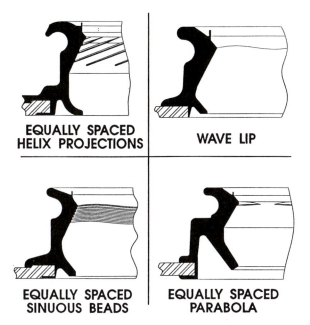

EQUALLY SPACED
HELIX PROJECTIONS | WAVE LIP

EQUALLY SPACED
SINUOUS BEADS | EQUALLY SPACED
PARABOLA

Figure 29 Hydrodynamic features. (Courtesy of Chicago Rawhide.)

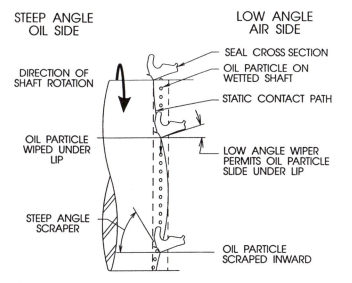

STEEP ANGLE
OIL SIDE

LOW ANGLE
AIR SIDE

DIRECTION OF
SHAFT ROTATION

SEAL CROSS SECTION

OIL PARTICLE ON
WETTED SHAFT

STATIC CONTACT PATH

OIL PARTICLE
WIPED UNDER
LIP

LOW ANGLE WIPER
PERMITS OIL PARTICLE
SLIDE UNDER LIP

STEEP ANGLE
SCRAPER

OIL PARTICLE
SCRAPED INWARD

Figure 30 Wavy seal edge provides birotational pumping. (Courtesy of Chicago Rawhide.)

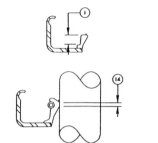
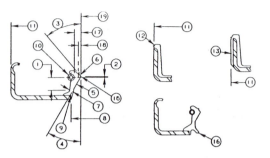

Item	Name	Typical Value	Remarks
1	Lip or beam length	2.5 to 5.0mm	Short element used for pressure, but increases wear. Long element increases flexibility and shaft followability.
2	"R" value (spring axial position)	0.25 to 0.75mm	Distance between centerline of spring and lip contact on shaft. Lip tip should always be toward oil side of spring centerline when sealing oil.
3	Oil side angle	40 to 70°	Must be larger than air side angle for proper sealing.
4	Air side or barrel angle	20 to 35°	Must be smaller than oil side angle for proper sealing.
5	Outside lip surface	--	Forms air side of lip.
6	Inside lip surface	--	Forms oil side of lip.
7	Flex section	0.63 to 1.5mm	Thick sections used for pressure, but increases wear. Thin section increases flexibility and shaft followability.
8	Case I.D.	--	Smallest diameter of metal case.
9	Heel section	--	Seal section that is bonded to the metal case.
10	Garter spring	--	Coiled wire spring used to maintain seal force.
11	Seal O.D.	--	Outer diameter of seal case/rubber O.D.
12	Nose gasket	--	Elastomeric material bonded to the face of the steel case. Forms a gasket when pressed into a bore.
13	Rubber O.D.	--	Elastomeric material that covers the steel case outer diameter.
14	Contact width	0.1 to 0.25mm (New)	Axial width of wear track when seal is installed on shaft. Will increase as seal wears.
15	Contact line	--	Line formed where the outside and inside lip surfaces meet to form the sealing edge.
16	Radial dirt lip	--	Short non-sprung lip to prevent entry of outside contaminants.
17	Head thickness	1.25 to 2.50mm	Radial distance between contact line and the inner diameter of the spring groove.
18	Interference	0.75 to 3.5mm	Diametrical difference between the seal lip I.D. and the shaft. Dependent upon shaft size.
19	Lip I.D.	--	Inner diameter of the primary sealing lip.

Figure 31   Sealing element geometry. (Courtesy of Chicago Rawhide.)

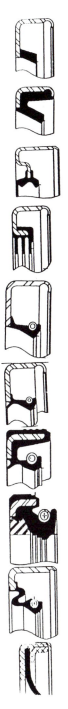

Seal Type	Typical Application
Non-spring loaded light duty	Retains grease at low speeds (less than 10m/sec). Used as excluder if primary lip faces atmosphere
Non-spring loaded	Rubber inside case prevents corrosion when used as a contaminant excluder
Non-spring loaded dual lip	Automotive wheel seal--retains grease while excluding contaminants
Heavy-duty excluder	Excludes contaminants in very dirty, low-speed applications (farm implements). Grease is packed between lips
Single lip spring loaded	Seal oil at speeds up to 25 m/sec
Dual lip spring loaded	Secondary lip excludes contaminants. Not effective at high speeds
Rubber O.D.	Rubber O.D. seals aluminum bores where thermal expansion can cause bore leaks. Also used if bores have minor imperfections
High pressure	Used to seal pressures up to 10,000 kPa in reciprocating applications (rack and pinion power steering)
High runout seal	Will withstand shaft dynamic runout up to 1.5 mm (diesel engine rear crankshaft)
Assembled seal	Wafer is tightly clinched between metal stampings to prevent rotation of wafer

Figure 32   General seal types. (Courtesy of Chicago Rawhide.)

runout, and other applications. Some seals are made with materials (i.e., PTFE) that cannot be bonded directly to the metal case. The lip material is clinched tightly between metal cases to prevent slippage of the wafer. Unitized seals incorporate a wear surface into the design that eliminates the need for the end user to grind the shaft surface (Fig. 33). Modern seal designs incorporate several components into a single package to provide added value to the customer (Fig. 34). Instead of several components to stock and install, only one is required.

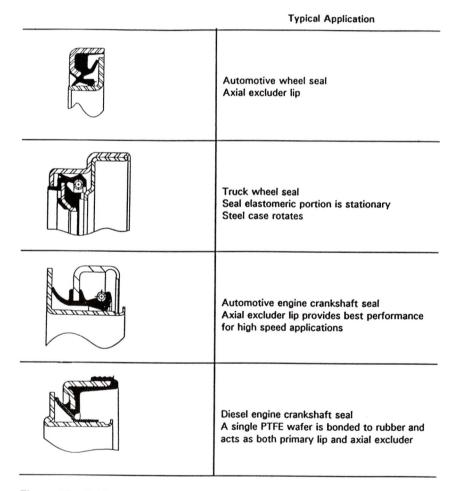

	Typical Application
	Automotive wheel seal Axial excluder lip
	Truck wheel seal Seal elastomeric portion is stationary Steel case rotates
	Automotive engine crankshaft seal Axial excluder lip provides best performance for high speed applications
	Diesel engine crankshaft seal A single PTFE wafer is bonded to rubber and acts as both primary lip and axial excluder

Figure 33   Unitized seal designs. (Courtesy of Chicago Rawhide.)

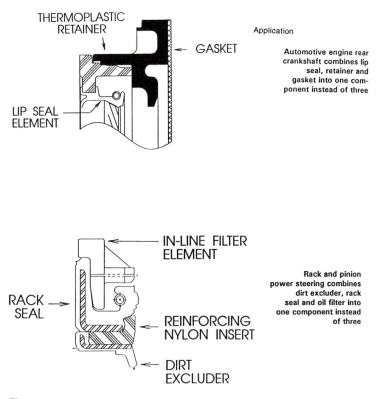

THERMOPLASTIC
RETAINER

Application

← GASKET

Automotive engine rear
crankshaft combines lip
seal, retainer and
gasket into one com-
ponent instead of three

LIP SEAL
ELEMENT

IN-LINE FILTER
ELEMENT

RACK
SEAL

Rack and pinion
power steering combines
dirt excluder, rack
seal and oil filter into
one component instead
of three

REINFORCING
NYLON INSERT

← DIRT
EXCLUDER

Figure 34 Value added seals. (Courtesy of Chicago Rawhide.)

C. Specifications

Radial lip seals will function properly only if the proper specifications for the shaft, bore, and seal are adhered to. Many seal manufacturers provide general-purpose seals to meet ISO and SAE (Society of Automotive Engineers) dimensional standards (Tables 28 and 29). Equipment designers should use these tables to allow the proper space for standard seals. Nonstandard seals will demand premium prices and may not be available when required.

Steel, cast iron, or stainless steel materials should be used for the shaft. If other materials are used, a steel sleeve should be pressed on the shaft to provide a wear surface for the seal.

The shaft should be hardened to Rockwell C-30 minimum. Harder shafts will not improve seal performance or prevent shaft scoring but will minimize shaft damage during handling. To prevent this damage, Rockwell C-45 is recommended. The shaft should be ground to provide pockets of lubrication to

Table 28 Dimensional Standards for Radial Lip Seals

DS	DB	W	DS	DB	W	DS	DB	W	DS	DB	W
6	16	7	25	52	7	45	65	8	120	150	12
6	22	7	28	40	7	50	68	8	130	160	12
7	22	7	28	47	7	50	72	8	140	170	15
8	22	7	28	52	7	55	72	8	150	180	15
8	24	7	30	42	7	55	80	8	160	190	15
9	22	7	30	47	7	60	80	8	170	200	15
10	22	7	30	52	7	60	85	8	180	210	15
10	25	7	32	45	8	65	85	10	190	220	15
12	24	7	32	47	8	65	90	10	200	230	15
12	25	7	32	52	8	70	90	10	220	250	15
12	30	7	35	50	8	70	95	10	240	270	15
15	26	7	35	52	8	75	95	10	260	300	20
15	30	7	35	55	8	75	100	10	280	320	20
15	35	7	38	55	8	80	100	10	300	340	20
16	30	7	38	58	8	80	110	10	320	360	20
18	30	7	38	62	8	85	110	12	340	380	20
18	35	7	40	55	8	85	120	12	360	400	20
20	35	7	40	62	8	90	120	12	380	420	20
20	40	7	42	55	8	95	120	12	400	440	20
22	35	7	42	62	8	100	125	12			
22	40	7	45	62	8	110	140	12			
22	47	7									
25	40	7									
25	47	7									

Dimensions in mm. DS = shaft diameter, DB = bore diameter, W = seal width.

prevent excessive lip wear during the initial hours of run-in. Excessive shaft roughness will cause premature lip wear and leakage. Plunge ground surfaces are recommended with a finish of 0.25 to 0.50 μm RA.

Shaft lead is miniature screw threads that result during machining operations. Grinding should remove these threads to prevent oil from being screwed out from the sump, which results in gross leakage. The lead angle specification is zero $\pm 0.05°$. The lead angle (Eq. (5)) is determined by measuring how far a loop of string placed on the shaft with a 25- to 30-g weight attached will advance during a measured time period when the shaft is slowly rotated.

$$\text{Lead angle} = \arctan \left[\frac{\text{string advance (mm)}}{\pi[\text{shaft dia (mm)}][\text{time of advance (min)}][\text{rpm}]} \right]$$

$$(5)$$

The shaft should have burr-free chamfers or a radius to protect the seal during installation. The chamfer should be 1.80 to 3.00 mm long with a lead-in angle of 15° to 30°. A radius of 3.00 mm can be used as an alternative method of providing a lead in at the shaft corners. The ISO and SAE shaft diameter tolerances appear in Table 30.

Table 29 Dimensional Standards for Radial Lip Seals

DS	DB	W	DS	DB	W	DS	DB	W	DS	DB	W
.500	.999	5/16	1.625	2.502	3/8*	2.750	3.500	7/16**	3.875	4.876	1/2
.500	1.124	5/16	1.625	2.623	3/8*	2.750	3.623	7/16**	3.875	4.999	1/2
.500	1.250	5/16	1.750	2.374	3/8*	2.750	3.751	7/16**	3.875	5.125	1/2
.625	1.124	5/16	1.750	2.502	3/8*	2.750	3.875	7/16**	3.875	5.251	1/2
.625	1.250	5/16	1.750	2.623	3/8*	2.875	3.623	7/16**	4.000	4.999	1/2
.625	1.375	5/16	1.750	2.750	3/8*	2.875	3.751	7/16**	4.000	5.125	1/2
.625	1.499	5/16	1.875	2.623	3/8*	2.875	3.875	7/16**	4.000	5.251	1/2
.750	1.250	5/16	1.875	2.750	3/8*	2.875	4.003	7/16**	4.000	5.375	1/2
.750	1.375	5/16	1.875	2.875	3/8*	3.000	3.751	7/16**	4.250	5.251	1/2
.750	1.499	5/16	1.875	3.000	3/8*	3.000	3.875	7/16**	4.250	5.375	1/2
.750	1.624	5/16	1.875	3.125	3/8*	3.000	4.003	7/16**	4.250	5.501	1/2
.875	1.375	5/16	2.000	2.623	3/8*	3.000	4.125	7/16**	4.500	5.501	1/2
.875	1.499	5/16	2.000	2.750	3/8*	3.125	4.125	1/2	4.500	5.625	1/2
.875	1.624	5/16	2.000	2.875	3/8*	3.125	4.249	1/2	4.500	5.751	1/2
.875	1.752	5/16	2.000	3.000	3/8*	3.125	4.376	1/2	4.750	5.751	1/2
1.000	1.499	5/16	2.000	3.125	3/8*	3.125	4.500	1/2	4.750	6.000	9/16
1.000	1.624	5/16	2.125	2.750	7/16**	3.250	4.249	1/2	5.000	6.000	9/16
1.000	1.752	5/16	2.125	2.875	7/16**	3.250	4.376	1/2	5.000	6.250	9/16
1.000	1.874	5/16	2.125	3.000	7/16**	3.250	4.500	1/2	5.000	6.375	9/16
1.125	1.624	5/16	2.125	3.125	7/16**	3.250	4.626	1/2	5.250	6.250	9/16
1.125	1.752	5/16	2.125	3.251	7/16**	3.375	4.249	1/2	5.250	6.375	9/16
1.125	1.874	5/16	2.250	3.000	7/16**	3.375	4.376	1/2	5.250	6.500	9/16
1.125	2.000	5/16	2.250	3.125	7/16**	3.375	4.500	1/2	5.250	6.625	9/16
1.250	1.752	5/16	2.250	3.251	7/16**	3.375	4.626	1/2	5.500	6.500	9/16
1.250	1.874	5/16	2.250	3.371	7/16**	3.500	4.376	1/2	5.500	6.625	9/16
1.250	2.000	5/16	2.375	3.125	7/16**	3.500	4.500	1/2	5.500	6.750	9/16
1.250	2.125	5/16	2.375	3.251	7/16**	3.500	4.626	1/2	5.500	6.875	9/16
1.375	2.000	3/8	2.375	3.371	7/16**	3.500	4.751	1/2	5.750	6.750	9/16
1.375	2.125	3/8	2.375	3.500	7/16**	3.625	4.626	1/2	5.750	6.875	9/16
1.375	2.250	3/8	2.500	3.251	7/16**	3.625	4.751	1/2	5.750	7.000	9/16
1.375	2.374	3/8	2.500	3.371	7/16**	3.625	4.876	1/2	5.750	7.125	9/16
1.500	2.125	3/8	2.500	3.500	7/16**	3.625	4.999	1/2	6.000	7.125	9/16
1.500	2.250	3/8	2.500	3.623	7/16**	3.750	4.626	1/2	6.000	7.500	9/16
1.500	2.374	3/8	2.625	3.371	7/16**	3.750	4.751	1/2			
1.500	2.502	3/8	2.625	3.500	7/16**	3.750	4.876	1/2			
1.625	2.250	3/8	2.625	3.623	7/16**	3.750	4.999	1/2			
1.625	2.374	3/8	2.625	3.751	7/16**						

Dimensions in inches. DS = shaft diameter, DB = bore diameter, W = seal width.
Source: Society for Automotive Engineers (J946).
*3/8 in. without inner cup; 1/2 in. with inner cup.
**7/16 in. without inner cup; 1/2 in. with inner cup.

Table 30 Shaft Diameter Tolerances

Shaft diameter (mm)	Tolerance	Shaft diameter (in.)	Tolerance
To and including 100	±0.08	To and including 4.000	±0.003
100.01 through 150	±0.10	4.001 through 6.000	±0.004
150.01 through 250	±0.13	6.001 through 10.000	±0.005

Table 31 Bore Specifications

	Seal width (W)	Housing bore depth	Chamfer length	Max. housing bore corner radius	Surface finish
Dimensions in mm	$W \leqslant 10$	$W + 0.9$	0.70–1.00	0.50	$< 3.2 \ \mu m$ RA
	$W > 10$	$W + 1.2$	1.20–1.50	0.75	
Dimensions in inches	All	$W + 0.016$	0.06–0.09	0.047	$< 125 \ \mu m$ RA

Source: Society for Automotive Engineers (J946).

    Seal bores are machined in the housings of the equipment and the seal is pressed into the bore. Proper press fit must be maintained to prevent leakage. Seal cases are usually made of steel, and no leakage will occur when a metal O.D. seal is properly installed into a properly toleranced ferrous bore (Tables 31, 32). If the housing is made of a nonferrous material, differential thermal expansion between the steel case and the nonferrous housing can result in O.D. leakage. Rubber O.D. seals are recommended to prevent this leakage (Table 33).

Table 32 Metal O.D. Seal Specifications

	Seal O.D. (D_o)	Bore tolerance	Nominal pressfit	Seal O.D. tolerance	Seal O.D. OOR
Dimensions in mm	$D_o \leqslant 50$	$-0.0 \ +0.039$	$+0.20/+0.08$	0.18	
	$50 < D_o \leqslant 80$	$-0.0 \ +0.046$	0.14	$+0.23/+0.09$	0.25
	$80 < D_o \leqslant 120$	$-0.0 \ +0.054$	0.15	$+0.25/+0.10$	0.30
	$120 < D_o \leqslant 180$	$-0.0 \ +0.063$	0.17	$+0.28/+0.12$	0.40
	$180 < D_o \leqslant 300$	$-0.0 \ +0.075$	0.21	$+0.35/+0.15$	0.25% of
	$300 < D_o \leqslant 440$	$-0.0 \ +0.84$	0.28	$+0.45/+0.20$	outside diameter
Dimensions in inches	$D_o \leqslant 1$	± 0.001	0.004	± 0.002	0.005
	$1 < D_o \leqslant 3$	± 0.001	0.004	± 0.002	0.006
	$3 < D_o \leqslant 4$	± 0.0015	0.005	± 0.002	0.007
	$4 < D_o \leqslant 6$	± 0.0015	0.005	$+0.003/-0.002$	0.009
	$6 < D_o \leqslant 8$	± 0.002	0.006	$+0.003/-0.002$	0.012
	$8 < D_o \leqslant 9$	± 0.002	0.007	$+0.004/-0.002$	0.015
	$9 < D_o \leqslant 10$	± 0.002	0.008	$+0.006/-0.002$	0.002 in./in.
	$10 < D_o \leqslant 20$	$-0.004 \ +0.002$	0.008	$+0.006/-0.002$	of seal OD
	$20 < D_o \leqslant 40$	$-0.006 \ +0.002$	0.008	$+0.008/-0.002$	
	$40 < D_o \leqslant 60$	$-0.010 \ +0.002$	0.008	$+0.010/-0.002$	

Source: Society for Automotive Engineers (J946).

Table 33 Rubber O.D. Seal Specifications (bore tolerance same as Table 32)

	Seal O.D. (D_o)	Nominal pressfit	Seal O.D. tolerance	Seal O.D. OOR
Dimensions in mm	$D_o \leqslant 50$	0.20	+0.30/+0.15	0.25
	$50 < D_o \leqslant 80$	0.25	+0.35/+0.20	0.35
	$80 < D_o \leqslant 120$	0.25	+0.35/+0.20	0.50
	$120 < D_o \leqslant 180$	0.32	+0.45/+0.25	0.65
	$180 < D_o \leqslant 300$	0.34	+0.50/+0.25	0.80
	$300 < D_o \leqslant 440$	0.38	+0.55/+0.30	1.00
Dimensions in inches	$D_o \leqslant 1$	0.009	±0.003	0.010
	$1 < D_o \leqslant 3$	0.011	±0.003	0.014
	$3 < D_o \leqslant 4$	0.012	±0.003	0.020
	$4 < D_o \leqslant 9$	0.012	±0.004	0.025
	$9 < D_o \leqslant 10$	0.014	±0.004	0.031
	$10 < D_o \leqslant 20$	0.017	±0.005	0.039
	$20 < D_o \leqslant 40$	0.018	±0.006	0.045
	$40 < D_o \leqslant 60$	0.030	±0.007	0.050

Source: Society for Automotive Engineers (J946).

The seal dimensions have specifications that must be adhered to for good performance. The seal O.D. tolerance, the nominal press fit required between the seal O.D. and the housing and the maximum allowable out of round (OOR) appear in Table 32 for metal O.D. seals and 33 for rubber O.D. seals. Seal O.D. is determined by the average of a minimum of three equally spaced measurements. The out of round is the difference between the largest and smallest of these readings. The tolerances for the seal primary lip and seal width appear in Tables 34 and 35. The radial wall dimension is defined to be the radial dis-

Table 34 Primary Lip Inner Diameter Tolerances

	Shaft diameter (DS)	Tolerance
Dimensions in mm	$DS \leqslant 75$	±0.05
	$75 < DS \leqslant 150$	±0.65
	$150 < DS \leqslant 250$	±0.75
Dimensions in inches	$DS \leqslant 3.000$	±0.020
	$3.000 < DS \leqslant 6.000$	±0.025
	$6.000 < DS \leqslant 10.000$	±0.025

Source: Society for Automotive Engineers (J946).

Table 35 Seal Width Tolerances

	Seal width (W)	Tolerance
Dimensions in mm	$W < 10$	± 0.3
	$W > 10$	± 0.4
Dimensions in inches	$W < 0.400$	± 0.015
	$W > 0.400$	± 0.020

tance from the seal O.D. to the primary lip. The radial wall variation is the difference between the maximum and minimum values of the radial wall measurements. The maximum values for radial wall variation appear in Table 36. The load that the seal exerts on the shaft after installation can be measured and toleranced (Table 37).

Garter springs (Fig. 35) are used on some seals to provide an additional lead that will compensate for radial load changes that occur when the rubber lip is exposed to heat and oil. The spring should be designed to pull the primary lip in about 0.25 to 0.50 mm until the garter spring coils touch each other. The spring is manufactured by coiling carbon steel (SAE 1050 to 1095) or stainless steel (SAE 30302 to 30304) wire with a degree of back winding to give a high initial tension (Fig. 36). The initial tension should be 50% to 80% of the desired spring load when the seal is installed on the shaft. A high initial tension will result in a low spring rate and less variation in total load. Tolerances for the spring assembled inside diameter and other parameters appear in Table 38.

D. Installation Procedures

Premature leakage is often caused by improper installation procedures. Automatic or semiautomatic hydraulic equipment is recommended in high-

Table 36 Specifications for Radial Wall Variation

Shaft diameter (DS)	Max. radial wall variation (RWV)	Shaft diameter (DS)	Max. radial wall variation (RWV)
Dimensions in mm		Dimensions in inches	
$DS \leqslant 75$	0.6	$DS \leqslant 3$	0.025
$75 < DS \leqslant 150$	0.8	$3 < DS \leqslant 6$	0.030
$150 < DS \leqslant 250$	1.0	$6 < DS \leqslant 10$	0.040

Source: Society for Automotive Engineers (J946).

Table 37 Radial Load Specification

Shaft diameter (DS)	Radial load tolerance (range)	Shaft diameters (DS)	Radial load tolerance (range)
Dimensions in mm		Dimensions in inches	
DS ⩽ 75	Nominal load ±45%	DS ⩽ 3	Nominal load ±45%
75 < DS ⩽ 250	Nominal load ±40%	3 < DS ⩽ 10	Nominal load ±40%

Source: Society for Automotive Engineers (J946).

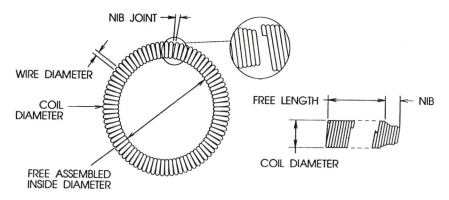

Figure 35 Garter spring features. (Courtesy of Chicago Rawhide.)

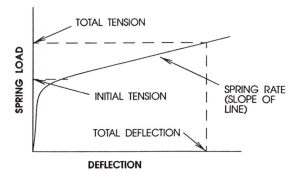

Figure 36 Spring load curve. (Courtesy of Chicago Rawhide.)

Table 38 Spring Tolerances

Spring load	20% of nominal load or ±0.14 N (which ever is largest)
Coil diameter	±0.13 mm
Wire diameter	Reference dimension (±0.03 mm when specified)
Coil-to-wire diameter ratio	5 or larger

Assembled inner diameter (AID) tolerances			
Wire diameter	AID tolerance	Wire diameter	AID tolerance
Dimensions in mm		Dimensions in inches	
0.15–0.28	±0.20	0.006–0.011	±0.008
0.30–0.48	±0.30	0.012–0.019	±0.012
0.50–0.76	±0.40	0.020–0.030	±0.015
0.80–1.40	±0.50	0.031–0.055	±0.020

Courtesy: Society for Automotive Engineers (J946).

volume production assembly operations. When replacing a seal, an arbor press should be used. The seal should be examined before installation to be sure it is the proper part for the application and no handling damage has occurred to the lip or O.D. The housing and shaft should be checked to be sure there are no burrs or scratches that will cut the lip or damage the seal O.D. during installation. A new seal should always be used and the lip should be prelubricated with oil from the application. The press fit or installation tool should be designed slightly smaller than the seal O.D. to press near the outside, strongest portion of the seal to prevent crushing. Assembly cones are used to prevent seal lip damage when the shaft edge is badly burred or if the shaft has keyways or splines (Fig. 37). The seal should not be cocked during installation and the installation tool should bottom the seal in the bore. If the seal is installed in a through bore, then the installation tool should bottom out on a surface that has

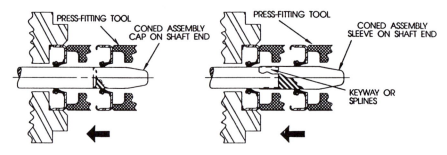

Figure 37 Assembly cones. (Courtesy of Chicago Rawhide.)

been machined perpendicular to the shaft axis. The tool can bottom on the shaft face or the front of the housing (Fig. 38). After installation, the seal area should be marked to protect the seal during painting operations. Paint baking times and temperatures should be minimized to protect the seal material. Caustic cleaning fluids should not be allowed to contact the lip material.

E. Troubleshooting Lip Seal Problems

Leakage can often result from system problems and is often interpreted as a defective seal. It is recommended that the area around the seal be carefully inspected before the seal is removed. A fluorescent dye is often placed in the oil and a black light is used to pinpoint where the leakage originates. After the seal is removed, the shaft and bore should be closely inspected to be sure there are no nicks, scratches, excessive runout or end play (bad bearings), or any other condition that could cause leakage. Other causes for lip seal leakage and recommended corrective action appear in Table 39.

IX. DYNAMIC SEALS: MECHANICAL FACE

A. Function and General Seal Designs

Mechanical face seals prevent leakage of fluids in rotating shaft applications that have requirements that exceed the capabilities of packings and elastomeric

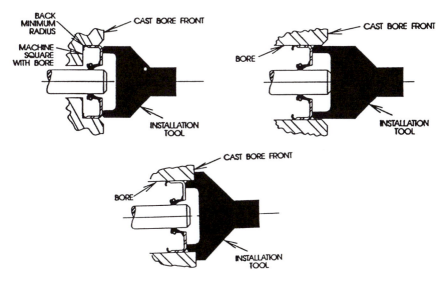

Figure 38 Preventing seal cocking during installation. (Courtesy of Chicago Rawhide.)

Table 39 Seal System Failure Analysis

Symptom	Possible cause	Corrective action
Early lip leakage	Nicks, tears or cuts in seal lip	Examine shaft. Eliminate burrs and sharp edges. Use correct mounting tools to protect seal lip from splines, keyways or sharp shoulders. Handle seals with care. Keep seals packaged in storage and in transit.
	Rough shaft	Finish shaft to 0.25 to 0.50 μm RA or smoother
	Scratches or nicks on surface	Protect shaft after finishing.
	Lead on shaft	Plunge-grind shaft surface.
	Excessive shaft whip or runout	Locate seal close to bearings. Ensure good, accurate machining practices.
	Cocked seal	Use correct mounting tools and procedures.
	Paint on shaft or seal element	Mask seal and adjacent shaft before painting.
	Turned-under lip	Check shaft chamfer for roughness. Machine chamfer to 0.8 μm RA or smoother, blend into shaft surface. Check shaft chamfer for steepness. Use recommended lead chamfer. Use correct mounting tools and procedures.
	Damaged or "popped out" spring	Use correct mounting tools and procedures to apply press-fit force uniformly. Protect seals in storage and transit.
	Damaged or distorted case	Use correct mounting tools and procedures to apply press-fit force uniformly. Protect seals in storage and transit.
	OD seal on shaft or lip element	Use care in applying OD sealant. Purchase precoated seals.
Lip leakage, intermediate life	Excessive lip wear	Check seal cavity for excessive pressure. Provide vents to reduce pressure. Provide proper lubrication for seal. Check shaft finish. Make sure finish meets specifications.
	Element hardening and cracking	Reduce sump temperature if possible. Upgrade seal material. Provide proper lubrication for seal. Change oil frequently. Change seal design.
	Element corrosion and reversion	Check material-lubricant compatibility. Change material.
	Excessive shaft wear	Check shaft hardness. Harden to Rockwell C30 minimum. Change oil frequently to remove contaminants. Use dust lip in dirty atmosphere.
O.D. leakage	Scored seal O.D.	Check housing machining. Use 3.2 μm RA or less. Check edges on housing bore. Use recommended chamfer. Remove burrs.
	Damaged seal case	Use correct mounting tools and procedures to apply press-fit uniformly. Protect seals in storage and transit.

Source: Chicago Rawhide.

radial lip seals. They provide long life without shaft wear. The shaft finish, hardness, direction or rotation, and material is not critical for seal function. The seal will function in a variety of fluids (acids, salts, oils, gases) with large pressure (up to 20,000 kPa), speed (up to 50,000 rpm), and temperature excursions (-200 to $650°C$). Seals can be mounted in tandem to accommodate higher-pressure requirements. They provide positive sealing for food processing, hazardous chemicals, and radioactive fluids. The disadvantages include high initial cost, larger space requirements than radial lip seals, and an inability to handle end play.

Mechanical seal design employs a spring-loaded seal head that is pressed against a stationary mating face. The faces are lapped to a finish of 0.08 to 0.4 μm RA and must have flatness to within three light bands of helium (0.87 μm). Secondary seals (O ring, etc.) are used to prevent leakage between components and prevent slippage between the seal components and the shaft or housing. Pins are sometimes employed at high speeds to lock the seal components in place. The simplest design configuration employs rotating heads and stationary seats and is used for speeds up to 6000 rpm. At higher speeds, stationary heads and rotating seats are used to provide better balance. Methods of providing a flush or buffer liquid is sometimes employed to cool the seal area, protect the seal faces from contaminants in the sealed fluid and to serve as a barrier to prevent leakage of dangerous fluids (Fig. 39).

The springs, seal metal components, and housing are often made with noncorrosive materials such as stainless steel or plated mild steels.

B. Spring/Bellows Design

Springs or bellows are used to control the loading of the two faces and to apply a closing force at all times. The spring load is especially important at low or zero pressures to keep the primary face properly seated on the mating ring. Typical designs and the advantages/disadvantages appear in Fig. 40.

The single-metallic-coil spring is the simplest design and offers low cost. It is the simplest to assemble with fewer parts than multiple coils. It is resistant to blockage and corrosion. A single coil requires more axial space than other designs and may become difficult to manufacture and compress as seal size increases; centrifugal force may tend to unwind or tighten the spring as shaft speed increases. This tendency to unwind or tighten depends upon the direction of the coil windings (left hand or right hand). The single coil can also provide uneven loading on the faces, which will result in uneven face wear.

Multiple coils are used to provide more even face loads. They require less axial space, are used for large diameter shafts, and are insensitive to centrifugal forces developed at high shaft rpm. A multiple-coil spring package has more parts than a single coil, is more difficult to assemble, and is more expensive. They also are less corrosion resistant and more prone to clog with contaminants than a single-coil spring.

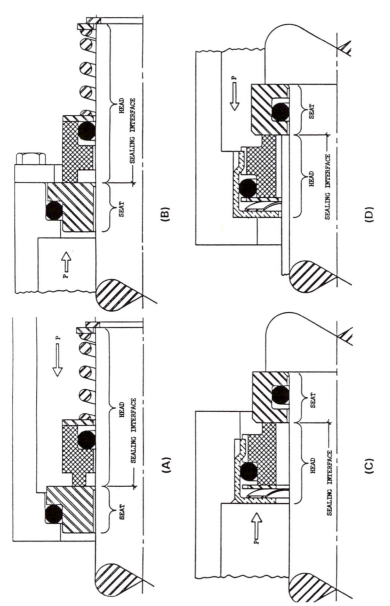

Figure 39 Typical mechanical seal configurations: (A) rotating internally mounted seal head—pressure on outside diameter of faces; (B) rotating externally mounted seal head—pressure on inside diameter of faces; (C) stationary internally mounted seal head—pressure on inside diameter of faces; (D) stationary externally mounted seal head—pressure on outside diameter of faces.

Spring Type	Advantages	Disadvantages
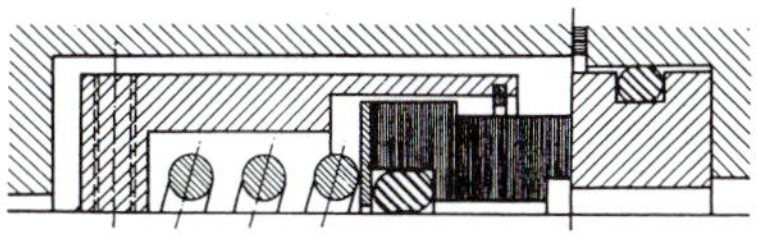 Single coil	Corrosion, blockage resistant. Low stress levels. Low cost. Greater axial tolerance	Uneven loading. Requires more axial space. Difficult to compress as size increases. May unwind/tighten at high speeds
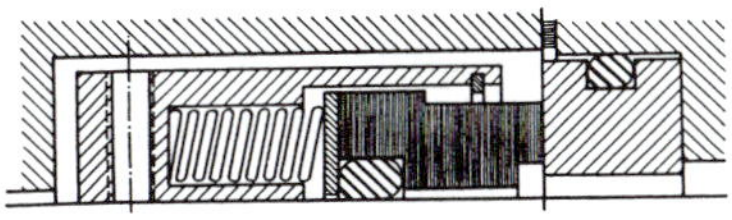 Multiple coils	Less axial space required. Even face loads. Resists high speeds.	Less corrosion/blockage resistance. High stress levels. More cost.
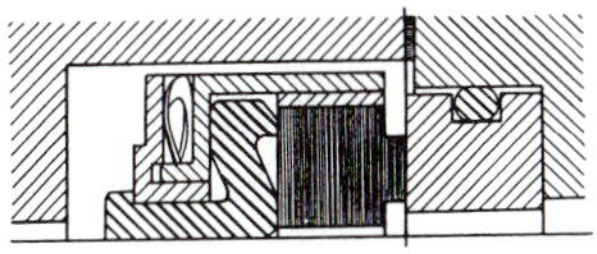 Wave spring/Belleville washer	Saves space	High spring rate. High cost.
 Elastomer bellows	Also provides secondary seal. Relatively inexpensive.	Cannot be used in all fluids. Has temperature limitations
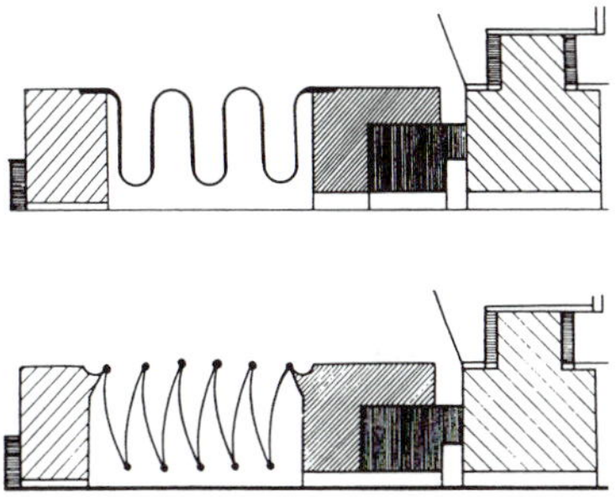 Corrugated or welded metal bellows	Provides secondary seal. Corrosion resistant. High temperature. High controlled rate	Expensive. Requires more space than coil strings.

Figure 40 Spring/bellows designs.

Wave springs or Bellville washers are used when axial space is limited. These springs typically have a high spring rate and a higher cost than coil springs.

Elastomeric molded springs also act as a secondary seal. Two functions are thus accomplished with a single component, which often results in a lower cost. Unfortunately, elastomers cannot be used at high temperatures and may be attacked by the fluids that are being sealed. Table 3 must be used to ensure the correct polymer type is selected.

Corrugated or welded metal bellows are used in high-temperature applications and function as a secondary seal as well as a spring. Corrosion-resistant materials can be selected to seal chemically active fluids. These designs are very expensive and require more space than coil springs.

No firm design rules exist to determine the proper nominal spring load. The choice is usually based on experience and may range from 0.035 to 0.35 MPa. Typically, low spring loads are selected for an unlubricated gas seal, and high loads will be used for a seal that is used in low-pressure application or for a balanced seal.

C. Secondary Seals

Elastomeric secondary seals are made from polymers which are selected to meet the temperature requirements of the application and to be compatible with the application fluid (Table 40). The most common and least expensive secondary seal design is the elastomeric O ring. Other designs such as the chevron, the U cup, and the wedge can be molded from elastomers and are pressure activated (Fig. 41). As the seal faces wear, the internal pressure and the spring push the secondary seals along the shaft. These seals are commonly called "pusher seals." The shaft must be in good condition with no corrosion, pitting, or solid deposits to inhibit seal movement.

When corrugated or welded metal bellows are used to provide face loading, they also serve as a secondary seal that does not contact the shaft.

D. Face Materials

The rotating face of a mechanical seal can be bronze, ceramic, carbon babbitt, tungsten carbide, or, most commonly, a carbon/graphite. Carbon/graphite faces have excellent dry running characteristics with good thermal conductivity. The proportions of carbon and graphite are controlled during the manufacturing process. The powder is compressed into a rough shape and then heated to 1000–2000°C. The material is then machined to the final shape and the contacting face is lapped. The carbon/graphite is often impregnated with metal or resin to add strength, wear resistance, and a lubricating film. Counterfaces are made of a variety of materials. A selection chart for face materials in various media appears in Table 40.

E. Seal Balance

When the internal pressure increases, the force pushing the seal faces together can also increase which will accelerate wear and friction. Changing face design by adding a step will balance the internal face pressure and reduce face loading

Table 40 Selection Chart for Face Materials in Various Media

Water	Salt solution	Sea water	Acids	Gasoline	Hydrocarbons
CG vs $\begin{cases} \text{B} \\ \text{NIR} \\ \text{C} \\ \text{TC} \\ \text{SC} \end{cases}$ TC vs SC	CG vs $\begin{cases} \text{SS} \\ \text{C} \\ \text{MO} \end{cases}$ CB vs PB	CB vs $\begin{cases} \text{AB} \\ \text{SC} \end{cases}$ TC vs SC B vs LP	CG vs $\begin{cases} \text{SS} \\ \text{C} \\ \text{CPTFE} \\ \text{HABC} \end{cases}$ C vs $\begin{cases} \text{GPTFE} \\ \text{PTFE} \end{cases}$	CG vs $\begin{cases} \text{NIR} \\ \text{C} \\ \text{SF} \\ \text{SC} \\ \text{TC} \end{cases}$	CG vs $\begin{cases} \text{NIR} \\ \text{SF} \\ \text{TC} \\ \text{SC} \end{cases}$

Rotating Face Materials
B, bronze
CB, carbon-babbitt
CG, carbon-graphite
TC, tungsten carbide
C, ceramic

Counterface Materials
AB, aluminum-bronze
B, bronze,
C, ceramic
MO, monel
NIR, nickel-resist
TC, tungsten carbide
SC, silicon carbide
PB, phosphor-bronze
LP, laminated plastic
CPTFE, carbon-filled Teflon (nonoxidizing acids)
GPTFE, glass-filled Teflon (oxidizing acids)
PTFE, Teflon
SS, stainless steel
HABC, Hasteloy, A, B, or C
SF, stellite hard facing on stainless steel

(Fig. 42). The balance ratio (Eq. (6)) is defined as the ratio of the hydraulic loading area to the seal interface area. When the balance ratio decreases, the load due to the internal pressure decreases.

$$B = \frac{\text{Hydraulic load area}}{\text{Seal interface area}} = \frac{\frac{1}{4}\pi(D_o^2 - D_s^2)}{\frac{1}{4}\pi(D_o^2 - D_i^2)} = \frac{D_o^2 - D_s^2}{D_o^2 - D_i^2} \tag{6}$$

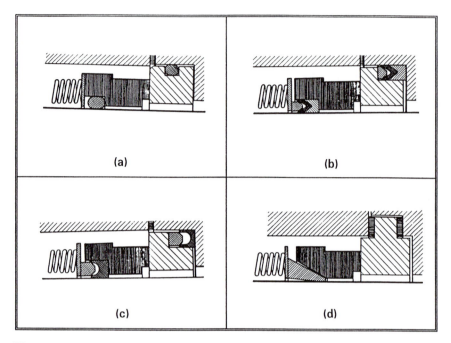

Figure 41   Secondary seal configurations: (a) O ring; (b) chevron; (c) U cup; (D) wedge.

where

B = balance ratio

D_o = interface outside diameter, mm

D_i = interface inside diameter, mm

D_s = shaft diameter, mm

Unbalanced seals are used for low-pressure applications (less than 1000 kPa) and have balance ratios of 1.1 to 1.2. Balanced seals for higher pressures have a balance ratio of 0.65 to 0.85 to reduce face loading while maintaining stability. Seals with balance ratios less than 0.65 can be hydraulically unstable if the pressure fluctuates and are more expensive to manufacture.

F.   PV Values

The PV factor (Eq. (7)) is obtained by multiplying the internal system pressure drop across the seal face pressure in kPa by the rubbing velocity in m/s. It is used as a guide in estimating the upper operating limits of various seal materials and design combinations (Table 41). The average rubbing velocity is often

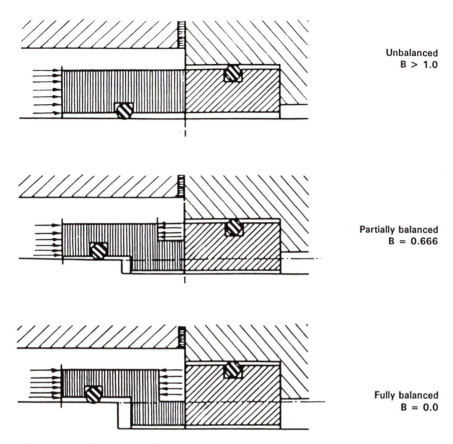

Unbalanced
B > 1.0

Partially balanced
B = 0.666

Fully balanced
B = 0.0

Figure 42 Mechanical seal balance.

Table 41 PV Limits (kPa m/s) for Various Seal Face Material Combinations

Face materials	Water and aqueous solutions		Other fluids	
	Unbalanced seals	Balanced seals	Unbalanced seals	Balanced seals
Stainless steel carbon	550	—	3,000	—
Lead bronze carbon	2,300	—	3,500	—
Stellite/carbon	5,000	8,500	5,200	58,000
Chrome oxide/carbon	7,000	45,000	—	—
Alumina ceramic/carbon	3,500	25,000	8,800	42,000
Tungsten carbide/tungsten carbide	4,500	50,000	7,000	42,000
Tungsten carbide/carbon	7,000	42,000	8,800	122,000

Note: Multiply PV value in kPa m/s by 23.5 to obtain PSI ft/min.

estimated using the shaft diameter, but a more precise calculation will use the average diameter of the interface area.

$$PV = \Delta P \frac{\pi}{1000} \frac{D_o + D_i}{2} \frac{N}{60} \tag{7}$$

where
ΔP = pressure drop across seal face, kPa
N = shaft speed, rpm
PV = PV factor, kPa · m/s

Seal capability is low if the PV factor is less than 700 kPa · m/s and high if PV is greater than 10000 kPa · m/s. Balanced seals will have higher PV values than unbalanced seals since the pressure acting on the seal faces is reduced.

G. Troubleshooting Mechanical Face Seals

Table 42 provides guidance when determining the cause of premature seal leakage.

X. DYNAMIC SEALS: NONCONTACTING
A. Bushings

A bushing seal has a small clearance gap between the bushing inner diameter and the rotating shaft that allows a controlled amount of leakage. The leakage can be minimized by reducing the gap clearance or increasing the bushing length. Bushing materials are chosen that provide low friction and will wear smoothly to accommodate shaft-to-bore misalignment or shaft dynamic runout. Babbitt is a good material for low-temperature applications. Bronze, aluminum alloys, carbon graphite, PTFE mixed with molybdenum disulfide, or graphite is used at higher temperatures (Table 43). The simplest design is a sleeve that is fixed in the housing, but clearances and leakage must be fairly large to accommodate shaft misalignment, cocking, and eccentricity (Fig. 43). This disadvantage is overcome by allowing the bushing ring to float in the radial direction. An axial spring forces the face of the ring against the housing to prevent leakage around the seal assembly. Pins are used to prevent rotation of the seal. The force between the seal face and the housing can be increased by the system pressure. By balancing the seal, this force can be controlled much like a mechanical seal.

Some seals have segmented, interlocking rings that float. A garter spring provides a radial force that pushes the rings together and tightly around the shaft to reduce the clearance and leakage.

Other seal designs consist of a series of bushings. Each one reduces the pressure until the final stage vents to the atmosphere. This is done to replace a single long bushing with a series of short ones to prevent rubbing damage if the shaft is not perfectly aligned with the housing.

Fixed bushings are used primarily to throttle the leakage of fluids and are not used to seal gases. The large clearance of fixed bushings and low viscosity of gases would result in unacceptable leakage rates. Gases are sealed with floating bushings that have very low clearances. Leakage rates can be estimated with the equations in Table 44.

B. Labyrinth Seals

Labyrinth seals are an extension of the bushing seal concept and are primarily used to seal gases in compressors and steam turbines. A series of teeth and expansion chambers presents a long tortuous path for the gas to follow as it escapes the chamber. The simplest design is the straight-through design that generates high gas velocities at the throat of each constriction. The energy is dissipated when the gas expands into the chamber beyond the throat. The efficiency of the labyrinth is improved by alternating grooves or teeth, but assembly is more complicated. The gases continually change directions as they flow from one chamber to another and the energy losses are greater than a straight labyrinth. Stepped labyrinths are also more effective than straight labyrinths, and the assembly is much simpler than labyrinths with alternating teeth (Fig. 44). Combinations of labyrinth types can be found in most turbine applications. If the process gas is dangerous, the labyrinth can be buffered by introducing an inert gas or air at a slightly higher pressure than the process gas. A portion of the buffer gas will leak into the labyrinth. A portion of the inert gas will also leak into the atmosphere (Fig. 45). The leakage of steam from a labyrinth can be estimated from Eq. (8).

$$G = \frac{11.34KA(P_i/V_i g[1 - (P_o/P_i)^2])^{1/2}}{N_T - \text{Ln}(P_o/P_i)} \tag{8}$$

where

A = clearance area = $2\pi R_s h$, cm^2

R_s = shaft radius, cm

h = radial clearance between shaft and labyrinth tooth, cm

P_i = upstream pressure, g/cm^2

P_o = downstream pressure, g/cm^2

N_T = number of teeth in labyrinth

g = gravitational constant, 980 cm/s^2

K = experimental flow coefficient: straight labyrinth, 60 to 120; inclined teeth, staggered teeth, 30 to 65

G = flow rate of steam, kg/h

V_i = specific volume of steam at P_i, cm^3/g

Table 42 Troubleshooting Face Seals

Symptoms	Appearance/observations	Causes
Seal leaks steadily whether shaft is stationary or rotating	Full 360° contact patterns on mating ring, little or no measurable wear	Secondary seal leakage caused by • Nicked, scratched or porous seal surfaces • O-ring compression set • Chemical attack due to improper choice of elastomer • O-ring extrusion due to excessive pressure, temperature or elastomeric swell • Corroded nickel or pitted secondary seal counter-surface • Corroded, cracked or porous housing material
	Noise from flashing or popping. High wear or thermal distress on mating ring. High wear and carbon deposits on primary ring. Possible edge chipping on primary ring. Thermal distress over 1/3 of mating ring, located 180° from inlet of seal flush. High wear and possible carbon deposits on primary ring. Thermal distress at 2–6 locations on mating ring. High wear and possible carbon deposits on primary ring.	Sealed liquid vaporizing at seal interface, caused by • Low suction or stuffing box pressure • Improper running clearance between shaft and primary ring • Insufficient cooling • Circumferential flush groove in gland plate missing or blocked • Overloaded seal • PV value too high
	High wear and grooving on mating ring	• Poor lubrication from sealed fluid • Abrasives in fluid • PV value too high

Symptom	Observation	Cause
Steady leakage at low pressure, little or no leakage at high pressure	Damaged mating ring. Eccentric contact pattern, although width equals that of primary ring. Possible cracks on mating ring.	Misaligned mating ring caused by • Improper clearance between gland plate and stuffing box • Lack of concentricity between shaft O.D. and stuffing box I.D.
Seal leaks steadily when shaft is rotating, little or no leakage when shaft is stationary	Two large contact spots, pattern fades away between spot. Contact through about 270°, pattern fades away at low spot. Contact spots at each bolt location.	Mechanical distortion caused by • Overtorqued bolts • Out-of-square clamping parts • Out-of-flat stuffing box faces • Nicked or burred gland surface
Steady leakage when shaft is rotating, no leakage when shaft is stationary	Heavy contact on mating ring O.D., fades to no visible contact at I.D. Possible edge chipping on primary ring O.D.	Deflection of primary ring from overpressurization. Seal faces not flat because of improper lapping.
	Heavy contact pattern on mating ring I.D., fades to no visible contact at O.D. Possible edge chipping on primary ring I.D.	Thermal distortion of seal faces. Seal faces not flat because of improper lapping.
	Contact pattern on mating ring is slightly larger than primary ring width. Possible high spot opposite drive pinhole.	Out-of-square mating caused by • Nicked or burred gland surfaces • Improper drive pin extension • Misaligned shaft • Piping strain on pump casing causing distortion • Bearing failure • Shaft whirl

Table 43 Floating-Bushing Ring-Seal Materials and Environment Combinations

Environment	Seal ring material	Shaft material
Oil	Babbitt	Hardened steel
	Bronzes	Shafting
	Aluminum	Chrome plate
	Carbon graphite	Nitrided steels
Water	Bronzes	440-C
	416 stainless steel	Chrome plated
	Stellite	
	Carbon graphite	
	Ceramics	Chrome plate
Gas (Air, CO_2),	Carbon graphite	Tool steels, hardened
H_2, He, N_2, O_2		
		Tungsten carbide plate
		Ceramic plate
		Chrome carbide
		Chrome plate
		Stainless steel (300)
		Stainless steel (400)
		Stainless steels 50 Rockwell C

The leakage rate is directly proportional to the radial clearance between the labyrinth tooth inner diameter and the shaft. It is inversely proportional to the number of teeth. To minimize leakage, the clearance h must be minimized and the number of teeth should be as many as practical. The teeth are typically designed to taper from a wide root to a narrow tip width (0.15 to 0.50 mm) with a taper angle of 8° to 12°. The tip of the tooth is made from materials that will wear or bed in if initial rubbing contact is made with rotating parts. Typical materials chosen are brass, bronze, soft aluminum, or any other soft materials that will wear evenly and cleanly and withstand the temperature without corrosion in the process gas. Leaded nickel-tin bronze will function very well in steam up to 500°C. If air or oxygen is present, this material will corrode rapidly at 500°C and its use is restricted to temperatures of 230°C or less. Some aircraft designs use a honeycomb, with abradable material that will fit almost line to line after the initial bedding or wear in. A multitude of labyrinth designs are used with other seal types to exclude contaminants.

C. Visco Seals

Some seal designs have a screw thread machined on the shaft or in the housing that provide a positive pumping or screwing action to keep the fluid from

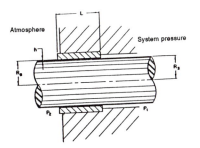

Fixed Bushing

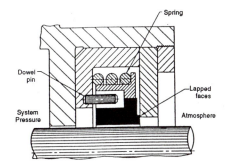

Floating Bushing Balanced

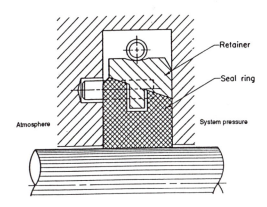

Segmented Floating Bushing

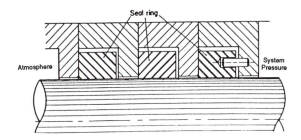

Multi-Stage Bushing Seal

Figure 43 Bushing seals.

Table 44   Estimated Bushing Seal Leakage

	Incompressible (cm^3/s)	Compressible (g/s)
Laminar-flow leakage rate (concentric)	$Q_L = \dfrac{\pi R h^3 g \Delta P}{6 \nu p L}$	$G_L = \dfrac{\pi R h^3 g}{12 \nu L}\left[\dfrac{P_1^2 - P_0^2}{P_s}\right]\dfrac{T_s}{T_1}$
Laminar-flow leakage rate (eccentric)	$Q_{Le} = Q_L\left[1 + \dfrac{3}{2}n^2\right]$	$G_{Le} = G_L\left[1 + \dfrac{3}{2}n^2\right]$
Turbulent-flow leakage rate (concentric)	$Q_T = 2\pi R\left[\dfrac{h^3 g \Delta P}{0.0665 \nu^{1/4} \rho L}\right]^{4/7}$	$G_T = 2\pi R h \alpha \sqrt{g P_i \rho_i}$
Turbulent-flow leakage rate (eccentric)	$Q_{Te} 1.315 Q_T$	$G_{Te} 1.5 G_T$

Note:

Q_L	= laminar concentric flow rate	cm^3/s
Q_{Le}	= laminar eccentric flow rate	cm^3/s
Q_T	= turbulent concentric flow rate	cm^3/s
Q_{Te}	= turbulent eccentric flow rate	cm^3/s
G_L	= leakage rate of laminar concentric	g/s
G_{Le}	= leakage rate of laminar eccentric	g/s
G_T	= leakage rate of turbulent concentric	g/s
G_{Te}	= leakage rate of turbulent eccentric	g/s
R	= mean radius of annulus	cm
h	= mean radial clearance	cm
ν	= kinematic viscosity	cm^2/s
ρ	= fluid density	g/cm^3
ρ_i	= gas density at the upstream condition	g/cm^3
Δp	= pressure drop	g/cm^2
P_i	= upstream pressure	g/cm^2
P_o	= downstream pressure	g/cm^2
P_s	= standard pressure	g/cm^2
T_s	= standard temperature	K
T_i	= upstream temperature	K
α	= flow resistance coefficient	empirical
n	= eccentricity/radial clearance	e/h
g	= gravitation constant	$980\ cm/s^2$
e	= shaft to bushing eccentricity	cm

flowing out of the sump. These seals are effective only when the shaft is rotating and must have an auxiliary seal to prevent leakage at low speeds or when the shaft has stopped.

The seal is effective only if the shaft rotates in one direction. If the shaft is reversed, the fluid will be screwed out of the sump into the atmosphere. These

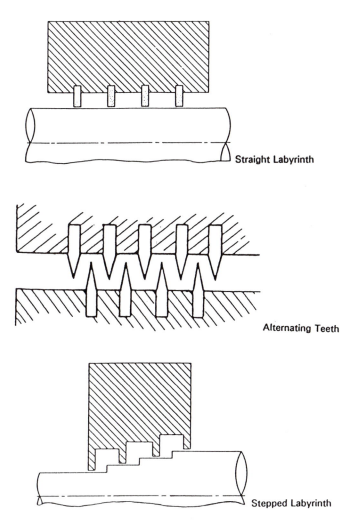

Straight Labyrinth

Alternating Teeth

Stepped Labyrinth

Figure 44 Labyrinth seals: (a) straight labyrinth; (b) alternating teeth; (c) stepped labyrinth.

seals have no contacting parts, so they have long life with high reliability. They are often used in applications where leakage of process fluid is unacceptable. Geometry factors for visco seals are defined in Fig. 46. Recommended values for these geometry factors appear in Table 45.

D. Magnetic Seals

A typical magnetic seal consists of a series of gaps created by machining triangular grooves or serrations onto a ferrous shaft or more typically a ferrous

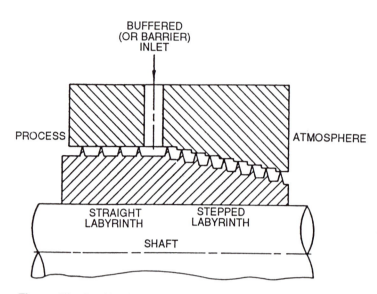

Figure 45 Combination labyrinth with buffer gas.

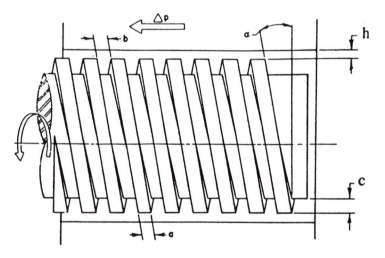

Figure 46 Visco seal geometry.
a = axial land width, cm
a = axial groove width, cm
a = groove depth, cm
a = radial clearance, cm
α = helix angle, °
γ = $b/(a + b)$
β = $(h + c)/h$

Table 45 Recommended Geometry Factors for Visco Seal

	Laminar	Turbulent
α	15–20°	10–15°
y	0.5	0.62
β	3.6–4.1	4.1–6.5

sleeve that is pressed on a shaft. A ring magnet is placed between pole pieces that creates a magnetic field between the gaps and the pole pieces (Fig. 47). The fluid film prevents leakage of internal gases and the entrance of light contamination. Magnetic seals are not used to seal liquids since the liquid splash tends to wash away and dilute the magnetic fluids. Leakage will eventually result. Magnetic seals have no contacting parts that result in low friction and no wear. The only friction results from the shear of the magnetic fluid in the gaps. This can be minimized by selecting low-viscosity magnetic fluids. Since there is no wear of components, long life is expected and there is no specific requirements for shaft surface finish. Upper shaft speed is limited only by the ability to remove the heat generated by the fluid shear and thus prevent fluid breakdown in the gaps. Shaft speeds up to 120,000 rpm have been achieved. The typical radial clearance in the gap is 0.05 to 0.125 mm. The seal must be installed close to bearings to minimize shaft-to-bore misalignment and dynamic runout to prevent contact of the rotating shaft and the pole pieces. The upper limit for the gap radial clearance is about 0.25 mm. Larger gaps will significantly decrease the ability to seal pressure. The magnetic fluid consists of a carrier fluid that contains a colloidal suspension of microscopic ferrous parti-

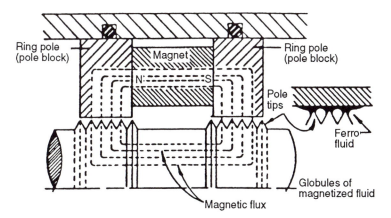

Figure 47 Magnetic seal. (Courtesy of Ferrofluidics Corporation.)

cles. The average size of the particles is 100 Å. The fluid becomes magnetized in the presence of a magnetic field. The viscosity will rise slightly as the external magnetic field is intensified. When the fluid becomes magnetically saturated, the viscosity will rise abruptly. The strength of the external magnetic field is designed to prevent the fluid from reaching saturation and high viscosities. Typical values range from 15,000 to 20,000 Oe. For a single serration or stage, the theoretical pressure differential that can be withheld is calculated from Eq. (9).

$$\Delta P = \frac{M_s H}{40,000\pi} \qquad (9)$$

where

ΔP = pressure differential, kPa

M_s = saturation level for magnetic fluid, G

H = strength of external magnetic fluid, Oe

For example, a fluid with a saturation level of 500 G will theoretically sustain a pressure drop of 60 kPa in an external magnetic field of 15,000 Oe. A conservative design limit would be about 35 kPa for a single stage. Multiple stages are used to extend the pressure range. Normal machining practices allow one serration or stage per millimeter. For typical fluids, this results in about 35 kPa/mm or stage. Stages are added until the seal can withstand the desired pressure drop.

Magnetic seals cannot be used to seal reciprocating shafts; they are expensive, the fluid needs to be replaced periodically, and they are limited to pressures of 7000 kPa or less. They cannot be used to seal fluids and the magnetic fluid may evaporate in a vacuum or at high temperatures. Typical magnetic fluids and their properties are given in Table 46.

Table 46 Magnetic Fluid Properties

Carrier fluid	Saturation magnetization (G.)	Viscosity (cP at 27°C)	Density (g/ml)	Pour point (°C)
Ester	300	75	1.258	−54
Ester	450	100	1.440	−63
Hydrocarbon	200	200	1.080	−51
Synthetic hydrocarbon	300	120	1.195	−38
Fluorocarbon	300	3500	2.245	−27
Polyphenyl ether	450	4500	1.665	−12

Source: Ferrofluidics Corporation.

The required axial space increases as the pressure increases and may be excessive at high pressures. Two hundred stages (200 mm) are required to seal 7000 kPa. Several sets of ring magnets and pole pieces may be required to insure that field strength is maintained over the length of the seal.

E. Troubleshooting Noncontacting Dynamic Seals

Many noncontacting seals are designed to have a small amount of leakage. Other designs (magnetic, visco) will run for thousands of hours without leakage. Guidelines for correcting excessive leakage of noncontacting dynamic seals appear in Table 47.

XI. ECONOMICS OF SEAL SELECTION

There are many varieties of seal types and materials to choose from for a given application. Many will function equally well initially, and the user is often tempted to select the seal that has the lowest initial cost. Many times selecting a

Table 47 Troubleshooting Noncontacting Dynamic Seals

Problem	Causes	Corrective action
Excessive leakage at start-up	Excessive radial clearance	Decrease radial clearance
	Excessive pressure	Improve or add vent
	Excessive lubricant fill	Decrease fill or increase cavity
	Excessive temperature	Decrease fill or increase cavity
	Excessive vibration	Increase lube cavity
	Excessive end play	Increase lube cavity
	Radial contact	Fill scratches and voids
	Bypass leakage	Replace damaged secondary seals
Excessive leakage after a period of time	Loss of lubricant viscosity	Improve lubricant
	Excessive relubrication	Decrease relubrication
	Contaminant ingress	Add contaminant shield
	Loss of magnetic fluid	Replace magnetic fluid
Water ingress	Static leakage	Add static seal
	Partial vacuum	Clean or add vent
	No lube leakage	Increase lube fill
	Excessive wet environment	Add water shield
Dirt ingress	Excessive radial clearance	Decrease radial clearance
	Excessively dirty environment	Add dirt seal
	Partial vacuum	Clean or add vent
	No lube leakage	Increase lube fill

seal with a low initial cost is the most expensive solution. Inexpensive materials for O rings, packings, gaskets, etc., may fail early because the application temperature exceeds the material capabilities or because the sealed fluid is aggressive. The low initial price will be quickly offset by expensive repairs, loss of internal fluid, and machine downtime. A higher initial investment in a more expensive material will result in longer life with less downtime and fewer repair bills. A complete analysis of the true cost of a sealing system should include the initial operating costs and other miscellaneous costs that accrue during the life of the equipment.

Initial costs include the purchase price of the seal, special machining operations on the housing and shaft to accommodate the seal, and the time required to assemble, inspect, and test the sealing system integrity. Operating costs include maintenance, replacement parts, fluid loss, and, in some cases, power consumed and the cost of maintaining an inventory of spare parts. Some seals wear the shaft surface and/or the housing walls. These parts may require refinishing or replacement before a new seal is installed. Some seal systems may require buffer fluids, pumps, and filters. Operating costs should include the loss of buffer fluid, the power required to drive the pumps and the cost to replace the filters. Miscellaneous costs cover the loss of production when the machinery is idle while the seal is being replaced. Leaking seals can cause damage to other components if a fire results or if lubricant levels of harmful fluids can create health hazards or damage the environment.

Rotary shaft applications can be sealed with soft packings, elastomer radial lip seals, and mechanical seals. For a 100-mm shaft, the initial costs of a heavy-duty balanced mechanical seal is over 100 times that of packings or an elastomeric radial lip seal (Table 48). Packings will require frequent inspection, gland tightening, and frequent replacement. Some leakage will also be present. Spring-loaded radial lip seals will minimize or eliminate leakage. Some shaft wear will result and may require repair when the seals are

Table 48 Relative Purchase Costs for Various Seal Types

Shaft diameter (mm)	Soft packing	Elastomeric radial lip seals	Mechanical		
			Simple	Heavy-duty unbalanced	Heavy-duty balanced
25	0.4	0.6	10	60	75
100	1.0	1.0	40	95	105
200	2.0	2.50	75	110	250
250	3.0	5.5	100	300	500

Source: Chicago Rawhide.

replaced. Lip seals require less inspection, no gland tightening, and replacement is less frequent than packings.

The mechanical seal will give the longest life and damage to components are essentially nonexistent. The user should estimate the total costs for each seal type under consideration. The total costs should be compared before a final decision can be made. Many times, environmental and health concerns are strong enough to override financial considerations. A foolproof sealing system must be selected without concerns for the initial costs.

NOMENCLATURE

Abrasion resistance: The resistance of a rubber composition to wearing away by contact with a moving abrasive surface.

Aeration: Air (or gas) bubbles entrained or accumulated in a liquid.

Aging: Change in characteristics of rubbers with time specifically influenced by environmental factors (e.g., light, heat, etc.).

Air side: The side of a seal which faces outward or toward the atmosphere, as opposed to the fluid being sealed.

Aniline point: General indication of the aromatic content of an oil fluid determined as the lowest temperature at which the oil is miscible with an equal volume of aniline.

Antiextrusion ring (device): A ring or similar device assembled with a seal to prevent the seal extruding into the clearance space.

Antioxidant: Additive in a rubber mix to resist oxidation.

Antiozonant: Additive in a rubber to resist degradation caused by ozone.

Automatic seal: General term applied to describe seal designs that are pressure-energized. More specifically, it is used to classify certain types of flexible lip seals.

Axial clearance: Clearance between a sealing element and the inside face of the cover.

Axial interference: Clearance or dimensional difference between the I.D. or O.D. of a seal and the assembled rod (or shaft) or housing diameter respectively.

Back (of seal): The side of a seal facing outward or opposite to that facing the fluid being sealed.

Backup ring: *See* antiextrusion ring.

Backup washer: *See* antiextrusion ring.

Bedding in: *See* run-in.

Bellows seal: Face-type seal with contact pressure generated by a bellows.

Blister: A raised cavity or sac that deforms the surface of a seal material.

Board: A thick (greater than 1.5 mm) and rigid nonmetallic material in sheet or strip form commonly used to make gaskets.

Bond: The adhesion established by vulcanization between two cured elastomer surfaces or between one cured elastomer surface and one nonelastomer surface.

Bore: A cylindrical surface in the machine housing which mates with the outside diameter of the seal.

Braid: Hollow or solid structure of round, square, or polygonal section constructed from interlocking filaments or yarn strand laid obliquely to the axis of the braid.

Braid-over-braid: A braid produced by more than one pass through a multiple-carrier binding machine.

Breakaway friction: Frictional force to be overcome to initiate movement, specifically, static friction.

Brittle point: Temperature at which an elastomer becomes brittle.

Bull ring: A type of piston ring. A rigid or semirigid ring employed at one end or both ends of a packing to exclude extrusion of the packing into the clearance space.

Case: Metal component of a seal to which the sealing element is bonded, clamped, or otherwise contained.

Centring ring (centering ring): An extension of a gasket designed to locate the gasket centrally on a flange.

Checking: Cracking or crazing of the surface of an elastomer due to the action of sunlight.

Chevron seal: Seal ring (or ring set) of V-shaped cross section.

Chrysotile: Fibrous magnesium silicate, or "white" asbestos mineral.

Clearance: Dimensional difference between sealing element and related component.

Coaxial seal: A composite seal in the form of two (or more) coaxial ring members.

Collar seal: Characteristic type of flexible lip rod seal, also known as a hat seal or hat ring.

Composite seal: Seal comprising two (or more) separate materials, usually bonded together.

Compression modulus: Ratio of compression stress to resulting compression strain expressed as percentage of the original dimension.

Compression packing: Resilient sealing material, usually in plaited form, for fitting in a gland or stuffing box and compressed to expand radially by tightening of the gland cover to produce a seal.

Compression seal: Seal working on the principle of being compressed to fill the clearance space.

Compression set: Permanent deformation of rubber after subscription to compression for a period of time. Specifically determined as the ratio of dimensional change to compression strain.

Conductivity: Elastomers are considered conductive when they possess a direct current resistivity of less than 10^5 ohm/cm.

Contact force (contact load): Total interface pressure between a seal and the adjacent surface.

Controlled gap seal: A seal designed to maintain constant clearance with shaft.

Copolymer: A polymeric material comprising molecules of two or more different kinds.

Cover: Member of casing protecting or strengthening a seal element.

Creep: A transient stress-strain condition in which the strain or deformation increases as the load stress remains constant. (This condition is approached in flat-face gasketing joints in which the bolt undergoes a high elongation relative to any creep that might take place in the gasket.)

Creep relaxation: A transient stress-strain condition in which the strain increases concurrently with the decay of stress. (This is the most common condition existing in flat-face gasketing assemblies in which the bolt exhibits a relatively large amount of elongation.)

Crocidolite: Fibrous iron silicate, or "blue" asbestos.

Crown height: The height of a (gasket) sealing element above the surface of a retainer.

Cup-specific type of piston seal defined by its geometry: (i.e., cup-shaped); but also utilized as a description of a (seal) case.

Cure: Vulcanization process applied to rubbers.

Diametral clearance: Difference between I.D. of a seal and the shaft or rod diameter; or the I.D. of a seal and its housing.

Die-formed ring: A packing ring mechanically compacted into an (apparently) homogeneous form.

Dielectric strength: The voltage required to puncture a sample of known thickness and is expressed as volts per millimeter of thickness.

Double-acting seal: Seal for reciprocating movements capable of sealing with both directions of movement.

Dry running: Rubbing contact without any liquid being present at the interface.

Durometer hardness: Arbitrary measurement of hardness related to the resistance to penetration of an indentor point on a durometer.

Dynamic friction: Friction generated when relative movement takes place between two contacting surfaces.

Dynamic runout: Twice the distance the center of the shaft is displaced from the center of rotation and expressed in TIR. The runout to which the seal lip is subjected due to the outside diameter of the shaft not rotating in a true circle (DRO).

Dynamic seal: A seal capable of working with relative movement (either reciprocating or rotary) between components being sealed.

Eccentricity, shaft: *See* shaft-to-bore misalignment.

Elasticity: Property inherent in elastomeric materials of readily returning to its original form when released from a deforming load.

Elastomer: Rubbers or rubberlike materials possessing elasticity.

Elastomeric lip seal: Device which uses an elastomer to form a lip that affects a seal.

Elongation: Strain defined as the extension between benchmarks produced by a tensile force applied to specimen and is expressed as a percentage of the original distance between the marks. Ultimate elongation is the elongation at the moment of rupture.

End play: A measure of axial movement encountered or allowed, with reference to the shaft on which the seal lip contacts.

Extrusion: Permanent displacement of part of a seal into a gap under the action of fluid pressure.

Face: Front surface of a seal (where appropriate).

Face seal: Seal embodying two faces in rubbing contact in a plane at right angles to the axis of the seal.

Filler: A solid compounding ingredient which may be added, usually in finely divided form, in relatively large proportions, to a polymer.

Filler ring: Elastic ring assembled with a U ring or V ring to consolidate the section.

Finger spring: A spring form with flexible fingers.

Flange: The rigid members of a gasketed joint that contact the sides or edges of the gasket.

Flanged joint: *See* gasketed joint, which is a preferred term.

Flash: Thin extensions of the elastomer formed by extrusion at the parting lines in the mold cavity or vent points.

Flashing: A rapid change in fluid state, from liquid to gaseous. In a dynamic seal, this can occur when frictional energy is added to the fluid as the latter passes between the primary sealing faces, or when fluid pressure is reduced below the fluid's vapor pressure because of a pressure drop across the sealing faces.

Flash line: Exaggerated degree of flash due to clearance or gap between mold parts.

Flex cracking: Surface cracks resulting from repeated flexual cycling.

Flex fatigue resistance: The ability to withstand fatigue resulting from repeated distortion by bending, extension, or compression.

Flexibility: Flexibility of an elastomeric material during exposure to a predetermined temperature for a specific length of time.

Flexual modulus: Stress at a certain strain—not a ratio and not a constant, but merely the coordinate of a point on the stress-strain curve.

Flinger: Washer-type ring mounted next to a gland or gland follower for directing any leakage away from the shaft.

Flinger ring: Secondary seal element in the form of a ring generating "windback" action.

Flow line: Imperfection in a molding due to imperfect flow of material during molding.

Flow mark: Imperfection in a molding due to incomplete flow of material in the mold.

Fluid side: That side of the seal which in normal use faces toward the fluid being sealed.

Fluoroelastomer: A saturated polymer in which hydrogen atoms have been replaced with fluorine. It is characterized by excellent chemical and heat resistance.

Follower: *See* gland follower.

Front: The side of a seal facing the fluid to be sealed. Specifically applied to rotary shaft in seal descriptions.

Garter: Pseudostatic exclusion seal, usually in the form of elastomeric bellows.

Garter spring: Helical wire spring of circular geometry fitted to a lip seal (specifically an oil seal) to enhance lip contact pressure.

Gasket: Static seal made from deformable sheet material sandwiched and compressed between two mating plane surfaces.

Gasketed joint: The collective total of all members used to effect a gasketed seal between two separate items.

Gland: General description of housing or cavity for accommodating compression packings or sealing rings.

Gland cover: Fixed gland member fitted on the nonpressure side of a gland to retain the seal against the action of pressure.

Gland follower: Adjustable gland member which can be tightened to compress and expand radially the packing in a gland.

Hard face: A facing of high hardness applied to softer materials.

Hardness: The resistance to indentation. Measured by the relative resistance of the material to an indentor point of any one of a number of standard hardness testing instruments.

Hat ring: *See* collar seal.

Head: Portion of a seal carrying the sealing edge.

Header: Ring of hard material used in conjunction with seal ring(s) to locate the seal(s) and eliminate axial movement.

Heart seal: Solid elastomeric ring seal with a heart-shaped cross section.

Heat buildup: The temperature rise in a rubber body resulting from hysteresis.

Heel: The part of the seal cross section adjacent to the clearance gap on the nonpressurized side or adjacent to the shaft on the back of an oil seal.

Helix angle: Angle between screw threads on a seal or shaft and a plane perpendicular to the shaft axis.

H-ring: Solid elastomeric seal ring of H-shaped cross section.

Housing: A rigid structure which supports and locates the seal assembly with respect to the shaft. Annular recess into which a shaft or rod seal is assembled.

Hydraulic packing: A packing specifically designed for the sealing of hydraulic fluids.

Hydroseal: A sealing system having helically disposed elements formed on the shaft surface.

Hysteresis: The percent energy lost per cycle of deformation, or 100% minus the resilience percentage. Results from internal friction and is manifest by the conversion of mechanical energy into heat.

Inclusion: Foreign matter included in the seal material.

Interface: The region between the static and dynamic sealing surfaces in which there is contact, or which experiences the closest approach and effects the primary seal.

Interference: Negative dimensional difference between a seal I.D. or O.D. and the final seal assembly diameter.

International rubber hardness degrees (IRHD): A standard unit used to indicate the relative hardness of elastomeric materials, where 0 represents a material having a Young's modulus of 0, and 100 represents a material of infinite Young's modulus.

Junk ring: Alternative name for an antiextrusion ring.

Knit line: A blemish of the sealing element created by premature curing during molding operation.

Lantern ring: Ring with radial ports located at an intermediate position in a gland to allow coolant or lubricant to be introduced.

Lead-in (chamfer): Chamber introduced in component(s) to facilitate assembly or seal on to a rod or shaft or into a cylinder or housing.

Leak: The passage of matter through interfacial openings or passageways, or both, in the gasket or seal.

Leakage rate: The quality of fluid passing through a seal or gasket in a given period of time.

Lip: The part or edge of a seal which forms the sealing surface.

Lip, auxiliary: *See* lip, secondary; lip seal.

Lip, dirt: *See* lip, secondary; lip seal.

Lip, dust: *See* lip, secondary; lip seal.

Lip, molded: A type of seal lip which requires no trimming to form the contact line.

Lip, primary: The normally flexible elastomeric component of a lip seal assembly, which rides against the rotating surface and effects the seal.

Lip, secondary: A short, non-spring-loaded lip, located at the outside seal face of a radial lip seal to prevent ingress of atmospheric contaminants. Synonymous: lip, dirt; lip, auxiliary; lip, dust.

Lip, spring-retaining: The portion of the primary lip that restricts the axial movement of the extension spring from a predetermined position.

Lip, static: That section of the helix seal lip incorporating the contact line.

Lip force: The radial force exerted by an extension spring and/or lip of a seal on the mating shaft. Lip force is expressed as force per unit of shaft circumference.

Lip insert: A material such as PTFE bonded onto a lip of an elastomeric seal to provide improved sealing and/or exclusion performance not attainable with the elastomeric material alone.

Lip interference: *See* interference.

Lip opening pressure (LOP): The pressure necessary for flowing air at 10,000 cm^3/m between the contact surface of a radial lip seal and a shaft-size mandrel under the following conditions: the seal case outer diameter clamped to be concentric with the mandrel and the pressurized air applied to the outside lip surface.

Lip seal: A seal where the sealing surface is in the form of a flexible lip.

Load: Actual pressure at sealing face of a seal, normally the sum of the interference load and fluid pressure acting on the seal.

Lubricant starvation: Lack of proper lubrication at the seal interface, which may cause premature wear and early failure.

Maintenance factor: The factor that provides the additional preload capability in the flange fasteners to maintain sealing pressure on a gasket after internal pressure is applied to the joint.

Migration: Degradation products removed from an elastomeric sealing element and escaping to other parts of the system.

Mix: General description of a rubber compound formulation.

Modulus: Specifically the shape of the stress/strain curve for a material at a given elongation.

Modulus, rubber: The tensile stress at a specified elongation. A measure of resistance to deformation.

Modulus, Young's: The ratio of the stress to the resulting strain (the latter expressed as a fraction of the original heights or thickness in the direction of the force).

Modulus of elasticity: The ratio of stress to the strain produced by that stress when stress is proportional to strain. But in rubber, modulus measurements are made in comparison or shear, rather than in tension, and they are only valid for strains up to about 15%.

Mold: Device to shape materials and cool or cure them to produce the desired product.

Mold mark: Imperfection in a mold duplicating a surface defect on the mold itself.

Mold release or mold lubricant: The substance used to coat the surfaces of a mold to prevent the elastomer from adhering to the mold cavity surface during vulcanization.

Mold shrinkage: Loss of dimension of a molded product after removal from the mold and subsequent cooling.

Monomer: A single organic molecule usually containing carbon and capable of additional polymerization.

Multiple seal: A seal set comprising two or more seal rings of sealing elements.

Neck bush: Throttle bush fitted at the bottom of a stuffing box or gland.

Nib, spring: A short end section of an extension spring formed by a reduction in the coil diameter used to join the two ends in forming a garter spring.

Nick: A void created in the seal material after molding.

Nitrile: A general term for the copolymers of butadiene and acrylonitrile.

Nonfill: A void in the seal material.

Off-register: Step or break in the surface of a molded product due to faulty register of the mold.

Oil resistance: The measure of an elastomer's ability to withstand the deteriorating effect of oil on the mechanical properties.

Oil seal: A seal designed primarily for the retention of oil.

Oil side: The side of a seal facing the fluid being sealed.

Oil swell: The change in volume of a rubber material due to absorption of oil.

O ring: Solid elastomer ring seal of circular cross section.

Ozone cracking: Surface cracking of rubber due to the degrading effect of ozone.

Packing: A deformable material used to prevent or control the passage of matter between surfaces which move in relation to each other.

Panting: Movement between sealed surfaces of a static seal due to pressure fluctuations and insufficient clamping or tightening.

Pedestal ring: Support for a U-ring seal.

Permanent set: The amount of residual displacement in a rubber part after a distorting load has been removed.

Permeability: A measure of the ease with which a liquid or gas can pass through a rubber film.

Pitting: Surface voids produced by mechanical erosion (wear) or chemical action.

Plasticity: The degree or rate at which unvulcanized elastomer and elastomeric compounds will flow when subjected to forces of compression, shear, or extrusion.

Plasticizer: A material which, when incorporated in elastomer or a polymer, will change its hardness, flexibility, processibility, and/or plasticity.

Plunge ground: The surface texture of shaft or wear sleeve produced by presenting the grinding wheel perpendicular to the rotating shaft without axial motion.

Polyacrylate: A type of elastomer characterized by an unsaturated chain and being a copolymer of alkyl acrylate and some other monomer such as chlorethyl vinyl ether or vinyl chloracetate.

Polymer: Generic term for an organic compound of high molecular weight and consisting of recurrent structural groups. Materials with long-chain molecules, such as natural rubber and synthetic rubbers.

Polymerization: The ability of certain organic compounds to react together to form a single molecule of higher atomic weight.

Polytetrafluoroethylene (PTFE): A fluoropolymer with excellent thermal and chemical resistance and a low coefficient of friction. Usually compounded with fillers such as molybdenum disulfide, graphite, pigments, and glass fibers to improve wear characteristics and other properties.

Porosity: A multitude of minute cavities in the seal material.

Postcure: Second cure given to a material after the pressure or press cure is complete.

Precure (press cure) partial cure: A term frequently used to designate the first cure of a material that is given more than one cure in its manufacture.

Pseudostatic seal: An exclusion seal (e.g., bellows or garter) for excluding dust, dirt, etc., but also capable of accommodating relative movement between the components to which it is attached.

Pusher-type seal: A mechanical seal in which the secondary seal is automatically pushed along the shaft or sleeve to compensate for wear.

PV factor: An arbitrary term which is the product of face pressure and relative sliding velocity. The term is normally considered to provide some measure of severity of service or seal life.

Quad ring: Solid elastomeric ring seal of modified circular cross section giving four sealing ridges.

Quench: A neutral fluid introduced into a seal cavity to dilute fluid which may have leaked past the seal.

Radial clearance: Clearance between the shaft and internal diameter of an oil seal cover.

Radial interference: Negative dimensional difference between the radial dimension of a seal and its housing or space into which it is filled.

Radial load: Total load carried by the lip of a rotary shaft seal or rod seal.

Radial pressure: Also known as contact pressure. It is the average pressure exerted by a seal on a shaft. It is obtained by dividing the total lip force by the total lip contact area.

Radial wall dimension (RWD): The distance between the seal lip contact line and the seal outside diameter measured in a radial direction on a finished seal in the free state.

Radial wall variation (RWV): The difference between the minimum and maximum radial wall dimensions when measured around 360° of the seal lip.

Rectangular seal: Solid elastomeric ring seal of rectangular cross section.

Relaxation: Decrease in stress occurring with time under constant load or deformation.

Resilience: Energy recovery property of an elastomer under deformation cycles. Specifically the ratio of energy returned to energy input, per cycle of rapid deformation.

Resistance, cold: The ability of a seal or sealing material to withstand the effects of a low-temperature environment without loss of serviceability.

Resistance, heat: The ability of a seal or sealing material to resist the deteriorating effects of elevated temperatures.

Resistance, ozone: The ability of a material to withstand the deteriorating effects of ozone (surface cracking).

Rib: A long, narrow projection which is normally triangular or rectangular in cross section and which is molded into the outside lip surface of a helix seal. It is oriented at an angle to the shaft axis. One end of the rib forms part of the seal-lip contact surface. Also placed on shafts or housings to form a screw seal.

Rider ring: Wear or load-carrying ring associated with some form of ring seal (usually a metallic ring or piston ring).

Ring: Any circular seal or seal element.

Ring gasket: A flange gasket which lies wholly within the ring of bolts.

Rotary seal: Seal type specifically suitable for sealing rotating shafts or rotary motions.

Rotary seal ring: Driven or rotating face of a mechanical seal.

Rubber: Elastomeric substance, either natural or synthetic.

Rubber face: Rubber coating applied to a seal case to provide a sealing surface against the seal housing.

Run-in: Period of initial operation and high wear that develops the interface contact patterns.

Running friction: Friction generated by a seal under dynamic operating conditions.

R value: The axial distance between the seal contact line and the center line of the spring groove of a radial lip seal.

Scoop trim: A trimmed surface which is concave applies to elastomeric lip seal.

Scratch: A shallow discontinuity in the seal material often caused by improper handling on installation.

Scraper (scraper ring): Heavy-duty wiper seal to exclude grit and heavier contaminants with reciprocating rod movements.

Scorch: Premature curing of vulcanized rubber due to excessive heat.

Scuffing: Surface roughness produced by mechanical wear.

Seal, birotational: A rotary shaft seal which will seal fluid regardless of direction of shaft rotation.

Seal, dynamic: A seal which has rotating, oscillating, or reciprocating motion between it and its mating surface, in contrast to stationary-type seal, such as a gasket.

Seal, helix: An elastomeric hydrodynamic lip seal having helical ribs on the outisde lip surface.

Seal, hydrodynamic: A dynamic sealing device which utilizes the viscous shear and inertia forces of the fluid; imparted by a helically grooved or ribbed seal lip, shaft, or housing to generate a pressure differential that opposes fluid flow.

Seal, lip: An elastomeric seal which prevents leakage in dynamic and static applications by reason of controlled interference between the seal lip and the mating service.

Seal, mechanical: Any material or device that prevents or controls the passage of matter across the separable members of a mechanical assembly.

Seal, radial: A seal which exerts radial sealing pressure in order to retain fluids and/or exclude foreign matter.

Seal, radial lip: A type of seal which features a flexible sealing member referred to as a lip. The lip is usually of an elastomeric material. It exerts radial sealing pressure on a mating shaft in order to retain fluids and/or exclude foreign matter.

Seal, split: A seal which has its primary sealing element split, approximately parallel with the shaft axial centerline. Typically used where conventional installation methods are impractical or impossible.

Seal, unirotational: A seal designed for applications having a single direction or shaft rotation.

Seal, unitized: A seal assembly in which all components necessary for accomplishing the complete sealing function are retained in a single packet.

Sealer, case O.D.: A coating applied to the case O.D. to prevent leakage between the seal case and the housing bore.

Sealing edge: The extreme section of an oil seal which provides the actual seal.

Sealing element: Portion of the seal section or seal element covering the sealing edge.

Sealing land: Flat portion of sealing edge of an oil seal after prolonged contact with shaft.

Seal plate: Alternative description for a gland cover.

Seal width: The overall axial dimensions of a seal.

Secondary seal: O ring, bellows, or similar device which accommodates leakage from the primary seal of a mechanical face seal.

Separator: An intermediate ring of thin, stiff material which allows individual rings in a seal assembly to slide over one another.

Shaft eccentricity: The radial distance which the geometric center of a shaft is displaced from the axis of shaft rotation.

Shaft lead: Spiral grooves on a shaft surface caused by a relative axial movement of grinding wheel to shaft.

Shaft offset: The radial distance between the axis of the seal bore and axis of shaft rotation. (synonym: shaft-to-bore misalignment)

Shaft out of round: The deviation of the shaft cross section from true circle. Out of round is measured as the radial distance, on a polar chart recording, between concentric, circumscribed, and inscribed circles which just contain the trace and are so centered that the radial distance is minimized.

Shaft surface finish: A term used to describe the quality, appearance, or characteristic of the shaft surface resulting from operations, such as grinding, polishing, burnishing, etc.

Shell: Case of an oil seal.

Shore hardness: The relative hardness of an elastomer obtained with the use of a shore durometer instrument.

Silicone: An elastomer with a base polymer of dimenthyl polysiloxane, with various attached vinyl or phenyl groups.

Single acting seal: *See* seal, unirotational.

Slinger: *See* flinger.

Slipper seal: Coaxial seal comprising an O ring and a hard low-friction PTFE bearing ring (slipper ring).

Soft packing: Gland packing of soft resilient material.

Spew: Excess material forced from a mold during molding process.

Split-ring seal: Split rigid section ring (usually metallic) similar in form and principle to piston ring.

Split seal: Elastomeric seal ring split to facilitate assembly.

Spring groove: A depression formed in the head section of the seal. It is generally semicircular in form and serves to accommodate and locate the garter spring.

Spring index: The ratio of the mean coil diameter to the wire diameter of a garter spring.

Spring initial tension: The "preload" that has been wound into the coils of a spring during the coiling operation.

Spring rate: The force, independent of initial tension, which is required for extending the working length of a spring a unit distance.

Spring windup: The tendency of a garter spring with its ends assembled together to deform from a flat surface. Excessive spring windup results in the spring forming a "figure 8" configuration.

Squeeze: Deformation of a seal produced when assembled with an interference fit.

Static friction: Instantaneous or "holding" friction of a seal under static conditions.

Stationary seal ring: Static face of a mechanical seal.

Stick-slip: Jerky or irregular motion when a seal is operating under varying static-dynamic friction.

Stiction: Initial friction or breakaway friction when motion is started.

Strain: The deformation of a gasket specimen under the action of applied force or stress.

Stress relaxation: The loss in stress when an elastomer is held at a constant strain over a period of time.

Stress relaxation or creep: That characteristic of all elastomers to show gradual increase in deformation under constant load with passage of time. It is usually expressed as percent relative creep, which equals total deformation minus initial deformation divided by initial deformation times 100.

Stuffing box: Alternative name for a gland for containing packings or seal rings.

Surface speed: The linear velocity calculated from the shaft rotational speed using the nominal shaft diameter.

Swell: Increase in volume of a seal or elastomeric material when in contact with a fluid.

Synthetic rubber: Synthetic elastomers made by polymerization of one or more monomers.

Tear resistance: The force per unit of thickness required to propagate a nick or cut in a direction normal to the direction of the applied force or to initiate tearing in a direction normal to the direction of the stress.

Tensile strength: The force per unit of the original cross-sectional area which is applied at the time of the rupture of a specimen.

Tensile stress: More commonly called "modulus," the stress required to produce a certain elongation.

Tension set: Increase in normal (unstressed) length of an elastomeric specimen after initial stretching and release.

Thread stretch: Loosing of bolted assemblies or unions due to bedding down of threads.

Throttle bush: Restrictive bush fitted at the bottom of a stuffing box or gland; also descriptive of a bushing seal.

Toroidal seal: Alternative name for an O ring.

Torsional vibration: A vibration which has a circumferential or angular direction. It is often generated by a stick-lip action between mating seal faces.

Track: Mark made on a shaft by a rotary seal.

Trapped O ring: Static seal using an O ring in a special groove.

Trim: The removal of superfluous parts from a molded product, usually removal of parting line flash or feed sprues.

Trim diameter: Diameter of an oil seal, without spring.

Trim face: Front surface of a rotary oil seal trimmed to an angle to firm the sealing edge.

Trimming angle: Angle between the trimmed face of a seal lip and the seal axis.

Unbonded flash: Flash which does not properly adhere to the mating material to which it is intended to be bonded.

Undercure: A degree of cure less than desired.

Unit seal: A seal ring consisting of a single ring and not normally subject to axial compression.

U ring: Flexible lip seal of U-shaped cross section.

Volume swell: Increase in physical size caused by the swelling action of a liquid, generally expressed as a percent of the original volume.

V ring: Flexible lip seal of V-shaped cross section; also known as a chevron seal.

Vulcanization: Heat process treatment for rubber to stabilize and harden it.

Wear bond: Mark made by a rotary seal on a shaft.

Wear rate: The amount of seal contact surface wear per unit of time.

Wear sleeve: A ring of hard material associated with a seal or seal assembly and intended to take rubbing wear.

Weepage: Small amount of leakage from a seal, arbitrarily defined as leakage rate of less than one drop per minute.

Yield factor (minimum design seating stress): The factor that represents the pressure in megapascals (or pounds-force per square inch) over the contact area of the gasket that is required to provide a sealed joint, with no internal pressure in the joint.

7

GEAR DRIVES

Richard H. Ewert*

Consultant, Factory Cost Systems and Gearing, St. Paul, Minnesota

I. INTRODUCTION

This chapter provides practicing engineers, most of whom are not specialists, with the basics of gearing. Engineers working for companies that are gear users need to have some knowledge of gearing, but only as one element of the many things they must know. The material in this chapter will be helpful to them and others who want to know something, but not everything, about gearing.

This chapter by itself is not intended to tell the reader how to design gears. It does, however, discuss some of the pitfalls, and it tells where to look to find out all about gear design.

After a brief history, we give some definitions, then basic gear geometry, then relate this to some of the common gear tooth cutting methods and to gear tooth action. Initially we look primarily at spur gears, then discuss other types of gears, splines, gear measurement, gear materials, heat treatments, gear failures, and gear efficiencies. Then we talk about enclosed gear drives and lubrication, discuss some of the modern gear tooth cutting machines, and finally make some predictions about what to look for in the future.

Definitions for many of the common terms used in gear technology are provided in the text.

The subject of gearing is extensive, and the calculations required for gear design are complex, so the material in this chapter can only touch on the

*Past president, American Gear Manufacturers Association, Alexandria, Virginia; past president (retired), Sewall Gear Manufacturing Company, St. Paul, Minnesota

basics. It is suggested, however, that the material presented herein will be helpful to the relative newcomer to gearing, as well as to the nonspecialist practicing engineer, even if this is the most he or she ever learns. If there is desire to delve more deeply into gear engineering and design, then this material can be the stepping-stone to further study.

A word of caution: The study of gearing can become addictive. I've been told that many years ago a prominent gear engineer's wife gave birth to a baby girl. They couldn't agree on a name for her, so he named her Involute, the name of the curve used for the profiles of most gear teeth!

II. THE HISTORY OF GEARS*

Meshing gear teeth have been in use by mankind for well over 3000 years [1]. As a matter of fact we don't know where or when the first gears were used. Gears came out of the mists of antiquity along with other mechanical devices such as the potter's wheel, the inclined plane, pulleys and levers, and many others.

The earliest gears were probably made out of wood. The Greeks did make small gears for astronomical devices out of metal. The Romans used wooden gears in grist mills for power transmission, as well as iron and bronze gears in devices to control motion where very little power was needed.

All through the middle ages gears were used in water mills. The mechanical clock invented in the 1300s used small metal gears. Pendulum clocks that flourished in early times were full of gears. Some of these clocks in churches were quite large.

The first really extensive use of power gears occurred in Holland where thousands of windmills pumped seawater out of the land. These giant windmills used a train of wooden gears to take the power from a fast-running wind rotor at the top of the mill and deliver it to a slow-running Archimedes type of screw pump at the bottom of the mill.

As the Industrial Revolution got underway in the 1700s there was a need for all kinds of gear drives. Rather good gears were made in cast iron using patterns that were quite accurately made with wooden teeth, spokes, and hub. Soon even better gears were made, after the first gear-cutting machines were invented and used to mill gear tooth shapes in cast iron or steel gear blanks. In the nineteenth century people considered that a *normal* gear had cast teeth and a *precision* gear had cut teeth.

As the twentieth century began, the automobile and airplane came into being. Already in the late nineteenth century, steam-powered ships, electric-

---

*By Darle W. Dudley, P.E., Dudley Technical Group, Inc., San Diego, California.

powered trolley cars, and geared machine tools had come into widespread use. Gear drives came into general use in vehicles, mills, and all manner of mechanical devices. The development of gear engineering went hand-in-hand with the very widespread use of gears.

The technology of gearing required a complex array of machinery to make the needed gears, plus engineering rules and data to deal with things like tooth contact stresses, tooth bending stresses, lubrication of fast-meshing gear teeth, and metallurgical understanding of the metals (or other materials) used in gearing.

What was needed, of course, was a gear design that was practical to make with machines available and a size and quality good enough for the gear drive to last as long as needed in each application. And there was, of course, the usual complication that the gear drive had to be reasonably low in cost and not too large or too heavy!

All during the twentieth century there has been steadily accelerating progress in the gear art. The art of making gears has changed from hand operating machines making one part at a time to automatic machines, computer controlled, that can achieve production rates and quality levels never thought possible in the early part of the century.

The engineering design art for gears has blossomed to where a whole shelf of technical books is needed by the engineer to understand how to do all the geometric calculations applicable to design and have all the needed data on materials and processes. Then much more information is needed on trade standards plus data to cover all the failure hazards that must be avoided to get the required gear life and reliability.

To summarize what has happened in gear technology:

Over 90% of man's technical knowledge applicable to gearing has been developed since 1890—the last 100 years.

The last 100 years can be divided into four periods, which each period having approximately equal overall advancement in the gear art [2]:

1. 1890–1930: The initial period in which gear manufacturing and gear engineering became recognized and established.

2. 1930–1960: Gear manufacturing and gear engineering became broad based. The gear art matured substantially.

3. 1960–1980: An era of considerable advancement in gear technology. The gear art became very sophisticated.

4. 1980–the present: The old established gear art saw drastic changes as innovations in machinery design, computer control, new materials, and new lubricants came into being. The traditional way things used to be done in the gear shop or the gear design office have all tended to change substantially.

The gear art has now progressed to one of immense complexity, amazing capability, and a bewildering tendency to make the machinery and design rules of only 30 years ago obsolete! Gear technology is highly advanced now just like the technologies having to do with space vehicles, fax machines, plastics, lasers, and many other things.

III. DEFINITIONS

A *gear* is a machine element, the purpose of which is to transmit motion and power from its shaft, through another gear to its shaft, by means of gear teeth. A usual requirement is that the motion transmitted be uniform. That is, if the driving gear is rotating at a uniform angular velocity, the driven gear must rotate uniformly also. The smaller of two gears in mesh is called the *pinion,* the larger the *gear.*

A *spur gear* is a gear with the teeth cut parallel to its axis of rotation.

A *helical gear* (pronounced "hellical" by gear people) is a gear with teeth cut at an angle to its axis of rotation. The teeth are sections of a helix that spirals around the body of the gear, not straight as are spur gear teeth.

A *bevel gear* is a gear with teeth cut on the outside of a cone.

Worm gearing is a gearset of which the driving member (the worm) is of screw thread configuration. The axis of the worm is normally at 90° to the worm gear shaft.

An *external gear* is a gear with the teeth cut on the outside of a cylinder.

An *internal gear* is a gear with the teeth cut on the inside of a hollow cylinder.

A *face gear* is a gear with the teeth cut on the end of a cylinder.

Spiroid gears are a family of gears in which the tooth design is in an intermediate zone between bevel, worm and face gear design. The spiroid design is patented by the Spiroid division of Illinois Tool Works, Chicago.

Pitch circle: Envision two disks in contact with each other, one driving the other by friction without slipping. Relative motion would be transmitted from the shaft of one to the shaft of the other in inverse proportion to the diameters of the two disks (if the driven disk is twice the size of the driver, its speed will be half the speed of the driver). If these two disks are provided with gear teeth, the outside diameters of which would be larger and the root diameters of which would be smaller than the diameters of the disks, the circumference of each disk is called the *pitch circle* of the driver and driven gears.

Pitch diameter (PD), nominal, is the diameter of the pitch circle of a gear. This is the PD for the tooth cutting operation. The nominal pitch circles of two gears in mesh at standard centers will be tangent to each other.

The *operating* pitch circles of two gears when meshing at greater than standard centers, as when the pinion is oversize, will be greater than the nominal pitch circles and *they* will now be tangent to each other.

Pitch is a measure of tooth spacing and size, expressed as follows:

Diametral pitch (DP) is the number of teeth per inch of *pitch diameter* (PD). Therefore DP = PD/number of teeth, or PD = number of teeth/DP, so as DP decreases the tooth size increases. For example, 1 DP teeth are larger than 2 DP teeth.

Circular pitch (CP) is the distance between the center lines of two adjacent teeth as measured along the pitch circle (in practice the measurement is made between the corresponding flanks of two adjacent teeth). The relationship between CP and DP is, therefore, CP = 3.1416/DP. Circular pitch is therefore the circumference of the pitch circle divided by the number of teeth.

Module is a measure of tooth spacing and size in the metric system, and is millimeters of pitch diameter per tooth. As the module increases the tooth size also increases. The relationship between module and DP is M = 25.400/DP.

Speed ratio is the number of teeth on the gear divided by the number of teeth (T) on the pinion. Thus with a 50T gear driven by a 20T pinion, the speed ratio is 2.5 to 1.

IV. THE INVOLUTE CURVE

The requirement of uniform angular velocity dictates the use of mating tooth forms which are *conjugate* to each other ("reciprocally related and interchangeable as to properties, as two points, lines, etc."—Webster New World Dictionary, Second College Edition, 1986). Of those which meet this requirement the *involute* curve is the easiest to produce in the shop, and is, therefore, used as the profile for most gear teeth. So it is important to have an understanding of some of the properties of the involute curve.

The involute is the curve traced by the end of a taut string being unwound from a nonrotating circle. In gearing this circle is called the *base circle*. Figures 1a and 1b show elements of a single involute curve and construction of involute curves. The latter shows involutes from two base circles with radii R_{b1} and R_{b2} respectively and various positions of the string being unwound from each base circle, called *generating lines*. It is seen that the unwinding string is always normal to the involute curve it is generating and is tangent to its base circle at any position it is in.

The involute curves from both base circles in Fig. 1b are drawn so the two involutes are in contact with each other.

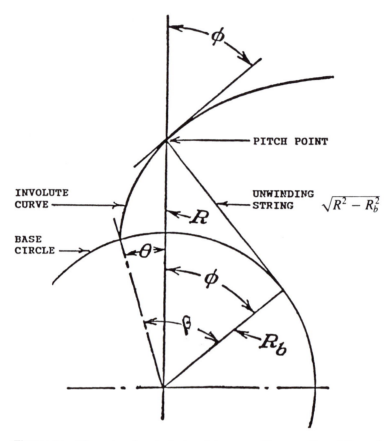

Figure 1a Elements of a single involute curve. R_b = radius of base circle, R = radius to involute curve (pitch radius), θ = angle from start of involute curve to radius r (vectorial angle), ϕ = pressure angle at radius R. (From Ref. 19.)

Note that the generating line from R_{b1}, when in position *AP*, continues on and is tangent also to R_{b2} at *D*. The intersection of this common tangent with the centerline of the two base circles defines the *pitch point P*, and by similar triangles it is seen that the pitch radii (R_1 and R_2) are directly proportional to the base circle radii.

If you will visualize what happens as these gears rotate, you will see that the point of tooth contact is always along the common tangent to the two base circles (line *AD*), and so this is called the *path of contact* or *line of action*.

The angle ϕ is the *pressure angle* (PA), so called because it is the angle through which the pressure, or power, is transmitted from the driving to the driven teeth.

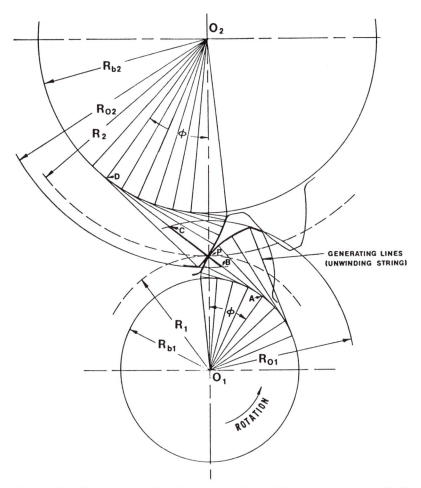

Figure 1b   Construction of involute curves. R_{b1} and R_{b2} = base circle radii, R_1 and R_2 = pitch radii, ϕ = pressure angle at R_1 and R_2. (From Ref. 19.)

We speak often of *contact ratio*. This is easy to define, and visualize, from Fig. 1b. If we draw the circles representing the outside radii (R_{01} and R_{02}) of these two gears, they intersect the line of action at points C and B. We see that contact can take place only between these two points. The *arc of action* is the arc through which one tooth travels from the time it first makes contact with its mating tooth until it ceases to be in contact, and is, therefore, the length BC measured at the radius of the base circle.

Referring to Fig. 1a, the involute curve is expressed mathematically* as follows:

Let

R_b = radius of base circle

R = radius to involute curve (pitch radius)

θ = angle from start of involute curve to radius r (vectorial angle)

β = angle through which string has been unwound

Then

$$\theta = \beta - \tan^{-1}\frac{\sqrt{R^2 - R_b^2}}{R_b}$$

The length of the generating line $\sqrt{R^2 - R_b^2}$ is the length of the circumference of the base circle which is subtended by the angle β.

Hence

$$\sqrt{R^2 - R_b^2} = R_b\beta \quad \text{or} \quad \beta = \frac{\sqrt{R^2 - R_b^2}}{R_b}$$

$$\therefore \quad \theta = \frac{\sqrt{R^2 - R_b^2}}{R_b} - \tan^{-1}\frac{\sqrt{R^2 - R_b^2}}{R_b}$$

This is the polar equation of the involute curve.

Since the first term = $\tan \phi$ and the second term = ϕ, this equation can be written

$$\theta = \tan \phi - \phi = \text{Inv}\,\phi, \quad \text{or} \quad \text{Inv}\,\phi = \tan \phi - \phi$$

If ϕ is the angle in degrees, then

$$\text{Inv}\,\phi = \tan \phi - \frac{\pi\phi}{180}$$

From Fig. 1a we also have the relationship between the pitch and base circles:

$$R_b = R \cos \phi$$

$$* \qquad * \qquad *$$

The term Inv ϕ is known as the *involute function*. Involute functions are tabulated in most gear books, one of which is *Gear Drive Systems* by Peter Lynwander. A portion of this tabulation is reproduced here.

Involute functions are used in various formulas relating to involute gear teeth. Three of these, which relate to tooth thickness, are shown on pp. 527, 528, 551, and 552.

Involute Functions: A Partial Tabulation

Angle (deg)	Cosine	Involute	Angle (deg)	Cosine	Involute
17.0	0.956305	0.009025	19.2	0.944376	0.013134
17.1	0.955793	0.009189	19.3	0.943801	0.013346
17.2	0.955278	0.009355	19.4	0.943223	0.013562
17.3	0.954761	0.009523	19.5	0.942641	0.013779
17.4	0.954240	0.009694	19.6	0.942057	0.013999
17.5	0.953717	0.009866	19.7	0.941471	0.014222
17.6	0.953191	0.010041	19.8	0.940881	0.014447
17.7	0.952661	0.010217	19.9	0.940288	0.014674
17.8	0.952129	0.010396	20.0	0.939693	0.014904
17.9	0.951594	0.010577	20.1	0.939094	0.015137
18.0	0.951057	0.010760	20.2	0.938493	0.015372
18.1	0.950516	0.010946	20.3	0.937889	0.015609
18.2	0.949972	0.011133	20.4	0.937282	0.015850
18.3	0.949425	0.011323	20.5	0.936672	0.016092
18.4	0.948876	0.011515	20.6	0.936060	0.016337
18.5	0.948324	0.011709	20.7	0.935444	0.016585
18.6	0.947768	0.011906	20.8	0.934826	0.016836
18.7	0.947210	0.012105	20.9	0.934204	0.017089
18.8	0.946649	0.012306	21.0	0.933580	0.017345
18.9	0.946085	0.012509	21.1	0.932954	0.017603
19.0	0.945519	0.012715	21.2	0.932324	0.017865
19.1	0.944949	0.012923	21.3	0.931691	0.018129

Source: Peter Lynwander, *Gear Drive Systems*. Marcel Dekker, New York, 1983.

Contact ratio is the length *BC* divided by the *base pitch*. Base pitch is the distance between two successive involutes measured along the circumference of the base circle, so it is the circumference of the base circle divided by the number of teeth. When the contact ratio is less than unity, which it happens to be in the illustration, it means that it is geometrically impossible for more than one tooth to be in contact at a time (for spur gears). For good operating characteristics, the contact ratio of a spur gearset should generally be about 1.4 or greater for relatively high-speed gearing (about 2000 ft/min *pitch line velocity* or greater), or gearing which must roll as smoothly as possible regardless of speed. A contact ratio of perhaps 1.2 for low-speed gearing might be acceptable but should, if possible, be greater than this.

By comparing layouts of this type it can be shown that contact ratio increases as pressure angle decreases. So in the pressure angles we usually talk about, 14½° and 20°PA, 14½°PA gears have a greater contact ratio than 20°, for a given number of teeth in pinion and gear, assuming no *undercut* (see Section VI). This means that 14½°PA gearing with no undercut would operate more quietly than 20°PA gearing, assuming both are made to exactly the same degree of accuracy.

Note that the radius of curvature of the involute increases as the pressure angle increases. If we compare 14½° and 20°, it is over 40% greater for 20°, and this greater radius of curvature of the 20°PA tooth enables it to withstand greater surface pressure than the 14½°PA tooth. At the same time the 20°PA tooth is thicker at the *root diameter* and will therefore withstand greater bending loads than the 14½°.

For most applications 20°PA has more advantages than 14½° and has therefore become more prevalent. In recent years 25°PA has gained acceptance for heavily loaded relatively low speed drives.

One might wonder why a pressure angle of 14½° became a standard in the first place. Why not 20°, which became a standard later, or 15° or maybe 18°? The reason is not very scientific or profound. In the old days coarse pitch gears had cast teeth. Laying out the teeth in the patterns was simpler for 14½° because the sine of 14½° is 0.25038, close enough to the fraction ¼ that the pattern maker could use the latter to lay out gear teeth.

V. GEAR CUTTING BY GENERATING PROCESSES

It is interesting to examine how these facts are used in the production and application of gearing. First, in cutting gear teeth by *generating*: As the base circle gets larger and larger its radius of course approaches infinity, and the involute curve approaches a straight line and eventually becomes a straight-sided *basic rack* tooth. A rack tooth will, therefore, roll with any gear tooth of the same pitch and pressure angle (the teeth are conjugate to each other). If, therefore, the cutting tool is made to the basic rack form, it is capable of generating gear teeth in any diameter gear.

One such cutting tool is the *hob,* a tool of screw thread configuration with flutes milled normal to the thread, or threads, to provide the cutting edges. Its cutting action is that of continuous milling, the gear blank rotating while the hob, also rotating, is moved across the face of the gear, or in hobbing worm gears moved radially or tangentially (see Section XII.C.2). A single-thread hob cutting a 50-tooth gear would therefore have to rotate 50 times faster than the gear it is cutting.

A typical hob for cutting spur or helical gears and a worm gear hob are illustrated in Figs. 2 and 3.

A Fellows shaper cutter is another type of tool. Its configuration is that of a gear, usually about 4 to 6 in. diameter, with the outside diameter (O.D.) and teeth tapered to provide clearance for the cutting edges (see Figs. 4–6). Its tooth form is the involute curve corresponding to the number of teeth and pressure angle of the cutter. Its cutting action is that of shaping as it reciprocates across the gear face, with both the cutter and the gear it is cutting rotating in the speed ratio of the number of teeth in the gear being cut divided by the number of teeth in the cutter.

Spur, helical, and double helical gears can be shaped. Cutters for shaping helicals must have the same helix angle as the gear being cut. The cutters for shaping external helicals must be opposite hand of the gear being cut, and for cutting internal helicals the same hand as the gear being cut. For internal gears the number of teeth in the cutter must not exceed that which would trim a portion of the involute (this becomes critical when the number of teeth in the internal is relatively small). For cutting helicals the gear shaper must be fitted with *guides* to impart rotary motion to the cutter as it reciprocates. Internal gears *must* be shaped, as well as gears whose teeth abut a shoulder or another gear. When the latter it is called a *cluster gear*. In such cases a relief groove must be provided in the gear blank for cutter and chip clearance.

Maag gear cutting machines cut teeth with a shaping action also, using cutters which are of basic rack configuration.

A common finishing process is *shaving,* used after hobbing or shaping to improve finish and accuracy. Gears to be shaved must be cut with *preshave* hobs or shaper cutters, which remove stock below the *active profile* of the teeth to provide clearance for the tips of the shaving cutter's teeth. Shaving cutters are essentially helical gears with serrated teeth to provide the cutting edges. The cutter drives the gear being shaved. A shaving cutter is illustrated in Fig. 6.

Another finishing process is grinding the profiles of the teeth, used to remove heat treat distortion on hardened gears. Grinding is required on hardened gears specified as AGMA quality 10 or better. Honing or lapping gear teeth can be used to improve finish and to remove minute profile errors on hardened gears. A honing machine works like a shaving machine, with the hone a nonmetallic helical gear impregnated with abrasive particles. Lapping of spur gear teeth by running two gears together with an abrasive lapping compound is a "last resort" method to remove hardening distortion, except for large gears

SPUR GEAR (OR HELICAL) HOB

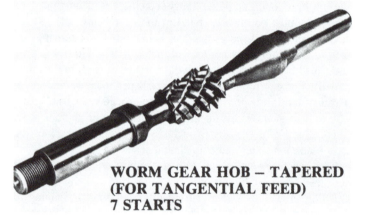

**WORM GEAR HOB – TAPERED
(FOR TANGENTIAL FEED)
7 STARTS**

Figure 2 Spur gear (or helical) hob. (Courtesy of Sewall Gear Manufacturing Company, St. Paul, MN.)

Figure 3 Worm gear hob—tapered (for tangential feed), 7 starts. This hob was used for hobbing the worm gear shown on page 548. (Courtesy of Sewall Gear Manufacturing Company, St. Paul, MN.)

and when performed by an expert. Excessive lapping of spurs will destroy the involute profile, as when spur gear teeth are in mesh rolling action takes place only at the pitch point and sliding action above and below the pitch point, so the stock removal occurs above and below the pitch point.

A straight-tooth Gleason bevel gear generating machine uses two single-shaper cutters or, in faster machines, two rotary interlocking cutters, with the

SHAPER CUTTER

FORM CUTTER

SHAVING CUTTER

Figure 4   Shaper cutter. (Courtesy of Sewall Gear Manufacturing Company, St. Paul, MN.)

Figure 5   Form cutter. (Courtesy of Sewall Gear Manufacturing Company, St. Paul, MN.)

Figure 6   Shaving cutter. (Courtesy of Sewall Gear Manufacturing Company, St. Paul, MN.)

cutting edges constrained to move in the path that the teeth of a mating gear would follow. Such cutters are illustrated in Figs. 7–9.

A spiral bevel gear generating machine uses a rotating face-milling cutter, fed into the workpiece, with cutter and work rotating together to provide the generating motion required. Such a cutter is illustrated in Fig. 10. Gleason

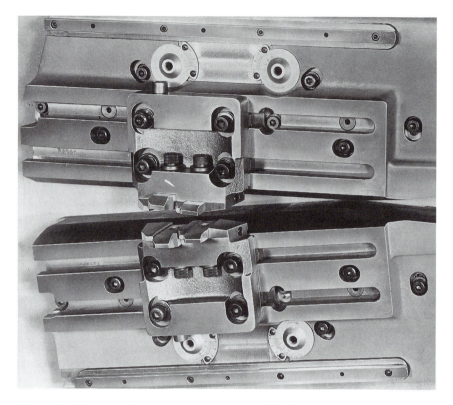

Figure 7   Gleason reciprocating cutters. For use with Gleason reciprocating tool generators nos. (14) 429 and (24) 434, for cutting straight-tooth Coniflex® bevel gears. (Courtesy of The Gleason Works, Rochester, NY.)

spiral bevel machines also cut Zerol bevels (spiral bevels with 0° spiral angle) and hypoid gears.

VI. GEAR TOOTH ACTION

A word about *undercut* and *modification*. For pinions with small numbers of teeth, the generating tool will cut away that portion of the involute which is near and below the base circle (R_{b1} in Fig. 1b). This cutaway portion is called undercut and it increases as the number of teeth becomes less. For 14½°PA gears with standard *full-depth* teeth, undercut occurs at about 32 (and fewer) teeth. For 20°PA full depth spur gears it occurs at about 18 (and fewer) teeth, and for 25°PA spur gears it occurs at about 12 (and fewer) teeth.

Figure 8 Gleason interlocking cutters, for use with Gleason interlocking tool generators nos. (104) 439 and (114) 442, for cutting straight-tooth Coniflex® bevels. (Courtesy
of The Gleason Works, Rochester, NY.)

Undercut pinions will therefore have a less-than-standard active tooth
profile, so contact ratio is reduced. Also, undercut reduces tooth strength as it
decreases tooth thickness at the root of the teeth. It is, therefore, an important
consideration in gear design. It is minimized, or sometimes eliminated altogether, by hobbing or shaping small pinions on oversize diameters. A limiting
factor in the amount of modification possible is the diameter at which the tooth
would come to a point.

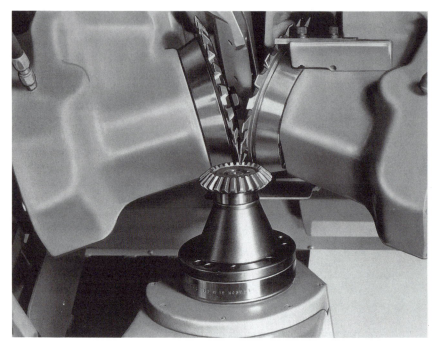

Figure 9 Interlocking cutters in operation. (Courtesy of The Gleason Works, Rochester, NY.)

Figure 10 Gleason face-mill cutter, for cutting spiral, Zerol bevel, and hypoid gears. (Courtesy of The Gleason Works, Rochester, NY.)

The pinion shown in Fig. 1b is a modified pinion. This is evident from the relatively long *addendum* shown ($R_{01} - R_1$), as compared to the *working depth* below the pitch line ($R_{02} - R_2$). On a nonmodified pinion, addendum and working depth below pitch line are equal. The distance from the pitch line to the bottom of the tooth space (the *root*) is the *dedendum*.

To determine the diameter at which the pinion tooth would come to a point, when cut on an oversize diameter, the following formula will do the trick. When the arc tooth thickness and pressure angle at a given radius are known, to find the radius at which the tooth becomes pointed, proceed as follows (see Fig. a):

$$\text{Inv}\,\phi_2 = \frac{T_1}{2r_1} + \text{Inv}\,\phi_1$$

$$r_2 = \frac{r_1 \cos \phi_1}{\cos \phi_2}$$

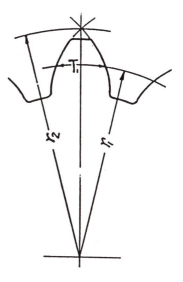

where

r_1 = given radius

r_2 = radius at which tooth becomes pointed

T_1 = arc tooth thickness at r_1

ϕ_1 = pressure angle at r_1

ϕ_2 = pressure angle at r_2

Figure a

Example. (For a 30T, 6DP, 14½°PA gear) $r_1 = 2.500$ in., $T_1 = 0.2618$, $\phi_1 = 14.500°$.

$$\text{Inv}\,\phi_1 = \tan \phi_1 - \frac{\pi\phi}{180} = 0.25862 - 0.25307 = 0.00555$$

$$\text{Inv}\,\phi_2 = \frac{0.2618}{2 \times 2.500} + 0.00555 = 0.05791$$

$$\phi_2 = 30.693° \qquad \cos \phi_2 = 0.85991 \qquad \cos \phi_1 = 0.96815$$

$$r_2 = \frac{2.500 \times 0.96815}{0.85991} = 2.8147$$

The formula to be used when it is necessary to determine the arc tooth thickness at a radius other than the radius at which the tooth thickness is known is (see Fig. b)

$$\cos \phi_2 = \frac{r_1 \cos \phi_1}{r_2}$$

$$T_2 = 2r_2 \left[\frac{T_1}{2r_1} + \text{Inv } \phi - \text{Inv } \phi_2 \right]$$

where

r_1 = given radius

ϕ_1 = pressure angle at r_1

T_1 = arc tooth thickness at r_1

r_2 = radius where tooth thickness is to be determined

ϕ_2 = pressure angle at r_2

T_2 = arc tooth thickness at r_2

This formula is useful in determining width of top land by substituting r_o (outside radius) for r_2, and then determining top land width for an oversize pinion. A rule of thumb for the latter might be that the minimum top land width should be about 20% of the top land width at r_o.

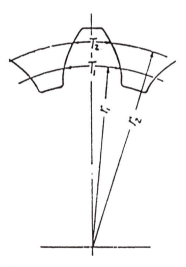

Figure b

For a nonmodified pinion and gear the full-depth addendum is 1/DP. The dedendum is 1.157/DP in the Brown & Sharpe system, 1.25/DP in the AGMA full depth system, 1.35/DP for full depth teeth to be shaved or ground, and 1.40/DP in the full depth fillet root system for teeth to be induction hardened by scanning. For stub tooth addenda and dedenda see later in this section.

Oversize modification of pinions is desirable for another reason also. It increases the *arc of recess* (length *PC* in Fig. 1b) and reduces the *arc of approach* (length *BP*) when the pinion is the driver. The tooth action through the arc of approach tends to scuff the tooth surfaces under heavy load, and the action through the arc of recess is a polishing action (*provided* the drive is properly lubricated). So properly lubricated *recess action* gearing will operate more smoothly and have longer life than gearing with standard tooth proportions. See Fig. 11 for illustrations of standard versus recess action gears.

Note, however, that if the gear drives the pinion, as it would in a speed increases, and if the pinion is oversize, the intended recess action would become *approach action*. Another case of gear driving the pinion is in a hoist drive when the hoist is letting the load descend. In general, gearing in such drives should have teeth of standard proportions, or if the pinion must be modified oversize to increase tooth strength, then the gear must be modified oversize also, to make the angles of approach and recess as nearly equal as possible.

When a gearset of standard proportions is to be replaced by a modified gearset, usually the standard center distance must be held. In such a case the gear is cut on an undersize diameter—undersize by the same amount the pinion diameter has been increased. When this is done we have *long-and-short addendum gears*.

A special case when modification of the driving pinion is almost always advisable is for nonmetallic pinions—nylon, fabric-impregnated bakelite, etc. Here tooth deflection under load is greater than with metal pinions, and modification minimizes the tendency for the tips of the gear teeth to gouge into the flanks of the pinion teeth as the latter deflect.

Sometimes it is best to avoid modification of small pinions and live with the undercut you get, or specify *stub teeth,* or perhaps even use *form cut* (also called *rotary cut*) pinion and gear (see Section VII.A). An example is open gearing which is lubricated by heavy grease applied with a stick. We have seen that pinion modification increases the arc of recess, and in this case recess action can be a problem as it is a wiping action of the pinion teeth on the gear teeth. It will tend to wipe away the grease lubricant faster than a nonmodified pinion, since on the latter the arcs of recess and approach are equal. So modification is not recommended for drives where the lubricant used is so heavy that it will not flow back quickly. For such drives the stub tooth is sometimes the answer. Undercut is minimized, or sometimes eliminated, because

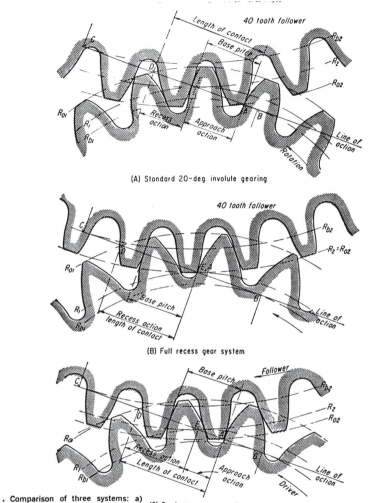

(A) Standard 20-deg. involute gearing

(B) Full recess gear system

. . Comparison of three systems: a) standard involute gears, b) full RA gears, and c) semi-RA gears. All are 20-deg pressure angle spur gears with a 20-tooth pinion (driver) and a 40-tooth follower. Note the directions of rotation. Teeth of the standard form first make contact at A; detrimental wear occurs during the entire approach action from A to E. Recess action E to D, on the other hand, is of a beneficial nature. In full RA system, all tooth contact occurs during recess action. But because the length of contact has been reduced, the minimum number of teeth permissible for the pinion has gone up to 20 teeth or more. Semi-RA gears may be the best compromise as they reduce approach action yet increase contact length.

(C) Semi recess gear system

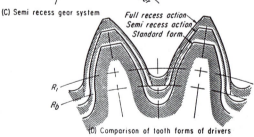

(D) Comparison of tooth forms of drivers

Figure 11   Standard versus recess action gears. (From Earle Buckingham, *Manual of Gear Design, Section 2, Spur and Internal Gears*. Revised by Eliot K. Buckingham, Buckingham Associates, Inc., Springfield, Vermont, 1980.)

the root diameter is greater, but contact ratio is less because a stub tooth has a shorter active profile. So unless the number of teeth in the gear is quite large stub teeth can be troublesome. Properly designed stub tooth gearing is good, however, for relatively low speed heavily loaded drives.

There are two stub tooth systems, both with 20°PA. The first was the *Fellows stub,* designated as a fraction, such as 4/5 DP, which says that the pitch of the gear is 4 DP with a 5 DP tooth depth. A later system is the *AGMA 20° stub* system. In this system the addendum is 0.8/DP and the dedendum is 1/DP. The AGMA system is interchangeable with the Fellows system and has largely replaced it.

VII. CROWNED TEETH

Crowned teeth are often specified for spur, helical, and straight-tooth bevel gears. They are thicker at or near the center of the tooth face than at their ends. Crowning is accomplished by crown hobbing, crown shaving, or crown grinding. On straight bevels the trade designation of crowned teeth is Coniflex (a Gleason Works trademark). The purpose of crowning is to minimize the chance of *end bearing* of the teeth when there is misalignment in assembly, distortion from heat treating, or deflection under load.

The amount of crown is commonly in the neighborhood of 0.0003 in./in. of face (0.0006 in. change in tooth thickness per inch of face). It can be varied but if assembly conditions call for more than this, the designer must be concerned with the compressibility of the tooth surfaces and tooth deflection. If the crown exceeds the amount the teeth will compress and deflect under full load to provide adequate tooth bearing area, the full face of the gearing cannot be used in determining the load rating.

VIII. FORM CUTTING OF SPUR GEARS

Something should be said about *form cutting* (sometimes called *rotary cutting*) of spur gears. Form cutters are milling cutters which have the form of a tooth space (see Fig. 5). They produce an approximate involute form. In general, gears produced by this method are not too satisfactory except for low pitch line velocity operation and in relatively heavy pitches, and in those cases of open gearing, referred to previously, for which modification should be avoided, and strength requirements dictate that undercut cannot be tolerated.

Standard form cutters are *range cutters,* so called because there are 8 cutters in a set, or 15 in a set for greater profile accuracy, each of which is designed to cut a range of numbers of teeth. Cutters for pinions in the undercut range are modified so they can mill below the base circle radius. Gear cutters are modified to produce gear teeth which will clear this lower portion of the pinion tooth profile. Since the profiles of form cut teeth are made up of more than one curve they are called *composite* tooth forms. The involute portion of the composite form is correct for the minimum number of teeth in each range. Range cutters are listed in the Appendix (Table 12). When greater accuracy of profile than can be obtained with range cutters is required, cutters designed for cutting specific numbers of teeth are used.

In general, form cut gearing is not interchangeable with generated tooth gearing, except in sets.

IX. GEAR RACK

Gear rack in pitches heavier than about 3 DP is usually form cut, because most rack generating machines are not designed to cut the heavier pitches. It is best to use single-purpose rack cutters to cut true involute teeth (straight sides on rack teeth), with some *tip relief* to minimize the possibility of tip interference. A rack and pinion is pictured in Fig. 12.

Some manufacturers of medium to heavy pitch rack use #1 range cutters (135 teeth to rack). As noted this produces a tooth profile which a 135-tooth gear would have, so it is not correct for a rack. The practice of making all rack with *basic rack form* teeth ensures smoother rolling with generated tooth pinions.

Finer pitch rack (finer than about 3 DP) can be form cut but usually is generated. Rack shapers are used for generating gear rack teeth, using the same type of reciprocating cutters used for cutting spur or helical gears by shaping.

X. LARGE-DIAMETER GEARS

An interesting application of heavy pitch gear rack is in the manufacture of large-diameter gears, larger than can be produced on conventional hobbing machines. One application for such gears is their use as the final drive to transmit low-speed, high-torque motion to the parabolic-shaped reflectors of large radiotelescopes and satellite tracking antennas. Such structures with reflectors of from 30- to 150-ft diameters and larger have been built. These final drive gears must be large in diameter to provide a high reduction ratio and to keep gear tooth loads resulting from inertia, momentum, and wind forces within practical limits. The largest antenna drive gear known to the writer has 1980 teeth and is 110-ft pitch diameter, a full circle gear which is the azimuth drive for a large antenna. Figures 13 and 14 show a typically tracking antenna and a closeup of the final drive gear, which in this case is the elevation drive and is a 120° sector of about 30-ft pitch radius. The azimuth drive gear for this antenna is a gear about 72 in. diameter, cut on the O.D. of a bearing. The reflector diameter of this antenna is about 60 ft.

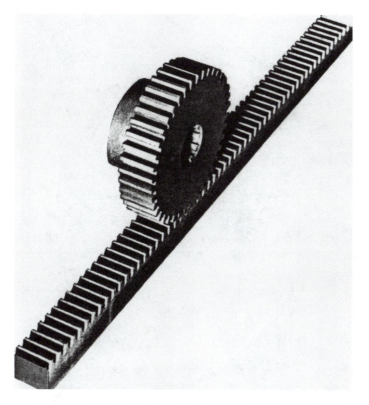

Figure 12 Gear rack and pinion (spur). (Courtesy of Sewall Gear Manufacturing Company, St. Paul, MN.)

Figure 13   A typical tracking antenna. (Courtesy of Harris Corporation, Melbourne, FL, and Sewall Gear Manufacturing Company, St. Paul, MN.)

Figure 14 Antenna final drive gear. (Courtesy of Harris Corporation, Melbourne, FL, and Sewall Gear Manufacturing Company, St. Paul, MN.)

These gears are made in segments cut as straight gear racks, then mechanically formed to the required radius. The cutting tools used are form cutters, with the pitch, pressure angle, and profile nonstandard, so that after the rack is formed to its specified radius its pitch becomes standard and its tooth profile becomes a true involute, conjugate to conventionally hobbed pinion teeth. Assembled with the aid of jigs to set the tooth spacing at the joints between the segments, the gearing operates with remarkable angular uniformity and with tooth-bearing patterns typical of generating tooth gearing.

XI. BACKLASH

Before proceeding to a discussion of some of the most common types of gears, it is important to understand that *all* types must have *backlash*. This is the gap between the profiles of meshing teeth when the driving side of a tooth is held tight against the driven at operating center distance. It can be measured with a feeler gauge.

A rule of thumb, useful as a guide, for determining the amount of backlash for various pitches is 0.030 to 0.050 in. divided by the diametral pitch. Using this rule, a 2 DP gearset would have 0.015 to 0.025 in. backlash, a 3 DP gearset 0.010 to 0.017 in., and so on. In practice, however, the designer must think in terms of tooth thicknesses and tooth thickness (TT) tolerances, and then decide what to specify for each gear. In addition, *runout* and center distance tolerances have to be considered, as each will cause the effective tooth thicknesses at the operating pitch circles to vary. Actual backlash, therefore, is what results from these variables after a gearset is assembled. It should not be specified as such until the designer has determined that the backlash he has in mind is consistent with the tolerances he has specified.

To illustrate, the backlash from applying data from AGMA Classification Manual [14], AGMA Standard 390.03—see Appendix), assuming Quality 8C, 2 DP 20°PA spur gearing (not modified), 12 in. PD pinion and 50 in. PD gear, can be determined as shown in Table 1.

The foregoing comments on backlash and Table 1 apply to industrial gearing up to about AGMA Quality 10. In addition, there are many applications where heat expansion and cold contraction must be considered, and relative expansion and contraction between gears and housings when they are of dissimilar materials.

One example is marine gearing—the propeller drive gears which reduce the speed of the prime movers (steam turbines or diesel engines) to the much lower speeds of the propellers. Most navy and many commercial ships are driven by steam turbines, which operate at much higher speeds than diesel engines. On these the first reduction gears get up to speed quickly at start-up, generating heat at the mesh which takes a while to dissipate through the lubricant and then through the housing. This heat will cause the first reduction gearing to expand,

Table 1 Determination of Backlash

	Pinion	Gear
Runout tolerance (total indicator reading)	0.0075″ (max.)	0.0105″ (max).
Corresponding effect on tooth thickness (TT)[a]	+0.0027″ −0.0027″	+0.0038″ −0.0038″
Specified TT tolerance (assumed)	+0.0000″ −0.0048″	+0.0000″ −0.0048″
Center distance tolerance (assumed)	+0.0050″ −0.0000″	
Corresponding effect on TT[a]	+0.0000″ −0.0018″	+0.0000″ −0.0018″

The resulting total effective tooth thickness variation, measured at the operating pitch circles, would be as follows:

	Pinion	Gear
Runout effect on TT	+0.0027″ −0.0027″	+0.0038″ −0.0038″
TT mfg. tolerance	+0.0000″ −0.0048″	+0.0000″ −0.0048″
Center distance effect on TT	+0.0000″ −0.0018″	+0.0000″ −0.0018″
Variation resulting from above	+0.0027″ −0.0093″	+0.0038″ −0.104″
TT reduction for runout	−0.0027″ −0.0027″	−0.0038″ −0.0038″
TT reduction for 0.015 in. min. (assumed) B/L	−0.0075″ −0.0075″	−0.0075″ −0.0075″
Total TT variation at pitch circle	−0.0075″ −0.0195″	−0.0075″ −0.0217″

Gearsets made to these specifications, with center distance tolerance as indicated, would therefore operate with blacklash of 0.015 inch minimum to 0.041 inch maximum.

[a] Variation in depth of mesh changes effective tooth thickness in the ratio of 1 to 0.73 for 20°PA.

often by enough to reduce backlash substantially. Therefore when these gears are manufactured, their teeth must be cut thinner than would be considered normal for industrial gearing.

Most ship propulsion gearing is double helical, some single helical, designed with finer pitches than would normally be used for industrial gearing of the same size. For example, a typical marine gear of 100 in. PD would probably be about 4 DP, where a typical industrial gear would be more like 2 DP. The reason for this is that the finer pitch gearing will operate more quietly than heavier pitch, important to the navy in its efforts to minimize sound which could be picked up by enemy vessels. Gear noise is important to operators of passenger ships as well. They do not want to hear complaints about gear noise from passengers in cabins located above the engine room.

An example where gears and housings are of dissimilar materials is gearing in army ground transportation vehicles which operate in a wide variety of temperatures, from the tropics to the arctic. Most of these gears are enclosed in aluminum or magnesium housings which have different coefficients of expansion than steel. Therefore the expansion coefficients of both gears and housings must be considered when specifying backlash.

Still another example is aerospace gearing, the most highly loaded of which is the gearing driving propellers and helicopter rotors. These are of necessity light weight, are made from premium steels, carburized and hardened and teeth finished by grinding. They are usually enclosed in aluminum or magnesium housings. Here again, therefore, the coefficients of expansion of both gears and the housings must be considered when specifying backlash.

Backlash is in no way related to gear quality. One example is the aerospace gearing described above. It is AGMA Q13 to Q15, but is required to have more backlash than most commercial quality gearing.

XII. GEAR TYPES

The foregoing comments were confined primarily to spur gearing, spur gear rack, and straight bevel gearing. The other types in common use are essentially variations of these, developed for applications in which the basic types would not perform satisfactorily. For example, helical and herringbone gears will operate more quietly than spurs made to the same degree of accuracy, and size for size they will transmit more power.

The various types of gearing in most common use can be categorized as follows:

Parallel shafts
    Spur
    Helical
    Herringbone

Each of these can be made with internal as well as external teeth, except that an internal herringbone must have a relief groove between helices for cutter and chip clearance. A gear rack is any one of the above with infinite pitch radius.

Intersecting shafts
    Straight bevel
    Spiral bevel
    Zerol
Nonintersecting nonparallel shafts
    Hypoid
    Worm gearing
    Crossed-axis helical

A. Parallel Shafts

1. Spur

Spur gears are gears with the teeth cut parallel to their axes of rotation. They are the simplest to design and make as compared to other gear types, and probably most gear designers start with spur gearing for a given application, changing to helicals or herringbones if found to be necessary. A typical commercial spur gearset and a combination internal-external spur are shown in Figs. 15 and 16.

2. Helical

Helical gears are gears with the teeth cut at an angle to their axes of rotation. The teeth are sections of a helix that spiral around the body of the gear, not straight as are spur gear teeth.

An important consideration in designing helical gears is to work from standard spur gear pitches. The advantages are twofold. The helix angle can be selected to provide adequate tooth overlap (a good rule of thumb is 1.2 or greater pitches overlap), and the designer is not limiting his sources of supply. With this approach the pitch of the spur gear hob or shaper cutter becomes the normal pitch of the helical gear. The circular pitch in the transverse direction (direction of rotation) is thus greater than the circular pitch of the hob or cutter. It equals the normal pitch divided by the cosine of the helix angle. Normal pitch helical gears are therefore larger in diameter than spur gears of the same pitch and number of teeth.

An alternative is to design to available *transverse pitch* hobs or cutters. This approach results in helical gears with the same numbers of teeth and pitch diameters they would have if designed as spur gears. Therefore transverse pitch helicals can be used to replace spurs when ratio and center distance must not be changed, if provision is made to handle the lateral thrust which occurs with helical gearing. The helix angles most commonly used for transverse pitch

SPUR GEAR & PINION

COMBINATION INTERNAL-EXTERNAL SPUR

15

16

HELICAL GEAR & PINION

INTERNAL GEAR & PINION (HELICAL)

17

18

Figures 15–18 (Courtesy of Sewall Gear Manufacturing Company, St. Paul, MN.)

helicals are 7½° and 15°, sometimes 23° and 30°, and 45° for crossed-axis helicals. As indicated, the disadvantages of transverse pitch helicals are that optimum helix angle for tooth overlap cannot be selected and sources of supply are limited to those who have these special hobs. In addition, the tooth strength is less since the normal tooth thickness of transversal pitch helical gearing is less than the normal tooth thickness of helicals designed from spur hobs or cutters.

Illustrations of commercial quality (AGMA Q8) helical gearing, external and internal, are shown in Figs. 17 and 18. The hands of the helices are opposite each other on the externals (the pinion is left hand and the gear is right hand) and on the internal gear and pinion in the illustration both are the same hand (left hand).

3. Herringbone

Herringbone gears are related to helicals in that they are, in effect, two helicals side by side with the hands of the helices opposed to each other. The advantages of herringbones over helicals are primarily in drives requiring very few teeth, such as found in steel mill roll drives and in geared pumps, and in drives where axial thrust cannot be tolerated.

Herringbone gears with a relief groove between the two helices are sometimes called *double helical gears.* They are usually considered to be more satisfactory than continuous-tooth herringbones for high-speed drives because the apex of the teeth on the latter can form an oil trap, causing noise, and at the higher speeds increase heat generation. Also relief grooves minimize tooth breakage due to axial runout, as continuous teeth will have stress concentration at the apex. With little or no axial runout stress concentration is less and continuous teeth are less apt to break than teeth with a relief groove. The most common helix angle for herringbones is 30°, and the pitch designation refers to the transverse direction.

The drives for most oilfield pumping units are herringbones. One reason for this is that one of the oldest and largest manufacturers of such equipment invested heavily in herringbone gear generating machines years ago. The gear reducers they designed and built therefore do not require thrust bearings, as would single helical reducers. Another reason is that the thrust bearings which would have to be used with single helicals would fail prematurely because of the reversing load direction, which in an oilfield pumping unit drive, occurs twice per revolution of the low-speed gear. What happens is that the load reversals loosen the thrust bearings, resulting in dynamic loads on them which they are not designed to handle.

An illustration of a typical oilfield pumping unit drive, with the top housing cover removed, is shown in Fig. 19, and a pumping unit is shown in Fig. 20.

Figure 19 Herringbone gear oil well pumping unit drive. (Courtesy of Lufkin Industries, Inc., Lufkin, TX.)

B. Bevel Gearing with Intersecting Shafts

1. Straight Bevel

Straight-tooth bevel gearing is gearing with the teeth cut on the outside of a cone and which, if extended, would intersect at the cone's apex. Since the diameter of a cone is greatest at its base the teeth will be thicker at the base and this is called the *heel* of the teeth. The thinner end of the teeth is called the *toe*. The *cone distance* of a bevel is the length from the base to the apex of the cone. The face of a bevel is the length from the heel to the toe and should *never* be greater than one-third the cone distance. The pitch of a bevel is measured at the heel. Two bevels with equal numbers of teeth, designed to run with each other, are called *miter gears*. The shaft angle of bevel gears is usually 90°, although bevel gearing can be designed to operate on any shaft angle.

Most straight-tooth bevels today are generated with crowned teeth. These are called Coniflex bevels, a Gleasaon Works trademark. The initial tooth bearing area of Coniflex teeth will be an elongated "bubble" shape, which should,

Figure 20 Oilfield pumping unit. (Courtesy of Lufkin Industries, Inc., Lufkin, TX.)

when assembled, be located nearer the toe than the heel of the teeth. Figure 21 shows a set of Coniflex straight bevel gears.

There are many machines still in operation which cannot cut Coniflex teeth. Bevels cut on these machines should be cut to have an initial toe bearing and assembled so as not to lose it, as the toe end of the teeth will deflect under load more than the heel, and will have better running-in characteristics.

2. Spiral Bevel

Spiral bevel gearing is the "helical" version of straight-tooth bevels. The teeth are generated on an angle to an element of the pitch cone, with the spiral angle selected (usually 35°) so the entering tooth comes into contact before the leaving tooth lets go. This feature of spiral bevels and helicals supplements the involute contact ratio referred to previously.

3. Zerol

Zerol bevels are spiral bevels with a 0° spiral angle. A possible advantage over Gleason Coniflex straight-tooth bevels is that each tooth has one convex and one concave surface, while a Coniflex tooth has two convex surfaces. The ini-

Figure 21   Straight-bevel gears. (Courtesy of The Gleason Works, Rochester, NY.)

tial tooth bearing area of Zerol gearing can therefore be greater than Coniflex gearing.

C.   Nonintersecting Nonparallel Shafts

1.   Hypoid

Hypoid gears are spiral bevels with the pinion shaft offset. Their most common use is in automotive rear axle drives.

Figures 22 and 23 show a spiral bevel gearset and a hypoid gearset.

Figure 22 Spiral bevel gears. (Courtesy of The Gleason Works, Rochester, NY.)

2. Worm Gearing

Worm gearing, properly designed and produced, provides a good trouble-free drive for smooth operation and permits a wide selection of ratios, but there are precautions which must be taken for optimum gear life. For example, with a high *lead* angle (say 30°), the gear should be tangentially hobbed as infeeding the hob will remove some of the active profile. Also it is desirable to specify higher pressure angles for the high lead angles (25° to 30°PA for lead angles over 30° or so). Selection of materials is critical. SAE 65 bronze with about 1% nickel (see Section XIV) is good for the gear, and the worm should be hardened steel with the threads ground after hardening. Care in assembly is critical, too. The worm gear should be adjusted laterally so the tooth bearing pattern is offset toward the "leaving" side of the teeth, as determined from the direction of rotation of the worm.

Finally, for optimum results, worm gearing should be run in after assembly, at no load to start, gradually increasing to full load, running until a polished area appears on the gear teeth. This may not be necessary if the initial tooth bearing area is satisfactory, but if it is less than about 60% of the face width

Figure 23   Hypoid gears. (Courtesy of The Gleason Works, Rochester, NY.)

there could be trouble without a run-in. The lubricant used for the run-in should be whatever is specified for worm gearing. Do *not* use lapping compound in an attempt to hurry the run-in process.

Hobs for cutting worm gears differ from spur and helical hobs in that they are usually designed to cut the *throat* of the worm gear (they are called *topping hobs*) as well as the teeth. They must have the same number of threads (or *starts*) as the mating worm, and the outside diameter must be somewhat larger than the O.D. of the mating worm, never the same or smaller. The pitch of worms and worm gears is generally measured in the transverse direction.

The capacity of a worm gear drive can be increased by designing it as recess-action worm gearing. The improved characteristics result from the increase in the recess (polishing) portion of the thread-tooth contact. This is achieved in worm gearing by hobbing the worm gear undersize, not by making the driving worm larger, as one might at first expect from the discussion on spur and helical gear modification. The reason for this is that the worm thread is the basic rack form no matter what its diameter, so increasing the worm diameter will change only the lead angle, not thread-tooth action.

A worm and gear set is illustrated in Fig. 24.

In *double-enveloping worm gearing* the worm is shaped like an hour-glass, so it will be in contact with more worm gear teeth than a conventional worm. This type of gearing will transmit more power than a conventional worm and gear set of the same size, but more heat will be generated in high-speed operation and assembly time will be longer, as worm alignment and position are more critical than with conventional type worm gearing.

3. Crossed-Axis Helical

Crossed-axis helical gears are helicals with helix angles of the same hand, usually designed to operate at a 90° shaft angle. These are sometimes called *spiral gears*. They have point contact and hence are limited to relatively light load applications. An illustration of a set of crossed-axis helicals appears in Fig. 25.

XIII. GEAR MEASUREMENT

It is common practice to check spur gears by taking measurements of diameters over pins, which are not pins at all, but are short sections of hardened and ground steel rods of various diameters. To measure a gear a pin is laid in tooth spaces opposite each other and diameter measurements are made over the tops of the pins. If the gear has an even number of teeth the pins are placed 180° apart, if an odd number the pins are placed as close as possible to 180° apart. The diameter of the pin to be used is selected to suit the pitch of the gear to be checked. It must be of such size so that it will rest in the tooth space, touching the tooth flanks at or slightly above the pitch line, stick out above the O.D. of the gear, and must not touch the root diameter. Standard pin sizes have been established, and gear makers and users can buy standard size pins for the commonly used pitches of teeth. There are published tables to enable determination of measurements over pins for the standard pitches without going through lengthy calculations.

Table 2 gives the standard pin sizes commonly used. The pin diameters change as the tooth size changes. For metric design *multiply* the pin diameter constant by the module. For English system design, *divide* the pin diameter constant by the diametral pitch.

Helical gears also can be checked over pins, although theoretically balls should be used instead of pins because a helical tooth is curved. The calculations should be made for ball measurement, but most gear manufacturers then use pins for the actual measurement. There is a slight difference between the calculated ball measurement and the actual pin measurement, the difference

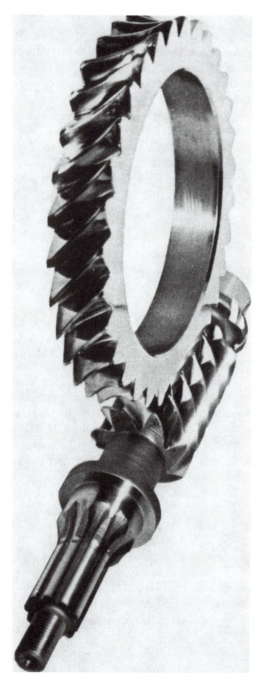

Figure 24 Worm gearing. (Photo courtesy of Sewall Gear Manufacturing Company, St. Paul, MN.)

Figure 25 Crossed-axis helicals. (Photo courtesy of Sewall Gear Manufacturing Company, St. Paul, MN.)

Table 2 Pin Sizes Used to Check the Tooth Thickness and Diameters of Spur Gears

Type of tooth	Pressure angle	Pin diameter constant
External, standard or near standard proportions	14½° to 25°	1.727
		1.920
		1.680
External, long-addendum pinion	14½° to 25°	1.920
Internal, standard proportions	14½° to 25°	1.680
		1.440

increasing as the helix angle increases, but is usually so slight that it can be ignored.

For heavy pitch gears, gear tooth vernier calipers are often used, instead of pins, to measure tooth thicknesses. They measure chordal tooth thickness, from which circular tooth thickness can be calculated if desired for determination of backlash.

For large-diameter gears it is necessary also to measure the chordal length over a number of teeth, from which tooth thickness and accuracy of tooth spacing can be determined. For such measurements a *span measuring tool* is used, set to touch the flanks of teeth at the ends of each span at or near the middle of the tooth height. A span measuring tool will normally be set by a vernier and will be equipped with a dial indicator to indicate any deviation from the theoretical chord length.

The pitch size categories of "fine," "medium," and "heavy" are usually considered to be as shown in Figure 26.

Category	DP	CP(DP/3.1416)	TT(Nominal) (CP/2)
Fine	20 DP & finer	0.1571 in. & finer	0.0785 in. & finer
Medium	19 DP to 4 DP	0.1653 to 0.7854 in.	0.0826 to 0.3927 in.
Heavy	3½DP to 1 DP	0.8976 to 3.1416 in.	0.4488 to 1.5708 in.

Figure 26 Pitch size categories

Gear tooth verniers measure the chordal tooth thickness, so this must be calculated from the known circular, or arc, tooth thickness. The formulas are

$$\text{arc } \beta = \frac{T}{2r} \text{ rad}$$

$$T_c = 2r \sin \beta$$

See Fig. c.

where

T = arc tooth thickness at r

T_c = chordal tooth thickness at r

Example. (For a 30T, 6DP, 14½°PA gear).

$r = 2.500$ in. and $T = 0.2618$ in.

$$\text{Arc } \beta = \frac{0.2618}{5.000} = 0.05236 \text{ rad} = 3.000°$$

$$\sin \beta = 0.05234$$

$$T_c = 2 \times 2.500 \times 0.05234 = 0.2617 \text{ in.}$$

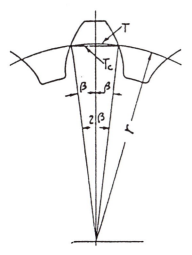

Figure c

It is then necessary also to find the height of the chordal addendum. The formulas are

$$\text{arc}\,\beta = \frac{T}{2r}$$

$$a_c = r_0 - \cos\beta$$

See Fig. d.

where

   T = arc tooth thickness at r

   T_c = chordal tooth thickness at r

   r = given radius

   r_0 = outside radius

   a_c = chordal addendum

Example.   (For a 30T, 6DP, 14½° PA gear).

$r = 2.500$, in., $r_0 = 2.6667$ in., and $T = 0.2618$ in.

$$\text{Arc}\,\beta = \frac{0.2618}{5.000} = 0.05236\,\text{rad} = 3.000°$$

$$\cos\beta = 0.99863$$

$$a_c = 2.6667 - 2.500 \times 0.99865 = 0.1701\,\text{in.}$$

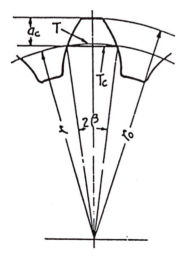

Figure d

Measurement of bevel gearing is accomplished by mounting gear and pinion on a Gleason bevel gear checking machine and rolling them together, driving the gear by the pinion, with a brake on the gear shaft to provide a load. Red lead is applied to the gear teeth, enabling examination of tooth bearing area and position and measuring of backlash and runout.

Worms are checked over pins. Worm gears are checked by meshing with a mating worm on a fixture and checked for tooth bearing area and position and for backlash. For repeat or production jobs it is desirable to use a master worm, made to a high degree of accuracy.

XIV. GEAR MATERIALS
A. Steel

In steels, the low carbons are carburizing grades and the .40 to .50 carbons are through-treating grades. Typical examples of such steels are 8620 and 4140. Higher strength steels often specified are 4320 and 4340. The last two digits in the numbers represent the points of carbon present, so 8620 is low carbon and 4140 is medium carbon (a "point" of carbon is 0.01%, so 4140 steel contains 0.40% of carbon). The first two digits indicate the type of alloying element in the steel. Tables 3 and 4 list typical gear steels, the first giving the common heat treat practice and hardenability characteristics of each, the second the chemical composition of each. If the first two digits are 10, there are no alloying elements except manganese, and these are called "carbon steels."

The .40 to .50 carbon grades are suitable for flame or induction hardening as well as through-hardening. It should be noted that usually induction hardening is preferable to flame hardening, particularly for tooth-to-tooth hardening of the heavier pitches, because with the induction process closer control of hardness pattern and depth is possible. When hardening of the roots as well as the flanks of the teeth is required, gears to be induction hardened should be cut with hobs that produce full fillet roots. Large steel gears to be induction hardened should be weldments with rolled steel rims, not castings. Castings can, and often do, have inclusions which cause cracks when quenched.

An important thing to consider when specifying steel to be used for making gears is its cleanliness—the absence of impurities.

Alloy steel for barstock and forgings for gears is produced in electric furnaces, is commonly vacuum-degassed, inert-atmosphere-shielded, and bottom-poured to improve cleanliness and reduce the content of hydrogen, oxygen, and nitrogen. Improved cleanliness results in improved transverse ductility and impact strength but can reduce machinability. An example would be if sulfur content is reduced to less than 0.015%.

Vacuum-degassed steel can be further refined by vacuum arc remelting or electroslag remelting. These processes further reduce inclusion size and con-

Table 3 Typical Gear Materials—Wrought Steel

Common alloy steel grades	Common heat treat practice[a]	General remarks/application
1045	T-H, I-H, F-H	Low hardenability
4130	T-h	Marginal hardenability
4140	T-H, T-H&N, I-H, F-H	Fair hardenability
4145	T-H, T-H&N, I-H, F-H	Medium hardenability
8640	T-H, T-H&N, I-H, F-H	Medium hardenability
4340	T-H, T-H&N, I-H, F-H	Good hardenability in heavy sections
Nitralloy 135 mod.	T-H&N	Special heat treatment
Nitralloy G	T-H&N	Special heat treatment
4150	I-H, F-H, T-H, TH&N	Quench crack sensitive Good hardenability
4142	I-H, F-H, T-H&N	Used when 4140 exhibits marginal hardenability
4350[b]	T-H, I-H, F-H	Quench crack sensitive, excellent hardenability in heavy sections
1020	C-H	Very low hardenability
4118	C-H	Fair core hardenability
4620	C-H	Good case hardenability
8620	C-H	Fair core hardenability
4320	C-H	Good core hardenability
8822	C-H	Good core hardenability in heavy sections
3310[b]	C-H	Excellent hardenability (in heavy sections for all three grades)
4820	C-H	
9310	C-H	

[a] C-H = carburize harden; F-H = flame harden; I-H = induction harden; T-H = through harden; T-H&N = through harden then nitride.
[b] Recognized, but not current standard grade.
Source: Extracted from AGMA 2004-B89, *Gear Materials and Heat Treatment Manual*, with the permission of the publisher, the American Gear Manufacturers Association, 1500 King Street, Suite 201, Alexandria, VA 22314.

tent, and gas, for improved fatigue strength, to produce the highest quality steel for critical gear applications.

Two obvious "critical gear applications" are aircraft engine gearing and helicopter transmission gearing.

B. Cast Iron

Cast iron is a good gear material if shock loading is minimal. When it is used, usually the driving pinion is steel, or perhaps a nonmetallic material if noise is

Table 4 Standard Alloy Steels—Chemical Composition Limits (%)

AISI or SAE Number	C	Mn	P Max	S Max	Si	Ni	Cr	Mo
1045	0.43/0.50	0.60/0.90	0.040	0.050	—	—	—	—
4130	0.28/0.33	0.40/0.60	0.035	0.040	0.15/0.35	—	0.80/1.10	0.15/0.25
4140	0.38/0.43	0.75/1.00	0.035	0.040	0.15/0.35	—	0.80/1.10	0.15/0.25
4145	0.43/0.48	0.75/1.00	0.035	0.040	0.15/0.35	—	0.80/1.10	0.15/0.25
8640	0.38/0.43	0.75/1.00	0.035	0.040	0.15/0.35	0.40/0.70	0.40/0.60	0.15/0.25
4340	0.38/0.43	0.60/0.80	0.035	0.040	0.15/0.35	1.65/2.00	0.70/0.90	0.20/0.30
4150	0.48/0.53	0.75/1.00	0.035	0.040	0.15/0.35	—	0.80/1.10	0.15/0.25
4142	0.40/0.45	0.75/1.00	0.035	0.040	0.15/0.35	—	0.80/1.10	0.15/0.25
4350	0.48/0.53	0.60/0.80	0.035	0.040	0.15/0.35	1.65/2.00	0.70/0.90	0.20/0.30
1020	0.18/0.23	0.30/0.60	0.040	0.050	—	—	—	—
4118	0.18/0.23	0.70/0.90	0.035	0.040	0.15/0.35	—	0.40/0.60	0.08/0.15
4620	0.17/0.22	0.45/0.65	0.035	0.040	0.15/0.35	1.65/2.00	—	0.20/0.30
8620	0.18/0.23	0.70/0.90	0.035	0.040	0.15/0.35	0.40/0.70	0.40/0.60	0.15/0.25
4320	0.17/0.22	0.45/0.65	0.035	0.040	0.15/0.35	1.65/2.00	0.40/0.60	0.20/0.30
8822	0.20/0.25	0.75/1.00	0.035	0.040	0.15/0.35	0.40/0.70	0.40/0.60	0.30/0.40
3310	0.08/0.13	0.45/0.60	0.025	0.025	0.15/0.30	3.25/3.75	1.40/1.75	—
4820	0.18/0.23	0.50/0.70	0.035	0.040	0.15/0.35	3.25/3.75	—	0.20/0.30
9310	0.08/0.13	0.45/0.65	0.025	0.025	0.15/0.35	3.00/3.50	1.00/1.40	0.08/0.15

Table 5 Minimum Hardness and Tensile Strengths for Gray Cast Iron

ASTM[a] class no.	Brinell hardness no.	Tensile strength (KSI)
20	155	20
30	180	30
35	205	35
40	220	40
50	250	50
60	285	60

[a] See ASTM A48 for additional information.

a factor and if loads are not too great. Cast iron and the nonmetallic materials are seldom used as gear materials in the enclosed drives produced by speed reducer manufacturers.

Cast iron is a generic term for the family of high-carbon and silicon iron alloys, classified by the following categories.

1. Gray Iron

Gray iron contains carbon—usually over about 3.0%—which is present as graphite flakes. Minimum hardness requirements are tabulated in Table 5.

2. Ductile Iron

Ductile iron, sometimes referred to as nodular iron, is characterized by the spheroidal shape of the graphite in the metal. This is produced by the introduction of magnesium and rare earth elements. Table 6 gives the mechanical properties of ductile iron.

Table 6 Mechanical Properties of Ductile Iron

ASTM[a] grade	Heat treatment	Brinell hardness	TS, min. (KSI)	YS, min. (KSI)	Elong. in 2″(min. %)
60-40-18	Anneal	170 max.	60	40	18.0
65-45-12	As cast or anneal	156–207	65	45	12.0
80-55-06	Normalize	187–255	80	55	6.0
100-70-03	Quench and temper	241–302	100	70	3.0
120-90-02	Quench and temper	As specified	120	90	2.0

[a] See ASTM A536 or SAE J434 for further information.

C. Powdered Metal

Powdered metal parts are formed by compressing metal powders in a die cavity and heating (sintering) it to bond the particles metallurgically. Die costs are high so large quantities are necessary to amortize die costs. The powdered metal usually specified for gear applications is alloy steel.

D. Gear Bronzes

Gear bronzes are used mostly for worm gears and are selected for their wear resistance when running with a hardened and ground steel worm. These are the phosphor or tin bronzes, manganese bronzes, aluminum bronze, and silicon bronzes. The bearing characteristics of aluminum bronze are inferior to the phosphor bronzes but better than for manganese bronze. The silicon bronzes main use is for lightly loaded gearing.

There are many other bronze alloys but relatively few are suitable for worm gear application. In addition to those listed above there is one which works well for many applications. It is a copper-zinc-manganese alloy, leaded, available as drop forgings or in bars (Mueller 6731 or 6730). The most common of the cast bronzes is phosphor bronze, SAE 65 (now SAE C90700), which is improved if about 1% nickel is added and chill cast, or centrifugally cast if the configuration permits.

For best results in worm gear drives the worm must be carburized and hardened, with the threads ground. Also, as noted previously, for optimum results worm gearing should be run in after assembly, at no load to start, increasing to full load after a reasonable time, or until a polished area appears on the gear teeth.

E. Nonmetallic Materials

There are several nonmetallic gear materials, the most common of which are the phenolic resins. One of these is canvas or linen-impregnated bakelite. Usually gears made from this material are machined. This material, called *thermosetting laminates,* is available in sheets of various thicknesses, square or round rods, and tubes. It is produced by coating the fabric sheets with liquid phenol-formaldehyde resin, cutting and stacking between metal plates, and bonding under pressures of 1000 to 2500 psi and temperatures of 270 to 350°F. One common use for this material is in automobile timing gears.

The nylons, a family of *thermoplastic polymers,* usually do not have fabric reinforcing materials but can be compounded with them. Gears made from this material are either machined or made with molded teeth if the quantities are large enough to justify the cost of the molds.

In drives which must operate quietly and are not too heavily loaded, the pinion is often specified to be nonmetallic. As noted in Section V, tooth deflection under load is greater than for metal pinions, so such pinions should be modified oversize (long addendum) to minimize the tendency of the gear teeth to gouge into the flanks of the pinion teeth as the latter deflect.

XV. INSPECTION OF MATERIALS

A. Hardness Testing

Materials and finished gears in the machinable hardness range are usually checked on a Brinell hardness machine, which presses a hardened steel ball into the surface to be checked. The diameter of the impression is a measure of the hardness, and is calibrated to give what is called the *Brinell hardness number* (BHN).

A Rockwell test machine can also be used to check materials in the machinable hardness range. This machine forces a diamond penetrator into the surface at given loads, with the depth of penetration calibrated to give the *Rockwell hardness number*. For hardnesses over about 400 BHN a Rockwell machine must be used. Rockwell C scale (RC) is the most common, for which the load is 150 kg and the hardness range is about 25 to 68 RC. Other scales, loads and ranges are as follows:

Scale	Load	Use
Rockwell A (RA)	60 kg	Small gears, medium shallow case.
Rockwell 30-N (R30-N)	30 kg	Shallow depth case-hdnd. parts.
Rockwell 15-N (R15-N)	15 kg	Very shallow depth case parts.

Another device for checking hardness is the scleroscope. This drops a hardened ball from a given height onto the surface being checked. The height of rebound is a measure of the surface hardness.

For all hardness checks the surface must be relatively smooth (say 64 μin. max.) and flat. The relationship between Brinell and Rockwell hardnesses and tensile strength can be obtained from steel warehouse catalogs. One of these tables, reproduced from a Central Steel & Wire Co. catalog, appears as Table 7.

Another device for checking hardness is the *Tukon*. This is a laboratory instrument used for determining case depths in test bars placed in the carburizing furnace with the gears being carburized and hardened. The hardened test bar is sectioned, the section polished to a mirror finish, and readings taken from the surface inward every 0.002 in. More accurate results are obtained by sectioning a few of the gears from each batch after hardening if the gears are

Table 7 Hardness Conversion for Carbon and Alloy Steels (all values are approximate)

Brinell Hardness Number (Carbide Ball)	Rockwell Hardness Numbers					Tensile Strength	
	C Scale	A Scale	15N Scale Superficial	B Scale	30T Scale Superficial	ksi	MPa
—	66	84.5	92.5	—	—	—	—
722	64	83.4	91.8	—	—	—	—
688	62	82.3	91.1	—	—	—	—
654	60	81.2	90.2	—	—	—	—
615	58	80.1	89.3	—	—	—	—
577	56	79.0	88.3	—	—	313	2160
543	54	78.0	87.4	—	—	292	2010
512	52	76.8	86.4	—	—	273	1880
481	50	75.9	85.5	—	—	255	1760
455	48	74.7	84.5	—	—	238	1640
443	47	74.1	83.9	—	—	229	1580
432	46	73.6	83.5	—	—	221	1520
421	45	73.1	83.0	—	—	215	1480
409	44	72.5	82.5	—	—	208	1430
400	43	72.0	82.0	—	—	201	1390
390	42	71.5	81.5	—	—	194	1340
381	41	70.9	80.9	—	—	188	1300
371	40	70.4	80.4	—	—	182	1250
362	39	69.9	79.9	—	—	177	1220
353	38	69.4	79.4	—	—	171	1180
344	37	68.9	78.8	—	—	166	1140
336	36	68.4	78.3	—	—	161	1110
327	35	67.9	77.7	—	—	156	1080
319	34	67.4	77.2	—	—	153	1050
311	33	66.8	76.6	—	—	149	1030
301	32	66.3	76.1	—	—	146	1010
294	31	65.8	75.6	—	—	141	970
286	30	65.3	75.0	—	—	138	950
279	29	64.6	74.5	—	—	135	930
271	28	64.3	73.9	—	—	131	900
264	27	63.8	73.3	—	—	128	880
258	26	63.3	72.8	—	—	125	860
253	25	62.8	72.2	—	—	123	850
247	24	62.4	71.6	—	—	119	820
243	23	62.0	71.0	—	—	117	810
240	—	—	—	100	83.1	116	800
234	—	—	—	99	82.5	114	785
222	—	—	—	97	81.1	104	715
210	—	—	—	95	79.8	100	690
200	—	—	—	93	78.4	94	650
195	—	—	—	92	77.8	92	635
185	—	—	—	90	76.4	89	615
176	—	—	—	88	75.1	86	590
169	—	—	—	86	73.8	83	570
162	—	—	—	84	72.4	81	560
156	—	—	—	82	71.1	77	530
150	—	—	—	80	69.7	72	495
144	—	—	—	78	68.4	69	475
139	—	—	—	76	67.1	67	460
135	—	—	—	74	65.7	65	450
130	—	—	—	72	64.4	63	435
125	—	—	—	70	63.1	61	420
121	—	—	—	68	61.7	59	405
117	—	—	—	66	60.4	57	395
114	—	—	—	64	59.0	—	—

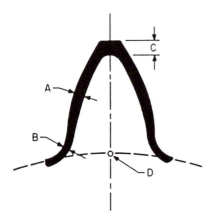

Figure 27 Carburized case pattern. Shaded area is all 50 RC in hardness or higher. Unshaded area is less than 50 RC. A, profile case thickness; B, root case thickness; C, tip case thickness; D, core hardness taken here.

small enough, or for larger and heavier pitch gears, sectioning individual gear teeth, then polishing the sectioned areas and taking readings as above. Since this is a destructive test, it is practical only when the quantities are large enough to warrant it.

Illustrations (schematic) showing sectioned gear teeth, one a carburized case pattern and one an induction-hardened (by scanning) case pattern, are shown as Figs. 27 and 28. By etching the polished surfaces the hardened area can be plainly seen.

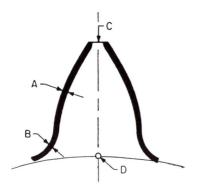

Figure 28 Induction-hardened case pattern, by scanning method of heating. A, profile case thickness; B, root case thickness; C, top land, no case at center; D, core hardness taken here.

The Tukon instrument can be used to check any hardness, so it is used also to determine core hardness of case hardened parts, induction as well as carburized, and in determining hardness penetration in through-hardening steels. The effective carburized case depth is usually specified to be about one-sixth the tooth thickness at the pitch line and is the depth from the surface to where the hardness is 50 RC.

B. Inspection for Internal Defects—Ultrasonic

Ultrasonic inspection is a nondestructive test method to determine the internal soundness and cleanliness of material by passing sound (ultrasound) through it. Very short sound waves of a frequency greater than 20,000 cycles per second (audible limit) are voltage generated and transmitted into the part by a transducer. In the method most often used, returning sound waves are transformed into voltage and monitored on an oscilloscope screen. (This discussion is from ANSI/AGMA Standard 2004-B89 [5], with permission of AGMA.)

C. Surface Finish

Surface finish requirements are designed by the symbol ^{xx}, with the numerical value xx in microinches. Table 8 defines the various degrees of finish.

D. Inspection for Surface Defects

1. Dye Penetrant

A red dye is applied to the surface, wiped off, then a white powder in suspension is applied. If there are surface cracks, the dye will have penetrated into them and will bleed through the white coating.

Table 8 Symbols for Surface Finish

Symbol	Type of finish
1000	Extremely rough, like a sand casting
500	Very rough, not suitable for mating surfaces
250	Rough, heavy toolmarks
125	Fine, even toolmarks
64	Fine, no objectionable toolmarks
32	Smooth, toolmarks barely visible
16	Very smooth, ground finish
8	Near mirrorlike, polished surface
4	Mirrorlike, no marks

2. Magnetic Particle

Magnetic particle inspection is a nondestructive testing method for locating surface and near surface discontinuities in ferromagnetic material. When a magnetic field is introduced into the part, discontinuities laying approximately transverse to the magnetic field will cause a leakage field. Finely divided ferromagnetic particles, dry or in a water or oil suspension, are applied over the surface of the material under test. These particles will gather and hold at the leakage field, making the discontinuities visible to the naked eye.

XVI. HEAT TREATING OF STEEL GEARS

Almost all steel gears require some sort of heat treatment, either before or after machining. An explanation of what happens when steel is heated and then cooled follows.

First it should be said that the amount of carbon present determines the steel's response to heat treatment more than any other element, in alloy as well as plain carbon steels. In alloy steels, the various alloying elements (manganese, nickel, chromium, molybdenum, and vanadium) are added in various combinations to improve core properties in carburizing grades, and hardness depth and toughness in through-treating grades. Relatively small amounts of phosphorus, silicon, and sulfur appear in all steels.

A. Through-Hardening Steels

We will talk first about heat treating the through-hardening grades of steel (0.30% or more carbon). The unmachined steel, semifinished or finished gear or part is heated to what is called the critical range of temperatures. This is usually 1450 to 1650°F, which is varied with the carbon content and the alloying elements in the steel. Some examples follow:

Steel	Critical range (°F)
1030	1575 to 1650
1045	1450 to 1550
4140	1525 to 1625
4340	1475 to 1525

The steel, gears, or parts are held at the critical long enough to ensure that they are heated throughout, then cooled rapidly (quenched) by submerging in water or oil. Water is a more severe quench medium (faster cooling) than oil and is used in quenching the carbon steels. Oil must be used in quenching the alloy steels as they will harden at a slower cooling rate and would be likely to crack if quenched in water.

To explain what happens when steel is heated and then quenched we must look at its structure, which can be likened to crystalline rocks composed of different materials. These mineral-like constituents which may occur in steel are ferrite, cementite, austenite, martensite, troostite, sorbite, and pearlite, which can be identified with the aid of a microscope after specimens have been highly polished and then etched with dilute acids which attack, in varying degree, the different constituents. These constituents are defined as follows:

1. Ferrite

Ferrite is composed largely of pure iron with small amounts of other elements in solution in the solid state. It is soft and ductile, is strongly magnetic, has a tensile strength of about 40,000 psi and an elongation of about 40%. In its free state, ferrite is found in all carbon steels with less than 0.85% carbon when these steels have been cooled slowly from temperatures above or within the critical range.

2. Cementite

Cementite is a chemical compound having the formula Fe_3C, containing 93.333% iron and 6.666% carbon by weight. Other elements also combine with carbon to form cementite. This is the hardest of all constituents occurring in steel. It is found normally in all slowly cooled steels containing *over* 0.85% carbon.

3. Austenite

Austenite is a solid solution of carbon in iron produced at temperatures at and above the critical range. It is transformed into other constituents as the steel is cooled through the critical range. It is nonmagnetic and can be detected because of this property. *Retained austenite* in hardened steel causes the surface hardness, in the as-quenched condition, to be less than expected. It can be transformed into martensite (the hard constituent) by deep freezing, from room temperatures, by submerging in a dry ice and alcohol mixture. If this doesn't work, other reasons for a too-soft surface could be decarburization, deteriorating quenchant or malfunctioning quenchant agitators.

4. Martensite

Martensite is the hard constituent, obtained when steel is cooled rapidly from within or above the critical range. The lower the carbon content of the steel the more rapidly it must be cooled to obtain martensite. The hardness of martensite increases with the carbon content.

5. Troostite

Troostite is formed by cooling less rapidly than to obtain martensite, or by reheating steel in which martensite has occurred. Steels which contain troostite are difficult to machine.

6. Sorbite

Sorbite is a mixture of small unoriented particles of ferrite and cementite. It is produced by regulating the rate of cooling from temperatures above the critical range or by tempering steel after hardening.

7. Pearlite

Pearlite is a mixture of ferrite and cementite in alternate layers. All steels contain some pearlite after cooling slowly from within or above the critical range. Normally pearlite has a tensile strength of about 125,000 psi and elongation of about 10%.

Figure 29 shows the temperatures at which these various constituents are formed, and the schematic appearance of each when etched and examined under a microscope.

B.   Review of Constituents[a]

Martensite, troostite, sorbite, and pearlite are transition products resulting from the decomposition of austenite. At temperatures above the critical range all steels are composed entirely of grains of austenite. The rates of cooling determine the constituents which are retained in the steel at normal temperatures. The first transition product resulting from the most rapid cooling of steel is martensite which is the strongest, hardest, and least ductile constituent. Slower rates of cooling produce in turn troostite, sorbite, and pearlite, with increasing ductility and decreasing strength and hardness. When martensite is obtained by very rapid cooling, its transformation to troostite or sorbite is brought about by slowly reheating the steel to temperatures below the critical range. It is therefore possible to produce varying degrees of strength, hardness, and ductility in a given steel by controlling the rate of cooling and the temperatures of reheating.

XVII.   PRINCIPAL HEAT TREATING PROCESSES

A.   Annealing

Annealing is accomplished by heating to the critical range of temperatures and cooling slowly, usually in the furnace. The constituents present in annealed

---

[a] From Harry L. Campbell, *The Working, Heat Treating, and Welding of Steel*, Wiley, New York.

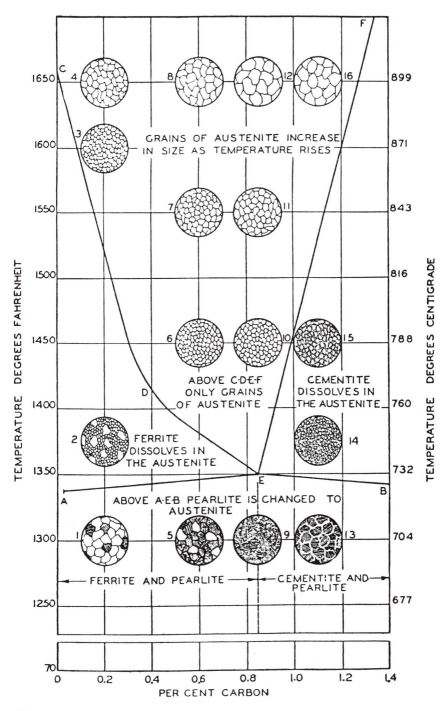

Figure 29   Critical range diagram for steels.

steel are ferrite and pearlite, the proportion of ferrite to pearlite increasing as the percentage of carbon increases (they are about equal at 0.45% carbon). The purposes of annealing are primarily to improve machinability, relieve internal stresses and refine the crystalline structure.

B. Normalizing

Normalizing is accomplished by heating to about 100°F above the critical range and cooling in still air at room temperature (say about 70°F). The resultant structure is composed largely of sorbite. The purposes of normalizing are primarily to refine the grain structure prior to hardening, produce a homogeneous structure throughout, and reduce segregation in castings and forgings.

C. Tempering

Tempering requires reheating after hardening to temperatures below the critical range, followed by the desired rate of cooling. The purpose of tempering is to relieve stresses caused by quenching and to reduce the hardness and brittleness which result from quenching steels containing more than about 0.40% carbon. The higher the tempering temperature (probably a maximum of about 100°F below the critical) the lower will be the final hardness. So tempering temperature is specified according to the final hardness required.

Figure 30 indicates the various temperature ranges we have discussed for hardening, annealing, normalizing, and tempering.

D. Case Hardening

1. Carburizing and Hardening

First we will discuss *carburizing* and *hardening*. This is achieved by heating the low-carbon steels in a carbon-rich atmosphere, most often at 1650 to 1750°F. The steel absorbs carbon to a depth determined by the time it is in the furnace. Steels for carburizing usually contain 0.15% to 0.20% carbon. After carburizing the case will contain about 1.0% carbon at the surface, diminishing to 0.40% to 0.45% carbon at the depth where a hardness of 50 RC is required. This is known as the effective case thickness (see Section XV.A, Tukon inspection procedure).

Some of the steels frequently requiring carburizing and hardening are 1015, 1020, 8620, and 4320. The first two are carbon steels, the second two alloys. The core properties of the carbon steels will be affected very little by the carburizing and hardening process, but the core properties of the alloys will be improved as they have hardenability characteristics of their own. This is why alloy low-carbon steels are selected for applications requiring tooth strength as well as tooth surface hardness.

To harden, the parts are quenched from the carburizing temperature, then, for best results, reheated at 1400 to 1450°F and again quenched. If the surface

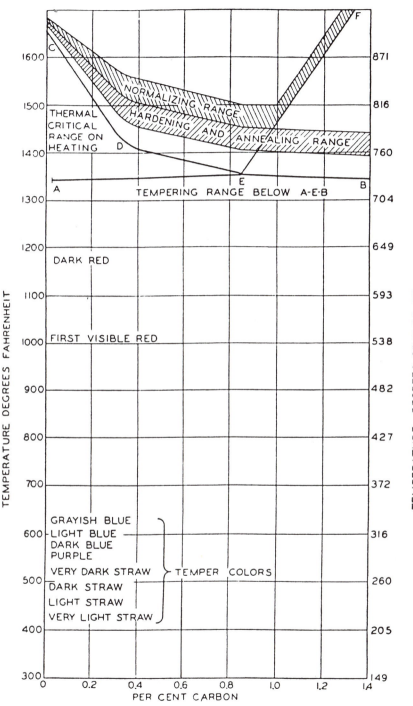

Figure 30 Temperature ranges used in heat treating steels.

hardness is less than expected, retained austenite could again be the culprit, which as stated for through-hardening steels can be eliminated by deep-freezing in a mixture of dry ice and alcohol.

2. Nitriding

The steels most often used are a special alloy called Nitralloy, usually containing about 0.40% carbon and pretreated to the desired core properties. Other alloys such as 4140 can also be nitrided. This process is accomplished by heating to relatively low temperatures (about 950°F) in an atmosphere of dry ammonia gas for 10 to 40 hours depending on part size and required case depth, then cooled in air. The case is harder than obtained with carburizing and hardening, and much shallower. Because of the low temperature used in nitriding and because the part is not quenched, distortion is less than occurs in carburizing and hardening.

3. Flame Hardening

Flame hardening is a case hardening process when tooth-to-tooth hardening is used for the heavier pitches. Through-hardening steels are required, so when finer pitch gears are spin hardened the teeth are hardened throughout.

In tooth-to-tooth flame hardening an oxyfuel burner, or flame head, is used, shaped to heat the flanks of the tooth. Theoretically the flame head can be shaped to harden the roots as well, but it is difficult to achieve consistency in root hardness and depth. The flame head is followed by a quench spray nozzle as the flame head traverses the length of each tooth. The quench medium is usually water.

4. Induction Hardening

Induction hardening of through-hardening steels, tooth to tooth, heats the tooth flanks, and the roots when required, by inducing an electric current into the surfaces to be hardened. This is accomplished by passing high-frequency alternating electric current through an inductor block, which in turn produces a high-frequency magnetic field which induces a current in the surfaces of the teeth, heating them to the critical temperature range. The depth of heating is controlled by varying the speed of scanning, the frequency, and the power. The inductor is followed by a quench spray nozzle having a shape similar to the inductor. The quench medium is water with an additive to reduce the severity of the quench to approximate an oil quench. Oil cannot be used as it would clog the spray nozzle and because of danger of fire. The inductor, which must not touch the workpiece surfaces, is shaped to harden the flanks of each tooth when the roots are not to be hardened, or the flanks of adjacent teeth *and* the root and fillets when they are. The latter is recommended as tooth bending strength is increased. *It is important* to make up a two- or three-tooth section, harden it by scanning, then check to make sure the depth, pattern, and hardness are as specified before hardening a complete gear.

Finer pitch gears—finer than about 5 DP—are spin induction hardened by use of a wrap-around coil. This hardens the teeth throughout. Illustrations of the various types of coils, inductor blocks or flame heads used, and hardness patterns obtained, are shown schematically as Fig. 31. Induction hardening is preferred over flame hardening as hardness and depth are more accurately controlled.

There are induction hardening machines available which do their work with the gears being hardened submerged in water, or water with an additive. It is said that distortion is less when such machines are used. Flame or induction hardening is specified when the size and configuration of the gear make it impractical to carburize and harden.

Sometimes it is necessary to case harden the teeth only, leaving the rest of the gear soft for finish machining after the teeth are hardened. This of course is accomplished in induction of flame hardening. It is accomplished in carburizing by painting the surfaces to remain soft with a copper powder suspended paint or by copper plating (which requires masking the teeth). The painting or plating prevents carbon penetration in the painted or plated surfaces.

A recent development in induction hardening is called the Micropulse system, developed by Contour Hardening (Indianapolis). This system will rapidly harden gear teeth and other irregular shapes which require a high degree of accuracy. In this system a gear is preheated by radiofrequency pulses, then given a final pulse to heat it to the required critical temperature, from which it is quenched by a water-base solution applied through the coil. It is said that the cycle time is fast and that distortion is minimal.

5. Shot Peening

Shot peening is the bombardment of stress concentration surfaces, such as the roots and fillets of gear teeth, with hardened steel or glass shot. The diameters of the shot used are related to the fillet radius of the gear teeth to be shot peened. Usually the flanks of the teeth are not directly peened. Shot peening produces minute indentations in the surfaces at which the peening device is aimed. The pellets are propelled by centrifugal force from a rotating wheel or by air pressure. These indentations produce residual compressive stress in the peened surfaces. This increases the bending strength of gear teeth as under load the usual tensile stress from bending has to first overcome the compressive stress before the teeth start to bend. The result is that shot peened teeth can have as much as 25% greater strength, both bending and fatigue, than teeth which have not been shot peened.

Probably the greatest benefit from shot peening is for hardened gears as the quenching of them can cause surface tensile stresses which shot peening can overcome. It is important that any heat treating be done prior to shot peening as heating after peening will draw the peened surfaces, eliminating the desired residual compression stress.

SPIN HARDENING

FLANK HARDENING

INDUCTOR OR FLAME HEAD   INDUCTOR OR FLAME HEAD

FLANK AND ROOT HARDENING

INDUCTOR OR FLAME HEAD

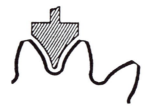

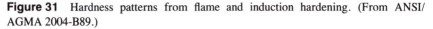

Figure 31 Hardness patterns from flame and induction hardening. (From ANSI/ AGMA 2004-B89.)

It is recommended that gears which are expected to be highly stressed be subjected to shot peening, with automated equipment, operated by experienced people, in shops which have adequate checking capabilities.

Shot peening is not to be confused with blasting for cleaning.

XVIII. SPLINES

Splines are a series of teeth, or slots and keys, on a shaft and in the bore of a gear, sprocket, sheave, etc., used to transmit power from one to the other. There are two types of splines, involute and parallel side.

A. Involute Splines

Involute splines have the same form as involute gear teeth except the teeth are one-half the depth of standard full depth gear teeth and the pressure angle is 30°. The spline is designated by a fraction, similar to the Fellows stub for gear teeth, in which the numerator is the diametral pitch and the denominator is always twice the numerator. There are 14 standard pitches, as follows: 2.5/5, 3/6, 4/8, 5/10, 6/12, 8/16, 10/20, 12/24, 16/32, 20/40, 24/48, 32/64, 40/80, and 48/96. The minimum number of teeth is 6 and the maximum is 60. There are both flat root and fillet root types.

There are three types of fits: (1) *major diameter,* a fit controlled by varying the major diameter of the external spline; (2) *sides of teeth,* a fit controlled by varying the tooth thickness, customarily used for fillet root splines; (3) *minor diameter,* a fit controlled by varying the minor diameter of the internal spline. Each type of fit is further divided into three classes: (a) *sliding,* clearance at all points; (b) *close,* close on either major or minor diameter or sides of teeth; (c) *press,* interference on either the major or minor diameter, or sides of teeth.

There are also 37.5° and 45°PA standard involute splines, but these are less common than 30°PA splines. Involute spline proportions, dimensions, fits, and tolerances are given in detail in Ref. 7, from which the foregoing data were taken.

B. Parallel Side (or "Straight-Sided") Splines

SAE Standard parallel side splines are illustrated in Fig. 32 and Tables 9 and 10, with dimensions and proportions. They are equally spaced keyways. External splines can be milled on a milling machine equipped with an indexing mechanism, or hobbed if the special hobs are available. Internal parallel side splines can be cut on a keyseater equipped with an indexing mechanism, or broached if broaches are available.

Some of the advantages of involute over parallel side splines are that standard gear inspection procedures can be used, self-centering equalizes stresses and they have maximum strength at the roots of the teeth.

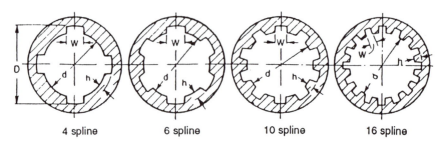

4 spline   6 spline   10 spline   16 spline

Figure 32   Parallel side splines.

Table 9   Dimensions of Spline Fittings (SAE standard, all dimensions in inches)

Nominal diam	4-spline for all fits		6-spline for all fits		10-spline for all fits		16-spline for all fits	
	D max[a]	W max[b]	D max[a]	W max[b]	D max[a]	W max[b]	D max[a]	W max[b]
¾	0.750	0.181	0.750	0.188	0.750	0.117		
7/8	0.875	0.211	0.875	0.219	0.875	0.137		
1	1.000	0.241	1.000	0.250	1.000	0.156		
1 1/8	1.125	0.271	1.125	0.281	1.125	0.176		
1¼	1.250	0.301	1.250	0.313	1.250	0.195		
1 3/8	1.375	0.331	1.375	0.344	1.375	0.215		
1½	1.500	0.361	1.500	0.375	1.500	0.234		
1 5/8	1.625	0.391	1.625	0.406	1.625	0.254		
1¾	1.750	0.422	1.750	0.438	1.750	0.273		
2	2.000	0.482	2.000	0.500	2.000	0.312	2.000	0.196
2¼	2.250	0.542	2.250	0.563	2.250	0.351		
2½	2.500	0.602	2.500	0.625	2.500	0.390	2.500	0.245
3	3.000	0.723	3.000	0.750	3.000	0.468	3.000	0.294
3½	...	...	...	...	3.500	0.546	3.500	0.343
4	...	...	...	...	4.000	0.624	4.000	0.392
4½	...	...	...	...	4.500	0.702	4.500	0.441
5	...	...	...	...	5.000	0.780	5.000	0.490
5½	...	...	...	...	5.500	0.858	5.500	0.539
6	...	...	...	...	6.000	0.936	6.000	0.588

[a] Tolerance allowed of −0.001 in. for shafts ¾ to 1¾ in., inclusive; of −0.002 for shafts 2 to 3 in., inclusive; −0.003 in. for shafts 3½ to 6-in., inclusive; for 4-, 6-, and 10-spline fittings; tolerance of −0.003 in. allowed for all sizes of 16-spline fittings.

[b] Tolerance allowed of −0.002 in. for shafts ¾ in., to 1¾ in., inclusive; of −0.003 in. for shafts 2 to 6 in., inclusive, for 4-, 6-, and 10-spline fittings; tolerance of −0.003 allowed for all sizes of 16-spline fittings.

Table 10 Spline Proportions

No. of splines	W for all fits	Permanent fit		To slide when not under load		To slide under load	
		h	d	h	d	h	d
4	0.241D	0.075D	0.850D	0.125D	0.750D		
6	0.250D	0.050D	0.900D	0.075D	0.850D	0.100D	0.800D
10	0.156D	0.045D	0.910D	0.070D	0.860D	0.095D	0.810D
16	0.098D	0.045D	0.910D	0.070D	0.860D	0.095D	0.810D

XIX. COSTS

Selection of materials, heat treatments, gear tooth data, and tolerances all will affect costs. Attention to the purely geometrical means to improve gear tooth action, discussed in this chapter, will help keep costs to a practical minimum. For example, it costs no more to hob or shape an oversize pinion than a standard one. Recess-action worm gearing is no more costly than conventional. Crowned teeth are little if any more expensive to produce than straight teeth.

Costs are affected most drastically, probably by the tolerances specified. In this regard it is recommended that a painstaking study be made of Appendix Tables 13 and 14, reproduced from Ref. 8, which tabulate tolerances for AGMA Q3 through Q15. There are later standards (ANSI/AGMA 2000-A88 [9] for spur and helical gears and AGMA 390.03A [4] for bevel, hypoid, fine pitch worm gearing and racks), but the writer used 390.03 because these tables are easier to understand than the latter tables, and the values are the same. Intelligent use of these tables still requires study of the AGMA publications indicated above.

These and other AGMA standards which could be of interest to an aspiring gear designer are listed in the Appendix.

Now we define "costs": The cost of a gear (or any machined part) is made up of the costs of material, subcontract costs, labor and shop overhead (or "burden"). For most gears the major costs are labor and burden. To determine what these are it is necessary, first, to measure, and then record, the machine time for each machining operation, then convert the time into dollars. Determination of material and subcontract costs is easy. Determination of labor and burden costs is not easy, and must be done properly to avoid trouble. It must be remembered that in a free market competitive environment, cost control can determine whether one gets or loses business, and if successful in getting, whether one keeps it.

We start with a couple of definitions:

1. *Direct labor cost.* The wages paid to employees who work directly on the product, excluding fringe benefits (these, as well as indirect labor costs, are part of factory burden).
2. *Burden costs.* The total of all manufacturing, material handling, factory services, and general factory expenses.

A. Allocation of Costs

It is no problem to accumulate total direct labor and factory burden costs over a period of time, totaling each at the close of each accounting period, and presenting them as part of the typical P&L statement. It *is* a problem to allocate, equitably, direct labor and factory burden costs to individual job orders. This must be done in some manner before a job becomes an order, in estimating the cost of each job preparatory to submitting a price quotation. For control purposes, after a job becomes an order, it is important that factory costs be recorded in the same "language" used in the estimate. There are a number of ways to allocate labor and burden costs to cost estimates and individual job orders. Two of the most common follow.

1. Direct Labor plus Shop Average Burden Percentage

A common procedure is to record direct labor times on job order sheets, convert times to dollars, then multiply the direct labor cost by a percentage to cover burden costs, having determined the percentage to use from historical data. This system possibly is satisfactory for a captive operation, where only overall costs are the primary concern of management. It has pitfalls, however, particularly in a shop engaged in jobbing work, as it ignores the fact that some machines cost more to operate, excluding the cost of direct labor, than others. This results in reported "costs" on some machines being too high, others too low. Also there is the problem of reporting direct labor costs on multiple operated equipment as the machines-per-man ratio will vary from day to day in the average shop. This presents a problem in developing meaningful standards of performance, as the reported cost on a job cannot be directly compared to the original estimate. It presents a further problem in estimating costs, as a machines-per-man ratio must be assumed in developing the estimate that could turn out to be too high a ratio when the job is run.

2. The Machine-Hour Rate System

The machine-hour rate method of allocating costs permits both the estimating and reporting of costs on the basis of machine time for each machining operation. It is particularly useful in job shop estimating and costing, where, as in the typical gear or general machine shop, investment in equipment is large and varies greatly from machine to machine, and where many machines are multiple operated (one man operates two or more machines).

With this system a rate is developed for each machine which represents each machine's proportionate share of direct labor and factory burden costs. Proportionate direct labor costs are determined from the top labor rate for each machine divided by its machines-per-man ratio. Proportionate factory burden costs are determined by first projecting such costs, then allocating them equitably to each machine, then dividing by the expected operating hours of each machine.

The procedure for developing a machine-hour rate cost system is too extensive to be covered in this chapter. For more information on this system, see Ewert [10].

XX. GEAR FAILURES

Almost every gear user will sooner or later be confronted with gear failures. Some of the most common are illustrated in Figs. 33 to 40.

A. Pitting

Pitting is a surface fatigue failure which takes a while to occur. It is designated as "micro" (small pits) or "macro" (large pits). Usually micropitting occurs first. Pitting is caused by load concentration from misalignment or errors in profile or tooth spacing, by bad design or by lubrication problems.

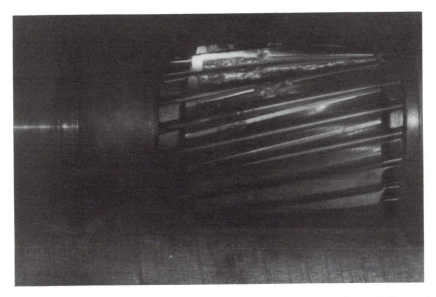

Figure 33 A case-hardened helical pinion with teeth broken and shattered, the result of severe surface damage from pitting. (Courtesy of Dudley Engineering Group, Inc., San Diego, CA.)

Figure 34 Simple bending fatigue on case-hardened spiral bevel gear teeth. Overload at heel of teeth was due to misfit with pinion. (Courtesy of Dudley Engineering Group, Inc., San Diego, CA.)

Figure 35 Severe pitting on case-hardened helical pinion teeth. Continued running would have resulted in tooth breakage. (Courtesy of Dudley Technical Group, Inc., San Diego, CA.)

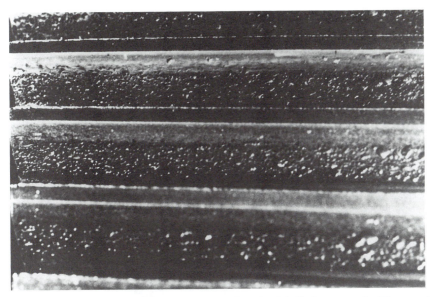

Figure 36 Micro- and macropitting on medium-hard helical gear teeth. The speed was not fast enough to develop a good elastohydrodynamic oil film and tooth loading was too heavy. (Courtesy of Dudley Technical Group, Inc., San Diego, CA.)

Figure 37 Serious "ledge"/on medium-hard helical gear teeth due to micro- and macropitting. Tips of teeth now carry most of the load. With continued running, ultimate failure will probably be tooth breakage. (Courtesy of Dudley Technical Group, Inc., San Diego, CA.)

Figure 38 Enlarged view of spur gear tooth to show light scoring and patches of micropitting. This gear was running hot with a thin oil and had a poor surface finish. (Courtesy of Dudley Technical Group, Inc., San Diego, CA.)

B. Breakage

Breakage is caused by extensive macropitting, which weakens the teeth, or most often by fatigue, with a crack usually starting at the tooth fillet. Fatigue results from repetitive stressing and can be detected by a relatively smooth surface from where the crack started, then a jagged surface where the break occurred. Breakage from sheer overload will have an all-jagged surface at the break.

C. Scoring

Scoring results, usually, because there is a lubrication problem. It can result also from bad design, poor workmanship, poor surface finish, involute error, and other things as well. The two degrees of scoring are called "initial scoring" and "severe scoring." Initial scoring often stops after a while, and severe scoring keeps going until enough material is removed to weaken the tooth, resulting in breakage. Initial scoring will be relatively smooth, and severe scor-

Figure 39 Case-hardened ground-tooth helical gear with pronounced waves in the ground finish. There is micropitting on most of the wave crests. (Courtesy of Dudley Technical Group, Inc., San Diego, CA.)

ing will cause a rough surface. If initial scoring does not stop, it will become severe.

There are several more ways a gear tooth can be damaged and more kinds of tooth breakage. In this chapter we talk only about the most common. To determine the exact causes of a failure a gear expert should be consulted, and it is important to do this so steps can be taken to avoid trouble in the future. For

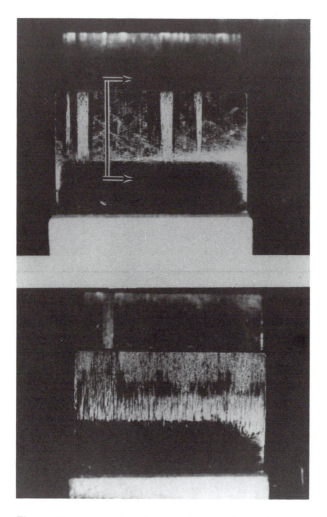

Figure 40   Upper view shows a closeup of a tooth flank with just four radial score marks. The lower view shows a tooth flank completely covered with scoring. These were case-hardened and ground teeth, heavily loaded and run with an oil without strong chemical additives to prevent scoring. (Courtesy of Dudley Technical Group, Inc., San Diego, CA.)

example, a change in the type of lubricant could be all that is necessary, having first determined that misalignment, backlash, and runout are within acceptable limits and that the hardness of pinion and gear are as specified.

If there is tooth profile error, the gears probably should be replaced. Such error can be detected by observing the wear pattern on the teeth. If the wear is greater at the tip of the pinion teeth, for example, this would indicate that the pinion was cut with a hob that generated a plus involute at the tips or provided insufficient tip relief, or both.

XXI. EFFICIENCY OF GEARS

This discussion is about gear tooth mesh and worm thread and worm gear tooth mesh efficiencies, from which the heat, or power, losses can be determined (power loss = 100 − efficiency).

The tooth mesh efficiencies and corresponding power losses of spur, helical, herringbone, straight and spiral bevel, and Zerol gears all will average about 0.98 to 0.99 (98% to 99%). The heat or power losses are therefore about 1.0% to 2.0%. While these power losses seem small and are often ignored by designers, they are important because they must be considered, along with other losses which occur in a drive, when looking at the amount of heat from friction which must be dissipated.

Hypoid gears, because they have higher tooth rubbing velocities than spiral bevels, will have a tooth mesh efficiency of about 0.90 (90%) or a power loss of about 10%.

Worm gear drives are different because their efficiency varies with the worm's lead angle, the pressure angle, the position of the pitch plane of the worm and other factors as well. The thread-tooth action is primarily sliding. To illustrate we will have a look at the worm and gear drives described as follows: first, single thread; second, triple thread; third, sextuple thread.

Worm hardened steel with the threads ground, 3.000 in. PD, 0.500 in. pitch (axial), 30°PA (normal).
Worm gear phosphor bronze (SAE 65).
Shaft angle 90°
Worm speed 1000 rpm

Let
P = axial pitch, in.

L = lead, in.

R_1 = pitch radius of worm, in.

λ = lead angle

ϕ_n = normal pressure angle

V = velocity of worm at R_1, ft/min

V_s = sliding velocity, ft/min

n = rpm of worm

f = coefficient of friction

Example 1. Single thread (1 start) (0.500 in lead), $P = 0.500$, $R_1 = 1.500$ in., $L = 0.500$ in.

$$\tan \lambda = \frac{L}{2} \times \pi \times R_1 = \frac{0.500}{2} \times \pi \times 1.500 = 0.053052$$

$$\lambda = 3°02'13''$$

$$\sin 2\lambda = 0.10581$$

$$\cos \lambda = 0.99859$$

$$\cos \phi_n = 0.86603$$

$$V = 3.000 \times \pi \times \frac{1000}{12} = 785.4 \text{ ft/min}$$

$$V_s = \frac{V}{\cos \lambda} = \frac{785.4}{0.99859} = 786.5 \text{ ft/min}$$

$$f = 0.030$$

Note that f varies with V_s (sliding velocity) approximately as follows:

V_s	f
200	0.05
400	0.04
800	0.03
1000	0.02

The values of f are experimental values for the worm and gear materials indicated herein. They are from Dudley [11].

$$\text{Efficiency } (E) = \frac{\cos \phi_n \times \sin 2\lambda}{\cos \phi_n \times \sin 2\lambda + 2f}$$

$$= \frac{0.86603 \times 0.10581}{0.86603 \times 0.10581 + 0.06}$$

$$= \frac{0.09163463}{0.15163463} = 0.60 \quad (60\%)$$

Power loss $= 100 - E = 40\%$

The formula for E is from Buckingham [12, p. 416].

Example 2. Same worm specs as Example except triple thread (3 starts) (1.500 in. lead), $L = 1.500$ in.

$$\tan \lambda = \frac{1.500}{2} \times \pi \times 1.500$$

$$= 0.15915$$

$$\lambda = 9°02'34''$$

$$\sin 2\lambda = 0.31044$$

$$\cos \lambda = 0.98758$$

$$\cos \phi_n = 0.86603$$

$$V_s = \frac{V}{\cos \lambda} = \frac{785.4}{0.98758} = 795.3 \text{ ft/min}$$

$$E = \frac{0.86603 \times 0.31044}{0.86603 \times 0.31044 + 0.06} = 0.82 \quad (82\%)$$

Power loss $= 100 - E = 18\%$

Example 3. Same worm specs as in Example 1 except sextuple thread (6 starts) (3.000 in. lead), $L = 3.000$ in.

$$\tan \lambda = \frac{3000}{2} \times \pi \times 1.500$$

$$= 0.31831$$

$$\lambda = 17°39'24''$$

$$\sin 2\lambda = 0.57804$$

$$\cos \lambda = 0.95289$$

$$\cos \phi_n = 0.86603$$

$$V_s = \frac{V}{\cos \lambda} = \frac{785.4}{0.95289} = 824.2 \text{ ft/min}$$

$$E = \frac{0.86603 \times 0.57804}{0.86603 \times 0.57804 + 0.06} = 0.89 \quad (89\%)$$

Power loss $= 100 - 89 = 11\%$

As you see, the power loss is reduced substantially as the lead angle increases. This is true up to a lead angle of about 30°, after which there is little change in power loss.

XXII. ESTIMATING GEAR SIZE (FOR SPUR GEARS)

To estimate spur pinion and gear diameters, center distance, pitch, and face width, follow the steps in this example, which is an open gearset driven by an electric motor through a speed reducer (the desired overall ratio is too high for a single gearset).

We start by making the following assumptions:

1. Medium accuracy (AGMA Q8)
2. Pinion hardness 225 BHN, gear hardness 210 BHN, 20°PA, not modified
3. 20-hp motor, 1750 rpm, coupled to input shaft of a 20:1 ratio speed reducer, with pinion mounted on reducer output shaft
4. Desired speed of gear shaft: 30 rpm (approx.)
5. Efficiency of reducer: 90%

Then

Pinion speed = 1750/20 = 87.5 rpm

$$\text{Ratio} = \frac{87.5}{30} = 2.917:1 \ (\text{use } 3:1).$$

$$\text{Actual speed of gear shaft} = \frac{87.5}{3} = 29.17 \text{ rpm}$$

To determine tangential tooth load (W_t), proceed as follows. The relationship between horsepower and torque is

$$\text{hp} = \frac{2\pi NT}{33,000} \quad \text{or} \quad T = \frac{33,000 \times \text{hp}}{2\pi N}$$

where

N = motor rpm

T = torque = pinion pitch radius $W_t = (d/2)W_t$

Now

33,000 ft/lb/min × 12 = 396,000 in.-lb/min = 1 hp   (by definition.)

Substituting assumed and known values in the equation for torque, we get

$$T = \frac{396,000 \times 20}{2 \times \pi \times 1750} \times 20 \times 0.90 = 12,965 \text{ in.-}lb$$

$$W_t = \frac{12,965}{d/2} = \frac{25,930}{d} \text{ lb}$$

Since gears in the hardness range in this example are more apt to fail from surface stresses than tooth breakage, we will work from Hertz stresses, the formulas for which are derived from the work of H. Hertz of Germany. He used two steel cylinders, pressed them together, and measured the width of the deformed area. From his formula the formula for compressive stress is derived and is

$$s_c = \sqrt{\frac{0.70}{2/E \cos \phi \sin \phi}} \sqrt{\frac{W_t}{Fd} \frac{m_g + 1}{m_g}}$$

where

s_c = compressive stress

E = modulus of elasticity, which is stress/strain as determined from test bars. For steel, $E = 30,000,000$

ϕ = pressure angle, 20° in this example

W_t = tangential tooth load

F = face width, in.

d = pitch diameter of pinion, in.

m_g = speed ratio, 3:1 in this example

So

$$s_c = \sqrt{\frac{0.70}{(2/30,000,000) \times 0.93969 \times 0.34202}} \sqrt{\frac{W_t}{Fd} \left[\frac{m_g + 1}{m_g} \right]}$$

$$= 5715 \sqrt{\frac{W_t}{Fd} \left[\frac{m_g + 1}{m_g} \right]}$$

It is convenient to call the term under the radical the "K factor," so the above becomes

$$s_c = 5715\sqrt{K}$$

In our example,

$$K = \frac{25,930/d}{Fd} \left[\frac{3 + 1}{3} \right] = \frac{25930}{Fd^2} \times 1.333 = \frac{34,565}{Fd^2}$$

$$s_c = 5715 \sqrt{\frac{34,565}{Fd^2}} \qquad s_c^2 = \frac{32,661,225 \times 34,565}{Fd^2}$$

$$Fd^2 = \frac{32,661,225 \times 34565}{s_c^2}$$

s_{ac} = allowable compressive stress = 100,000 psi (approx.) for steel in the assumed hardness range

$$Fd^2 = \frac{32{,}661{,}225 \times 34565}{(100{,}000)^2} = 112.9$$

If $F = 3$, $d^2 = 37.63$, $d = 6.134$ in.
If $F = 2.5$, $d^2 = 45.16$, $d = 6.720$ in.

A 20°PA nonmodified pinion must have at least 18 teeth to avoid undercut, so it looks like the pitch we need is 3DP. We choose a 3 in. face in case we need the added tooth bending strength over what it would be with a 2.5 in. face. Therefore,

$$N_p = 6.134 \times 3.000 = 18.4 \text{ (use 19)}$$

$$N_g = 19 \times 3 = 57$$

where

N_p = no. of teeth in pinion

N_g = no. of teeth in gear

$$\text{Pinion PD} = \frac{19}{3.000} = 6.333 \text{ in.}$$

$$\text{Gear PD} = \frac{57}{3.000} = 19.000 \text{ in.}$$

$$\text{CD} = \frac{6.333 + 19.000}{2} = 12.667 \text{ in.}$$

$$W_t = \frac{T}{d/2} = \frac{12{,}965 \times 2}{d} = \frac{25{,}930}{6.333} = 4094 \text{ lb}$$

To determine the approximate bending stress at the root of the teeth, we will use the original Lewis formula, derived by Wilfred Lewis in 1893 from the formula for stress at the base of a cantilever beam loaded at the end. The Lewis formula is

$$s_t = \frac{W_t \times \text{DP}}{F \times Y}$$

(Note: in final design calculations use the more accurate AGMA modified Lewis formula)
where

s_t = tensile stress

W_t = tangential tooth load

DP = diametral pitch

F = face width, in.

Y = tooth form factor, or "Y factor" = 0.287 (from table)

Hence,

$$s_t = \frac{4094 \times 3.000}{3 \times 0.287} = 14,265 \text{ psi}$$

The yield point of steel in the 200 BHN range is about 60,000 psi. Gear teeth have a root fillet stress concentration factor of close to 2 (Lewis did not use a stress concentration factor in his formula, but he allowed for it by using low design stresses). We will divide 60,000 by 2 for stress concentration and then by 2 for a margin of safety. This gives an approximate "Lewis" stress of 15,000 psi, which says that we are in the ball park on allowable bending stress.

There are, however, many more things to consider before a final design is reached. Some of these are:

1. The characteristics of the driven machine. Is it apt to stall on occasion? If so the stall torque of the motor must be considered in calculating the breaking strength of the teeth. Does it vibrate? If so, the dynamic load on the gear teeth will be increased.
2. The foregoing calculations assume full contact across the face. This almost never happens because of shaft misalignment, shaft deflection under load, helix angle error, etc.
3. Allowances must be made for the low pitch line velocity in this example (about 145 ft/min). This results in a marginal oil film thickness between meshing teeth, even when a relatively heavy oil is used. This is what happens with plain bearings—at low speeds the oil film is not properly distributed, but is improved as the speed increases.

All the above things require that the gears be derated.

4. The number of hours per day the machine is expected to operate. Say that this is 8 hours per day, 5 days per week. The rpm of the pinion is 87.5, so the tooth contacts per minute of any one tooth are therefore 87.5. Tooth contacts per week = $87.5 \times 60 \times 60 = 210,000$, and for a 48-week year, 10,080,000 cycles per year. The life of such a gearset, properly designed and assembled, is typically assumed to be about 10^8 cycles, so this gearset should be good for almost 10 years $(100,000,000/10,080,000 = 9.92 \text{ years})$.

The trick will be to complete the design to achieve a life of about 10 years, now following all the steps and gear trade criteria outlined in Dudley [13, Chapter 3, pp. 3.9–3.153] and in applicable AGMA standards. It is suggested, however, that only experienced gear design engineers be entrusted with developing the final specifications of the gears.

If this gearset is to become a production item, then a few sample sets should be made up to the final specifications, assembled, and operated for a long enough time to determine if the gears are, in fact, properly designed.

XXIII. ENCLOSED GEAR DRIVES

Enclosed gear drives are called *speed reducers* (or *speed increasers*) when the enclosure, called the *housing,* contains the bearings, seals, shafts, etc., as well as the gears, as opposed to enclosures placed over open gearing for safety reasons or protection from dust. Speed reducer housings are usually cast iron for standard units and steel weldments for low-quantity special units.

Most of the standard units listed by manufacturers of such drives are speed reducers, not increasers, although some of those listed could be used as speed increasers if none of its pinions are modified oversize, or if they are, their mating gears are oversize by the same amount so that angles of approach and recess are approximately equal (see Section V).

The most common types of speed reducers are discussed here.

A. Parallel Shaft

Parallel shaft reducers are single reduction (one gearset), double reduction (two gearsets), and triple reduction (three gearsets). The input shaft is driven by a motor through a V-belt drive or direct connected through a coupling. The output shaft has either a spur pinion cut integral with it, or pressed on and keyed, driving a large gear pressed on and keyed to the shaft of the machine being driven, or is connected to the shaft to be driven by one of several types of couplings. The gearing in these units is helical or herringbone. In triple reduction units the low-speed gear and pinion are sometimes spurs, with the gear induction hardened and the pinion carburized and hardened.

B. Shaft Mounted, Parallel Shaft

Shaft-mounted, parallel shaft reducers are single or double reduction with input shaft and no output shaft. In this type of reducer the low-speed gear bearings are mounted on the outside diameter of its hubs and its bore slides on to the shaft to be driven, driving by means of a key, often held in place by a Dodge (or equivalent) Taper-Lock bushing. Shaft-mounted reducers must be restrained from rotating. This is accomplished by connecting a rod, called the *torque arm,* to the reducer and a rigid part of the machine being driven, or by flange mounting. The gearing in these units is helical.

C. In-Line

In-line drives are double reduction, with the gearing arranged so that the input and output shafts are in line with each other. A common use for this type of drive is in a *gearmotor,* in which an electric motor is connected to the reducer by flange mounting. The gearing in these units is helical in integral horsepower drives. In fractional horsepower in-line drives the gearing is usually spur, fine pitch, and narrow face.

D. Right-Angle Drives

There are three basic types of right-angle drives: (1) worm gear reducers, comprised usually of one worm and one worm gear (2) combination worm gear and helical reducers, in which the initial reduction is usually a worm and worm gear; and (3) combination spiral bevel and helical reducers, in which the initial reduction is usually a spiral bevel pinion and gear.

All of these reducer types are catalogued by several gear and reducer manufacturers. To determine who these are write American Gear Manufacturers Association, 1500 King St., Alexandria, VA 22314, or call 703-684-0211. Catalogs containing illustrations, dimensions, load ratings, etc., can be obtained from the reducer manufacturers or their distributors. AGMA can provide the names of unmounted gear manufacturers as well.

E. Other Enclosed Drives

There are many other enclosed gear arrangements. A few of the most common are:

1. Epicyclic Gearing

Epicyclic gearing is a family of gear arrangements which includes two of the most common: *planetary* parallel shaft drives and the bevel gear *differential*. The planetary parallel shaft drive is composed of the *sun gear* in the center, meshing with three or more *planet gears* which also mesh with an internal gear, called the *annulus* or *cage*. The planets must be equally spaced (at 120° when there are three). Depending on which member is fixed, which is input, and which is output, the terms used to describe each are "planetary," "star," and "solar." These are illustrated in Fig. 41a–c. Table 11 provides data for each arrangement. Planetary drive applications include automatic-shift automobile transmissions, aircraft propellor drives, and many others.

Table 11 Single (Simple) Epicyclic Gear Data

Arrangement	Fixed member	Input member	Output member	Overall ratio	Range of ratios normally used
Planetary	Ring	Sun	Cage	$N_r/N_s + 1$	3:1–12:1
Star	Cage	Sun	Ring	N_r/N_s	2:1–11:1
Solar	Sun	Ring	Cage	$N_s/N_r + 1$	1.2:1–1.7:1

To assemble: $N_r = N_s + 2N_p$

$\dfrac{N_r + N_s}{\text{No. of planets}}$ must equal a whole number

N_s = number of sun teeth N_p = number of planet teeth N_r = number of ring (annulus) teeth

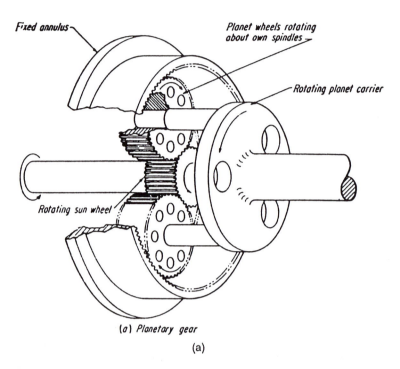

(a) Planetary gear

(a)

Figure 41   Simple epicyclic gears. (a) Planetary, (b) star gear, (c) solar gear. (From *Gear Handbook,* Darle W. Dudley, ed., McGraw-Hill Book Company, Inc., 1962, with permission.)

2.   Bevel Gear Differential

The bevel gear differential's most common application is in the rear axle drive of automobiles. Figure 42 is a schematic drawing of such a drive. In the illustration, S is the drive shaft, D is the hypoid pinion, E the hypoid gear (bore not keyed), H and K are the bevel side gears which are keyed to the wheel axles, one to the left wheel, the other to the right wheel, T are the studs which provide the bearings for the pinions R and by which the power from gear E is transmitted to the differential assembly as seen in the illustration. There are four differential pinions, spaced 90° apart.

When the car is going straight ahead, D drives E and the differential gears revolve as a unit, with no gear tooth action between the side gears and the four pinions. Whenever the car starts to turn a corner, say to the right, the left wheel will have to travel further, so the left axle B must turn faster than C.

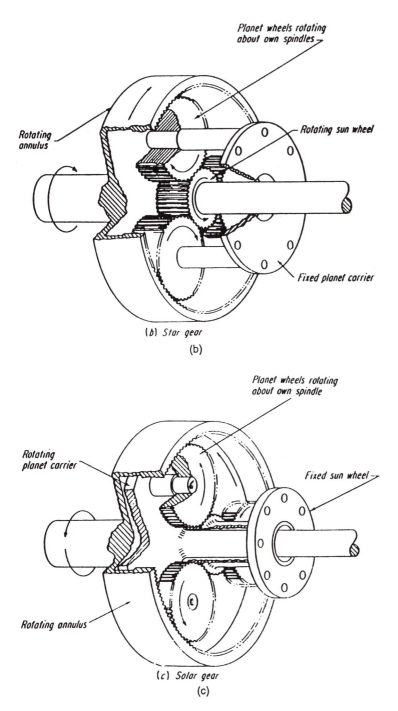

(b) Star gear

(b)

(c) Solar gear

(c)

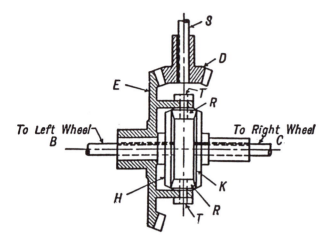

Figure 42   Bevel gear differential. (From Ref. 18.)

Then the differential gears begin to move relative to each other, the action being that of an epicyclic gear train.

XXIV. TRANSMISSION GEARING

Probably one of the most widespread uses of gearing is in automotive, truck, off-road vehicles, etc., transmissions. These gears are helical, in *constant mesh* the shifting from low to higher speeds and reverse being accomplished by toothed clutches sliding between each gearset. Illustrations of such transmissions are shown in Figs, 43 and 44. These are manual transmissions. Figure 45 shows an automatic transmission.

An interesting aside is a story told to me by an old friend and mentor, the late Professor Earle Buckingham of MIT. When Ford Motor Company built their first Model A gear shift transmission, it was noisy. Ford called in Professor Buckingham to solve the problem. In those days (about 1928), improvement in gear accuracy could have helped, but Ford had achieved about the most accurate hardened gearing possible at that time.

Professor Buckingham spent a couple of days at Ford and in a few weeks sent Ford his recommendations. He had changed the number of teeth in each gear so that the mesh in each gearset was a harmonic with others in the transmission. The noise level did not change, but the sound was musical and that ended the complaints.

Figure 43   Five-speed manual transmission. (From *Gear Handbook,* Darle W. Dudley, ed., McGraw-Hill Book Company, Inc., 1962, with permission.)

At about the time of the preceding solution for Ford's problem, Buckingham had written a book, *Spur Gears,* published in 1928. In Chapter VI there is a section entitled "Music of the Gears," in which he uses an automobile transmission as an example, stating that: "By keeping the numbers of teeth in these gears to harmonious ratios, we can avoid certain unpleasant sounds." I have been told that in addition to solving Ford's problem by this approach he solved the same problem for several other automobile manufacturers as well.

There are many more types and sizes of enclosed gear drives, too numerous to cover in this chapter, the purpose of which is to give you the basics and tell you where to look for more detailed information.

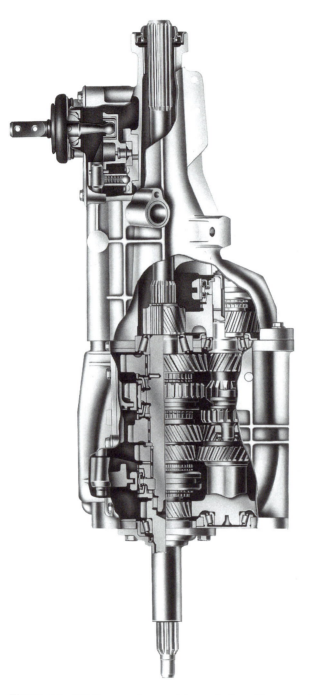

Figure 44 T-5 five-speed manual transmission. (Courtesy of Borg Warner Automotive Corporation, Muncie, IN.)

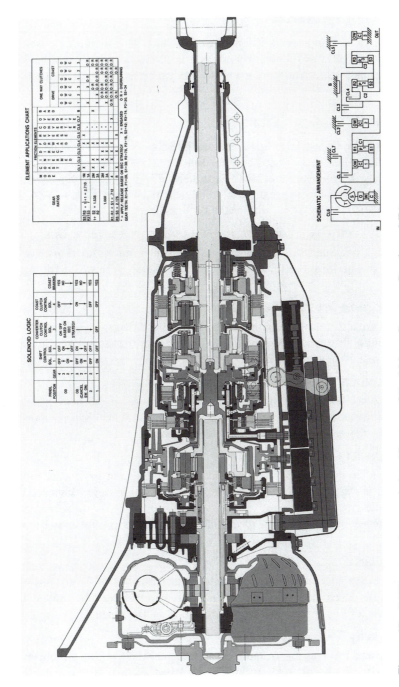

Figure 45 E40D automatic transmission. (Courtesy of Ford Motor Company, Dearborn, MI.)

XXV. LUBRICATION

Improperly lubricated gears will not last very long. This is true, of course, for any part subjected to friction. This section is divided into two parts, the lubrication of industrial open gearing and the lubrication of industrial enclosed gear drives.

A. Industrial Open Gearing

First we will discuss lubrication of open gearing—gears that operate with no enclosure or that are enclosed only for safety reasons or protection from dust, and whose bearings are lubricated separately. The gear types this covers are spur, helical, herringbone and bevel. Lubricants recommended in AGMA Standard 251.02 [14] are listed, with methods of application.

1. Rust and Oxidation Inhibited Gear Oils (R & O Gear Oils)

These are petroleum based and include chemical additives to resist rust and oxidation. Specifications for these are listed in Tables 21 and 23 in the Appendix.

2. Extreme Pressure Lubricants (EP Oils)

EP oils are also petroleum based, with chemical additives which form a film to withstand high surface pressures. Specifications for these are listed in Tables 16, 22, and 24 in the Appendix.

3. Residual Compounds—Diluent Type

Residual compounds are straight mineral or EP oils in the heavier grades, which contain a volatile solvent to make application easier. Specifications for these are shown in Table 16 (note 2) in the Appendix.

4. Special Compounds

These and some greases are available. AGMA does not have standards for these. It is suggested that recommendations for their use can be obtained from those who produce them.

There are several methods of application of lubricants. The most common follow.

5. Splash

Splash is the simplest method of applying lubricants to relatively low-speed gearing, in which the gear or an idler in mesh with the gear dips into the lubricant. Oils for this system are shown in Table 18 in the Appendix.

6. Gravity Feed or Forced Dip

Gravity feed involves one or more oilers or a pan which allows oil to drip into the gear mesh at a set rate, again for low-speed gearing. Recommended oils and rates of application are shown in Tables 19 and 20 in the Appendix.

7. Intermittent Mechanical Spray Systems

Heavy oil is used which will stay on the gear teeth through several revolutions, activated automatically at timed intervals, usually once every revolution of the gear. See Table 19 in the Appendix.

8. Continuous Pressure Lubrication

Continuous pressure lubrication uses a circulating system with a pump, producing a continuous spray to the gear teeth. See Tables 18 and 19 in the Appendix.

9. Application by a Brush, Stick, or Hand

This is for the heavier grades of lubricants, heavy pitch, and very low speed gears. An example of such a drive is the antenna final drive gear shown in Fig. 14. To determine how often application must be made, close observation is required.

B. Industrial Enclosed Gear Drives

Now we discuss lubrication of industrial enclosed gear drives—speed reducers and increasers—in which the enclosure contains the bearings, seals, shafts, etc., as well as the gears, and which have been designed and rated in accordance with applicable AGMA Standards. Covered are helical, herringbone, straight or spiral bevel, and spur gearing, operating at or less than 5000 ft/min pitch line velocity. Also covered are worm gear drives operating at or less than 2400 rpm worm speed or no more than 2000 ft/min rubbing speed. Oil sump temperature is limited to 200°F with ambient temperatures from 15° to 125°F.

The first two types of lubricants recommended in AGMA Standard 250.04 [15] are the same as the first two for industrial open gearing in Section XXV.A. The third and fourth types are given here.

1. Compounded Gear Oils

Compounded gear oils are a blend of petroleum-based oil with 3% to 10% fatty oils or synthetic fatty oils, used frequently in worm gear drives.

2. Synthetic Gear Lubricants

Synthetic gear lubricants are diesters, polyglycol, and synthetic hydrocarbons used in enclosed drives for special operating conditions. They will operate over a wider temperature range than mineral oils, but have disadvantages such as incompatibility with other lube system components. Good results with synthetics have been obtained with worm gear drives.

It is important to change oil in enclosed drives periodically. For new units the first change should be made after 500 operating hours or four weeks, whichever occurs first. Subsequent changes should be made every 2500 operating hours or six months, whichever occurs first. These recommended oil change periods are for normal operating conditions. When the old oil is drained, the unit should be flushed with flushing oil which is compatible with the lube oil. In general, solvents should not be used for flushing. See Tables 23–27 in the Appendix. For more detailed information on gear lubrication refer to the AGMA Standards indicated herein, or contact any of the major oil companies, several of which have conducted extensive research on this subject.

XXVI. GEAR TECHNOLOGY DEVELOPMENTS, 1960 TO THE PRESENT AND BEYOND

As pointed out at the start of this chapter, the advances in gear technology and manufacture matured in the period 1930–1960, then advanced greatly in the period 1960–1980, and even more greatly in the 1980s. So first we'll talk about the important developments during the 1960s and the 1980s.

A. The 1960s

In the 1960s AGMA further developed standards for the rating of surface durability and strength of spur, helical, and bevel gear teeth. Metallurgical quality gained importance in gear design considerations during this period. Gear honing machines were developed to provide better finish than grinding, required on high-speed highly loaded gears in the aerospace field. New designs of cutting tools and special materials for cutting tool teeth made possible the cutting of harder materials. One of these was the "skiving" hob which can hob gears up to about 60 RC hardness. EP lubricants were improved and synthetic oils came into use for high-temperature applications.

B. The 1980s

The development of "CBN" grinding for high-volume work. CBN stands for cubic boron nitride, an abrasive used as a coating on metal wheels. Higher speed grinding with vitreous wheels. A new cast iron called "austempered ductile iron" (ADI) was being developed. This material has good fatigue and toughness characteristics, is 10% lighter in weight then steel, has improved noise and vibration dampening, and has improved wear and scuffing resistance. Gear lubricants were further improved. More developments of honing techniques to provide better finishes than grinding. Development of CNC (computer numerical control) hobbing and shaping machines which, among other advantages, reduces setup times substantially. Earlier numerical control machines (NC) were controlled by tape. Development of titanium coating for hobs, shaper cutters, and other cutting tools to prolong their life and permit faster cutting speeds.

C. The Early 1990s

There were developments in cutting tools. Two examples were developed by Pfauter-Maag Cutting Tools, Loves Park, IL. One is called the Wafer shaper cutter, the other the Wafer hob (the hob has no wafers but is so-named as it was developed for the same purpose as the shaper cutter—to be thrown away when worn out, not resharpened). See Figs. 46–49 taken from Ashcroft and Cluff [16,17].

The Wafer shaper cutter uses a thin disk with teeth to provide the cutting edges. It will last longer than a single sharpening of a conventional cutter. When it becomes dull it is thrown away and replaced with a new wafer. No changes in machine settings are required, as is the case with a conventional cutter every time it is resharpened.

The Wafer hob is longer than a conventional hob. Instead of resharpening it is shifted to a new cutting position, and after several shifts it is thrown away. It has many more gashes than a conventional hob, hence has much thinner teeth which cannot be resharpened. The additional gashes produce a smoother tooth surface than can be obtained with a conventional hob.

There are limitations in application of these hobs. They are, however, an attractive solution to new production applications in relatively fine pitches of 10 DP and finer.

D. Current Developments

It is always difficult to predict what might happen in the future. It is safe to say, however, that we can expect more developments. Some of these will be

Further improvement in gear rating formulas.
New gear materials, both metallic and nonmetallic, will come into use.
Gear lubricant technology will make further progress, both for the oil used and for the additives.
There will be further improvement in gear quality and methods of quality control.
For high-production gears the trend will continue toward full automation of all machining and inspection operations.
Honing, peening, plating, laser bombardment, etc., can be expected to expand further in the next 20 or so years.

To summarize, in the words of Dudley [2]

Gears are here to stay. All manner of machinery on land, in water, in the air and in outer space require gears. The weight, cost, reliability and excellence of performance (in terms of noise and reliability) are of prime importance. The development of gear design and the development of ways to make gears will go on indefinitely. Man's progress on earth and in the whole cosmos beyond the earth requires a continued evolution of the gear art.

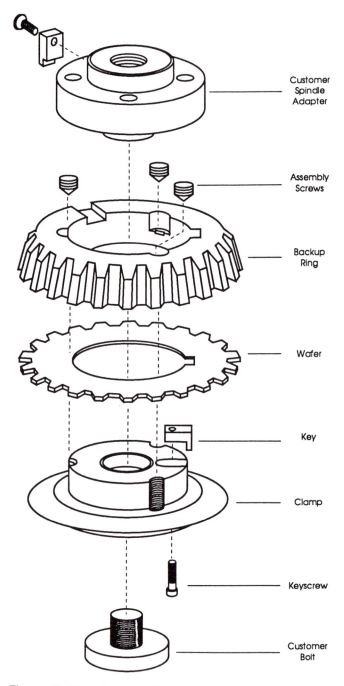

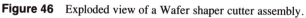

Figure 46   Exploded view of a Wafer shaper cutter assembly.

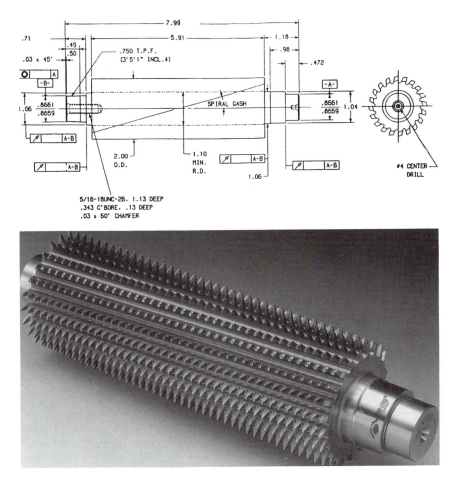

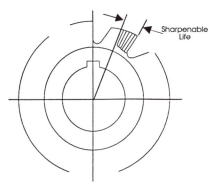

Figure 47 Design of a helically gashed shank-type solid Wafer hob for 10DP or finer.

Figure 48 Sharpenable life of a hob.

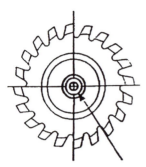

Figure 49 Nonresharpenable Wafer hob.

XXVII. RECENT DEVELOPMENTS IN GEAR CUTTING MACHINES

A few of the recent developments in gear cutting machines are briefly described and illustrated in this section. Fellows Corporation has developed and is producing a line of CNC Hydrostroke gear shapers. The heart of these machines is a hydromechanical system that applies the entire cutting force concentric to the cutting spindle, eliminating the side loads which occur in crank-type stroking systems. They also feature "rapid return," the noncutting portion of a stroke cycle, which increases production. The CNC system provides faster setup times and permits virtually unattended operation (Fig. 50). Fellows has developed many other new features on the equipment they produce.

The Gleason Works has developed and is now producing a new line of spiral bevel and hypoid generators and grinding machines, called the Phoenix. These machines use six-axis CNC to produce spiral bevel and hypoid gear geometries with no mechanical adjustments and make faster setups.

Gleason also now manufactures a line of hobbing machines, called the G-Tech machines. These are CNC machines, eliminating the need for index and differential change gears, manual setup of hob swivel angles, hob position, and hob feed. Also crown and taper hobbing can be performed without the use of cams (Figs. 51 and 52).

National Broach & Machine Co. has further developed a number of their Red Ring line of machines. One is the model GFD single-axis CNC rotary gear shaving machine, which can be conventional, diagonal, plunge, or tangential shave, with crowning, taper, or straight attachments. Another is the Red Ring CNC five-axis GF-300 gear finisher. This machine can be configured as a shaver, hard honer, roll finisher, and CBN (cubic boron nitride) form grinder. Still another is the Red Ring SF-900 CNC precision gear grinder. This machine features versatility in choosing the best grinding wheel for the job,

Figure 50   Fellows CNC hydrostroke gear shaper. (Courtesy of Fellows Corporation, Springfield, VT.)

fast setups, automatic wheel size compensation, and stock dividing. The GF-300 and SF-900 machines are illustrated in Figs. 53 and 54.

Other National Broach products include broaching machines and many others, all of which have been improved and/or redesigned over recent years.

American Pfauter has developed and is producing their latest vertical hobbing machines, the PE150 and PE300. Both are CNC machines. They feature shorter machining times, faster hob change, lower tool costs, shorter idle times (when one man is operating more than one machine), greater versatility, and faster setup. Both machines are available with automatic loading and unloading systems for production jobs. The PE150, illustrated in Fig. 55, is equipped with a loading and unloading system.

There are, of course, others of gear machine manufacturers who have done a lot of development work. There isn't space to mention them all here. Here again AGMA can tell you who most of them are.

There are many small- to medium-size gear manufacturers who are making good gears with old but well-maintained equipment. In general, these are the

Figure 51 Gleason Phoenix spiral bevel and hypoid generator. (Courtesy of The Gleason Works, Rochester, NY.)

Figure 52 Gleason G-Tech hobbing machine. (Courtesy of The Gleason Works, Rochester, NY.)

Figure 53 National Broach Red Ring CNC GF-300 gear finisher. (Courtesy of National Broach & Machine Company, Mt. Clemens, MI.)

Figure 54   National Broach Red Ring SF-900 precision gear grinder. (Courtesy of National Broach & Machine Company, Mt. Clemens, MI.)

Figure 55   American Pfauter PE 150 CNC hobbing machine. (Courtesy of American Pfauter, Loves Park, IL.)

people who are specialists in producing relatively small lots in the medium to heavy pitch range. They would do better, of course, with the newer more sophisticated machines described herein, but can't afford them, except maybe one at a time over a period of years. But there will always be such people—somebody has to be able to make the small lots, and to make just one gearset in a hurry when required to replace gears which failed, and often can't be obtained from the original equipment manufacturer.

Perhaps, in retrospect, it should be to the little guy that this chapter should be dedicated—to those who hope some day to achieve the ultimate but find that it takes time to do it. But it can be done with determination and persistence.

ACKNOWLEDGMENTS

The writer is indebted to Darle W. Dudley for his contributions to this chapter. In addition to writing the section on the history of gears, he proofread the manuscript, offering suggestions for revisions and additions which were most helpful. E. J. Wellauer, Vice President/Engineering (Retired), Falk Corpora-

tion, Milwaukee, also offered a number of suggestions which were very helpful, as did E. R. (Ned) Sewall, my successor at Sewall Gear. In addition, I
received some suggestions from a "non-gear man," Gregory P. Struve, P.E.,
Vice President of Robert Muir Company (Minnesota Office). He suggested
some clarifications which he felt would be helpful for readers who have only a
minimal knowledge of gearing.

I am indebted also to the late Professor Earle Buckingham of MIT. I have
used a number of his definitions (paraphrased) and excerpts from notes taken in
discussions with him throughout.

NOTATION (listed in approximate order of appearance)

T	Number of teeth
F	Face width, in.
PD	Pitch diameter, in. (T/DP)
DP	Diametral pitch (PD/T)
CP	Circular pitch, in. (3.1416/DP)
M	Module, mm (25.400/DP)
R_b	Radius of base circle, in.
r	Pitch radius, in. (Fig. 1a)
R	Pitch radius, in. (Fig. 1b)
ϕ	Pressure angle, deg
θ	Vector angle, deg (angle from start of involute curve to pitch point)
P	Pitch point of spurs
R_O	Outside radius, in.
OD	Outside diameter, in.
CD	Center distance, in.
TT	Tooth thickness, in.
BHN	Brinell hardness number
RC	Rockwell hardness number (C scale)
RA	Rockwell hardness number (A scale)
R30-N	Rockwell hardness number (30-N scale)
R15-N	Rockwell hardness number (15-N scale)
TS	Tensile strength, psi
YP	Yield point, psi
R of A	Reduction of area, %
El	Elongation, %
π	3.1416: circumference of a circle divided by its diameter, an irrational number. Determination of the circumference can be accomplished mathematically by assuming that a polygon of infinite number of sides is inscribed within a circle with a radius of unity, which is a calculus problem. The resulting value of pi is 3.14159265 . . . and on and on. For most engineering calculations the value 3.1416 is used. (The reader interested in learning more about pi can read *A History of Pi,* by Petr Beckman, St. Martin's Press, New York, 1976.)

P	Axial pitch of a worm, in.
L	Lead of a worm, in. ($P \times$ no. of starts)
λ	Lead angle of a worm, deg
ϕ_n	Normal pressure angle, deg
V	Velocity, rpm
V_s	Sliding velocity of a worm, ft/min
n (or N)	Speed, rpm
W_t	Tangential tooth load, lb
f	Coefficient of friction
E	Efficiency
hp	Horsepower $= 2\pi NT/33{,}000$
T	Torque, in.-lb
s_c	Compressive stress, psi
E	Modulus of elasticity, stress/strain (30,000,000 for steel)
M_g	Speed ratio
s_{ac}	Allowable compressive stress, psi
s_t	Tensile stress, psi
Y	Tooth form factor

Note: All dimensions are in the English system except the reference to module (M), which is metric. The relationship between English and metric is 1 in. = 2.54 cm or 25.4 mm.

REFERENCES

1. Dudley, Darle W., *The Evolution of the Gear Art*, American Gear Manufacturers Association, 1969.
2. Dudley, Darle W., "Gear Technology—Past, Present and Future," a paper presented at the International Conference on Gearing, Zhengzhow, Henan, China, 1988.
3. Buckingham, Earle, *Manual of Gear Design*, Section 2, revised by Eliot K. Buckingham, Buckingham & Associates, Springfield, VT, 1980.
4. AGMA Standard 390.03A, *Gear Handbook, Gear Classification, Materials and Measuring Methods for Bevel, Hypoid, Fine Pitch Worm Gearing and Racks, only as Unassembled Gears*, 1980.
5. ANSI/AGMA Standard 2004-B89, *Gear Materials and Heat Treatment Manual*, 1989.
6. Campbell, Harry L., *The Working, Heat Treating and Welding of Steel*, Wiley, New York.
7. Society of Automotive Engineers, Standard ANSI B92.1—1970, *Involute Splines and Inspection*, 7th Printing, 1988.
8. AGMA Standard 390.03, *Gear Handbook*, Vol. 1, *Gear Classification, Materials and Measuring Methods for Unassembled Gears*, 1971.
9. ANSI/AGMA Standard 2000-A88, *Gear Classification and Inspection Handbook, Tolerances and Measuring Methods for Unassembled Spur and Helical Gears*, 1988.

10. Ewert, Richard H., "Costs, Their Determination, Allocation and Control," a consolidation of several papers presented at American Gear Manufacturers Association meetings, 1963 through 1985. (A complete copy can be obtained by contacting the author at 1064 Lombard Ave., St. Paul, MN 55105-3255.)

11. Dudley, Darle W. (Editor), *Gear Handbook,* McGraw-Hill, New York, 1962.

12. Buckingham, Earle, *Analytical Mechanics of Gears,* McGraw-Hill, New York, 1949.

13. Dudley, Darle W., *Handbook of Practical Gear Design,* McGraw-Hill, New York, 1984.

14. AGMA Standard 251.02, *Lubrication of Industrial Open Gearing,* 1974.

15. AGMA Standard 250.04, *Lubrication of Industrial Enclosed Gear Drives,* 1981.

16. Ashcroft, Geoffrey, "Disposable Cutting Tools for Gear Manufacturing," a paper presented at AGMA Mfg. Symposium, Chicago, 1991.

17. Ashcroft, Geoffrey and Cluff, Brian W., "High Efficiency Gear Hobbing," a paper presented at AGMA Fall Meeting, Detroit, 1991.

18. Schwamb, Merrill & James, *Elements of Mechanism,* 4th ed., Wiley, New York, 1930.

19. Ewert, Richard H., *Gearing: Basic Theory and Its Applications*—A Primer, Sewall Manufacturing Company, St. Paul, MN, 1980.

APPENDIX

AGMA Standards, a Partial list
Range Cutters for Cutting Spur Gears (Table 12)
Medium to Coarse Pitch Gear Tolerances for Spur, Helical, and Herringbone Gears (Table 13)
Tooth Thickness Tolerances for Spur, Helical, and Herringbone Gears (Table 14)
Applications and Suggested Quality Numbers (Table 15)
Lubrication of Industrial Open Gearing (Tables 16–22)
Lubrication of Industrial Enclosed Gear Drives (Tables 23–27)

A. AGMA Standards, a Partial List

This is a partial list of AGMA Standards,* selected for use by those who wish to pursue gear design procedures, or those whose responsibility is maintenance of existing equipment.

Number	Title
201.02	Tooth Proportions for Course Pitch Involute Spur Gears
250.04	Specification—Lubrication of Industrial Enclosed Gear Drives
251.02	Specification—Lubrication of Industrial Open Gearing
341.02	Design of General Industrial Course Pitch Cylindrical Worm Gearing
390.03a	Handbook—Gear Classification, Matrials, and Measuring Methods for Bevel, Hypoid, Fine Pitch Worm Gearing and Racks, Only as Unassembled Gears
908.B89	Geometry Factors for Determining the Pitting Resistance and Bending Strength of Spur, Helical and Herringbone Teeth
1012-F90	Gear Nomenclature, Definitions of Terms, with Symbols
2000-A88	Gear Classification and Inspection Handbook—Tolerances and Measuring Methods for Unassembled Spur and Helical Gears
2001-B88	Fundamental Rating Factors and Calculation Methods for Involute Spure and Helical Gear
2002-B88	Tooth Thickness Specification and Measurement
2003-A86	Rating the Pitting Resistance and Bending Strength of Generated Straight Bevel, Zerol Bevel and Spiral Bevel Gear Teeth
2004-B89	Gear Materials and Heat Treatment Manual
2005-B88	Design Manual for Bevel Gears
2008-B90	Standard for Assembling Bevel Gears

*A complete list of Standards can be obtained from AGMA at no charge. Address: 1500 King St., Suite 201, Alexandria, VA 22314

Table 12   Range Cutters for Cutting Spur Gears

Cutter number	Range of number of teeth, 8 cutter set	Range of number of teeth, 15 cutter set
1	135 to rack	135 to rack
1½	—	80 to 134
2	55 to 134	55 to 79
2½	—	42 to 54
3	35 to 54	35 to 41
3½	—	30 to 34
4	27 to 34	26 to 29
4½	—	23 to 25
5	21 to 26	21 and 22
5½	—	19 and 20
6	17 to 20	17 and 18
6½	—	15 and 16
7	14 to 16	14
7½	—	13
8	12 and 13	12

Note: Cutters for pinions in the undercut range are modified so they can mill below the base circle radius. Gear cutters are modified to produce gear teeth that will clear this lower portion of the modified pinion tooth profile. Since the profiles of form cut teeth are made up of more than one curve they are called composite tooth forms. The involute portion of the composite form is correct for the minimum number of teeth indicated for each range.

Table 13 Medium to Coarse Pitch Gear Tolerances for Spur, Helical, and Herringbone Gears (tolerances in ten-thousandths of an inch)

AGMA Quality Number	Normal Diametral Pitch	RUNOUT TOLERANCE — PITCH DIAMETER (INCHES)										PITCH TOLERANCE +/– — PITCH DIAMETER (INCHES)									
		3/4	1½	3	6	12	25	50	100	200	400	3/4	1½	3	6	12	25	50	100	200	400
3	1/2					788.2	938.6	1106.9	1305.5	1539.6	1815.7										
3	1				477.8	563.5	671.1	791.4	933.4	1100.8	1298.1										
3	2			289.7	341.6	402.9	479.8	565.9	667.4	787.1	928.2										
3	4			207.1	244.3	288.1	343.1	404.6	477.2	562.7	663.7										
3	8			148.1	174.7	206.0	245.3	289.3	341.2	402.4	474.5										
4	1/2					563.0	670.4	790.7	932.5	1099.7	1297.0										
4	1				341.3	402.5	479.3	565.3	666.7	786.3	927.3										
4	2			208.6	244.0	287.8	342.7	404.2	476.7	562.2	663.0										
4	4			147.9	174.5	206.8	245.0	289.0	340.8	402.0	474.1										
4	8			105.8	124.8	147.1	175.2	206.6	243.7	287.4	338.9										
5	1/2					402.1	478.9	564.8	666.1	785.5	926.4										
5	1				243.8	287.5	342.4	403.8	476.2	561.6	662.4										
5	2			147.8	174.3	205.6	244.8	288.7	340.5	401.6	473.6										
5	4		89.6	105.7	124.6	147.0	175.0	206.4	243.5	287.1	338.6										
5	8		64.1	75.6	89.1	106.1	126.1	147.6	174.1	205.3	242.1										
6	1/2					287.2	342.1	403.4	475.8	561.1	661.7					38.4	43.7	49.4	55.9	63.2	71.4
6	1				174.1	205.4	244.6	288.4	340.2	401.2	473.1				29.1	32.9	37.4	42.3	47.8	54.1	61.1
6	2			106.8	124.5	146.8	174.9	206.2	243.2	286.8	338.3			22.0	24.9	28.1	32.0	36.2	41.0	46.3	52.3
6	4		64.0	75.5	89.0	105.0	125.0	147.4	173.9	205.1	241.9		16.7	18.9	21.3	24.1	27.4	31.0	35.1	39.6	44.8
6	8	38.8	45.8	54.0	63.6	75.1	89.4	105.4	124.3	146.6	172.9	12.6	14.3	16.1	18.2	20.6	23.5	26.6	30.0	33.9	38.4
6	12	31.9	37.6	44.4	52.3	61.7	73.5	86.6	102.2	120.5	142.1	11.5	13.0	14.7	16.7	18.8	21.5	24.3	27.4	31.0	35.0
6	20	24.9	29.4	34.6	40.8	48.2	57.4	67.7	79.8	94.1	111.0	10.3	11.6	13.1	14.9	16.8	19.1	21.6	24.5	27.6	31.3
7	1/2					205.2	244.3	288.1	339.8	400.8	472.7				20.5	27.0	30.8	34.8	39.3	44.5	50.3
7	1				124.4	146.7	174.7	206.0	243.0	286.6	337.9				17.5	23.1	26.4	29.8	33.7	38.1	43.1
7	2			75.4	88.9	104.9	124.9	147.3	173.7	204.9	241.6			15.5	16.0	19.8	22.6	25.5	28.8	32.6	36.9
7	4		45.7	53.9	63.6	75.0	89.3	105.3	124.2	146.5	172.8		11.7	13.3	14.5	17.0	19.3	21.8	24.7	29.9	31.6
7	8	27.7	32.7	38.5	45.5	53.6	63.9	75.3	88.8	104.7	123.5	8.9	10.1	11.4	12.9	14.5	16.5	18.7	21.1	23.9	27.0
7	12	22.8	26.9	31.7	37.4	44.1	52.5	61.9	73.0	86.1	101.5	8.1	9.2	10.4	11.7	13.3	15.1	17.1	19.3	21.8	24.7
7	20	17.8	21.0	24.7	29.2	34.4	41.0	48.3	57.0	67.2	79.3	7.2	8.2	9.3	10.5	11.8	13.5	15.2	17.2	19.5	22.0

(continued)

Table 13 Continued

AGMA Quality No.	Normal Diametral Pitch	Runout 3/4	Runout 1¼	Runout 3	Runout 6	Runout 12	Runout 25	Runout 50	Runout 100	Runout 200	Runout 400	Pitch 1¼	Pitch 3	Pitch 6	Pitch 12	Pitch 25	Pitch 50	Pitch 100	Pitch 200	Pitch 400	Profile 3/4	Profile 1¼	Profile 3	Profile 6	Profile 12	Profile 25	Profile 50	Profile 100	Profile 200	Profile 400	Lead 1&Less	Lead 2	Lead 3	Lead 4	Lead 5
8	1/2					146.5	174.5	205.8	242.2	286.3	337.6				19.0	21.7	24.5	27.7	31.3	35.4					42.6	47.7	53.1	59.1	65.7	73.1					
8	1				88.8	104.8	124.8	147.2	173.8	204.7	241.6			14.4	16.3	18.6	21.0	23.7	26.8	30.3				29.3	31.5	35.3	39.3	43.7	48.6	54.1					
8	2			53.9	63.5	74.9	89.2	105.2	124.1	146.3	172.6		10.9	12.3	14.0	15.9	18.0	20.3	23.0	26.0			18.8	21.0	23.3	26.1	29.0	32.3	36.0	40.0	5	8	11	13	16
8	4		32.7	38.5	45.4	53.6	63.8	75.2	88.7	104.6	123.4	8.3	9.3	10.6	11.9	13.6	15.4	17.4	19.7	22.2		12.5	13.9	15.5	17.2	19.3	21.5	23.9	26.6	29.6					
8	8	19.8	23.3	27.5	32.5	38.3	45.6	53.8	63.4	74.8	88.2	7.1	8.0	9.0	10.2	11.7	13.2	14.9	16.8	19.0	9.3	10.3	11.5	12.8	14.3	15.9	17.7	19.7	21.9						
8	12	16.3	19.2	22.6	26.7	31.5	37.5	44.2	52.1	61.5	72.5	6.5	7.3	8.3	9.3	10.6	12.0	13.6	15.4	17.4	7.8	8.6	9.6	10.7	12.0	13.3	14.8	16.5	18.4						
8	20	12.7	15.0	17.7	20.8	24.6	29.3	34.5	40.7	48.0	56.6	5.8	6.5	7.4	8.3	9.5	10.7	12.1	13.7	15.5	6.2	6.9	7.7	8.6	9.6	10.7	11.9	13.2	14.7						
9	1/2					104.7	124.7	147.0	173.4	204.5	241.2				13.4	15.3	17.3	19.5	22.1	24.9					30.4	34.1	37.9	42.2	46.9	52.2					
9	1				63.5	74.8	89.1	105.1	124.0	146.2	172.4			10.2	11.5	13.1	14.8	16.7	18.9	21.4				20.2	22.5	25.2	28.1	31.2	34.7	38.6					
9	2			38.5	45.4	53.7	63.7	75.2	88.6	104.5	123.3		7.7	8.7	9.8	11.2	12.7	14.3	16.2	18.3			13.5	15.0	16.7	18.6	20.7	23.1	25.7	28.6	4	7	9	11	13
9	4		23.3	27.5	32.4	38.3	45.6	53.7	63.4	74.7	88.1	5.9	6.7	7.6	8.6	9.7	10.9	12.3	13.8	15.7		8.9	10.0	11.1	12.3	13.8	15.3	17.1	19.0	21.1					
9	8	14.1	16.7	19.6	23.2	27.4	32.6	38.4	45.3	53.4	63.0	4.6	5.1	5.8	6.6	7.5	8.5	9.6	10.8	12.2	6.6	7.4	8.2	9.1	10.2	11.4	12.6	14.1	15.6						
9	12	11.6	13.7	16.2	19.1	22.5	26.8	31.6	37.2	43.9	51.8	3.8	4.3	4.8	5.4	6.1	6.9	7.8	8.8	9.9	5.5	6.2	6.9	7.6	8.6	9.5	10.6	11.8	13.1						
9	20	9.1	10.7	12.6	14.9	17.6	20.9	24.7	29.1	34.3	40.4	3.0	3.3	3.7	4.2	4.8	5.4	6.0	6.7	7.6	4.4	4.9	5.5	6.1	6.8	7.6	8.5	9.4	10.5						
10	1/2					74.8	89.0	105.0	123.8	146.1	172.3				9.4	10.8	12.2	13.7	15.5	17.6					21.7	24.3	27.1	30.1	33.5	37.3					
10	1			45.3	53.5	63.7	75.1	88.5	104.4	123.1				7.2	8.1	9.2	10.4	11.8	13.3	15.0				14.5	16.1	18.0	20.0	22.3	24.8	27.6					
10	2			27.5	32.5	38.4	45.3	53.7	63.3	74.7	88.1		5.4	6.1	6.9	7.9	8.9	10.1	11.4	12.9			9.6	10.7	11.9	13.3	14.8	16.5	18.3	20.4	3	5	7	9	10
10	4		16.7	19.6	23.2	27.3	32.5	38.4	45.3	53.4	63.0	4.1	4.6	5.2	5.9	6.7	7.6	8.6	9.8	11.0		6.4	7.1	7.9	8.8	9.9	11.0	12.2	13.6	15.1					
10	8	10.1	11.9	14.0	16.6	19.5	23.3	27.4	32.4	38.2	45.0	3.5	4.0	4.5	5.1	5.8	6.6	7.4	8.3	9.4	4.7	5.3	5.9	6.5	7.3	8.1	9.0	10.0	11.2						
10	12	8.3	9.8	11.5	13.6	16.1	19.1	22.6	26.6	31.4	37.0	2.8	3.2	3.6	4.1	4.6	5.3	6.0	6.8	7.6	4.0	4.4	4.9	5.5	6.1	6.8	7.6	8.4	9.4						
10	20	6.5	7.6	9.0	10.6	12.5	14.9	17.6	20.8	24.5	28.9	2.3	2.6	2.9	3.3	3.7	4.2	4.8	5.4	6.1	3.2	3.5	3.9	4.3	4.8	5.4	6.0	6.7	7.5						
11	1/2					53.4	63.4	75.0	88.5	104.3	123.0				6.9	7.9	8.9	10.0	11.4	12.9					15.5	17.4	19.3	21.5	24.0	26.7					
11	1			32.4	38.2	45.5	53.6	63.2	74.6	88.0			5.1	5.8	6.5	7.4	8.3	9.4	10.6			10.3	11.5	12.9	14.3	15.9	17.7	19.7						
11	2		19.6	23.1	27.3	32.5	38.3	45.2	53.3	62.9		3.8	4.3	4.8	5.4	6.1	6.9	7.8	8.8			7.4	8.2	9.2	10.2	11.4	12.7	14.1		3	4	6	7	8
11	4		11.9	14.0	16.6	19.5	23.2	27.4	32.3	38.1	45.0	2.8	3.2	3.6	4.1	4.6	5.2	5.9	6.6	7.8	4.9	5.5	5.6?	6.3	7.0	7.8	8.7	9.7	10.8					
11	8	7.2	8.5	10.0	11.8	14.0	16.6	19.5	23.1	27.3	32.2	2.4	2.7	3.0	3.4	3.8	4.3	4.9	5.4	6.1	3.6	3.8	4.2	4.9	5.4	6.0	6.4	7.2	8.0						
11	12	5.9	7.0	8.2	9.7	11.5	13.7	16.1	19.0	22.4	26.4	2.0	2.2	2.5	2.8	3.2	3.6	4.1	4.7	5.4	3.0	3.3	3.5	3.9	4.3	4.9	5.4	6.0	6.7						
11	20	4.6	5.5	6.4	7.6	9.0	10.7	12.6	14.8	17.5	20.6	1.8	2.0	2.3	2.6	2.9	3.3	3.7	4.2	4.8	2.5	2.8	2.8	3.1	3.5	3.9	4.3	4.8	5.4						
12	1/2					38.1	45.4	53.6	63.2	74.5	87.9				4.7	5.3	6.0	6.8	7.7	8.7					11.1	12.4	13.8	15.4	17.1	19.0					
12	1		14.0	16.5	19.5	23.1	27.3	32.3	38.1	44.9		3.5	4.0	4.6	5.2	5.8	6.6	6.8	7.7		7.4	8.2	9.2	10.2	11.4	12.7	14.1						
12	2		8.5	10.0	11.8	13.9	16.6	19.5	23.1	27.2	32.1	2.7	3.0	3.4	3.9	4.4	5.0	5.6	6.4	32.1			5.5	6.1	6.8	7.6	8.4	9.4	10.4	2	3	5	6	7
12	4	5.2	6.1	7.2	8.5	10.0	11.8	13.9	16.6	19.6	23.0	2.0	2.3	2.6	2.9	3.3	3.8	4.3	4.8	5.5	3.6	4.0	4.5	5.0	6.8	7.6	8.4	9.4						
12	8	4.2	5.0	5.9	6.9	8.2	9.8	11.5	13.6	16.0	18.9	1.8	2.0	2.3	2.6	2.9	3.3	3.8	4.3	4.8	3.0	3.3	3.1	3.9	4.3	4.6	5.1	5.7						
12	12	3.3	3.9	4.6	5.4	6.4	7.6	9.0	10.6	12.5	14.7	1.6	1.7	2.0	2.3	2.6	3.0	3.3	3.8	4.3	2.4	2.5	2.2	2.8	3.1	4.6	6.2	5.1	4.8						
12	20									13.0	15.0	1.4	1.3	1.8	2.0	2.3	2.6	3.0	3.4	3.8	2.0	2.2	2.0	2.5	2.8	3.1	3.5	4.3	3.8						

LEAD TOLERANCE — FACE WIDTH (INCHES): 1 and Less, 2, 3, 4, 5 (AGMA 8); 4, 7, 9, 11, 13 (AGMA 9); 3, 5, 7, 9, 10 (AGMA 10); 3, 4, 6, 7, 8 (AGMA 11); 2, 3, 5, 6, 7 (AGMA 12)

AGMA QUALITY NUMBER	NORMAL DIAMETRAL PITCH	RUNOUT TOLERANCE — PITCH DIAMETER (INCHES)										PITCH TOLERANCE +/- — PITCH DIAMETER (INCHES)										PROFILE TOLERANCE — PITCH DIAMETER (INCHES)									LEAD TOLERANCE — FACE WIDTH (INCHES)				
		3/4	1¼	3	6	12	25	50	100	200	400	3/4	1¼	3	6	12	25	50	100	200	400	1¼	3	6	12	25	50	100	200	400	1 and Less	2	3	4	5
13	1/2					27.2	32.4	38.3	45.1	53.2	62.8					3.3	3.8	4.2	4.8	5.4	6.1			5.3	7.9	8.9	9.9	11.0	12.2	13.6	2	3	4	4	5
	1				16.5	19.5	23.2	27.4	32.3	38.1	44.9				2.5	2.8	3.2	3.6	4.1	4.6	5.3		3.5	3.9	5.9	6.6	7.3	8.1	9.0	10.1					
	2			10.0	11.8	13.9	16.6	19.6	23.1	27.2	32.1			1.9	2.1	2.4	2.8	3.1	3.4	4.0	4.5	2.3	2.6	2.9	4.3	4.9	5.4	6.0	6.7	7.4					
	4		6.1	7.2	8.4	10.0	11.9	14.0	16.5	19.5	22.9		1.4	1.6	1.8	2.1	2.4	2.7	3.0	3.4	3.8	1.7	2.1	2.7	3.1	3.5	4.0	4.4	4.9	5.5					
	8	3.7	4.3	5.1	6.0	7.1	8.5	10.0	11.8	13.9	16.4	1.1	1.3	1.6	1.8	2.0	2.3	2.7	3.0	3.4	3.8	1.4	1.8	2.1	2.3	2.7	3.0	3.3	3.7	4.1					
	12	3.0	3.6	4.2	5.0	5.9	7.0	8.2	9.7	11.4	13.5	1.0	1.2	1.4	1.6	1.8	2.0	2.3	2.6	2.9	3.3	1.3	1.5	1.7	2.0	2.2	2.5	2.8	3.1	3.4					
	20	2.4	2.8	3.3	3.9	4.6	5.4	6.4	7.6	8.9	10.5	0.8	1.0	1.1	1.3	1.5	1.7	1.9	2.1	2.4	2.7	1.1	1.4	1.6	1.6	1.8	2.0	2.2	2.5	2.7					
14	1/2					19.5	23.0	27.3	32.2	38.0	44.8					2.3	2.6	3.0	3.4	3.8	4.3			3.8	5.7	6.3	7.1	7.8	8.7	9.7	1	2	3	4	
	1				11.8	13.9	16.6	19.5	23.0	27.2	32.1				1.7	1.9	2.1	2.5	2.8	3.3	3.7		2.7	2.8	4.2	4.7	5.2	5.8	6.5	7.2					
	2			7.2	8.4	9.9	11.8	14.0	16.5	19.4	22.9			1.3	1.5	1.7	1.9	2.2	2.5	2.8	3.2	2.0	2.1	2.1	3.1	3.5	3.9	4.3	4.8	5.3					
	4		4.3	5.1	6.1	7.1	8.5	10.0	11.8	13.9	16.4		1.0	1.1	1.3	1.5	1.7	1.9	2.1	2.4	2.7	1.5	1.5	1.8	2.3	2.6	2.9	3.2	3.5	3.9					
	8	2.6	3.1	3.6	4.3	5.1	6.0	7.1	8.4	9.9	11.7	0.8	0.9	1.0	1.1	1.3	1.5	1.7	1.8	2.1	2.3	1.1	1.3	1.5	1.7	1.9	2.0	2.3	2.6	2.9					
	12	2.2	2.5	3.0	3.5	4.2	5.0	5.9	6.9	8.2	9.6	0.7	0.8	0.9	1.0	1.1	1.3	1.4	1.6	1.9	2.1	1.0	1.1	1.3	1.4	1.6	1.8	2.0	2.2	2.4					
	20	1.7	2.0	2.3	2.8	3.3	3.9	4.6	5.4	6.4	7.5	0.6	0.7	0.8	0.9	1.0	1.2	1.3	1.4	1.7	1.9	0.9	1.0	1.1	1.3	1.4	1.5	1.6	1.8	2.0					
15	1/2					13.9	16.6	19.5	23.0	27.2	32.0					1.8	2.0	2.2	2.4	2.7	3.0			2.7	4.0	4.5	5.0	5.6	6.2	6.9	1	2	2	3	3
	1				8.4	10.0	11.8	14.0	16.5	19.4	22.9				1.2	1.4	1.5	1.8	2.0	2.3	2.6		1.8	2.0	3.0	3.3	3.7	4.1	4.6	5.1					
	2			5.1	6.0	7.1	8.5	10.0	11.8	13.9	16.4			1.0	1.1	1.2	1.4	1.5	1.7	2.0	2.2	1.2	1.5	1.5	2.2	2.5	2.8	3.1	3.4	3.8					
	4		3.1	3.7	4.3	5.1	6.1	7.1	8.4	9.9	11.7		0.9	0.8	0.9	1.0	1.2	1.3	1.5	1.7	1.9	1.3	1.0	1.5	1.6	1.8	2.0	2.3	2.5	2.8					
	8	1.9	2.2	2.6	3.1	3.6	4.3	5.1	5.8	6.9	8.4	0.6	0.7	0.7	0.8	0.9	1.0	1.1	1.3	1.4	1.6	0.9	1.0	1.1	1.2	1.4	1.5	1.7	1.9	2.1					
	12	1.5	1.8	2.1	2.5	3.0	3.6	4.2	4.9	5.8	6.9	0.5	0.6	0.6	0.7	0.8	0.9	1.0	1.1	1.3	1.5	0.7	0.9	0.9	1.0	1.1	1.3	1.4	1.6	1.7					
	20	1.2	1.4	1.7	2.0	2.3	2.8	3.3	3.9	4.6	5.4	0.4	0.5	0.6	0.6	0.7	0.8	0.9	1.0	1.2	1.3	0.6	0.7	0.7	0.8	0.9	1.0	1.1	1.3	1.4					

Source: Extracted from AGMA *Gear Handbook* (AGMA 390.03, 1971) with the permission of the publisher, American Gear Manufacturers Association, 1500 King St., Suite 201, Alexandria, VA 22314.

Table 14 Tooth Thickness Tolerance Classes for Spur, Helical, and Herringbone Gearing (all tolerance values in inches)

Quality Number	Diametral Pitch	Class				
		A	B	C	D	E
3	.5	.074				
	1.2	.031				
	2.0	.019				
	3.2	.012				
	5.0	.0075				
4	.5	.074				
	1.2	.031				
	2.0	.019				
	3.2	.012				
	5.0	.0075				
5	.5	.074				
	1.2	.031				
	2.0	.019	.0093			
	3.2	.012	.006			
	5.0	.0075	.0037			
	8.0	.005	.0025			
6	.5	.074				
	1.2	.031				
	2.0	.019	.0093			
	3.2	.012	.006			
	5.0	.0075	.0037			
	8.0	.005	.0025			
	12.0	.003	.0018			
	20.0	.0024	.0012	.0006		
	32.0	.0016	.0008	.00043		
7 and higher	.5	.074				
	1.2	.031				
	2.0	.019	.0093	.0048		
	3.2	.012	.006	.003		
	5.0	.0075	.0037	.0019		
	8.0	.005	.0025	.00125	.00063	
	12.0	.003	.0018	.0009	.00044	
	20.0	.0024	.0012	.0006	.0003	.00016
	32.0	.0016	.0008	.00043	.0002	.0001
	50	.0012	.0006	.0003	.00014	.00007
	80	.0008	.00045	.00022	.00011	.000055
	120	.00067	.00034	.00017	.00009	.000045
	200	.0005	.00025	.00013	.00006	.00003

Source: Extracted from AGMA *Gear Handbook* (AGMA 390-03, 1971) with the permission of the publisher, American Gear Manufacturers Association, 1500 King St., Suite 201, Alexandria, VA 22314.

Table 15 Applications and Suggested Quality Numbers

This is a tabulation of many industrial and end use applications for spur, helical, herringbone, bevel, and hypoid gearing, racks and fine-pitch worms and worm gearing. A typical AGMA quality number range is shown for each of the many industries and applications in this table. When selecting a quality number for an industry or an application which is not shown, use a similar industry or application as a guide. The AGMA quality number shown opposite each item of equipment identifies the quality of gearing generally used. There may be certain designs or operating conditions that would justify specifying gears to a lower or higher quality number. Locate the application desired and obtain the AGMA quality number suggested. In the interest of economy, use the lower quality number shown, unless some of the conditions of the equipment of its operation indicate the use of a higher quality number.

Note: Quality numbers selected from this table should be preceded by the letter "Q" when shown in specifications or on drawings.

Application	Quality number[a]	Application	Quality number[a]
Aerospace		Brewing industry	
Actuators	7–11	Agitator	6–8
Control gearing	10–12	Barrel washer	6–8
Engine accessories	10–13	Cookers	6–8
Engine power	10–13	Filling machines	6–8
Engine starting	10–13	Mash tubs	6–8
Loading hoist	7–11	Pasteurizer	6–8
Propeller feathering	10–13	Racking machine	6–8
Small engines	12–13	Brick-making industry	5–7
Agriculture		Bridge machinery	5–7
Baler	3–7	Briquette machines	5–7
Beet harvester	5–7	Cement industry	
Combine	5–7	(Quarry operation)	
Corn picker	5–7	Conveyor	5–6
Cotton picker	5–7	Crusher	5–6
Farm elevator	3–7	Diesel-electric locomotive	8–9
Field harvester	5–7	Electric dragline (cast gear)	3
Peanut harvester	3–7	(cut gear)	6–8
Potato digger	5–7	Electric locomotive	6–8
Air compressor	10–11	Electric shovel (cast gear)	3
Automotive industry	10–11	(cut gear)	6–8
Bailing machine	5–7	Elevator	5–6
Bottling industry		Locomotive crane (cast gear)	3
Capping	6–7	(cut gear)	5–6
Filling	6–7	(Plant operation)	
Labeling	6–7	Air separator	5–6
Washer, sterilizer	6–7	Ball mill	5–7

Application	Quality number[a]	Application	Quality number[a]
[Cement industry]		[Computing and accounting machines]	
Comped mill	5–6	Typewriter	8
Cooler	5–6	Construction equipment	
Elevator	5–6	Backhoe	6–8
Feeder	5–6	Crames, open gearing	3–6
Filter	5–6	enclosed gearing	6–8
Kiln	5–6	Ditch digger	3–8
Kiln slurry agitator	5–6	transmission	6–8
Overhead crane	5–6	Drag line	5–8
Pug, rod, and tube mills	5–6	Dumpster	6–8
Pulverizer	5–6	Paver, loader	3
Raw and finish mill	5–6	transmission	8
Rotary dryer	5–6	mixer	3–5
Slurry agitator	5–6	swing gear	3–5
Chewing gum industry		mixing bucket	3
Chicle grinder	6–8	Shaker	8
Coater	6–8	Shovels, open gearing	3–6
Mixer-kneader	6–8	enclosed gearing	6–8
Molder-roller	6–8	Stationary mixer, transmission	8
Wrapper	6–8	drum gears	3–5
Chocolate industry		Stone crusher, transmission	8
Glazer, finisher	6–8	conveyor	6
Mixer, mill	6–8	Truck mixer, transfer case	9
Molder	6–8	drum gears	3–5
Presser, refiner	6–8	Cranes	
Tempering	6–8	Boom hoist	5–6
Wrapper	6–8	Gantry	5–6
Clay working machinery	5–7	Load hoist	5–7
Commercial meters		Overhead	5–6
Gas	7–9	Ship	5–7
Liquid, water, milk	7–9	Crushers	
Parking	7–9	Ice, feed	6–8
Computing and accounting machines		Portable and stationary	6–8
Accounting—billing	9–10	Rock, ore, coal	6–8
Adding machine—calculator	7–9	Dairy industry	
Addressograph	7	Bottle washer	6–7
Bookkeeping	9–10	Homogenizer	7–9
Cash register	7	Separator	7–9
Comptometer	6–8	Dams and locks	
Computing	10–11	Tainter gates	5–7
Data processing	7–9	Dishwasher	
Dictating machine	9	Commercial	5–7

Application	Quality number[a]	Application	Quality number[a]
Distillery industry		[Flour mill industry]	
Agitator	5–7	Grain cleaner	7–8
Bottle filler	5–7	Grinder	7–8
Conveyor, elecator	6–7	Hulling	7–8
Grain pulverizer	6–8	Milling, scouring	7–8
Mash tub	5–7	Polisher	7–8
Mixer	5–7	Separator	7–8
Yeast tub	5–7	Foundry industry	
Electric furnace		Conveyor	5–6
Tilting gears	5–7	Elevator	5–6
Electronic instrument control		Ladle	5–6
and guidance systems		Molding machine	5–6
Accelerometer	10–12	Overhead cranes	5–6
Airborne temperature recorder	12–13	Sand mixer	5–6
Aircraft instrument	12	Sand slinger	5–6
Altimeter-stabilizer	9–11	Tumbling mill	5–6
Analog computer	10–12	Home appliances	
Antenna assembly	7–9	Blender	6–8
Antiaircraft detector	12	Mixer	7–9
Automatic pilot	9–11	Timer	8–10
Digital computer	10–12	Washing machine	8–10
Gun-data computer	12–13	Machine tool industry	
Gyro caging mechanism	10–12	Hand motion	
Gyroscope-computer	12–13	(other than Indexing and Positioning)	6–9
Pressure transducer	12–13	Feed drives	8 and up
Radar, sonar, tuner	10–12	Speed drives	8 and up
Recorder, telemeter	10–12	Multiple spindel drives	8 and up
Serve system component	9–11	Power drives, 0–800 fpm	6–8
Sound detector	9	800–2000 fpm	8–10
Transmitter, receiver	10–12	2000–4000 fpm	10–12
Engines		Over 4000 fpm	12 and up
Diesel, semidiesel		Indexing and positioning—	
and internal combustion		approximate positioning	6–10
Engine accessories	10–12	Accurate indexing and positioning	12 and up
Supercharger	10–12	Marine industry	
Timing gearings	10–12	Anchor hoist	6–8
Transmission	8–10	Cargo hoist	7–8
Farm equipment		Conveyor	5–7
Milking machine	6–8	Davit gearing	5–7
Separator	8–10	Elevator	6–7
Sweeper	4–6	Small propulsion	10–12
Flour mill industry		Steering gear	8
Bleacher	7–8	Winch	5–8

Application	Quality number[a]	Application	Quality number[a]
Metal working		[Paper and pulp]	
Bending roll	5–7	Box machines	6–8
Draw bench	6–8	Building paper	6–8
Forge press	5–7	Calendar	6–8
Punch press	5–7	Chipper	6–8
Roll lathe	5–7	Coating	6–8
Mining and preparation		Digester	
Agitator	5–6	Envelope machines	6–8
Breaker	5–6	Food container	6–8
Car dump	5–7	Glazing	6–8
Car spotter	7–8	Grinder	
Centrifugal drier	7–8	Log conveyor—elevator	5–7
Clarifier	7–8	Mixer, agitator	6–8
Classifier	7–8	Paper, machine	
Coal digger	6–10	auxiliary	8–9
Concentrator	5–6	main drive	10–12
Continuous miner	6–7	Press, couch, drier rolls	6–8
Cutting machine	6–10	Save-all	
Conveyor	5–7	Slitting	10–12
Drag line, open gearing	3–6	Steam drum	6–8
enclosed gearing	6–8	Varnishing	6–8
Drills	5–6	Wall paper machine	6–8
Drier	5–6	Paving industry	
Electric locomotive	6–8	Aggregate drier	5–7
Elevator	5–6	Aggregate spreader	5–7
Feeder	6–8	Asphalt mixer	5–7
Flotation	5–6	Asphalt spreader	5–7
Grizzly	5–6	Concrete batch mixer	5–7
Hoists, skips	7–8	Photographic equipment	
Loader (underground)	5–8	Aerial	10–12
Rock drill	5–6	Commercial	8–10
Rotary/car dump	6–8	Printing industry	
Screen (rotary)	7–8	Press, book	9–11
Screen (shaking)	7–8	Flat	9–11
Separator	5–6	Magazine	9–11
Sedimentation	5–6	Newpaper	9–11
Shaker	6–8	Roll reels	6–7
Shovel	3–8	Rotary	9–11
Triple gearing	5–7	Book binding	
Washer	6–8	Pump industry	
Paper and pulp		Liquid	10–12
Bag machines	6–8	Rotary	6–8
Bleacher, decker			

Application	Quality number[a]	Application	Quality number[a]
[Pump industry]		[Steel industry]	
Slush-duplex-triplex	6–8	blooming-mill side guard	5–6
Vacuum	6–8	car haul	5–6
Quarry industry		coil conveyor	5–6
Conveyor-elevator	6–7	edger drives	5–6
Crusher	5–7	electrolyic line	6–7
Rotary screen	7–8	flange-machine ingot buggy	5–6
Shovel-electric-diesel		leveler	6–7
Radar and missile		magazine pusher	6–7
Antenna elevating	8–10	mill shear drives	6–7
Data gear	10–12	mill table drives	
Launch pad azimuth	8	(under 800 ft/min)	5–6
Ring gear	9–12	mill table drives	
Rotating drive	10–12	(over 800–1800 ft/min)	6–7
Railroads		mill table drives	
Construction hoist	5–7	(over 1800 ft/min)	8
Wrecking crane	6–8	nail and spike machine	5–6
Rubber and plastics		piler	5–6
Boot and shoe machines	6–8	plate mill rack and pinion	5–6
Drier, press	6–8	plate mill side guards	5–6
Extruder, strainer	6–8	plate turnover	5–6
Mixer, tuber	6–8	preheat furnace pusher	5–6
Refiner, calender	5–7	processor	6–7
Rubber mill, scrap cutter	5–7	pusher rack and pinion	5–6
Tire building	6–8	rotary furnace	5–6
Tire chopper	5–7	shear depress table	5–6
Washer, banbury mixer	5–7	slab squeezer	5–6
Small power tools		slab-squeezer rack and pinion	5–6
Bench grinder	6–8	slitter, side trimmer	6–7
Drills-saws	7–9	tension reel	6–7
Hair clipper	7–9	tilt, table, upcoiler	5–6
Hedge clipper	7–9	transfer car	5–6
Sander, polisher	8–10	wire drawing machine	6–7
Sprayer	6–8	Blast furnace, coke plant	
Space navigation		Open-hearth and soaking pits	
Sextant and star tracker	13 and up	Miscellaneous drives	
Steel industry		Bessemer tilt-car dump	5–6
Auxiliary and miscellaneous drives		coke pusher, distributor	5–6
annealing furnaces	5–6	conveyor, door lift	5–6
bending roll	5–6	electric-furnace tilt	5–6
blooming-mill manipulator	5–6	hot metal car tilt	5–6
blooming-mill rack and pinion	5–6	hot metal charger	5–6
		jib hoist, dolomite machine	5–6

Application	Quality number[a]	Application	Quality number[a]
[Steel industry]		Rod mills	
larry car	5-6	Roughing	6-7
mixing bin, mixer tilt	5-6	Intermediate	7-8
ore crusher, pig machine	5-6	Finishing	10-12
pulverizer, quench car	5-6	High speed	12-14
shaker, stinter conveyor	5-6	Skelp mills	
stinter machine, skip hoist	5-6	Roughing	6-7
slag crusher, slip hoist	5-6	Intermediate	7-8
Primary and secondary rolling mill drives		Finishing	7-9
Blooming and plate mill	5-6	Structural and rail mills	
Heavy-duty hot mill drives	5-6	heavy	
Slabbing and strip mill	5-6	Reversing rougher	5-6
Hot mill drives		Finishing	5-6
Sendzimer-Stekel	7-8	Light	
Tandem-temper-skin	6-7	Roughing	5-6
Cold mill drives		Finishing	5-6
Bar, merchant, rail, road	5-6	Overhead cranes	
Structural, tube	5-6	Billet charger, codd mill	5-6
Mill gearing		Bucket handling	5-6
Billet mills		Car repair shop	5-6
Free roughing	5-6	Cast house, coil storage	5-6
Tandem roughing	5-6	Charging machine	5-6
Finishing	5-6	Cinder yard, hot top	5-6
Cold mills		Coal and ore bridges	5-6
Reversing	7-8	Electric furnace charger	5-6
Tandem	7-8	Hot metal, ladle	5-6
Temper	7-8	Hot mill, ladle house	5-6
Foil	7-8	Jib crane, motor room	5-6
Hot mills		Mold yard, rod mill	5-6
Blooming and slabbing mills	5-6	Ore unloader, stripper	5-6
Continuous hot strip mills	5-6	Overhead hoist	5-6
Tandem roughing		Pickler building	5-6
(including scalebreaker)	5-6	Pig machine, sand house	5-6
Finishing	6-7	Portable hoist	5-6
Merchant mills		Scale pit, shipping	5-6
Roughing	6-7	Scrap balers and shears	5-6
Intermediate	7-8	Scrap preparation	5-6
Finishing	7-9	Service shops	5-6
Plate mills		Skull cracker	5-6
Reversing roughing	5-6	Slab handling	5-6
Unidirectional roughing	5-6	Precision gear drives	
Unidirectional finishing	5-6	Diesel electric gearing	8-9
		Flying shear	9-10

Application	Quality number[a]	Application	Quality number[a]
[Steel industry]		[Miscellaneous]	
Shear timing gears	9–10	Fishing reel	6
High-speed reels	8–9	Gauges	8–10
Locomotive timing gears	9–10	IBM card puncher, sorter	8
Pump gears	8–9	Metering pumps	7–8
Tube reduction gearing	8–9	Motion picture equipment	8
Turbine	9–10	Popcorn machine, commercial	6–7
		Pumps	5–7
Miscellaneous		Sewing machine	8
Clocks	6	Slicer	7–8
Counters	7–9	Vending machine	6–7

Source: Extracted from AGMA *Gear Handbook* (AGMA 390-03, 1971) with the permission of the publisher, American Gear Manufacturers Association, 1500 King St., Suite 201, Alexandria, VA 22314.
[a]Quality numbers are inclusive, from lowest to highest numbers shown.

B. AGMA Standards: Lubrication of Industrial Open Gearing[a]

Table 16 Viscosity Ranges for AGMA Open Gear Lubricants (AGMA standards: lubrication of industrial open gearing[a]

R and O Gear Oils	Viscosity Ranges ASTM System[1]	Extreme Pressure Gear Oils
AGMA Lubricant No.	SSU at 100° F	AGMA Lubricant No.
4	626 to 765	4 EP
5	918 to 1,122	5 EP
6	1,335 to 1,632	6 EP
7	1,919 to 2,346	7 EP
8	2,837 to 3,467	8 EP
9	6,260 to 7,650	9 EP
10	13,350 to 16,320	10 EP
11	19,190 to 23,460	11 EP
12	28,370 to 34,670	12 EP
13	850 to 1,000 at 210°F[3]	13 EP
Residual Compounds[2]	Viscosity Ranges[2]	Metric Equivalent Viscosity Ranges
AGMA Lubricant No.	SSU at 210° F	cSt at 210°F (98.9°C)
14R	2,000 to 4,000	428.5 to 857.0
15R	4,000 to 8,000	857.0 to 1714.0

[1] "Viscosity System for Industrial Fluid Lubricants," ASTM D-2422. Also British Standards Institute, B.S. 4231.

[2] Residual compounds-diluent type, commonly known as solvent cutbacks, are heavy bodied oils containing a volatile, nonflammable, diluent for ease of application. The diluent evaporates leaving a thick film of lubricant on the gear teeth. Viscosities listed are for the base compound without diluent. **Caution:** These lubricants may require special handling and storage procedures. Diluents can be toxic or irritating to the skin. Consult lubricant supplier's instructions.

[3] Viscosities of AGMA lubricant numbers 13 and above are specified at 210°F as measurement of Saybolt viscosities of these heavy lubricants at 100°F would not be practical.

[a] Tables 16 through 22 (AGMA Tables 1 through 6) were extracted from AGMA Specifications—Lubrication of Industrial Open Gearing (AGMA 251.02), with the permission of the publisher, the American Gear Manufacturers Association, 1500 King St., Suite 201, Alexandria, VA 22314.

Table 17 Equivalent Viscosities of Other Systems (for Reference Only)

AGMA Lubricant No.	Equivalent ASTM Grade	Metric Equivalent Viscosity Ranges[1] cSt at 37.8°C (100°F)	Approximate Viscosity SSU at 210°F
4,4EP	S700	135 to 165	
5,5EP	S1000	198 to 242	
6,6EP	S1500	288 to 352	
7,7EP	S2150	414 to 506	
8,8EP	S3150	612 to 748	160
9,9EP	S7000	1350 to 1650	225
10,10EP	S15,000	2880 to 3520	300
11,11EP	S21,500	4140 to 5060	400
12,12EP	S31,500	6120 to 7480	500
13,13EP	S150,000	25,600 to 38,400	900

[1] At the time of printing of this standard new ASTM and ISO Standards just issued reflect a changeover from 37.8°C (100°F) to 40°C (104°F) as the accepted base temperature for viscosity measurement. During the time interval normally required by the petroleum industry to make such a change across all product lines, lubricant falling within the viscosity tolerances at either temperature shall be accepted as meeting AGMA viscosity requirements for a particular grade. It is expected however, that within five years all AGMA Lubricant Standards will specify 40°C as the only acceptable base temperature for lubricant viscosity measurements.

Table 18 Recommended AGMA Lubricants[5] for Continued Methods of Application

Ambient Temperature[1] in degrees Fahrenheit	Character of Operation	Pressure Lubrication Pitch Line Velocity		Splash Lubrication Pitch Line Velocity		Idler Immersion Pitch Line Velocity
		Under 1000 Ft./Min.	Over 1000 Ft./Min.	Under 1000 Ft./Min.	1000 to 2000 Ft./Min.	Up to 300 Ft./Min.
15-60[2]	Continuous	5 or 5 EP	4 or 4 EP	5 or 5 EP	4 or 4 EP	8 - 9 8 EP - 9 EP
	Reversing or Frequent "Start Stop"	5 or 5 EP	4 or 4 EP	7 or 7 EP	6 or 6 EP	8 - 9 8 EP - 9 EP
50-125[2]	Continuous	7 or 7 EP	6 or 6 EP	7 or 7 EP	6 or 6 EP	11 or 11 EP
	Reversing or Frequent "Start Stop"	7 or 7 EP	6 or 6 EP	9 - 10[3] 9 EP - 10 EP	8 - 9[4] 8 EP - 9 EP	11 or 11 EP

[1] Temperature in vicinity of the operating gears

[2] When ambient temperatures approach the lower end of the given range, lubrication systems must be equipped with suitable heating units for proper circulation of lubricant and prevention of channeling. Check with lubricant and pump suppliers.

[3] When ambient temperature remains between 90 and 125°F at all times, use 10 or 10 EP.

[4] When ambient temperature remains between 90 and 125°F at all times, use 9 or 9 EP.

[5] AGMA viscosity number recommendations listed above refer to gear lubricants shown in Table 15. Although both R & O and EP oils are listed, the EP is preferred.

Table 19   Recommended AGMA Lubricants[2] for Intermittent Methods of Application Limited to 1500 Feet per Minute Pitch Line Velocity[1]

Ambient Temperature In Degrees Fahrenheit[3]	Mechanical Spray Systems[5]		Gravity Feed or Forced Drip Method Using Extreme Pressure Lubricant
	Extreme Pressure Lubricant	Residual Compound[4]	
15 to 60	– –	14R	– –
40 to 100	12 EP	15R	12 EP
70 to 125	13 EP	15R	13 EP

[1] Feeder must be capable of handling lubricant selected.
[2] AGMA viscosity number recommendations listed above refer to gear oils shown in Table 16.
[3] Ambient temperature is temperature in vicinity of the gears.
[4] Diluents must be used to facilitate flow through applicators.
[5] Greases are sometimes used in mechanical spray systems to lubricate open gearing. A general-purpose EP grease of number 1 consistency (NGL1) is preferred. Consult gear manufacturer and spray system manufacturer before proceeding.

Table 20   Recommended Quantities of Lubricant for Intermittent Methods of Application Where Pitch Line Velocity Does Not Exceed 1500 Feet per Minute for Automotive, Semiautomotive, Hand Spray, Gravity Feed, or Forced Drip Systems

Gear Diameter In Feet	[1] Ounces Per Application At Intervals Of:														
	¼ Hour					1 Hour					4 Hours[2]				
	Face Width In Inches					Face Width In Inches					Face Width In Inches				
	8	16	24	32	40	8	16	24	32	40	8	16	24	32	40
10	.2	.3	.4	.5	.6	.8	1.2	1.6	2.0	2.4	5.0	6.0	8.0	10.0	12.0
12	.3	.3	.4	.5	.6	1.2	1.4	1.8	2.2	2.6	6.0	7.0	9.0	11.0	13.0
14	.3	.4	.5	.6	.7	1.4	1.6	2.0	2.4	2.8	7.0	8.0	10.0	12.0	14.0
16	.4	.5	.6	.7	.8	1.6	2.0	2.4	2.8	3.2	8.0	10.0	12.0	14.0	16.0
18	.5	.6	.7	.8	.9	2.0	2.4	2.8	3.2	3.6	10.0	12.0	14.0	16.0	18.0
20	.6	.7	.8	.9	1.0	2.4	2.8	3.2	3.6	4.4	12.0	14.0	16.0	18.0	20.0
22	.7	.8	.9	1.0	1.1	2.8	3.2	3.6	4.0	4.8	14.0	16.0	18.0	20.0	22.0
24	.8	.9	1.0	1.1	1.2	3.2	3.6	4.0	4.4	5.2	16.0	18.0	20.0	22.0	24.0
26	.9	1.0	1.1	1.2	1.3	3.6	4.0	4.4	4.8	5.6	18.0	20.0	22.0	24.0	26.0
28	1.0	1.1	1.2	1.3	1.4	4.0	4.4	4.8	5.2	6.0	20.0	22.0	24.0	26.0	28.0

[1] The spraying time should equal the time for one and preferably two revolutions of the gear to insure complete coverage. Periodic inspections should be made to insure that sufficient lubricant is being applied to give proper protection.
[2] Four hours is the maximum interval permitted between applications of lubricant. *More frequent application of smaller quantities is preferred.* However, where diluents are used to thin lubricants for spraying, intervals must not be so short as to prevent diluent evaporation.

Table 21 Specifications for R & O Gear Oils

Property	Test Procedure	Criteria for Acceptance
Viscosity	ASTM D88	Must be specified in Table 1
Viscosity Index	ASTM D2270	*90 min
Oxidation Stability	ASTM D943	*Hours to reach a neutralization number of 2.0 *AGMA Grade* *Hours (min)* 1, 2 1500 3, 4 750 5, 6, 7, 8 500
Rust Protection	ASTM D665	AGMA Lubricant Nos. 4 thru 8 – No rust after 24 hours with synthetic sea water AGMA Lubricant Nos. 9 thru 13 – No rust after 24 hours with distilled water
Corrosion Protection	ASTM D130	*#1 strip after 3 hours at 250°F
Foam Suppression	ASTM D892	*Must be within the limits specified in paragraph 1 of Appendix
Demulsibility	ASTM D2711	*Must be within the limits specified in paragraph 2 of Appendix
Cleanliness	None	Must be free from grit and abrasives

*Applicable to AGMA lubricant nos. 4 through 8 only.

Table 22 Specification for Extreme Pressure Gear Lubricants

Property	Test procedure	Criteria for acceptance
Viscosity	ASTM D88	Must be specified in Table 16
Viscosity index	ASTM D2270	90 min[a]
Oxidation stability	ASTM D2893	Increase in kinematic viscosity of oil sample at 210°F should not exceed 10%[a]
Rust protection	ASTM D665	No rust after 24 h with distilled water
Corrosion protection	ASTM D130	#1 strip after 3 h at 212°F[a]
Foam suppression	ASTM D892	Must be within the limits specified in paragraph 3 of Appendix[a]
Cleanliness	None	Must be free from grit and abrasives
EP property	Timken test FZG test	An oil which passes either a 45 lb Timken OK load or 9 stages on the FZG machine is considered acceptable.
Additive solubility	None	Must be filtered to 100 microns (wet or dry) without loss of EP additive

[a] Applicable to AGMA lubricant nos. 4EP through 8EP only.

C. AGMA Standards: Lubrication of Industrial Enclosed Gearing [a]

Table 23 Specification for R&O Gear Oils (including compounded gear lubricants)

Property	Test Procedure	Criteria for Acceptance
Viscosity	ASTM D88	Must be as specified in Table 26
Viscosity Index	ASTM D2270	90 min.
Oxidation Stability	ASTM D943	*Hours to reach a neutralization number of 2.0 AGMA Grade Hours (minimum) 1, 2 1500 3, 4 750 5, 6 500
Rust Protection	ASTM D665	No rust after 24 hours with synthetic sea water
Corrosion Protection	ASTM D130	#1 strip after 3 hours at 120°C (250°F)

Foam Suppression	ASTM D892	Must be within these limits:
		
		                               Max Volume of Foam (mL.) After:
		            Temperature 5 Minute Blow 10 Minute Rest
		Sequence I 24°C (75°F) 75 10
		Sequence II 93.5°C (200°F) 75 10
		Sequence III 24°C (75°F) 75 10

Demulsibility	ASTM D2711	*Must be within these limits: Max percent water in the oil after 5-hour test 0.5% Max cuff after centrifuging 2.0 mL. Min total free water collected during entire test 30.0 mL.
Cleanliness	None	Must be free from grit and abrasives

Note: The criteria for acceptance indicated for Oxidation Stability and Desirability is not applicable to Compounded Gear Oils.

[a] Tables 23 through 27 (AGMA Tables 1 through 5) were extracted from AGMA Specifications—Lubrication of Industrial Enclosed Gear Drives (AGMA 250.04), with the permission of the publisher, the American Gear Manufacturers Association, 1500 King St., Suite 201, Alexandria, VA 22314

Table 24 Specification for Extreme Pressure Gear Lubricants

Property	Test Procedure	Criteria for Acceptance
Viscosity	ASTM D88	Must be as specified in Table 27
Viscosity Index	ASTM D2270	90 min.
Oxidation Stability	ASTM D2893	Increase in kinematic viscosity of oil sample at 95°C (210°F) should not exceed 10%
Rust Protection	ASTM D665	No rust after 24 hours with distilled water
Corrosion Protection	ASTM D130	#1 strip after 3 hrs. at 100°C (212°F)
Foam Suppression	ASTM D892	Must be within these limits:

		Temperature	Max Volume of Foam (mL) After:	
			5 Minute Blow	10 Minute Rest
		Sequence I 24°C (75°F)	75	10
		Sequence II 93.5°C (200°F)	75	10
		Sequence III 24°C (75°F)	75	10

Demulsibility	ASTM D2711 (Modified for 90 mL. water)	Must be within these limits:		
			AGMA Grades	
			2 EP to 6 EP	7 EP, 8 EP
		Max percent water in the oil after 5 hour test	1.0%	1.0%
		Max cuff after centrifuging	2.0 mL	4.0 mL
		Min total free water collected during entire test (start with 90 mL. of water)	60.0 mL	50.0 mL

Property	Test Procedure	Criteria for Acceptance
Cleanliness	None	Must be free from grit and abrasives
EP Property	ASTM D2782 (Timken Test) DIN 51-354 (FZG Test)	An oil must pass both a 60 lb. Timken OK load, and 11 stages on the FZG machine with A/8.3/90°C parameters for acceptance.
Additive Solubility	None	Must be filterable to 25 μm (microns) (wet or dry) without loss of EP additive

Table 25 AGMA Lubricant Number Recommendations for Enclosed Helical, Herringbone, Straight Bevel, Spiral Bevel, and Spur Gear Drives

Type of Unit[a]	AGMA Lubricant Number[b, c]	
	Ambient Temperature[d, e]	
Low Speed Center Distance	−10°C to +10°C (15°F to 50°F)	10°C to 50°C (50°F to 125°F)
Parallel Shaft, (single reduction)		
Up to 200 mm (to 8 in.)	2 – 3	3 – 4
Over 200 mm, to 500 mm (8 to 20 in.)	2 – 3	4 – 5
Over 500 mm (over 20 in.)	3 – 4	4 – 5
Parallel Shaft, (double reduction)		
Up to 200 mm (to 8 in.)	2 – 3	3 – 4
Over 200 mm (over 8 in.)	3 – 4	4 – 5
Parallel Shaft, (triple reduction)		
Up to 200 mm (to 8 in.)	2 – 3	3 – 4
Over 200 mm, to 500 mm (8 to 20 in.)	3 – 4	4 – 5
Over 500 mm (over 20 in.)	4 – 5	5 – 6
Planetary Gear Units, (housing diameter)		
Up to 400 mm (to 16 in.) O.D.	2 – 3	3 – 4
Over 400 mm (over 16 in.) O.D.	3 – 4	4 – 5
Straight or Spiral Bevel Gear Units		
Cone distance to 300 mm (to 12 in.)	2 – 3	4 – 5
Cone distance over 300 mm (over 12 in.)	3 – 4	5 – 6
Gearmotors and Shaft Mounted Units	2 – 3	4 – 5
High Speed Units[f]	1	2

[a] Drives incorporating overrunning clutches as backstopping devices should be referred to the gear drive manufacturer as certain types of lubricants may adversely affect clutch performance.

[b] Ranges are provided to allow for variations in operating conditions such as surface finish, temperature rise, loading, speed, etc.

[c] AGMA viscosity number recommendations listed above refer to R&O gear oils shown in Table 26. EP gear lubricants in the corresponding viscosity grades may be substituted where deemed necessary by the gear drive manufacturer.

[d] For ambient temperatures outside the ranges above, consult the gear manufacturer. Some synthetic oils have been used successfully for high- or low-temperature applications. (See 3.4 of text)

[e] Four point of lubricant selected should be at least 5°C (9°F) lower than the expected minimum ambient starting temperature. If the ambient starting temperature approaches lubricant pour point, oil sump heaters may be required to facilitate starting and insure proper lubrication.

[f] High-speed units are those operating at speeds above 3600 rpm or pitch line velocities above 25 m/s (5000 fpm) or both. Refer to Standard AFGMA 421. *Practice for High Speed Helical and Herringbone Gear Units,* for detailed lubrication recommendations.

Table 26 AGMA Lubricant Number Recommendations for Enclosed Cylindrical and Double-Enveloping Worm Gear Drives

Type, Worm Gear Drive	Worm Speed[c] Up to (rpm)	AGMA Lubricant Numbers[a] Ambient Temperature[b]		Worm Speed[c] Above (rpm)	AGMA Lubricant Numbers[a] Ambient Temperature[b]	
		−10°C to +10°C (15° to 50°F)	10°C to 50°C (50° to 125°F)		−10°C to +10°C (15° to 50°F)	10°C to 50°C (50° to 125°F)
Cylindrical Worm[d]						
Up to 150 mm (to 6 in.)	700	7 Comp, 7 EP	8 Comp, 8 EP	700	7 Comp, 7 EP	8 Comp, 8 EP
Over 150 mm, to 300 mm (6 to 12 in.)	450	7 Comp, 7 EP	8 Comp, 8 EP	450	7 Comp, 7 EP	7 Comp, 7 EP
Over 300 mm, to 450 mm (12 to 18 in.)	300	7 Comp, 7 EP	8 Comp, 8 EP	300	7 Comp, 7 EP	7 Comp, 7 EP
Over 450 mm, to 600 mm (18 to 24 in.)	250	7 Comp, 7 EP	8 Comp, 8 EP	250	7 Comp, 7 EP	7 Comp, 7 EP
Over 600 mm (over 24 in.)	200	7 Comp, 7 EP	8 Comp, 8 EP	200	7 Comp, 7 EP	7 Comp, 7 EP
Double-Enveloping Worm[d]						
Up to 150 mm (to 6 in.)	700	8 Comp	8A Comp	700	8 Comp	8 Comp
Over 150 mm, to 300 mm (6 to 12 in.)	450	8 Comp	8A Comp	450	8 Comp	8 Comp
Over 300 mm, to 450 mm (12 to 18 in.)	300	8 Comp	8A Comp	300	8 Comp	8 Comp
Over 450 mm, to 600 mm (18 to 24 in.)	250	8 Comp	8A Conp	250	8 Comp	8 Comp
Over 600 mm (over 24 in.)	200	8 Comp	8A Comp	200	8 Comp	8 Comp

[a] Both EP and compounded oils are considered suitable for cylindrical worm gear service. Equivalent grades of both are listed in the table. For double-enveloping worm gearing, EP oils in the corresponding velocity grades may be substituted only where deemed necessary by the worm gear manufacturer.

[b] Pour point of the oil used should be less than the minimum ambient temperature expected. Consult gear manufacturer on lube recommendations for ambient temperatures below −10°C (14°F).

[c] Worm gears of either type operating at speeds above 2400 rpm or 10 m/s (200 fpm) rubbing speed may require force feed lubrication. In general, a lubricant of lower viscosity than recommended in the above table shall be used with a force feed system.

[d] Worm gear drives may also operate satisfactorily using other types of oils. Such oils should be used, however, only upon approval by the manufacturer.

Table 27   Viscosity Ranges for AGMA Lubricants

Rust and Oxidation Inhibited Gear Oils	Viscosity Range[a]	Equivalent ISO Grade[b]	Extreme Pressure Gear Lubricants[c]	Viscosities of Former AGMA System[d]
AGMA Lubricant No.	mm²/s (cSt) at 40°C		AGMA Lubricant No.	SSU at 100°F
1	41.4 to 50.6	46		193 to 235
2	61.2 to 74.8	68	2 EP	284 to 347
3	90 to 110	100	3 EP	417 to 510
4	135 to 165	150	4 EP	626 to 765
5	198 to 242	220	5 EP	918 to 1122
6	288 to 352	320	6 EP	1335 to 1632
7 Comp[e]	414 to 506	460	7 EP	1919 to 2346
8 Comp[e]	612 to 748	680	8 EP	2837 to 3467
8A Comp[e]	900 to 1100	1000	8A EP	4171 to 5098

[a] *Viscosity System for Industrial Fluid Lubricants,* ASTM 2422. Also British Standards Institute, B.S. 4231.
[b] *Industrial Liquid Lubricants—ISO Viscosity Classifications.* International Standard, ISO 3448.
[c] Extreme pressure lubricants should be used only when recommended by the gear drive manufacturer.
[d] AGMA 250.03, May, 1972 and AGMA 251.02, November, 1974.
[e] Oils marked Comp are compounded with 3% to 10% fatty or synthetic fatty oils.
Viscosity ranges for AGMA lubricant members will henceforth be identical to those of ASTM 2422.

8

CHAIN DRIVES

David W. South

Consultant, Madison Heights, Michigan

I. INTRODUCTION

Chains are manufactured to various degrees of precision ranging from unfinished castings or forgings to chains having certain accurately machined parts. It may be stated that a chain, composed of a series of links pinned together, is a form of flexible gear connecting two toothed sprocket wheels mounted on parallel shafts. Their main advantages are that they are as flexible as belts, they are as positive as gears, they provide excellent design flexibility, convenience, and resistance to shock loading, they operate satisfactorily in adverse surroundings, they can be manufactured in various special steels to resist specific environments, and they have unlimited shelf life, simplicity of installation and general economy. Mechanical efficiency, in comparison to gear and belt drives, is favorable. In other words, chains operating under ideal conditions can have efficiencies as high as 97 to 99 percent.

There are three principal types of chains: roller, silent, and engineering steel (class) chains. In terms of applications, their major uses are: (1) power transmission drives, (2) conveyors, (3) bucket elevators, (4) tension linkages.

When standard chain components are modified to adapt the chain for use in conveying, elevating, and timing operations, or to push, pull, or clamp as necessary, what are known as "attachments" must be employed. The components usually modified are (1) the link plates—which are provided with extended lugs that may be straight or bent and (2) the chain pins—which are extended in length so as to project substantially beyond the outer surface of the pin link plates. Practically all chains can be readily equipped with attachments for whatever specific function is required. Since there are thousands of attachment designs (several standardized also), it would be beyond the scope of this chapter to cover them. However, at the end of each section reference to some of the more commonly used types is given (if applicable).

Like other standards and development organizations the American National Standards Institute (ANSI) has established certain standards that relate to the various types of sprocket-engagement chains in general use. For the primary purpose of interchangeability between manufacturers, this family of standards is the result of over 50 years of relentless standardization activity. There are currently 17 standards, all of which are covered in this chapter.

Various manufacturers have individually developed *nonstandard roller chains* (prior to ANSI standard adoption) that are similar in form and construction to standard roller chains but do not conform "dimensionally" to standard chains. Various types and sizes are still available from the originating manufacturers, primarily for replacement on existing drives. Since their manufacture is rapidly being discontinued, they are not recommended for new installations.

A. Historical Background

The ideal of utilizing a flexible, yet strong machine element capable of transmitting rotative power, over relatively long distances if required, to drive machinery, to operate a military mechanism, or to transport materials along conveyors and up elevators is a fundamental concept known for centuries.

We learn from a thesis written by Philo of Byzantium around 200 B.C. that the developmental process for a sprocket wheel–flexible chain interaction already existed. This primitive, yet effective chain drive was used as a form of bucket elevator (waterwheel) in which the chains were fashioned with buckets (attachments) to lift water to higher elevations. It is also understood from Philo's thesis that a reciprocating chain drive, incorporated as a form of "tension linkage," was used, perhaps to modernize repeating catapults. Most likely, these early chain drives employed a simple round-link type of chain.

Other people made significant contributions to the very early development of chains. While working at the Arsenal at Rhodes, Dionysius of Alexandria designed a flat-link type of chain—a concept that was undoubtedly a major advancement over the cruder round-link design. Leonardo da Vinci in the 1500s made sketches of several interesting chain styles, some of which bear remarkable similarities to modern chains. Also in the 1500s, Ramelli designed a waterwheel-driven rotary pump, in which a chain drive was used for power transmission.

Undoubtedly, many other early examples may be found of uses of chain for power of transmission or the conveyance of materials. However, it was not until the 1800s (the advent of the Industrial Revolution) that the history of modern chain development actually began.

In the early 1800s, the cog chain was developed to transmit power of motion between the shafts of treadmills to water elevators, weaving looms, and harvesting machinery. This chain was constructed of rectangular cast links connected by looped and riveted iron bands. Because the links contacted the sprocket tooth, or cog as it was called, it became known as *cog chain*.

Although this chain made a strong contribution to the mechanization of farm implements, it broke easily and was troublesome to repair in the field. These problems were overcome in 1873 with the development of *cast detachable chain*. This was a simple cast metal chain composed of identical links that could easily be coupled or uncoupled by hand. Thus the mechanization of agricultural equipment soon became a practical reality, bringing with it a need for the positive transmission of power and accurate timing of motions that only a chain-and-sprocket drive could provide. This mechanization rapidly spread to other major industries such as metalworking and textile manufacture.

This basic detachable chain design is the earliest of the industrial chain concepts that have come down to us today virtually unchanged. Malleable iron detachable chain, and its fabricated steel counterparts, made with highly sophisticated and automated equipment, is still in use today and is available from manufacturers and their distributors.

Also in the 1800s, the cast basic detachable links had been complemented with integral cast *appendage* or attachment links that could be installed at intervals in the basic chains. By the 1890s, designs of bucket elevators were being offered as a standardized machine by many manufacturers.

For conveying of bulk materials horizontally or up mild machines, slats or *flights* rather than buckets were bolted to the cast attachments, this becoming the modern apron conveyor. To convey bulk materials up fairly steep inclines as well as horizontally, *drag* conveyors using different styles of attachments and scraper flights in a trough were developed.

The next important development after the cast detachable chain was the cast *pintle chain*. This type of chain, the direct ancestor of both standard roller chain and engineering-class steel chains, had a "closed-barrel" design, heavier link sections, and steel pins or "pintles." It evolved from the cast detachable chain link design to meet the requirements of heavier loads, higher speeds, and more severe operating conditions in both drives and conveying applications. Modifications were soon made to the original pintle chain design to meet specialized needs. For example, wear plates or "shoes" were added to provide greater resistance to sliding wear. Even though not all earlier innovations were so successful, this design is still widely used today.

Later on, both pitch and width of cast pintle links were expanded to provide *drag chains* of high efficiency and dependability. Thus by the 1900s the cast pintle chain had been developed to the point that industry, in the United States and abroad, had an efficient and dependable product available for power transmission, conveying, and elevating.

About a decade after cast detachable chain was introduced, a chain made entirely of steel components was introduced for use in driving bicycles. This was the beginning of the precision roller chain market. By the early 1900s, roller chain was in use to drive wheels of the "safety bicycle," the newly invented automobile, and the propellors of the Wright Brothers' airplane flown at Kitty Hawk. Thus the public demand for large quantities of chain to be installed on bicycles and soon, on motorcycles, automobiles, and trucks, sparked organized methods of manufacturing that led to the size and success of the modern chain industry.

Engineering steel chains were first developed in the 1880s for about the same reasons that steel roller chain was developed—to meet the needs for higher strength, speeds, and shock resistance and for better dimensional control than was obtainable from the materials and methods of manufacturing cast links.

The early engineering steel chains were a direct takeoff from cast pintle chains in that they were designed with equivalent pitches and dimensions similar to those of cast chains so that the steel chains could be fitted to identical sprockets. This procedure meant that higher-strength chains could be substituted for cast chains when necessary without needing to change to sprockets of a different design.

Precision inverted-tooth chain, popularly known as "silent chain," was developed in the early part of the nineteenth century. Its advent was spectacular; very few products have ever appeared in such a large version in a highly publicized application and performed with the effectiveness of products manufactured over 100 years later.

Silent chain did not see much growth in the United States, in spite of its auspicious European beginning, until the 1920s. By the 1930s silent chain was used as a high-speed, quiet alternative to precision roller chain.

B. Standard Roller Chain Construction

Roller chain is a series of connected journal bearings, assembled with alternate *roller links* and pin links (see Fig. 1). Each roller link consists of two bushings, rollers, and link plates (side plates). Rollers are slipped onto the bushings; then the bushings are press-fitted into the side plates to complete a single roller link section. The roller has sufficient side clearance to permit free movement and lubricant access. These same roller links are used for *single-* and *multiple-stranded* roller chains.

Figure 1   An exploded view of roller chain construction. (Courtesy of Browning Manufacturing, Maysville, KY.)

A multiple strand roller chain is basically an assembly of two or more (generally no more than six) single-strand chains, placed side by side, and they have roller links identical to those of ordinary single-strand chains.

The pin link plates carry pins that extend through the entire width of the chain assembly, and thus maintain uniform alignment of the individual strands. However, link plates of pin links for multiple-strand roller chains are not necessarily identical to those used in single-strand roller chain. Therefore, pin links of roller chains should always be supplied on a unit-link basis rather than attempting to make up such links from miscellaneous parts.

Pin links have two pins press-fitted into two link plates, and when assembled, the two pins slip into bushings of adjacent roller links. The side plate (link plate) on a pin link section is held in place by way of the pin being riveted (mushroomed out) on the ends. Generally, it is standard manufacture on chain sizes No. 25 through No. 60 for the pins to be of riveted construction. The "cottered" style is usually standard on chains No. 80 and larger. This type has a hole in one end of the pin that extends beyond the side plate on one side so as to receive the cotter.

C. Standard Roller Chain Dimensions

The principal dimensions of standard roller chain are (1) pitch, (2) roller width, and (3) roller diameter (see Fig. 2). The *pitch* is the distance in inches between the centers of the adjacent joint members in a taut chain. In other words, in a roller chain the pitch is the distance between the centers of the bushings or rollers. *Roller width,* or chain width, is the distance between the "inner faces" of the roller link plates (that is, chain width is not the width of the chain over the link plates or the pins).

ANSI standard roller chains range in pitch from 0.25 to 3 in. Also included in this category are the *heavy series* chains, identified by the suffix "H," and starting with the 0.75-in. pitch (No. 60) and larger. The only difference between the standard series chains and the heavy series chains is the thickness of the link plates. Heavy series chains have link plates as thick as the next larger-sized standard chain. They are commonly used where space and weight limitations prohibit use of larger chain sizes. In effect, they have a higher yield strength than ANSI standard and will withstand greater shock loads. Multiple stranded heavy series chains are normally made to order and require special sprockets.

D. Standard Roller Chain Numbers

The standard roller chain numbering (designation) system is very easy to understand. It is merely a shorthand device that provides complete identification of a particular roller chain. Chain numbers consist of at least two digits. The very last number, or the *right-hand digit* designates chain style: 0 = standard roller chain of usual proportions; 1 = narrow width, lightweight roller chain; 2 = carrier (large-diameter) roller chain (for double-pitch conveyor chains only); 5 = rollerless (bushing only) chain. Also, for multiple-stranded roller chains, add a dash (hyphen) and the appropriate "suffixed" number of strands. Number(s) to the left of the right-hand digit indicate chain pitch, in 1/8-in. units or increments. For example, a No. 40 chain = 4 × 1/8 or a ½-in. pitch chain of basic proportions; a No. 35 chain = 3 × 1/8 or a 3/8-in. pitch rollerless bushing chain; and a No. 140 chain = 14 × 1/8 or a 1¾-in. pitch chain of basic proportions. Standard chain numbers and general chain dimensions are shown in Table 1.

Figure 2 The principal dimensions of a roller chain. (Courtesy of American Chain Association, 1982.)

Table 1 Standard Chain Numbers (Sizes) and General Chain Dimensions (in inches), ANSI Standard Roller Chain

ANSI Standard Chain No.		Pitch P	Max. Roller Diameter Dr	Width W	Pin Diameter Dp	Link Plate Thickness LPT		Meas. Load, Lb.
Std.	Heavy					Std.	Heavy	
25*	. . . H	$^1/_4$	0.130*	$^1/_8$	0.0905	0.030	. . . .	18
35*	. . . H	$^3/_8$	0.200*	$^3/_{16}$	0.141	0.050	. . . .	18
41†	. . . H	$^1/_2$	0.306	$^1/_4$	0.141	0.050	. . . .	18
40	. . . H	$^1/_2$	$^5/_{16}$	$^5/_{16}$	0.156	0.060	. . . .	31
50	. . . H	$^5/_8$	0.400	$^3/_8$	0.200	0.080	. . . .	49
60	60H	$^3/_4$	$^{15}/_{32}$	$^1/_2$	0.234	0.094	.125	70
80	80H	1	$^5/_8$	$^5/_8$	0.312	0.125	.156	125
100	100H	$1^1/_4$	$^3/_4$	$^3/_4$	0.375	0.156	.187	195
120	120H	$1^1/_2$	$^7/_8$	1	0.437	0.187	.219	281
140	140H	$1^3/_4$	1	1	0.500	0.219	.250	383
160	160H	2	$1^1/_8$	$1^1/_4$	0.562	0.250	.281	500
180	180H	$2^1/_4$	$1^{13}/_{32}$	$1^{13}/_{32}$	0.687	0.281	.312	633
200	200H	$2^1/_2$	$1^9/_{16}$	$1^1/_2$	0.781	0.312	.375	781
240	240H	3	$1^7/_8$	$1^7/_8$	0.937	0.375	.500	1125

*Without rollers.
†Lightweight Chain.

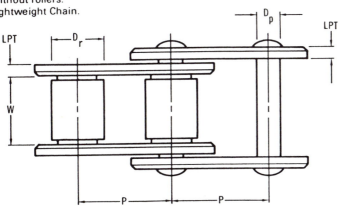

Source: American Chain Association, 1982.

In the multiple-stranded roller chain series, for example, a chain designation of "60-2" indicates a common parallel assembly of two strands of No. 60 chain, "60-3" designates a triple strand, "60-4" a quadruple-stranded assembly, and so forth.

II. STANDARD PRECISION ROLLER CHAINS

A. Single or Base Pitch Roller Chains (ANSI B29.1)

Out of all the many styles of chains that are available, the most widely used for basic transmission of power are the *precision roller chains*. Their operation is of fundamental consideration. The rollers (shock absorbers that reduce the effects of impact) engage and bear against the sprocket teeth. Bushings are accurately press-fitted into the link plates so that they do not turn. When engaging sprockets, the rollers roll in to mesh with the teeth and set at the bottom of the space between the teeth to provide a positive gripping action. The load is distributed over a number of teeth to give smooth, uninterrupted transmission of power.

Single or base pitch chains are available in 13 pitch sizes, ranging from ¼ to 3 in. (see Table 1). Additionally, unitized multiple-strand assemblies are available and are commonly purchased in widths up to eight strands. General dimensions and horsepower ratings are shown in Tables 2 to 15. Also, if properly outfitted, these chains can provide unlimited light-duty conveying applications. A wide variety of attachments are available.

B. Double-Pitch Power Transmission Roller Chains (ANSI B29.3)

Double-pitch, or "extended-pitch" as they are sometimes called, precision roller chains are particularly applicable on power drives where speeds are slow to moderate, loads are generally medium-type, or center distances between sprockets are relatively long. Since the side plates on ANSI B29.3 chains consist of typically narrow-waisted (or "figure 8") contouring, they are commonly referred to as *power transmission series* chains.

ANSI standard double-pitch roller chains are manufactured to the same standards as ANSI standard series (base-pitch) chains. In other words, in this design the bushings, pins, and rollers of a specific base-pitch chain are assembled with link plates that are twice the base-pitch dimension.

Again, double-pitch transmission series chains have twice the distance between rollers of corresponding standard roller chains, and nomenclature is as follows: on drive series add 2000 to the base ANSI standard number. For example, a No. 2050 chain = base number 50 or 5/8-in. pitch × 2 = 10/8 or 1¼-in. pitch. Since double-pitch chains contain only half as many rollers,

Table 2 General Dimensions and Horsepower Capacity, No. 25 Roller Chain (ANSI B29.1) 1/4-in. Pitch

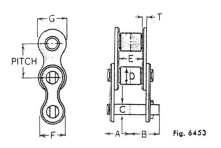

Fig. 6453

Specifications and dimensions

Chain number	Average ultimate strength, pounds	Joint bearing area, sq. in.	Weight per foot, pounds	A	B	C
25	875	.017	.08	.15	.19	.091

Ratings

Number of teeth in small sprocket	Maximum bore, inches	Horsepower for single strand chain▲ RPM of small sprocket																		
		100	500	900	1200	1800	2500	3000	3500	4000	4500	5000	5500	6000	6500	7000	7500	8000	8500	9000	10000
11	.313	0.05	0.23	0.39	0.50	0.73	0.98	1.15	1.32	1.38	1.16	0.99	0.86	0.75	0.67	0.60	0.54	0.49	0.45	0.41	0.35
12	.375	0.06	0.25	0.43	0.55	0.80	1.07	1.26	1.45	1.57	1.32	1.12	0.97	0.86	0.76	0.68	0.61	0.56	0.51	0.47	0.40
13	.438	0.06	0.27	0.47	0.60	0.87	1.17	1.38	1.58	1.77	1.49	1.27	1.10	0.96	0.86	0.77	0.69	0.63	0.57	0.53	0.45
14	.563	0.07	0.30	0.50	0.65	0.94	1.27	1.49	1.71	1.93	1.66	1.42	1.23	1.08	0.96	0.86	0.77	0.70	0.64	0.59	0.50
15	.563	0.08	0.32	0.54	0.70	1.01	1.36	1.61	1.85	2.08	1.84	1.57	1.36	1.20	1.06	0.95	0.86	0.78	0.71	0.65	0.56
16	.563	0.08	0.34	0.58	0.76	1.09	1.46	1.72	1.98	2.23	2.03	1.73	1.50	1.32	1.17	1.05	0.94	0.86	0.78	0.72	0.61
17	.625	0.09	0.37	0.62	0.81	1.16	1.56	1.84	2.11	2.38	2.22	1.90	1.64	1.44	1.28	1.14	1.03	0.94	0.86	0.79	0.67
18	.750	0.09	0.39	0.66	0.86	1.24	1.66	1.96	2.25	2.53	2.42	2.07	1.79	1.57	1.39	1.25	1.12	1.02	0.93	0.86	0.73
19	.813	0.10	0.41	0.70	0.91	1.31	1.76	2.07	2.38	2.69	2.62	2.24	1.94	1.70	1.51	1.35	1.22	1.11	1.01	0.93	0.79
20	.875	0.10	0.44	0.74	0.96	1.38	1.86	2.19	2.52	2.84	2.83	2.42	2.10	1.84	1.63	1.46	1.32	1.20	1.09	1.00	0.86
21	.875	0.11	0.46	0.78	1.01	1.46	1.96	2.31	2.66	2.99	3.05	2.60	2.26	1.98	1.76	1.57	1.42	1.29	1.17	1.08	0.92
22	.938	0.11	0.48	0.82	1.07	1.53	2.06	2.43	2.79	3.15	3.27	2.79	2.42	2.12	1.88	1.69	1.52	1.38	1.26	1.16	0.99
23	1.000	0.12	0.51	0.86	1.12	1.61	2.16	2.55	2.93	3.30	3.50	2.98	2.59	2.27	2.01	1.80	1.62	1.47	1.35	1.24	1.06
24	1.063	0.13	0.53	0.90	1.17	1.69	2.27	2.67	3.07	3.46	3.73	3.18	2.76	2.42	2.15	1.92	1.73	1.57	1.44	1.32	1.12
25	1.188	0.13	0.56	0.94	1.22	1.76	2.37	2.79	3.21	3.61	3.96	3.38	2.93	2.57	2.28	2.04	1.84	1.67	1.53	1.40	1.20
28	1.250	0.15	0.63	1.07	1.38	1.99	2.68	3.15	3.62	4.09	4.54	4.01	3.47	3.05	2.70	2.42	2.18	1.98	1.81	1.66	1.42
30	1.313	0.16	0.68	1.15	1.49	2.15	2.88	3.40	3.90	4.40	4.89	4.45	3.85	3.38	3.00	2.68	2.42	2.20	2.01	1.84	1.57
32	1.500	0.17	0.73	1.23	1.60	2.30	3.09	3.64	4.18	4.72	5.25	4.90	4.25	3.73	3.30	2.96	2.67	2.42	2.21	2.03	1.73
35	1.688	0.19	0.80	1.36	1.76	2.53	3.41	4.01	4.61	5.20	5.78	5.60	4.86	4.26	3.78	3.38	3.05	2.77	2.53	2.32	1.98
40	1.875	0.22	0.92	1.57	2.03	2.93	3.93	4.64	5.32	6.00	6.68	6.85	5.93	5.21	4.62	4.13	3.73	3.38	3.09	2.83	2.42
Lubrication type■			A				B									C					

▲ Ratings are based on a service factor of 1.
The ratings tabled above apply directly to lubricated, single strand, standard roller chains.

■ Type A: Manual or drip (Maximum chain speed 500 FPM)
Type B: Bath or disc (Maximum chain speed 3500 FPM)
Type C: Forced (pump)

Source: Rexnord Corporation, Link-Belt Roller Chain Operation, Indianapolis, IN.

Table 3 General Dimensions and Horsepower Capacity, No. 35 Roller Chain (ANSI B29.1) 3/8-in. Pitch

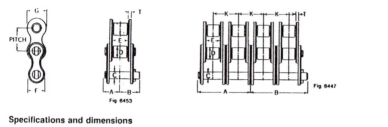

Fig 6453 / Fig 6447

Specifications and dimensions

Chain number	Chain width, number of strands	Average ultimate strength, pounds	Joint bearing area, sq. in.	Weight per foot, pounds	Dimensions, inches								
					A	B	C	D	E	F	G	K	T
35	Single	2,150	.041	.22	.24	.27							
35-2	Double	4,300	.082	.42	.43	.46							
35-3	Triple	6,450	.123	.62	.63	.67							
35-4	Quadruple	8,600	.164	.82	.83	.86	.141	.200△	.19	.31	.35	.399	.05
35-5	Quintuple	10,750	.205	1.06	1.03	1.06							
35-6	Sextuple	12,900	.246	1.27	1.23	1.26							

Ratings

Horsepower for single strand chain ▲

RPM of small sprocket

Number of teeth in small sprocket	Maximum bore, inches	100	500	900	1200	1800	2500	3000	3500	4000	4500	5000	5500	6000	6500	7000	7500	8000	8500	9000	10000
11	.563	0.18	0.77	1.31	1.70	2.45	3.30	2.94	2.33	1.91	1.60	1.37	1.18	1.04	0.92	0.82	0.74	0.67	0.62	0.57	0.48
12	.625	0.20	0.85	1.44	1.87	2.70	3.62	3.35	2.66	2.17	1.82	1.56	1.35	1.18	1.05	0.94	0.85	0.77	0.70	0.64	0.55
13	.688	0.22	0.93	1.57	2.04	2.94	3.95	3.77	3.00	2.45	2.05	1.75	1.52	1.33	1.18	1.06	0.95	0.87	0.79	0.73	0.62
14	.813	0.24	1.01	1.71	2.21	3.18	4.28	4.22	3.35	2.74	2.30	1.96	1.70	1.49	1.32	1.18	1.07	0.97	0.88	0.81	0.69
15	.875	0.25	1.08	1.84	2.38	3.43	4.61	4.68	3.71	3.04	2.55	2.17	1.88	1.65	1.47	1.31	1.18	1.07	0.98	0.90	0.77
16	.938	0.27	1.16	1.97	2.55	3.68	4.94	5.15	4.09	3.35	2.81	2.40	2.08	1.82	1.62	1.45	1.30	1.18	1.08	0.99	0.85
17	1.063	0.29	1.24	2.10	2.73	3.93	5.28	5.64	4.48	3.67	3.07	2.62	2.27	2.00	1.77	1.58	1.43	1.30	1.18	1.09	0.93
18	1.188	0.31	1.32	2.24	2.90	4.18	5.61	6.15	4.88	3.99	3.35	2.86	2.48	2.17	1.93	1.73	1.56	1.41	1.29	1.18	1.01
19	1.250	0.33	1.40	2.37	3.07	4.43	5.95	6.67	5.29	4.33	3.63	3.10	2.69	2.36	2.09	1.87	1.69	1.53	1.40	1.28	1.10
20	1.313	0.35	1.48	2.51	3.25	4.68	6.29	7.20	5.72	4.68	3.92	3.35	2.90	2.55	2.26	2.02	1.82	1.65	1.51	1.39	1.18
21	1.438	0.37	1.56	2.64	3.42	4.93	6.63	7.75	6.15	5.03	4.22	3.60	3.12	2.74	2.43	2.17	1.96	1.78	1.62	1.49	1.27
22	1.563	0.38	1.64	2.78	3.60	5.19	6.97	8.21	6.59	5.40	4.52	3.86	3.35	2.94	2.61	2.33	2.10	1.91	1.74	1.60	1.37
23	1.688	0.40	1.72	2.92	3.78	5.44	7.31	8.62	7.05	5.77	4.83	4.13	3.58	3.14	2.79	2.49	2.25	2.04	1.86	1.71	1.46
24	1.750	0.42	1.80	3.05	3.96	5.70	7.66	9.02	7.51	6.15	5.15	4.40	3.81	3.35	2.97	2.66	2.40	2.17	1.99	1.82	1.56
25	1.813	0.44	1.88	3.19	4.13	5.95	8.00	9.43	7.99	6.54	5.48	4.68	4.05	3.56	3.16	2.82	2.55	2.31	2.11	1.94	1.65
28	2.125	0.50	2.12	3.61	4.67	6.73	9.05	10.7	9.47	7.75	6.49	5.55	4.81	4.22	3.74	3.35	3.02	2.74	2.50	2.30	1.96
30	2.281	0.54	2.29	3.89	5.03	7.25	9.74	11.5	10.5	8.59	7.20	6.15	5.33	4.68	4.15	3.71	3.35	3.04	2.77	2.55	2.17
32	2.500	0.58	2.45	4.17	5.40	7.77	10.4	12.3	11.6	9.47	7.93	6.77	5.87	5.15	4.57	4.09	3.69	3.35	3.06	2.81	0
35	2.781	0.64	2.70	4.59	5.95	8.56	11.5	13.6	13.2	10.8	9.08	7.75	6.72	5.90	5.23	4.68	4.22	3.83	3.50	3.21	0
40	3.250	0.73	3.12	5.30	6.87	9.89	13.3	15.7	16.2	13.2	11.1	9.47	8.21	7.20	6.39	5.72	5.15	4.68	0		
Lubrication type ■		A				B							C							

▲ Ratings are based on a service factor of 1.
The ratings tabled above apply directly to lubricated, single strand, standard roller chains. For multiple strand chains, apply the factors shown in the table at right.

■ Type A: Manual or drip (Maximum chain speed 370 FPM)
Type B: Bath or disc (Maximum chain speed 2800 FPM)
Type C: Forced (pump)

Multiple strand factors

Number of strands	Multiple strand factor
2	1.7
3	2.5
4	3.3
5	4.1
6	5.0
7 or more	Consult Link-Belt Chain Division

Source: Rexnord Corporation, Link-Belt Roller Chain Operation, Indianapolis, IN.

Table 4 General Dimensions and Horsepower Capacity, No. 40 Roller Chain (ANSI B29.1) 1/2-in. Pitch

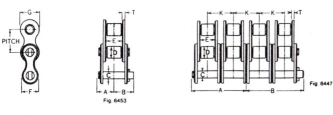

Fig. 6453   Fig. 6447

Specifications and dimensions

Chain number	Chain width, number of strands	Average ultimate strength, pounds	Joint bearing area, sq. in.	Weight per foot, pounds	A	B	C	D	E	F	G	K	T
40	Single	4,000	.068	.39	.32	.38							
40-2	Double	8,000	.136	.79	.60	.67							
40-3	Triple	12,000	.204	1.18	.89	.95	.156	.313	.31	.41	.47	.566	.06
40-4	Quadruple	16,000	.272	1.57	1.17	1.23							
40-5	Quintuple	20,000	.340	1.97	1.45	1.52							
40-6	Sextuple	24,000	.408	2.36	1.73	1.80							

(Dimensions, inches — columns A B C D E F G K T)

Ratings

Horsepower for single strand chain▲ — RPM of small sprocket

Number of teeth in small sprocket	Maximum bore, inches	50	100	200	300	400	500	700	1000	1200	1400	1600	1800	2400	3000	3500	4000	5000	6000	7000	8000
11	.750	0.23	0.43	0.80	1.16	1.50	1.83	2.48	3.42	4.03	4.63	5.22	4.66	3.03	2.17	1.72	1.41	1.01	0.77	0.61	0.50
12	.844	0.25	0.47	0.88	1.27	1.65	2.01	2.73	3.76	4.43	5.09	5.74	5.31	3.45	2.47	1.96	1.60	1.15	0.87	0.69	0.57
13	1.000	0.28	0.52	0.96	1.39	1.80	2.20	2.97	4.10	4.83	5.55	6.26	5.99	3.89	2.79	2.21	1.81	1.29	0.98	0.78	0.64
14	1.188	0.30	0.56	1.04	1.50	1.95	2.38	3.22	4.44	5.23	6.01	6.78	6.70	4.35	3.11	2.47	2.02	1.45	1.10	0.87	0.71
15	1.250	0.32	0.60	1.12	1.62	2.10	2.56	3.47	4.78	5.64	6.47	7.30	7.43	4.82	3.45	2.74	2.24	1.60	1.22	0.97	0.79
16	1.375	0.35	0.65	1.20	1.74	2.25	2.75	3.72	5.13	6.04	6.94	7.83	8.18	5.31	3.80	3.02	2.47	1.77	1.34	1.07	0.87
17	1.500	0.37	0.69	1.29	1.85	2.40	2.93	3.97	5.48	6.45	7.41	8.36	8.96	5.82	4.17	3.31	2.71	1.94	1.47	1.17	0.96
18	1.625	0.39	0.73	1.37	1.97	2.55	3.12	4.22	5.82	6.86	7.88	8.89	9.76	6.34	4.54	3.60	2.95	2.11	1.60	1.27	0
19	1.750	0.42	0.78	1.45	2.09	2.71	3.31	4.48	6.17	7.27	8.36	9.42	10.5	6.88	4.92	3.91	3.20	2.29	1.74	1.38	0
20	1.875	0.44	0.82	1.53	2.21	2.86	3.50	4.73	6.53	7.69	8.83	9.96	11.1	7.43	5.31	4.22	3.45	2.47	1.88	1.49	0
21	2.063	0.46	0.87	1.62	2.33	3.02	3.69	4.99	6.88	8.11	9.31	10.5	11.7	7.99	5.72	4.54	3.71	2.66	2.02	1.60	0
22	2.188	0.49	0.91	1.70	2.45	3.17	3.88	5.25	7.23	8.52	9.79	11.0	12.3	8.57	6.13	4.87	3.98	2.85	2.17	1.72	0
23	2.250	0.51	0.96	1.78	2.57	3.33	4.07	5.51	7.59	8.94	10.3	11.6	12.9	9.16	6.55	5.20	4.26	3.05	2.32	1.84	0
24	2.250	0.54	1.00	1.87	2.69	3.48	4.26	5.76	7.95	9.36	10.8	12.1	13.5	9.76	6.99	5.54	4.54	3.25	2.47	1.96	0
25	2.281	0.56	1.05	1.95	2.81	3.64	4.45	6.02	8.30	9.78	11.2	12.7	14.1	10.4	7.43	5.89	4.82	3.45	2.63	0	
28	2.625	0.63	1.18	2.20	3.18	4.11	5.03	6.81	9.39	11.1	12.7	14.3	15.9	12.3	8.80	6.99	5.72	4.09	3.11	0	
30	2.750	0.68	1.27	2.38	3.42	4.43	5.42	7.33	10.1	11.9	13.7	15.4	17.2	13.6	9.76	7.75	6.34	4.54	3.45	0	
32	3.000	0.73	1.36	2.55	3.67	4.75	5.81	7.86	10.8	12.8	14.7	16.5	18.4	15.0	10.8	8.54	6.99	5.00	0		
35	3.563	0.81	1.50	2.81	4.04	5.24	6.40	8.66	11.9	14.1	16.2	18.2	20.3	17.2	12.3	9.76	7.99	5.72	0		
40	3.781	0.93	1.74	3.24	4.67	6.05	7.39	10.0	13.8	16.3	18.7	21.1	23.4	21.0	15.0	11.9	9.76	6.99	0		
Lubrication type ■		A				B								C							

▲ Ratings are based on a service factor of 1.
The ratings tabled above apply directly to lubricated, single strand, standard roller chains. For multiple strand chains, apply the factors shown in the table at right.

■ Type A: Manual or drip   (Maximum chain speed 300 FPM)
Type B: Bath or disc   (Maximum chain speed 2300 FPM)
Type C: Forced (pump)

Multiple strand factors

Number of strands	Multiple strand factor
2	1.7
3	2.5
4	3.3
5	4.1
6	5.0
7 or more	Consult Link-Belt Chain Division

Source: Rexnord Corporation, Link-Belt Roller Chain Operation, Indianapolis, IN.

Table 5   General Dimensions and Horsepower Capacity, No. 41 Roller Chain (ANSI B29.1) 1/2-in. Pitch

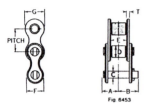

Fig. 6453

Specifications and dimensions

Chain number	Average ultimate strength, pounds	Joint bearing area, sq. in.	Weight per foot, pounds	Dimensions, inches							
				A	B	C	D	E	F	G	T
41	2,000	.049	.27	.27	.32	.141	.306	.25	.32	.38	.05

Ratings

Number of teeth in small sprocket	Maximum bore, inches	Horsepower for single strand chain▲ RPM of small sprocket																			
		50	100	200	300	400	500	700	1000	1200	1400	1600	1800	2400	3000	3500	4000	5000	6000	7000	8000
11	.875	0.13	0.24	0.44	0.64	0.82	1.01	1.37	1.88	1.71	1.36	1.11	0.93	0.61	0.43	0.34	0.28	0.20	0.15	0.12	0.10
12	.969	0.14	0.26	0.49	0.70	0.91	1.11	1.50	2.07	1.95	1.55	1.27	1.06	0.69	0.49	0.39	0.32	0.23	0.17	0.14	0.11
13	1.125	0.15	0.28	0.53	0.76	0.99	1.21	1.63	2.25	2.20	1.75	1.43	1.20	0.78	0.56	0.44	0.36	0.26	0.20	0.16	0.13
14	1.250	0.16	0.31	0.57	0.83	1.07	1.31	1.77	2.44	2.46	1.95	1.60	1.34	0.87	0.62	0.49	0.40	0.29	0.22	0.17	0.14
15	1.313	0.18	0.33	0.62	0.89	1.15	1.41	1.91	2.63	2.73	2.17	1.77	1.49	0.96	0.69	0.55	0.45	0.32	0.24	0.19	0.16
16	1.438	0.19	0.36	0.66	0.95	1.24	1.51	2.05	2.82	3.01	2.39	1.95	1.64	1.06	0.76	0.60	0.49	0.35	0.27	0.21	0.17
17	1.563	0.20	0.38	0.71	1.02	1.32	1.61	2.18	3.01	3.29	2.61	2.14	1.79	1.16	0.83	0.66	0.54	0.39	0.29	0.23	0.19
18	1.688	0.22	0.40	0.75	1.08	1.40	1.72	2.32	3.20	3.59	2.85	2.33	1.95	1.27	0.91	0.72	0.59	0.42	0.32	0.25	0
19	1.813	0.23	0.43	0.80	1.15	1.49	1.82	2.46	3.40	3.89	3.09	2.53	2.12	1.38	0.98	0.78	0.64	0.46	0.35	0.28	0
20	1.875	0.24	0.45	0.84	1.21	1.57	1.92	2.60	3.59	4.20	3.33	2.73	2.29	1.49	1.06	0.84	0.69	0.49	0.38	0.30	0
21	2.063	0.26	0.48	0.89	1.28	1.66	2.03	2.74	3.78	4.46	3.59	2.94	2.46	1.60	1.14	0.91	0.74	0.53	0.40	0.32	0
22	2.188	0.27	0.50	0.93	1.35	1.74	2.13	2.89	3.98	4.69	3.85	3.15	2.64	1.71	1.23	0.97	0.80	0.57	0.43	0.34	0
23	2.250	0.28	0.53	0.98	1.41	1.83	2.24	3.03	4.17	4.92	4.11	3.37	2.82	1.83	1.31	1.04	0.85	0.61	0.46	0.37	0
24	2.250	0.29	0.55	1.03	1.48	1.92	2.34	3.17	4.37	5.15	4.38	3.59	3.01	1.95	1.40	1.11	0.91	0.65	0.49	0.39	0
25	2.313	0.31	0.57	1.07	1.55	2.00	2.45	3.31	4.57	5.38	4.66	3.81	3.20	2.08	1.49	1.18	0.96	0.69	0.53	0	
28	2.625	0.35	0.65	1.21	1.75	2.26	2.77	3.74	5.16	6.08	5.52	4.52	3.79	2.46	1.76	1.40	1.14	0.82	0.62	0	
30	2.813	0.38	0.70	1.31	1.88	2.44	2.98	4.03	5.56	6.55	6.13	5.01	4.20	2.73	1.95	1.55	1.27	0.91	0.69	0	
32	3.125	0.40	0.75	1.40	2.02	2.61	3.20	4.33	5.96	7.03	6.75	5.52	4.63	3.01	2.15	1.71	1.40	1.00	0		
35	3.563	0.44	0.83	1.54	2.22	2.88	3.52	4.76	6.57	7.74	7.72	6.32	5.29	3.44	2.46	1.95	1.60	1.14	0		
40	3.875	0.51	0.96	1.78	2.57	3.33	4.07	5.50	7.59	8.94	9.43	7.72	6.47	4.20	3.01	2.39	1.95	1.40	0		
Lubrication type■		A					B							C							

▲ Ratings are based on a service factor of 1.
The ratings tabled above apply directly to lubricated, single strand, standard roller chains.

■ Type A: Manual or drip   (Maximum chain speed   300 FPM)
Type B: Bath or disc   (Maximum chain speed 2300 FPM)
Type C: Forced (pump)

Source: Rexnord Corporation, Link-Belt Roller Chain Operation, Indianapolis, IN.

Table 6 General Dimensions and Horsepower Capacity, No. 50 Roller Chain (ANSI B29.1) 5/8-in. Pitch

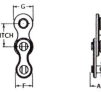

Fig 6453

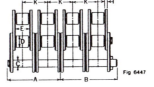

Fig 6447

Specifications and dimensions

Chain number	Chain width, number of strands	Average ultimate strength, pounds	Joint bearing area, sq. in.	Weight per foot, pounds	Dimensions, inches A	B	C	D	E	F	G	K	T
50	Single	6,250	.108	.70	.41	.48							
50-2	Double	12,500	.216	1.39	.76	.83							
50-3	Triple	18,750	.324	2.09	1.12	1.19	.200	.400	.38	.52	.59	.713	.08
50-4	Quadruple	25,000	.432	2.76	1.48	1.54							
50-5	Quintuple	31,250	.540	3.15	1.84	1.90							
50-6	Sextuple	37,500	.648	3.77	2.19	2.25							

Ratings

Number of teeth in small sprocket	Maximum bore inches	Horsepower for single strand chain▲ RPM of small sprocket 50	100	300	500	900	1000	1200	1400	1600	1800	2100	2400	2700	3000	3500	4000	4500	5000	5500	6000
11	.969	0.45	0.84	2.25	3.57	6.06	6.66	7.85	8.13	6.65	5.58	4.42	3.62	3.04	2.59	2.06	1.68	1.41	1.20	1.04	0.92
12	1.125	0.49	0.92	2.47	3.92	6.65	7.31	8.62	9.26	7.58	6.35	5.04	4.13	3.46	2.95	2.34	1.92	1.61	1.37	1.19	1.04
13	1.313	0.54	1.00	2.70	4.27	7.25	7.97	9.40	10.4	8.55	7.16	5.69	4.65	3.90	3.33	2.64	2.16	1.81	1.55	1.34	0
14	1.438	0.58	1.09	2.92	4.63	7.86	8.64	10.2	11.7	9.55	8.01	6.35	5.20	4.36	3.72	2.95	2.42	2.03	1.73	1.50	0
15	1.625	0.63	1.17	3.15	4.99	8.47	9.31	11.0	12.6	10.6	8.88	7.05	5.77	4.83	4.13	3.27	2.68	2.25	1.92	1.66	0
16	1.750	0.67	1.26	3.38	5.35	9.08	9.98	11.8	13.5	11.7	9.78	7.76	6.35	5.32	4.55	3.61	2.95	2.47	2.11	1.83	0
17	1.875	0.72	1.34	3.61	5.71	9.69	10.7	12.6	14.4	12.8	10.7	8.50	6.96	5.83	4.98	3.95	3.23	2.71	2.31	2.01	0
18	2.063	0.76	1.43	3.83	6.07	10.3	11.3	13.4	15.3	13.9	11.7	9.26	7.58	6.35	5.42	4.30	3.52	2.95	2.52	0	
19	2.250	0.81	1.51	4.07	6.44	10.9	12.0	14.2	16.3	15.1	12.7	10.0	8.22	6.89	5.88	4.67	3.82	3.20	2.73	0	
20	2.375	0.86	1.60	4.30	6.80	11.5	12.7	15.0	17.2	16.3	13.7	10.8	8.88	7.44	6.35	5.04	4.13	3.46	2.95	0	
21	2.563	0.90	1.69	4.53	7.17	12.2	13.4	15.8	18.1	17.6	14.7	11.7	9.55	8.01	6.84	5.42	4.44	3.72	3.18	0	
22	2.688	0.95	1.77	4.76	7.54	12.8	14.1	16.6	19.1	18.8	15.8	12.5	10.2	8.59	7.33	5.82	4.76	3.99	3.41	0	
23	2.813	1.00	1.86	5.00	7.91	13.4	14.8	17.4	20.0	20.1	16.9	13.4	11.0	9.18	7.84	6.22	5.09	4.27	0		
24	2.875	1.04	1.95	5.23	8.29	14.1	15.5	18.2	20.9	21.4	18.0	14.3	11.7	9.78	8.35	6.63	5.42	4.55	0		
25	2.906	1.08	2.03	5.47	8.66	14.7	16.2	19.0	21.9	22.8	19.1	15.2	12.4	10.4	8.88	7.05	5.77	4.83	0		
28	3.250	1.23	2.30	6.18	9.79	16.6	18.3	21.5	24.7	27.0	22.6	18.0	14.7	12.3	10.5	8.35	6.84	5.73	0		
30	3.563	1.33	2.48	6.66	10.5	17.9	19.7	23.2	26.6	30.0	25.1	19.9	16.3	13.7	11.7	9.26	7.58	0			
32	3.750	1.42	2.66	7.14	11.3	19.2	21.1	24.9	28.6	32.2	27.7	22.0	18.0	15.1	12.9	10.2	8.35	0			
35	4.125	1.57	2.93	7.86	12.5	21.1	23.2	27.4	31.5	35.5	31.6	25.1	20.6	17.2	14.7	11.7	9.55	0			
40	5.125	1.81	3.38	9.08	14.4	24.4	26.8	31.6	36.3	41.0	38.7	30.7	25.1	21.0	18.0	14.3	0				
Lubrication type■		A		B								C									

▲ Ratings are based on a service factor of 1.
The ratings tabled above apply directly to lubricated, single strand, standard roller chains.
For multiple strand chains, apply the factors shown in the table at right.

■ Type A: Manual or drip (Maximum chain speed 250 FPM)
  Type B: Bath or disc (Maximum chain speed 2000 FPM)
  Type C: Forced (pump)

Multiple strand factors

Number of strands	Multiple strand factor
2	1.7
3	2.5
4	3.3
5	4.1
6	5.0
7 or more	Consult Link-Belt Chain Division

Source: Rexnord Corporation, Link-Belt Roller Chain Operation, Indianapolis, IN.

Table 7 General Dimensions and Horsepower Capacity, No. 60 Roller Chain (ANSI B29.1) 3/4-in. Pitch

Fig. 6453

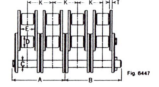

Fig. 6447

Specifications and dimensions

Chain number	Chain width, number of strands	Average ultimate strength, pounds	Joint bearing area, sq. in.	Weight per foot, pounds	Dimensions, inches							
					A	B	C	D	E	F	G	K	T
60	Single	9,000	.162	1.02	.50	.60							
60-2	Double	18,000	.324	2.00	.95	1.05							
60-3	Triple	27,000	.486	3.00	1.40	1.50							
60-4	Quadruple	36,000	.648	3.83	1.84	1.95	.234	.469	.50	.60	.71	.897	.09
60-5	Quintuple	45,000	.810	5.02	2.30	2.40							
60-6	Sextuple	54,000	.972	6.02	2.69	2.85							

Ratings

Number of teeth in small sprocket	Maximum bore, inches	Horsepower for single strand chain▲																			
		RPM of small sprocket																			
		25	50	100	200	300	500	700	900	1000	1100	1200	1400	1600	1800	2000	2500	3000	3500	4000	4500
11	1.250	0.41	0.77	1.44	2.69	3.87	6.13	8.30	10.4	11.4	12.5	11.9	9.4	7.70	6.45	5.51	3.94	3.00	2.38	1.95	1.63
12	1.344	0.45	0.85	1.58	2.95	4.25	6.74	9.12	11.4	12.6	13.7	13.5	10.7	8.77	7.35	6.28	4.49	3.42	2.71	2.22	1.86
13	1.500	0.50	0.92	1.73	3.22	4.64	7.34	9.94	12.5	13.7	14.9	15.2	12.1	9.89	8.29	7.08	5.06	3.85	3.06	2.50	0
14	1.750	0.54	1.00	1.87	3.49	5.02	7.96	10.8	13.5	14.8	16.2	17.0	13.5	11.1	9.26	7.91	5.66	4.31	3.42	2.80	0
15	1.938	0.58	1.08	2.01	3.76	5.41	8.57	11.6	14.5	16.0	17.4	18.8	15.0	12.3	10.3	8.77	6.28	4.77	3.79	3.10	0
16	2.125	0.62	1.16	2.16	4.03	5.80	9.19	12.4	15.6	17.1	18.7	20.2	16.5	13.5	11.3	9.66	6.91	5.26	4.17	3.42	0
17	2.313	0.66	1.24	2.31	4.30	6.20	9.81	13.3	16.7	18.3	19.9	21.6	18.1	14.8	12.4	10.6	7.57	5.76	4.57	3.74	0
18	2.500	0.70	1.31	2.45	4.58	6.59	10.4	14.1	17.7	19.5	21.2	22.9	19.7	16.1	13.5	11.5	8.25	6.28	4.98	4.08	0
19	2.688	0.75	1.39	2.60	4.85	6.99	11.1	15.0	18.8	20.6	22.5	24.3	21.4	17.5	14.6	12.5	8.95	6.81	5.40	4.42	0
20	2.813	0.79	1.47	2.75	5.13	7.38	11.7	15.8	19.8	21.8	23.8	25.7	23.1	18.9	15.8	13.5	9.66	7.35	5.83	0	
21	3.063	0.83	1.55	2.90	5.40	7.78	12.3	16.7	20.9	23.0	25.1	27.1	24.8	20.3	17.0	14.5	10.4	7.91	6.28	0	
22	3.250	0.87	1.63	3.05	5.68	8.19	13.0	17.5	22.0	24.2	26.4	28.5	26.6	21.8	18.2	15.6	11.1	8.48	6.73	0	
23	3.438	0.92	1.71	3.19	5.96	8.59	13.6	18.4	23.1	25.4	27.7	29.9	28.4	23.3	19.5	16.7	11.9	9.07	7.19	0	
24	3.625	0.96	1.79	3.35	6.24	8.99	14.2	19.3	24.2	26.6	29.0	31.3	30.3	24.8	20.8	17.8	12.7	9.66	7.67	0	
25	3.750	1.00	1.87	3.50	6.52	9.40	14.9	20.1	25.3	27.8	30.3	32.7	32.2	26.4	22.1	18.9	13.5	10.3	8.15	0	
28	4.188	1.13	2.12	3.95	7.37	10.6	16.8	22.8	28.5	31.4	34.2	37.0	38.2	31.3	26.2	22.4	16.0	12.2	0		
30	4.500	1.22	2.28	4.26	7.94	11.4	18.1	24.5	30.8	33.8	36.8	39.8	42.4	34.7	29.1	24.8	17.8	13.5	0		
32	4.750	1.31	2.45	4.56	8.52	12.3	19.4	26.3	33.0	36.3	39.5	42.7	46.7	38.2	32.0	27.3	19.6	14.9	0		
35	5.500	1.44	2.68	5.03	9.38	13.5	21.4	29.0	36.3	39.9	43.5	47.1	53.4	43.7	36.6	31.3	22.4	17.0	0		
40	6.250	1.67	3.11	5.81	10.8	15.6	24.7	33.5	42.0	46.1	50.3	54.4	62.5	53.4	44.7	38.2	27.3	0			
Lubrication type■		A			B								C								

▲ Ratings are based on a service factor of 1.
The ratings tabled above apply directly to lubricated, single strand, standard and heavy series roller chains.

■ Type A: Manual or drip (Maximum chain speed 220 FPM)
Type B: Bath or disc (Maximum chain speed 1800 FPM)
Type C: Forced (pump)

Multiple strand factors

Number of strands	Multiple strand factor
2	1.7
3	2.5
4	3.3
5	4.1
6	5.0
7 or more	Consult Link-Belt Chain Division

Source: Rexnord Corporation, Link-Belt Roller Chain Operation, Indianapolis, IN.

Table 8 General Dimensions and Horsepower Capacity, No. 80 Roller Chain (ANSI B29.1) 1-in. Pitch

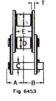

Fig 6453

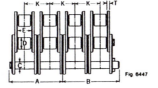

Fig 6447

Specifications and dimensions

Chain number	Chain width, number of strands	Average ultimate strength, pounds	Joint bearing area, sq. in.	Weight per foot, pounds	Dimensions, inches A	B	C	D	E	F	G	K	T
80	Single	16,000	.275	1.67	.63	.74							
80-2	Double	32,000	.550	3.31	1.21	1.30							
80-3	Triple	48,000	.825	4.97	1.78	1.87	.313	.625	.63	.75	.91	1.153	.13
80-4	Quadruple	64,000	1.100	6.76	2.35	2.44							
80-5	Quintuple	80,000	1.375	8.21	2.92	3.03							
80-6	Sextuple	96,000	1.650	9.84	3.50	3.61							

Ratings

Number of teeth in small sprocket	Maximum bore, inches	Horsepower for single strand chain▲ — RPM of small sprocket																			
		25	50	100	200	300	400	500	700	900	1000	1200	1400	1600	1800	2000	2200	2400	2700	3000	3400
11	1.625	0.97	1.80	3.36	6.28	9.04	11.7	14.3	19.4	23.0	19.6	14.9	11.8	9.69	8.12	6.93	6.01	5.27	4.42	3.77	1.70
12	1.750	1.06	1.98	3.69	6.89	9.93	12.9	15.7	21.3	26.2	22.3	17.0	13.5	11.0	9.25	7.90	6.85	6.01	5.04	4.30	0
13	2.000	1.16	2.16	4.03	7.52	10.8	14.0	17.1	23.2	29.1	25.2	19.2	15.2	12.5	10.4	8.91	7.72	6.78	5.68	4.85	0
14	2.250	1.25	2.34	4.36	8.14	11.7	15.2	18.6	25.1	31.5	28.2	21.4	17.0	13.9	11.7	9.96	8.63	7.57	6.35	5.42	0
15	2.563	1.35	2.52	4.70	8.77	12.6	16.4	20.0	27.1	34.0	31.2	23.8	18.9	15.4	12.9	11.0	9.57	8.40	7.04	6.01	0
16	2.875	1.45	2.70	5.04	9.41	13.5	17.6	21.5	29.0	36.4	34.4	26.2	20.8	17.0	14.2	12.2	10.5	9.25	7.76	6.62	0
17	3.125	1.55	2.88	5.38	10.0	14.5	18.7	22.9	31.0	38.9	37.7	28.7	22.7	18.6	15.6	13.3	11.5	10.1	8.49	7.25	0
18	3.375	1.64	3.07	5.72	10.7	15.4	19.9	24.4	33.0	41.4	41.1	31.2	24.8	20.3	17.0	14.5	12.6	11.0	9.25	7.90	0
19	3.688	1.74	3.25	6.07	11.3	16.3	21.1	25.8	35.0	43.8	44.5	33.9	26.9	22.0	18.4	15.7	13.6	12.0	10.0	8.57	0
20	3.813	1.84	3.44	6.41	12.0	17.2	22.3	27.3	37.0	46.3	48.1	36.6	29.0	23.8	19.9	17.0	14.7	12.9	10.8	0	
21	4.125	1.94	3.62	6.76	12.6	18.2	23.5	28.8	39.0	48.9	51.7	39.4	31.2	25.6	21.4	18.3	15.9	13.9	11.7	0	
22	4.438	2.04	3.81	7.11	13.3	19.1	24.8	30.3	41.0	51.4	55.5	42.2	33.5	27.4	23.0	19.6	17.0	14.9	12.5	0	
23	4.625	2.14	4.00	7.46	13.9	20.1	26.0	31.8	43.0	53.9	59.3	45.1	35.8	29.3	24.6	21.0	18.2	15.9	13.4	0	
24	4.688	2.24	4.19	7.81	14.6	21.0	27.2	33.2	45.0	56.4	62.0	48.1	38.2	31.2	26.2	22.3	19.4	17.0	14.2	0	
25	4.750	2.34	4.37	8.16	15.2	21.9	28.4	34.7	47.0	59.0	64.8	51.1	40.6	33.2	27.8	23.8	20.6	18.1	15.1	0	
28	5.375	2.65	4.94	9.23	17.2	24.8	32.1	39.3	53.2	66.7	73.3	60.6	48.1	39.4	33.0	28.2	24.4	21.4	0		
30	5.750	2.85	5.33	9.94	18.5	26.7	34.6	42.3	57.3	71.8	78.9	67.2	53.3	43.6	36.6	31.2	27.1	23.8	0		
32	6.313	3.06	5.71	10.7	19.9	28.6	37.1	45.4	61.4	77.0	84.6	74.0	58.7	48.1	40.3	34.4	29.8	26.2	0		
35	7.750	3.37	6.29	11.7	21.9	31.6	40.9	50.0	67.6	84.8	93.3	84.7	67.2	55.0	46.1	39.4	34.1	0			
40	9.375	3.89	7.27	13.6	25.3	36.4	47.2	57.7	78.1	99.0	108	103	82.1	67.2	56.3	48.1	20.0	0			
Lubrication type■			A				B								C						

The dotted line indicates the point at which pin and bushing galling is likely to begin. When the desired selection falls within the shaded area, consult Link-Belt Chain Division for guidance.

▲ Ratings are based on a service factor of 1.
The ratings tabled above apply directly to lubricated, single strand, standard and heavy series roller chains. For multiple strand chains, apply the factors shown in the table at right.

■ Type A: Manual or drip (Maximum chain speed 170 FPM)
Type B: Bath or disc (Maximum chain speed 1500 FPM)
Type C: Forced (pump)

Multiple strand factors

Number of strands	Multiple strand factor
2	1.7
3	2.5
4	3.3
5	4.1
6	5.0
7 or more	Consult Link-Belt Chain Division

Source: Rexnord Corporation, Link-Belt Roller Chain Operation, Indianapolis, IN.

Table 9 General Dimensions and Horsepower Capacity, No. 100 Roller Chain (ANSI B29.1) 1¼-in. Pitch

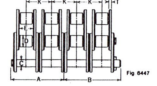

Fig. 6453

Fig 6447

Specifications and dimensions

Chain number	Chain width, number of strands	Average ultimate strength, pounds	Joint bearing area, sq. in.	Weight per foot, pounds	Dimensions, inches								
					A	B	C	D	E	F	G	K	T
100	Single	26,500	.401	2.72	.76	.89							
100-2	Double	53,000	.802	5.19	1.46	1.59							
100-3	Triple	79,500	1.203	7.67	2.16	2.29	.375	.750	.75	.97	1.13	1.408	.16
100-4	Quadruple	106,000	1.604	10.1	2.86	2.99							
100-5	Quintuple	132,500	2.005	12.7	3.56	3.69							
100-6	Sextuple	159,000	2.406	15.2	4.26	4.39							

Ratings

Number of teeth in small sprocket	Maximum bore, inches	Horsepower for single strand chain ▲ — RPM of small sprocket																			
		25	50	100	200	300	400	500	600	700	800	900	1000	1200	1400	1600	1800	2000	2200	2400	2600
11	2.000	1.85	3.45	6.44	12.0	17.3	22.4	27.4	32.3	37.1	32.8	27.5	23.4	17.8	14.2	11.6	9.71	8.29	7.19	6.31	1.29
12	2.250	2.03	3.79	7.08	13.2	19.0	24.6	30.1	35.5	40.8	37.3	31.3	26.7	20.3	16.1	13.2	11.1	9.45	8.19	7.18	0
13	2.500	2.22	4.13	7.72	14.4	20.7	26.9	32.8	38.7	44.5	42.1	35.3	30.1	22.9	18.2	14.9	12.5	10.6	9.23	8.10	0
14	2.813	2.40	4.48	8.36	15.6	22.5	29.1	35.6	41.9	48.2	47.0	39.4	33.7	25.6	20.3	16.6	13.9	11.9	10.3	9.05	0
15	3.250	2.59	4.83	9.01	16.8	24.2	31.4	38.3	45.2	51.9	52.2	43.7	37.3	28.4	22.5	18.4	15.5	13.2	11.4	10.0	0
16	3.500	2.77	5.17	9.66	18.0	26.0	33.6	41.1	48.4	55.6	57.5	48.2	41.1	31.3	24.8	20.3	17.0	14.5	12.6	11.1	0
17	3.813	2.96	5.52	10.3	19.2	27.7	35.9	43.9	51.7	59.4	63.0	52.8	45.0	34.3	27.2	22.3	18.7	15.9	13.8	0.79	0
18	4.188	3.15	5.88	11.0	20.5	29.5	38.2	46.7	55.0	62.3	68.6	57.5	49.1	37.3	29.6	24.2	20.3	17.4	15.0	0	
19	4.563	3.34	6.23	11.6	21.7	31.2	40.5	49.5	58.3	67.0	74.4	62.3	53.2	40.5	32.1	26.3	22.0	18.8	16.3	0	
20	4.875	3.53	6.58	12.3	22.9	33.0	42.8	52.3	61.6	70.8	79.8	67.3	57.5	43.7	34.7	28.4	23.8	20.3	17.8	0	
21	5.250	3.72	6.94	13.0	24.2	34.8	45.1	55.1	65.0	74.6	84.2	72.4	61.8	47.0	37.3	30.6	25.8	21.9	19.0	0	
22	5.625	3.91	7.30	13.6	25.4	36.6	47.4	58.0	68.3	78.5	88.5	77.7	66.3	50.4	40.0	32.8	27.5	23.4	20.3	0	
23	5.813	4.10	7.66	14.3	26.7	38.4	49.8	60.8	71.7	82.3	92.8	83.0	70.9	53.9	42.8	35.0	29.4	25.1	7.74	0	
24	6.000	4.30	8.02	15.0	27.9	40.2	52.1	63.7	75.0	86.2	97.2	88.5	75.6	57.5	45.6	37.3	31.3	26.7	0		
25	6.125	4.49	8.38	15.6	29.2	42.0	54.4	66.6	78.4	90.1	102	94.1	80.3	61.1	48.5	39.7	33.3	28.4	0		
28	7.000	5.07	9.47	17.7	33.0	47.5	61.5	75.2	88.6	102	115	112	95.2	72.4	57.5	47.0	39.4	33.7	0		
30	7.625	5.47	10.2	19.0	35.5	51.2	66.3	81.0	95.5	110	124	124	106	80.3	63.7	52.2	43.7	10.0	0		
32	8.250	5.86	10.9	20.4	38.1	54.9	71.1	86.9	102	118	133	136	116	88.5	70.2	57.5	45.2	0			
35	9.125	6.46	12.0	22.5	42.0	60.4	78.3	95.7	113	130	146	156	133	101	80.3	65.8	55.1	0			
40		7.46	13.9	26.0	48.5	69.8	90.4	111	130	150	169	188	163	124	98.1	80.3	0				
Lubrication type ■		A			B								C								

The dotted line indicates the point at which pin and bushing galling is likely to begin. When the desired selection falls within the shaded area, consult Link-Belt Chain Division for guidance.

▲ Ratings are based on a service factor of 1.
The ratings tabled above apply directly to lubricated, single strand, standard and heavy series roller chains. For multiple strand chains, apply the factors shown in the table at right.

■ Type A: Manual or drip (Maximum chain speed 150 FPM)
Type B: Bath or disc (Maximum chain speed 1300 FPM)
Type C: Forced (pump)

Multiple strand factors

Number of strands	Multiple strand factor
2	1.7
3	2.5
4	3.3
5	4.1
6	5.0
7 or more	Consult Link-Belt Chain Division

Source: Rexnord Corporation, Link-Belt Roller Chain Operation, Indianapolis, IN.

Table 10 General Dimensions and Horsepower Capacity, No. 120 Roller Chain (ANSI B29.1) 1½-in. Pitch

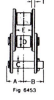
Fig 6453

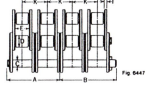
Fig. 6447

Specifications and dimensions

Chain number	Chain width, number of strands	Average ultimate strength, pounds	Joint bearing area, sq. in.	Weight per foot, pounds	Dimensions, inches								
					A	B	C	D	E	F	G	K	T
120	Single	37,000	.606	3.72	.96	1.13							
120-2	Double	74,000	1.212	7.38	1.84	2.02							
120-3	Triple	111,000	1.818	11.0	2.74	2.91							
120-4	Quadruple	148,000	2.424	14.7	3.63	3.81	.438	.875	1.00	1.13	1.38	1.789	.19
120-5	Quintuple	185,000	3.030	18.4	4.52	4.70							
120-6	Sextuple	222,000	3.636	22.0	5.42	5.59							

Ratings

Number of teeth in small sprocket	Maximum bore, inches	Horsepower for single strand chain▲																			
		RPM of small sprocket																			
		10	25	50	100	200	300	400	500	600	700	800	900	1000	1100	1200	1300	1400	1600	1800	2000
11	2.438	1.37	3.12	5.83	10.9	20.3	29.2	37.9	46.3	54.6	46.3	37.9	31.8	27.1	23.5	20.6	18.3	16.4	13.4	11.2	9.59
12	2.750	1.50	3.43	6.40	11.9	22.3	32.1	41.6	50.9	59.9	52.8	43.2	36.2	30.9	26.8	23.5	20.9	18.7	15.3	12.8	10.9
13	3.188	1.64	3.74	6.98	13.0	24.3	35.0	45.4	55.5	65.3	59.5	48.7	40.8	34.9	30.2	26.5	23.5	21.0	17.2	14.4	12.3
14	3.625	1.78	4.05	7.56	14.1	26.3	37.9	49.1	60.1	70.8	66.5	54.4	45.6	39.0	33.8	29.6	26.3	23.5	19.2	16.1	8.94
15	4.000	1.91	4.37	8.15	15.2	28.4	40.9	53.0	64.7	76.3	73.8	60.4	50.6	43.2	37.4	32.9	29.1	26.1	21.3	17.9	0
16	4.438	2.05	4.68	8.74	16.3	30.4	43.8	56.8	69.4	81.8	81.3	66.5	55.7	47.6	41.2	36.2	32.1	28.7	23.5	19.7	0
17	4.688	2.19	5.00	9.33	17.4	32.5	46.8	60.6	74.1	87.3	89.0	72.8	61.0	52.1	45.2	39.6	35.2	31.5	25.8	21.6	0
18	5.188	2.33	5.32	9.92	18.5	34.6	49.8	64.5	78.8	92.9	97.0	79.4	66.5	56.8	49.2	43.2	38.3	34.3	28.1	23.5	0
19	5.563	2.47	5.64	10.5	19.6	36.6	52.8	68.4	83.6	98.5	105	86.1	72.1	61.6	53.4	46.8	41.5	37.2	30.4	25.5	0
20	5.938	2.61	5.96	11.1	20.7	38.7	55.8	72.2	88.3	104	114	92.9	77.9	66.5	57.6	50.6	44.9	40.1	32.9	27.5	0
21	6.375	2.75	6.28	11.7	21.9	40.8	58.8	76.2	93.1	110	122	100	83.8	71.6	62.0	54.4	48.3	43.2	35.4	29.6	0
22	6.500	2.90	6.60	12.3	23.0	42.9	61.8	80.1	97.9	115	131	107	89.9	76.7	66.5	58.4	51.8	46.3	37.9	16.6	0
23	6.688	3.04	6.93	12.9	24.1	45.0	64.9	84.0	103	121	139	115	96.1	82.0	71.1	62.4	55.3	49.5	40.5	0	
24	7.000	3.18	7.25	13.5	25.3	47.1	67.9	88.0	108	127	146	122	102	87.4	75.8	66.5	59.0	52.8	43.2	0	
25	7.250	3.32	7.58	14.1	26.4	49.3	71.0	91.9	112	132	152	130	109	92.9	80.6	70.7	62.7	56.1	45.9	0	
28	8.375	3.76	8.57	16.0	29.8	55.7	80.2	104	127	150	172	154	129	110	95.5	83.8	74.3	66.5	54.4	0	
30	9.250	4.05	9.23	17.2	32.1	60.0	86.4	112	137	161	185	171	143	122	106	92.9	82.4	73.8	42.4	0	
32	10.000	4.34	9.90	18.5	34.5	64.3	92.6	120	146	173	199	188	158	135	117	102	90.8	81.3	0		
35		4.78	10.9	20.3	38.0	70.9	102	132	162	190	219	215	180	154	133	117	104	92.9	0		
40		5.52	12.6	23.5	43.9	81.8	118	153	187	220	253	263	220	188	163	143	127	59.5	0		
Lubrication type■		A		B			C														

The dotted line indicates the point at which pin and bushing galling is likely to begin. When the desired selection falls within the shaded area, consult Link-Belt Chain Division for guidance.

▲ Ratings are based on a service factor of 1.
The ratings tabled above apply directly to lubricated, single strand, standard and heavy series roller chains. For multiple strand chains, apply the factors shown in the table at right.

■ Type A: Manual or drip (Maximum chain speed 130 FPM)
Type B: Bath or disc (Maximum chain speed 1200 FPM)
Type C: Forced (pump)

Multiple strand factors

Number of strands	Multiple strand factor
2	1.7
3	2.5
4	3.3
5	4.1
6	5.0
7 or more	Consult Link-Belt Chain Division

Source: Rexnord Corporation, Link-Belt Roller Chain Operation, Indianapolis, IN.

Table 11   General Dimensions and Horsepower Capacity, No. 140 Roller Chain (ANSI B29.1) 1¾-in. Pitch

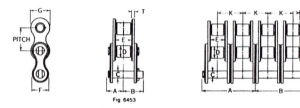

Fig 6453   Fig 6447

Specifications and dimensions

Chain number	Chain width, number of strands	Average ultimate strength, pounds	Joint bearing area, sq. in.	Weight per foot, pounds	Dimensions, inches								
					A	B	C	D	E	F	G	K	T
140	Single	48,500	.726	4.69	1.02	1.21							
140-2	Double	97,000	1.452	9.25	1.98	2.17							
140-3	Triple	145,500	2.178	13.8	2.94	3.13	.500	1.000	1.00	1.31	1.56	1.924	.22
140-4	Quadruple	194,000	2.904	18.4	3.90	4.09							
140-5	Quintuple	242,500	3.630	22.9	4.86	5.05							
140-6	Sextuple	291,000	4.356	27.4	5.82	6.01							

Ratings

Number of teeth in small sprocket	Maximum bore, inches	Horsepower for single strand chain ▲ — RPM of small sprocket																			
		10	25	50	100	150	200	250	300	350	400	450	500	550	600	700	800	1000	1200	1400	1600
11	2.750	2.12	4.83	9.02	16.8	24.2	31.4	38.4	45.2	52.0	58.6	65.2	71.6	75.2	66.0	52.4	42.9	30.7	23.3	18.5	15.2
12	3.125	2.33	5.31	9.91	18.5	26.6	34.5	42.2	49.7	57.1	64.4	71.6	78.7	85.7	75.2	59.7	48.9	35.0	26.6	21.1	17.3
13	3.625	2.54	5.79	10.8	20.2	29.0	37.6	46.0	54.2	62.2	70.2	78.0	85.8	93.5	84.8	67.3	55.1	39.4	30.0	23.8	19.5
14	4.063	2.75	6.27	11.7	21.8	31.5	40.8	49.8	58.7	67.4	76.0	84.5	93.0	101	94.8	75.2	61.6	44.1	33.5	26.6	21.8
15	4.688	2.96	6.76	12.6	23.5	33.9	43.9	53.7	63.2	72.7	81.9	91.1	100	109	105	83.4	68.3	48.9	37.2	29.5	0
16	5.063	3.18	7.24	13.5	25.2	36.3	47.1	57.5	67.8	77.9	87.8	97.7	107	117	116	91.9	75.2	53.8	41.0	32.5	0
17	5.625	3.39	7.73	14.4	26.9	38.8	50.3	61.4	72.4	83.2	93.8	104	115	125	127	101	82.4	59.0	44.9	35.6	0
18	5.938	3.61	8.23	15.4	28.6	41.3	53.5	65.3	77.0	88.5	99.8	111	122	133	138	110	89.8	64.2	48.9	38.8	0
19	6.250	3.82	8.72	16.3	30.4	43.7	56.7	69.3	81.6	93.8	106	118	129	141	150	119	97.4	69.7	53.0	42.1	0
20	6.688	4.04	9.22	17.2	32.1	46.2	59.9	73.2	86.3	99.1	112	124	137	149	161	128	105	75.2	57.2	45.4	0
21	7.125	4.26	9.72	18.1	33.8	48.7	63.1	77.2	91.0	104	118	131	144	157	170	138	113	80.9	61.6	48.9	0
22	7.438	4.48	10.2	19.1	35.6	51.3	66.4	81.2	95.6	110	124	138	151	165	178	148	121	86.8	66.0	52.4	0
23	8.125	4.70	10.7	20.0	37.3	53.8	69.7	85.2	100	115	130	145	159	173	187	158	130	92.8	70.6	56.0	0
24	8.625	4.92	11.2	20.9	39.1	56.3	72.9	89.2	105	121	136	151	166	181	196	169	138	98.9	75.2	59.7	0
25	8.813	5.14	11.7	21.9	40.8	58.8	76.2	93.2	110	126	142	158	174	189	205	180	147	105	80.0	63.5	0
28	10.000	5.81	13.3	24.7	46.2	66.5	86.2	105	124	143	161	179	197	214	232	213	174	125	94.8	0	
30		6.26	14.3	26.7	49.7	71.6	92.8	113	134	154	173	193	212	231	249	236	193	138	105	0	
32		6.71	15.3	28.6	53.3	76.8	99.5	122	143	165	186	206	227	247	267	260	213	152	116	0	
35		7.40	16.9	31.5	58.7	84.6	110	134	158	181	205	227	250	272	295	297	243	174	130	0	
40		8.54	19.5	36.4	67.9	97.7	127	155	182	210	236	263	289	315	340	363	297	213	0		
Lubrication type ■		A				B												C			

The dotted line indicates the point at which pin and bushing galling is likely to begin. When the desired selection falls within the shaded area, consult Link-Belt Chain Division for guidance.

▲ Ratings are based on a service factor of 1.
The ratings tabled above apply directly to lubricated, single strand, standard and heavy series roller chains. For multiple strand chains, apply the factors shown in the table at right.

■ Type A: Manual or drip (Maximum chain speed 110 FPM)
Type B: Bath or disc (Maximum chain speed 1100 FPM)
Type C: Forced (pump)

Multiple strand factors

Number of strands	Multiple strand factor
2	1.7
3	2.5
4	3.3
5	4.1
6	5.0
7 or more	Consult Link-Belt Chain Division

Source: Rexnord Corporation, Link-Belt Roller Chain Operation, Indianapolis, IN.

Table 12 General Dimensions and Horsepower Capacity, No. 160 Roller Chain (ANSI B29.1) 2-in. Pitch

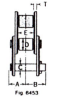
Fig 6453

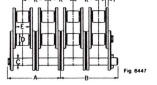
Fig 6447

Specifications and dimensions

Chain number	Chain width, number of strands	Average ultimate strength, pounds	Joint bearing area, sq. in.	Weight per foot, pounds	Dimensions, inches								
					A	B	C	D	E	F	G	K	T
160	Single	68,000	.991	6.12	1.23	1.41							
160-2	Double	136,000	1.982	12.5	2.38	2.56							
160-3	Triple	204,000	2.973	18.6	3.52	3.71	.563	1.125	1.25	1.56	1.81	2.305	.25
160-4	Quadruple	272,000	3.964	24.8	4.67	4.86							
160-5	Quintuple	340,000	4.955	31.9	5.82	6.00							
160-6	Sextuple	408,000	5.946	38.3	6.97	7.15							

Have dimensions certified for installation purposes.
Available in riveted and cottered construction.

Ratings

Number of teeth in small sprocket	Maximum bore, inches	Horsepower for single strand chain▲																		
		RPM of small sprocket																		
		10	25	50	100	150	200	250	300	350	400	450	500	550	600	700	800	900	1000	1100	1200
11	3.250	3.07	7.01	13.1	24.4	35.2	45.6	55.7	65.6	75.4	85.0	94.5	96.6	83.7	73.5	58.3	47.7	40.0	34.1	29.6	26.0
12	3.500	3.38	7.70	14.4	26.8	38.6	50.1	61.2	72.1	82.8	93.4	104	110	95.4	83.7	66.4	54.4	45.6	38.9	33.7	29.6
13	4.000	3.68	8.40	15.7	29.2	42.1	54.6	66.7	78.6	90.3	102	113	124	108	94.4	74.9	61.3	51.4	43.9	38.0	33.4
14	4.563	3.99	9.10	17.0	31.7	45.6	59.1	72.3	85.2	97.8	110	123	135	120	105	83.7	68.5	57.4	49.0	42.5	37.3
15	5.000	4.30	9.80	18.3	34.1	49.2	63.7	77.9	91.7	105	119	132	145	133	117	92.8	76.0	63.7	54.4	47.1	41.4
16	5.625	4.61	10.5	19.6	36.6	52.7	68.3	83.5	98.4	113	127	142	156	147	129	102	83.7	70.2	59.9	51.9	45.6
17	6.313	4.92	11.2	20.9	39.1	56.3	72.9	89.1	105	121	136	151	166	161	141	112	91.7	76.8	65.6	56.9	49.9
18	6.500	5.23	11.9	22.3	41.6	59.9	77.6	94.8	112	128	145	161	177	175	154	122	99.9	83.7	71.5	62.0	54.4
19	7.063	5.55	12.7	23.6	44.1	63.5	82.2	101	118	136	153	171	188	190	167	132	108	90.8	77.5	67.2	59.0
20	7.250	5.86	13.4	25.0	46.6	67.1	86.9	106	125	144	162	180	198	205	180	143	117	98.1	83.7	72.6	63.7
21	7.750	6.18	14.1	26.3	49.1	70.7	91.6	112	132	152	171	190	209	221	194	154	126	105	90.1	78.1	68.5
22	8.125	6.50	14.8	27.7	51.6	74.4	96.3	118	139	159	180	200	220	237	208	165	135	113	96.6	83.7	0
23	8.750	6.82	15.6	29.0	54.2	78.0	101	124	146	167	189	210	231	251	222	176	144	121	103	89.5	0
24	9.625	7.14	16.3	30.4	56.7	81.7	106	129	152	175	197	220	241	263	237	188	154	129	110	95.4	0
25		7.46	17.0	31.8	59.3	85.4	111	135	159	183	206	229	252	275	252	200	164	137	117	101	0
28		8.43	19.2	35.9	67.0	96.5	125	153	180	207	233	259	285	311	298	237	194	162	139	120	0
30		9.08	20.7	38.7	72.2	104	135	165	194	223	251	279	307	336	331	263	215	180	154	0	
32		9.74	22.2	41.5	77.4	111	144	176	208	239	269	300	329	359	365	289	·237	198	169	0	
35		10.7	24.5	45.7	85.2	123	159	194	229	263	297	330	363	395	417	331	271	227	180	0	
40		12.4	28.3	52.8	98.5	142	184	225	265	304	343	381	419	457	494	404	331	257	0		
Lubrication type■		A		B									C							

The dotted line indicates the point at which pin and bushing galling is likely to begin. When the desired selection falls within the shaded area, consult Link-Belt Chain Division for guidance.

▲ Ratings are based on a service factor of 1.
The ratings tabled above apply directly to lubricated, single strand, standard and heavy series roller chains. For multiple strand chains, apply the factors shown in the table at right.

■ Type A: Manual or drip (Maximum chain speed 100 FPM)
Type B: Bath or disc (Maximum chain speed 1000 FPM)
Type C: Forced (pump)

Multiple strand factors

Number of strands	Multiple strand factor
2	1.7
3	2.5
4	3.3
5	4.1
6	5.0
7 or more	Consult Link-Belt Chain Division

Source: Rexnord Corporation, Link-Belt Roller Chain Operation, Indianapolis, IN.

Table 13 General Dimensions and Horsepower Capacity, No. 180 Roller Chain (ANSI B29.1) 2¼-in. Pitch

  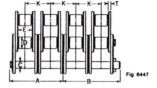

Fig. 6453

Fig. 6447

Specifications and dimensions

Chain number	Chain width, number of strands	Average ultimate strength, pounds	Joint bearing area, sq. in.	Weight per foot, pounds	Dimensions, inches							
					A	B	C	D	E	F	G	K	T
180	Single	86,000	1.364	9.06	1.39	1.56							
180-2	Double	172,000	2.728	17.6	2.69	2.86							
180-3	Triple	258,000	4.092	26.9	3.98	4.15							
180-4	Quadruple	344,000	5.456	35.8	5.28	5.45	.688	1.406	1.41	1.85	2.14	2.592	.28
180-5	Quintuple	430,000	6.820	44.7	6.57	6.74							
180-6	Sextuple	516,000	8.184	53.6	7.87	8.04							

Have dimensions certified for installation purposes.
Available in riveted or cottered construction.

Ratings

Number of teeth in small sprocket	Maximum bore, inches	Horsepower for single strand chain▲																			
		RPM of small sprocket																			
		10	25	50	100	150	200	250	300	350	400	450	500	550	600	650	700	800	900	1000	1100
11	3.375	4.24	9.68	18.1	33.7	48.6	62.9	76.9	90.6	104	117	124	106	92.0	80.7	71.6	64.1	52.4	46.9	37.5	32.5
12	4.000	4.66	10.6	19.8	37.0	53.4	69.1	84.5	99.6	114	129	142	121	105	92.0	81.6	73.0	59.7	50.1	42.8	37.1
13	4.500	5.08	11.6	21.6	40.4	58.2	75.4	92.1	109	125	141	156	136	118	104	92.0	82.3	67.4	56.5	48.2	0
14	5.375	5.51	12.6	23.4	43.7	63.0	81.6	99.8	118	135	152	169	152	132	116	103	92.0	75.3	63.1	53.9	0
15	5.875	5.93	13.5	25.3	47.1	67.9	88.0	108	127	146	164	182	169	146	129	114	102	83.5	70.0	59.7	0
16	6.250	6.36	14.5	27.1	50.5	72.8	94.3	115	136	156	176	196	188	161	142	126	112	92.0	77.1	65.8	0
17	6.688	6.79	15.5	28.9	54.0	77.7	101	123	145	167	188	209	204	177	155	138	123	101	84.4	72.1	0
18	7.375	7.22	16.5	30.8	57.4	82.7	107	131	154	177	200	222	222	193	169	150	134	110	92.0	78.5	0
19	7.750	7.66	17.5	32.6	60.8	87.6	114	139	164	188	212	236	241	209	183	163	145	119	99.8	85.2	0
20	8.250	8.10	18.5	34.5	64.3	92.6	120	147	173	199	224	249	260	226	198	176	157	129	108	92.0	0
21	9.313	8.53	19.5	36.3	67.8	97.6	126	155	182	209	236	262	280	243	213	189	169	138	116	99.0	0
22		8.97	20.5	38.2	71.3	103	133	163	192	220	248	276	300	260	228	203	181	148	124	0	
23		9.41	21.5	40.1	74.8	108	140	171	201	231	260	290	318	278	244	216	194	159	133	0	
24		9.86	22.5	42.0	78.3	113	146	179	210	242	273	303	333	296	260	231	206	169	142	0	
25		10.3	23.5	43.9	81.8	118	153	187	220	253	285	317	348	315	277	245	220	180	151	0	
28		11.6	26.6	49.6	92.5	133	173	211	249	286	322	358	394	374	328	291	260	213	178	0	
30		12.5	28.6	53.4	99.6	144	186	227	268	308	347	386	424	414	364	322	289	236	198	0	
32		13.4	30.7	57.2	107	154	199	244	287	330	372	414	455	456	401	355	318	260	0		
35		14.8	33.8	63.1	118	170	220	268	316	363	410	456	501	522	458	406	364	291	0		
40		17.1	39.0	72.9	136	196	254	310	365	420	473	526	579	575	524	465	398	244	0		
Lubrication type■		A		B							C										

The dotted line indicates the point at which pin and bushing galling is likely to begin. When the desired selection falls within the shaded area, consult Link-Belt Chain Division for guidance.

▲ Ratings are based on a service factor of 1.
The ratings tabled above apply directly to lubricated, single strand, standard and heavy series roller chains. For multiple strand chains, apply the factors shown in the table at right.

■ Type A: Manual or drip (Maximum chain speed 95 FPM)
Type B: Bath or disc (Maximum chain speed 950 FPM)
Type C: Forced (pump)

Multiple strand factors

Number of strands	Multiple strand factor
2	1.7
3	2.5
4	3.3
5	4.1
6	5.0
7 or more	Consult Link-Belt Chain Division

Source: Rexnord Corporation, Link-Belt Roller Chain Operation, Indianapolis, IN.

Table 14 General Dimensions and Horsepower Capacity, No. 200 Roller Chain (ANSI B29.1) 2½-in. Pitch

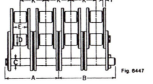

Fig 6453

Fig. 6447

Specifications and dimensions

Chain number	Chain width, number of strands	Average ultimate strength, pounds	Joint bearing area, sq. in.	Weight per foot, pounds	A	B	C	D	E	F	G	K	T
200	Single	106,000	1.681	10.9	1.54	1.89							
200-2	Double	212,000	3.362	21.0	2.96	3.31							
200-3	Triple	318,000	5.043	31.5	4.38	4.73	.781	1.563	1.50	1.94	2.31	2.817	.31
200-4	Quadruple	424,000	6.724	43.2	5.80	6.14							
200-5	Quintuple	530,000	8.405	53.9	7.22	7.56							
200-6	Sextuple	636,000	10.086	64.6	8.64	8.98							

(Dimensions C through T — .781, 1.563, 1.50, 1.94, 2.31, 2.817, .31 — apply to all rows.)

Have dimensions certified for installation purposes.
Available in riveted or cottered construction.

Ratings

Horsepower for single strand chain▲ — RPM of small sprocket

Number of teeth in small sprocket	Maximum bore, inches	10	15	20	30	40	50	70	100	150	200	250	300	350	400	450	500	550	600	650	700
9	2.625	4.54	6.54	8.47	12.2	15.8	19.3	26.1	36.0	51.9	67.3	82.2	96.9	111	119	100	85.4	74.1	65.0	57.6	0
10	3.125	5.08	7.32	9.49	13.7	17.7	21.6	29.3	40.4	58.2	75.4	92.1	109	125	140	117	100	86.7	76.1	67.5	0
11	3.750	5.64	8.12	10.5	15.1	19.6	24.0	32.5	44.8	64.5	83.5	102	120	138	156	135	115	100	87.8	77.9	0
12	4.625	6.19	8.92	11.6	16.6	21.6	26.4	35.7	49.2	70.8	91.8	112	132	152	171	154	132	114	100	0	
13	5.250	6.75	9.72	12.6	18.1	23.5	28.7	38.9	53.6	77.2	100	122	144	166	187	174	148	129	113	0	
14	5.875	7.31	10.5	13.6	19.7	25.5	31.1	42.1	58.1	83.7	108	132	156	179	202	194	166	144	126	0	
15	6.125	7.88	11.3	14.7	21.2	27.4	33.5	45.4	62.6	90.1	117	143	168	193	218	215	184	159	140	0	
16	6.688	8.45	12.2	15.8	22.7	29.4	36.0	48.7	67.1	96.6	125	153	180	207	234	237	203	176	154	0	
17	7.688	9.02	13.0	16.8	24.2	31.4	38.4	52.0	71.6	103	134	163	193	221	249	260	222	192	169	0	
18	8.375	9.59	13.8	17.9	25.8	33.4	40.8	55.3	76.2	110	142	174	205	235	265	283	242	209	184	0	
19	9.125	10.2	14.6	19.0	27.3	35.4	43.3	58.6	80.8	116	151	184	217	249	281	307	262	227	199	0	
20	9.750	10.7	15.5	20.1	28.9	37.4	45.8	61.9	85.4	123	159	195	229	264	297	331	283	245	0		
21		11.3	16.3	21.1	30.5	39.5	48.2	65.3	90.0	130	168	205	242	278	313	348	305	264	0		
22		11.9	17.2	22.2	32.0	41.5	50.7	68.7	94.6	136	177	216	254	292	330	366	327	283	0		
23		12.5	18.0	23.3	33.6	43.5	53.2	72.0	99.3	143	185	226	267	307	346	384	349	303	0		
24		13.1	18.9	24.4	35.2	45.6	55.7	75.4	104	150	194	237	279	321	362	402	372	323	0		
25		13.7	19.7	25.5	36.8	47.6	58.2	78.8	109	156	203	248	292	335	378	421	396	343	0		
26		14.3	20.6	26.6	38.4	49.7	60.7	82.2	113	163	212	259	305	350	395	439	420	364	0		
Lubrication type■		A						B							C						

The dotted line indicates the point at which pin and bushing galling is likely to begin. When the desired selection falls within the shaded area, consult Link-Belt Chain Division for guidance.

▲ Ratings are based on a service factor of 1.
The ratings tabled above apply directly to lubricated, single strand, standard and heavy series roller chains. For multiple strand chains, apply the factors shown in the table at right.

■ Type A: Manual or drip (Maximum chain speed 85 FPM)
Type B: Bath or disc (Maximum chain speed 900 FPM)
Type C: Forced (pump)

Multiple strand factors

Number of strands	Multiple strand factor
2	1.7
3	2.5
4	3.3
5	4.1
6	5.0
7 or more	Consult Link-Belt Chain Division

Source: Rexnord Corporation, Link-Belt Roller Chain Operation, Indianapolis, IN.

Table 15 General Dimensions and Horsepower Capacity, No. 240 Roller Chain (ANSI B29.1) 3-in. Pitch

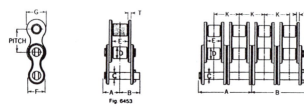

Fig. 6453

Fig. 6447

Specifications and dimensions

Chain number	Chain width, number of strands	Average ultimate strength, pounds	Joint bearing area, sq. in.	Weight per foot, pounds	Dimensions, inches								
					A	B	C	D	E	F	G	K	T
240	Single	152,200	2.488	16.4	1.85	2.20							
240-2	Double	304,400	4.976	32.2	3.58	3.93							
240-3	Triple	456,600	7.464	49.4	5.31	5.75							
240-4	Quadruple	608,800	9.952	65.7	7.04	7.38	.938	1.875	1.88	2.44	2.81	3.458	.38
240-5	Quintuple	761,000	10.440	82.0	8.76	9.11							
240-6	Sextuple	913,200	14.928	98.4	10.50	10.84							

Have dimensions certified for installation purposes.
Available in riveted and cottered construction.

Ratings

Number of teeth in small sprocket	Maximum bore, inches	Horsepower for single strand chain ▲ — RPM of small sprocket																			
		5	10	15	20	25	30	40	50	60	80	100	125	150	175	200	250	300	350	400	450
9	3.250	3.92	7.31	10.5	13.6	16.7	19.6	25.4	31.1	36.7	47.5	58.1	71.0	83.6	96.1	108	132	156	169	138	116
10	4.000	4.39	8.19	11.8	15.3	18.7	22.0	28.5	34.9	41.1	53.2	65.0	79.5	93.7	108	121	148	175	198	162	136
11	4.500	4.86	9.08	13.1	16.9	20.7	24.4	31.6	38.6	45.5	59.0	72.1	88.1	104	119	136	164	194	223	187	156
12	5.500	5.34	9.97	14.4	18.6	22.7	26.8	34.7	42.4	50.0	64.8	79.2	96.8	114	131	148	181	213	245	213	0
13	6.250	5.83	10.9	15.7	20.3	24.8	29.2	37.9	46.3	54.5	70.6	86.4	106	124	143	161	197	232	267	240	0
14	7.875	6.31	11.8	17.0	22.0	26.9	31.7	41.0	50.1	59.1	76.5	93.6	114	135	155	175	213	251	289	268	0
15	8.813	6.80	12.7	18.3	23.7	28.9	34.1	44.2	54.0	63.6	82.4	101	123	145	167	188	230	271	311	297	0
16	9.688	7.29	13.6	19.6	25.4	31.0	36.6	47.4	57.9	68.2	88.4	108	132	156	179	202	247	290	334	328	0
17		7.78	14.5	20.9	27.1	33.1	39.0	50.6	61.8	72.9	94.4	115	141	166	191	215	263	310	356	359	0
18		8.28	15.4	22.3	28.8	35.2	41.5	53.8	65.8	77.5	100	123	150	177	203	229	280	330	379	377	0
19		8.78	16.4	23.6	30.6	37.4	44.0	57.0	69.7	82.2	106	130	159	187	215	243	297	350	402	393	0
20		9.28	17.3	24.9	32.3	39.5	46.5	60.3	73.7	86.8	112	138	168	198	228	257	314	370	423	407	0
21		9.78	18.2	26.3	34.1	41.6	49.0	63.5	77.7	91.5	119	145	177	209	240	270	331	390	439	421	0
22		10.3	19.2	27.6	35.8	43.8	51.6	66.8	81.7	96.2	125	152	186	220	252	284	348	410	454	435	0
23		10.8	20.1	29.0	37.6	45.9	54.1	70.1	85.7	101	131	160	195	230	265	298	365	430	469	448	0
24		11.3	21.1	30.4	39.3	48.1	56.7	73.4	89.7	106	137	167	205	241	277	312	382	450	483	0	
25		11.8	22.0	31.7	41.1	50.3	59.2	76.7	93.8	110	143	175	214	252	290	327	399	470	496	0	
26		12.3	23.0	33.1	42.9	52.4	61.8	80.0	97.8	115	149	183	223	263	302	341	416	491	509	0	
Lubrication type ■		A							B								C				

The dotted line indicates the point at which pin and bushing galling is likely to begin. When the desired selection falls within the shaded area, consult Link-Belt Chain Division for guidance.

▲ Ratings are based on a service factor of 1.
   The ratings tabled above apply directly to lubricated, single strand, standard and heavy series roller chains. For multiple strand chains, apply the factors shown in the table at right.

■ Type A: Manual or drip (Maximum chain speed 75 FPM)
   Type B: Bath or disc (Maximum chain speed 800 FPM)
   Type C: Forced (pump)

Multiple strand factors

Number of strands	Multiple strand factor
2	1.7
3	2.5
4	3.3
5	4.1
6	5.0
7 or more	Consult Link-Belt Chain Division

Source: Rexnord Corporation, Link-Belt Roller Chain Operation, Indianapolis, IN.

bushings, and pins, they have lighter weight and offer greater economy than comparable standard roller chains.

In this series, six standard pitch sizes are currently available, the smallest being of 1-in. pitch, "2040" derived from base-pitch No. 40, and the largest in the series being of 3-in. pitch, "2120" derived from the base-pitch chain No. 120. Some sizes are offered in the "heavy" series but not in the standard. General dimensions and horsepower ratings are shown in Tables 16 to 19.

Most manufacturers offer this chain in multiple-stranded assemblies, however, they are not available in standard form. Also, since these chains are more applicable for power drive systems than for conveyance, a limited number of chain attachments are readily available.

C. Double-Pitch Conveyor Roller Chains (ANSI B29.4)

Conveyor series precision roller chains are also manufactured or derive many of their principal dimensions from the same standards as ANSI B29.1—base-pitch roller chains. As stated, drive or transmission series standard roller chains are provided with figure-8-styled link plates. On the other hand, since the *conveyor series* standard roller chains are primarily intended for conveying applications, they are provided with "straight-edge" or straight parallel-sided link plates.

The conveyor series roller chain is available in two styles: those with standard or regular diameter rollers, in which both pin link plates and roller link plates are of consistent height, so that when they are employed in horizontal conveyor runs, they can be supported, by either sliding contact or rails or tracks (this is the chief reason for the straight-sided parallel link plates). In other words, it is assumed that standard rollers will be supplied in a conveyor application that is expected to slide or that is supported by other means. In the second style, those with large-diameter or "carrier" rollers, rollers larger in diameter than the height of the link plates are advantageous, so as to provide rolling action or support along a desired conveyor path, generally horizontal. In theory, rolling friction is typically half, or even less, than the level of sliding friction. Hence, it is practical to use a large-diameter roller series chain, so as to minimize chain *working load* on long or rough conveyor paths.

Double-pitch conveyor roller chains are currently available in seven standard pitch sizes and are offered with the choice of regular or large-diameter rollers in each pitch size. Again, in the same manner as in the double-pitch transmission series, pitch proportions are a direct derivative of base-pitch standard chain numbers. The smallest size is the 1-in. pitch, derived from base-pitch chain number 40, and the largest size is the 4-in. pitch, which is derived from base-pitch chain number 160. Additionally, these chains are offered in the heavy link plate series. Regular or standard-thickness link plates are manufactured in 1-in. and 1¼-in. pitch sizes, while 1½-in. through 4-in.

Table 16 General Dimensions and Horsepower Capacity, No. 2040 Double-Pitch Power-Transmission Roller Chain (ANSI B29.3) 1-in. Pitch

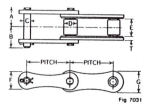

Fig 7031

Specifications and dimensions

Chain number	Average ultimate strength, pounds	Joint bearing area, sq. in.	Weight per foot, pounds	Dimensions, inches							
				A	B	C	D	E	F	G	T
2040	4,000	.068	.28	.32	.38	.156	.313	.31	.39	.45	.06

Have dimensions certified for installation purposes.
Available in riveted or cottered construction.

Ratings

Number of effective teeth in small sprocket	Maximum bore, inches	Horsepower▲ RPM of small sprocket																
		25	50	100	150	200	250	300	400	500	600	700	800	900	1000	1100	1200	1300
11	2.188	.202	.379	.687	.958	1.19	1.41	1.59	1.89	2.14	2.32							
12	2.250	.223	.419	.766	1.07	1.34	1.58	1.81	2.16	2.46	2.71	2.88						
13	2.281	.243	.458	.842	1.18	1.48	1.76	2.00	2.44	2.79	3.08	3.31	3.48					
14	2.625	.263	.497	.914	1.28	1.63	1.93	2.20	2.67	3.09	3.44	3.70	3.91	4.10				
15	2.750	.283	.535	.989	1.39	1.76	2.09	2.40	2.93	3.38	3.77	4.08	4.32	4.52	4.67			
16	3.000	.303	.572	1.06	1.49	1.89	2.25	2.59	3.17	3.67	4.09	4.44	4.73	4.96	5.13			
17	3.438	.322	.611	1.13	1.59	2.02	2.41	2.77	3.41	3.95	4.41	4.80	5.10	5.38	5.57	5.72		
18	3.625	.342	.648	1.20	1.70	2.15	2.57	2.94	3.63	4.21	4.71	5.13	5.48	5.76	5.97	6.15		
19	3.750	.361	.687	1.27	1.80	2.28	2.72	3.14	3.86	4.49	5.02	5.48	5.85	6.17	6.41	6.61	6.70	
20	3.781	.380	.720	1.34	1.90	2.40	2.87	3.29	4.07	4.72	5.29	5.76	6.17	6.50	6.77	6.98	7.13	
21	4.000	.399	.758	1.41	1.99	2.52	3.01	3.47	4.27	4.97	5.57	6.07	6.50	6.86	7.13	7.35	7.50	
22	4.500	.419	.794	1.48	2.08	2.64	3.15	3.63	4.48	5.20	5.83	6.37	6.81	7.18	7.48	7.71	7.87	
23		.437	.829	1.54	2.18	2.76	3.30	3.79	4.68	5.42	6.09	6.64	7.11	7.49	7.80	8.04	8.21	8.30
24		.456	.866	1.60	2.27	2.88	3.44	3.96	4.87	5.67	6.35	6.92	7.40	7.80	8.12	8.37	8.54	8.63
25		.475	.902	1.67	2.36	3.00	3.58	4.11	5.07	5.90	6.60	7.19	7.73	8.10	8.42	8.67	8.84	8.94
30		.568	1.076	1.99	2.81	3.56	4.24	4.86	5.95	6.93	7.76	8.40	8.90	9.38	9.72	9.95	10.09	10.15
35		.657	1.247	2.30	3.24	4.09	4.86	5.56	6.81	7.86	8.71	9.42	9.99	10.43	10.72	10.93	10.97	
40		.748	1.413	2.60	3.65	4.59	5.44	6.22	7.57	8.67	9.60	10.31	10.86	11.23	11.49	11.61		
Lubrication type■		1					2						3					

▲ Ratings based on service factor of 1.

■ Type 1: Manual drip (4 to 10 drops per minute), or splash.
Type 2: Rapid drip (20 drops per minute minute minimum); splash, or disc.
Type 3: Disc or forced.

Source: Rexnord Corporation, Link-Belt Roller Chain Operation, Indianapolis, IN.

Table 17   General Dimensions and Horsepower Capacity, No. 2050 Double-Pitch Power-Transmission Roller Chain (ANSI B29.3) 1¼-in. Pitch

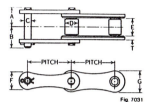

Fig. 7031

Specifications and dimensions

Chain number	Average ultimate strength, pounds	Joint bearing area, sq. in.	Weight per foot, pounds	Dimensions, inches							
				A	B	C	D	E	F	G	T
2050	6,250	.108	.41	.41	.48	.200	.400	.38	.48	.55	.08

Have dimensions certified for installation purposes.
Available in riveted or cottered construction.

Ratings

Number of effective teeth in small sprocket	Maximum bore, inches	Horsepower▲ RPM of small sprocket															
		25	50	100	150	200	250	300	350	400	450	500	550	600	700	800	900
11	2.625	.385	.72	1.29	1.78	2.19	2.56	2.85	3.12	3.33	3.53						
12	2.875	.428	.80	1.44	1.99	2.48	2.90	3.26	3.58	3.86	4.10	4.31					
13	2.938	.467	.87	1.59	2.20	2.74	3.23	3.65	4.03	4.36	4.66	4.91	5.11	5.30			
14	3.250	.506	.95	1.73	2.41	3.01	3.55	4.02	4.45	4.84	5.17	5.48	5.73	5.96			
15	3.563	.544	1.02	1.87	2.61	3.27	3.86	4.39	4.88	5.31	5.68	6.02	6.31	6.57	6.94		
16	3.750	.582	1.09	2.00	2.81	3.52	4.16	4.74	5.26	5.73	6.16	6.55	6.87	7.19	7.61		
17	3.875	.620	1.16	2.14	2.99	3.77	4.46	5.09	5.66	6.17	6.63	7.05	7.42	7.75	8.24	8.62	
18	4.313	.658	1.23	2.27	3.19	4.01	4.75	5.41	6.03	6.58	7.09	7.54	7.94	8.31	8.84	9.28	
19	4.500	.696	1.31	2.41	3.39	4.25	5.05	5.76	6.42	7.00	7.55	8.04	8.46	8.87	9.42	9.90	
20	5.125	.732	1.38	2.54	3.56	4.48	5.32	6.07	6.75	7.38	7.95	8.46	8.92	9.35	9.97	10.49	
21	5.500	.769	1.45	2.66	3.75	4.70	5.59	6.38	7.10	7.77	8.37	8.90	9.39	9.84	10.50	11.06	11.44
22	5.875	.806	1.52	2.79	3.92	4.92	5.86	6.69	7.45	8.14	8.76	9.33	9.84	10.31	11.01	11.59	12.00
23		.842	1.58	2.91	4.09	5.16	6.12	6.98	7.78	8.50	9.15	9.74	10.27	10.76	11.50	12.10	12.52
24		.879	1.65	3.05	4.27	5.37	6.38	7.28	8.10	8.85	9.54	10.16	10.70	11.21	11.97	12.59	13.03
25		.914	1.72	3.17	4.45	5.59	6.62	7.58	8.42	9.20	9.91	10.55	11.12	11.64	12.42	13.05	13.50
30		1.092	2.06	3.77	5.28	6.63	7.84	8.93	9.92	10.82	11.62	12.35	12.99	13.57	14.39	15.06	15.48
35		1.267	2.38	4.35	6.07	7.59	8.96	10.18	11.27	12.26	13.14	13.92	14.59	15.17	16.00	16.62	16.94
40		1.44	2.70	4.91	6.82	8.51	10.00	11.33	12.51	13.57	14.49	15.28	15.95	16.57	17.29	17.78	
Lubrication type■		1			2					3							

▲ Ratings based on service factor of 1.
■ Type 1: Manual drip (4 to 10 drops per minute), or splash.
Type 2: Rapid drip (20 drops per minute minimum); splash, or disc.
Type 3: Disc or forced.

Source: Rexnord Corporation, Link-Belt Roller Chain Operation, Indianapolis, IN.

Table 18 General Dimensions and Horsepower Capacity, No. 2060 Double-Pitch Power-Transmission Roller Chain (ANSI B29.3) 1½-in. Pitch

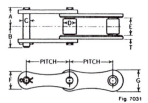

Fig 7031

Specifications and dimensions

Chain number	Average ultimate strength, pounds	Joint bearing area, sq. in.	Weight per foot, pounds	Dimensions, inches							
				A	B	C	D	E	F	G	T
2060	9,000	.162	.65	.50	.60	.234	.469	.50	.60	.71	.13
2060H	9,000	.162	.79	.56	.65	.234	.469	.50	.60	.71	.13

Have dimensions certified for installation purposes.
Available in riveted or cottered construction.

Ratings

Number of effective teeth in small sprocket	Maximum bore, inches	Horsepower▲ RPM of small sprocket															
		25	50	75	100	125	150	200	250	300	350	400	450	500	550	600	650	700
11	3.125	.66	1.21	1.70	2.15	2.54	2.93	3.58	4.12	4.56	4.93							
12	3.563	.73	1.34	1.90	2.41	2.85	3.30	4.05	4.70	5.24	5.71	6.08						
13	3.813	.79	1.48	2.09	2.65	3.15	3.65	4.52	5.27	5.91	6.46	6.92	7.32					
14	4.125	.86	1.60	2.27	2.90	3.45	4.00	4.97	5.79	6.54	7.17	7.72	8.18	8.58				
15	4.500	.92	1.72	2.45	3.14	3.74	4.34	5.39	6.32	7.14	7.86	8.48	9.01	9.48				
16	4.688	.99	1.85	2.64	3.36	4.01	4.66	5.82	6.82	7.73	8.52	9.21	9.80	10.34	10.77			
17	5.250	1.05	1.97	2.82	3.59	4.28	4.98	6.22	7.32	8.29	9.14	9.91	10.56	11.14	11.64	12.06		
18	5.563	1.12	2.10	2.99	3.82	4.56	5.31	6.63	7.82	8.85	9.78	10.60	11.31	11.96	12.50	12.97		
19	5.938	1.18	2.23	3.17	4.05	4.83	5.62	7.03	8.29	9.42	10.41	11.29	12.08	12.76	13.35	13.87	14.30	
20	6.188	1.25	2.34	3.34	4.26	5.09	5.93	7.41	8.74	9.92	10.97	11.91	12.74	13.46	14.08	14.64	15.10	
21	6.438	1.31	2.46	3.51	4.49	5.36	6.24	7.80	9.19	10.43	11.55	12.52	13.40	14.14	14.83	15.42	15.90	
22	7.000	1.37	2.58	3.67	4.70	5.62	6.54	8.16	9.62	10.93	12.08	13.13	14.04	14.84	15.55	16.15	16.67	
23		1.44	2.69	3.83	4.90	5.86	6.83	8.53	10.06	11.42	12.62	13.71	14.67	15.49	16.22	16.87	17.38	17.83
24		1.50	2.80	4.00	5.11	6.11	7.12	8.90	10.47	11.90	13.16	14.28	15.27	16.14	16.89	17.56	18.11	18.57
25		1.56	2.92	4.17	5.32	6.36	7.41	9.27	10.89	12.37	13.67	14.84	15.86	16.76	17.53	18.21	18.79	19.24
30		1.86	3.48	4.96	6.32	7.55	8.78	10.94	12.76	14.55	16.05	17.38	18.54	19.53	20.38	21.11	21.70	22.16
35		2.16	4.03	5.73	7.29	8.67	10.06	12.52	14.67	16.54	18.17	19.61	20.80	21.88	22.73	23.40	23.99	
40		2.45	4.55	6.46	8.20	9.70	11.31	13.99	16.33	18.35	20.08	21.57	22.84	23.86	24.64	25.42		
Lubrication type■		1						2						3				

▲ Ratings based on service factor of 1.

■ Type 1: Manual drip (4 to 10 drops per minute), or splash.
Type 2: Rapid drip (20 drops per minute minute minimum); splash, or disc.
Type 3: Disc or forced.

Source: Rexnord Corporation, Link-Belt Roller Chain Operation, Indianapolis, IN.

Table 19 General Dimensions and Horsepower Capacity, No. 2080 Double-Pitch Power-Transmission Roller Chain (ANSI B29.3) 2-in. Pitch

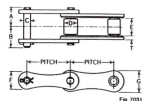

Fig. 7031

Specifications and dimensions

Chain number	Average ultimate strength, pounds	Joint bearing area, sq. in.	Weight per foot, pounds	Dimensions, inches							
				A	B	C	D	E	F	G	T
2080	16,000	.275	1.10	.63	.74	.313	.625	.63	.75	.91	.13

Have dimensions certified for installation purposes.
Available in riveted or cottered construction.

Ratings

Number of effective teeth in small sprocket	Maximum bore, inches	Horsepower▲															
		RPM of small sprocket															
		10	20	30	40	50	60	70	80	90	100	150	200	250	300	350	400	450
11	4.250	.66	1.24	1.78	2.26	2.76	3.20	3.60	3.99	4.38	4.78	6.36	7.60					
12	4.625	.72	1.37	1.96	2.52	3.08	3.56	4.03	4.48	4.92	5.36	7.20	8.68	9.82				
13	4.938	.79	1.49	2.15	2.77	3.36	3.91	4.44	4.95	5.45	5.93	8.02	9.73	11.08				
14	5.250	.85	1.62	2.33	3.01	3.66	4.26	4.85	5.42	5.96	6.49	8.82	10.75	12.29	13.60			
15	5.688	.91	1.74	2.52	3.25	3.95	4.60	5.25	5.86	6.45	7.03	9.60	11.74	13.46	14.94			
16	6.188	.98	1.87	2.70	3.48	4.24	4.94	5.64	6.29	6.93	7.56	10.36	12.70	14.59	16.24	17.65		
17	7.375	1.04	1.99	2.88	3.71	4.52	5.28	6.02	6.72	7.40	8.09	11.10	13.63	15.69	17.50	19.04		
18	7.938	1.11	2.11	3.05	3.94	4.80	5.61	6.40	7.14	7.87	8.60	11.82	14.53	16.76	18.72	20.38	21.77	
19	8.438	1.17	2.23	3.23	4.17	5.09	5.94	6.77	7.56	8.33	9.10	12.52	15.40	17.80	19.90	21.67	23.18	
20	9.313	1.23	2.35	3.40	4.40	5.36	6.26	7.13	7.98	8.78	9.60	13.20	16.25	18.81	21.04	22.91	24.52	
21	9.563	1.29	2.47	3.57	4.62	5.62	6.58	7.49	8.39	9.23	10.09	13.87	17.08	19.79	22.14	24.11	25.80	
22	9.813	1.36	2.58	3.74	4.84	5.90	6.89	7.84	8.79	9.67	10.57	14.53	17.90	20.74	23.20	25.27	27.03	
23		1.42	2.70	3.90	5.06	6.16	7.20	8.19	9.18	10.10	11.05	15.18	18.71	21.66	24.23	26.40	28.22	
24		1.48	2.82	4.05	5.27	6.43	7.51	8.54	9.56	10.53	11.52	15.82	19.51	22.55	25.23	27.50	29.38	30.98
25		1.54	2.93	4.20	5.48	6.69	7.81	8.89	9.94	10.95	11.98	16.45	20.30	23.42	26.20	28.57	30.52	32.16
30		1.84	3.50	5.02	6.54	7.96	9.29	10.59	11.74	12.97	14.23	19.46	23.91	27.52	30.70	33.56	35.52	37.26
35		2.14	4.07	5.82	7.56	9.19	10.71	12.21	13.48	14.92	16.35	22.26	27.23	31.21	34.65	37.57	39.66	
40		2.43	4.61	6.60	8.55	10.38	12.09	13.76	15.17	16.80	18.36	24.88	30.28	34.52	38.09	40.96	43.07	
Lubrication type■		1										2			3			

▲ Ratings based on service factor of 1.
■ Type 1: Manual drip (4 to 10 drops per minute), or splash.
Type 2: Rapid drip (20 drops per minute minute minimum); splash, or disc.
Type 3: Disc or forced.

Source: Rexnord Corporation, Link-Belt Roller Chain Operation, Indianapolis, IN.

pitch sizes are offered with the option of heavy series link plates. General dimensions and chain sizes are shown in Tables 20 and 21. Nomenclature for these chains is as follows: conveyor series same as drive series, except use prefix "C" (e.g., C2040). Add right-hand digit "2" to designate carrier roller chain (e.g., C2042). Add suffix "H" to designate heavy series (e.g., C2060H).

Carrier roller chains are identical to standard conveyor series double-pitch roller chains, with the exception of having larger-diameter rollers. This is what classifies them as carrier roller chain. These chains come with standard steel rollers but are also available in thermoplastic-type rollers.

In this design, the standard steel roller is replaced by a thermoplastic roller. The roller diameter and width are the same as those of the steel roller it replaces. Nomenclature is as follows: use ANSI standard double-pitch number, make right-hand digit 2, and add suffix "D" (for Delrin, a registered trademark thermoplastic) (e.g., C2042D). The suffix "D," is not an ANSI standard desig-

Table 20　General Dimensions, Double-Pitch Conveyor Roller Chain (ANSI B29.4)

Standard diameter steel rollers

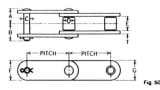

Fig. 5074

Carbon steel

Chain number	Chain pitch, inches	Average ultimate strength, pounds	Joint bearing area, square inches	Weight per foot, pounds	Dimensions, inches								Sprocket Data, Pages
					A	B	C	D	E	F	G	T	
C 2040	1.000	4,000	.068	.32	.32	.38	.156	.313	.31	.47	.47	.06	E-10
C 2050	1.250	6,250	.108	.51	.41	.48	.200	.400	.38	.59	.59	.08	E-12
C 2060H	1.500	9,000	.176	1.05	.56	.65	.234	.469	.50	.68	.68	.13	E-14
C 2080H	2.000	16,000	.295	1.67	.69	.81	.313	.625	.63	.91	.91	.16	E-16
C 2100H	2.500	26,500	.427	2.55	.83	.95	.375	.750	.75	1.13	1.13	.19	E-18
C 2120H	3.000	37,000	.636	3.57	1.02	1.19	.438	.875	1.00	1.38	1.38	.22	E-20
C 2160H	4.000	68,000	1.014	6.18	1.30	1.48	.562	1.125	1.25	1.87	1.87	.28	E-22

Stainless steel

Chain number	Chain pitch, inches	Average ultimate strength, pounds	Joint bearing area, square inches	Weight per foot, pounds	Dimensions, inches								Sprocket Data, Pages
					A	B	C	D	E	F	G	T	
C 2040 SN	1.000	3,000	.068	.32	.32	.38	.156	.313	.31	.45	.45	.06	E-10
C 2050 SN	1.250	4,700	.108	.51	.41	.48	.200	.400	.38	.55	.55	.08	E-12
C 2060H SN	1.500	6,750	.176	1.05	.56	.65	.234	.469	.50	.71	.71	.13	E-14
C 2080H SN	2.000	12,000	.295	1.67	.69	.81	.313	.625	.63	.91	.91	.16	E-16
C 2100H SN	2.500	18,750	.427	2.55	.83	.95	.375	.750	.75	1.13	1.13	.19	E-18
C 2120H SN	3.000	27,000	.636	3.57	1.02	1.19	.438	.875	1.00	1.38	1.38	.22	E-20
C 2160H SN	4.000	47,000	1.014	6.18	1.30	1.48	.562	1.125	1.25	1.87	1.87	.28	E-22

Source: Rexnord Corporation, Link-Belt Roller Chain Operation, Indianapolis, IN.

Table 21 General Dimensions, Double-Pitch Conveyor Roller Chain (ANSI B29.4)

Large diameter steel or Delrin rollers

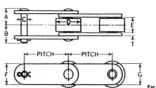

Carbon steel

Fig. 4383

Chain number	Chain pitch, inches	Average ultimate strength, pounds	Joint bearing area, square inches	Weight per foot, pounds	Dimensions, inches							
					A	B	C	D	E	F	G	T
C 2042	1.000	4,000	.068	.56	.32	.38	.156	.625	.31	.47	.47	.06
C 2052	1.250	6,250	.108	.83	.41	.48	.200	.750	.38	.59	.59	.08
C 2062H	1.500	9,000	.176	1.55	.56	.65	.234	.875	.50	.68	.68	.13
C 2082H	2.000	16,000	.295	2.35	.69	.81	.313	1.125	.63	.91	.91	.16
C 2102H	2.500	26,500	.427	3.85	.83	.95	.375	1.563	.75	1.13	1.13	.19
C 2122H	3.000	37,000	.636	6.22	1.02	1.19	.438	1.750	1.00	1.38	1.38	.22
C 2162H	4.000	68,000	1.014	9.34	1.30	1.48	.562	2.250	1.25	1.87	1.87	.28

Stainless steel

Chain number	Chain pitch, inches	Average ultimate strength, pounds	Joint bearing area, square inches	Weight per foot, pounds	Dimensions, inches							
					A	B	C	D	E	F	G	T
C 2042 SN	1.000	3,000	.068	.56	.32	.38	.156	.625	.31	.47	.47	.06
C 2052 SN	1.250	4,700	.108	.83	.41	.48	.200	.750	.38	.59	.59	.08
C 2062H SN	1.50	6,750	.176	1.55	.56	.65	.234	.875	.50	.68	.68	.13
C 2082H SN	2.000	12,000	.295	2.35	.69	.81	.313	1.125	.63	.91	.91	.16
C 2102H SN	2.500	18,750	.427	3.85	.83	.95	.375	1.563	.75	1.13	1.13	.19
C 2122H SN	3.000	27,000	.636	6.22	1.02	1.19	.438	1.750	1.00	1.38	1.38	.22
C 2162H SN	4.000	47,000	1.014	9.34	1.30	1.48	.562	2.250	1.25	1.87	1.87	.28

Source: Rexnord Corporation, Link-Belt Roller Chain Operation, Indianapolis, IN.

nation but rather a typical manufacturer's coding system. It is mentioned only for identification purposes.

Carrier roller chains using thermoplastic rollers are designed to be used: (1) where lubrication cannot be tolerated or is impractical (generally because of limited maintenance conditions); (2) where smooth, quiet operation is desired, because thermoplastic absorbs shock, reduces pulsations, and rolls quietly in conveyor runways and when engaging sprocket teeth; and (3) where long life is required—thermoplastic rollers will have less build up of foreign materials on them.

The weight of these chains is generally one-third less than that of comparable "all-steel" carrier roller chains. This means easier handling and, on some applications, will reduce the weight of the conveyor equipment and possibly the horsepower required.

General maintenance requirements for maximum roller life, include lubricating the chains during the *break-in* period (see manufacturer's recommendations). Also, they are not recommended where the temperature is below 0°F, or continuously above 180°F.

Multiple-strand assemblies are not available in standard form. A broad range of conveyance attachments is readily available for assembly into double-pitch conveyor roller chains. Of course, no "horsepower ratings" tables are available for these chains, since they are primarily intended for conveyor applications rather than power drive usage.

III. ENGINEERING-CLASS CHAINS

A. Steel Detachable-Link Chain (ANSI B29.6)

Detachable-link chains have been used for many years for drive, conveyor, and elevator applications where loads are light and without intermittent shocks, speeds are uniform, and conditions of operation are generally nonabrasive. Undoubtedly, the largest users of this type of chain throughout the years have been the agricultural and lumber industries.

The design of this chain is a fundamental consideration. Steel detachable-link chain is an assembly of identical one-piece formed-steel links that hook and interfit together, so as to form a continuous chain. Essentially, this design allows the end bar of one link to freely articulate inside the hook of the adjacent link (see Table 22).

Detachable chain is easy to assemble and to disassemble (detach). Individual links are inserted from one side, at the proper angle (generally perpendicular). When fully inserted, the link is lowered to the same plane as the rest of the

Table 22   General Dimensions, Steel Detachable-Link Chain (ANSI B29.6)

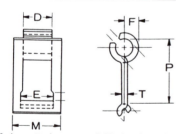

Various attachments available for these sizes

CHAIN NUMBER	NO. OF LINKS IN 10 FEET	MINIMUM TENSILE POUNDS	DIMENSIONS — INCHES		P † Pitch	T
25	133	760	.422	.180	⁴⁹⁄₆₄	.904	.073
32	104	1320	.594	.230	¹⁵⁄₁₆	1.157	.090
32W	104	1320	.594	.232	1 ⁷⁄₁₆	1.157	.095
42	87	1680	.781	.265	1 ⁷⁄₃₂	1.375	.105
50H	87	2240	.781	.280	1 ⁷⁄₃₂	1.375	.125
51	106	1680	.703	.233	1 ³⁄₃₂	1.133	.100
52	80	2160	.844	.303	1¹³⁄₃₂	1.508	.120
55	74	2240	.796	.320	1 ⁹⁄₃₂	1.630	.125
62	73	3520	.984	.335	1 ⁹⁄₁₆	1.654	.148
62H	73	3600	.984	.342	1 ⅝	1.654	.155
67H	52	4400	1.093	.448	1 ⅞	2.313	.185
67W	52	3600	1.093	.428	2 ¾	2.313	.155
67XH	52	5000	1.093	.448	1 ⅞	2.313	.200

Source: Drives Inc., Fulton, IL.

chain and becomes an interlock chain segment until it is again raised to the proper angle and detached.

Detachable chain operates with the closed side of the hook riding next to the sprocket wheel. For drive applications, the "direction of travel" is in the direction of the hook; for conveyor and elevator applications, the direction of travel is in the direction of the end bar.

There are currently 15 standard sizes available, ranging in pitch size from 0.904 to 2.313 in. A variety of attachments are available for the most popular sizes.

B. Leaf Chain (ANSI B29.8)

Leaf chain, sometimes called "cable" or "balance" chain, generally has greater tensile strength than standard roller chains and is therefore ideally used for *tension linkage* applications. Leaf chain is constructed of figure-8-shaped interlaced link plates held together by riveted pins. There are no bushings or rollers incorporated in this design. Link plates are generally medium-carbon steel, heat-treated, and pins are case-hardened for toughness. Leaf chain is, undoubtedly, a by-product derivative of base-pitch B29.1 precision roller chains.

Leaf chain standard numbering system consists of both odd and even lacings (plate configurations). Its nomenclature indicates type, pitch, and lacing: A "BL" prefix (type B, heavy series) is used followed by three of four digits. The last two digits indicate the lacing combination (2 × 2, 3 × 4, 6 × 6, etc.), and the digit(s) to the left indicates pitch in 1/8-in. increments. For example, a "BL534" leaf chain would be a heavy series 5/8-in. pitch with alternate links having three- and four-link plates. There are currently 10 pitch sizes available, ranging from ½ to 2½-in. (see Table 23). In the past there was an "AL" or light series, which consisted of link plates having regular thickness. However, in the 1977 revision of ANSI B29.8 this light series was dropped (perhaps because of current trends in leaf chain usage and new designs), and now all lacing patterns are of the BL type.

Leaf chain is not usually applied in endless strands, as it has no provision for sprocket tooth engagement (the only type of chain in the B29 series) but rather its changes of direction articulate over plain sheaves rollers. Hence, force is applied to the ends through *clevises* or other means of attachment through the chain pins.

They are designed for practically any application requiring a strong, flexible linkage for transmitting reciprocating motion (motion back and forth alternately) or lift, rather than rotative power. Leaf chains are especially well suited for use on hydraulic fork-life trucks, as counterweight chains for machine tools, elevators, and oven doors, and other similar lifting or balancing applications.

Table 23 General Dimensions, Leaf Chain (ANSI B29.8)

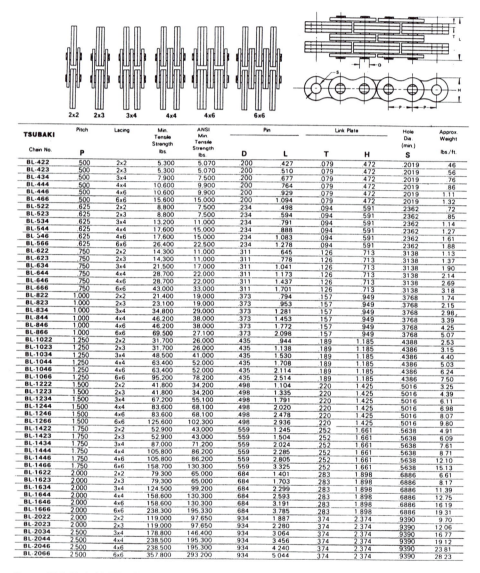

TSUBAKI Chain No.	Pitch P	Lacing	Min. Tensile Strength lbs.	ANSI Min. Tensile Strength lbs.	Pin D	Pin L	Link Plate T	Link Plate H	Hole Dia. (min.) S	Approx. Weight lbs./ft.
BL-422	.500	2x2	5.300	5.070	.200	.427	.079	.472	.2019	.46
BL-423	.500	2x3	5.300	5.070	.200	.510	.079	.472	.2019	.56
BL-434	.500	3x4	7.900	7.500	.200	.677	.079	.472	.2019	.76
BL-444	.500	4x4	10.600	9.900	.200	.764	.079	.472	.2019	.86
BL-446	.500	4x6	10.600	9.900	.200	.929	.079	.472	.2019	1.11
BL-466	.500	6x6	15.600	15.000	.200	1.094	.079	.472	.2019	1.32
BL-522	.625	2x2	8.800	7.500	.234	.498	.094	.591	.2362	.72
BL-523	.625	2x3	8.800	7.500	.234	.594	.094	.591	.2362	.85
BL-534	.625	3x4	13.200	11.000	.234	.791	.094	.591	.2362	1.14
BL-544	.625	4x4	17.600	15.000	.234	.888	.094	.591	.2362	1.27
BL-546	.625	4x6	17.600	15.000	.234	1.083	.094	.591	.2362	1.61
BL-566	.625	6x6	26.400	22.500	.234	1.278	.094	.591	.2362	1.88
BL-622	.750	2x2	14.300	11.000	.311	.645	.126	.713	.3138	1.13
BL-623	.750	2x3	14.300	11.000	.311	.778	.126	.713	.3138	1.37
BL-634	.750	3x4	21.500	17.000	.311	1.041	.126	.713	.3138	1.90
BL-644	.750	4x4	28.700	22.000	.311	1.173	.126	.713	.3138	2.14
BL-646	.750	4x6	28.700	22.000	.311	1.437	.126	.713	.3138	2.69
BL-666	.750	6x6	43.000	33.000	.311	1.701	.126	.713	.3138	3.18
BL-822	1.000	2x2	21.400	19.000	.373	.794	.157	.949	.3768	1.74
BL-823	1.000	2x3	23.100	19.000	.373	.953	.157	.949	.3768	2.15
BL-834	1.000	3x4	34.800	29.000	.373	1.281	.157	.949	.3768	2.98
BL-844	1.000	4x4	46.200	38.000	.373	1.453	.157	.949	.3768	3.39
BL-846	1.000	4x6	46.200	38.000	.373	1.772	.157	.949	.3768	4.25
BL-866	1.000	6x6	69.500	27.100	.373	2.098	.157	.949	.3768	5.07
BL-1022	1.250	2x2	31.700	26.000	.435	.944	.189	1.185	.4388	2.53
BL-1023	1.250	2x3	31.700	26.000	.435	1.138	.189	1.185	.4386	3.15
BL-1034	1.250	3x4	48.500	41.000	.435	1.530	.189	1.185	.4386	4.40
BL-1044	1.250	4x4	63.400	52.000	.435	1.708	.189	1.185	.4386	5.03
BL-1046	1.250	4x6	63.400	52.000	.435	2.114	.189	1.185	.4386	6.24
BL-1066	1.250	6x6	95.200	78.200	.435	2.514	.189	1.185	.4386	7.50
BL-1222	1.500	2x2	41.800	34.200	.498	1.104	.220	1.425	.5016	3.25
BL-1223	1.500	2x3	41.800	34.200	.498	1.335	.220	1.425	.5016	4.39
BL-1234	1.500	3x4	67.200	55.100	.498	1.791	.220	1.425	.5016	6.11
BL-1244	1.500	4x4	83.600	68.100	.498	2.020	.220	1.425	.5016	6.98
BL-1246	1.500	4x6	83.600	68.100	.498	2.478	.220	1.425	.5016	8.07
BL-1266	1.500	6x6	125.600	102.300	.498	2.936	.220	1.425	.5016	9.80
BL-1422	1.750	2x2	52.900	43.000	.559	1.245	.252	1.661	.5638	4.91
BL-1423	1.750	2x3	52.900	43.000	.559	1.504	.252	1.661	.5638	6.09
BL-1434	1.750	3x4	87.000	71.200	.559	2.024	.252	1.661	.5638	7.61
BL-1444	1.750	4x4	105.800	86.200	.559	2.285	.252	1.661	.5638	8.71
BL-1446	1.750	4x6	105.800	86.200	.559	2.805	.252	1.661	.5638	12.10
BL-1466	1.750	6x6	158.700	130.300	.559	3.325	.252	1.661	.5638	15.13
BL-1622	2.000	2x2	79.300	65.000	.684	1.401	.283	1.898	.6886	6.61
BL-1623	2.000	2x3	79.300	65.000	.684	1.703	.283	1.898	.6886	8.17
BL-1634	2.000	3x4	124.500	99.200	.684	2.299	.283	1.898	.6886	11.39
BL-1644	2.000	4x4	158.600	130.300	.684	2.593	.283	1.898	.6886	12.75
BL-1646	2.000	4x6	158.600	130.300	.684	3.191	.283	1.898	.6886	16.19
BL-1666	2.000	6x6	238.300	195.330	.684	3.785	.283	1.898	.6886	19.31
BL-2022	2.000	2x2	119.000	97.650	.934	1.887	.374	2.374	.9390	9.70
BL-2023	2.000	2x3	119.000	97.650	.934	2.280	.374	2.374	.9390	12.06
BL-2034	2.500	3x4	178.800	146.400	.934	3.064	.374	2.374	.9390	16.77
BL-2044	2.500	4x4	238.500	195.300	.934	3.456	.374	2.374	.9390	19.12
BL-2046	2.500	4x6	238.500	195.300	.934	4.240	.374	2.374	.9390	23.81
BL-2066	2.500	6x6	357.800	293.200	.934	5.044	.374	2.374	.9390	28.23

Source: U.S. Tsubaki, Wheeling, IL.

C. Heavy-Duty Offset-Sidebar Roller Chain (ANSI B29.10)

Offset drive chains are designed for the heaviest applications requiring power transmission such as construction machinery, conveyors, and power drives; and they operate under the most severe conditions at moderately high speeds. Offset-sidebar power-transmission roller chain is an assembly of identical interfitting links with heavily walled or extra-thick sidebars, which are formed so as to provide one wide end and one narrow end (offset) for each link (see Fig. 3). Some manufacturers offer this chain in a "solid bushed" design rather than the typical roller/bushing design, which, in effect, allows the pin diameter to be increased, providing a larger cross section for higher ultimate strength and an increase in bearing area for greater working load.

These rugged, all-steel chains are suitable for a wide variety of drive applications where difficult operating conditions prevail. Hence, they are commonly used for the propel and crowd drives on cranes and power shovels, for drum drives on transit mixer trucks, etc., and, depending upon which chain size is selected, are quite capable of transmitting as high as 500 horsepower. See Tables 24 to 31 for horsepower ratings. There are currently eight standard pitch sizes available, ranging from 2½ to 7 in. Multiple-strand assemblies are not available as standard products.

D. Combination Chain (ANSI B29.11)

These chains are constructed of an alternating assembly of cast one-piece center links, or inner block links (generally malleable iron), and outer links, or pin links, that are made of sturdy steel sidebars, or link plates and pins. Hence, the term *combination chain* is appropriately applied.

An interesting optional feature (not in the standard) that most manufacturers offer in some or all chain sizes is the employment of accurately proportioned,

Figure 3 Typical heavy-duty offset-sidebar roller chain. (Courtesy of American Chain Association, 1982.)

Table 24 Horsepower Capacity, No. 2010 (ANSI B29.10) 2.500-in. Pitch

No. of Teeth	Horsepower capacity RPM												
	2	3	7	10	20	30	40	100	200	250	350	450	600
9	1.1	1.4	2.7	3.9	7.7	11.6	15.4	38.6	77.2	96.5	135.1	100.1	65.0
10	1.1	1.5	3.0	4.3	8.6	12.9	17.2	42.9	85.8	107.3	150.2	117.2	76.1
11	1.2	1.7	3.3	4.7	9.4	14.2	18.9	47.2	94.4	118.0	165.2	135.2	87.8
12	1.3	1.8	3.6	5.1	10.3	15.4	20.6	51.5	103.0	128.7	180.2	154.1	100.1
13	1.4	1.9	3.9	5.6	11.2	16.7	22.3	55.8	111.5	139.4	195.2	173.7	112.8
14	1.5	2.0	4.2	6.0	12.0	18.0	24.0	60.1	120.1	150.2	210.2	194.2	126.1
15	1.5	2.1	4.5	6.4	12.9	19.3	25.7	64.4	128.7	160.9	225.2	215.3	139.9
16	1.6	2.2	4.8	6.9	13.7	20.6	27.5	68.6	137.3	171.6	240.3	237.2	154.1
17	1.7	2.3	5.1	7.3	14.6	21.9	29.2	72.9	145.9	182.3	255.3	259.8	168.8
18	1.8	2.4	5.4	7.7	15.4	23.2	30.9	77.2	154.5	193.1	270.3	283.1	183.9
19	1.9	2.5	5.7	8.2	16.3	24.5	32.6	81.5	163.0	203.8	285.3	307.0	–
20	1.9	2.6	6.0	8.6	17.2	25.7	34.3	85.8	171.6	214.5	300.3	331.5	–
21	2.0	2.7	6.3	9.0	18.0	27.0	36.0	90.1	180.2	225.2	315.3	356.7	–
22	2.1	2.8	6.6	9.4	18.9	28.3	37.8	94.4	188.8	236.0	330.4	382.5	–
23	2.1	3.0	6.9	9.9	19.7	29.6	39.5	98.7	197.4	246.7	345.4	405.3	–
24	2.2	3.1	7.2	10.3	20.6	30.9	41.2	103.0	205.9	257.4	360.4	414.4	–
	Manual lubrication					Oil bath				Oil stream lubrication			

Source: Jeffrey Chain Corporation, Morristown, TN.

Table 25 Horsepower Capacity, No. 2512 (ANSI B29.10) 3.067-in. Pitch

No. of Teeth	Horsepower capacity												
	RPM												
	1	3	6	10	20	40	100	150	200	250	300	350	400
9	1.0	2.4	4.0	6.4	12.7	25.5	63.7	95.6	127.4	159.3	191.1	171.8	140.6
10	1.1	2.6	4.3	7.1	14.2	28.3	70.8	106.2	141.6	177.0	212.4	198.9	164.7
11	1.2	2.7	4.7	7.8	15.6	31.1	77.9	116.8	155.7	194.7	231.3	215.5	190.0
12	1.3	2.9	5.1	8.5	17.0	34.0	85.0	127.4	169.9	212.4	248.6	231.5	216.5
13	1.4	3.1	5.5	9.2	18.4	36.8	92.0	138.0	184.1	230.1	265.3	247.0	232.3
14	1.4	3.3	5.9	9.9	19.8	39.6	99.1	148.7	198.2	247.8	281.4	262.1	246.4
15	1.5	3.5	6.4	10.6	21.2	42.5	106.2	159.3	212.4	265.5	296.9	276.6	260.0
16	1.6	3.7	6.8	11.3	22.7	45.3	113.3	169.9	226.5	283.2	312.0	290.6	273.2
17	1.7	3.8	7.2	12.0	24.1	48.1	120.3	180.5	240.7	300.9	326.5	304.1	285.9
18	1.7	4.0	7.6	12.7	25.5	51.0	127.4	191.1	245.9	318.6	340.5	317.1	–
19	1.8	4.2	8.1	13.5	26.9	53.8	134.5	201.8	269.0	336.3	354.0	329.7	–
20	1.9	4.3	8.5	14.2	28.3	56.6	141.6	212.4	283.2	354.0	367.1	341.9	–
21	1.9	4.5	8.9	14.9	29.7	59.5	148.7	223.0	297.3	371.7	379.2	353.0	–
22	2.0	4.7	9.3	15.6	31.1	62.3	155.7	233.6	311.5	389.4	391.7	364.8	–
23	2.1	4.9	9.8	16.3	32.6	65.1	162.8	244.2	325.6	407.1	403.4	375.7	–
24	2.2	5.1	10.2	17.0	34.0	68.0	169.9	254.9	339.8	424.8	414.6	386.1	–
	Manual lubrication					Oil bath				Oil stream lubrication			

Source: Jeffrey Chain Corporation, Morristown, TN.

Table 26 Horsepower Capacity, No. 2814 (ANSI B29.10) 3.500-in Pitch

No. of Teeth	RPM												
	1	3	6	10	20	35	60	100	125	150	200	250	300
9	1.4	3.3	5.5	8.8	17.6	30.8	52.8	88.1	110.1	132.1	176.1	178.7	170.8
10	1.5	3.5	6.0	9.8	19.6	34.2	58.7	97.8	122.3	146.8	145.7	196.1	187.4
11	1.6	3.8	6.5	10.8	21.5	37.7	64.6	107.6	134.5	161.4	215.2	213.0	203.6
12	1.8	4.1	7.0	11.7	23.5	41.1	70.4	117.4	146.8	176.1	234.8	229.5	219.4
13	1.9	4.3	7.6	12.7	25.4	44.5	76.3	127.2	159.0	190.8	254.4	245.6	234.7
14	2.0	4.6	8.2	13.7	27.4	47.9	82.2	137.0	171.2	205.5	273.9	261.2	249.6
15	2.1	4.8	8.8	14.7	29.4	51.4	88.1	146.8	183.4	220.1	292.1	276.3	264.1
16	2.2	5.1	9.4	15.7	31.3	54.8	93.9	156.5	195.7	234.8	307.7	291.1	278.2
17	2.3	5.3	10.0	16.6	33.3	58.2	99.8	166.3	207.9	249.5	322.8	305.5	–
18	2.4	5.5	10.6	17.6	35.2	61.6	105.7	176.1	220.1	264.2	337.6	319.4	–
19	2.5	5.8	11.2	18.6	37.2	65.1	115.5	185.9	232.4	278.8	351.9	333.0	–
20	2.6	6.0	11.7	19.6	39.1	68.5	117.4	195.7	244.6	293.5	365.8	346.1	–
21	2.7	6.2	12.3	20.5	41.1	71.9	123.3	205.5	256.8	308.2	379.3	358.9	–
	Manual lubrication						Oil bath				Oil stream lubrication		

Source: Jeffrey Chain Corporation, Morristown, TN.

Table 27 Horsepower Capacity, No. 3315 (ANSI B29.10) 4.073-in. Pitch

No. of Teeth	Horsepower capacity RPM												
	1	3	6	10	20	30	40	65	80	100	125	150	200
9	2.0	4.7	8.0	12.8	25.5	38.3	51.1	83.0	102.1	127.7	159.6	168.2	166.3
10	2.2	5.1	8.7	14.2	28.4	42.6	56.7	92.2	113.5	141.8	177.3	185.0	182.9
11	2.4	5.5	9.4	15.6	31.2	46.8	62.4	101.4	124.8	156.0	195.0	201.5	199.2
12	2.5	5.9	10.2	17.0	34.0	51.1	68.1	110.6	136.2	170.2	212.8	217.6	215.1
13	2.7	6.3	11.1	18.4	36.9	55.3	73.8	119.9	147.5	184.4	230.5	233.4	230.7
14	2.9	6.6	11.9	19.9	39.7	59.6	79.4	129.1	158.9	198.6	248.2	248.8	246.0
15	3.0	7.0	12.8	21.3	42.6	63.8	85.1	138.3	170.2	212.8	265.9	263.9	261.0
16	3.2	7.3	13.6	22.7	45.4	68.1	90.8	147.5	181.6	227.0	280.7	278.7	275.6
17	3.3	7.7	14.5	24.1	48.2	72.3	96.5	156.7	192.9	241.1	295.3	293.2	289.9
18	3.5	8.0	15.3	25.5	51.1	76.6	102.1	165.0	204.3	253.3	309.5	307.3	303.9
19	3.6	8.4	16.2	27.0	53.9	80.9	107.8	175.2	215.6	269.5	323.5	321.2	317.6
20	3.8	8.7	17.0	28.4	56.7	85.4	113.5	184.4	227.0	283.7	337.1	334.7	–
21	3.9	9.0	17.9	29.8	59.6	89.4	119.2	193.6	238.3	297.9	350.5	347.9	–
	Manual lubrication				Oil bath					Oil stream lubrication			

Source: Jeffrey Chain Corporation, Morristown, TN.

Table 28 Horsepower Capacity, No. 3618 (ANSI B29.10) 4.500-in. Pitch

No. of Teeth	Horsepower capacity RPM												
	1	3	6	10	20	30	35	50	65	80	100	125	150
9	2.6	6.0	10.2	16.3	32.6	48.9	57.0	81.5	105.9	130.4	153.8	156.6	158.8
10	2.8	6.5	11.1	18.1	36.2	54.3	63.4	90.5	117.7	144.9	169.5	172.5	175.0
11	3.0	7.0	12.0	19.9	39.8	59.8	69.7	99.6	129.5	159.4	184.8	188.1	190.8
12	3.3	7.5	13.0	21.7	43.5	65.2	76.1	108.7	141.3	173.9	199.8	203.4	206.3
13	3.5	8.0	14.1	23.5	47.1	70.6	82.4	117.7	153.0	188.3	214.6	218.4	221.6
14	3.7	8.5	15.2	25.4	50.7	76.1	88.7	126.8	164.8	202.8	229.1	233.2	236.6
15	3.9	8.9	16.3	27.2	54.3	81.5	95.1	135.8	176.6	217.3	243.4	247.7	251.3
16	4.1	9.4	17.4	29.0	58.0	86.9	101.4	144.9	188.3	231.8	257.4	261.9	265.7
17	4.2	9.8	18.5	30.8	61.6	92.4	107.8	153.9	200.1	246.3	271.1	275.9	279.9
18	4.4	10.2	19.6	32.6	65.2	97.8	114.1	163.0	211.9	260.8	284.6	289.6	293.8
19	4.6	10.7	20.6	34.4	68.8	103.2	120.4	172.0	223.7	275.3	297.8	303.1	307.5
20	4.8	11.1	21.7	36.2	72.4	108.7	126.8	181.1	235.4	289.8	310.7	316.3	320.9
21	5.0	11.5	22.8	38.0	76.1	114.1	133.1	190.1	247.2	304.2	323.5	329.2	334.0
	Manual lubrication					Oil bath					Oil stream lubrication		

Source: Jeffrey Chain Corporation, Morristown, TN.

Table 29 Horsepower Capacity, No. 4020 (ANSI B29.10) 5.000-in. Pitch

No. of Teeth	Horsepower capacity												
	RPM												
	0.5	1.0	3	6	10	20	30	35	50	65	80	100	125
9	2.0	3.4	7.8	13.3	21.1	42.2	63.3	73.8	105.5	139.9	139.3	145.3	151.6
10	2.2	3.7	8.5	14.4	23.4	46.9	70.3	82.0	117.2	147.6	153.6	160.2	–
11	2.3	3.9	9.1	15.5	25.8	51.6	77.4	90.3	128.9	161.2	167.7	174.9	–
12	2.5	4.2	9.7	16.9	28.1	56.3	84.4	98.5	140.7	174.5	181.6	189.4	–
13	2.6	4.5	10.3	18.3	30.5	61.0	91.4	106.7	152.4	187.7	195.2	203.7	–
14	2.8	4.7	10.9	19.7	32.8	65.6	98.5	114.9	164.1	200.6	208.7	217.7	–
15	2.9	5.0	11.5	21.1	35.2	70.3	105.5	123.1	175.8	213.4	222.0	231.6	–
16	3.1	5.2	12.1	22.5	37.5	75.0	112.5	131.3	187.5	225.9	235.0	245.2	–
17	3.2	5.5	12.7	23.9	39.9	79.7	119.6	139.5	199.3	238.2	247.8	258.6	–
18	3.4	5.7	3.3	25.3	42.2	84.4	126.6	147.7	211.0	250.4	260.5	271.7	–
	Manual lubrication						Oil bath				Oil stream lubrication		

Source: Jeffrey Chain Corporation, Morristown, TN.

Table 30 Horsepower Capacity, No. 4824 (ANSI B29.10) 6.000-in. Pitch

No. of Teeth	Horsepower capacity RPM												
	0.5	1.0	3	6	10	20	30	35	40	45	50	60	70
9	3.1	5.3	12.2	20.7	30.3	66.0	96.1	101.5	106.3	110.8	115.0	122.6	129.0
10	3.4	5.7	13.2	22.4	36.6	73.3	106.2	112.1	117.5	122.5	127.1	135.5	–
11	3.6	6.2	14.2	24.2	40.3	80.6	116.1	122.6	128.5	133.9	139.0	148.2	–
12	3.9	6.6	15.2	26.4	44.0	87.9	126.0	133.0	139.4	145.3	150.8	160.8	–
13	4.1	7.0	16.2	28.6	47.6	95.3	135.7	143.2	150.1	156.5	162.4	173.2	–
14	4.4	7.4	17.1	30.8	51.3	102.6	145.3	153.4	160.8	167.6	173.9	185.4	–
15	4.6	7.8	18.0	33.0	55.0	109.9	154.8	163.4	171.3	178.5	185.3	197.5	–
16	4.8	8.2	18.9	35.2	58.6	117.3	164.2	173.3	181.6	189.3	196.5	209.5	–
17	5.1	8.6	19.8	37.4	62.3	124.6	173.4	183.1	191.9	200.0	207.6	221.3	–
18	5.3	9.0	20.7	39.6	66.0	131.9	182.6	192.7	202.0	210.6	218.5	233.0	–
	Manual lubrication						Oil bath					Oil stream	

Source: Jeffrey Chain Corporation, Morristown, TN.

Table 31 Horsepower Capacity, No. 5628 (ANSI B29.10) 7.000-in. Pitch

No. of Teeth	Horsepower capacity RPM													
	0.1	0.5	1	2	4	6	10	15	20	25	30	35	40	45
9	1.3	4.6	7.7	13.1	22.2	30.2	48.1	67.1	76.7	85.0	92.5	99.4	105.7	.0
10	1.4	4.9	8.4	14.2	24.0	32.7	53.5	74.2	84.8	94.0	102.3	109.0	.0	.0
11	1.6	5.3	9.0	15.2	25.9	35.3	58.8	81.2	92.8	103.0	112.0	120.3	.0	.0
12	1.7	5.7	9.6	16.3	27.6	38.5	64.2	88.2	100.8	111.8	121.7	130.7	.0	.0
13	1.8	6.0	10.2	17.3	29.4	41.7	69.5	95.1	108.7	120.6	131.2	140.9	.0	.0
14	1.9	6.4	10.8	18.3	31.1	44.9	74.8	102.0	116.5	129.2	140.6	151.1	.0	.0
15	2.0	6.7	11.4	19.3	32.7	48.1	80.2	108.8	124.3	137.8	150.0	161.1	.0	.0
16	2.1	7.1	12.0	20.3	34.4	51.3	85.5	115.5	132.0	146.4	159.3	171.1	.0	.0
17	2.2	7.4	12.5	21.2	36.4	54.5	90.9	122.2	139.6	154.8	168.5	180.9	.0	.0
18	2.3	7.7	13.1	22.2	38.5	57.7	96.2	128.8	147.1	163.2	177.5	190.7	.0	.0

Manual lubrication   Oil bath

Source: Jeffrey Chain Corporation, Morristown, TN.

solid-type "elliptical" barrels, which in some applications, may be advantageous. In this design extra metal is added to the critical areas where the sprocket-to-chain contact causes most chain wear. Additionally, a type "W" (other manufacturers may have a different coding) combination chain, designed for heavy-duty drag conveyor service, is also offered by some manufacturers. Their principal use is in the forest products industry, for woodyard conveying systems. In this design, the cast center links have wear shoes added to the sides, so as to provide extra surface area for increased resistance to abrasive and sliding wear normally encountered in these types of environments. For more severe conditions, a heavier center link design may be furnished. This gives more surface area (improved sliding resistance) on the link barrels to retard wear on the return run. These chains may be identified by the suffix "W1" or "W2," respectively, or they may not. Therefore, refer to a manufacturers' catalog for proper identification. *Note*: The two examples just given, elliptical barrel and type W drag chain, have been taken out of context. They are not outlined by ANSI standards, but rather by "increased variation developments" designed by various manufacturers and are mentioned in this section for identification purposes only.

Combination chains are used extensively for conveyor and elevator applications where a wide variety of abrasive and nonabrasive materials must be handled, and they are also being used in general industrial assembly conveyors. Because link surfaces are ample enough to resist sliding wear, they are also widely used in drag conveyor applications. Hence, because the inherent design of combination chains involves chain joints that are well protected and have generous pin bearing surfaces, they are widely used on bucket elevators and conveyors handling rough materials, and are ideal for applications in the cement, chemical, paper and pulp, fertilizer, quarrying, mining, and other related industries. Although they are not normally recommended for drive applications, some of the smaller sizes have proven to be quite effective, and they are sometimes used for power transmission service.

There are currently seven standard pitch sizes available, ranging from 1.631 to 6.050 in. They are identified by their common prefix "C," which is in the standard. A variety of attachments are available, for both cast center links and outer sidebars. General dimensions and chain sizes are shown in Table 32.

E. Steel-Bushed Rollerless Chain (ANSI B29.12)

Steel-bushed rollerless chain also called "knuckle chain," is of an all-steel construction, available with both straight and offset sidebar types, the latter not being in the standard. These chains are used exclusively in conveying and elevating operations, rather than in power drives. These rugged, steel chains are ideally used for heavy-duty service on bucket elevators (especially where

Table 32   General Dimensions, Combination Chains (ANSI B29.11)

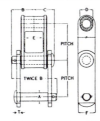

Chain number *	Average pitch, inches	Allow-able chain pull, pounds ▲	Average ultimate strength, pounds ■	Links in approx. 10 feet	Weight per foot, pounds ✕	Dimensions, inches							Attachments available ∗	
						A	B	C	D	E	F	T	Center link attachments	Steel sidebar attachments
C 55☐	1.631	1,110	9,000	74	2.0	.375	1.08	.98	.72	.69	.75	.19	G19, K1	A22
C 77	2.308	1,400	11,000	52	2.2	.435	1.12	1.03	.72	.69	.88	.19	F2, K1	
C 188	2.609	1.950	14,000	46	3.3	.500	1.43	1.23	.88	.94	1.13	.25	A22, F2, G6, G19, K1, K2	G27, K1, K2
C 60	2.307	2,180	19,000	52	3.5	.500½	1.57	1.38	.75	.94	1.00	.25	K1	
C 131	3.075	3,220	24,000	39	6.5	.625	1.90	1.63	1.24	1.14	1.50	.38	F2, G6, G19, K1, K2	K1, K2
C 102B	4.000	4,000	24,000	30	6.3	.625	2.33	2.04	.98	1.92	1.50	.38	K2	K2, S1
C 110	6.000	4,000	24,000	20	6.2	.625	2.33	2.04	1.26	1.92	1.50	.38	K2	K2
C 102½	4.040	5,550	36,000	30	9.7	.750	2.48	2.28	1.38	2.00√P	1.75	.38	F2, G6, K3	K3, S1
C 111	4.760	5,950	36,000	25½	9.8	.750	2.55	2.34	1.42	2.42	1.75	.38	F2, K2	K2, S1
C 132	6.050	8,330	50,000	20	14.5	1.000	3.27	3.06	1.73	3.04	2.00	.50	K2, RF12	K2, S1

C Combination Chains (Type W)

Available in riveted or
cottered construction.
Cottered construction shown.
Riveted construction furnished¹
unless otherwise specified.

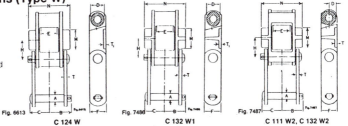

Fig. 6613   C 124 W   Fig. 7486   C 132 W1   Fig. 7487   C 111 W2, C 132 W2

Chain number ∗ +	Average pitch, inches	Allowable chain pull, pounds ▲	Average ultimate strength, pounds ■	Links in approx. 10 feet	Weight per foot, pounds ✕	Dimensions, inches											Attach-ments available ∗
						A	B	C	D	E	F	H	M	N	T	T₁	
C 111 W2	4.760	5,950	36,000	25½	11.8	.750	2.55	2.34	.72	2.42	1.75	2.38	2.50	5.13	.38	.44	S1
C 124 W☐	4.063	6,300	60,000	29½	15.4	.875	2.56	2.31	1.75	1.69	2.25	2.03	1.50	4.44	.50	.38	
C 132 W1	6.050	8,330	50,000	20	15.6	1.000	3.27	3.06	1.73	3.04	2.00	3.03	3.50	6.25	.50	.44	S1
C 132 W2	6.050	8,330	50,000	20	16.0	1.000	3.27	3.06	.87	3.04	2.00	3.03	3.50	6.25	.50	.44	S1

Have dimensions certified for installation purposes.

Source: Rexnord Corporation, Link-Belt Engineered Chain Operation, Morgantown, NC.

centers are widely spaced) or conveyors that must operate under extremely abrasive, gritty, or other unfavorable surroundings.

This class of chain is essentially based on the same dimensional framework or proportions as the combination chains of B29.11, and are therefore very similar in appearance. Like the combination chain, they consist of an alternating assembly of inner links, but with solid bushings; and outer links, or pin links, in which sidebars are fabricated from bar steel, and thus they offer greater ultimate strength than their counterpart. Hence, they are often used to replace combination chains when systems are modified to handle heavier loads or to operate under more severe conditions.

There are currently six standard pitch sizes available, ranging from 2.609 to 6.050 in. A variety of attachment links are available. General dimensions and chain sizes are shown in Table 33.

F. Mill Chain—"H" Type (ANSI B29.14)

Essentially, mill chains are a derivative by-product of the basic cast pintle chain. Therefore, "H"-class mill chain is sometimes referred to as "H"-type pintle chain, depending upon the manufacturer (class and type may be used interchangeably). H-class mill chain is an extremely strong, serviceable chain originally designed for heavy drives and transfer conveyor purposes in sawmills and the paper and pulp industry. In addition to their principal use in the forest products industry, they are also used in many other industrial applications where a heavy, rugged sliding chain is required to function in generally abrasive atmospheres.

The chains are of malleable iron, in which the sidebars of the links are cast with broad, reinforcing wearing shoes, so as to minimize wear and stiffen the links for sliding in troughs or over floors and runways.

Mill chain of this class may operate in two directions. As drive chain, it travels in the direction of the closed barrel; for elevator or conveyor chain applications, it should travel toward the open ends of the links. H-class mill chain is available in pitch sizes ranging from 2.308 to 4.000 in. Various regular and specialized attachments are available. General dimensions and chain sizes are shown in Table 34.

G. Mill Chain—Welded Type (ANSI B29.16)

Welded-steel mill chains are similar to fabricated steel chains except that links are made as integral weldments rather than being held together by means of tight fits and locking surfaces. More precisely, mill chain is a consecutive assembly of identical welded-steel offset links connected together with steel pins.

Welded-steel chains provide an excellent alternative to similar cast chains where increased strength, wear life, and dimensional accuracy are demanded.

Table 33 General Dimensions, Steel-Bushed Rollerless Chains (ANSI B29.12)

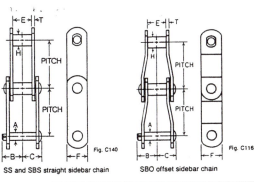

SS and SBS straight sidebar chain   SBO offset sidebar chain

Chain number	Average pitch, inches	Allowable chain pull, pounds	Average ultimate strength, pounds	Links in approx. 10 feet	Weight per foot, pounds	Dimensions, inches							Attachments available
						A	B	C	E	F	H	T	
SS 152	1.506	1,230	9,000	80	2.2	.312	.95	.86	.81	.88	.63	.16	D6, D7
SBS 188	2.609	2,750	25,000	46	3.8	.500	1.43	1.23	1.07	1.13	.88	.25	A22, G19, K1, K2
SBS 131	3.075	4,500	40,000	39	7.4	.625	1.90	1.62	1.29	1.50	1.25	.38	K2
SBS 1972	3.075	4,900	70,000	39	9.2	.625	1.87	1.69	1.50	1.75	1.25	.38	BM55
SBO 2103	3.075	5,000	40,000	39	5.6	.750	1.70	1.33	1.38	1.50	1.25	.25	F29
SBS 2162	3.075	5,600	60,000	39	9.4	.750	1.89	1.69	1.38	1.75	1.25	.38	D4
SBS 110	6.000	6,300	40,000	20	6.3	.625	2.33	2.04	2.13	1.50	1.25	.38	K2
SBS 102B	4.000	6,500	40,000	30	6.9	.625	2.33	2.04	2.13	1.50	1.00	.38	G19, K2
SBS 102½	4.040	7,900	50,000	30	9.4	.750	2.32	2.12	2.25	1.75	1.25	.38	K3
SBS 111	4.760	8,850	50,000	25½	10.2	.750	2.55	2.34	2.63	2.00	1.44	.38	K2
SBS 844	6.000	9,200	70,000	20	10.4	.750	2.72	2.47	2.50	2.00	1.19	.50	K2
SS 2136	•	9,900	170,000	34	21.1	.937	2.56	2.39	1.91	2.38	1.75	.56	D3, BM55
SBS 2236	4.000	9,900	170,000	33	19.2	.936	2.54	2.36	1.91	2.38	1.75	.56	D2
SBS 856	6.000	14,000	100,000	20	13.5	1.000	3.08	2.90	3.00	2.50	1.75	.50	K2, K3, K6, K24, K35
SBS 2857	6.000	14,000	130,000	20	21.0	1.000	3.08	2.90	3.00	3.25	1.75	.50	K44
SBS 150 Plus	6.050	15,100	100,000	20	16.6	1.000	3.29	3.07	3.35	2.50	1.75	.50	K2, K3
SBS 4871	9.000	15,300	130,000	13.3	14.6	1.000	3.23	2.98	3.38	3.00	1.75	.50	D3
SBS 850 Plus	6.000	16,100	200,000	20	25.3	1.313	3.23	2.95	2.25	3.25	2.00	.63	
SBO 850 Plus	6.000	16,100	200,000	20	24.6	1.313	3.23	2.95	2.25	3.25	2.00	.63	
SS 1654	6.000	18,300	250,000	20	35.4	1.497	3.19	3.19	2.25	4.00	2.50	.63	
SBS 2859	6.000	21,800	200,000	20	34.0	1.241	3.97	3.65	3.75	4.00	2.37	.63	K44
SBS 2864	7.000	21,800	200,000	17	31.0	1.241	3.97	3.65	3.75	4.00	2.37	.63	K443
SBS 2865	7.000	25,000	200,000	17	32.6	1.241	4.16	3.84	4.50	4.00	2.37	.63	K3
SBS 2866	7.000	27,000	200,000	17	35.4	1.360	4.35	4.08	4.50	4.00	2.37	.63	K3
SBO 6065	6.000	27,600	600,000	20	51.7	1.750	3.820	3.460	3.000	4.750	3.000	.75	

Source: Rexnord Corporation, Link-Belt Engineered Chain Operation, Morgantown, NC.

Hence, they are frequently used as replacement chains for cast pintle and combination chains when load and other application factors are increased. Undoubtedly, their primary function or design is for heavy-duty conveyor and elevator service, although in the smaller sizes they can readily be used for slow-speed power drives.

Welded-steel mill chains are commonly available in two series of material combinations: (1) those in which the regular-type steel links (with offsets) are supplied, but coupled together by heat-treated pins; and (2) those with both welded-steel links and pins heat-treated. This series is preferred when additional resistance to abrasion is required, so as to provide increased strength and to retard wear.

Table 34   General Dimensions, Mill Chains ("H" Type) (ANSI B29.14)

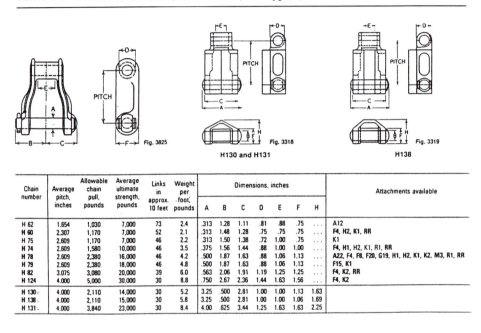

H130 and H131   H138

Chain number	Average pitch, inches	Allowable chain pull, pounds	Average ultimate strength, pounds	Links in approx. 10 feet	Weight per foot, pounds	Dimensions, inches							Attachments available
						A	B	C	D	E	F	H	
H 62	1.654	1,030	7,000	73	2.4	.313	1.28	1.11	.81	.88	.75	...	A12
H 60	2.307	1,170	7,000	52	2.1	.313	1.48	1.28	.75	.75	.75	...	F4, H2, K1, RR
H 75	2.609	1,170	7,000	46	2.2	.313	1.50	1.38	.72	1.00	.75	...	K1
H 74	2.609	1,580	10,000	46	3.5	.375	1.56	1.44	.88	1.00	1.00	...	F4, H1, H2, K1, R1, RR
H 78	2.609	2,380	16,000	46	4.2	.500	1.87	1.63	.88	1.06	1.13	...	A22, F4, F8, F20, G19, H1, H2, K1, K2, M3, R1, RR
H 79	2.609	2,380	18,000	46	4.8	.500	1.87	1.63	.88	1.06	1.13	...	F15, K1
H 82	3.075	3,080	20,000	39	6.0	.563	2.06	1.91	1.19	1.25	1.25	...	F4, K2, RR
H 124	4.000	5,000	30,000	30	8.8	.750	2.67	2.36	1.44	1.63	1.56	...	F4, K2
H 130	4.000	2,110	14,000	30	5.2	3.25	.500	2.81	1.00	1.00	1.13	1.63	
H 138	4.000	2,110	15,000	30	5.8	3.25	.500	2.81	1.00	1.00	1.06	1.69	
H 131	4.000	3,840	23,000	30	8.4	4.00	.625	3.44	1.25	1.63	1.63	2.25	

Source: Rexnord Corporation, Link-Belt Engineered Chain Operation, Morgantown, NC.

There are currently eight standard pitch sizes available, ranging from 2.609 to 6.050 in. The prefix "W" is employed in the designation. A variety of attachments are available. General dimensions and chain sizes are shown in Table 35.

H.   Hinge-Type Flat-Top Conveyor Chain (ANSI B29.17)

These chains are suitable (recommended) for straight-running conveyance applications of all kinds. As the term *hinge-type* implies, this design of conveyor chain utilizes a very basic concept, and its construction is of fundamental consideration. In this style, interfitting curled joints similar to a typical door hinge are used. Thus, the chain is a consecutive assembly of identically formed steel flat-top links coupled together by steel pins. Hence, these chains are held together by means of tight fits and interlocking surfaces, creating a totally nonwelded construction. Sprocket wheel engagement occurs when the curled hinges, which protrude on the bottom side of the flat-top conveying surface, come in contact with the teeth.

Table 35 General Dimensions, Mill Chains (Welded Type) (ANSI B29.16)

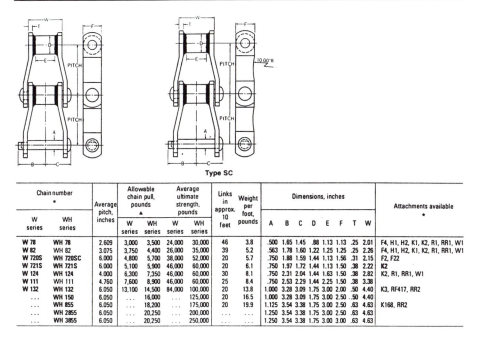

Type SC

Chain number *		Average pitch, inches	Allowable chain pull, pounds ▲		Average ultimate strength, pounds		Links in approx. 10 feet	Weight per foot, pounds	Dimensions, inches								Attachments available *
W series	WH series		W series	WH series	W series	WH series			A	B	C	D	E	F	T	W	
W 78	WH 78	2.609	3,000	3,500	24,000	30,000	46	3.8	.500	1.65	1.45	.88	1.13	1.13	.25	2.01	F4, H1, H2, K1, K2, R1, RR1, W1
W 82	WH 82	3.075	3,750	4,400	26,000	35,000	39	5.2	.563	1.78	1.60	1.22	1.25	1.25	.25	2.26	F4, H1, H2, K1, K2, R1, RR1, W1
W 720S	WH 720SC	6.000	4,800	5,700	38,000	52,000	20	5.7	.750	1.88	1.59	1.44	1.13	1.56	.31	2.15	F2, F22
W 721S	WH 721S	6.000	5,100	5,900	46,000	60,000	20	6.1	.750	1.97	1.72	1.44	1.13	1.50	.38	2.22	K2
W 124	WH 124	4.000	6,300	7,350	46,000	60,000	30	8.1	.750	2.31	2.04	1.44	1.63	1.50	.38	2.82	K2, R1, RR1, W1
W 111	WH 111	4.760	7,600	8,900	46,000	60,000	25	8.4	.750	2.53	2.29	1.44	2.25	1.50	.38	3.38	K3, RF417, RR2
W 132	WH 132	6.050	13,100	14,500	84,000	100,000	20	13.8	1.000	3.28	3.09	1.75	3.00	2.00	.50	4.40	K168, RR2
...	WH 150	6.050	...	16,000	...	125,000	20	16.5	1.000	3.28	3.09	1.75	3.00	2.50	.50	4.40	
...	WH 855	6.050	...	18,200	...	175,000	20	19.9	1.125	3.54	3.38	1.75	3.00	2.50	.63	4.63	
...	WH 2855	6.050	...	20,250	...	200,000	...	...	1.250	3.54	3.38	1.75	3.00	2.50	.63	4.63	
...	WH 3855	6.050	...	20,250	...	250,000	...	...	1.250	3.54	3.38	1.75	3.00	3.00	.63	4.63	

Source: Rexnord Corporation, Link-Belt Engineered Chain Operation, Morgantown, NC.

There is only one common pitch size for standard chain models that fall under ANSI B29.17—the 1½-in. pitch size. However, five top plate widths are offered as standard, ranging from 3.25 to 7.50 in. Standard chain models are available in carbon steel 24C26 and in austenitic stainless steel 24A26. Note that manufacturers who do not offer this chain in the standard, commonly refer to them as *815 Series flat-top chain* for their identification purposes. Attachments are not available. General dimensions are shown in Table 36.

I. Drag Chain—Welded Type (ANSI B29.18)

Out of all the chains that are manufactured, drag chains have the widest body, some with a breadth of as much as 16 in., and rank second or third in pitch size. These chains are specifically designed for the loads and operating conditions frequently imposed by drag conveyor service, and they are not intended for power drive usage. They are generally operated in troughs and are used for moving sawdust, wood chips, logs, mill refuse, ashes, coal, and other bulk materials. Welded-steel drag chains have sufficiently high tensile strength to

Table 36   General Dimensions, Hinge-Type Flat-Top Conveyor Chains (ANSI B29.17)

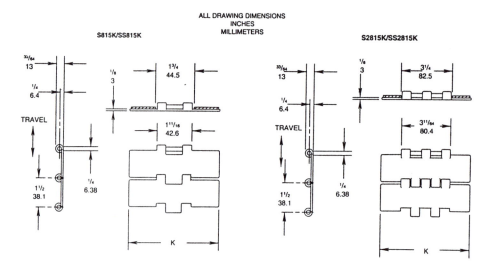

Table No. 1   SERIES 815 (1½ PITCH - 38.1 MM)

Material		K Std. Width		Weight	
Hardened Steel	Stainless Steel	Inches	MM	Lbs. per Ft.	Kg. per Meter
S815K 2¼	SS815K 2¼	2¼	57.2	1.43	2.13
S815K 2⅝	SS815K 2⅝	2⅝	66.7	1.60	2.38
S815K 3¼	SS815K 3¼	3¼	82.6	1.84	2.74
S815K 3½	SS815K 3½	3½	88.9	2.02	3.00
S815K 4	SS815K 4	4	101.6	2.14	3.19
S815K 4½	SS815K 4½	4½	114.3	2.34	3.48
S815K 6	SS815K 6	6	152.4	2.94	4.38
S815K 7½	SS815K 7½	7½	190.5	3.54	5.27

DOUBLE HINGE SERIES 2815 (1½ PITCH - 38.1 MM)

S2815K 7½	SS2815K 7½	7½	190.5	3.75	5.58

Source: Morse Industrial, Ithaca, NY.

absorb shock loading impacts and the stresses imposed by heavy loads. Hence, they are well suited for the many service requirements of drag conveyors in sawmills, pulp and paper mills, and related industries.

Drag chains consist of a consecutive assembly of identical rugged, all-steel, welded offset links (jig welded), connected together by large-diameter steel pins. This design is very practical. The all-steel construction facilitates reworking in the field if distortion or damage occurs during operation. The barrels of these chains are provided with a shaped cross section that creates an optimal pushing and/or scraping action while in use. In other words, the front or leading side of the barrel in each link is square or vertical to facilitate effective pushing or scraping. This flat or vertical surface is important because it reduces any tendency for the chain to "ride over" the material being conveyed. The back or trailing side of the barrel is rounded, which assures proper contact with the sprocket.

Several manufacturers offer two or three design versions of the basic standard welded-steel design. They include:

H-malleable drag chain: A consecutive assembly of identical straight-sided malleable iron links with thick, wide wearing shoes that are contoured (vertical ribs) to stiffen the link, to prevent snagging or interlocking of adjacent strands when used in a multiple assembly, and to protect chain pins and other subsequent damage to chain or trough.

Combination drag chain: An alternate assembly of malleable iron center links and an outer link with heavy, straight-sided medium carbon steel sidebars. They are designed with heavier sections than comparable sizes of H-drag chain, thus providing higher working load ratings. They too are provided with broad wearing shoes to prevent damage.

Extra-heavy-duty drag chain: A consecutive assembly of heavy-duty cast steel, rather than malleable iron links, joined together by heavy steel, T-headed cottered pins that are held securely by heavy, protective lugs. Broad wearing shoes are also cast into the sides of each link.

As with welded mill chains, all models are available in two material series or combinations: those with only pins heat-treated and those with both pins and welded links heat-treated. There are currently nine standard chain models available, ranging in pitch size from 5 to 8 in. Special attachments are available but have to be welded to the steel links. General dimensions and chain sizes are shown in Table 37.

J. Pintle Chain—Class 700 (ANSI B29.21)

Manufacturers offer several different types of pintle chain (e.g., class 400, class 800, class 900 and H pintle). However, class 700 pintle chain, commonly referred to as *chain for water and sewage treatment plants,* is the only group of pintle chains covered under an American standard, specifically B29.21. These chains are totally suited for sewage plant applications as well as other conveying and elevating uses. Thus, they are widely used for settling tank conveyors in sewage treatment plants and are sometimes used on bucket elevators.

Class 700 pintle chain is a consecutive assembly of identical cast malleable iron, slightly offset links joined together by removable steel pins (T-head), which are locked in place by lugs, cast on the sidebars to prevent rotation.

Table 37   General Dimensions, Drag Chains (Welded Type) (ANSI B29.18)

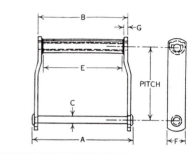

Chain number	Average pitch, inches	Allowable chain pull, pounds	Average ultimate strength, pounds	Links in approx. 10 feet	Weight per foot, pounds	Dimensions, inches						Attachments available
						A	B	C	E	F	G	
WD 102	5.000	8,500	51,000	24	11.2	9.23	7.75	.750	6.50	1.50	.38	C1, C4, RR1, W1
WD 104	6.000	8,500	51,000	20	8.4	6.86	5.38	.750	4.13	1.50	.38	C1, C4, RR1, W1
WD 110	6.000	8,500	51,000	20	11.9	12.04	10.56	.750	9.31	1.50	.38	C1, C3, C4, RR1, W1
WD 112	8.000	8,500	51,000	15	9.9	12.04	10.56	.750	9.31	1.50	.38	C1, C4, RR1, W1
WD 116	8.000	8,500	51,000	15	14.4	15.86	14.38	.750	13.13	1.75	.38	C1, C4, RR1, W1
WD 480	8.000	11,500	70,000	15	18.7	14.60	12.75	.875	11.25	2.00	.50	C3, C4, RR1, W1

Source: Rexnord Corporation, Link-Belt Engineered Chain Operation, Morgantown, NC.

Riveted chain construction is recommended for sewage application. There are two standard model types: a parallel or straight-sidebar type which is intended to be run with plain sprockets; and a thick, curved-sidebar type of heavy cross section to provide a support surface for contact with sprocket rims or flanges.

There are currently five standard chain models available, all of which have a common pitch length of 6 in. A variety of attachment links are available. General dimensions and chain sizes are shown in Table 38.

K.   Drop-Forged Rivetless Chain (ANSI B29.22)

Drog-forged rivetless chain is generally regarded as one of the strongest chains ever developed. Their construction design is a fundamental consideration. This chain is made up of an alternating assembly of a single inner or center loop-type link and two separate outer links or sidebars, coupled together by symmetrically forged T-headed pins that lock into the sidebars to prevent rotation. Hence, the simplicity of design permits easy assembly or dismantling of the chain without the use of tools.

Rivetless chain is ideally adapted to general service in long conveyor runs, such as in retarding and assembly conveyors, where chain weight is of prime consideration, and for flight conveyors operating in materials that would tend to pack in fully enclosed chain joints (such materials do not tend to pack in its open structure). Also, its design permits extensive use for overhead trolley

Table 38   General Dimensions, Pintle Chains (Class 700) (ANSI B29.21)

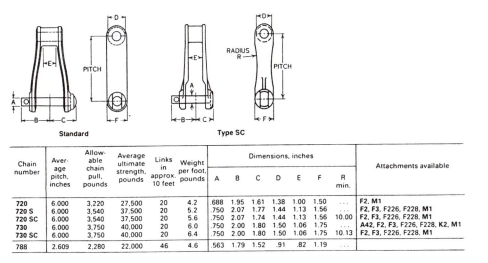

Standard   Type SC

Chain number	Average pitch, inches	Allowable chain pull, pounds	Average ultimate strength, pounds	Links in approx. 10 feet	Weight per foot, pounds	Dimensions, inches							Attachments available
						A	B	C	D	E	F	R min.	
720	6.000	3,220	27,500	20	4.2	.688	1.95	1.61	1.38	1.00	1.50	...	F2, M1
720 S	6.000	3,540	37,500	20	5.2	.750	2.07	1.77	1.44	1.13	1.56	...	F2, F3, F226, F228, M1
720 SC	6.000	3,540	37,500	20	5.6	.750	2.07	1.74	1.44	1.13	1.56	10.00	F2, F3, F226, F228, M1
730	6.000	3,750	40,000	20	6.0	.750	2.00	1.80	1.50	1.06	1.75	...	A42, F2, F3, F226, F228, K2, M1
730 SC	6.000	3,750	40,000	20	6.4	.750	2.00	1.80	1.50	1.06	1.75	10.13	F2, F3, F226, F228, M1
788	2.609	2,280	22,000	46	4.6	.563	1.79	1.52	.91	.82	1.19	...	

Source: Rexnord Corporation, Link-Belt Engineered Chain Operation, Morgantown, NC.

conveyor applications where service over irregular routes, both horizontally and vertically, demands that the chain flex in a transverse or multiplane operation. These conveyor services are driven by sprockets or a caterpillar drive. The caterpillar drive employs a strand of power chain (precision roller chain) fitted with "driving dogs," which interengage with the drop-forged rivetless chain (see Fig. 4).

Standard rivetless chain is available in two styles: (1) Conventional or regular type chains are recommended for general applications. Usually furnished with only the T-head pins heat-treated, they are identified by three or four righthand digits (no prefix). (2) Type X or heavy duty chains are designed to have greater ultimate strength (transverse strength), increased flexibility, and better operating performance. Hence, they are especially recommended for trolley conveyor service. They are furnished with all components heat-treated. Type X chains have smoothly curved contours on their center links where they come in contact with the sidebars so as to permit free vertical flexure. They are identified by the prefix "X" and followed by three or four digits. Additionally, a *nonstandard* variation of the drop-forged rivetless chain is offered by some manufacturers. This type is known as "steel barloop chain" or "S-bar link chain," depending upon the manufacturer, and is usually identified by the prefix "S." Again, this is followed by three or four righthand digits. S- (steel) bar link chain is used for conveying applications on sliding surfaces where ser-

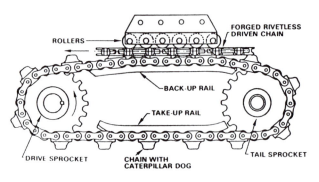

Figure 4 Typical forged rivetless chain overhead trolley conveyor application employing a caterpillar-type drive. (Courtesy of U.S. Tsubaki, Inc., Union Chain Division, Sandusky, OH.)

vice is generally severe. These chains consist of an alternating assembly of rivetless chain center links and two outer fabricated steel sidebars joined together by steel rivets. All components are hardened for maximum strength and wear resistance. Also, since the loop link is the same forging as is used on rivetless chain, sprocket interchangeability between the two types of chains is practical.

There are currently 15 standard models of drop-forged rivetless chain available, ranging in pitch size from 2 to 9 in. Various attachments are available for both rivetless chain and S-bar link chain. In the latter, attachments can be welded to the steel sidebars, or rivetless chain attachments can be bolted to the forged center links. General dimensions and chain sizes are shown in Table 39.

IV. SPECIALTY CHAINS

A. Self-Lubricating Roller Chains

Self-lubricating chain is built for service comparable to that of ANSI roller chain and is interchangeable with any ANSI chain of the same type. The only difference is that they are self-lubricating, and in some cases their roller width may be slightly less than those of standard roller chains. However, they will still run on sprockets manufactured to ANSI standards. Like other standard chains, they are produced in base-pitch, single- and multiple-width (strands), double-pitch drive series and are also available in the hollow-pin design from most manufacturers.

Construction of this chain is fundamental to its use. The heavy-walled, oil-impregnated, sintered-steel bushings replace the conventional bushings of ANSI roller chain. In effect, this porous bushing material functions somewhat as a sponge, with the lubricant retained in the voids. These voids (pockets) usually make up about 20%, if not more, of the volume of the bushing.

Table 39 General Dimensions, Drop-Forged Rivetless Chains (ANSI B29.22)

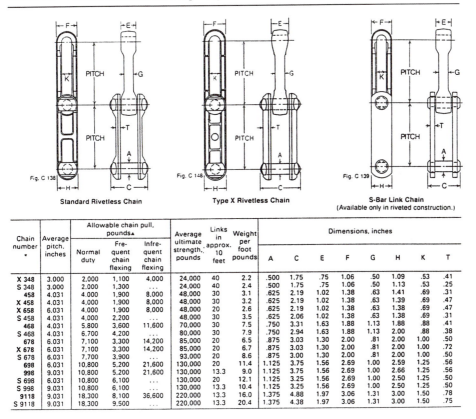

Standard Rivetless Chain — Fig. C 138   Type X Rivetless Chain — Fig. C 148   S-Bar Link Chain — Fig. C 139 (Available only in riveted construction.)

Chain number	Average pitch, inches	Allowable chain pull, pounds			Average ultimate strength, pounds	Links in 10 feet approx.	Weight per foot pounds	Dimensions, inches							
		Normal duty	Frequent chain flexing	Infrequent chain flexing				A	C	E	F	G	H	K	T
X 348	3.000	2,000	1,100	4,000	24,000	40	2.2	.500	1.75	.75	1.06	.50	1.09	.53	.41
S 348	3.000	2,000	1,300	...	24,000	40	2.4	.500	1.75	.75	1.06	.50	1.13	.53	.25
458	4.031	4,000	1,900	8,000	48,000	30	3.1	.625	2.19	1.02	1.38	.63	1.41	.69	.31
X 458	4.031	4,000	1,900	8,000	48,000	30	3.2	.625	2.19	1.02	1.38	.63	1.39	.69	.47
X 658	6.031	4,000	1,900	8,000	48,000	20	2.6	.625	2.19	1.02	1.38	.63	1.38	.69	.47
S 458	4.031	4,000	2,200	...	48,000	30	3.5	.625	2.06	1.02	1.38	.63	1.38	.69	.31
468	4.031	5,800	3,600	11,600	70,000	30	7.5	.750	3.31	1.63	1.88	1.13	1.88	.88	.41
S 468	4.031	6,700	4,200	...	80,000	30	7.9	.750	2.94	1.63	1.88	1.13	2.00	.88	.38
678	6.031	7,100	3,300	14,200	85,000	20	6.5	.875	3.03	1.30	2.00	.81	2.00	1.00	.50
X 678	6.031	7,100	3,300	14,200	85,000	20	6.7	.875	3.03	1.30	2.00	.81	2.00	1.00	.72
S 678	6.031	7,700	3,900	...	93,000	20	8.6	.875	3.00	1.30	2.00	.81	2.00	1.00	.50
698	6.031	10,800	5,200	21,600	130,000	20	11.4	1.125	3.75	1.56	2.69	1.00	2.59	1.25	.56
998	9.031	10,800	5,200	21,600	130,000	13.3	9.0	1.125	3.75	1.56	2.69	1.00	2.66	1.25	.56
S 698	6.031	10,800	6,100	...	130,000	20	12.1	1.125	3.25	1.56	2.69	1.00	2.50	1.25	.50
S 998	9.031	10,800	6,100	...	130,000	13.3	10.4	1.125	3.25	1.56	2.69	1.00	2.50	1.25	.50
9118	9.031	18,300	8,100	36,600	220,000	13.3	16.0	1.375	4.88	1.97	3.06	1.31	3.00	1.50	.78
S 9118	9.031	18,300	9,500	...	220,000	13.3	20.4	1.375	4.38	1.97	3.06	1.31	3.00	1.50	.75

Source: Rexnord Corporation, Link-Belt Engineered Chain Operation, Morgantown, NC.

The self-lubricating technique is of basic importance. While in operation, the lubricant (a special nongumming type) contained within the oil-impregnated bushing flows over all vital (frictional) surfaces. When the drive stops, the lubricant is reabsorbed by the bushing. This cycle continues throughout the chain's service life.

Self-lubricating chain has three main advantages not found in conventional chain:

1. *Oil-impregnated bushing*: A protective film of oil completely lubricates the live bearing area between pin and bushing, minimizing wear by reducing metal-to-metal contact and internal rusting of pin and bushing surfaces. The oil film on the exterior surface of the bushing provides constant lubrication between sprocket teeth and chain. Thus, there is

built-in lubrication at three critical points—pin, link plate, and sprocket contact surface.

2. *Extended bushing*: Bushing extends beyond surface of the plates and acts as a lubricated thrust bearing to eliminate plate galling and seizing, thus reducing friction and heat between link plates.
3. *Positive clearance*: The positive clearance between pin link plate and roller link plate creates a self-cleaning action to eliminate build-up of dust, dirt, and corrosion. This positive clearance also acts as a barrier to keep abrasive particles out of the chain joint. This extends chain life under dirty, dusty, and mildly corrosive conditions.

Self-lubricating chains have unlimited practical applications. These chains have been used successfully in the following areas: timing and indexing drives, farm equipment, food processing, baking, printing, textile machinery, chemical processing, materials handling, packing machinery, and many similar applications where lubrication cannot be tolerated or is impractical.

In summary, regular self-lubricating chain is recommended for operation in ambient temperatures in the 0–200°F range. Of course, chains with special lubricants for a much wider temperature range (− 50–450°F) are available, but on a made-to-order basis.

Note that these special self-lubricating chains may be selected in accordance with the selection procedures outlined near the end of this chapter. However, since manufacturers publish separate or special horsepower rating tables for these chains outside their general catalog, it is recommended that a company representative be referred to for accurate chain design calculations. Also, most manufacturers advise that the use of their "Quick Selector" chart be omitted. In other words, rating values for self-lubricating chains vary slightly, and the standard horsepower rating tables shown in this text should be used *only* for conventional roller chain. Standard and special attachments are available for required applications.

B. Extra-Clearance Roller Chain

Extra-clearance chain, sometimes referred to as "side-bow" roller chain, is designed to permit side flexing and twisting not tolerated by conventional roller chains (chain joints). These chains are constructed of ANSI standard roller links, but are combined with special pin links designed to allow greater clearance between the pins and bushings, and between the roller link and pin link side plates.

Generally, side plates are treated for wear and shock resistance, and pins may be case-hardened alloy steel and zinc-plated, so as to minimize joint seizure. Extra joint and transverse clearances allow the chain to accommodate several inches of lateral displacement, and are also well suited to transmit loads operating in a straight line. In other words, this flexible roller chain permits twists of 8 degrees, and sidebars of 4 in., approximately. (See Figure 5.)

Extra-clearance chains are primarily designed for curved conveyor applications, such as bottling, packaging, canning, etc., in which the wide use of "live-roll" conveyors is employed, and to transmit power where there may be abnormal chain twist, as on transit mixers and crawler drives for shovels and cranes.

These chains are primarily designed to operate on sprockets cut for ANSI standard roller chains of the same pitch if applied on the "straightway" system. However, if applied on a curve or twist section type system, sprockets with *reduced* tooth thickness are recommended. Custom-made sprockets with a *zero-degree pressure angle* are recommended for tangential meshing with chain-type systems. It is suggested that a guide rail or track be used to help guide the chain around a curved system.

In summary, extra-clearance chains are specialty chains and do not follow an ANSI standard numbering system. Therefore, a manufacturer's catalog will have to be referred to when determining a specific chain number. Normally, they are only available in the single-pitch, single-strand series. These chains may be selected in accordance with the selection procedures described near the end of this chapter. However, it is suggested that a manufacturers' representative be consulted regarding horsepower ratings, since they are not given in a general catalog. The horsepower rating tables in this text should be used *only* for standard or conventional roller chain. Some attachments are available.

C. Hollow-Pin Roller Chains

Hollow-pin chains allow unusual flexibility for conveyor applications. They are identical to ANSI standard roller chain in pitch, bushing diameter, roller width,

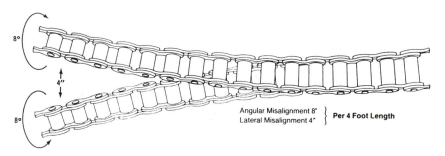

8°

4″

8°

Angular Misalignment 8°
Lateral Misalignment 4″ } Per 4 Foot Length

Figure 5 Extra-clearance chains are designed to allow unusual side-flexing and twisting not possible with standard roller chains. (Courtesy of Morse Industrial, Ithaca, NY.)

and roller diameter. They are available in the single-pitch series with standard rollers or rollerless. They are also available in the double-pitch conveyor series with standard rollers, rollerless, or with large-diameter carrier rollers constructed of steel or thermoplastic. Single and double-pitch chains of rollerless design have bushing diameters that are the same as comparable chain rollers.

This "hollow-pin" feature provides unlimited design versatility. Many types of crossroads, pins, rollers, and custom attachments may be inserted at any point along the chain. The connecting link will not accept an attachment unless the chain is removed from the drive system. (See Figure 6.) Attachments are held in place by using retaining clips.

Since there is no ANSI standard numbering system for hollow-pin chain, you will need to refer to a manufacturer's catalog regarding their specific numbering system. Again, selection procedures previously discussed are applicable here.

D. Corrosion-Resistant Roller Chains

Roller chains that must operate in conditions involving water, acids, alkalis, or high temperatures are known as noncorrosive and heat-resistant roller chains. These include roller chains that are an assembly of stainless steel (almost all components), and those that are an assembly of plated components.

These durable chains (stainless and plated) are manufactured to the same dimensional specifications as ANSI standard roller chain, and therefore will operate over the same sprockets.

Stainless steel chains are made from a combination of materials to give long service life. Such materials include 300, 302, 303, 316, and 400 series stainless steels, 18-8 stainless steel, and 17-4 Ph, 17-7 Ph chrome–nickel steels, which are generally reserved for round parts (bushings, pins, and rollers). Some of

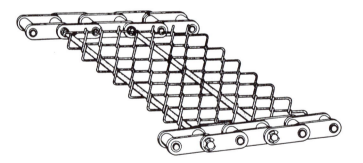

Figure 6 Typical "cross-rod" example. A wire mesh conveyor, using cross-rod attachments assembled to parallel strands of double-pitch hollow pin chain. (Courtesy of Rexnord Corporation, Link-Belt Roller Chain Operation, Indianapolis, IN.)

these materials possess varying degrees of magnetic properties, whereas others exhibit none. For utmost corrosion resistance, chains made with all parts of 302 and 303 series stainless steel are recommended, whereas "round parts" from the 400 series stainless steel are generally preferred when maximum wear resistance is required.

Plated roller chains are preferred when chains operate in mildly corrosive environments (generally not in acids or alkalis) or for decorative applications. They provide greater corrosion resistance than their counterpart carbon-steel chain, and they are lower in cost than stainless steel. Manufacturers apply these platings to individual chain components prior to assembly, to insure that all vital joint wearing areas receive the benefit of uniform plating. Such plating types include nickel, tin–nickel, zinc, electroless nickel, cadmium, hard chrome and zinc–phosphate.

The corrosion resistance of roller chain is primarily dependent on the inherent resistance to corrosion of the materials used in its manufacture. The "corrosion resistance data" that appears in manufacturers' catalogs are generally obtained under controlled conditions and should be regarded as indicative of the corrosion resistance to be expected from chain under similar conditions and not as a guarantee of performance in commercial or industrial service. Hence, suitability of a chain for operation under specific adverse conditions should be determined by testing, using realistic service conditions. Stainless steel chain parts are not ordinarily heat-treated, since this reduces corrosion resistance. Resistance to wear cannot therefore be compared with that of case-hardened standard steel roller chain. (See Table 40.)

Stainless and plated roller chains are available in base-pitch single and multiple-strand series, and double-pitch conveyor series; with standard and large-sized rollers. Additionally, the *average ultimate strength* of stainless and plated roller chains must be taken into consideration for proper design and selection procedures. For example, because of the ductility of stainless chains, a slight permanent set may be measured at approximately 50% of published average ultimate strengths. The average ultimate strength of plated chain is approximately 10% less than that of unplated chain. A manufacturers' catalog or representative should be consulted for appropriate average ultimate strength listings. Again, these chains may also be selected in accordance with this chapter. Approved attachments are available.

V. ROLLER CHAIN SPROCKETS

Roller chain sprockets may be described as toothed wheels whose teeth are shaped to mesh with roller chain. Theory is a fundamental consideration. Chains and sprockets interact with each other to convert linear motion to rotary motion or vice versa, since the chain moves in an essentially straight line between sprockets and moves in a circular path while engaged with each

Table 40 Corrosion Resistance of Stainless Steel Chains

	Concentration	Temperature	SS Series	NS Series	AS Series
Nitric acid	5%	70°F	A	A	B
	20%	70°F	A	A	B
	50%	Boiling	A	A	B
	Concentrated	Boiling	D	D	D
Sulphuric acid	5%	70°F	C	B	C
	5%	Boiling	E	C	E
	50%	70°F	D	C	
	50%	Boiling	E	D	E
	Concentrated	70°F	A	A	
	Concentrated	Boiling	D	D	E
Hydrochloric acid		70°F	E	E	E
Phosphoric acid	1%	70°F	--A	--A	A
	5%	70°F	C	A	
	10%	70°F	C	A	
Oxalic acid	5%	70°F-Boiling	A	A	D
	10%	Boiling	D	G	E
Acetic acid	5-10%	70°F	A	A	A
	20-100%	70°F	A	A	A
	50%	Boiling	C	B	D
Formic acid	5%	70~140°F	B	A	D
Lactic acid	5%	70°F	A	A	A
	5%	150°F	A	A	
	10%	Boiling	B	A	D
Butyric acid	5%	70~150°F	A	A	B
Citric acid	5%	70~150°F	A	A	A
	15%	Boiling	A	A	
Chromic acid	5%	70°F	A	A	A
	10%	Boiling	C	B	
Iodine			E	D	E
Fluorine		70°F	E	E	E
Chlorine gas	Dry	70°F	C	B	E
	Wet	70°F	D	C	E
Bromine		70°F	E	D	E
Carbon disulfide		70°F	A	A	A
Carbon tetrachloride		70°F	A	A	A
	5-10%	70°F	*C	*B	
Carbolic acid		70°F	A	A	
Tartar acid		70°F	A	A	A
Oleic acid		70°F	*A	A	
Malic acid		210°F	B	A	
Aqua ammonia		70°F	A	A	A
Ammonia gas		120°F		D	

	Concentration	Temperature	SS Series	NS Series	AS Series
Hydroxide calcium	10-20%	Boiling	A	B	A
	50%	Boiling	C	B	
Calcined soda			A	A	B
Sodium carbonate	5%	70~150°F	A	A	A
Sodium dicarbonate	Saturated	70°F	A	A	A
Sodium thiosulfate	5-10%	70~150°F	A	A	
Ammonium sulfate	1-5%	70°F	A	A	
Sodium chloride	5-20%	70~150°F	*A	A	B
	Saturated	Boiling	B	A	
Zinc chloride	5%	70°F	*A	*A	
Zinc sulphate	5%-Saturated	70°F	A	A	B
Ferric chloride	1%	70°F	*.*B	*A	
	5%	70°F	*.*D	*C	D
Ethyl alcohol		70°F-Boiling	A	A	
Methyl alcohol		70°F	A	A	A
		150°F	*C	B	
Chloroform		70°F	A	A	A
Soda (carbonated water)			A	A	
Vinegar		70°F	A	A	
Sea water			*A	*A	C
Milk		150°F	A	A	
Syrup			A	A	
Volatile oil			A	A	A
Fruit juice			A	A	
Mayonnaise		70°F	*A	A	
Gelatine			A	A	
Glycerine			A	A	A
Ketchup		70°F	*A	A	
Coffee		Boiling	A	A	
Beer			A	A	

A: Excellent (Corrosion rate below 0.00035 inch/month)
B: Satisfactory (Corrosion rate 0.00035-0.0035 inch/month)
C: Marginal (Corrosion rate 0.0035-0.01 inch/month)
D: Unsatisfactory (Corrosion rate 0.01-0.035 inch/month)
E: Not recommended. (Corrosion rate over 0.035 inch/month)

++: Presence of nitric acid activates corrosion.
* : Pitting is activated. Should be kept dry after exposure.

ANTI-CORROSIVE/HEAT RESISTANT CHAIN

Source: U.S. Tsubaki, Wheeling, IL.

sprocket. Although a number of tooth designs (profiles) have been developed throughout the years, the three general principles or prerequisites for any chain sprocket tooth-form construction are still the same:

1. Provide smooth engagement and disengagement with the moving chain.
2. Provide equal distribution of the transmitted load so that each tooth engaged with the chain will carry its share of the load.
3. Provide accommodation for normal changes in chain length (within reasonable limits) as the chain elongates as a result of wear during its service life.

Additionally, all chain sprockets must be manufactured with reasonable static balance if smooth operation is to be obtained. Because the overall performance of a drive installation depends largely on sprocket–chain interaction, choosing the right sprocket is an important as choosing the right chain. Hence, it is essential that the design or selection of chain and sprocket be compatible.

1. Sprocket Terminology (Diameters)

Although there are many similarities between terms for gears and sprockets, their meanings may not be exactly identical. Therefore, the following nomenclature shall be used with reference to standard roller chain sprockets.

Pitch diameter is the diameter of the pitch circle that passes through the centers of the link pins as the chain is wrapped on the sprocket. Since the chain pitch is measured on a straight line between the centers of adjacent pins, the chain pitch lines form a series of chords of the sprocket pitch circle. Therefore, it is essentially a relationship or function of the chain pitch and the number of teeth in the sprocket.

Bottom diameter is the diameter of a circle tangent to the curve (called the seating curve) at the bottom of the tooth gap. It allows the chain roller centers to follow the sprocket pitch circle. It equals the pitch diameter minus the roller diameter. This dimension is used for measuring the diameter on sprockets having even numbers of teeth.

Caliper diameter is the distance from the bottom of one tooth gap to that of the nearest opposite tooth gap. This dimension is used for measuring roller chain sprockets with an odd number of teeth. On the other hand, the caliper diameter is the same as the bottom diameter, but for a sprocket with an even number of teeth. In other words, the caliper diameter for a sprocket with an even number of teeth is equal to or the same as the bottom diameter.

Outside diameter is the diameter over the "tips" of the sprocket teeth.

Minimum through bore (hub) length must be sufficient to allow a long enough key to withstand the torque transmitted by the shaft. This also assures stability of the sprocket on the shaft. See Fig. 7 for general diameter measurements.

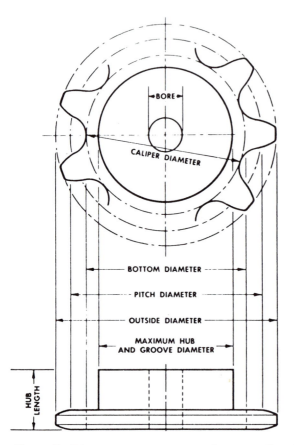

Figure 7 Diameter measurements made on a roller chain sprocket. (Courtesy of American Chain Association, 1982.)

2. Sprocket System Analysis

Basic Geometry and Relative Motions.   The sprocket layout is based on the pitch circle, the diameter of which is such that the circle passes through the center of each of the chain's joints when that joint is engaged with the sprocket. Since each chain link is rigid, the engaged chain forms a polygon whose sides are equal in length to the chain's pitch. The *pitch circle* of a sprocket, then, is a circle that passes through each corner, or vertex, of the pitch polygon.

The action of the moving chain as it engages with the rotating sprocket is one of consecutive engagement. Each link must articulate, or swing through a specific angle, to accommodate itself to the pitch polygon, and each link must be completely engaged, or seated, before the next in succession can begin its articulation.

As depicted in Fig. 8, with a sprocket tooth at top-dead-center position, link *AB* is completely (properly) seated; with rotation of one-half a link, so that the space or rollers between two teeth is also at top-dead-center position. Hence, as roller *A* approaches the sprocket, it tends to follow the chordal line to the position of roller *B*. However, engagement with the sprocket forces it to follow the arc of the pitch circle. When roller *A* has advanced to the position shown in the bottom part of Fig. 8, it has been raised through the distance *PR–r*, as shown. Continued rotation will bring the next tooth to top-dead-center position, completing the articulation of that particular link through the angle required (or 360° per number of teeth) to become completely seated again. The cycle will repeat with the link that follows, and so on.

Since the center-to-center dimension of each engaged link forms a chord relative to the pitch circle of the sprocket, this angular swing of each link as each joint articulates is commonly referred to as *chordal action*. Chordal action, or polygonal effect, is the variation of the velocity of the driven sprocket that results because the actual instantaneous velocity ratio is not constant but a function of the sprocket angular position. In other words, generally the sprockets of a chain drive do not have a constant angular velocity ratio, and the linear velocity of the chain is not constant. This is because the chain does not wrap around the sprocket in the form of a pitch circle, but rather in the form of a pitch polygon.

Thus, the direct result of each link's chordal action is referred to as chordal *rise and fall*. In other words, the incoming straight-chain strand(s) position has a tendency to move slightly up or down (a slight whipping action). This is an unwanted effect and is dependent upon whether a chain joint or a tooth is at top-dead-center position. Again, because of chordal rise and fall, there are slight variations in the linear velocity of the incoming chain strand, which in turn produces dynamic loads, vibrations, and premature chain wear. Essentially, the analysis of chordal action is made under the assumption that the chain strand acts like a connecting rod in a four-bar linkage. Experiments at low speeds yield excellent agreement with the theoretical results. At higher speeds the chain vibrations are sufficient to alter the static behavior of the chordal action. Nevertheless, this action remains a main source of excitation.

Thus, as the number of teeth in a sprocket increases, the angle of articulation (wrap), and therefore the chordal action, decreases. Hence, chordal action is more pronounced with sprockets having a small number of teeth. For example, in a sprocket with 10 teeth on its periphery, the velocity variation is 5.15 percent; for 20 teeth, 1.25 percent; and at 30 teeth, velocity variation is only 0.55 percent. Therefore, it is a considerable advantage in having as many teeth as possible on the small sprocket. However, when it becomes necessary to use small sprockets, those with an odd number of teeth are usually selected. Because, when the chain velocity is increasing on one side it is decreasing on the other, and inertia forces on the two sides tend to balance each other. Chor-

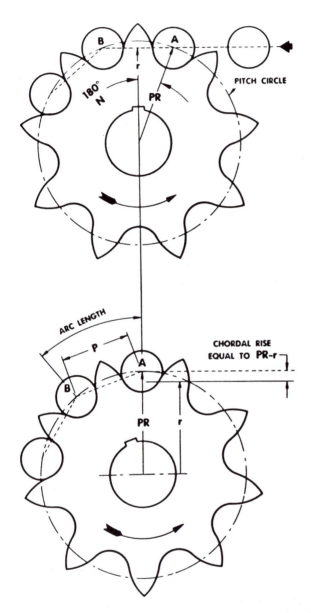

Figure 8 Chordal action of chain. (Courtesy of American Chain Association, 1982.)

dal action may generally be ignored if the minimum number of sprocket teeth is 15.

Force Distribution on the Sprocket Teeth.   The seating of a roller chain on a typical sprocket is shown in Fig. 9. When the chain is new the rollers seat at the bottom of the tooth spaces. As the chain wears, the pitch of each link is increased, but because wear is random, it is difficult to predict the pressure angles. The tension in the chain decreases very rapidly but never becomes zero unless the pressure angle is zero. Once the chain has reached an elongated state, the roller engages the top curve of the teeth farther and farther away from the working curve, leading to an undesirable situation. However, to help remedy this, the tooth profiles are made so that the force of the sprocket on the roller has a radially outward component. This causes a worn chain to take a position on the sprocket teeth that corresponds to the worn pitch, thus distributing the load among all the teeth that are in contact with the chain. Generally, a maximum of 3 percent has been admitted to current ratings for 15,000 hours of service. Increased elongation indicates that the hardened layers of pins and bushings are worn away and the soft cores reached. A dangerous rapid wear increase follows at this stage.

Roller Impact Loading.   Theoretically, correct gear teeth come into contact without impact. Because of inaccuracies in machining and deflections there is always some impact. This can be reduced by using gears made with a high degree of accuracy, but in roller chains, impact loading is inherent in the design. A sprocket having a small number of teeth, running at high speeds, leads to roller breakage. Lubricant viscosity also plays an important part in roller fatigue because the rollers must squeeze the lubricant before they are completely seated, and squeezing provides good damping. The presence of impact loading is one reason why chains with large pitches cannot be operated at high speeds.

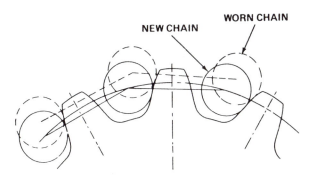

Figure 9   Diagram showing wear on new and worn chain (chain seating on a sprocket). (Courtesy of American Chain Association, 1982.)

3. Sprocket Materials

The most commonly used sprocket materials are iron and steel. Where exposure to a corrosive environment is of concern, stainless steel or bronze may be employed for added resistance. For large sprockets cast iron is widely used, especially in drives with large speed ratios, since the teeth of the larger sprocket are subjected to fewer chain engagements in a given time. For severe service, cast steel or steel plate are preferred.

The smaller sprockets of a drive are usually made of steel. With this material the body of the sprocket can be heat-treated to produce toughness for shock resistance, and the tooth surfaces can be hardened to resist wear. Additionally, many small sprockets are now made by powder metallurgy, and in some light-duty sprocket applications, they are being manufactured by injection molding of plastic materials.

The two basic factors that determine what type of materials are to be used in making sprockets are the application (job to be done) and the size of the sprocket. Generally, most standard sprockets are made of low-carbon steel.

Hardening of Teeth. In certain applications hardened teeth may be desirable to increase sprocket strength and equalize wear between small and large sprockets. Hardened teeth substantially increase sprocket life and their use is recommended under the following conditions: (1) when pinion or drive speed ratios are 4:1 or greater; (2) when the chain drive is operating in dirty, dusty, or other unusually abrasive conditions; (3) with slow speed, heavily loaded drives, usually at chain speeds of 100 fmp or less; (4) with moderate speed drives where sprockets have 25 teeth or fewer.

As to the degree of hardness, this is governed by conditions prevailing in each application. For most operating conditions a hardness range of Rockwell C35 to C50 is entirely satisfactory for sprockets that are induction-hardened (teeth only).

Sprockets may also have their teeth flame-hardened. This is generally found on large-diameter, low-carbon steel sprockets. A hardness range of Rockwell C20 to C40 is suitable for most applications. Either hardening method prevents possible plate distortion that may arise in overall heat treatment or furnace carburizing. Basically, the diameter and pitch of the sprocket govern the method used.

4. Sprocket Manufacture

Sprockets are manufactured in various ways and from many types of materials. Manufacturing methods and materials used are reflected in the precision of the teeth and chain contact surfaces of the sprockets and in the surface finishes. Roller chain sprockets are manufactured to the tight, and accepted, standards of ANSI B29.1, which defines the dimensions, tolerances, and tooth forms used in such manufacture. These may be referred to as *cut-tooth sprockets* for roller chain, both single and double-pitch.

Basically, standard sprockets may be classified as either semiprecision (cast tooth or flame-cut tooth) or precision (machine-cut tooth). However, today's technology also permits manufacture of cast or flame-cut semiprecision sprockets with an accuracy rapidly approaching that of machine-cut teeth.

The types of sprockets used in any installation are really a function of the type of chain used (or identified in terms of contact and operation with the chain). Hence, sprocket tooth form designs will vary according to the type of chain with which they are to operate.

It is critical that the tooth form be accurately machine-cut (precision) for roller chains and silent chains, which are normally made from highly finished steel or cast iron. Cast roller chain sprockets have cut teeth; and the rim, hub, face, and bore are machined. The smaller sprockets are generally cut from steel bar stock and are finished all over. Sprockets cut from bar stock are limited in size by the maximum diameter of the commercially available stock. Sprockets are often made from forgings or forged bars. The extent of finishing depends on the particular specifications that are applicable. Additionally, many sprockets are made by welding a steel hub(s) to a steel plate, so as to produce a one-piece sprocket of desired proportions and one that can be heat-treated if required.

Engineering-class chains are normally operated over cast tooth or flame-cut (semiprecision) sprockets. However, this is a generalization and exceptions are not hard to find. For example, some sprockets for engineering steel offset drive chains (ANSI B29.10) are being employed with teeth that are machine-cut, particularly for use in multiple-strand sprockets. This is to ensure proper alignment of the stranded assemblies while they are under heavy loads. Additionally, for cast-link, combination, and welded-link chains, cast-to-shape and flame-cut sprockets are widely used. However, regardless of how the sprocket teeth may have been formed, the most important criterion of sprocket selection is that the chain and sprockets have been designed to operate with one another. Again, hardness of tooth surfaces and toughness of body are factors important to sprocket life. These requirements can readily be met by the use of proper materials and heat treatment.

Machine-cut sprockets can be produced by milling, hobbing, or shaping. Milling cutters and hobs that will produce the standard tooth form are readily available from various tool manufacturers. For example, cutters for Fellows gear shapers are available from the Fellows Company or their distributors. As previously stated, many of the smaller sprockets are currently being made using powder metallurgy techniques, and in the lighter-duty types plastic injection molding is being featured. Thus, sprocket selection is simply a matter of choosing the right chain, with the appropriate (compatible) sprocket for the required task to be performed. Sprockets with machine-cut teeth are smooth and quiet in operation and are commonly used in high-speed drives. Sprockets with cast or flame-cut teeth are economical and give adequate life when used with engineering-class chains.

5. Types of Sprockets

According to the American National Standards Institute (ANSI B29.1), there are four standard types of roller chain sprockets (except for the tooth forms, silent chain sprocket types are similar also): Type A—these are plain (flat plate) sprockets without a hub projection; Type B—these have a hub on one side only; Type C—these have a hub on both sides; Type D—these are sprockets that employ a detachable hub. These four standard sprocket types are shown in Fig. 10.

Between the hub and the rim, a sprocket may be one of several forms, generally according to the size. The smaller sprockets are usually solid, except for the shaft bore, since their weight is not excessive. For medium sizes, which are usually cast, webbed forms are used, having cored holes (generally referred to as "lightning holes") between the hub and the rim to help reduce the weight. The larger sprockets are usually cast and made in the form of spoked wheels to obtain the maximum possible weight reduction.

Detachable Hub Sprockets. To the maintenance worker or installer these "quick-disconnect" sprockets, as they are commonly referred to, can be a real life-saver. Rather than it being necessary to remove the whole assembly from the shaft (which can be very troublesome and time-consuming), the toothed wheel itself can quickly and easily be detached by removing the bolts or set screws.

These demountable hub-type sprockets are available from all manufacturers, and come in various forms or combinations. This particular style may be referred to as a taper-lock or split bushing sprocket. These sprockets are

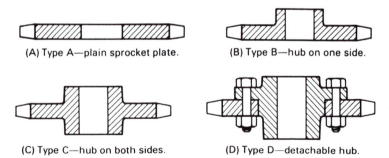

(A) Type A—plain sprocket plate.   (B) Type B—hub on one side.

(C) Type C—hub on both sides.   (D) Type D—detachable hub.

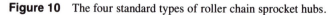

Figure 10 The four standard types of roller chain sprocket hubs.

manufactured in both the B and C sprocket types and combine a tapered split bushing with a tapered bore sprocket. This design provides for a positive grip fit to the shaft and a choice of finished bore sizes within a single sprocket selection.

Additionally, some type D sprockets consist of a solid detachable hub and a split plate, or a solid hub and a solid plate. Others are made with both the hub and the plate in split form. These "split form" sprockets are most advantageous where the sprocket must be mounted between bearings, so as to permit easy installation or removal without disturbing the shafting. Note that in the installation of any split sprocket, caution must be exercised to prevent any variation of the pitch at the "split line." Hence, special dowel pins incorporated in this design come with the sprocket and are inserted on the split line for this purpose.

Shear Pin and Slip Clutch Sprockets. These special-purpose or optional-accessory-type sprockets (relatively inexpensive and valuable) are designed to prevent damage to the drive or to other equipment, caused by overloads or stalling. This is their only purpose.

A shear pin sprocket consists of a hub, keyed to the shaft, and a sprocket that is free to rotate either on the shaft or on the hub when the *shear pin* breaks (see Fig. 11). The shear pin is designed to transmit up to a predetermined amount of torque and to shear when overloaded. When the pin shears, the sprocket is free to rotate either on the shaft or on the hub, which is keyed to the shaft. The shear pin is made of a material with known shearing strength, and its failure (fatigue) point can be established with reasonable accuracy by machining it with a groove of a calculated diameter at the shear plane. Hence, the assembly can be returned on-line and in working condition by simply inserting a new shear pin.

Slip or jaw clutches are generally used where a low-speed device is required for positively connecting or disconnecting two rotating machine elements, and where engagements or disengagements are not too frequent. Essentially, these devices may be classified as square jaw type or spiral (sawtooth) type.

Square jaw construction is recommended where engagement or disengagement in motion and under load is not necessary. Thus, engagement is possible only when the two halves are in position so that jaws of the shifter half can enter the mating pockets (notches) of the stationary half. Square jaw types transmit power in both directions.

Spiral jaw clutches are made either "right" or "left" hand. A right-hand clutch drives when turning clockwise as viewed from the plain end of the driving half. Spiral jaws permit shifting more readily, and are sometimes used where the clutch must be engaged when in motion. Such applications should be limited to low speeds and loads. Spiral jaw clutches transmit power in one direction only, according to which type of "hand" is purchased.

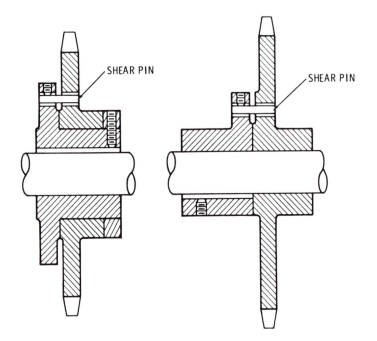

Figure 11   Typical shear pin hub sprockets. (Courtesy of American Chain Association, 1982.)

Regardless of which style is chosen, the sprocket half of a clutch is free to turn on the shaft. Square and spiral jaw clutches are generally manufactured from gray iron, with cast jaws ground smooth. For heavier workloads, it is recommended that clutches with machined jaws be used to provide uniform contact over larger surfaces. For more severe service, it is suggested that cast steel jaw clutches be supplied.

Multiple-Strand Roller Chain Sprockets.   Sprockets for multiple-strand roller chains are produced as integral wheels with correct spacing for the strands and alignment of teeth for which they were intended.

Double-Pitch Roller Chain Sprockets.   Sprockets for double-pitch roller chain are made with different pitch diameters and bottom diameters than sprockets of the corresponding base chain of the ANSI series. However, ANSI base chain sprockets having 35 or more teeth may be used with double-pitch chain, since at this point the slight differences become negligible.

There are basically two types of designs for double-pitch chain sprockets, and these are commonly designated as:

1. *Single-duty* sprockets (single-cut tooth form), having a single tooth for each chain pitch (i.e., all teeth are effective, or engaging), during every revolution. Obviously, the number of effective teeth on a single-cut sprocket is equal to the number of actual teeth.
2. *Double-duty* sprockets (double-cut tooth form), having two teeth per chain pitch, so that the chain is engaged by every other tooth form (i.e., only half of the actual teeth function as effective teeth in each revolution. (See Figs. 12 and 13).

Again, it is advantageous when using double-duty sprockets to employ those with an odd number of teeth, so that each tooth will engage the chain only on every other revolution, thus practically doubling the life of the sprocket. With an even number of actual teeth, the same effect can be obtained by periodically advancing the chain by one tooth on the sprocket. Additionally, this "double-duty action" will also result when single-pitch chain sprockets are used with double-pitch chains. But again, since there are slight differences in pitch and bottom diameters, this system is not suggested for full-time service, but rather

Figure 12 Single-duty sprocket for double-pitch chain. (Courtesy of American Chain Association, 1982.)

Figure 13 Double-duty sprocket for double-pitch chain. (Courtesy of American Chain Association, 1982.)

in an emergency situation. Note that a double-pitch conveyor series chain which has large-diameter (carrier) rollers on it should not be operated with double-duty sprockets, as insufficient space exists between the rollers and such a chain will not be accommodated. Special sprockets are made for this purpose.

VI. CHAIN LENGTH AND CENTER DISTANCE

Chain length is a function or relationship of the number of teeth in both sprockets and of the center distance. The final chain dimension arrived at cannot have a fractional part of a pitch, so if the calculated length is not an integer, the designer must adjust it to the next higher whole number. Hence, for the roller and silent chains, the length should, if possible, be adjusted (calculated) to the nearest even number of pitches, to avoid using an offset link. However, if the centers of the drive cannot be adjusted to accommodate an even number of pitches, use of an offset link is inevitable and it may be selected.

Unquestionably, the simplest chain arrangements that can readily accommodate easy chain length and center distance calculations are those of the two-sprocket layout design. Chain length should be between 70 and 160 pitches. A maximum chain length of 200 pitches can be used without excessive catenary tension or the danger of the chain jumping the sprocket teeth. Generally, a long chain is recommended over a short chain because the rate of chain elongation is lower and a longer chain has a greater shock-absorbing capability. When fixed center distances are long and drive ratios are less than 2:1, large-diameter sprockets should be selected to prevent the tight and slack strands from striking each other after significant normal wear occurs.

Obviously, when both sprocket are the same size, for a 1:1 ratio, both strands will be parallel, and chain length is simply twice the center distance plus the number of teeth in one sprocket. However, in most layout designs, one sprocket has more teeth than the other, which means that it is not possible for the straight chain strands between the two sprockets to run in a perfectly parallel plane. Although various methods are used for calculating length or center distance, they all are based on the same assumption: the number of teeth in the two sprockets is used to take the theoretical angle between the two strands into account.

Using N for the number of teeth in the larger sprocket, n for the number of teeth in the smaller sprocket, and with F (sometimes K) as a factor varying in value with the magnitude of $N - n$, the following simple chain length calculation shown in Table 41 may be employed (chain length, in pitches, when center distance, in pitches, is known). The chain length in pitches can be multiplied by the chain pitch to give the length in *inches*. The actual chain length should be larger than the calculated value by 1/64-in./ft of length. Of course this is only an approximate value for chain length. Additionally, some manufacturers reproduce in their catalog's what is known as quick selector "nomograms" (charts in which a straightedge is laid across two outside columns or scales on their respective points, and the intersection of the straightedge with the center column or scale gives the desired information). These nomograms are to be used only for quick approximations.

In general, typical drives with adjustable center distances of 30 to 50 times the chain pitch between sprocket shafts is most desirable, and the distance must be greater than half the sum of the outside diameters of the sprockets. This is the *preferred center distance*. Extremely short center distances should be avoided, if possible, especially for ratios greater than 3:1, because it is desirable to have at least 120° of wrap (angle of wrap) in the arc of contact on a power sprocket. This is to provide sufficient tooth clearance between the teeth of the two sprockets. For ratios of 3:1 or less, the wrap will be 120° or more for any center distance or numbers of teeth. In order to obtain a wrap of 120° or more for ratios greater than 3½:1, the center distance must not be less than the difference between the outside diameters of the two sprockets. This may be referred to as *minimum center distance*. A center distance equivalent to approx-

Table 41   Chain Length Formula and Constants

Chain Length for Two Sprocket Drives
L = 2C + ½N + ½n + F

L = Chain Length, Pitches   C = Center Distance, Pitches
N = No. of Teeth, Large Sprocket   n = No. of Teeth, Small Sprocket

Condensed Table of Values for F in Chain Length Equation

CENTER DISTANCE IN PITCHES

N-n	10	15	20	25	30	35	40	45	50	55	60	65	70	75	80
10	.25	.17	.12	.10	.08	.07	.06	.06	.05	.05	.04	.04	.04	.03	.03
11	.31	.20	.15	.12	.10	.08	.07	.07	.06	.06	.05	.05	.04	.04	.03
12	.37	.24	.18	.14	.12	.10	.09	.08	.07	.07	.06	.06	.05	.05	.04
13	.43	.28	.21	.17	.14	.12	.10	.09	.08	.08	.07	.07	.06	.06	.05
14	.50	.33	.25	.20	.16	.14	.12	.11	.10	.09	.08	.08	.07	.07	.06
15	.57	.38	.28	.23	.18	.16	.14	.12	.11	.10	.09	.09	.08	.08	.07
16	.65	.43	.32	.26	.21	.18	.16	.14	.13	.12	.11	.10	.09	.09	.08
17	.73	.49	.36	.29	.24	.20	.18	.16	.14	.13	.12	.11	.10	.10	.09
18	.82	.55	.41	.33	.27	.23	.20	.18	.16	.15	.14	.13	.12	.11	.10
19	.92	.61	.46	.37	.30	.26	.22	.20	.18	.16	.15	.14	.13	.12	.11
20	1.02	.68	.51	.41	.34	.29	.25	.22	.20	.18	.17	.16	.15	.14	.13
21	1.13	.75	.56	.45	.37	.32	.28	.25	.22	.20	.18	.17	.16	.15	.14
22	1.24	.82	.61	.49	.41	.35	.31	.28	.24	.22	.20	.19	.18	.16	.15
23	1.36	.90	.67	.53	.45	.38	.34	.30	.26	.24	.22	.20	.19	.17	.16
24	1.48	.98	.73	.58	.49	.42	.37	.33	.29	.26	.24	.22	.21	.19	.18
25	1.61	1.06	.79	.63	.53	.45	.40	.35	.31	.28	.26	.24	.22	.20	.19
26	1.74	1.15	.86	.68	.57	.49	.43	.38	.34	.31	.29	.26	.24	.22	.21
27	1.88	1.24	.93	.73	.61	.53	.46	.41	.37	.33	.31	.28	.26	.24	.23
28	2.02	1.33	1.00	.78	.66	.57	.50	.45	.40	.36	.33	.30	.28	.26	.25
29	2.27	1.43	1.07	.84	.71	.61	.53	.48	.43	.39	.35	.32	.30	.28	.27
30	2.32	1.54	1.15	.91	.76	.65	.57	.51	.46	.42	.38	.35	.32	.30	.29
31	2.48	1.64	1.22	.97	.81	.69	.61	.54	.49	.44	.40	.37	.34	.32	.30
32	2.65	1.75	1.30	1.04	.87	.74	.65	.58	.52	.47	.43	.40	.37	.34	.32
33	2.82	1.86	1.38	1.10	.92	.79	.69	.61	.55	.50	.46	.42	.39	.36	.34
34	3.00	1.97	1.47	1.17	.98	.84	.73	.65	.59	.53	.49	.45	.42	.39	.36
35	3.19	2.09	1.56	1.24	1.04	.89	.77	.69	.62	.56	.52	.48	.44	.41	.38
36	3.38	2.22	1.65	1.32	1.10	.94	.82	.73	.66	.60	.55	.51	.47	.44	.41
37	3.58	2.34	1.74	1.39	1.16	.99	.87	.77	.69	.63	.58	.53	.49	.46	.43
38	3.78	2.47	1.84	1.47	1.22	1.05	.92	.81	.73	.66	.61	.56	.52	.49	.46
39	3.99	2.60	1.94	1.55	1.29	1.10	.96	.85	.77	.70	.64	.59	.55	.51	.48
40	4.21	2.74	2.04	1.63	1.34	1.16	1.01	.90	.81	.74	.68	.63	.58	.54	.51
41	4.43	2.88	2.15	1.71	1.41	1.22	1.06	.94	.85	.77	.71	.66	.61	.56	.53
42	4.66	3.03	2.26	1.80	1.49	1.28	1.12	.99	.89	.81	.74	.69	.64	.59	.56
43	4.90	3.18	2.37	1.89	1.56	1.34	1.17	1.04	.93	.85	.77	.72	.67	.62	.58
44	5.14	3.33	2.49	1.98	1.64	1.41	1.23	1.09	.98	.89	.81	.76	.70	.65	.61
45	5.39	3.49	2.60	2.07	1.72	1.47	1.28	1.14	1.02	.93	.85	.79	.73	.68	.64
46	5.65	3.65	2.72	2.16	1.80	1.54	1.34	1.19	1.07	.97	.89	.82	.76	.71	.67
47	5.92	3.81	2.84	2.25	1.88	1.60	1.40	1.24	1.12	1.01	.93	.86	.79	.74	.70
48		3.98	2.96	2.35	1.96	1.67	1.46	1.30	1.17	1.06	.97	.89	.83	.77	.73
49		4.15	3.08	2.45	2.04	1.74	1.52	1.35	1.22	1.10	1.01	.93	.86	.80	.76
50		4.32	3.21	2.55	2.13	1.82	1.59	1.41	1.27	1.15	1.05	.97	.90	.84	.79
51		4.50	3.34	2.65	2.21	1.89	1.65	1.46	1.32	1.20	1.09	1.01	.93	.87	.82
52		4.68	3.47	2.76	2.30	1.96	1.72	1.52	1.37	1.25	1.14	1.05	.97	.91	.85
53		4.87	3.61	2.87	2.39	2.04	1.78	1.58	1.42	1.30	1.18	1.09	1.01	.94	.88
54		5.06	3.75	2.98	2.48	2.12	1.85	1.64	1.48	1.35	1.23	1.13	1.05	.98	.92
55		5.26	3.89	3.09	2.57	2.20	1.92	1.70	1.53	1.40	1.27	1.17	1.09	1.02	.95
56		5.46	4.04	3.21	2.67	2.28	1.99	1.77	1.59	1.45	1.32	1.22	1.13	1.06	.99
57		5.67	4.19	3.33	2.77	2.36	2.06	1.83	1.65	1.50	1.37	1.26	1.17	1.10	1.02
58		5.88	4.34	3.45	2.87	2.45	2.13	1.90	1.71	1.55	1.42	1.31	1.21	1.14	1.06
59		6.09	4.49	3.57	2.97	2.53	2.21	1.96	1.77	1.60	1.47	1.35	1.25	1.18	1.10
60		6.31	4.65	3.70	3.07	2.62	2.29	2.03	1.83	1.66	1.52	1.40	1.30	1.22	1.14
61		6.53	4.81	3.82	3.17	2.71	2.37	2.10	1.89	1.71	1.57	1.45	1.34	1.26	1.18
62		6.76	4.97	3.95	3.27	2.80	2.45	2.17	1.95	1.77	1.62	1.50	1.39	1.30	1.22
63		6.99	5.13	4.08	3.37	2.89	2.53	2.24	2.01	1.83	1.67	1.55	1.43	1.34	1.26
64		7.23	5.30	4.21	3.48	2.98	2.61	2.31	2.08	1.89	1.73	1.60	1.48	1.38	1.30

Source: Diamond Chain Company, Indianapolis, IN.

imately 80 times the pitch may be considered an approved or acceptable *maximum center distance* (the normal center distance is 40 times the chain pitch). Very long center distances will only result in excessive catenary tension in the chain and may cause the chain to jump sprocket teeth. If such distances are unavoidable, the chain should be supported by guides or by idlers. Better yet, one may consider the possibility that two or more drives be substituted for a single drive, or that double-pitch chain be considered in conditions of slow speed and moderate loading. Of course, this is always recommended for extremely long center distances.

Drives may be installed with either adjustable or fixed center distances. However, experience indicates that adjustable center distances make for good, sound design practices (at least one of the shafts should be mounted in such a way that it can be moved outward from the theoretically exact dimension). Hence, adjustable centers simplify the control of chain slack. the shaft centers adjustment design provisions should allow for at least 1½ pitches. This will provide a means to adjust for chain-wear elongation as it occurs during the service life of the chain. Beyond this point, the chain length can be shortened by removal of links and the centers adjusted inward to compensate. On the other hand, a little slack is needed to allow the chain to position itself on the sprocket without inducing tension in the chain, which could cause excessive wear. For chain drives with nonadjustable (fixed) center distances, an idler or "shoe" may be employed to control incurred blacklash from normal chain wear. A center distance of approximately 30 pitches is preferred, and the center distance should be selected to produce an initial chain tightness.

The center distance between sprockets is calculated by assuming (theoretically exact value) that the chain is wrapped snugly around both sprockets at exact pitch diameter, and chain at exact length. Since sprocket manufacturing tolerances are *minus* values only and chain length tolerance is *plus* values only, the calculated center distance will usually provide a satisfactory amount of slack in the chain's return strand. The center distance in chain pitches can be calculated approximately by the formula shown in Fig. 14. Again, using N for the larger sprocket and n for the smaller sprocket, L for the chain length in pitches; and where W is a factor that varies in value from 0.20264 to 0.22302, depending on the magnitude of the expression $(L - n)/(N - n)$ from 12.00 down to 1.008. Note that for $(L - n)/(N - n)$ greater than 12.00, use 0.20264, and for this less than 1.008, use 0.22302 (see Table 42). The sprocket center distance in inches is calculated by multiplying the center distance in pitches by the pitch in inches.

For *multisprocket* or *serpentined* drive arrangements, when all sprockets are of the same size and all are "inside" the chain, the chain length in pitches is equal to the sum of all consecutive center distances in pitches, plus the number of teeth on one sprocket.

The chain cannot have a fractional part of a pitch, so if the calculated length is not an integer, the designer must adjust it to the next higher whole number. For roller chain or inverted tooth chain, the length should, if possible, be adjusted to the nearest even number, to avoid using an offset link.

$$C = \tfrac{1}{4}\left\{L - \frac{N+n}{2} + \sqrt{\left[L - \frac{N+n}{2}\right]^2 - W(N - n)^2}\right\}$$

where:

C = Shaft Center Distance in pitches

L = Length of chain in pitches,

N = Number of teeth in larger sprocket,

n = Number of teeth in smaller sprocket,

The value of W in the formula is related to the ratio $(L-n)/(N-n)$. The value of W ranges from 0.20264 to 0.22302 as the ratio $(L-n)/(N-n)$ varies from 12 to 1.008. When a close approximation of center distance is satisfactory, as when center-distance adjustment if provided, W may be taken as $(2)/(\pi)^2$, or 0.20264. For fixed-center drives, where you need an accurate dimension, calculate the ratio and determine the true value of W from Table 5.

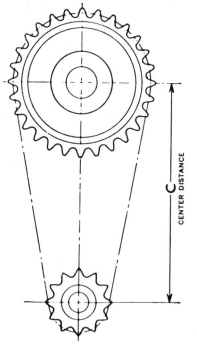

CENTER DISTANCE

C

Figure 14 Center distance formula—calculation of shaft centers. (Courtesy of Jeffrey Chain Corporation, Whitney Chain Operations, Morristown, TN.)

Table 42 Center Distance—Values of the Variable W

$\frac{L \cdot n}{N \cdot n}$	W	$\frac{L \cdot n}{N \cdot n}$	W	$\frac{L \cdot n}{N \cdot n}$	W	$\frac{L \cdot n}{N \cdot n}$	W	$\frac{L \cdot n}{N \cdot n}$	W
12.00	0.20264	2.70	0.20297	1.33	0.20574	1.130	0.20929	1.034	0.21648
11.00	0.20265	2.60	0.20301	1.32	0.20584	1.120	0.20967	1.032	0.21681
10.00	0.20266	2.50	0.20306	1.31	0.20595	1.110	0.21011	1.030	0.21715
9.00	0.20267	2.40	0.20311	1.30	0.20607	1.100	0.21061	1.028	0.21751
8.00	0.20268	2.30	0.20318	1.29	0.20619	1.090	0.21119	1.026	0.21789
7.00	0.20269	2.20	0.20325	1.28	0.20632	1.080	0.21184	1.024	0.21828
6.00	0.20271	2.10	0.20334	1.27	0.20645	1.070	0.21258	1.022	0.21869
5.00	0.20272	2.00	0.20345	1.26	0.20659	1.060	0.21343	1.020	0.21914
4.80	0.20272	1.90	0.20358	1.25	0.20674	1.058	0.21360	1.019	0.21937
4.60	0.20273	1.80	0.20374	1.24	0.20689	1.056	0.21378	1.018	0.21961
4.40	0.20273	1.70	0.20396	1.23	0.20705	1.054	0.21397	1.017	0.21987
4.20	0.20274	1.60	0.20424	1.22	0.20721	1.052	0.21416	1.016	0.22014
4.00	0.20276	1.50	0.20463	1.21	0.20738	1.050	0.21437	1.015	0.22042
3.80	0.20277	1.40	0.20517	1.20	0.20755	1.048	0.21458	1.014	0.22072
3.60	0.20279	1.39	0.20523	1.19	0.20774	1.046	0.21481	1.013	0.22104
3.40	0.20281	1.38	0.20530	1.18	0.20793	1.044	0.21505	1.012	0.22137
3.20	0.20284	1.37	0.20538	1.17	0.20815	1.042	0.21531	1.011	0.22173
3.00	0.20288	1.36	0.20546	1.16	0.20839	1.040	0.21558	1.010	0.22212
2.90	0.20291	1.35	0.20555	1.15	0.20865	1.038	0.21586	1.009	0.22255
2.80	0.20294	1.34	0.20564	1.14	0.20895	1.036	0.21616	1.008	0.22302

Source: Jeffrey Chain Corporation, Whitney Chain Operations, Morristown, TN.

For other types of multisprocket drives, an accurate drafter's layout (to large scale) should be made. Hence, for extreme accuracy in chain length calculations involving these parameters, the following procedure is recommended.

1. Adjust dividers to the exact scale pitch, verifying the adjustment by stepping off exactly 10 to 20 pitches, total scale measurement, to provide a summation check.
2. Check the accuracy of each pitch circle, after drawing it in lightly, by stepping off the exact number of pitches.
3. Draw the tangents to represent the pitch line of the chain.
4. Starting on a tangent line, near a point of tangency, step off the arcs of chain contact on the pitch circle with the dividers. The last step with the dividers should be on the text tangent line.
5. All arcs of contact and tangents, and use the next greater whole number of pitches for the chain length. Should the layout indicate an objectionable amount of slack chain, this slack may be eliminated by changing the numbers of sprocket teeth or the shaft locations or by using an idler sprocket (see Fig. 15).

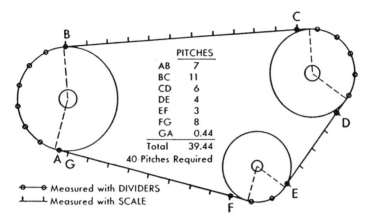

The table within the figure:

PITCHES	
AB	7
BC	11
CD	6
DE	4
EF	3
FG	8
GA	0.44
Total	39.44

40 Pitches Required

●——● Measured with DIVIDERS
⌐——⌐ Measured with SCALE

Figure 15 Multi-sprocket chain length calculation method. (Courtesy of Diamond Chain Company, Indianapolis, IN.)

VII. ROLLER CHAIN INSTALLATION

For trouble-free operation, careful and accurate installation is essential. Before installing sprockets, make sure the shafts are parallel and level. If for some reason the shafts must be set at a decline or incline from level position, be certain both shafts are at exactly the same angle.

By using a "torpedo" level or some other type of small level, horizontal shafts may easily be checked for levelness. And by using a tape measure, the distance between shafts on both sides should be made equal (square). (See Fig. 16.) Shafts are now both level and aligned. On shafts operating at angles, it is best to use a direct-reading "angle finder" to get the correct angle of both shafts so they can be matched.

After you have made sure the shafts are parallel and level, locate the sprockets as near as possible to the shaft bearings. Sprocket alignment after mounting can be checked with a straightedge held against the side of the

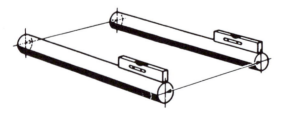

Figure 16 Common method used for leveling and aligning (squaring) shafts. (Courtesy of American Chain Association, 1982.)

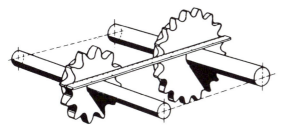

Figure 17   Common method used for sprocket alignment. (Courtesy of American Chain Association, 1982.)

sprockets. (See Fig. 17.) Keys and set screws should then be checked for tightness.

For drives with adjustable shafts, decrease the center distance for easier chain installation. With the chain already assembled, wrap it around the teeth of one sprocket, then stretch it across to the next and slip the chain over the sprocket. Do not "run" the chain onto the sprocket. After the chain is properly engaged in the teeth of the sprockets, tighten the drive to the proper tension. Alignments should then be rechecked.

For drives with nonadjustable (fixed) shafts, the center distance will have to be at the specified dimension. Of course, the chain will have to be apart, since there is no way the chain can be slipped over both sprockets.

To help accomplish this, what is known as a *connecting link* will have to be used, so the "free ends" of the chain may be attached. A connecting link, commonly referred to as a "master link" in the plant or shop, is a special type of pin link used to provide easy installation or removal of a roller chain with an *even number* of pitches. There are two types of connecting links for roller chains: drive fit and slip fit.

The *drive fit* type is similar to a standard pin link, in that each pin is riveted to the link plate at each end. The other end of the pin has somewhat of an interference fit in its link plate. The free link plate is retained by either a spring clip, cotter pin, or roll pin. (See Fig. 18.)

The *slip fit* type (not shown) is similar in design with each pin riveted to the link plate at one end. However, the other end of the pin is a slip-fit (a light press fit) in its link plate. This facilitates easy removal of the link plate for disassembly or assembly. Note that for greatest security, drive-fit connecting links are recommended. On less critical applications where speed is slow and maximum convenience for coupling and uncoupling is desired, slip-fit styles may be used.

If an "odd" number of pitches is required (avoid this if possible), a roller link at one end of the open chain will have to be substituted with what is known as an *offset link;* this is connected to the roller link on the other end, so as to

SPRING CLIP CONNECTING LINK   COTTER PIN CONNECTING LINK

Figure 18   The "free" link plate on a connecting link is held securely in place by either a spring clip or cotter pins. Sometimes roll pins replace cotters. (Courtesy of Browning Manufacturing, Maysville, KY.)

make the chain endless. (See Fig. 19.) An offset link is a combination of a roller link and a pin link and is used *only* when there is an odd number of pitches in a chain strand. These are also available for double-pitch roller chain (with standard rollers or carrier rollers). This offset link offers high fatigue strength and long life. Also, when an offset link is used, no connecting link is required.

Still another method of providing for an odd number of pitches in a chain strand is use of an *offset section*. (See Fig. 20.) They offer greater stability than their counterpart, the single offset link. A riveted offset section should be used when severe operating conditions exist. Since they are of riveted constructed, a standard connecting link will have to be used as the point of attachment.

Now you are ready to install the chain. Fit the chain on both sprockets (making sure the chain is properly seated on the teeth), bringing the free chain ends together on one sprocket. Next, insert the pins of the connecting link into the available holes of the two roller links of the chain (see Fig. 21); then place

OFFSET LINK

Figure 19   On a chain with an odd-number of pitches, an offset link is used to connect chain ends. (Courtesy of Browning Manufacturing, Maysville, KY.)

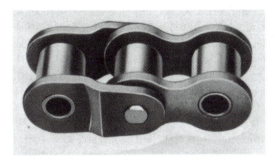

Figure 20 If more stability is required when connecting chain ends on an odd-number of pitches, an offset section is recommended. (Courtesy of American Chain Association, 1982.)

Figure 21 A connecting link is used to keep the chain together. (Courtesy of Rexnord Corporation, Link-Belt Roller Chain Operation, Indianapolis, IN.)

the free side into the protruding pins of the connecting link. Last, depending on which style is used, secure the side plate with the cotters, roll pins, or spring clip that came in the package. Once the fasteners has been installed, do not turn on the drive until you have checked the free side plate, making sure it is not squeezing against the sprocket teeth. This could lead to interference with the free-flexing of the chain joint.

If you experience difficulty when trying to insert the connecting link, a special tool called a *chain puller* may be helpful. Normal procedure is to first rotate the drive or rewrap the chain, so that the free ends of the chain are away from the sprocket teeth; then the chain puller can be properly hooked around the end rollers, pulling them together so you can install the connecting link quickly and easily.

A. Roller Chain Tensioning

For fixed shaft and adjustable shaft drives, recommended chain tensioning shall be as follows: The return strand of a chain generally has some slack. This slack results in a sag called *catenary sag* of the chain. *Catenary* refers to the curve theoretically formed by a perfectly flexible, uniformly dense, inextensible cable (in this case chain strand) suspended from two points. This sag must be of the correct amount if the chain is to operate properly. On a horizontal drive this "flexure," measured from a straight line, should be approximately 3 percent, but not less than 2 percent, of the center distance between shafts. For 45 to 90° drives, slack (sag) should be 0.5 to 1 percent. Generally, the preferred drive arrangement is to have the return side on the bottom of the drive; however, this is not always the case. When a chain is being tensioned with the slack side on top of the drive, this same catenary method can be used, but the straightedge must be placed as shown in Fig. 22.

To determine the amount of sag (at the top or bottom of the drive), pull one side of the chain taut, allowing all the excess chain to accumulate in the opposite span. As illustrated, place a straightedge over the slack span, and pushing down at the center of the span using a steel ruler, measure the amount of sag. If necessary, adjust drive centers for proper sag that will result in correct chain tension.

Never operate a drive with both sides of the chain taut, or excessive chain and bearing wear will result. On the other hand, if slack side is too loose, vibration and unwanted chain whipping will again lead to excessive wear or cause loss of power. Additionally, when using chain guards or casings, be sure to provide sufficient clearance for the slack strand.

VIII. ROLLER CHAIN MAINTENANCE

It is important that regular preventative maintenance schedules be set up for roller chain drives. Chain drives are no different from other pieces of equip-

Figure 22   The preferred method for measuring the catenary sag of a chain operating on a horizontal drive with slack side on top. Sag should be about 2 to 3 percent of the horizontal center distance. (Courtesy of Rexnord Corporation, Link-Belt Roller Chain Operation, Indianapolis, IN.)

ment in operation. Wherever mechanical contact is present, wear and fatigue is inevitable. Usually when performing periodic maintenance duties, a quick visual inspection of the drive components is sufficient to catch any initial faults before they develop into major problems. Basically, what you are looking for when performing this visual inspection is improper chain tension, insufficient or improper lubrication, apparent misalignments of shafts and sprockets, broken or damaged components, and evidence of wear.

Perhaps one of the first things to do when carrying out a maintenance inspection is to check the chain. A dead giveaway for misalignment problems is that unusual wear can be seen on the inner surface of the roller link side plates. It may also be noticed that a groove has been worn into the side of the sprockets, just below the tooth form. This type of wear also generally accompanies misalignment problems.

Even a drive that has had proper tension maintained and that has operated under normal conditions will eventually display excessive tooth wear. This generally will cause a "hook-shaped" appearance on the tooth form and produce a sharp edge on the end of the tooth. When sprockets are in such a condition, their life may be prolonged by the simple, yet effective method of reversing them on their shaft, so as to expose the less-worn tooth surfaces on the opposite side to the chain. Double-duty sprockets with odd numbers of teeth have a particular tooth engage the chain only every other revolution, thus automatically doubling sprocket tooth life. On double-duty sprockets with an even number of teeth, sprocket life can be doubled simply by the chain being manually advanced one tooth periodically. Of course, these sprockets can also be reversed but over a much longer period.

Sometimes repair work will have to be performed on chains because of elongation (stretching) from normal wear or broken sections, thus requiring the proper removal of certain sections from a chain. The term "cutting the chain" is commonly used when referring to link or section removal. The chain is not actually cut, but instead the two pins of a link are driven or forced out of the link plate. It is important that while the pins are being taken out, the plates of adjacent links not be deformed.

To help simplify the task of removing chain links it is recommended that tools specially designed for this purpose be employed. These tools are called *chain breakers, chain cutters,* and *chain detachers.* Regardless of the style of tool utilized, they all serve one basic function: to properly extract the pins from a section and to prevent possible damage to adjacent components.

The chain detacher is a handy bench tool that is usually kept in the shop, rather than taken out to the job site when maintenance repairs on chains are necessary. Chain detachers generally consist of two parts: a fork and an anvil block or simply two hardened steel jaws that have provisions for clamping. Essentially, chain detachers are a special type of bench vise. Once the chain section has been installed, the pins can easily be driven squarely out of the side plate.

In order to remove these pins correctly, a hammer and pin-punch will be required. If the chain has pins that are riveted, it is important that their heads be ground off flush, before they are driven through the side plate. This will give added protection against possible damage to the pin-punch.

When using chain breakers and chain cutters, normally it will not be necessary to grind off the heads on riveted pins. This is because they make use of mechanical advantages, such as the lever and the screw. Both are convenient, portable tools for taking out in the field when a link or section removal is necessary. Thus, there is no need for hammer and punch.

IX. ROLLER CHAIN LUBRICATION

Proper lubrication of a roller chain is of extreme importance if long life and satisfactory service are to be achieved. Unless properly applied, the best lubricant cannot perform its functions. For slow speeds and light loads, simple methods of application give satisfactory results, but as speeds and loads increase, more precise methods are required. There are three basic systems for lubricating chains, referred to as type A, manual or drip lubrication; type B, bath or slinger/disc lubrication; and type C, oil stream lubrication.

Manual lubrication: Oil is applied liberally with a brush or spout can. The frequency of application is governed by local conditions and the chain speed.

Drip lubrication: A continuous supply of oil drops are directed between the link plate edges from a sight-feed lubricator, feeding through a wick-packed distributing pipe. Oil delivery rates should be regulated to assure complete lubrication without flooding.

Bath lubrication: The lower strand of the chain runs through a sump of oil in the drive housing. The oil level should reach the pitch line of the chain at its lowest point while operating.

Slinger/disc lubrication: The chain operates above a sump of oil. A rapidly rotating disc picks up the oil and "slings" it against a collection plate, from which oil drips into a trough and onto the chain.

Oil stream lubrication: The oil stream method is the most satisfactory of the three types of lubrication for chain drives running at relatively high speeds and loads. A pump delivers oil under pressure to nozzles that direct an oil spray onto the chain. (See Fig. 23.)

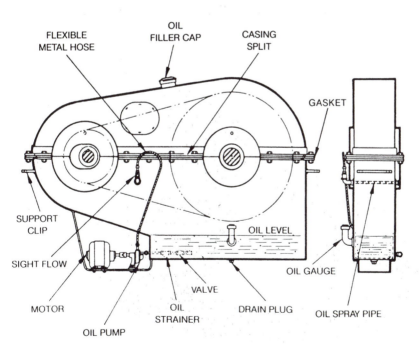

Figure 23 Type C—oil steam lubrication, also referred to as Type III—forced feed lubrication. Lubricant is supplied in a continuous stream by a circulating pump and distribution pipe or hose. (Courtesy of Ramsey Products Corporation, Charlotte, NC.)

X. STANDARD PRECISION INVERTED-TOOTH CHAIN

A. Silent Chain Drives (ANSI B29.2)

Silent chain, also called inverted-tooth chain, is used mainly for the transmission of mechanical power between rotating parallel shafts and for timing applications requiring high speed, no slip, and quiet operation. In addition to power transmission, silent chain is also available for conveying applications. For this purpose flat-back links are normally used.

Silent chain drives are renowned for their economical and efficient transmission (it may be as high as 99 percent) of power over a wide range of loads and speeds. These versatile chain drives combine the flexibility and quietness of a belt, the positive action and durability of gears, and the convenience and efficiency of chain. Basically, the "inverted-teeth" on these chains are designed so that they engage the cut-tooth wheels (sprockets) in a manner similar to the way a rack engages a gear. A silent chain drive is illustrated in Fig. 24.

The driving force behind the silent chain may best be described by the following characteristics: The chain passes over the face of the wheel like a belt and the wheel teeth do not project through it; the chain engages the wheel by means of teeth extending across the full width of the underside, with the excep-

Figure 24 Typical silent chain drive. (Courtesy of Rexnord Corporation, Link-Belt Roller Chain Operation, Indianapolis, IN.)

tion of those chains having a central guide link; the chain teeth and wheel teeth are of such a shape (in profile) that as the chain pitch increases through wear at the joints, the chain shifts outward upon the teeth, thus engaging the wheel on a pitch circle of increasing diameter. The result of this action is that the pitch of the wheel teeth increases at the same rate as the chain pitch. Another distinguishing feature of the silent chain is that the power is transmitted by and to all the teeth in the arc of contact, irrespective of the increasing pitch due to elongation. In other words, the silent chain and sprockets are designed so that when the chain pitch elongates (stretches) as a result of normal joint wear, the chain engages the sprocket on a correspondingly larger pitch circle. The load is therefore evenly distributed over the sprocket teeth engaging the chain.

The links of silent chain engage the sprocket with less sliding action and less impact than occurs in other types of chain drives. This results in longer sprocket life and the much quieter operation from which silent chain gets its name. Reduced impact on engagement also makes higher operating speeds possible.

1. Silent Chain Versus Other Methods of Mechanical Power Transmission

Each method of power transmission has features that for some applications make it more desirable than another method. There are overlapping areas of application but some of the comparative advantages of silent chain are as follows:

Silent chain drives compared with belts: (1) No slippage; (2) greater efficiency; (3) larger ratios possible; (4) require less space; (5) can withstand heavier overloads; (6) can be used with short center distance at higher ratios; (7) less affected by extremes in temperature or humidity or by tramp lubricants; (8) lower bearing loads.

Silent chain drives compared with gears: (1) Quieter than spur gears; (2) center distance much less restricted; (3) shaft location and parallelism tolerances are broader; (4) lower bearing loads; (5) no end thrust as with helical gears; (6) greater elasticity to absorb shock.

Silent chain drives compared with roller chain: (1) Much higher maximum operating speeds; (2) much quieter; (3) smoother transmission of power; (4) less loading impact and sliding as the chain engages the sprocket; (5) higher efficiency; (6) longer sprocket life.

2. Silent Chain Construction

Silent chain is manufactured to specification for "inverted-tooth (silent) chains and sprocket teeth," standard B29.2. It should be mentioned that this standard is intended primarily to provide for *interchangeability* between chains and sprockets of different manufacturers; that is, *does not* provide for a standardization of "joint" components and "link plate" contours, which differ in each manufacturers' design. However, all manufacturers' links are contoured to engage the standard sprocket tooth so that joint centers lie on the pitch diameter of the sprocket. Figure 25 shows the style of joint design used by one manufacturer, Ramsey Products Corporation (Charlotte, NC).

Silent chain is constructed of a series of pinned leaf links connected together with joint components designed to minimize friction and wear when the chain flexes as it engages and disengages the sprocket. The link plates of the chain are designed with cylindrical bushings that are interference-fit into the link plate to prevent rotation; they form rows of straight-sided teeth extending across the width of the inside of the chain which engage straight flank teeth on the sprockets.

Perhaps the most critical part of the chain is the pinned connection, which is not discussed in the ANSI standard. Detailed designs have been developed to reduce wear in these connections. The successful designs reduce sliding and promote "rolling contact" in the joints and between the chain and sprocket teeth. Several types of chain joints are manufactured to provide various degrees of quiet operation and long service life: round pin with bushing, round pin with segmented bushing, rocker pin, and rolling pins.

Just like their counterpart, the positive drive (timing belt), silent chains have the inherent characteristic of wanting to slide or ride off the drive while in motion. To help remedy this problem silent chain is retained on the sprockets by guide links, also termed side flange, or by flanges on at least one sprocket. Practically all silent chains are constructed of link plates with guide plates added to the chain to prevent this "lateral" movement. It should be mentioned that flanged sprockets are generally found on older drive systems and are seldom used today. However, a no-guide or nonguide chain is still available for replacement on these existing drives.

Silent chains are constructed as *center-guide, side-guide, duplex,* and *reversible.* Most methods employ chain guides that ride in a sprocket groove. Also, the majority of silent chain drives are of the center-guide type, where one or two guide links in the width of the chain engage the "guide grooves" cut in the sprocket teeth. A less frequently used construction is the side-guide type,

Figure 25 Typical silent chain or inverted-tooth chain. Center or middle-guide design (most widely used style). (Courtesy of Ramsey Products Corporation, Charlotte, NC.)

assembled with guide links at the outer edges of the chain that straddle the sprocket face.

The duplex assembly has teeth that project on both sides of the chain in every link. This allows drives that require sprocket teeth to contact both sides of the chain, as in serpentine drives or drives requiring idler sprockets where the chain must flex in either direction. The duplex chain uses a sprocket that has grooved teeth to allow for clearance of the reversed link plate. The reversible chain assembly is the same as the duplex chain except the teeth project on both sides every other pitch. Every other row of sprocket teeth must be grooved to allow the reversed link plates to clear the sprocket teeth.

3. Standard Silent Chain Numbers

The standard silent chain number or designation in the United States consists of a combined letter-and-number symbol: (1) a two-letter symbol SC; (2) one or two numerical digits indicating the *pitch* in eighths of an inch; and (3) two or three numerical digits indicating the chain *width* in quarter-inches.

For example, "SC302" designates a silent chain of 3/8-in. pitch and ½-in. in width. "SC1012" designates a silent chain of 1¼-in. pitch (10/8) and 3 in. wide (12/4). Link plates of silent chain manufactured to the ANSI Standard B29.2 are stamped with a symbol indicating the pitch. For example, SC-6, or simply 6, indicates a chain with a ¾-in. pitch (6/8).

Link plates of silent chains *not conforming* to the ANSI B29.2 standard are generally stamped with a number of the manufacturers' choice. Standard silent chain pitches are listed in Table 43. Also included in the standard is a 3/16-in. pitch small pitch silent chain, covered by ANSI B29.9. This standard covers the sprocket tooth form, standard sprocket diameters, general chain proportions, sprocket face dimensions and horsepower ratings. Standard horsepower ratings for the different chain sizes are shown in Tables 44 through 52 and are based upon *per inch of chain width*.

Table 43 Silent Chain Identification Stampings

CHAIN NUMBER	CHAIN PITCH	STAMP
SC3 (Width in ¼ in.)	$^3/_8$	SC3 or 3
SC4 "	$^1/_2$	SC4 or 4
SC5 "	$^5/_8$	SC5 or 5
SC6 "	$^3/_4$	SC6 or 6
SC8 "	1	SC8 or 8
SC10 "	1¼	SC10 or 10
SC12 "	1½	SC12 or 12
SC16 "	2	SC16 or 16

Table 44   Horsepower Capacity per Inch of Width, 3/16-in. Pitch Silent Chain (ANSI B29.2) (Standard Widths: 5/32, 7/32, 11/32, 13/32, 15/32, 17/32, 19/32, 21/32, 23/32, 25/32, 27/32, 29/32, 31/32)

Number of teeth	Revolutions per minute											
	500	600	700	800	900	1200	1800	2000	3500	5000	7000	9000
15	0.28	0.33	0.38	0.43	0.47	0.60	0.80	0.90	1.33	1.66	1.94	1.96
17	0.33	0.39	0.44	0.50	0.55	0.70	0.96	1.05	1.60	2.00	2.40	2.52
19	0.37	0.43	0.50	0.55	0.61	0.80	1.10	1.20	1.80	2.30	2.76	2.92
21	0.41	0.48	0.55	0.62	0.68	0.87	1.22	1.33	2.03	2.58	3.12	3.35
23	0.45	0.53	0.60	0.68	0.75	0.96	1.35	1.47	2.25	2.88	3.50	3.78
25	0.49	0.58	0.66	0.74	0.82	1.05	1.47	1.60	2.45	3.13	3.80	4.10
27	0.53	0.62	0.71	0.80	0.88	1.15	1.58	1.72	2.63	3.35	4.06	4.37
29	0.57	0.67	0.76	0.86	0.95	1.21	1.70	1.85	2.83	3.61	4.40	4.72
31	0.60	0.72	0.81	0.91	1.01	1.30	1.81	1.97	3.02	3.84	4.66	5.00
33	0.64	0.75	0.86	0.97	1.07	1.37	1.90	2.08	3.17	4.02	4.85	
35	0.68	0.80	0.92	1.03	1.14	1.45	2.03	2.21	3.41	4.27	5.16	
37	0.71	0.84	0.96	1.08	1.19	1.52	2.11	2.30	3.48	4.39	5.24	
40	0.77	0.91	1.04	1.16	1.29	1.64	2.28	2.50	3.77	4.76		
45	0.86	1.02	1.15	1.30	1.43	1.83	2.53	2.75	4.15	5.21		
50	0.95	1.12	1.27	1.37	1.58	2.00	2.78	3.02	4.52	5.65		
	Type I						Type II			Type III		

Source: American Chain Association, 1982.

Table 45   Horsepower Capacity per Inch of Width, 3/8-in. Pitch Silent Chain (ANSI B29.2) (Standard Widths: ½, ¾, 1, 1¼, 1½, 1¾, 2, 2¼, 2½, 2¾, 3, 3½, 4, 5, 6, 8)

No. of Teeth Small Sprocket	Revolutions Per Minute – Small Sprocket												
	100	500	1000	1200	1500	1800	2000	2500	3000	3500	4000	5000	6000
* 17	.46	2.1	4.6	4.9	5.3	6.5	6.9	7.9	8.5	8.8	8.8	–	–
* 19	.53	2.5	4.8	5.4	6.5	7.4	7.9	9.1	9.9	10	11	9.8	–
21	.58	2.8	5.1	6.0	7.3	8.3	9.0	10	11	12	12	12	10
23	.63	3.0	5.6	6.6	8.0	9.3	10	12	13	14	14	14	12
25	.69	3.3	6.1	7.3	8.8	10	11	13	14	15	15	15	14
27	.74	3.5	6.8	7.9	9.5	11	12	14	15	16	18	18	16
29	.80	3.8	7.3	8.5	10	12	13	15	16	18	19	19	18
31	.85	4.1	7.8	9.1	11	13	14	16	18	19	20	20	19
33	.90	4.4	8.3	9.8	12	14	15	18	19	21	21	21	20
35	.96	4.6	8.8	10	13	15	16	19	20	23	23	23	21
37	1.0	4.9	9.1	11	14	15	16	20	21	24	24	24	–
40	1.1	5.3	10	12	15	16	18	21	24	25	26	26	–
45	1.3	6.0	11	13	16	19	20	24	26	28	29	–	–
50	1.4	6.6	13	15	18	20	23	26	29	30	–	–	–
	Type I			Type II				Type III					

* For best results, smaller sprocket should have at least 21 teeth.

Source: American Chain Association, 1982.

Table 46 Horsepower Capacity per Inch of Width, ½-in. Pitch Silent Chain (ANSI B29.2)
(Standard Widths: ½, ¾, 1, 1¼, 1½, 1¾, 2, 2¼, 2½, 2¾, 3, 3½, 4, 5, 6, 8)

No. of Teeth Small Sprocket	Revolutions Per Minute – Small Sprocket										
	100	500	700	1000	1200	1800	2000	2500	3000	3500	4000
* 17	.83	3.8	5.0	6.3	7.5	10	11	11	11	11	–
* 19	.93	3.8	5.0	7.5	8.8	11	13	14	14	14	–
21	1.0	5.0	6.3	8.8	10	14	14	15	16	16	–
23	1.1	5.0	7.5	10	11	15	16	18	19	19	18
25	1.2	5.0	7.5	10	13	16	18	20	21	21	20
27	1.3	6.3	8.8	11	13	18	19	21	24	24	23
29	1.4	6.3	8.8	13	14	19	21	24	25	25	25
31	1.5	7.5	10	13	15	21	23	25	28	28	28
33	1.6	7.5	10	14	16	23	24	28	29	30	29
35	1.8	7.5	11	15	18	24	25	29	31	31	30
37	1.9	8.8	11	16	19	25	26	30	33	33	–
40	2.0	8.8	13	18	20	28	29	33	35	35	–
45	2.5	10	14	19	23	30	30	36	39	–	–
50	2.5	11	15	21	25	34	36	40	–	–	–

Type I   Type II   Type III

* For best results, smaller sprocket should have at least 21 teeth.

Source: American Chain Association, 1982.

Table 47 Horsepower Capacity per Inch of Width, 5/8-in. Pitch Silent Chain (ANSI B29.2)
(Standard Widths: 1, 1¼, 1½, 1¾, 2, 2½, 3, 4, 5, 6, 7, 8, 10)

No. of Teeth Small Sprocket	Revolutions Per Minute – Small Sprocket									
	100	500	700	1000	1200	1800	2000	2500	3000	3500
* 17	1.3	6.3	7.5	10	11	14	15	14	–	–
* 19	1.4	6.3	8.8	13	14	16	18	18	–	–
21	1.6	7.5	10	13	15	19	20	20	20	–
23	1.8	7.5	11	15	16	21	23	24	23	–
25	1.9	8.8	11	16	19	24	25	26	26	24
27	2.0	10	13	18	20	26	28	29	29	26
29	2.1	10	14	19	21	28	30	31	31	29
31	2.4	11	15	20	23	30	31	34	34	31
33	2.5	11	16	21	25	33	34	36	36	34
35	2.6	13	16	23	26	34	36	39	39	35
37	2.8	13	18	24	28	36	39	43	41	–
40	3.0	14	19	26	30	39	41	44	–	–
45	3.4	16	21	29	34	44	46	–	–	–
50	3.8	18	24	33	38	48	50	–	–	–

Type I   Type II   Type III

* For best results, smaller sprocket should have at least 21 teeth.

Source: American Chain Association, 1982.

Table 48 Horsepower Capacity per Inch of Width, ¾-in. Pitch Silent Chain (ANSI B29.2) (Standard Widths: 1, 1¼, 1½, 1¾, 2, 2½, 3, 3½, 4, 5, 6, 7, 8, 10, 12)

No. of Teeth Small Sprocket	Revolutions Per Minute – Small Sprocket								
	100	500	700	1000	1200	1500	1800	2000	2500
* 17	1.9	8.1	11	14	15	16	18	18	–
* 19	2.0	9.3	13	15	18	20	21	21	–
21	2.3	10	14	18	20	23	24	25	24
23	2.5	11	15	20	23	25	28	28	28
25	2.8	13	16	21	25	29	31	31	30
27	2.9	14	18	24	28	31	34	35	35
29	3.1	15	20	26	30	34	36	38	38
31	3.4	15	21	28	31	36	40	41	41
33	3.6	16	23	30	34	39	43	44	44
35	3.8	18	24	31	36	41	45	46	46
37	4.0	19	25	34	39	44	48	49	49
40	4.4	20	28	36	41	48	51	53	53
45	4.9	23	30	40	46	53	56	58	–
50	5.4	25	34	45	51	58	61	–	–
	Type I		Type II			Type III			

* For best results, smaller sprocket should have at least 21 teeth.

Source: American Chain Association, 1982.

Table 49 Horsepower Capacity per Inch of Width, 1-in. Pitch Silent Chain (ANSI B29.2) (Standard Widths: 2, 2½, 3, 4, 5, 6, 7, 8, 9, 10, 12, 14, 16)

No. of Teeth Small Sprockets	Revolutions Per Minute – Small Sprocket										
	100	200	300	400	500	700	1000	1200	1500	1800	2000
* 17	3.8	6.3	8.8	11	14	18	21	23	–	–	–
* 19	3.8	7.5	10	13	15	20	25	26	28	–	–
21	3.8	7.5	11	15	18	23	29	31	33	33	–
23	3.8	8.8	13	16	19	25	31	35	38	38	–
25	5.0	8.8	14	18	21	28	35	39	41	41	41
27	5.0	10	15	19	24	30	39	43	46	46	45
29	5.0	11	16	20	25	33	41	46	50	51	50
31	6.3	11	16	23	28	35	45	50	54	55	54
33	6.3	13	18	24	29	38	49	54	59	59	58
35	6.3	13	19	25	30	40	51	56	61	63	61
37	6.8	14	20	26	33	43	54	60	65	66	–
40	7.5	15	23	29	35	45	59	65	70	–	–
45	8.8	16	25	31	39	51	65	71	76	–	–
50	10	19	28	35	43	56	71	78	–	–	–
	Type I			Type II				Type III			

* For best results, smaller sprocket should have at least 21 teeth.

Source: American Chain Association, 1982.

Table 50 Horsepower Capacity per Inch of Width, 1¼-in. Pitch Silent Chain (ANSI B29.2) (Standard Widths: 2½, 3, 4, 5, 6, 7, 8, 9, 10, 12, 14, 16, 18, 20)

1¼" Pitch

No. of Teeth Small Sprockets	Revolutions Per Minute – Small Sprocket										
	100	200	300	400	500	600	700	800	1000	1200	1500
* 19	5.6	10	15	20	24	26	29	31	34	35	–
21	6.3	11	18	23	26	30	33	36	40	41	–
23	6.9	13	19	24	29	34	36	40	45	46	46
25	7.5	14	20	26	31	36	40	44	50	53	53
27	8.0	15	23	29	35	40	44	49	54	58	58
29	8.6	16	24	31	38	43	48	53	59	63	64
31	9.3	18	26	34	40	46	51	56	64	68	69
33	9.9	19	28	35	43	49	55	60	69	73	74
35	11	20	29	38	45	53	59	64	73	78	78
37	11	21	30	40	48	55	63	68	76	81	–
40	12	24	34	44	53	60	68	74	83	88	–
45	13	26	38	49	59	68	75	81	91	–	–
50	15	29	43	54	65	74	83	90	100	–	–
	Type I			Type II				Type III			

* For best results, smaller sprocket should have at least 21 teeth.

Source: American Chain Association, 1982.

Table 51 Horsepower Capacity per Inch of Width, 1½-in. Pitch Silent Chain (ANSI B29.2) (Standard Widths: 3, 4, 5, 6, 7, 8, 9, 10, 12, 14, 16, 18, 20, 22, 24)

1½" Pitch

No. of Teeth Small Sprockets	Revolutions Per Minute – Small Sprocket										
	100	200	300	400	500	600	700	800	900	1000	1200
* 19	8.0	15	21	28	31	35	39	40	41	43	–
21	8.8	16	24	30	36	40	44	46	49	49	–
23	10	19	26	34	40	45	49	53	55	56	55
25	10	20	29	38	44	50	55	59	61	65	64
27	11	23	31	40	48	54	60	64	68	70	70
29	13	24	34	44	51	59	65	70	74	75	76
31	14	25	36	46	55	64	70	75	79	81	83
33	14	28	39	50	59	68	75	80	85	88	89
35	15	29	41	53	63	71	79	85	90	93	94
37	16	30	44	59	66	76	84	90	96	99	–
40	18	33	48	66	73	83	90	98	105	–	–
45	19	38	54	68	81	93	101	108	113	–	–
50	21	41	59	75	89	101	111	118	–	–	–
	Type I		Type II					Type III			

* For best results, smaller sprocket should have at least 21 teeth.

Source: American Chain Association, 1982.

Table 52 Horsepower Capacity per Inch of Width, 2-in. Pitch Silent Chain (ANSI B29.2) (Standard Widths: 4, 5, 6, 7, 8, 10, 12, 14, 16, 18, 20, 22, 24, 30)

2'' Pitch

No. of Teeth Small Sprocket	Revolutions Per Minute – Small Sprocket								
	100	200	300	400	500	600	700	800	900
* 19	14	26	36	44	50	54	56	–	–
21	16	29	40	50	53	63	65	–	–
23	17	33	45	55	64	70	74	75	–
25	18	35	49	61	70	78	83	85	85
27	20	38	54	66	78	85	91	94	94
29	21	41	58	73	84	93	99	103	103
31	23	44	63	78	90	100	106	110	110
33	25	46	66	83	96	106	114	118	118
35	26	50	71	88	103	114	121	125	125
37	28	53	75	93	110	124	128	131	–
40	30	58	81	101	118	129	138	141	–
45	34	64	90	113	131	144	151	–	–
50	38	71	100	125	144	156	–	–	–

Type I   Type II   Type III

* For best results, smaller sprocket should have at least 21 teeth.

Source: American Chain Association, 1982.

4. Silent Chain Sprockets

Silent chain sprockets, also called wheels by some manufacturers, have teeth that are precision machined (cut-tooth) to conform to established standards. The dimension, tooth form, and tolerances are specified under ANSI B29.2. Except for tooth form, silent chain sprocket design parallels the general design practice for standard roller chain sprockets, as previously covered.

Sprockets for American National Standard silent chains have teeth with straight-line (flank) working faces. The tops of the teeth may be rounded or square. Bottom clearance below the working face is not specified but must be sufficient to clear the chain teeth. The standard tooth form is designed to mesh with link plate contours having an included angle of 60°. Sprocket teeth may be cut by either a straddle cutter or a hob.

Similar to the roller chain sprocket, silent chain sprockets or wheels are manufactured in three basic types identified by their hub arrangement. These are classified again, as types A, B, and C. The type A wheels, also called plate or plane wheels, do not have any hub projections. Type B wheels have a hub on one side only. Type C wheels have hubs on both sides. The hub projections are usually equidistant from the center line of the wheel. Also, type C wheels may be divided into three groups: *tapered bore*—for a tapered shaft or a split tapered bushing; *split sprocket*—for installation between bearings or other obstructive components; and *clamp hub*—for secure mounting of large sprockets on large-diameter shafts. In other words, when required these special features available on type C wheels are used to facilitate installation or replacement.

5. Silent Chain Installation

Correct installation of silent chain drives requires that the shafts and the sprockets (wheels) be accurately aligned. Shafts must be set level, or, if inclined from a level position, both shafts must be at exactly the same angle. Shafts must also be positioned parallel within very close limits. It cannot be emphasized enough that accurate alignment of silent chain drives is an important prerequisite to maximize chain-and-sprocket life. The same basic alignment, chain tension, and lubrication steps covered in detail for roller chains are applicable and should be followed for silent chain drives.

Horizontal shaft alignment may be obtained with the aid of a small level (machinist's or torpedo). This will help tell when both shafts are in exact horizontal position. Then shafts may be adjusted for parallel alignment, using any suitable measuring device such as calipers, feeler bars, or verniers. The distance between shafts on both sides of the sprockets must be equal (shafts must be squared). For drives with fixed shafts the center distance must be set at the exact specified dimension. And for adjustable shaft drives, the distance should be less than final operating distance for easier chain installation.

It is equally important that silent chain sprockets be in axial alignment for correct chain-and-sprocket tooth engagement. To accomplish this, apply a suitable straightedge to the machined side surfaces of both sprockets. This procedure is shown in Fig. 26. Next, tighten set-screws in sprocket hubs to guard against lateral movement and to hold the key in place.

Before installing the chain recheck all the preceding adjustments and correct any that may have been disturbed. Check the new chain to be sure it is free from contamination (dirt, grit, etc.). If a used chain is being reinstalled, check also to be sure it is free from abrasive, tramp contaminates. In either case, clean and relubricate if necessary. Next, bring the free ends of the chain together on one sprocket and install the necessary connecting hardware. It is important to keep in mind that since joint components vary with each manufacturer, the method of assembly will also vary. In general, after the components are assembled the pin is retained by a cotter, or is "peened" over the washer to rivet it in place. Figure 27 shows a typical connection procedure used on a Ramsey silent chain.

When it is not convenient to couple the chain on a sprocket, the use of a coupling tool (chain puller) may be beneficial. As mentioned, this device provides a convenient means for drawing the ends of the chain together, which otherwise would be very difficult, especially with a heavy chain.

Chain life will be shortened by both running too tight and running too loose. A chain that is too tight has an unnecessary additional load imposed upon it. The slack strand of chain should be free of tension when running. However, a

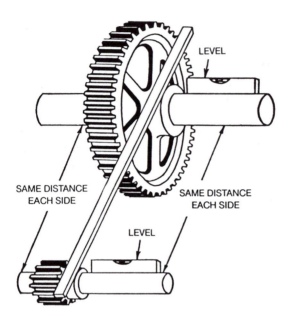

Figure 26 Common method used to align shafts, both horizontally and vertically (shaft parallelism should be checked before installing sprockets), and to align sprockets (sprockets should be aligned on the shafts so there is little or no lateral offset between sprocket faces). (Courtesy of Ramsey Products Corporation, Charlotte, NC.)

chain that is loose enough to whip or surge is subjected to additional joint wear, shock loads, and repeated stresses that contribute to fatigue failure. A chain must therefore be properly tensioned at installation and checked periodically.

With the drive stationary, determine the amount of sag in the horizontal drive by pulling one side of the chain (usually the slack or return side) taut, allowing all the excess chain to accumulate in the opposite span. Next, place a straightedge over the newly accumulated slack span, and pulling the chain down from the center, measure for proper sag.

Therefore, on drives where a line between shaft centers is horizontal, or inclined no more than 60°, the chain should be installed and maintained with a sag in one span equal to approximately 2 percent of the sprocket center distance. In summary, if there is no means for adjusting the chain tension, the drive should be designed so that the chain is taut at installation. Normal seating of the chain parts will provide sufficient sag.

6. Silent Chain Maintenance

Like any other machine component a silent chain drive should be periodically inspected and maintained if necessary. The time spent making such an inspec-

Making The Connection

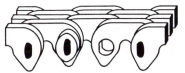

1. Lace ends of chain together so holes line up.

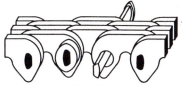

2. Insert longer pin through chain.

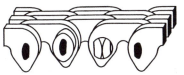

3. Insert short pin next to long pin. Note relative pin position.

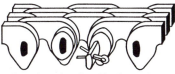

4. Put washer on long pin and install cotter pin.

A Set Of Typical Connecting Parts

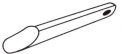

Long pin, headed one end, drilled other end (furnished headed one end, annealed other end for some sizes of chain)

Cotter pin

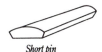

Short pin

Washer

Figure 27 Ramsey Products' version of making the "final connection." (Courtesy of Ramsey Products Corporation, Charlotte, NC.)

tion and necessary adjustments will be more than repaid by increased service life. The first inspection should be after 100 hours of operation. This is generally enough time to find and correct any improper conditions before serious problems can develop. Thereafter, most drives may be inspected at 500-hour intervals. However, drives subjected to shock loads or severe operating conditions should be inspected at 200-hour intervals. At each inspection the following items should be checked and corrected, if necessary:

1. *Check lubrication*: A chain that does not receive sufficient lubrication will wear prematurely. An early indication is the appearance of a reddish

brown iron oxide deposit on the chain. If this occurs, the method of lubricating the drive should be improved immediately. After the first 100 hours of operation, the lubricant in the drive should be drained and refilled with fresh oil, and each 500 hours hereafter. If at any time the oil is found to be contaminated, make preparations to change fluid immediately. If drive casings are used, they should be properly flushed and the chain carefully cleaned.

2. *Check chain tension*: Check chain tension and adjust as needed to maintain the proper sage in the slack span. This is an important check during the initial 100 hours of operation, since some slack will be apparent as a result of the seating of the chain joints. To correct this, adjust sprocket centers to take up accumulated slack and provide proper chain tension.

3. *Check chain wear*: It is a fact that progressive joint wear elongates chain pitch, causing the chain to lengthen and ride higher on the sprocket teeth. In other words, during operation, chain pins and bushings slide against each other as the chain engages, wraps, and disengages from its sprockets. No matter how well lubricated these chain parts are, it is inevitable that some metal-to-metal contact will occur, and these components eventually will wear.

The number of teeth in the large sprocket determines the amount of joint wear that can be tolerated before the chain jumps or rides over the ends of the sprocket teeth. When this critical degree of elongation is reached, the chain must be replaced.

The recommended measuring procedure is to remove the chain and suspend it vertically. Measure the chain wear elongation and if elongation exceeds functional limits, or is greater than 3 percent (0.36 in. in 1 foot) replace the entire chain.

Chain manufacturers have established tables of maximum elongation to aid in determining when wear has reached a critical point and replacement is necessary. This is determined by placing a certain number of chain pitches under tension, so that elongation can be measured.

Do not connect a new section of chain to a worn chain, because it may run rough and damage the drive. Do not continue to run a chain worn beyond 3 percent elongation, because the chain will not engage the sprockets properly and may damage the sprockets.

4. *Check sprocket tooth wear*: Check for roughness or binding when the chain engages or disengages from the sprocket. Inspect the sprocket teeth for reduced tooth section and "hooked" tooth tips. If these conditions are present, the sprocket teeth are excessively worn and the sprocket should be replaced. It is never advisable to operate a new chain on badly worn sprockets, but it is sometimes possible to reverse the sprockets so that the relatively unworn tooth surfaces contact the chain. Conversely, do not run a worn chain on new sprockets as it will cause the new sprockets to wear rapidly.

5. *Check sprocket alignment*: If noticeable wear is apparent on only one side of the guide links or guide grooves, there is misalignment. Realign the sprockets as previously outlined to prevent further abnormal chain and sprocket wear.

B. Chain Drive Selection Procedures

Basically, the design of a roller chain drive consists primarily of the selection of the chain and sprocket sizes; and although horsepower and speed are the prime considerations in selecting a drive, it is not quite as simple as this. There are a number of rules of good design practice that will enable users to obtain the best results from their roller chain drive. Basic design data or design factors needed for drive system component selection require knowledge of the following:

Required preliminary information for drive selection: (1) Source or type of input power (electric motor, internal combustion engine, turbine, etc.); (2) type of equipment to be driven; (3) horsepower to be transmitted; (4) size (diameter) and speed of driving shaft; (5) size (diameter) and speed of driven shaft; (6) approximate center distance between shafts; (7) relative position or arrangement of shafts; (8) space limitations; (9) desired life; (10) proposed method of lubrication; (11) corrosive environment, if any; (12) safety requirements.

The preceding items should not necessarily be considered in this order. Additionally, close attention must be paid to the horsepower ratings tables given throughout this chapter. They will be referred to frequently, since they are essential for the proper analysis of this section.

With these suggested guidelines in mind, the following step-by-step procedure can be used for an appropriate selection of roller chain and sprockets, as outlined in the following paragraphs:

Step 1: Determine class of driven load. These are typical applications and are classified according to the "usual" conception of the character of the load in question. Load characteristics should always be determined by consideration of the "actual" operating conditions. Typical driven load classifications may be determined from Table 53. If the specified type of load that you are looking for is not listed in this table, classify it by its similarity to a listed item.

Step 2: Determine service factor and strand factor. Service factors are selected for various applications after first determining the prime mover or power source type. Again, see Table 53 (right column). Hence, a service factor accounts for the peak loads caused by the power source and the type of driven equipment. Additionally, for multiple-strand chains, a "correction factor" (shown at the bottom right-hand corner of the horsepower ratings tables) *must*

Table 53   Service Factors Used for Determining Design Horsepower

Driven equipment	Service Factors		
	Input power		
	Internal combustion engine with hydraulic drive	Electric motor or turbine	Internal combustion engine with mechanical drive
Agitators, liquid stock	1.0	1.0	1.2
Beaters	1.2	1.3	1.4
Blowers, centrifugal	1.0	1.0	1.2
Boat propellers	1.4	1.5	1.7
Compressors			
centrifugal	1.2	1.3	1.4
reciprocating, 3 or more cylinders	1.2	1.3	1.4
reciprocating, singular, 2 cylinders	1.4	1.5	1.7
Conveyors			
uniformly loaded or fed	1.0	1.0	1.2
not uniformly loaded or fed.	1.2	1.3	1.4
reciprocating	1.4	1.5	1.7
Cookers, cereal	1.0	1.0	1.2
Crushers	1.4	1.5	1.7
Elevators, bucket			
uniformly loaded or fed	1.0	1.0	1.2
not uniformly loaded or fed.	1.2	1.3	1.4
Fans, centrifugal	1.0	1.0	1.2
Feeders			
rotary table	1.0	1.0	1.2
apron, belt, screw, rotary vane	1.2	1.3	1.4
reciprocating	1.4	1.5	1.7
Generators	1.0	1.0	1.2
Grinders	1.2	1.3	1.4
Hoists	1.2	1.3	1.4
Kettles, brew	1.0	1.0	1.2
Kilns and dryers, rotary	1.2	1.3	1.4
Lineshafts			
light or normal service.	1.0	1.0	1.2
heavy service	1.2	1.3	1.4
Machinery			
uniform load, nonreversing	1.0	1 0	1.2
moderate pulsating load, nonreversing	1.2	1.3	1.4
severe impact or variable load, reversing	1.4	1.5	1.7
Mills			
ball, pebble and tube	1.2	1.3	1.4
hammer, rolling	1.4	1.5	1.7
Pumps			
centrifugal	1.0 ·	1.0	1.2
reciprocating, 3 or more cylinders	1.2	1.3	1.4
Screens, rotary, uniformly fed	1.2	1.3	1.4
Basis for service factors: Uniform load	1.0	1.0	1.2
Moderate shock load	1.2	1.3	1.4
Heavy shock load.	1.4	1.5	1.7

Source: Rexnord Corporation, Link-Belt Roller Chain Operation, Indianapolis, IN.

must be applied to the horsepower capacity values given in this chapter, since these values are based upon single-strand chains. Hence, the proper horsepower capacity of a multiple-strand chain is equal to that of a single-strand chain of the same pitch, multiplied by the strand factor as set forth in these tables. Note that other selection factors may be applicable and must be applied to the factors just discussed. Most specialty chains have special or separate factor values that must be used to determine an accurate horsepower capacity rating.

Step 3: Computation of design horsepower. The design horsepower equals the horsepower to be transmitted multiplied by the appropriate service factor, or other applicable factor values (DH = HP × SF).

Step 4: Select tentative chain size. To help narrow down the range of chain sizes to be checked, tentatively select the chain pitch size required from the "quick selector" pitch selection chart shown in Fig. 28 (single-pitch) and 29 (double-pitch). How it works: First, locate the design horsepower (from step 3) on the *vertical axis* by reading up the strand columns—single, double, etc.—in order until the design horsepower is located. The number of strands indicated at the top of the column in which the design horsepower is *first* located is the normal recommendation. Second, locate the RPM of the small sprocket on the *horizontal axis* of both charts. Third, the intersection of the two lines (DH and RPM) will be in an area designated with the recommended chain pitch. If the intersection is near the borderline of the pitch area, the pitches on both sides of the line should be evaluated to obtain the most suitable selection.

Step 5: Final selection of chain and size of small sprocket. If a single-strand roller chain has been selected, refer directly to the appropriate horsepower ratings tables. What to do: Follow down the column headed the "RPM of small sprockets" and find the value nearest the design horsepower (the column just below). Then, follow this line horizontally *to the left* to find the required number of teeth for the small sprocket. Pay close attention to the column headed the "Maximum bore, inches." Check the compatibility of the maximum sprocket bore selected with the shaft size already established. Obviously, the sprocket (bore) must be large enough to be mounted on the shaft. However, if the sprocket selected will not accommodate the shaft, use a larger sprocket or make a new sprocket and chain selection from the same tables for the next larger chain size.

Step 6: Determine the required ratio. The ratio of the sprocket sizes is determined by the desired speed ratio. Normal maximum single-stage speed ratios should be 7:1 or less for speeds above 650 fpm. For lower speeds, ratios up to 10:1 have been used successfully, but with proper design. For greater than a 7:1 speed reduction, a double-stage reduction (two drives) drive is preferable and will give longer life, usually at a lower cost. Hence, the practical single reduction limit is affected by the minimum size of the small sprocket, the maximum size of the large sprocket, and the need for sufficient wrap on the small sprocket. Thus, the ratio = RPM high-speed shaft (driver)/RPM slow-speed shaft (driven).

Step 7: Select the number of teeth for the large sprocket. To compute the number of teeth for the large sprocket use the selected small sprocket and the specified shaft speed ratio—or simply, $N = n$/ratio (round off to nearest whole

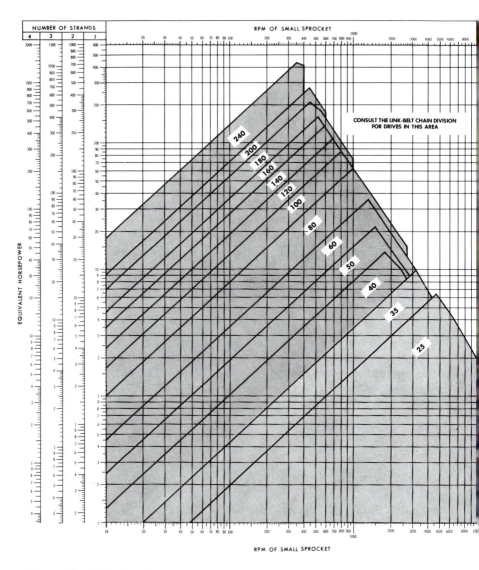

Figure 28 Trial chain length nomograph, single-pitch roller chains. (Courtesy of Rexnord Corporation, Link-Belt Roller Chain Operation, Indianapolis, IN.)

Chart D · Trial selection of double-pitch roller chains

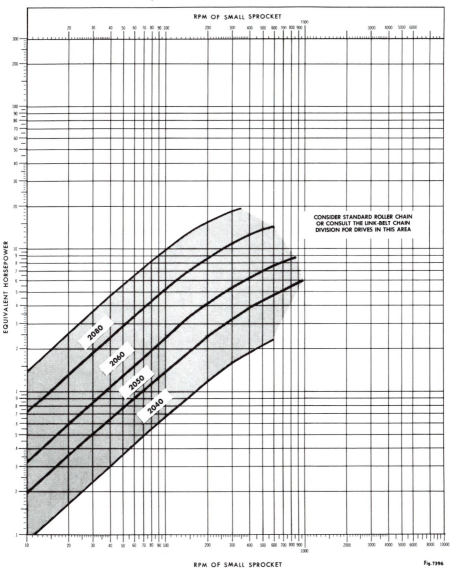

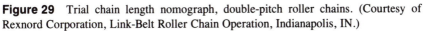

Figure 29 Trial chain length nomograph, double-pitch roller chains. (Courtesy of Rexnord Corporation, Link-Belt Roller Chain Operation, Indianapolis, IN.)

number). If the required accuracy of speed ratio cannot be obtained using an integral number of teeth on the large sprocket, try a larger size for the small sprocket. If this is done, it is advisable to go back and recheck the small sprocket size and the chain pitch. Additionally, if space limitations are specified in the design, it is advisable to keep this in mind and check for space clearances.

Step 8: Determine chain length. Follow the equation previously outlined in Section IV.F. The computed length may be fractional, but the actual chain must be an integral number of pitches. Again, if at all possible, the chain length should be an even number of pitches. An odd number of pitches would necessitate the use of an offset link, which is expensive and will reduce the chain's capacity.

Step 9: Determine final center distance between sprockets. Follow the equation previously outlined in Section IV.F. Sprocket centers must be greater than one-half the sum of the sprocket outside diameters to avoid any possible tooth interference. A suggested *minimum* center distance would equal the outside diameter of the large sprocket and one-half the outside diameter of the smaller sprocket. Drives so proportionately designed would assure that the minimum suggested "chain wrap" of 120° would be included on the small sprocket. The shortest practical center distance (as short as 20 pitches for pulsating drives) is recommended for high-speed drives to avoid chain whipping and potential drive damage. If the center distance is fixed and the adopted chain length results in any objectionable slack, a idler may and should be used, or a sprocket combination selected, that will alleviate objectionable sag and provide a relatively snug fit. Additionally, idler sprockets should be no smaller than the driver, have a minimum of 17 teeth, and should be located adjacent to the driving sprocket so that at least three teeth are in full engagement with the *slack* (non-load-carrying) span of chain. When idler sprockets must be installed in the "taut" span of chain, it must be kept in mind that the service life of the chain will be shorter because of the additional joint articulation under load. If possible, provide enough adjustment of the chain tightener to permit removal of two pitches of chain.

Step 10: Determine method of lubrication. The methods of lubrication shown in the "horsepower ratings tables" (indicated by "stair-step" lines) are based primarily on chain speed; however, the relative positions of driving and driven shafts often influence the method of lubrication. Recommendations and complete lubrication information are given in Section IV.I and should be referred to when necessary.

For convenience, a chain drive "checklist" is shown and may be used in conjunction with the procedures just described (see Fig. 30).

ITEM	DATA TO CHECK	SUGGESTED ALTERNATIVES
1	Small sprocket (Driver Sprocket) should have 17 or more teeth.	(a) Use next smaller chain pitch.
2	Large sprocket (Driven Sprocket) should have less than 120 teeth.	(a) Use next larger chain pitch. (b) Use more chain strands. (c) Speed ratio too large — divide into two drives.
3	Speed ratio should be 7 to 1 or less — 10 to 1 is a maximum.	(a) Divide into two drives.
4	With speed ratios over 3 to 1, the center distance between shafts should not be less than the outside diameter of the large sprocket less the outside diameter of the small sprocket to provide minimum recommended chain wrap of 120° on the small sprocket.	(a) Increase center distance. (b) Divide into two drives. (c) Use more chain strands. (d) Use next larger chain pitch. (e) Use next smaller chain pitch.
5	Center distance must be greater than ½ of the sum of the outside diameters of both sprockets to prevent interference of sprocket teeth.	(a) Increase center distance. (b) Use more chain strands. (c) Use next larger chain pitch. (d) Use next smaller chain pitch.
6	Number of teeth in large sprocket (Driven Sprocket) should conform to listed stock sprocket sizes.	(a) Select the closest stock sprocket and calculate the number of teeth in the small sprocket (Driver Sprocket) as follows: $$\frac{\text{NO. OF TEETH}}{\text{IN DRIVER SPROCKET}} = \frac{\text{RPM of DriveN} \times \text{No. of Teeth in DriveN}}{\text{RPM of DriveR}}$$
7	Number of teeth in small sprocket (Driver Sprocket) should conform to listed stock sprocket sizes.	(a) Select the closest stock sprocket and calculate the number of teeth in the large sprocket (Driven Sprocket) as follows: $$\frac{\text{NO. OF TEETH}}{\text{IN DRIVEN SPROCKET}} = \frac{\text{RPM of DriveR} \times \text{No. of Teeth in DriveR}}{\text{RPM of DriveN}}$$
	NOTE: If stock sprockets and the speed ratio desired do not balance out and the ratio cannot be changed, non-stock or special sprockets can be made to order.	
8	Sprockets selected must accommodate the specified shafts.	(a) Select the closest size sprockets which will accept the shafts.
9	The drive selected should fit the available space.	(a) Use next larger chain pitch. (b) Use more chain strands. (c) Use next smaller chain pitch.
10	The shaft center distance should be less than 80 pitches of chain.	(a) Install guides or idlers.
11	The center distance should be equal to or greater than the minimum recommended in Minimum Center Distances Table below.	(a) Use next smaller chain pitch. (b) Use more chain strands.

MINIMUM CENTER DISTANCES

PITCH	⅜"	½"	⅝"	¾"	1"	1¼"	1½"	1¾"	2"	2½"	3"
Minimum Center Distance	6"	9"	12"	15"	21"	27"	33"	39"	45"	57"	66"

ITEM	DATA TO CHECK	SUGGESTED ALTERNATIVES
12	The center distance should be within the optimum range of 30-50 chain pitches.	(a) Use next larger chain pitch. (b) Use next smaller chain pitch. (c) Use more chain strands.
13	The final drive design should have adequate capacity to handle the required horsepower rating for the chain pitch as calculated in Step 3 of Selection Procedure.	(a) Make new selection or contact the WHITNEY District Office for recommendations.
14	For sprockets under 24 teeth, speeds over 600 RPM, ratios over 4 to 1, and heavy loading...	(a) Harden sprocket teeth.

Figure 30 Drive chain selection procedure checklist. (Courtesy of Jeffrey Chain Corporation, Whitney Chain Operations, Morristown, TN.)

XI. DEFINITIONS OF TECHNICAL TERMS

The purpose of this glossary is to provide definitions of *selected terms* that are most significant or applicable to this chapter. Although this glossary cannot be deemed all-inclusive, it is largely representative of the chain industry and includes many important terms found in chain catalogs and engineering manuals. Included are not only the standard definitions but also terms the user is most likely to encounter in such catalogs, drawings, and specifications.

Angle of flex. The total angle of chain joint articulation as a chain enters or leaves a sprocket or "wheel." The angle is equal to 360° divided by the number of teeth in the sprocket.

Articulation. The action occurring at each chain joint in flexing from the straight to an angle and back to the straight (engaging and disengaging), as the joint passes around a sprocket or other path, causing it to flex. The angle of articulation is inversely proportional to the number of teeth in the sprocket.

Attachment. A modification or extension of standard chain links or pins to adapt the chain for purposes such as conveying, elevating, or timing.

Backlash. Movement (if any) of the chain along the pitch line of the sprocket when the direction of chain is reversed.

Block chain. An alternative name used by some manufacturers for "bar-link chain" or for certain styles of leaf chains. Basically, it is a type of chain in which the inner link is either a solid block or a block built up of laminations.

Block link. An inside link, usually used with cast chain links but sometimes applied to engineering steel chain inside links.

Bottom diameter. The diameter of a circle tangent to the seating curve ("pocket" configuration) at the bottom of each tooth space of a sprocket. *See also* Root Diameter.

Break-in. The elongation occurring during the first several hours of operation of a new chain. *See* Run-in.

Caliper diameter. The distance measured across the bottoms of two opposite tooth gaps in a sprocket having an even number of teeth, and measured between one tooth gap and the nearest opposite gap for a sprocket with an odd number of teeth.

Catenary effect. The catenary is the curve a length of chain assumes between its suspension points. Thus, catenary tension is the tensile load imposed on the chain due to the weight of the chain and the amount of sag between suspension points.

Center distance. The dimension between the centers of the shafts of a chain drive.

Chain casing. A sheet metal enclosure or housing around a chain drive which retains the lubricating system and protects the chain and the lubricant from contamination.

Chain guard. Generally an open-type structure made of sheet metal, expanded metal, or similar construction around a chain drive.

Chain length. The distance between the joint centers at each end of a taut chain strand. This distance is usually expressed in feet and/or inches or in pitches.

Chain width. Defined somewhat differently for various chains, but usually the inside width of the chain, between roller link plates for roller chain or sidebars of inside links for engineering steel chains. *Note*: This is not the "overall" or clearance width of the chain. Conversely, the width of silent chain is the overall width, although the width between guides will also be given when appropriate.

Chordal effect (action). The effect produced by the chain joint centers being forced to follow arcs instead of chords of the sprocket pitch circle.

Chordal pitch. The length of one side of the polygon formed by the lines between the joint centers as the chain is wrapped on the sprocket. It is a chord of the sprocket pitch circle and is equal to the chain pitch.

Circumferential stress. Stresses that act around the circumference of a bushing or cylinder.

Clevis. A block dimensioned to receive and anchor the end of a chain (e.g., leaf chain) length used in a tension linkage or hoisting application.

Connecting link. For a roller chain, a special pin link, or outer link made with one link plate, readily detactable to facilitate connecting or disconnecting the ends of a length of chain.

Crotch height. For a silent chain, the height of the link crotch above the pitch line of the link.

Cutter. A special cutting tool that machines the necessary tooth and tooth gap contours in a blank for a cut-tooth sprocket.

Cycle. Change in load level as a chain passes around a system. Usually the change is from negligible load to a load peak on a regular basis as the chain undergoes a complete cycle or operation.

Design horsepower. A calculated value arrived at by multiplying the specified horsepower by a service factor which reflects various operating conditions. It is the necessary value used to select the chain size for the drive.

Double-cut sprocket. For double-pitch (extended pitch) roller chains, a sprocket having two sets of effective teeth. Tooth spaces for the second set are located midway between those of the first set. Hence, such teeth make contact with the chain only on alternate revolutions of the sprocket.

Double-pitch roller chain. A roller chain having double or twice the pitch dimension of a standard, or base-pitch, roller chain, but otherwise having standard (identical) pins, bushings, and standard or oversized rollers. Also commonly referred to as *extended pitch chain.*

Effective teeth. In double-cut sprockets, the number of sprocket teeth engaging with the chain in one revolution. The term normally applies to sprockets for double-pitch roller chain.

Engineering (roller) chain. Manufacturers' term for nonstandard roller chain, usually double-pitch style but an odd pitch, intended for a specific special application.

Galling. Developing a condition (joint wear) on the live bearing surface of a pin or bushing of a chain where excessive friction between high spots result in "localized welding" with subsequent tearing and a further roughening of the contact surfaces.

Hub and groove diameter. The outside diameter of the hub, or the diameter at the base of a groove cut in the hub to provide clearance for link plates sidebars.

Idler sprocket. In a drive system, a sprocket that does not transmit power, but is positioned in the system's layout to control the chain's travel path, either to avoid contact with the machine's structure or to take up slack in the return strand of chain.

Included angle. For silent chains, the angle included between the outer surfaces of the link plate contours. This angle affects the layout of the sprocket tooth form.

Knuckle. Manufacturers' term for the bushings used in a steel-bushed rollerless chain. Such bearings are usually of a shouldered design.

Knuckle chain. Manufacturers' catalog term for steel-bushed rollerless chain.

Link. The portion of a chain's structure that connects adjacent pitch centers.

Link plate. For roller chains, any of the individual side plates that make up the chain's outer, inner, or offset links. For a silent chain, any one of the plates of which an assembled chain is composed. Also referred to as *sidebars.*

Load classification. A classification for drive loads based on the intensity of shock that is imposed on the drive.

Longitudinal stresses. Stresses that act along the span of a bushing or cylinder.

Measuring load. The magnitude of the specified standard load imposed on a length of chain when it is to be stretched taut for measuring purposes.

Multiple-strand chain. A roller chain (or other chain) made up of two or more individual strands, assembled into a unitized structure by means of special pins running or extending transversely through all rows.

Multiple-strand factor. A factor (value) by which the horsepower rating of a single-strand chain is multiplied to obtain the correct horsepower capacity of a chain with two or more strands.

Offset link. A specific type of chain link that utilizes bent, or offset, link plates and is assembled with a pin at one end and a bushing and roller at the other, so as to act as a combination link. Made for use in a straight link chain when an uneven number of links in the total strand is required. Also applies to silent chains. Can act as a form of "connecting link," where the pin is readily removable and replaceable.

Offset section. For a roller chain, a factory-assembled section, made up of a roller link and an offset link; for a silent chain, a factory-assembled section, made up of three or more links, one or more of which consists of offset plates. Offset sections are used to connect strands of chain having an odd number of pitches. Its use provides greater stability than its counterpart, the single offset link.

Overchaining. A drive is said to be "overchained" when it incorporates a chain of substantially higher rating than that indicated by normal selection procedures to have been necessary.

Over-gauge diameter. The measurement over gauge pins of specified diameter inserted in opposite tooth gaps of a sprocket. For an odd number of teeth, the pins are placed in the nearest opposite gaps.

Pitch. The fundamental dimension in chain and sprocket measurement. For chain (any type), it is the center-to-center dimension or distance between chain joints. For a sprocket, it is the chordal dimension between the centers of rollers bedded against the bottoms of adjacent tooth spaces.

Pitch circle. The circle passing through the chain joint centers in a chain wrapped around a sprocket of correct theoretical dimensions.

Pitch diameter. The diameter of the sprocket pitch circle.

Pocket curve. The curve or configuration of a sprocket at the bottom of the tooth gap, tangent to the bottom diameter circle.

Proof loading. The practice of subjecting new chain to a tensile loading to some predetermined percentage of the chain's rated strength. The purpose is to prove that each chain is free of substandard parts that might degrade its capability to withstand the specific load.

Pulsations. Fluctuations of a cyclic nature in load or speed.

Roller flex path. The path taken by the roller (or barrel) as it flexes onto or off of the sprocket.

Root diameter. For roller chains, same as bottom diameter. For silent chains, it is the diameter of the circle that would pass through the root line surfaces developed by the cutters. Hence, the theoretical bottom diameter of a sprocket, equal to the pitch diameter minus the chain roller or barrel diameter.

Run-in. The initial period of operation of any mechanism, during which the component parts "seat" themselves. *See* Break-in.

Scoring. Marring or scratching of pin or bushing caused by metallic debris being picked up in the contact surfaces of one of the parts.

Seating curve. A specific term for the pocket curve of a roller chain sprocket.

Seizing. Stiffening (or "freezing") of a chain joint as a result of roughness and high friction caused by galling.

Service factor. A factor (compensation value) by which the specified horsepower or working load of a chain is multiplied to calculate design horsepower for selection purposes. Hence, service factors take into account the surrounding operating conditions and various types of power sources.

Sidebar. Alternative term for link plate, more commonly used in connection with the tension member of an engineering steel chain.

Single-cut sprocket. For double-pitch roller chains, a sprocket having one set of effective teeth.

Take-up. Provision, in a drive or conveyor layout, for compensating for chain elongation, either by increasing center distance (as with a shaft take-up) or by increasing the chain's travel path (as with an idler sprocket, or shoe, on an adjustable mount).

Taper-Lock Sprocket. Manufacturers' term for sprockets equipped with a split tapered bushing (bore) for rigid mounting on a shaft. Generally facilitates installation or removal.

Tensioner. A take-up device that uses springs, weights, or air or hydraulic pressure to provide constant tension in the slack strand despite changing chain elongation caused by wear or thermal expansion and contraction.

Tension linkage. A chain application primarily transmitting motion (tensile force rather than horsepower) at low or intermittent speeds through short distances or holding a load in place when it is not moving. Hence, the motion is usually reciprocal rather than continuously in one direction.

Thermoplastic. A plastic material that can be softened and resoftened without significant changes in chemical composition.

Thimble. Specific manufacturers' term for the bushings used in certain engineering steel chain (roller type).

Tooth flange thickness. The width, in profile, of a roller chain sprocket tooth.

Tooth form. The shape of the sprocket tooth from the bottom of the seating curve up through the "working faces" (surface) to the tip of the tooth.

Tooth gap. The space (pocket) between two sprocket teeth.

Tooth profile. The outline (cross section) of a sprocket tooth as projected on a plane through the sprocket axis and the center of the tooth.

Topping curve. The curve of the outer portion of the tooth form. It is shaped to guide the roller (if any) smoothly into mesh with the sprocket.

Torque. The expression of load on a shaft or sprocket in terms of torsional moment. Usually expressed in *inch-pounds,* calculated as the product of chain tension or pull (in pounds) and sprocket pitch radius (in inches).

Transverse pitch. The lateral, or transverse, dimension between the center lines of adjacent strands of a multiple-strand chain assembly. For sprockets, the dimension between the center lines of adjacent tooth flanges, or profiles.

Underchaining. A drive is said to be underchained when it incorporates a chain of substantially lower rating than that indicated to be needed from normal selection procedures.

Working face. That portion of the face of the sprocket tooth contacted by the barrel, roller or bushing, or equivalent.

BIBLIOGRAPHY

Acme Chain, *Product Catalog,* Holyoke, MA, 1987.

American Chain Association, *Chains for Power Transmission and Material Handling: Design and Applications Handbook,* Marcel Dekker, New York, 1982.

Browning Manufacturing, *Catalog No. 11,* Maysville, KY, 1991.

Diamond Chain Company, *Product Guide,* Indianapolis, IN, 1987.

Jeffrey Chain Corporation, Catalog PT-213A, Whitney Chain Operations, Morristown, TN, 1986.

Jeffrey Chain Corporation, Catalog PT-222A, Morristown, TN, 1986.

Morse Industrial, *Catalog PT-93,* Ithaca, NY, 1986.

Ramsey Products Corporation, Catalog No. 490, Charlotte, NC.

Rexnord Corporation, Bulletin R100, Milwaukee, WI, 1992.

Rexnord Corporation, Bulletin 7010, Link-Belt Roller Chain Operation, Indianapolis, IN, 1993.

Rexnord Corporation, Bulletin 7500, Link-Belt Engineered Chain Operation, Morgantown, NC, 1993.

U.S. Tsubaki, Inc., *Bulletin UC8005* and *Bulletin 5300-B,* Union Chain Division, Sandusky, OH.

U.S. Tsubaki, Inc., *General Catalog, North American Version,* Wheeling, IL, 1991.

9

BELT DRIVES

James D. Shepherd

The Gates Rubber Company, Denver, Colorado

I. INTRODUCTION

A. History

Power transmission belts have played an important role in the industrial development of the world for more than 200 years. Flat belts of plied-up leather and regular cotton or hemp rope running in V-grooves (sheaves) were predominant in early industrial history. These belts transmitted power from large steam engines or water wheels to various types of production machinery, usually through a series of line shafts.

The first V-belt was developed in 1917 and patented the same year. V-ribbed belts were introduced in the 1950s and were a "cross" between a flat belt and a V-belt. This belt was introduced under the trade name Poly-V. During the same period (early 1950s) the first synchronous (timing) belt was marketed. The synchronous belts extended the power transmission market through the use of positive engagement of teeth on the belt with corresponding teeth on the pulley.

B. Advantages of Belt Drives

Belts have many advantages over other power transmission methods. One of the most important is overall economics. The efficiency and reliability of belts as a power transmission medium is well recognized. Belts are clean and require no lubrication. Belts can transmit power between shafts spaced widely apart with a wide selection of speed ratios. V-belts can be used for special design purposes such as clutching, speed variation and transmitting power in more than one plane (i.e., mule drives/quarter-turn drives, etc.). V- and V-ribbed belts are capable of handling large load fluctuations and have excellent shock-absorbing abilities.

C. Where Belts Are Used

Many methods of power transmission are used today to drive industrial, agricultural, and automotive machinery. When the desired speed of the driveN shaft is the same as the driveR, direct connection is the most common and economic method, sometimes through flexible couplings. When the speed of the driveN shaft must be in exact relationship with the driveR, gears or chains are most frequently used. However, synchronous belts have gained wide usage in this relatively limited field—primarily due to economics, lack of lubrication requirements, and very high power density of modern curvilinear belts.

By far the most numerous type of drive is where synchronization is not required, driveR and driveN speeds are different, and the prime mover and driveN machinery are separated by some distance. This type of drive lends itself ideally to the use of the various kinds of belts that exist today.

As discussed later, belts have an extremely broad application range (torque, speed, ambient conditions, etc.) that make them ideal for virtually any application.

II. CLASSIFICATION OF BELTS

A. Industry Belt and Pulley Standards

In order to understand belt types and how they are used in the marketplace, it is helpful to understand the domestic and international standards that relate to these products.

1. Purpose of Standardization

Standardization has been in existence since the turn of the century. For example, the American Society of Agricultural Engineers (ASAE) was formed in 1910 and the predecessor of the American National Standards Institute (ANSI) was established in 1918. Even in these early years, it was recognized that there was a need for standardization in the industry in order to increase parts interchangeability, improve manufacturing processes, and generally serve the using public.

The United States has participated in development of both national and international standards in order to promote interchangeability of power transmission belts in various applications and to provide availability of replacement parts on an international basis. With the increase of multinational companies, the international aspect of standardization has become increasingly important in the last decade. International standards activity for industrial belts is organized by the International Organization for Standardization (ISO). The activities of this group will be described later.

2. Main Standardization Bodies in the United States

American National Standards Institute. Virtually all domestic standards-writing organizations are governed by ANSI. The following groups are the actual administrators of belt and pulley power transmission standards in the United States:

1. Rubber Manufacturers Association (RMA)
2. Mechanical Power Transmission Association (MPTA)
3. Society of Automotive Engineers (SAE)
4. American Society of Agricultural Engineers (ASAE)
5. American Petroleum Institute (API)

Each of these groups has, as their main participants, the belt and pulley manufacturers. However, the users are invited to participate and in some cases, are quite active in the drafting of power transmission standards.

Rubber Manufacturers Association. The major writer of industrial belt power transmission standards for the United States is the Rubber Manufacturers Association. The drafting group for RMA is the Power Transmission Belt Technical Committee (PTBTC) and has, as its main membership, each of the domestic manufacturers of power transmission belts. Each of the standards are written such that all manufacturers contribute to and agree upon the contents of each standard.

In many cases, the PTBTC of RMA contributes to the drafting of standards administered by other organizations listed above. In addition, they also review each of the standards written by the other organizations to ensure compatibility with RMA and ISO standards.

PTBTC also works very closely with the Mechanical Power Transmission Association in drafting their standards since this is the association dealing with the V-belt sheaves and synchronous belt pulleys to which the power transmission belts are applied. Further, the RMA group works closely with the Rubber Association of Canada (RAC) in drafting standards that are compatible with Canadian procedures and manufacturing processes.

A description of the RMA published standards is shown in Table 1. These standards are reviewed on a five-year basis to assure that they reflect current industry knowledge and practice.

Table 1 RMA Standards

Publication number	Description
IP-20	Specifications: Classical Multiple V-Belts (A, B, C, D, and E Cross Sections)
IP-21	Specifications: Double-V (Hexagonal) Belts (AA, BB, CC, DD Cross Sections)
IP-22	Specifications: Narrow Multiple V-Belts (3V, 5V, and 8V Cross Sections) (1983)
IP-23	Specifications: Single V-Belts (2L, 3L, 4L, and 5L Cross Sections)
IL-24	Specifications: Synchronous Belts (MXL, XL, L, M, XH, and XXH Belt Sections)
IP-25	Specifications: Variable Speed V-Belts
IP-26	Specifications: V-Ribbed Belts (H, J, K, L, and M Cross Sections)

In addition to the published standards, RMA also publishes several bulletins pertaining to various functions and design methods of power transmission belts. Current bulletins are listed in Table 2.

Mechanical Power Transmission Association. The Mechanical Power Transmission Association is the standards organization for manufacturers and users of V-belt sheaves and synchronous belt pulleys/sprockets. A working

Table 2 Current RMA Bulletins

Publication number	Description
IP-3-1	Heat Resistance and Low Temperature Properties
IP-3-2	Oil Resistance and Chemical Resistance
IP-3-3	Static Conductive V-Belts
IP-3-4	Storage of Power Transmission Belts
IP-3-6	Use of Idlers with Power Transmission Belt Drives
IP-3-7	V-Flat Drives
IP-3-8	High-Modulus V-Belts
IP-3-9	Joined V-Belts
IP-3-10	V-Belt Drives with Twist and Non-Alignment Including Quarter Turn
IP-3-13	Mechanical Efficiency of Power Transmission Belt Drives
IP-3-14	A Drive Design Procedure for Variable Pitch V-Belt Drives

Table 3 SAE Standards

Number	Name
J636	V-Belts and Pulleys
J637	Automotive V-Belt Drives—Recommended Practice
J1278	SI (Metric Synchronous Belts and Pulleys
J1313	Automotive Synchronous Belt Drives
J1459	V-Ribbed Automotive Belts
J1596	Automotive V-Ribbed Belt Drives and Test Methods

relationship exists between MPTA and RMA—each reviews the other's standards for compatibility. In general, at least one joint meeting is held each year to review the active projects of each organization.

Society of Automotive Engineers. SAE is an engineering society "dedicated to advancing mobility" and includes automotive applications as well as related fields including aerospace, trucks and buses, and farm machinery. An SAE belt committee exists to guide the development of standards relating to application and testing of power transmission belts used in automotive applications. Each of the individual standards is written by a subcommittee consisting of manufacturer and user members. The standards written by SAE are listed in Table 3.

American Society of Agricultural Engineers. ASAE is an engineering society "created to advance the theory and practice of engineering in agricultural and the allied sciences and arts." As with previous societies, the various committees are staffed by both manufacturers and users. The standard for this market is S211: V-belt drive for agricultural machines.

American Petroleum Institute. As is the case with SAE and ASAE, API maintains a belt standard (API 1B) because of the special needs of this industry.

3. International Standardization Activity

The International Organization for Standardization. Although many nations have their own standardization groups, ISO is the only international organization which is responsible for and supported by most manufacturing nations in the world.

Belt power transmission activity is the function of ISO Technical Committee 41. Under this technical committee are two subcommittees dealing with power transmission belts:

Subcommittee 1: V-belt and V-ribbed standards
Subcommittee 4: Synchronous belt standards

The belt standards presently published by ISO are listed in Table 4. Each of these standards is formally reviewed on a five-year cycle.

Although none of the ISO standards is officially "used" in the United States, the U.S. representatives have been diligent in efforts to guide the ISO standards into practices consistent with U.S. technology and procedures. In general, none of the standards have significant conflicts with U.S. procedure. However, several cross sections are standardized by ISO which are not consistent with cross sections used in the United States. This has not presented a major problem with the exception of retrofitting some specialized equipment imported from outside the United States.

B. Friction Versus Positive Belt Drives

All power transmission belts are either friction drive or positive drive. Friction drive belts rely on coefficient of friction to transmit the power and require tension to maintain the coefficient of friction. Flat belts are the purest form of a friction drive while V-belts and V-ribbed belts rely on coefficient of friction but have a multiplying effect through the V or wedging characteristic.

Positive drive belts rely on the engagement of teeth on the belt with grooves on the pulley. There can be no real slip with this kind of belt, short of an actual ratcheting action.

Within these two categories of friction and positive drives there are various types of belts recognized in the industry. These are described in the following section.

C. Types of Belt Cross Sections

Figure 1 depicts various belt cross sections used on industrial drives. The heavy-duty industrial belts (classical and narrow) are used on large industrial drives, usually in matched sets. Variable-speed drives are used on devices to vary the driveN speed in relationship to the driveR speed. Synchronous belts re used where exact speed relationship must be maintained between the driveR and driveN shafts (they are also used for their low maintenance requirements and for energy efficiency benefits, as discussed later). The V-ribbed belts are typically used where high speed ratio and small diameters are encountered, such as on clothes dryers. The light-duty belts are used in the lawn and garden implement market and to drive a multitude of small machinery.

In addition to the industrial belts shown in Fig. 1, there are special cross sections used in the agricultural and automotive markets. There are also special cross sections such as flat and round belts that are used on special applications.

Each belt type is briefly discussed and reference will be made to RMA/MPTA standards wherever possible.

Table 4 International (ISO) Standards

Number	Name
ISO 22	Belt Drives—Flat Transmission Belts and Corresponding Pulleys—Dimensions and Tolerances
ISO 155	Belt Drives—Pulleys—Limiting Values for Adjustment of Centres
ISO 254	Belt Drives—Pulleys—Quality. Finish and Balance
ISO 255	Belt Drives—Pulleys for V-Belts (System Based on Datum Width)—Geometrical Inspection of Grooves
ISO 1081	Drives Using V-Belts and Grooved Pulleys—Terminology
ISO 1604	Belt Drives—Endless Wide V-Belts for Industrial Speed-Changers and Groove Profiles for Corresponding Pulleys
ISO 1813	Antistatic Endless V-Belts—Electrical Conductivity—Characteristic and Method of Test
ISO 2790	Belt Drives—Narrow V-Belts for the Automotive Industry and Corresponding Pulleys—Dimensions
ISO 3410	Agricultural Machinery—Endless Variable-Speed V-Belts and Groove Sections of Corresponding Pulleys
ISO 4183	Belt Drives—Classical and Narrow V-Belts—Grooved Pulleys (System Based on Datum Width)
ISO 4184	Classical and Narrow V-Belts—Lengths
ISO 5287	Narrow V-Belt Drives for the Automotive Industry—Fatigue Test
ISO 5289	Endless Hexagonal Belts for Agricultural Machinery and Groove Sections of Corresponding Pulleys
ISO 5290	Grooved Pulleys for Joined Narrow V-Belts—Groove Sections 9J, 15J, 20J and 25J
ISO 5291	Grooved Pulleys for Joined Classical V-Belts—Groove Sections AJ, BJ, CJ and DJ (Effective System)
ISO 8370	V- and Ribbed V-Belts—Dynamic Test to Determine Pitch Zone Location
ISO 8419	Narrow Joined V-Belts—Lengths in Effective System
ISO 9608	V-Belts—Uniformity of Belts—Centre Distance Variation—Specifications and Test Method
ISO 9980	Belt Drives—Grooved Pulleys for V-Belts (System Based on Effective Width)—Geometrical Inspection of Grooves
ISO 9981	Belt Drives—Pulleys and V-Ribbed Belts for the Automotive Industry—Dimensions—PK Profile
ISO 9982	Belt Drives—Pulleys and V-Ribbed Belts for Industrial Applications—Dimensions—PH, PJ, PK, PL, and PM Profiles

1. V-Belt (Trapezoidal)

The most familiar and widely used type of belt in the industry is the V-belt or trapezoidal belt. These belts have their name from the general trapezoidal or V-shaped as shown in Fig. 1. There are three major categories of industrial trapezoidal belts: classical (A, B, C, and D), narrow (3V, 5V, 8V), and light-duty

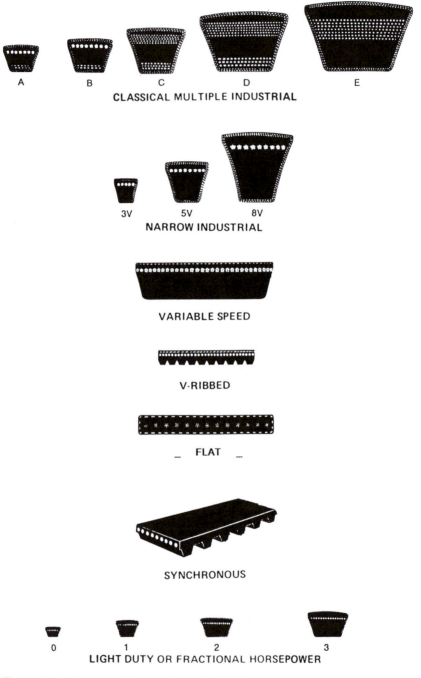

Figure 1 Belt cross sections used on industrial drives.

or fractional horsepower (2L, 3L, 4L, 5L). Each of these belts are described in RMA/MPTA Standards IP-20, IP-22, IP-23, respectively.

2. Joined Industrial V-Belts

Classical and narrow V-belts are available in the joined configuration, where several belt strands are connected by a tie band across the top of the belts (see Fig. 2). The tie band improves lateral stability and solves the problems of belts turning over or coming off the sheaves. The tie band rides above the sheave, so it does not interfere with the wedging action of the individual belt strands. These belts are also described in RMA/MPTA standards IP-20 and IP-22.

3. V-Ribbed Belts

Shallow ribs on the underside of V-ribbed belts mate with corresponding grooves in the pulley. This mating guides the belt and makes it more stable than a flat belt. Some designs of V-ribbed belts result in the complete filling of the pulley grooves. Because of the relatively small ribs, often combined with filled grooves, V-ribbed belts do not have as much wedging action as V-belts and, consequently, must operate at higher belt tensions.

Figure 2 Joined industrial belt.

V-ribbed belts have higher lateral stability than flat belts because of the mating grooves. In fact, they have the stability advantages approaching that of joined V-belts on drives where individual belts whip and turn over or come off the sheave. The tensile section of V-ribbed belts rides outside of the pulley grooves and the shallow ribs make V-ribbed belts more susceptible to jumping grooves on misaligned drives.

While V-ribbed belts may not be as flexible as flat belts, they perform well on small-diameter pulleys. V-ribbed belts are frequently used in clothes dryers, for example, because a speed ratio of 30:1 is required between the motor and drum; thus requiring an extremely small motor pulley. V-ribbed belts are described in RMA/MPTA Standard IP-26.

4. Flat Belts

Flat belts are the simplest and generally least expensive type of belt. They are made of leather, plastic, solid rubber, or rubber reinforced with fabric tensile cords. Reinforced flat belts are capable of transmitting high power—up to 500 hp if wide, flat belts are used on large pulleys. Such drives are cumbersome, however, and flat-belt drives have largely been replaced by more compact V-belt and synchronous drives.

Modern flat-belt drives are generally used for applications where high speed is more important than high-power transmission. At higher speeds, flat belts develop less centrifugal tension (which reduces the power transmission capabilities of the larger belts).

With their thin sections, flat belts also develop less bending tension as they flex around the pulley, especially on small-diameter pulleys. Because bending tension detracts from drive tension, flat belts offer advantages on drives with pulleys that may be too small for V-belts. But flat belts depend on high tensions to maintain traction and must be designed with rugged shafts and bearings to prevent high wear rates or early failure. Alignment also must be controlled closely because flat belts have a tendency to track off their pulleys. Many belt manufacturers can provide data on crowned pulleys, which enhance proper belt tracking. No RMA standard exists.

5. Round Belts

Sewing machines were the first modern machines on which round belts were used. Round leather belting is still used to some extent for applications of this kind. Round belts can be made to length (endless), or the proper length can be cut from rolls of belting and spliced with a metal hook to form an endless product.

The present-day use of round belts is rather limited, primarily for agricultural machinery drives and some light-duty or appliance drives, such as vacuum cleaners.

While round belts are frequently used on misaligned drives, their perceived advantage is not fully realized as the modern construction uses rectangular tensile members (see Fig. 3).

Materials used in the various components are similar to those used in V-belts. These round belts are available in many fractional diameters from 3/16 to 1 1/16 in.

No industry standards exist for round belts, but there are certain popular lengths which are occasionally stocked by some belt manufacturers. Most of the time, round belts are built to special lengths for specific OEM (original equipment manufacturer) drives.

Sheaves used for round-belt drives are normally those of the 2L, 3L, 4L, 5L, or A, B, C, D lines. Round grooves do not afford wedging and are seldom used.

6. Variable-Speed V-Belts

Within certain limits, V-belts are well suited for applications where the driveN speed must vary in relationship to the driveR speed. The speed ratio on these drives is controlled by moving one sheave side wall relative to the other so that the belt rides at different diameters. For industrial applications that require significant speed variation, special variable-speed V-belts are available. These belts have wide, thin cross sections specifically designed to provide a wide range of speed variation. They transmit up to 100 hp per belt and provide speed variations up to 10:1. RMA/MPTA Standard IP-25 covers belts with top widths ranging from 0.875 to 3.0 in.

The mechanics of these systems are described in more detail later.

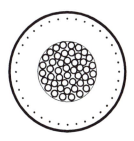

Old Style
Hand–Rolled
Construction

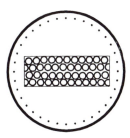

Modern
Machine–Made
Construction

Figure 3 Round-belt construction.

Figure 4   Segmented spliced belting.

7. Spliced Belting

Spliced belting comes in various forms and is usually purchased by the foot and spliced together mechanically or through heat bonding to form the proper length of belt. Figure 4 shows a type of belting which is made of several small segments linked together either mechanically or through intertwining the individual components to form a continuous V-belt. Figure 5 shows a conventional appearing V-belt linked with mechanical splices that are attached to each end of the belt. This belt actually is manufactured using many layers of a special fabric which will retain the splices.

Another type of spliceable belting is made of a thermoplastic material that is cut to length and then heat-spliced together.

For the most part, all of the above spliced belting is available in classical section belts (A, B, C, and D) and is used where it is difficult to replace belts or where fixed centers dictate that a belt must be made short and then stretched on a drive. In some cases the manufacturers indicate that the spliced belting has the same horsepower rating as normal endless belts, although the majority of the suppliers indicate that their belts have a reduced horsepower capacity.

Often spliced belting is used in an emergency, as it is relatively easy to "manufacture" and install a spliced belt on the drive to "keep the wheels turning" until a more permanent repair can be made.

8. Heavy-Duty Industrial Double V-Belts

Double V-belts (sometimes called hexagonal belts; see Fig. 6) are commonly used in multiple, and are used in applications where the drive geometry requires that power be transmitted from both sides of the belt. These belts operate in the same sheaves as regular classical V-belts. They are covered by RMA Standard IP-21.

Figure 5   Mechanically spliced belting.

Figure 6 Industrial double (hexagonal) V-belt.

9. Synchronous Belts

Synchronous belts (timing belts) were developed to transmit power in drives where exact synchronization was required. These belts transmit power through the positive engagement of belt teeth with pulley teeth (see Fig. 7). Thus, precise speed ratios and synchronization are possible in such applications as machine-tool indexing heads, automotive camshafts, office machines, and computer peripheral equipment.

Synchronous belt drives have an advantage over gears and chains in that they transmit reasonably high loads with relatively low noise and without lubrication. The shock-absorbing characteristics of the rubber teeth against the metal pulley also can be helpful. Because of their high tensile modulus, these belts require more accurate alignment than V-belts but are more tolerant than gears.

Synchronous belts have seen increasing use in application where V-belts and chain were used in the past. This is primarily due to three significant characteristics.

1. They require less maintenance. Both V-belts and chain stretch and require periodic takeups. In addition, chain requires lubrication.
2. They are extremely energy efficient. This is covered later.
3. Modern curvilinear belts have high "power density" (high horsepower ratings with relatively small package sizes).

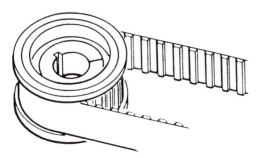

Figure 7 Synchronous belt and pulley.

It is important to remember, however, that the total tension in a synchronous belt increases as torque is applied. The interaction between belt teeth and pulley grooves teeth develops a radial force that increases running tensions. This is illustrated in Fig. 8A. As a result, dynamic bearing loads for synchronous belts and for V-belts are in the same general range at rated loads for a given application.

There are two basic types of synchronous belts: trapezoidal and curvilinear (see Fig. 8B). These are described in the following paragraphs.

Trapezoidal Synchronous Belts.   These have been used for about 30 years. They are common on many synchronized industrial drives and on automotive camshaft drives.

There are six standard belt sections: MXL, XL, L, H, XH, and XXH (0.080 to 1.250 in pitch). These are covered by RMA/MPTA Standard IP-24.

Curvilinear Synchronous Belts.   This category of belts has had the most dynamic development of any form of belt power transmission in the last two decades. Several manufacturers have developed products that are rated at up to three times more horsepower compared to trapezoidal belts.

The most prevalent belt in the curvilinear category is the HTD profile which was introduced in 1970. Since this development, other systems, generally called modified curvilinear profiles, have been introduced into the market. Table 5 lists the profiles most prevalent in North America, together with the primary patent numbers. Figure 9 describes the profile of each belt type. With the exception of HTD, the profiles are covered by unexpired patents and are carefully controlled by the manufacturers of their licenses. The development of these new profiles has focused on 8-mm and 14-mm pitches.

Unfortunately, this rapid growth of belt profiles has caused considerable confusion in the market. This is magnified by the fact that no national or international industry standard exists. This confusion is regrettable, as the products have superior performance capabilities and have excellent track records. It is recommended that the reader select a trusted vendor and rely on them to provide the best curvilinear belt for the application.

In addition to categorizing belts by *type,* it is helpful to categorize them by *application.* The following sections describe these applications.

D.   Heavy-Duty Industrial Belts

Heavy-duty industrial drives are typically higher hp application (above 5 hp) and run 8–24 h per day. Examples are large fans, industrial air compressors, and rock crushers.

Many industrial drives have load and reliability requirements that cannot be met by single V-belt installations. Therefore, 2 to 12 (or more) industrial V-belts are normally used together. Multiple-belt drives deliver up to several hun-

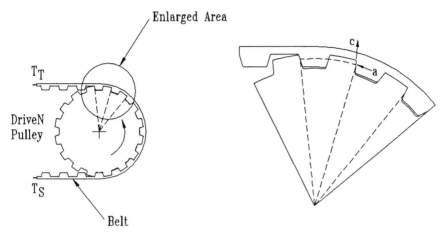

Figure 8A Synchronous belt and pulley forces. Belt must exert circumferential force *a* against pulley tooth to cause pulley rotation. Since tooth face is inclined, belt tries to slide up pulley tooth in direction *c*. This tends to elongate belt, increasing tension.

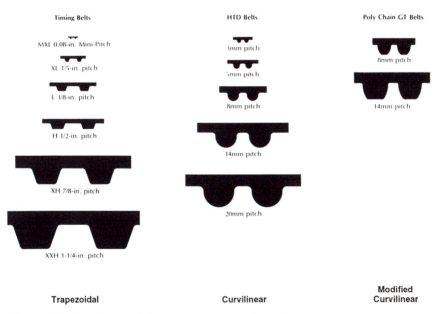

Figure 8B Synchronous belt types: common tooth profiles.

Table 5 Curvilinear Synchronous Profiles

Trade name	Patent or technology holder	Relative Hp rating	Pitches (mm)	Comments
HTD	Gates Rubber Co.	1.0	3,5,8,14,20	Original curvilinear profile. 8 mm × 14 mm patents have expired. Not interchangeable with other profiles.
STPD	Goodyear	1.0	8,14	Not widely used in U.S. Not stocked—OEM design only.
RPP	Pirelli	1.0	8,14	Gaining acceptance in U.S. Rated as interchangeable with HTD.
HPR	Pirelli	1.5	8,14	Same profile as RPP with new construction. Higher rating based on use in RPP pulleys.
HPPD	Goodyear	1.0	8,14	Tooth profile licensed from Pirelli. Construction technology per STPD patent. Rated as interchangeable with HTD pulleys.
GT	Gates Rubber Co.	2.0+	8,14	Trade name Poly Chain. Special urethane construction. Not interchangeable with other profiles.

This table reflects commercial practice in the North American market. While the companies listed developed the products, many other belt and power transmission component manufacturers are licensed to manufacture/ sell the products.

dred horsepower continuously and can absorb reasonable shock loads. Temperature limits range from −30 to +140°F.

There are two types of V-belts in this market: the narrow (3V, 5V, 8V) line and the classical (A, B, C, D). The two standards, published by the RMA/ MPTA are IP-22 for the narrow belt and IP-20 for the classical belt. Drives using these belts and which are properly designed and maintained can be expected to provide three to five years of service.

Heavy-duty industrial V-belts are also used in special applications:

V-flat drives
Variable-pitch drives
Crossed drives
Quarter- (and eighth-) turn drives
Vertical-shaft drives

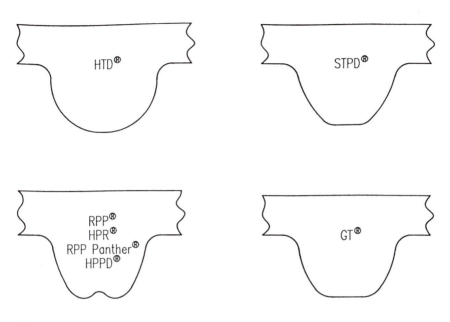

Figure 9 Curvilinear synchronous profiles.

Information on these special drives is given later in this chapter.

In addition to V-belts, larger profiles of V-ribbed and synchronous belts are used extensively on industrial applications.

E. Light-Duty Industrial V-Belts

Light-duty industrial applications are generally less than 5 hp in load and frequently run less than 8 h per day. Examples are small fans and machine tools.

Light-duty V-belts are similar to classical V-belts except that they are designed for more flexibility and lighter loads. Thus, they are more adaptable to the small sheave diameters found on light-duty, fractional-horsepower drives.

Most drives designed with light duty V-belts are single- rather than multiple-belt drives. Service requirements for these belts vary widely, from as little as 2 or 3 h per week for power lawnmowers to 40 or more hours per week for office machines. They are designated as 2L, 3L, 4L, and 5L sections and are described in RMA Standard IP-23.

Smaller profiles in V-ribbed and synchronous belts are also used extensively.

F. Automotive Belts

Automotive accessory drives have limited space available in engine compart-
ments. Not only must these belts transmit high horsepower, they also must do
so over relatively small-diameter sheaves while being subjected to ambient
temperatures of up to 300°F.

The SAE has standardized several cross sections for driving automotive
accessories. Prior to 1980, V-belts were predominantly used. These are described in SAE Standard J636. The most prevalent sizes were designated 0.380
and 0.500 (their approximate top widths).

In the 1980s domestic car manufacturers began converting to V-ribbed
belts. By the late 1980s most drives were driven by a single V-ribbed belt (K
section) and were tensioned by a spring-loaded idler. These belts are described
in SAE Standard J1459.

Automotive camshafts are driven by synchronous belts on some applications. Early camshaft drives used trapezoidal profiles (see synchronous section
above). These belts and pulleys are described in SAE Standard J1278. More
recently, the camshaft applications have been designed with special curvilinear
profiles. Because of the special requirements for this type of drive, the belt
supplier must always be involved with the design and testing of the drive.

G. Agricultural V-Belts

While most agricultural applications require high-horsepower-capacity belts,
the requirements are significantly different from those of most heavy-duty
industrial drives. For instance, industrial drives run at fairly constant loads,
whereas agricultural drives are subjected to a wide range of loads, including
intermittent large shock loads. Thus, agricultural V-belts tend to have a unique
construction to withstand the rigors of the application. In addition, most agricultural drives require the belt to bend in reverse over small-diameter idlers,
and are often subjected to very dirty ambient conditions, both of which require
special product design considerations.

Since load and speed requirements can vary widely from one moment to the
next, depending on the field conditions, a weighted average of the effects of
these variable conditions must be taken into account in the drive design. Also,
many agricultural drives have more than one driveN shaft, and many have one
or more idlers to accomplish the proper geometry and to apply the required
belt tension. These added sheaves complicate the drive design procedure. Some
belt manufacturers publish special design manuals or use computer programs
for these applications.

The ASAE publishes a standard covering cross sections for fixed- and
adjustable-speed agricultural V-belts. This standard, ASAE S211, describes the
classical sections (HA, HB, HC, HD, and HE), and double V-sections (HAA,

HBB, HCC, and HDD), and narrow sections (H3V, H5V, and H8V). The "H" designation is used to clarify that this unique market often uses a special construction belt. The standard was recently revised and now includes V-ribbed belts as well.

III. BELT CONSTRUCTION

Figure 10 shows the construction of a typical industrial V-belt and synchronous belts. Belts have up to five basic components:

1. Tensile cord
2. Overcord
3. Adhesion material (optional)
4. Undercord
5. Band/cover or tooth fabric (optional)

A. Tensile Cord

Since a belt is fundamentally a tension device, the tensile cord provides nearly all the tensile strength of the belt. The ideal tensile member must have combined qualities of fatigue strength, tensile strength, shock resistance, adhesion ability, stretch resistance, and resilience.

Dependent on the qualities described, the most frequently used fibers have included rayon, nylon, polyester, fiberglass, steel, and the relatively new aramid fibers. Polyester has virtually replaced rayon and nylon in V-belts because of its greater strength per unit volume and higher modulus. Thus, it provides greater potential load capacity with low stretch. Glass fibers, while exhibiting excellent strength and low stretch characteristics, are subject to compression failure on some twisted or shock-loaded applications. Steel has very high tensile strength and low stretch but has marginal fatigue resistance, high bending stiffness, and cutting problems during manufacturing. The new aramid fibers have an excellent blend of strength and modulus properties and are finding increasing use in belts today.

Synchronous belts require a high modulus material to minimize stretch so that tooth meshing is maintained. As a result, fiberglass, steel, and aramid are the only options of today's materials.

The tensile cord materials and some of their characteristics are listed in Table 6.

B. Overcord

The overcord material locates the tensile member in the belt and adds to the transverse support of the tensile member.

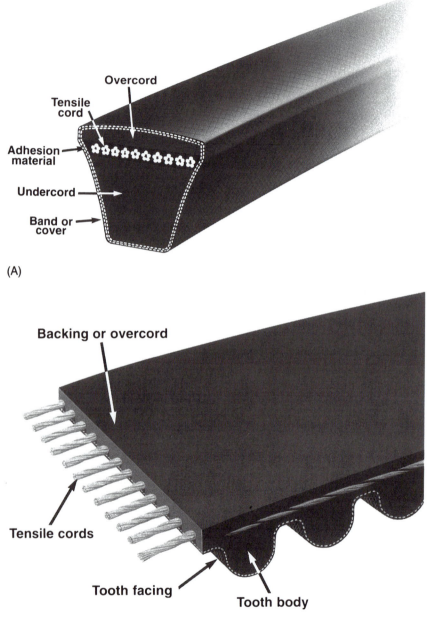

(A)

(B)

Figure 10   (A) Basic V-belt and (B) synchronous belt construction.

Table 6 Tensile Cord Materials

Type	Profiles
Nylon	High strength low modulus. Good for shock loads but needs frequent tension adjustment or automatic tensioner
Rayon	Medium strength and modulus. Minimal use today
Polyester	Good strength and modulus. Good fatigue resistance. Most prevalent cord in V- and V-ribbed belts. Is somewhat heat sensitive and will shrink at high temperatures
Fiberglass	Very high strength and modulus. Good fatigue properties when treated properly. Minimal usage in V and V-ribbed belt. Most prevalent cord in synchronous belts
Steel cable	Very high strength and modulus. Poor fatigue properties on small-diameter pulleys. Only used in made-to-order or specialized belts
Aramid	Very high modulus and tensile strength. Good fatigue propertise with special treatment. Used in some special belt lines and many made to order belts for the agricultural market

C. Adhesion Material

The adhesion material is used to ensure complete bonding of the tensile member to the other parts of the belt. It is important to the belt, since a breakdown of the bond in this area will result in premature belt failure. Depending on the construction and belt type, adhesion may be achieved with special compounding of the over- and undercord material (i.e., a special adhesion layer may not be required).

D. Undercord

The undercord has to support the tensile member or form the teeth which transmits the torque from the pulley to the tensile cord. In the case of V-type belts, it must be stiff enough to bridge the groove and keep the center tensile cords from sagging under load, yet flexible enough to bend over the sheave through millions of cycles without cracking. Notching or cogging of V-belts relieves bending stress in the undercord of smaller-diameter sheaves, but it can reduce the support that the tensile member vitally needs on longer diameters.

E. Band (Cover) or Tooth Fabric

The band or cover material (or tooth fabric) is used to protect the rest of the belt from oil, dust, and other destructive elements. The band is optional, and many modern V-belts do not include a cover material. In the case of synchronous belts, the tooth fabric resists the wear-in from the belt/pulley interface and adds strength to the tooth resisting shearing action.

Optimization of any belt's construction must take into consideration the specific characteristics of the application or market toward which the belt is directed. The selection of a particular tensile member and of undercord and overcord materials, and the choice between banded or nonbanded construction can mean success or failure in any given application. The belt suppliers can provide assistance in selecting the proper construction for unique applications.

IV. BELT POWER TRANSMISSION FUNDAMENTALS

A. Work and Power

In order to define the work being performed by a belt drive, a basic understanding of general power transmission is required.

Work can be defined as a force acting through a distance. A formula work is

$$W = Fl \tag{1}$$

where
W = work performed, ft · lb
F = force exerted, lb
l = distance moved, ft

Power, then, is defined as the rate at which work is performed, or work per unit time. The most frequently used English unit of power is the horsepower, which is defined as 33,000 ft · lb of work per minute. An equation for horsepower is

$$P = \frac{Fl}{33,000t} \tag{2}$$

where
P = power, hp
F = force exerted, lb
l = distance moved, ft
t = time, min

B. Belt Tensions

While considerable time could be spent discussing all the fundamentals of belt power transmission, several basic equations stand out as being most important.

The force that produces work with a belt drive acts on the rim of the pulley, causing it to rotate. When a drive is transmitting power, the belt pulls, or belt tensions, are not equal, as shown in Fig. 11.

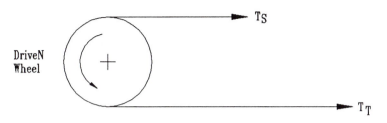

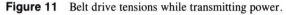

Figure 11 Belt drive tensions while transmitting power.

There is a tight-side tension T_T and a slack-side tension T_S. The difference between these two tensions $T_T - T_S$ is called effective pull or net pull. Substituting in Eq. (2) yields

$$T_T - T_S = \frac{33,000P}{V} \qquad (3)$$

where

T_T = tight-side tension, lb

T_S = slack-side tension, lb

P = power, hp

V = belt speed, ft/min

In this relationship, belt velocity V is defined as

$$V = \frac{D \times N}{3.82} \qquad (4)$$

where

V = belt speed, ft/min

D = pitch diameter, in.

N = rotational speed, r/min

If the horsepower to be transmitted and the belt speed are known, Eq. (3) can be used to determine the effective pull. A second relationship between T_T and T_S is required to find the value of each. The most commonly used relationship is the tension ratio.

1. Tension Ratio

The ratio of the tight-side to the slack-side tension is commonly referred to as the *tension ratio*. The higher the ratio between tight- and slack-side tension, the closer a given belt is to slip—the belt is loose. A low tension ratio means there is more slack-side tension in comparison to the tight-side tension, and the belt is less likely to slip. In the latter case, since the slack-side tension is the greater, the tight-side tension must also be proportionately greater in order to yield the same effective pull for transmitting the required power. The principle of tension ratio is illustrated in Fig. 12. The lower tension ratio creates a greater shaft pull although the effective pull is 40 lb in both cases.

The fundamental tension ratio formula for V-belts at the point of *impending slip* is

$$R = \frac{T_T}{T_S} = e^{W\mu\theta} \tag{5}$$

where

R = tension ratio

T_T = tight-side tension, lb

T_S = slack-side tension, lb

e = a constant, the base of natural logarithms, 2.718

μ = coefficient of friction between the belt and pulley surface

θ = arc of contact of the belt on the pulley, rad

W = wedging factor for V-belts; not present in the same formula for flat belts

Note that coefficient of friction, arc of contact, and wedging are the factors affecting tension ratio at the point of *impending slip*.

The wedging factor takes into account the multiplication of force between the belt and the groove surfaces that occurs because of the wedging action of the belt in the groove. The wedging factor increases the exponent of *e*, allowing V-belts to operate at higher average tension ratios than flat belts. The higher tension ratio translates directly into lower operating tensions and bearing loads for V-belts than for flat belts.

The same is true of the arc of contact (sometimes called belt wrap). At an increased arc of contact, a drive can operate at a higher tension ratio and lower operating tension.

There is a general relationship between tension ratio and belt slip. This can best be illustrated by Fig. 13. It should be noted that the shape of this curve varies greatly as a function of the belt construction, pulley diameter, ambient conditions, and belt/pulley angle system.

While some V-belts will operate satisfactorily at tension ratios of up to 20:1, the common V-belt *design* tension ratio of 5 (at 180° arc of contact) allows a reserve for nonideal conditions and tension loss that occurs as a result of belt stretch and belt/pulley wear. Design tension ratios of 9:1 are sometimes used with V-belt drives which are equipped with automatic tensioning devices. Flat belts are designed at a 2:1 to 2.5:1 tension ratio and V-ribbed belts are designed with a 4:1 tension ratio.

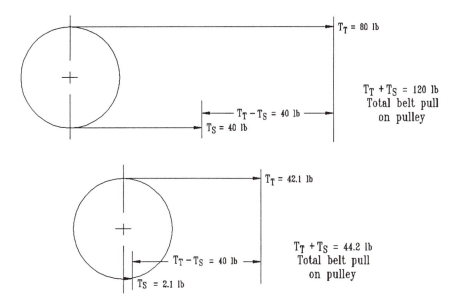

Figure 12   Effect of tension ratio on belt tension (tensions shown to scale).

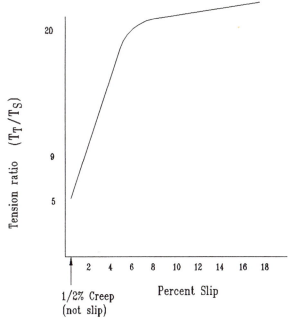

Figure 13   Tension ratio vs. percent slip.

While the preceding section describes tension ratios for friction type drives, it should not be ignored for synchronous drives. The typical synchronous drive design procedure does not use a tension ratio calculation or arc correction factor. Nevertheless, synchronous belts generate slackside tension because of the wedging of the belt and pulley tooth. For shaft load calculations the tensions should be calculated assuming a 5:1 tension ratio.

Equation (5) is generally not used in the actual belt applications design process. Indeed, the design tension ratio does not appear in most drive selection manuals. Rather, it is incorporated in a correction factor commonly derived from graphs or tables supplied in design manuals. The specific values of factor G for a two-sheave V-belt drive are shown in Table 7.

Using the arc-of-contact correction factor from Table 5, simplified formulas for tight-side and slack-side tension for V-belt drives may be derived.

For flat belts

$$T_T = \frac{66,000P}{GV} \tag{6}$$

$$T_S = \frac{33,000(2.0 - G)P}{GV} \tag{7}$$

For V-belts

$$T_T = \frac{41,250P}{GV} \tag{8}$$

$$T_S = \frac{33,000(1.25 - G)P}{GV} \tag{9}$$

where

G = arc-of-contact correction factor

V = belt speed, ft/min

P = power, hp

C. Stress Fatigue Process

The calculation of the working tension just described is fundamental to the design process. However, this is only one of three types of tension a belt actually "feels" in a single cycle of operation. The following section will discuss the source and effect of these tensions in a study of the process by which a belt fails in normal service—stress fatigue. V-belts will be used as an example throughout this section. However, most of the principles discussed will also apply to all power transmission belts.

In addition to the working tensions in the drive there are two other tensions developed when a belt is operating on a drive.

Table 7 Factor G for Arc-of-Contact Correction

$(D - d)/C$	Arc of contact (deg)	V-Belt factor G for belts in V-grooves	Flat belt factor G for belts on flat pulleys
0.0	180	1.00	1.00
0.1	174	0.99	0.98
0.2	169	0.97	0.96
0.3	163	0.96	0.93
0.4	157	0.94	0.91
0.5	151	0.93	0.88
0.6	145	0.91	0.86
0.7	139	0.89	0.83
0.8	133	0.87	0.80
0.9	127	0.85	0.77
1.0	120	0.82	0.74
1.1	113	0.80	0.71
1.2	106	0.77	0.67
1.3	99	0.73	0.63
1.4	91	0.70	0.59
1.5	83	0.65	0.55

1. Bending Tension

Bending tension occurs when a belt bends around a sheave or pulley. Since the top part of the belt being bent is in tension and the bottom part is in compression, compressive stresses also occur. But the effect of bending is most commonly evaluated in terms of tension introduced in the tensile member. The amount of tension incurred in a belt depends upon the radius of bend (diameter of sheave or pulley) and is described in the following equation:

$$T_B = \frac{C_B}{D} \tag{10}$$

where

T_B = bending tension, lb

C_B = a constant depending on the belt size and construction

D = sheave or pulley diameter, in.

2. Centrifugal Tension

Centrifugal tension occurs in a belt because of centrifugal force. The belt is rotating around the drive. It has weight and it tries to pull out away from the sheave or pulley. Tension in the belt results. This tension depends upon the belt speed and belt weight as shown in the following equation:

$$T_C = MV^2 \tag{11}$$

where

T_C = centrifugal tension, lb

M = a constant depending on the weight of the belt

V = belt speed, ft/min

Neither the bending nor centrifugal tensions are imposed on the sheaves, pulleys, shafts, or bearings—only upon the belts.

3. Tension Diagram for Stress Fatigue

A schematic tension diagram of the belt tensions imposed on the tensile member of the belt turning one revolution is shown in Fig. 14.

Starting at point A on the drive, the slack side tension, T_S, plus the centrifugal tension, T_C, are imposed on the belt. (The centrifugal tension is imposed equally in all parts of the drive.) When the belt enters the driveN sheave at point B, the bending tension, T_B, is imposed. As the belt goes from the slack side of the sheave to the tight side, the working tension increases from T_S to T_T. At point C, when the bending tension is removed, $T_T + T_C$ remain. The

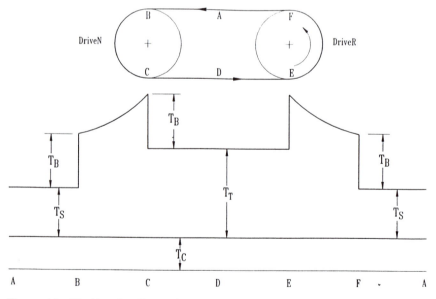

Figure 14 Working, bending, and centrifugal tensions in a belt drive.

opposite occurs over the driveR wheel—the bending tension is imposed at E and removed at F returning the belt to the original slack side plus centrifugal tensions.

Visualize this entire cycle occurring extremely fast—10 times per second is not an unusual application—and you get some idea of the stresses that a belt "feels" when transmitting power.

A belt wears out in normal service because of the above repetitive stresses incurred as it rotates around the drive. Proper analysis of these stresses should therefore make possible the prediction of life under given conditions. In order to use these stresses, they are expressed in terms of peak tension:

$$T_{\text{peak}} = T_T + T_B + T_C \tag{12}$$

where

T_{peak} = peak tension at a sheave, lb

T_T = tight-side belt tension, lb

T_B = bending tension, lb

T_C = centrifugal tension, lb

Peak tension is important because it directly relates to belt life. This relationship has been established through field testing of actual applications and lab testing. The correlation is illustrated in Fig. 15.

The curve on the right-hand portion of FIg. 15 is the empirical link between peak tensions in a belt and the service life it will give on a drive which imposes those peak tensions. The curves (each construction/cross section has the equivalent of sets of curves to describe the effects of the drive variables, such as diameter and tensions) are obtained by running hundreds of belts to failure under controlled laboratory conditions, and then checking the results with

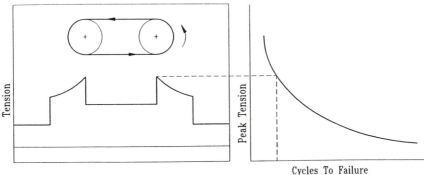

Cycles To Failure

Figure 15   Peak tensions versus cycles to failure.

actual controlled field tests. The value of cycles to failure can be converted to belt life by

$$\text{Life} = \frac{\text{Cycles}(L)}{1440V} \tag{13}$$

where

L = belt length, in.

V = belt speed, ft/min

Figure 15 and Eq. (13) can be combined to give a relationship for the load/life capacity of a belt. This data (usually in the form of formulas) can then be used to either

Establish a horsepower rating (load) for a belt on a given application to give a desired life, or

Predict how long a belt will last if a load on a given application is known.

For industrial applications the *horsepower rating method* is used to determine the proper drive. Other types of applications, such as automotive front end and agricultural, often use a drive design method that relates the load imposed on the belt to a measure of belt life. The belt manufacturers should be consulted on drives which are not standard industrial drives.

4. Illustration of Effect of Drive Variables

By using the previous information, the effect of changing drive conditions on life can be quickly illustrated (equal driveR and driveN diameters will be used in these illustrations for simplification). Assume first that the sheave diameters are reduced on a drive without changing belt speed, horsepower load, or belt length. Fig. 16 illustrates these changes.

Since the load, speed, and belt length do not change, T_T and T_C are the same and T_B is the only factor that changes. T_B increases, increasing the peak tension and thus decreasing the number of cycles to failure. This decreases belt life.

Next assume the horsepower load increases without changing diameters, belt speed, or belt length. Figure 17 illustrates the changes.

The working tensions, T_T and T_S, both increase because the load goes up. Thus the peak tension increases. The number of cycles to failure therefore decreases, decreasing belt life.

Next, assume that everything is the same except that belt length is reduced by reducing the center distance. Figure 18 illustrates the changes.

All tensions remain the same in this instance, and so peak tension is the same and the cycles to failure is the same. But belt life enters into the cycles-belt life Eq. (13) so that when length decreases, life decreases.

Finally, assume that belt speed increases. There are three effects:

Figure 16 Effect of reducing sheave diameters.

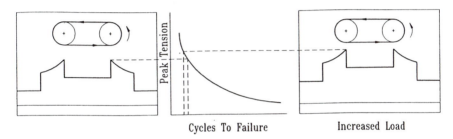

Figure 17 Effect of increasing horsepower load.

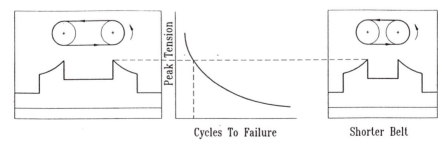

Figure 18 Effect of reducing belt length.

1. T_T decreases.
2. T_C increases.
3. Rate of imposed stresses increases.

The combined effect of these changes depends upon the original value of belt speed and on the amount of belt speed change. In general, increasing speed increases belt life (lower tensions). But, if the belt speed goes too high, the decrease in life caused by increasing T_C, plus the increased rate of imposed stresses, offsets the increase in life caused by reducing T_T. The overall effect is a decrease in belt life. This is illustrated in Fig. 19.

The above illustrations show the affect of drive parameter changes on belt life. These changes are taken into consideration in the horsepower ratings of the belts.

5. Rules of Thumb

The life curve illustrated in the previous figures is not a linear function. Therefore, it is difficult to generalize about the amount of change of belt life that will occur because of a change in one or more drive variables. It is necessary to substitute actual values into the design procedures. These drive design procedure take each one of the drive variables into account automatically.

But there are a few rules of thumb that give general indications of the effect of changes. These rules may help the user to estimate the affect of changing a drive variable:

1. When sheave diameters are reduced 10%, the belt life is reduced approximately half.

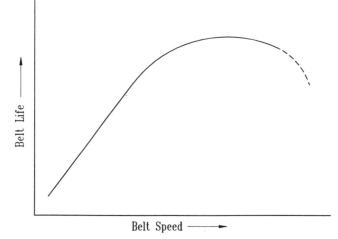

Figure 19 Belt life versus belt speed.

2. When horsepower load increases 10%, the belt life is reduced approximately half.
3. When belt length is cut in half, belt life is reduced approximately half—life is directly proportional to belt length.

V. BELT DYNAMICS

A. Belt Creep

Frictional power transmission belts are made from relatively low modulus materials. There is a slight change in the length of an increment of the belt as it progresses around the pulley from the slack side to the tight side. This change in length results in a slight change of the rotational speed of the driveN pulley relative to the driveR pulley. This is referred to as creep.

Since creep is dependent on the change in the belt length between the slack side and tight side of the drive, it is influenced by the tension ratio. The higher the tension ratio (the looser the belt), the greater the difference between the tight-side and slack-side tension and, therefore, the greater the amount of creep. For properly maintained V-belts drives, the belt creep is considered to be less than 0.5%.

Belt creep is not to be confused with belt slip which occurs because of an insufficient amount of tension in order to transmit the horsepower. Belt slip is often associated with a screeching noise and a significant heat buildup in the drive.

B. Belt Vibration

Belt vibration is a common phenomena that occurs in many belt drives. There are two fundamental vibration phenomena:

1. Plucking frequency
2. Torsional vibration

The plucking vibration of a belt is the same as vibration of a violin string, complicated by the fact that the belt is moving between the sheaves which forms its points of support, while the violin string has fixed points of support. Both the belt and the violin vibrate at their natural frequency because of some forcing function (movement of the bow in the case of the violin). In the case of the violin the vibration produces a pleasant sound because the frequency is in the audible range while belt frequency is generally in the very low, inaudible range.

The natural frequency of a span vibration has been described using the following equation:

$$f_{ni} = \frac{590iT}{hw\sqrt{\dfrac{T_{st}}{w} + 103.5s^2}} \tag{14}$$

where
  i = mode number: 1, 2, 3
  h = span length, in.
  T = belt tension, lb
  w = weight of belt, lb/ft
  s = belt speed, ft/min/1000

While there is less than perfect correlation between the above equation and the actual occurrence on drives, the equation can serve to illustrate the variables involved, and the changes that can be made to a drive to reduce the impact of vibration. In the case of the plucking frequency the two major variables are the span length and the belt tension. Obviously, the drive can be redesigned using different cross sections of belts or larger or smaller sheaves to impact the belt speed as well as the tension.

In the case of torsional vibration, it is well known that two discs connected by a slender shaft will have a natural frequency of torsional vibration. Picture one disc locked so that it cannot rotate. Twisting the second disc about the center of the shaft and then releasing it will produce an oscillating motion of the free disc (see Fig. 20A).

The natural frequency depends on the moment of the inertia of the mass of the nonlocked disc and the torsional spring constant of the shaft (torque required to produce a twist of 1 radian) of the shaft.

Not quite so well recognized is the fact that the V-belts on a drive are mechanically equivalent to a slender shaft coupling of the driveR and driveN machines and, as was the case with plucking frequency, is complicated by the rotation of the belt drive. The following equation has been used to describe torsional vibration:

$$f_n = 1.378\sqrt{K_o D_r^2 \left[\frac{N}{h}\right] \left[\frac{I_m R^2 + I_n}{I_m I_n}\right]} \tag{15}$$

where
  K_o = strain constant of belt
  D_r = driveR pitch diameter, in.
  N = number of belts

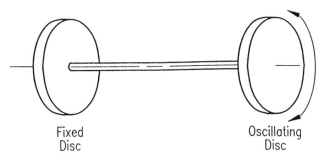

Fixed   Oscillating
Disc   Disc

Figure 20A   Torsional vibration of a two-disc system.

h = span length, in.

R = speed ratio (on speed down drives); on speed up drives use 1/speed ratio

I_m = moment of inertia of sheave and rotating parts of motor, slug ft^2

I_n = moment of inertia of sheave and rotating parts of driveN, slug ft^2

As was the case with the plucking frequency, the above equation does not have one to one correlation with actual practice but illustrates the variables involved. The major variables in a torsional vibration problem are the strain constant of the belt tensile member, inertias, number of belts, and span length.

Plucking and torsional vibration can result in extraordinary life reduction. Both can cause span vibration with amplitudes sufficient to cause belts to turn over in the grooves or come off the drive. In the case of torsional vibration, large amplitude vibration is not always present, but the oscillating tensions can still cause drastic life reduction. One illustration of this phenomena was an alternator drive and an experimental diesel engine. The drive had well over 10,000 h of calculated life, yet yielded less than 50 h at idle with only a 40-A load. By using high-resolution tachometers very high rate of acceleration and deceleration loads were measured. This resulted in very large acceleration loads. Taking these into account in the fatigue analysis revealed a calculated life of about 200 h—reasonable agreement with the actual results. It should be noted that very little span vibration amplitude was exhibited by this drive.

One possible solution to stability problems which result from large amplitude vibration, is to use joined belts. This will reduce the lateral vibration of the belt and maintain proper alignment as the belt enters the pulley minimizing the chance of belt turnover.

Another possible solution to stability problems is to design restraints to minimize the amplitude. These usually consist of pins located close to the belt or a trough against which the belt slide when the amplitude increases.

Because of the complexity of vibration analysis, it is recommended that significant vibration problems be referred to the belt supplier for further analysis.

C. Noise

As was the case with vibration, noise is an extremely complex subject and should be referred to the belt supplier if major problems exist. However, a basic knowledge of noise and its components will help the reader to understand the phenomena.

Audible sound is defined as any pressure *variation* perceptible by the human ear. The variation may occur in air, water, or other medium. These variations or waves have two components: (1) frequency and (2) amplitude or pressure.

The pressure or amplitude has received the most attention from governing organizations because it is the factor which produces damage to the ear. Sound pressure level is measured in decibels which is a dimensionless term. The "A" weighting scale is most frequently used as it weights the factors of the sound pressure levels to reflect the sensitivity of the human ear.

While it is obviously important to protect personnel from extremely high sound pressure levels, it is also important to protect them from *annoying* noise. This is often described as the quality of sound. The best example is scraping of fingernails against a chalkboard. While this will not damage your hearing, it is obviously annoying. This is where frequency enters the picture. Humans are most sensitive in the frequency range between 4000 and 10,000 Hz. Certain frequencies or combinations of frequencies in this area can be extremely annoying yet not damaging to the ear. Unfortunately, belt drives will occasionally emit frequencies in this range of sufficient sound pressure level to cause both annoyance and possible hearing damage given long-term exposure.

For the most part, V-belts are not considered to be producers of high sound pressure levels and thus only contribute (infrequently) to the annoyance factor. Generally, these problems can be cured with proper maintenance—primarily proper belt tension, cleanliness of the belt and pulleys, or replacement of belts damaged due to slippage or other causes.

Synchronous belts can produce sound pressure levels that exceed acceptable limits. In general, these drives are heavily loaded and high speed, usually in combination with wider belt widths. Many of today's newer generation curvilinear synchronous belts are designed to reduce sound pressure levels. For synchronous belt drives which exhibit excessive noise and/or objectionable frequencies, it is generally recommended that an inexpensive sound enclosure be built around the drive. While drive enclosure for V-belts and flat belts is not recommended because of significant heat build up, it is an acceptable practice for synchronous belts (which have minimal heat generation). A simple sound enclosure fabricated of plywood and foam rubber is quite effective, particularly in the frequency range objectionable to people.

Proper design practice and good maintenance generally result in acceptable noise levels. Problem drives must be referred to the supplier for detailed analysis.

D. Energy Efficiency/Heat Generation

A well designed and maintained belt drive is 95–98% efficient. For comparison, the published efficiencies of chain are in the 95–99% range and spur or worm gears are in the 92–96% range.

V-belt efficiency can be defined as

$$\text{Efficiency} = \frac{\text{horsepower out}}{\text{horsepower in}} \times 100$$

or

$$\text{Efficiency} = \frac{\text{torque out} \times \text{speed out}}{\text{torque in} \times \text{speed in}} \times 100 \tag{16}$$

Torque losses result from hysteresis caused by bending stresses and from friction at the belt-to-sheave interface. There may also be windage losses— losses caused by the belt and sheaves moving through the air.

Speed losses are the result of belt slip and belt creep (described earlier).

1. General Belt Design Considerations

Industrial belt drives operate most efficiently when designed according to recommended practices. Figure 20B shows the proper loss when belts are underloaded or overloaded. Efficiency losses become significant when a belt is run at 50% or less of its rated capacity.

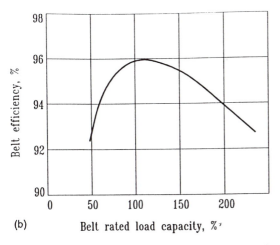

(b)

Figure 20B Belt efficiency versus rated load capacity—illustrates general tendency of all V-belts to operate most efficiently near their maximum rated load capacity.

The most important design variable is the use of proper sheave diameters. Sheaves should be selected by proper drive design methods, with consideration given to the National Electrical Manufacturers Association (NEMA) recommendations, space limitations, and efficiency. NEMA specifies the minimum recommended diameters for sheaves used on electric motors in NEMA Standard MG 1-14-42. These recommendations are designed to protect the motors by limiting bearing loads and shaft stresses. RMA also publishes information on recommended minimum sheave diameters for all belt cross sections.

In general, the larger the sheave, the greater the drive efficiency, because there is less hysteresis loss. Large diameters also result in less creep and reduced slip because of the lower operating tensions. If the drive requires a small-diameter sheave, the designer will sacrifice efficiency and allow for added stress on the motor and the driveN equipment.

Equally important, synchronous belts should be considered for drives where energy efficiency is important. Because of their thin section, they have very low hysteresis losses. But, the major benefit is that they have *no* speed loss. This can yield a gain of up to 5-8% over poorly maintained V-belt drives that are slipping due to improper tensioning. Many utility companies have accepted a 5% energy savings for synchronous drives (over V-belts) as a value for rebate programs.

Motors should be properly sized. Although optimization of the belt drive is important, it is only one component of the drive package. The plant engineer should examine each application carefully to make sure the motor has sufficient capacity to meet the horsepower requirements, yet is not overdesigned for the application—a condition that leads to electrical inefficiencies.

2. Improving Efficiency of Existing Drives

There are several options for improving efficiency of existing drives. Wrapped or fabric covered belts can be replaced with molded notch V-belts if the drive has sheave diameters smaller than RMA minimum recommendations. The notched version of the belt will reduce the bending stresses and the hysteresis losses. Since these belts cost more than the conventional belts, a cost analysis should be made to determine if the expected energy savings justify the conversion.

The efficiency loss that accompanies overbelting presents another opportunity. A drive designed years ago should be re-examined. For example, consider a 30-hp belt drive operating at 1160 rpm. The drive has two 9-in. pitch diameter sheaves and C144 belts. On the basis of a 1964 RMA rating of 9.31 hp per belt, this drive required four belts. The capacity of belts has improved since 1964, RMA ratings have changed twice since then. The RMA rating for this belt is now 12.72 hp. As a result, only three belts are required to do the job. Over-belting is eliminated, and efficiency may be slightly improved. **Caution:** *Using less than the full number of belts on an existing drive will result in* *uneven groove wear and may create problems when the next set of belts is installed.*

Often overlooked and neglected, proper maintenance can do more to help increase V-belt drive efficiency than anything else. Improper tensioning can cause efficiency losses of up to 10%.

3. Evaluating Belt Efficiency

Any well-designed drive that is operating efficiently will be only slightly warm to the touch immediately after shutdown, assuming ambient conditions are below 100°F. Since efficiency losses are in the form of heat generation, a drive that is uncomfortable to hold in your hand (approximately 140°F or greater), is an excellent candidate for further evaluation and holds promise for energy savings. **Caution:** *Observe all safety procedures when performing this inspection.*

VI. PULLEYS (SHEAVES AND SPROCKETS)

This industry uses three terms to describe the "discs" to which the belt is applied in order to transmit motion or power: pulleys, sheaves, sprockets.

In the United States the term "sheaves" is associated with V-belt; V-ribbed belts with pulleys; while "pulleys" and "sprockets" are associated with synchronous belts. In ISO the term "pulley" is applied to all belts. For the remainder of this chapter, the term "pulley" will be used to cover all types.

A. Materials

Various materials are used for power transmission pulleys. The most popular are listed in Table 8 and described below.

1. Cast Iron

Most industrial pulleys are made of gray or ductile cast iron. This is a cost effective material for medium to large volumes of pulleys. The majority cast pulleys are made with gray iron. The common material for gray cast iron is class 30B, ASTM A-48. This material permits casting of relatively complex shapes and thus provides design freedom at relatively low cost.

Ductile iron is also used for industrial pulleys but in a limited volume. It has excellent ductility and high modulus and tensile strength for heavier loading with lower deflection. Typical ductile cast iron would be 65-45-12, ASTM A-536.

Both gray and ductile iron have excellent wear characteristics as well as sufficient strengths for today's industrial environment. The tensile strength of gray cast iron limits its capability to 6500 feet per minute (fpm) rim speed while ductile cast iron can be run at speeds up to 8000 fpm.

Table 8 Pulley Materials

Type	Properties
Gray cast iron	Cost effective. Good wear resistance. Can be cast in complex shapes. Maximum rim speed: 6500 fpm
Ductile cast iron	Higher strength and modulus compared to gray cast iron. Good wear resistance. Maximum rim speed:[a] 8000 fpm[3]
Steel	High tensile strength. Can be cost effective for small pulleys. Depending upon alloy, maximum rim speed can exceed 15,000 fpm[3a]
Powdered metal	Cost effective for large quantities. Premium materials will result in properties similar to steel.
Aluminum	High strength-to-weight ratio. Medium to poor wear resistance. Depending upon the alloy, maximum rim speed can exceed 12,000 fpm[3]
Plastic	Low weight and good corrosion resistance. Low cost. Must be careful when applying to any drive that has moderate to heavy loads. Physical properties vary widely depending on base and filler material.

[a] Any pulley operating over 6500 fpm must be referred to the supplier for safety-related analysis.

2. Steel

Steel is used in limited quantities for smaller pulleys and prototype work. It is also used for certain special applications that require high tensile strength or high rim speeds. Depending on the alloying, the mechanical properties can be adjusted to meet special requirements. Typical grades of steel used in pulleys include 1018 and 1144. These materials, because of their higher tensile strength, have limiting rim speeds significantly greater than cast iron. For 1018 the rims speeds can be up to 12,000 fpm while 1144 permits rim speeds up to 16,000 fpm. Some stainless steel is used for special applications. Because of its high tensile strength it has a rim speed capability of up to 18,000 fpm.

3. Sintered Metal

Sintered/powdered metal parts are made by compacting the powered metal into the desired shape using rigid dies and then sintering the part to obtain the desired properties.

To meet required tolerances, additional coining or sizing operations are often needed. Although these terms are used interchangeably they are different procedures. Sizing involves small amounts of secondary metal deformation which improves the parts dimensional precision. Sizing slightly affects density and material properties. Coining is a final pressing operation used to change the upper and lower surface configuration of circular parts.

The sintered metal process has a wide range of applications. Part manufacturing capability is a function of the size of the press required to form the shape of the product. Because the dies used for the part are relatively expensive, this process and material is used primarily for high-volume parts usually in the range of 6 in. in diameter or less. The characteristics of the pulleys can vary significantly depending on the base material. In general, the properties are similar to steel.

4. Aluminum

Aluminum has a high strength-to-weight ratio. Because of its decreased wear resistance, compared to cast iron and steel, aluminum is typically used in non-power transmission and lightly loaded applications or where weight is an important factor. Aluminum's wear characteristics can be improved by hard anodizing or special hardening processes. Various aluminum alloys are available and can significantly change the tensile strength of the material. In general, the common materials permit pulley rim speeds up to 12,000–18,000 fpm, depending on the material chosen.

5. Plastic

Plastic's tensile and compressive strengths together with its stiffness properties are significantly lower than metal, especially steel. Its physical properties can be enhanced with stabilizers and fibrous reinforcements as well as highly compounded plastic materials. Plastic has low thermal conductivity. As a result, it will not dissipate the heat generated in a heavily loaded belt drive.

Plastic is affected by temperature and time under load which is commonly referred to as creep deformation. Because of its reduced properties, the rim speed and wear characteristics of this material is generally unacceptable for highly loaded industrial drives.

It is most commonly used on motion transmission and low horsepower applications, primarily where cost and corrosion can be a problem.

B. Mountings

1. Bored to Size

Some industrial pulleys are made with specific bore sizes integral to the pulley. For the most part, bored-to-size pulleys are relatively small in diameter or are specially made for a specific application with high volume.

Bushing Systems. By far, the majority of the industrial pulleys used today involve a bushing system. The bushing system provides a common tapered bore in the pulley that can be matched with a series of bushings which have a matching O.D. taper and various bore options. This system permits maximum flexibility of inventory with minimum parts.

There are three major bushing systems in the industrial market today:

1. Taper lock
2. QD (quick detach)
3. Split taper

2. Taper Lock

The taper lock system is shown in Fig. 21. This system is unique in that the bushing is completely contained within the pulley (there is no flange protruding from the bore). Because it is flush with the pulley, it requires a minimum amount of space on the shaft. The complete pulley installation is compact and neat in appearance.

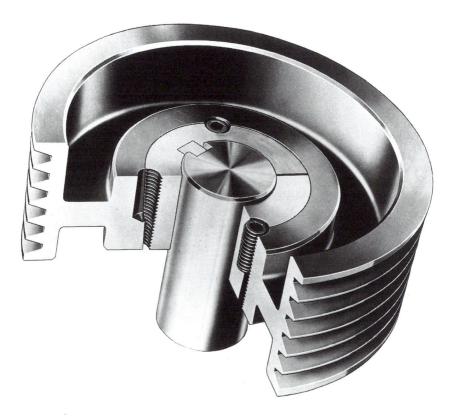

Figure 21   Taper lock bushing system.

3. Quick Detach

The QD system (quick detach) shown in Fig. 22 uses a flanged bushing to draw the bushing into the pulley. This bushing system is unique in that it can be mounted with the bushing flange next to the motor or bearing; as an alternative the pulley can be reversed along with the bushing and the bolts can be installed from the outside as well.

4. Split Taper

The last system is a split tapered system (Fig. 23) that looks similar to the QD although the bushing is not slit completely through the flange and there are two slits within the bushing to allow compression on the shaft. In addition, there is an external key between the OD of the bushing and the ID of the pulley.

Each of the above bushings are capable of transmitting full rated torque as well as "reasonable" horsepower spike conditions if properly designed and installed. The user should be careful to read the installation instructions. The installation procedure, while simple, must be followed carefully in order to receive full benefit of the product. None of the bushing systems are interchangeable.

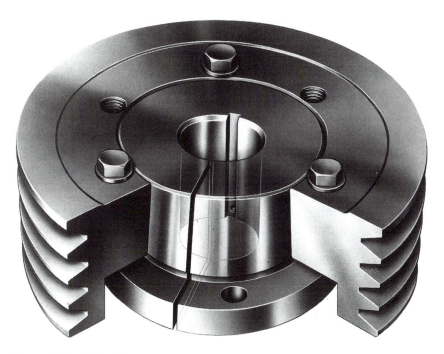

Figure 22   QD bushing system.

Figure 23   Split taper bushing.

C.   Design and Application Considerations

1.   *Balancing*

Commercially available cast iron and steel pulleys are manufactured and distributed with either static balancing or of a finished quality that permits vibration free operation within reasonable operation speeds. As a rule of thumb, for rim speeds less than 6500 fpm, no additional balancing is required. Some applications requiring critical smooth operation or with extremely wide pulleys should be referred to the manufacturer for possible additional balancing requirements, even if the rim speed is under 6500 fpm. For special applications or where high speed operation is anticipated, dynamic balancing should be specified when ordering the pulleys or done as a separate operation upon receipt of the pulleys. The dynamic balancing should be done according to the tolerances specified in the MPTA Bulletin SPB-86.

2. Corrosion Protection

For industrial cast iron and steel pulleys there are three basic methods of corrosion protection:

1. Painting
2. Plating
3. Dipping

The predominant protective method for cast iron pulleys is a painting process. Painting technology has improved significantly over the last several years and the corrosion resistance, particularly for shelf storage prior to installation is excellent with today's paints. The primary concern of painting is that the belt contact area be covered with minimal paint so as not to affect the belt operation during its early operating period.

For extreme corrosion resistant requirements, plating can be required. The primary plating materials are zinc, nickel, and chromium. As is the case with painting, care must be taken not to permit the plating to change the shape or surface finish of the belt contact area.

For storage purposes many manufacturers use a dipping process, usually in an oil based solution. This process offers good protection for most storage conditions. The dipping process does not distort the belt surface of the pulley provided that the dipping solution has a low viscosity.

For nonferrous materials such as aluminum, other processes can be used for corrosion protection. The primary method is anodizing. This is also used to enhance the wear characteristic of the aluminum surface through the use of hard anodizing.

3. Wear

Common industrial pulleys provide excellent wear resistance because of the characteristics of cast iron and steel. Under extremely abrasive conditions the wear characteristics can be enhanced through the use of hard chrome plating or the use of exotic materials such as stainless steel. In the case of aluminum pulleys, hard anodizing or hard chrome plating can reduce wear. In addition to enhancing the physical properties of the pulleys, it is common practice to increase the design service factor, thus decreasing the load that each belt carries resulting in lower pressure between the belt and pulley. Pulley and belt wear can also be reduced by carefully shielding the drive to minimize the abrasive material to which the drive is exposed.

4. Heat Dissipation

Heat dissipating characteristics of pulleys become important whenever a heat source becomes a significant factor in belt life reduction. The heat source can be the belt drive itself or some external heat source. In V-belts, it is general practice to use heat conductive pulley material to conduct the heat generated from the belt hysteresis away from the belt and dissipate it through radiation to the atmosphere. On highly loaded drives, such as snowmobile variable speed applications, it is common practice to design the pulley with a maximum radiating surface through the use of fins.

In the instances where the heat source is external to the belt drive, it may be desirable to isolate the belt from the heat source through the use of low-thermal-conductive materials. These applications are rare and should be referred to the belt supplier for further consideration.

VII. DRIVE DESIGN CONSIDERATIONS

A. Belt Application Range

Belts have been successfully applied over a wide range of speeds and loads. Because of the numerous belt cross sections available, virtually any power transmission application can be considered a potential candidate. While the normal speed range for belts is the typical motor speeds from 575 to 3450 rpm, applications for speeds from less than 10 to over 20,000 rpm are not uncommon. Similarly, drives have been designed for loads from less than 1/4 hp to well over 2000 hp. Often, even the large drives are space competitive with others forms of power transmission.

Belts are commonly used for nonpower transmission applications. Typically these involve conveying of materials, but other applications are possible. An example is bowling equipment, which consists of many belt drives that convey the ball and pins.

In addition to the horsepower/speed application range the designer should also take into consideration the ambient temperature and the ambient condition characteristics of the drive. Most of today's belts operate satisfactorily over a range from -30 to $140°F$. (See RMA Bulletin IP-3-1) Special belt constructions can extend this range but must be ordered at an added cost.

Belt drives are reasonably forgiving of most ambient conditions experienced in the typical factory application. These include moderate amounts of oil, dust, dirt, etc. Excessive amounts of oil or significant amounts of debris or abrasive dust can result in premature belt and pulley wear. Care should be taken to divert as much of the foreign material from the belt drive as possible and applications having excessive foreign material should have increased maintenance frequency.

B. Prime Mover and DriveN Characteristics

It is important to consider the characteristics of both the driveR and driveN unit. The torque curves for either the electric or internal combustion prime mover should be reviewed to determine if the design must account for abnormal peak loads or significant torque pulsations (i.e., one or two cylinder internal combustion engines). For the driveN, it is very important to understand the horsepower demands, together with any torque fluctuations that might be produced by the driveN. For example, fan horsepower requirement is a cubic relationship of the speed and thus small rpm changes can make significant differences in the driveN load. Heavy pulsating loads (i.e., rock crushers and piston pumps) can have significant impact on the belt's ability to deliver adequate service.

In all of the above cases, the belt supplier's design procedures permit the designer to take the driveR and driveN variables into consideration. This is done through the use of a *service factor* selected through a review process of both units.

Belt suppliers design manuals and engineering standards contain service factor tables. While detailed, they are easy to use and permit the designer to quickly review the characteristics of the drive components. Once the service factor is selected it is multiplied times the power requirement to arrive at the *design horsepower.*

It should be emphasized that service factor is not a safety factor. It is a correction factor to account for various drive conditions.

Sometimes an additional service factor is imposed on a drive when load variation cannot be easily evaluated (i.e., large acceleration or deceleration loads) or for ambient conditions (i.e., applications subjected to abrasive conditions or elevated ambient temperatures).

C. Single- Versus Multiple-Condition Duty Cycle

Most industrial drives can be properly designed using a single-design horsepower which is a combination of the horsepower requirement multiplied by the service factor described above. This is considered a single-condition duty cycle. In some cases it is necessary to analyze the drive in more detail using a multiple-condition duty cycle. An example of this is an automotive front-end accessory drive where the engine rpm can vary from 600 to over 6000 rpm and each of the accessories has varying torque requirements, depending on speed and other conditions. A proper design of an automotive front-end drive can encompass duty cycles of up to 30 conditions which take into account all of the engine speed and accessory load permutations. If an industrial drive is going to be subjected to significant load variations as a function of time, the designer should contact the belt supplier to determine if a special drive design procedure is required to optimize the belt design as well as the tension recommendation.

D. Accounting for Special Drive Conditions

In the preceding paragraphs methods of taking into account unusual operating conditions were described. This section will discuss two of the more frequently encountered special drive conditions which need to be considered.

1. Acceleration Loads

In addition to the normal load requirements, a belt drive is subjected to acceleration loads resulting from rapid speeding up or slowing down of the equipment. For the majority of drives, the acceleration loads are either small enough or occur seldom enough that they do not require special attention during design. In these cases, the extra loading is taken care of by the service factor selected.

When rapid speed changes occur or when acceleration and deceleration is a normal function of the application, it is best to determine if this extra loading affects the design. An example of this is an automobile engine. When the engine is "hot-rodded," the alternator is forced to accelerate quickly. If it is a fairly large alternator, with high inertia, the extra acceleration loads may equal or exceed the normal loading assigned for drive design purposes.

The extra loading due to high or frequent acceleration can be accounted for in the design by one of these methods:

An extra service factor, usually based on experience with the given application. This is less desirable than the following quantitative methods.

Calculating the acceleration load and adding it to the driveN machine's normal horsepower requirement. This is applicable when the extra load is truly imposed on top of the regular load for the major portion of the life of the drive.

Calculating the acceleration load and using it as a separate condition, as in multicondition automotive designs. This method is applicable when the times of acceleration do not coincide with other peak driveN loads, so that conditions must be treated separately.

Acceleration loads may be calculated from the following formula:

$$HP_a = (4.30 \times 10^{-9})WR^2 a\bar{N} \qquad (17)$$

where
HP_a = acceleration horsepower
WR^2 = inertia of rotating component, lb $\cdot$ in.^{2}
a = acceleration of rotating component, rev/sec
$\bar{N}$ = average speed of rotating component through the acceleration period, r/min

2. Unusual Ambient Conditions

Unusual ambient conditions are generally considered to be excessive heat, excessive cold, or excessive debris which will result in belt and pulley wear. This can also include misalignment which is inherent in the application and cannot be removed. The impact of each of these conditions can be minimized by reducing the load on the belt and thus the normal force between the belt and the pulley. In general, the practice of increasing the service factor by one or two points, perhaps more if the severity is great, is an acceptable practice. If an application deviates significantly from the belt application range described earlier in this section, the designer should contact the belt supplier for further guidance.

E. Speed Ratio Calculation

Before a discussion of speed ratio calculations is developed, it is necessary to describe the various diameter nomenclature that is used to describe the relationship of a belt in a pulley. In designing belt drives, two diameters are commonly used for a given belt/pulley combination: The pitch diameter and the outside diameter. This is illustrated in Fig. 24. Depending on the cross section, the outside diameters may not be a controlled value (i.e., variable speed belts). When this is the case, reference is made to an effective outside diameter which can be geometrically defined. The effective outside diameter (EOD) may coincide with the physical O.D.

Recently, a third diameter reference has been established for classical belts—datum diameter. It was necessary to establish this reference when the industry changed the relationship of O.D. and P.D. to reflect modern belt constructions. The datum system permitted belt and pulley suppliers to continue the same sizing nomenclature (previously based on the pitch diameter). The relationship between the outside/pitch/datum diameter is defined by the belt suppliers and standards organizations such as RMA, SAE, ASAE, etc.

Generally, the outside diameter (or alternatively the effective outside diameter) is used to determine belt length while the pitch diameter is used to determine belt tensions and belt speed.

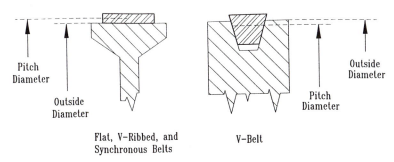

Figure 24   Pitch diameter and outside diameter.

For shaft speed calculations, the ratio between the RPM of the faster shaft and the RPM of the slower shaft is called speed ratio and is defined as follows:

$$\text{Speed ratio} = \frac{N \text{ of faster shaft, r/min}}{N \text{ of slower shaft, r/min}} \tag{18}$$

For design purposes the speed ratio of a belt drive is always considered to be equal to or greater than 1.

The individual RPM of the two pulleys is inversely proportional to their pitch diameters. In equation form this is

$$\text{Speed ratio} = \frac{N \text{ faster shaft}}{N \text{ slower shaft}} = \frac{\text{pitch diameter larger wheel}}{\text{pitch diameter smaller wheel}} \tag{19}$$

F. Belt Length Calculations

Belt length calculation for two-wheel drives is simple:

$$L = 2C + 1.57(D + d) + \frac{(D - d)^2}{4C} \tag{20}$$

where

L = belt length, in.

D = diameter of large sheave or pulley, in.

d = diameter of small sheave or pulley, in.

C = center distance, in.

Belt suppliers and engineering standards specify whether the belt calculation is made using effective outside, pitch, or datum diameter (it varies with each type of belt). For drives with more than two pulleys it is often necessary to make a drive layout to determine correct belt length. Procedures for this method are described in suppliers' manuals.

Once the correct belt length is determined the drive must be checked to make sure that provision has been made for installation and takeup of the belt. The values for installation and takeup are included in belt suppliers' literature or applicable standards.

G. Allowable Belt Speeds

Although, for each belt, there is a speed beyond which life or horsepower rating begins to decrease, generally speaking, the upper limit on belt speed is determined more by pulleys than by belts. This is especially true of today's smaller, more powerful belts, which develop less centrifugal stress because of their lower weight.

With respect to pulleys, both safety and balance are considerations in determining upper limits on belt speed.

Stock cast iron pulleys should not be run above 6500 ft/min belt speed without checking with the supplier for possible dynamic balancing or special material requirements.

V-belt drives are commonly designed (with steel pulleys) for belt speeds up to 9000 to 10,000 ft/min in automotive drives. Special belt designs can transmit loads at belt speeds greater than 20,000 ft/min.

At the other end of the range, low-belt-speed limitations are imposed only by economics. These applications usually have high torque requirements often requiring large belt sections and/or a large number of belts, and often larger-diameter pulleys.

H. Use of Idlers

Idlers are pulleys in a belt drive that are not loaded (do not transmit power). They are used for several reasons:

To provide takeup for fixed center drives
To clear obstructions
To turn corners (as in mule pulley idlers, Fig. 39)
To break up long spans where belt whip may be a problem
To provide automatic tensioning
To increase wrap on critical wheels (careful analysis is needed to make sure that the drive is actually benefitted)

Idlers may be run on the inside or outside of the drive (Fig. 25). An inside idler decreases arc of contact on adjacent wheels, while an outside idler increases arc. An outside idler, however, bends belts backward, and on some drives this nullifies the gain obtained with increased arc as far as belt life is concerned. Also, if an outside idler is used for takeup, the amount of takeup is limited by the belts on the opposite side of the drive.

If at all possible, idlers should be placed in the slack span of the drive to minimize their contribution to the fatigue of the belt. Spring-loaded, or weighted, idlers should always be located on the slack span because the spring force or weight can be much less. Such idlers should not be used on a drive where the belt's direction of rotation can be reversed, where the slack slide can become the tight side.

A flat idler pulley, either inside or outside, should be located as far as possible from the next pulley along the path of the belt. Belts move back and forth slightly across a flat pulley (due to misalignment or other factors) and there should be as much span length as possible so that the affects of tracking misalignment is minimized.

The diameter of inside idlers should be at least as large as the diameter of the smallest loaded wheel. The diameter of outside idlers should be at least one-third larger than the diameter of the smallest loaded wheel. (It should be

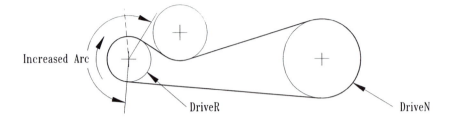

Idler Pulley on the Slack Side of a Drive

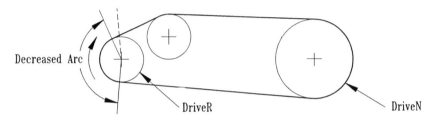

Figure 25 Effect of idler placement.

noted that some of today's modern V-belt and synchronous belt constructions are sensitive to back side idlers and may require special size limitations; contact the belt supplier for more information.)

Flat idlers should not be crowned (except on flat belt applications) but flanging of such idlers is good practice. In the case of flanged idlers, the inside bottom corners of the flange should not be rounded because the belt might then climb over the flange (see Fig. 26). The width of flat idlers must be greater than that of the belt (or the total width of the belts).

Brackets for idlers should be sturdily constructed. Frequently, the cause of drive problems described as belt stretch, belt instability, short belt life, belt roughness, belt vibration, etc., can be traced to weak idler backetry. RMA Bulletin IP-3-6 discusses idlers in belt systems.

I. Basic Drive Design Procedure

While the discussions in this chapter has concentrated on specific details of various complicated drives, the truth is that the majority of industrial belt drives are very simple to design and implement. The majority of these drives are two pulley applications with an electric motor and a driveN unit that usually does not have a complicated duty cycle. When this is the situation, any of the suppliers catalogs are very easy to use and the drive can be selected in less

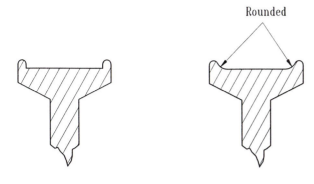

Figure 26　Flanging of flat idlers.

than 10 min. For drives that are more complicated, the designer has the option of contacting the belt supplier for design assistance or can use existing computer software which is discussed in the following section.

For the simple two pulley drive, the design out of a supplier's catalog is very straightforward. The following steps describe this belt selection process for a simple V-belt drive using an electric motor prime mover:

1. Find the design horsepower by selecting the proper service factor (using the selection table) and multiply the service factor times the horsepower requirement).

2. Select the proper belt section. Most suppliers' catalogs have selection charts to guide the designer to the best cross section for further analysis.

3. Using the proper drive selection table (see Table 9 for an example) find the motor rpm and desired driveN speed combination and then select a set of pulleys which meets the desired condition.

4. Proceeding to the right cross the line, select the belt length which meets the proper center distance. On the same line, determine the horsepower rating (per belt) for the combination of pulleys selected.

5. The horsepower rating determined by the pulley and belt combination must be adjusted using a correction factor determined by the color (or shading) of the belt length selected (usually a factor from 0.8 to 1.1 which accounts for the belt length affect on the rating).

6. Since the horsepower rating from the table is per belt, the number of belts required can be determined by dividing the design horsepower by the rated horsepower. The results must be rounded up to the next whole number.

Table 9 High-Power II V-Belt, PowerBand Belt, and Tri-Power Molded Notch V-Belt Drives

1160 RPM	1750 RPM	3450 RPM	Small Sheave	Large Sheave	Speed Ratio	A-26	A-29	A-31	A-33	A-35	A-37	A-40	A-42	A-43	A-46	A-49	A-53	A-57	A-60	A-65
1160	1750	3450	3.00	3.00	1.00	8.9	10.4	11.4	12.4	13.4	14.4	15.9	16.9	17.4	18.9	20.4	22.4	24.4	25.9	28.4
1160	1750	3450	3.20	3.20	1.00	8.6	10.1	11.1	12.1	13.1	14.1	15.6	16.6	17.1	18.6	20.1	22.1	24.1	25.6	28.1
1160	1750	3450	3.40	3.40	1.00	8.3	9.8	10.8	11.8	12.8	13.8	15.3	16.3	16.8	18.3	19.8	21.8	23.8	25.3	27.8
1160	1750	3450	3.60	3.60	1.00	8.0	9.5	10.5	11.5	12.5	13.5	15.0	16.0	16.5	18.0	19.5	21.5	23.5	25.0	27.5
1160	1750	3450	3.80	3.80	1.00	7.7	9.2	10.2	11.2	12.2	13.2	14.7	15.7	16.2	17.7	19.2	21.2	23.2	24.7	27.2
1160	1750	3450	4.00	4.00	1.00	7.4	8.9	9.9	10.9	11.9	12.9	14.4	15.4	15.9	17.4	18.9	20.9	22.9	24.4	26.9
1160	1750	3450	4.20	4.20	1.00	7.1	8.6	9.6	10.6	11.6	12.6	14.1	15.1	15.6	17.1	18.6	20.6	22.6	24.1	26.6
1160	1750	3450	4.60	4.60	1.00	6.4	7.9	8.9	9.9	10.9	11.9	13.4	14.4	14.9	16.4	17.9	19.9	21.9	23.4	25.9
1160	1750	3450	4.80	4.80	1.00		7.6	8.6	9.6	10.6	11.6	13.1	14.1	14.6	16.1	17.6	19.6	21.6	23.1	25.6
1160	1750	3450	5.00	5.00	1.00		7.3	8.3	9.3	10.3	11.3	12.8	13.8	14.3	15.8	17.3	19.3	21.3	22.8	25.3
1160	1750	3450	5.20	5.20	1.00		7.0	8.0	9.0	10.0	11.0	12.5	13.5	14.0	15.5	17.0	19.0	21.0	22.5	25.0
1160	1750	3450	5.60	5.60	1.00			7.4	8.4	9.4	10.4	11.9	12.9	13.4	14.9	16.4	18.4	20.4	21.9	24.4
1160	1750	3450	6.00	6.00	1.00				7.7	8.7	9.7	11.2	12.2	12.7	14.2	15.7	17.7	19.7	21.2	23.7
1160	1750	3450	6.40	6.40	1.00					8.1	9.1	10.6	11.6	12.1	13.6	15.1	17.1	19.1	20.6	23.1
1160	1750	3450	7.00	7.00	1.00							9.7	10.7	11.2	12.7	14.2	16.2	18.2	19.7	22.2
1115	1683	3317	4.60	4.80	1.04	6.3	7.8	8.8	9.8	10.8	11.8	13.3	14.3	14.8	16.3	17.8	19.8	21.8	23.3	25.8
1115	1683	3317	4.80	5.00	1.04		7.5	8.5	9.5	10.5	11.5	13.0	14.0	14.5	16.0	17.5	19.5	21.5	23.0	25.5
1115	1683	3317	5.00	5.20	1.04		7.1	8.1	9.1	10.1	11.1	12.6	13.6	14.1	15.6	17.1	19.1	21.1	22.6	25.1
1105	1667	3286	3.40	3.60	1.05	8.2	9.7	10.7	11.7	12.7	13.7	15.2	16.2	16.7	18.2	19.7	21.7	23.7	25.2	27.7
1105	1667	3286	3.60	3.80	1.05	7.8	9.3	10.3	11.3	12.3	13.3	14.8	15.8	16.3	17.8	19.3	21.3	23.3	24.8	27.3
1105	1667	3286	3.80	4.00	1.05	7.5	9.0	10.0	11.0	12.0	13.0	14.5	15.5	16.0	17.5	19.0	21.0	23.0	24.5	27.0
1105	1667	3286	4.00	4.20	1.05	7.2	8.7	9.7	10.7	11.7	12.7	14.2	15.2	15.7	17.2	18.7	20.7	22.7	24.2	26.7
1094	1651	3255	3.00	3.20	1.06	8.8	10.3	11.3	12.3	13.3	14.3	15.8	16.8	17.3	18.8	20.3	22.3	24.3	25.8	28.3
1094	1651	3255	3.20	3.40	1.06	8.5	10.0	11.0	12.0	13.0	14.0	15.5	16.5	17.0	18.5	20.0	22.0	24.0	25.5	28.0
1094	1651	3255	6.00	6.40	1.06					8.4	9.4	10.9	11.9	12.4	13.9	15.4	17.4	19.4	20.9	23.4
1084	1636	3224	5.20	5.60	1.07			7.7	8.7	9.7	10.7	12.2	13.2	13.7	15.2	16.7	18.7	20.7	22.2	24.7
1084	1636	3224	5.60	6.00	1.07			8.0	9.0	10.0	11.0	11.5	12.5	13.0	14.5	16.0	18.0	20.0	21.5	24.0
1074	1620	3194	4.60	5.00	1.08		7.6	8.6	9.6	10.6	11.6	13.1	14.1	14.6	16.1	17.6	19.6	21.6	23.1	25.6
1074	1620	3194	4.80	5.20	1.08		7.3	8.3	9.3	10.3	11.3	12.8	13.8	14.3	15.8	17.3	19.3	21.3	22.8	25.3
1064	1606	3165	4.20	4.60	1.09	6.7	8.2	9.2	10.2	11.2	12.2	13.7	14.7	15.2	16.7	18.2	20.2	22.2	23.7	26.2
1064	1606	3165	6.40	7.00	1.09						8.6	10.1	11.1	11.6	13.1	14.6	16.6	18.6	20.1	22.6
1055	1591	3136	3.60	4.00	1.10	7.7	9.2	10.2	11.2	12.2	13.2	14.7	15.7	16.2	17.7	19.2	21.2	23.2	24.7	27.2
1055	1591	3136	3.80	4.20	1.10	7.4	8.9	9.9	10.9	11.9	12.9	14.4	15.4	15.9	17.4	18.9	20.9	22.9	24.4	26.9
1045	1577	3108	3.40	3.80	1.11	8.0	9.5	10.5	11.5	12.5	13.5	15.0	16.0	16.5	18.0	19.5	21.5	23.5	25.0	27.5
1045	1577	3108	5.00	5.60	1.11			6.8	7.8	8.8	9.8	10.8	12.3	13.3	13.8	15.3	16.8	18.8	20.8	22.3
1036	1563	3080	3.00	3.40	1.12	8.6	10.1	11.1	12.1	13.1	14.1	15.6	16.6	17.1	18.6	20.1	22.1	24.1	25.6	28.1
1036	1563	3080	3.20	3.60	1.12	8.3	9.8	10.8	11.8	12.8	13.8	15.3	16.3	16.8	18.3	19.8	21.8	23.8	25.3	27.8
1036	1563	3080	4.60	5.20	1.12		7.5	8.5	9.5	10.5	11.5	13.0	14.0	14.5	16.0	17.5	19.5	21.5	23.0	25.5
1027	1549	3053	4.20	4.80	1.13	6.6	8.1	9.1	10.1	11.1	12.1	13.6	14.6	15.1	16.6	18.1	20.1	22.1	23.6	26.1
1018	1535	3026	4.00	4.60	1.14	6.9	8.4	9.4	10.4	11.4	12.4	13.9	14.9	15.4	16.8	18.4	20.4	22.4	23.9	26.4
1018	1535	3026	5.60	6.40	1.14				7.7	8.7	9.7	11.2	12.2	12.7	14.2	15.7	17.7	19.7	21.2	23.7
1009	1522	3000	5.20	6.00	1.15			7.3	8.3	9.3	10.4	11.9	12.9	13.4	14.9	16.4	18.4	20.4	21.9	24.4
1000	1509	2974	3.40	4.00	1.16	7.8	9.3	10.3	11.3	12.3	13.3	14.8	15.8	16.3	17.8	19.3	21.3	23.3	24.8	27.3
1000	1509	2974	3.60	4.20	1.16	7.5	9.0	10.0	11.0	12.0	13.0	14.5	15.5	16.0	17.5	19.0	21.0	23.0	24.5	27.0
1000	1509	2974	4.80	5.60	1.16		7.0	8.0	9.0	10.0	11.0	12.5	13.5	14.0	15.5	17.0	19.0	21.0	22.5	25.0
1000	1509	2974	6.00	7.00	1.16						8.9	10.4	11.4	11.9	13.4	14.9	16.9	18.9	20.4	22.5
991	1496	2949	3.20	3.80	1.17	8.1	9.7	10.7	11.7	12.7	13.7	15.2	16.2	16.7	18.2	19.7	21.7	23.7	25.2	27.7
991	1496	2949	7.00	8.20	1.17								9.7	10.7	11.7	13.2	15.2	17.2	18.7	21.2
983	1483	2924	3.00	3.60	1.18	8.5	10.0	11.0	12.0	13.0	14.0	15.5	16.5	17.0	18.5	20.0	22.0	24.0	25.5	28.0
983	1483	2924	4.20	5.00	1.18	6.4	7.9	8.9	9.9	10.9	11.9	13.4	14.4	14.9	16.4	17.9	19.9	21.9	23.4	25.9
975	1471	2899	4.00	4.80	1.19	6.7	8.2	9.2	10.2	11.2	12.2	13.7	14.7	15.2	16.7	18.2	20.2	22.2	23.7	26.2
975	1471	2899	5.00	6.00	1.19			7.5	8.5	9.5	10.5	12.0	13.0	13.5	15.0	16.5	18.5	20.5	22.0	24.5
967	1458	2875	3.80	4.60	1.20	7.0	8.5	9.5	10.5	11.5	12.5	14.1	15.1	15.6	17.1	18.6	20.6	22.6	24.1	26.6
959	1446	2851	4.60	5.60	1.21		7.1	8.1	9.1	10.1	11.1	12.6	13.6	14.1	15.6	17.1	19.1	21.1	22.6	25.1

Key to hp correction factors   0.7   0.8   0.9   [1.0]

| V-Belt No. and Center Distance | | | | | | | | | | | | Sheave Datum Diameters | | Rated HP per Bell (Including Allowance for Speed Ratio) | | | | | |
| | | | | | | | | | | | | | | Hi-Power II RPM of Small Sheave | | | Tri-Power Molded Notch RPM of Small Sheave | | |
A-68	A-71	A-75	A-81	A-85	A-90	A-96	A-100	A-105	A-112	A-120	A-128	Small Sheave	Large Sheave	1160 RPM	1750 RPM	3450 RPM	1160 RPM	1750 RPM	3450 RPM
29.9	31.4	33.4	36.4	38.4	40.9	43.9	45.9	48.4	51.9	55.9	59.9	3.00	3.00	1.62	2.13	3.01	2.16	2.82	3.99
29.6	31.1	33.1	36.1	38.1	40.6	43.6	45.6	48.1	51.6	55.6	59.6	3.20	3.20	1.87	2.50	3.61	2.39	3.14	4.49
29.3	30.8	32.8	35.8	37.8	40.3	43.3	45.3	47.8	51.3	55.3	59.3	3.40	3.40	2.13	2.86	4.20	2.62	3.45	4.97
29.0	30.5	32.5	35.5	37.5	40.0	43.0	45.0	47.5	51.0	55.0	59.0	3.60	3.60	2.38	3.21	4.77	2.84	3.76	5.44
28.7	30.2	32.2	35.2	37.2	39.7	42.7	44.7	47.2	50.7	54.7	58.7	3.80	3.80	2.63	3.57	5.33	3.07	4.07	5.90
28.4	29.9	31.9	34.9	36.9	39.4	42.4	44.4	46.9	50.4	54.4	58.4	4.00	4.00	2.88	3.92	5.87	3.29	4.37	6.34
28.1	29.6	31.6	34.6	36.6	39.1	42.1	44.1	46.6	50.1	54.1	58.1	4.20	4.20	3.13	4.26	6.40	3.50	4.67	6.77
27.4	28.9	30.9	33.9	35.9	38.4	41.4	43.4	45.9	49.4	53.4	57.4	4.60	4.60	3.62	4.94	7.40	3.94	5.25	7.58
27.1	28.6	30.6	33.6	35.6	38.1	41.1	43.1	45.6	49.1	53.1	57.1	4.80	4.80	3.86	5.28	7.88	4.15	5.53	7.96
26.8	28.3	30.3	33.3	35.3	37.8	40.8	42.8	45.3	48.8	52.8	56.8	5.00	5.00	4.10	5.61	8.34	4.36	5.81	8.32
26.5	28.0	30.0	33.0	35.0	37.5	40.5	42.5	45.0	48.5	52.5	56.5	5.20	5.20	4.34	5.94	8.78	4.57	6.09	8.67
25.9	27.4	29.4	32.4	34.4	36.9	39.9	41.9	44.4	47.9	51.9	55.9	5.60	5.60	4.81	6.59	9.60	4.98	6.63	9.32
25.2	26.7	28.7	31.7	33.7	36.2	39.2	41.2	43.7	47.2	51.2	55.2	6.00	6.00	5.28	7.22	10.3	5.38	7.16	9.90
24.6	26.1	28.1	31.1	33.1	35.6	38.6	40.6	43.1	46.6	50.6	54.6	6.40	6.40	5.74	7.84	11.0	5.78	7.68	10.4
23.7	25.2	27.2	30.2	32.2	34.7	37.7	39.7	42.2	45.7	49.7	53.7	7.00	7.00	6.42	8.73	†	6.36	8.42	†
27.3	28.8	30.8	33.8	35.8	38.3	41.3	43.3	45.8	49.3	53.3	57.3	4.60	4.80	3.68	5.03	7.57	3.99	5.32	7.72
27.0	28.5	30.5	33.5	35.5	38.0	41.0	43.0	45.5	49.0	53.0	57.0	4.80	5.00	3.92	5.37	8.05	4.20	5.60	8.10
26.6	28.1	30.1	33.1	35.1	37.6	40.6	42.6	45.1	48.6	52.6	56.6	5.00	5.20	4.16	5.70	8.51	4.41	5.88	8.46
29.2	30.7	32.7	35.7	37.7	40.2	43.2	45.2	47.7	51.2	55.2	59.2	3.40	3.60	2.19	2.95	4.37	2.67	3.52	5.11
28.8	30.3	32.3	35.3	37.3	39.8	42.8	44.8	47.3	50.8	54.8	58.8	3.60	3.80	2.44	3.30	4.94	2.89	3.83	5.58
28.5	30.0	32.0	35.0	37.0	39.5	42.5	44.5	47.0	50.5	54.5	58.5	3.80	4.00	2.69	3.66	5.50	3.12	4.14	6.04
28.2	29.7	31.7	34.7	36.7	39.2	42.2	44.2	46.7	50.2	54.2	58.2	4.00	4.20	2.94	4.01	6.04	3.34	4.44	6.48
29.8	31.3	33.3	36.3	38.3	40.8	43.8	45.8	48.3	51.8	55.8	59.8	3.00	3.20	1.68	2.22	3.18	2.21	2.89	4.13
29.5	31.0	33.0	36.0	38.0	40.5	43.5	45.5	48.0	51.5	55.5	59.5	3.20	3.40	1.93	2.59	3.78	2.44	3.21	4.63
24.9	26.4	28.4	31.4	33.4	35.9	38.9	40.9	43.4	46.9	50.9	54.9	6.00	6.40	5.34	7.31	10.5	5.43	7.23	10.0
26.2	27.7	29.7	32.7	34.7	37.2	40.2	42.2	44.7	48.2	52.2	56.2	5.20	5.60	4.43	6.07	9.04	4.62	6.16	8.81
25.5	27.0	29.0	32.0	34.0	36.5	39.5	41.5	44.0	47.5	51.5	55.5	5.60	6.00	4.90	6.72	9.86	5.03	6.70	9.46
27.1	28.6	30.6	33.6	35.6	38.1	41.1	43.1	45.6	49.1	53.1	57.1	4.60	5.00	3.71	5.07	7.66	4.04	5.40	7.87
26.8	28.3	30.3	33.3	35.3	37.8	40.8	42.8	45.3	48.8	52.8	56.8	4.80	5.20	3.95	5.41	8.14	4.25	5.69	8.25
27.7	29.2	31.2	34.2	36.2	38.7	41.7	43.7	46.2	49.7	53.7	57.7	4.20	4.60	3.25	4.43	6.74	3.60	4.82	7.06
24.1	25.6	27.6	30.6	32.6	35.1	38.1	40.1	42.6	46.1	50.1	54.1	6.40	7.00	5.86	8.01	11.3	5.88	7.83	10.7
28.7	30.2	32.2	35.2	37.2	39.7	42.7	44.7	47.2	50.7	54.7	58.7	3.60	4.00	2.50	3.38	5.11	2.94	3.91	5.73
28.4	29.9	31.9	34.9	36.9	39.4	42.4	44.4	46.9	50.4	54.4	58.4	3.80	4.20	2.75	3.74	5.67	3.17	4.22	6.19
29.0	30.5	32.5	35.5	37.5	40.0	43.0	45.0	47.5	51.0	55.0	59.0	3.40	3.80	2.25	3.03	4.54	2.72	3.60	5.29
26.3	27.8	29.8	32.8	34.8	37.3	40.3	42.3	44.8	48.3	52.3	56.3	5.00	5.60	4.22	5.78	8.68	4.46	5.96	8.61
29.6	31.1	33.1	36.1	38.1	40.6	43.6	45.6	48.1	51.6	55.6	59.6	3.00	3.40	1.74	2.30	3.35	2.26	2.97	4.28
29.3	30.8	32.8	35.8	37.8	40.3	43.3	45.3	47.8	51.3	55.3	59.3	3.20	3.60	1.99	2.67	3.95	2.49	3.29	4.78
27.0	28.5	30.5	33.5	35.5	38.0	41.0	43.0	45.5	49.0	53.0	57.0	4.60	5.20	3.74	5.11	7.74	4.04	5.40	7.87
27.6	29.1	31.1	34.1	36.1	38.6	41.6	43.6	46.1	49.6	53.6	57.6	4.00	4.60	3.27	4.48	6.83	3.60	4.82	7.06
27.9	29.4	31.4	34.4	36.4	38.9	41.9	43.9	46.4	49.9	53.9	57.9	4.00	4.60	3.02	4.14	6.30	3.44	4.59	6.77
25.2	26.7	28.7	31.7	33.7	36.2	39.2	41.2	43.7	47.2	51.2	55.2	5.60	6.40	4.95	6.81	10.0	5.13	6.85	9.75
25.9	27.4	29.4	32.4	34.4	36.9	39.9	41.9	44.4	47.9	51.9	55.9	5.20	6.00	4.48	6.16	9.21	4.72	6.31	9.10
28.8	30.3	32.3	35.3	37.3	39.8	42.8	44.8	47.3	50.8	54.8	58.8	3.40	4.00	2.27	3.08	4.63	2.77	3.67	5.40
28.5	30.0	32.0	35.0	37.0	39.5	42.5	44.5	47.0	50.5	54.5	58.5	3.60	4.20	2.52	3.43	5.20	2.99	3.98	5.87
26.5	28.0	30.0	33.0	35.0	37.5	40.5	42.5	45.0	48.5	52.5	56.5	4.80	5.60	4.00	5.50	8.31	4.30	5.75	8.39
24.4	25.9	27.9	30.9	32.9	35.4	38.4	40.4	42.9	46.4	50.4	54.4	6.00	7.00	5.42	7.44	10.7	5.53	7.38	10.3
29.2	30.7	32.7	35.7	37.7	40.2	43.2	45.2	47.7	51.2	55.2	59.2	3.20	3.80	2.04	2.76	4.13	2.54	3.36	4.92
22.7	24.2	26.2	29.2	31.2	33.7	36.7	38.7	41.2	44.7	48.7	52.7	7.00	8.20	6.59	8.99	†	6.51	8.64	†
29.5	31.0	33.0	36.0	38.0	40.5	43.5	45.5	48.0	51.5	55.5	59.5	3.00	3.60	1.79	2.39	3.53	2.31	3.04	4.42
27.4	28.9	30.9	33.9	35.9	38.4	41.4	43.4	45.9	49.4	53.4	57.4	4.20	5.00	3.30	4.52	6.92	3.65	4.89	7.20
27.7	29.2	31.2	34.2	36.2	38.7	41.7	43.7	46.2	49.7	53.7	57.7	4.00	4.80	3.05	4.18	6.39	3.44	4.59	6.77
26.0	27.5	29.5	32.5	34.5	37.0	40.0	42.0	44.5	48.0	52.0	56.0	5.00	6.00	4.27	5.87	8.86	4.51	6.03	8.75
28.1	29.6	31.6	34.6	36.6	39.1	42.1	44.1	46.6	50.1	54.1	58.1	3.80	4.60	2.80	3.83	5.85	3.22	4.29	6.33
26.6	28.1	30.1	33.1	35.1	37.6	40.6	42.6	45.1	48.6	52.6	56.6	4.60	5.60	3.79	5.20	7.92	4.09	5.47	8.01

Key to hp correction factors   1.0   1.1   1.2

† Rim speed higher than 6,500 feet per minute

7. Once the drive is selected, the motor pulley should be checked against the NEMA table to make sure that excessive shaft and bearing loads are not exceeded (most suppliers' design manuals contain these tables).
8. Depending on the above results, the designer may want to repeat the procedure several times to select various drive options.
9. Once the design is finalized, the designer will need to determine the installation and take up allowance and the required tension for the drive. Both of these are described in the design manuals furnished by the belt supplier.

J. Computer Programs Versus Design Manuals

All belt suppliers furnish design manuals with which the user can properly apply the product using rather straightforward procedures. In addition, some belt suppliers also have computer programs for either their own use or the customer's use.

For drive optimization or for drives having significant complications, computer belt design programs are of great benefit. Computer programs originate from three sources:

1. Belt/pulley supplier
2. Customer written
3. Third-party software

Each source has produced credible programs with virtually trouble-free drive design capability. Third-party software is generally written around industry standards which may not reflect an individual supplier's product ratings; in some cases their ratings are higher than industry standards. Third-party software tends to be rather generic in terms of the options afforded the user and do not cover proprietary products described in some belt suppliers' programs.

Customer developed software is often written to address a specific need—many variations of a similar application, ability to check inventory within a customer system, as a part of a CAD design system, etc. Provided that the software is carefully checked against the belt formulas for a wide variations of application, this software is often effective for customers with specific needs.

By far, the best alternative is to use software provided by the belt supplier. For the most part, these programs reflect state-of-the-art software technology and have been carefully checked by the designer of the software to assure comprehensive cross-checks which minimize the possibility of poor drive design.

VIII. AUTOMATIC TENSIONING METHODS

As the name implies, tension is maintained at a proper value by some automotive device, thus minimizing the amount of attention the drive requires. This is sometimes called the constant tension method of tensioning, even though the tension is not always constant. Automatic tensioning can be provided by the various means listed here and discussed in more detail later in this section:

Spring-loaded (or weighted) idlers
Spring-loaded prime mover bases
Gravity or reactive-torque motor bases
Torque-sensitive pulleys
Spring-loaded pulleys

Properly designed automatic tensioning systems can

Eliminate the need for manual takeup. This usually reduces maintenance costs, and is particularly important for drives which are relatively inaccessible.

Allow the drive to operate at minimum tension. Locked-center drives are slightly overtensioned initially so that adequate tension is provided for a longer time as wear and belt elongation occur. Automatic tensioning can eliminate this initial overtensioning and provide maximum belt and bearing life.

Provide the proper tension at all times. This eliminates the possibility of drive component damage from over- or undertensioning.

Some discretion must be used in applying most of the automatic tensioning devices when shock or pulsating loads occur—spring-loaded idlers and the various types of motor bases can encounter severe bounce and instability. This can result in belts being thrown off the drive or in drive component breakage. Although this bounce can sometimes be dampened out with shock absorbers, the best solution may be a locked-center drive. As mentioned earlier in this chapter, drives subjected to reversed rotation should not use spring-loaded idlers. However, with careful analysis, constant-tension motor bases may be an option.

Automatic tensioning should be used on friction type of drives only. Use of these systems on synchronous belt drives will usually result in belt ratcheting.

The following paragraphs deal with each of the various methods of automatic tensioning.

A. Spring-Loaded Idlers

These are used quite extensively on agricultural machinery. They are especially desirable on drives which encounter wide load variations along with high peak loads. A locked-center drive must be tensioned to prevent slip at the peak load, and as the load drops off, the total tension remains nearly constant at the higher value. On the other hand, a spring-loaded idler allows the total tension to decrease as the load drops off, thus increasing belt and bearing life.

This effect is best illustrated by an example. The V-belt drive in Fig. 27(a) uses a spring-loaded idler on the slack side. Arc of contact on the driveR pulley is 180°, and so the design tension ratio T_T/T_S is 5. Assume that the horsepower load is such that an effective pull $T_T - T_S$ of 40 lb is required. Thus the tight-side tension T_T is 50 lb and the slack-side tension T_S is 10 lb, so that the total tension $T_T + T_S$ is 60 lb. For a fixed-center (or fixed idler) drive the $T_T + T_S$ of 60 lb would be locked in the drive and the value would not change regardless of loads applied.

For the spring-loaded idler application, the idler spring and geometry must be designed, in this case, so that the idler force against the belt results in a 10-lb slack-side tension. This slack-side tension then remains at 10 lb regardless of whether the load increases or decreases, or even whether or not the drive is rotating.

If the load were taken completely off the driveN shaft, both T_T and T_S would be 10 lb and $T_T + T_S$ would be 20 lb. As load increases, T_T goes up while T_S stays at 10 lb. Assume that one-half the design horsepower is being transmit-

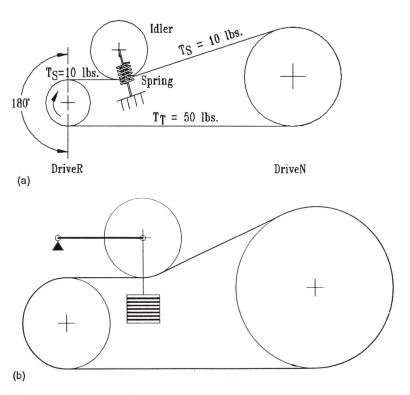

(a)

(b)

Figure 27 (a) Effect of spring-loaded idler placement; (b) Weighted tensioning idler.

ted, so that $T_T + T_S$ is 20 lb instead of 40 lb. Then, since T_S is a constant 10 lb, T_T would be 30 lb, and $T_T + T_S$ would be 40 lb, compared with 60 lb at full horsepower load.

It was previously mentioned that spring-loaded idlers should be located on the slack side of the drive. The idler force can be much less in this position. It was also mentioned that they should not be used on reversing drives—the drive will either slip (if the idler is located on the original slack side) or be severely overtensioned (if the idler is located on the original tight side). For example, if the drive in Fig. 27(a) were to have the driveR and driveN reversed, the spring-loaded idler would be in the tight span. But it imposes only 10 lb on that strand, and the tension on the new slack span would be even less. As a result, the drive would transmit only a fraction of the required horsepower without slipping. On the other hand, with the idler in this position, if the idler force were increased to 50 lb of span tension (so that $T_T - T_S$ would be 40 lb, and the load could be transmitted) and the drive reverted to the original driveR and driveN positions, T_T would be 90 lb ($T_T - T_S = 40$ lb and $T_S = 50$ lb) and $T_T + T_S$ would be 140 lb, a great deal overtensioned.

Weighted idlers, sometimes called gravity idlers, are a special case of the spring-loaded idler where the weight of the assembly is the idler force; there is no spring (see Fig. 27b). They were frequently used with flat belt drives. Weighted idlers are simple and the tension can be easily adjusted by adding or subtracting weight.

B. Spring-Loaded Prime Mover Bases

Spring-loaded motor bases are so called because they are used more with electric motors than with other types of prime movers. The principle on which they operate is sketched in Fig. 28.

Here, the spring force must be such that the motor is pulled back against the drive with a force equal and opposite to the resultant belt pull. The spring must have some extra force, however, to overcome the friction in the base, and, if the mounting is not horizontal, the effect of motor weight must be taken into account.

Spring-loaded motor bases, unlike spring-loaded idlers, provide a nearly constant total drive tension, $T_T + T_S$. Most of these bases are used in the industrial market; therefore, each individual drive is not always designed for the correct tension. Instead, standard bases which have an adjustment for the spring tension are available. A good way to set this spring is to install the drive and operate it, adjusting the spring tension until the drive does not slip under the highest normal load condition. Some spring force is lost as the belt seats in and elongates, and occasional spring adjustments may be necessary. Spring loaded motor bases can be used on drives which reverse direction of rotation unlike spring-loaded idlers.

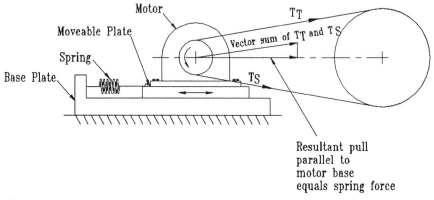

Figure 28   Spring-loaded motor base.

C.   Gravity Motor Bases

These use the weight of the motor to tension the drive, as shown in Fig. 29. The base plate is usually slotted so that the drive tension can be adjusted by moving the motor closer to or further from the pivot point.

1.   Reactive-Torque Motor Bases

These use the reactive torque of the motor to help tension the drive. When a motor is delivering power to a drive, its frame tries to rotate in the direction opposite to that of the pulley. Reactive torque is this tendency to rotate. See the illustration in Fig. 30.

By placing the motor in a cradle with a pivot point near the motor shaft, the motor can be allowed to move away from the driveN pulley, tensioning the

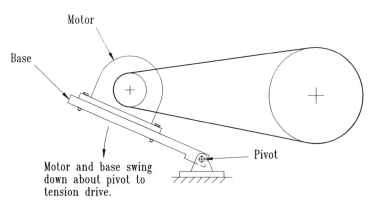

Figure 29   Gravity motor base.

Reactive Torque – motor frame tries to rotate in direction opposite to working torque.

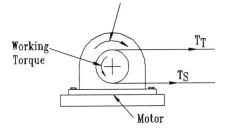

Figure 30 Reactive torque vs. working torque.

drive. As torque increases, tension increases (see Fig. 31). This reactive-torque base is similar to the gravity base except that the pivot point on the gravity base is quite far away from the motor. In that case, the reactive torque has little effect on drive tension.

The reactive-torque motor base must be installed so that the direction of rotation is correct. If the motor shaft in Fig. 31 were rotating in the opposite direction, the reactive torque would actually loosen the drive, rather than tightening it. Operating in the direction shown, both the reactive torque and the motor weight act to tension the drive.

Standard bases are usually available with slots so that the motor can be moved as needed to get the proper tension when the drive is installed. Although specification data usually shows a given setting (or a fixed based) for

Reactive torque forces motor and mounting plate to rotate to the left, tensioning drive.

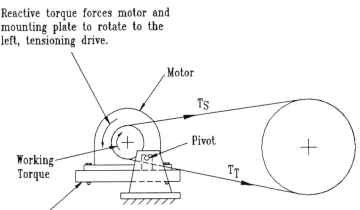

Figure 31 Reactive-torque motor base.

a certain NEMA motor frame size, it is good practice experimentally if the tension in the drive is correct after the drive is installed. This is done by imposing the highest load on the motor after the belts are run in, to see if the drive slips. The pivot point must be adjusted if the tension is not correct.

D. Torque-Sensitive Pulleys

These are made in separate halves, similar to variable-speed pulleys. In the past, threads or angled slots were used on one of the pulley halves so that it moved toward or away from the other half as torque demand increased or decreased. In more recent years, torque-sensitive variable-speed pulleys have included a combination of springs and cam surfaces which reacted to the torque demands placed on them by the drive (see Fig. 32). Like torque-reactive motor bases, torque-sensitive pulleys must be installed so that the direction of rotation is correct. If it is not, the pulleys will open up and the drive will not operate (it is possible to design a pulley with a double cam for use with reversing drives). If they are designed correctly, such pulleys can provide a system where the drive is properly tensioned for all load conditions (i.e., tension proportional to load).

The amount of total takeup that such pulleys can provide is limited by the total possible diameter change. This diameter change can be calculated:

$$\Delta D = \frac{TW}{\tan(\alpha/2)} - 2TH \tag{21}$$

where

ΔD = total possible change in effective outside diameter, in.

TW = top width of groove when closed, in.

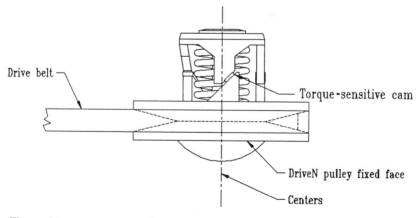

Figure 32   Torque-sensitive sheave.

α = groove angle, deg

TH = belt thickness, in.

If longer belts (which require greater takeup) or belts of small cross section (which have small ΔD) are used, provision for shaft movement for additional takeup must be made.

When torque-sensitive pulleys are used as the driveR on drives with high-inertia loads, problems of sporadic speed variation can occur as the drive comes up to speed. The pulleys keep opening and closing slightly as a result of overrun of the driveN unit.

Torque-sensitive pulleys in the light-duty or fractional-horsepower range are available as standard items from various manufacturers. They are occasionally built as a special item for other types of V-belt applications.

1. Spring-Loaded Pulleys

Spring-loaded pulleys are commonly used in variable-speed drives, but can be used to tension a drive if the resultant speed variation can be tolerated. They provide automatic tensioning even though their main function is to provide variable speed. These pulleys will be discussed later.

IX. SPECIAL DRIVE APPLICATIONS

A. Multiplane Drive Geometry

The usual drive design has the belts and pulleys essentially in one plane, as shown in Fig. 33. But there are many belt drives in which the belts and pulleys

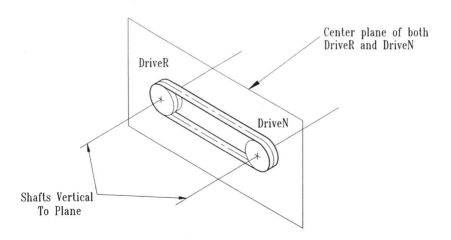

Figure 33   One-plane drive.

lie in more than one plane—a multiplane drive. One example is shown in Fig. 34.

This drive's functions include changing the direction of power transmission, as well as the normal functions of a belt drive. Such drives are used quite frequently on agricultural equipment.

Multiplane drives generally use V-belts. Flat and V-ribbed belts must be carefully designed as they are sensitive to the misalignment inherent in these kinds of drives. Synchronous belts are not recommended for multiplane drives although some applications have been developed through special coordination with the belt supplier.

There are two factors common to multiplane drives which must be considered in their design.

1. Twist of the belt (all multiplane drives)
2. Kink of the belt (most multiplane drives)

Twist of the belt reduces belt life because the edge of the belt must elongate slightly, as shown in Fig. 35. This elongation causes a higher tension in that part of the belt, reducing life.

Also, twist can cause V-belt turnover if it is too severe. Most belt supplier's design manuals give recommendations for minimum tangent lengths for a given amount of twist. (The tangent length is the length of the belt span between the points of contact on the adjacent pulleys.) These minimum tangent lengths are selected to keep the amount of twist to a minimum so that it does not affect the stability of the belt in the groove.

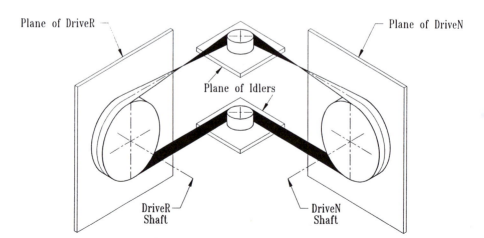

Figure 34 Multiplane drive.

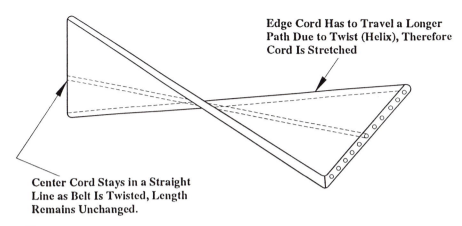

Edge Cord Has to Travel a Longer
Path Due to Twist (Helix), Therefore
Cord Is Stretched

Center Cord Stays in a Straight
Line as Belt Is Twisted, Length
Remains Unchanged.

Figure 35 Belt twist.

When a belt enters a pulley misaligned, kinking results, and the edge of the belt must elongate (see Fig. 36). This elongation creates additional tension in the belt, reducing life.

Kink of the belt reduces life even more than twist. Belts do not easily bend sideways, especially today's higher modulus constructions. A belt with a low tensile modulus of elasticity will undergo less increase in tension, and experience a smaller reduction in life than one with a high modulus. High-modulus steel cable belts, for example, can withstand very little twist or kink without significant life reduction.

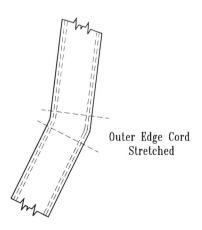

Outer Edge Cord
Stretched

Figure 36 Belt kink.

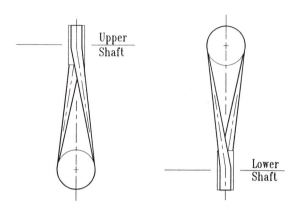

Figure 37   Quarter-turn drive.

Misalignment also causes increased wear of belt side walls from rubbing against the side of the pulleys during entry and exit. This wear can cause excessive tension loss. In severe cases of misalignment, belt stability in the groove can become a problem, causing belt turnover or belts coming off the drive. For this reason, deep-groove pulleys are commonly used for multiplane drives. Specially placed idlers to reduce inherent misalignment are also a possibility.

All multiplane drives should be referred to the belt supplier in order to ensure optimum design. In addition, RMA Bulletin IP-3-10 can serve as a valuable reference.

Three of the most popular multiplane drives are quarter-turn (Fig. 37), crossed belt (Fig. 38), and mule drive (Fig. 39). Each will be explained in more detail.

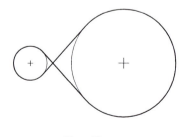

Preferred   More Wear
(a)   (b)

Figure 38   Crossed drives.

1. Cross-Belt Drives

The crossed drive is used when the driveR and driveN shafts must rotate in opposite directions. Two major factors must be considered when designing these drives: amount of belt twist and belt wear at the intersection.

On crossed drives when the twist occurs in too short a center distance, the belts can become unstable on the drive. Belt suppliers have developed recommendations for minimum center distance for various cross sections of belts.

The other problem that occurs with crossed drives is belt wear due to abrasion at the crossover point where belts touch. This can be minimized by design so that this crossover point is near the center of the drive. This results in less pressure between the spans (Fig. 38(a)). If the speed ratio is high, and the crossover point is near the small pulley, as shown in Fig. 38(b), both spans of belt try to occupy the same location more forcibly and wear is more severe.

An idler can be inserted in a crossed drive to prevent rubbing of the strands, as shown in Fig. 40. However, this arrangement causes misalignment of the drive and may result in reduced service. A thin, low-friction plate, inserted between the strands at the point of contact, can decrease the coefficient of friction and the relative velocity, thus reducing wear.

2. Quarter-Turn

Quarter-turn drives, which have the plane of the pulleys 90° apart (Fig. 37), were very popular with flat belt drives from tractor power take-offs to drive various implements during the early part of this century. A modern example is a vertical-shaft irrigation pump driven by a horizontal electric motor.

To design a quarter-turn or eighth-turn drive, first follow the standard procedure as if you were designing a regular drive to determine pulley sizes, number and type of belts, and center distance. When horsepower ratings cannot be read directly from tables, they must be determined from basic ratings with the various corrections applied. The arc-of-contact, or wrap, of V-belts on the pulley is greater on a quarter-turn than on a one-plane drive that has the same pulley diameters and center distance. Because of this fact, and because speed ratio is usually limited to 2.5:1, factor G, the arc-of-contact correction factor, can be taken as 1.0.

Where quarter-turn drives require a speed ratio greater than 2.5:1, a compound driver should be used. To design such a drive, the primary drive should be the quarter-turn (if possible) and the conventional drive should provide most of the speed ratio to optimize the drive. Use deep-grooved pulleys for individual V-belts to minimize the tendency of the belt to turn over in the groove (standard grooved pulleys must be used for joined V-belts).

3. Mule Drives

Mule pulley drives are those which use a driveR pulley, a driveN pulley, and two or more idlers for the purpose of changing the plane of pulleys. The first

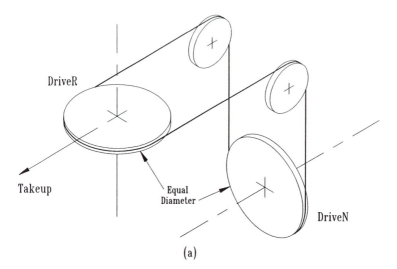

Figure 39   Various types of mule drives.

of such drives was used to "turn corners" with factory line shafts, as illustrated in Fig. 41.

There are various types of mule pulley drives. Three are shown in Fig. 39. Figure 39(a) shows a typical mule pulley drive, one in which the driveR and driveN pulleys are the same size and are at a 90° angle to each other. Figure 39(b) shows one which has unequal pulley sizes. For such a drive to be properly aligned, the idlers must be cocked to lie in a plane drawn through the tangent points on the driveR and driveN pulleys. Also, the tangent points on the idlers must be in the center planes of the driveR and driveN pulleys. Figure 39(c) shows a mule pulley drive where the four pulleys are mounted on only two shafts, which is sometimes used where the center distance is too short for a regular quarter-turn drive. The idlers are free to rotate on the shafts. Note that in Fig. 39(c) the driveR and driveN can be placed in any of the pulley locations to produce desired "direction" of power transmission relative to the driveR shaft.

4. Vertical-Shaft Drives

Vertical-shaft drives, while not multiple-plane drives, share some of the same design concerns. These are drives in which the driveR and driveN shafts are perpendicular to the ground. The primary concern for these drives is the effect of gravity which tends to force the belt out of the grooves or over the flange. With today's higher horsepower ratings and joined belts, the problems encountered with vertical shaft drives have been minimized. The higher ratings tend to

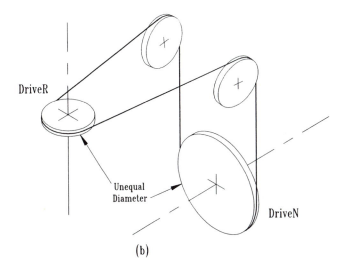

(b)

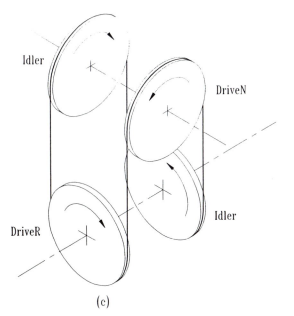

(c)

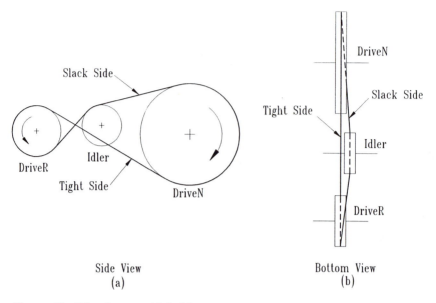

Figure 40   Idlers in crossed belt drives.

have higher tensions, and thus the belt has less sag between the driveR and driveN pulleys. This can be further minimized through the use of joined belts. For synchronous drives, the primary concern is making sure that both pulleys are flanged on the bottom so that the synchronous belt does not track off of either pulley.

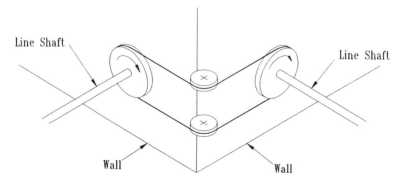

Figure 41   Mule drive.

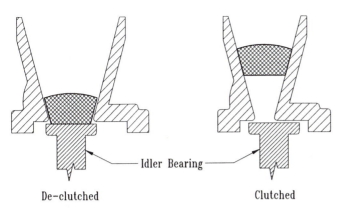

De-clutched   — Idler Bearing —   Clutched

Figure 42   Clutching sheave system.

B. Belt Clutching Drives

Belt clutches offer a simple and inexpensive way to disconnect a driveN shaft from a driveR shaft, if they are properly designed. Both flat belts and V-belts are used on clutching drives. Some V-ribbed belts have been used for clutching applications but the shallow grooves generally present significant design problems. Synchronous belts are obviously not suited to this application.

Clutching flat belt drives were used extensively in turn of the century factories when many different drives are powered from a line shaft, and were required to operate independently. V-belt clutching drives are used frequently in lawn and garden equipment and agricultural machinery.

Belt clutching mechanisms are of four types:

Clutching pulley, in which the side or sides of a V pulley move outward to declutch, allowing a V belt to run on a free-wheeling center "idler" as in Fig. 42.

Clutching idler, in which either an inside or outside idler is moved out of the drive, as in Fig. 43.

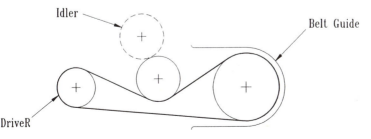

Figure 43   Clutching idler system.

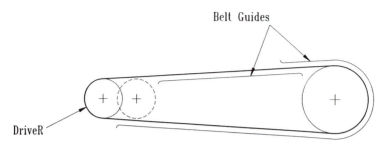

Figure 44   Clutching centers system.

Clutching centers, in which the center distance is shortened, as in Fig. 44.
Idle pulley, in which the belt is shifted (with a shifting fork) to a free-
   wheeling idle pulley (Fig. 45).

On clutching idler and clutching center drives, since the driveR wheel continues to rotate and belt velocity is zero, the belt must be kept from grabbing
the driveR wheel and causing the drive to operate sporadically. Under some
conditions, belt damage can occur. The path of the declutched belt can be controlled by belt guides, or guide pins, as shown in Fig. 46.

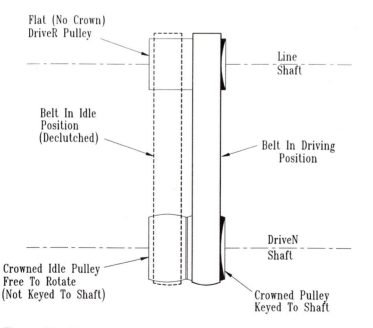

Figure 45   Clutching idle pulley.

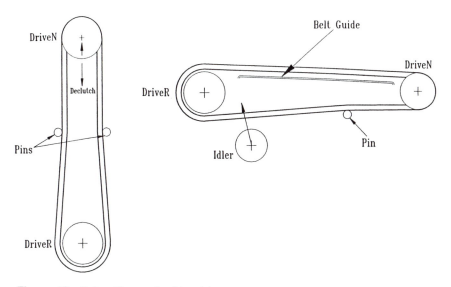

Figure 46   Belt guides on clutching drives.

Flanged pulleys are usually required when flat belts are clutched by clutching idler or clutching center methods, because belt slip during clutching can cause these belts to come off the pulley.

C.   Speed Variation by V-Belts

Many driveN machines require a change in speed. Some speed variation requirements are minimal (such as a ventilating fan that is adjusted only once after installation to allow for minor changes of the airflow requirements). Other machines, such as drill presses, may require changes in speed for different kinds of work. Still others require speed changes while operating.

DriveN speed changes can be accomplished by V-belt drives. Regardless of the type of V-belt used, the principle is always the same: the diameter at which the V-belt operates on one or more pulleys will vary. This is done by varying the width of the pulley groove. See Fig. 47.

The diameter range over which a V-belt operates in one or more pulleys gives a speed variation to the driveN shaft. Speed variation can be best expressed as a range, such as variation from 1750 to 875 rpm of the driveN shaft. Most often, however, it is expressed as a mathematical ratio, with the term "speed range" applied to it. For example, the variation can be expressed as 2.0 speed range.

Various industries use many types of V-belt and pulley combinations for driveN speed variation. Those most commonly referred to include variable-pitch drives and variable-speed drives.

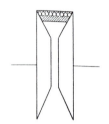 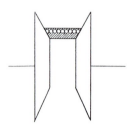

Belt Rides At Maximum Diameter.   Belt Rides At Minimum Diameter.

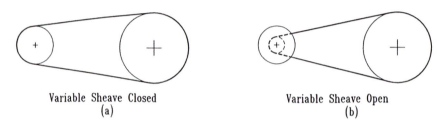

Variable Sheave Closed   Variable Sheave Open
(a)   (b)

Figure 47   Closed and opened variable sheave.

1. Variable-Pitch Drives

Standard lines of variable-pitch pulleys are used with both heavy-duty industrial belts (A, B, C, D, and 5V, 8V) and light-duty belts (3L, 4L, 5L). The
light-duty pulleys are usually single groove; the heavy-duty are usually multiple groove. These variable-pitch drives are used where required speed variation is not great.

In designing variable-pitch drives, either the driveR or the driveN can be
the variable pulley, or both, if more speed is required. One variable-pitch pulley gives about 20–40% variation, depending on cross section and diameter.
Two variable-pitch pulleys give about 40–100%.

When only one variable-pitch pulley is being used with multiple-belt drives,
a companion pulley with special groove spacing is required to minimize belt
misalignment. RMA Bulletin IP-3-14 contains more information on this type of
drive.

2. Variable-Speed Drives

For drives that require more speed variation than can be obtained with conventional (A, B, C, D or 5V, 8V or 3L, 4L, 5L) belts, variable-speed drives are
available. These drives use special wide, relatively thin belts. RMA Standard
IP-25 is a good reference.

Packaged units consisting of variable-speed belts and pulleys, combined with the motor and an output gear box (if specified), are available from about 1/2 to 100 hp. These will be either single- or dual-belt drives, depending on horsepower and supplier.

The speed range of variable-speed drives can be much greater than that of variable-pitch drives. Speed ranges up to 10 to 1 can be obtained on smaller units.

Spring-loaded or hand-adjustable motor pulleys are available for the standard variable-speed belts. These units permit customizing an application.

Industrial variable-speed drive design is often taken care of for the designer when he purchases off-the-shelf hardware for a given application. However, for special applications and OEM designs, the following discussion is offered for consideration. Because of the complexity of the variables, it is strongly recommended that the designer work with a belt supplier in order to achieve the best balance/compromise of variables.

The geometry considerations for a variable-speed drive are shown in Figs. 48 and 49. As can be seen from these figures, the major variables the designer can alter are the groove angle, belt top width, belt thickness, and sheave diameter. Each of these functions has a trade-off:

The smaller the angle the lower the cord support and the lower the belt's availability to carry the load.

Wider top width belts have less "beam strength" and thus tend to require more support (mostly increased belt thickness) for high horsepower.

The thickness of the belt helps to support the cord but also increases bending stresses and reduces the overall speed ratio available.

By using smaller diameters the actual pitch diameter change becomes a bigger percentage of the diameter and, thus, increases speed range available, but significantly reduces horsepower transmission capability.

Types of Adjustable-Speed Drives. The three basic types of adjustable speed drives are

1. Adjustable-center distance (one variable and one fixed pulley)
2. Fixed-center distance (two adjustable speed pulleys)
3. Variator (two fixed pulleys and one shifting variator pulley)

Figure 50 shows the adjustable-center-distance variable-speed drive. In this case the driveN is the fixed pulley and the variable speed is the driveR. This is the general form of this application and usually involves a spring-loaded motor pulley and a hand-adjustable screw-type adjustable motor base.

Figure 51 shows the fixed-center version of a variable-speed drive. This type of drive generally has a mechanically controlled driveR pulley and a spring-loaded (and sometimes torque-sensitive) driveN pulley. While this type

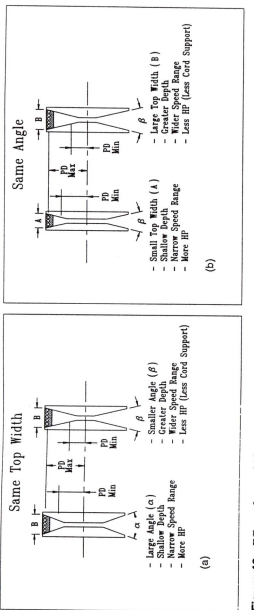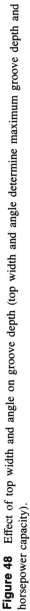

Figure 48 Effect of top width and angle on groove depth (top width and angle determine maximum groove depth and horsepower capacity).

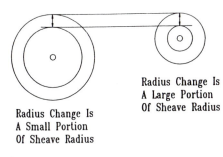

Radius Change Is
A Large Portion
Of Sheave Radius

Radius Change Is
A Small Portion
Of Sheave Radius

But: Smaller Sheave Has Less Arc of Contact
And: Causes Increased Internal Belt Flexure

Figure 49   Small sheaves provide maximum ratio change.

of drive is more complex, the speed variation available to the designer is virtually doubled that of the fixed-center drive.

The last style of variable-speed drive involves the variator geometry as shown in Fig. 52. This is a relatively simple device to set up and control but is rather complicated in its design procedure. The shifting occurs by moving the

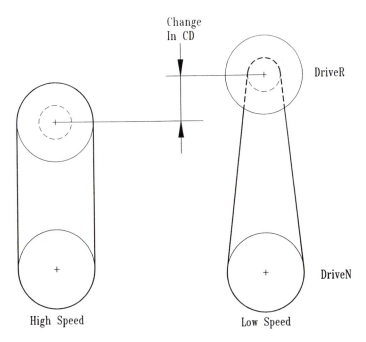

Change
In CD

DriveR

DriveN

High Speed   Low Speed

Figure 50   Speed ratio change using adjustable-center distance.

Speed Reduction

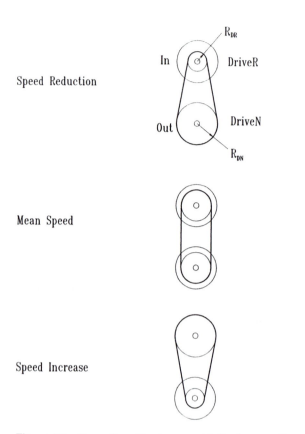

Mean Speed

Speed Increase

Figure 51   Two adjustable sheaves with fixed-center distance.

variator pulley (center pulley in Fig. 52) back and forth with a simple lever device. The disadvantages of this type of system is significant misalignment that occurs over most of the speed range and bearing wear in the free-floating section which cause cocking of the floating disk.

Of the industrial variable devices (complete package furnished by various suppliers), the most common drive is the fixed-center two-pulley adjustable drive. This is a proven reliable, efficient method of adjusting driveN speeds for various applications.

Agricultural machinery uses variable-speed drives of both the variable-pitch type (HA, HB, HC, etc.) and the wider, thinner belt type. The latter are often referred to as agricultural adjustable-speed drives. They use special construction belts with top widths from 1 to 2 1/2 in. Propulsion drives for combines (traction drives) and cylinder drives are examples of applications in which

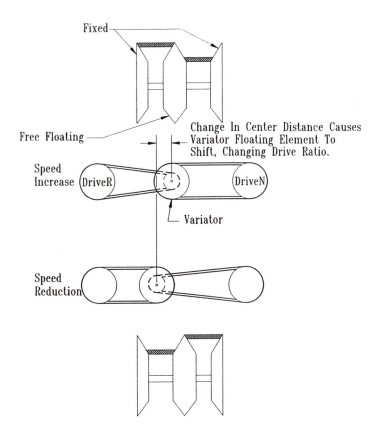

Figure 52 Three-sheave variator system.

adjustable-speed belts are used. Agricultural adjustable-speed belts are defined in ASAE Standard S211.

D. Torque Limiting

Many designers use belts for torque-limiting devices in order to protect the equipment. Synchronous belts are obviously not suitable for torque limiting and should not be considered as protective devices. V-belts, flat belts, and V-ribbed belts can be used for this application.

Because frictional belts slip and relieve peak loads, they frequently are excellent "equipment" protecting devices for unusual overload conditions. However, the designer *must not* rely on them to be safety devices when personnel safety is involved. Because the coefficient of friction, together with the wedging, varies considerably for various ambient conditions and belt construction, the ability to predict the point of breakaway slip is of concern. This is complicated by the user's tensioning method, which is generally quite variable. There are various design techniques that can help a designer minimize the extreme spread in peak torque power transmission capability, but these are beyond the scope of this chapter. It is strongly recommended that the designer use commercially available protective devices, particularly when personnel safety is involved.

E. V-Flat Drives

The typical V-flat drive is made up of a V-grooved driveR pulley and a large driveN pulley with a flat driving surface. The most practical drive is one with a large speed ratio in combination conversion of a flat belt drive to V-belts or V-ribbed belts, since it is not necessary to replace the large driveN pulley. These drives are also very practical for driving a large drum or cylinder where addition of V-grooves would be difficult and expensive (i.e., V-ribbed belts on a clothes dryer where the clothes drum is used as the driveN).

In addition to the normal V-flat drives which use V-belts or V-ribbed belts, variable-speed belts have been applied using a variable-pitch V-groove and flat driveN pulley.

The V-flat drive generally requires less tension than a flat belt drive. The design of V-flat drives is described in RMA Bulletin IP-3-7 and various belt suppliers' manuals.

X. BELT MAINTENANCE CONSIDERATIONS

An effective preventative maintenance program can save a significant amount of time and money. Inspection and replacement of belts and faulty drive components before they fail can reduce costly down time and production delays. In addition, proper maintenance can significantly impact the efficiency of belt drives, particularly friction belt drives.

A. Sources of Drive Problems

The sources of drive problems are shown generically in Fig. 53. While the absolute values of this chart may vary from belt type and application, the relationships are reasonable for most applications. By examining Fig. 53, the designer can quickly ascertain that improper drive maintenance is the overwhelming factor for drive problems. This section only deals in generalities with these sources of problems. Various belt suppliers have excellent preventive maintenance manuals.

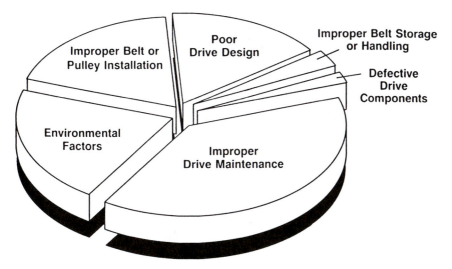

Figure 53 Sources of drive problems.

B. Belt Storage

While belt storage is a relatively small portion of the drive problems, it is worth a brief discussion. Under favorable storage conditions good-quality belts retain their initial service capabilities for up to eight years. The following are guidelines for proper storage:

Minimize exposure to sunlight and moisture.

Maintain storage area at temperatures not exceeding 100°F (except for short periods of time).

Avoid exposure to significant amounts of electrical machinery (the ozone by-product will degrade the belts).

Belt should be coiled and stored on racks that will not cause deformation of the belt over long storage periods.

An excellent source of further information is RMA Bulletin IP-3-4.

C. General Drive Inspection

As mentioned previously, most drive suppliers have preventive maintenance manuals. In general, these are excellent checklists for proper installation and maintenance of belts and pulleys. The primary elements of drive inspection and maintenance are those of common sense. These involve proper belt tension, proper drive alignment, replacement of worn parts (Fig. 54 shows the use of a

Figure 54   Sheave groove gauges.

set of groove gauges to measure wear on an industrial pulley), and normal machine inspection procedures. For the most part, the machine inspection procedures involve the use of good visual inspection, listening for signs of improper operation, and checking for excessive conditions such as vibration or excessive temperatures.

The well-trained mechanic can easily detect signs of impending belt failure. These signs include

Severe belt cracking
Excessive machine vibration
Excessive drive temperature (most belt drives are only slightly warm to the touch after the machine has been shutdown)
Excessive noise
Excessive belt debris/dust in the guard

D. Belt Tensioning

The two primary maintenance factors which will yield the biggest drive improvements ad optimum service are proper tensioning and drive alignment.

Proper drive tension recommendations can be furnished by the belt supplier or calculated using any of the supplier's drive design manuals. The process of tensioning the drive is fairly easy and the tools required are generally available in most maintenance cribs.

Figures 55 and 56 show two methods of tensioning a belt drive (both methods apply to both friction- and synchronous-type belts). Figure 55 shows a standard deflection method using a simple gauge easily obtained from a belt supplier. The process is to measure the amount of belt deflection for a specified deflection force. Normal procedure is to set the deflection at 1/16 in./in. of span length and measure the actual deflection force. These values can be calculated from the design manuals or furnished by the belt supplier. Figure 56 shows the use of a commercial tensioning gauge. For the most part, these gauges read directly in static tension. Commercial gauges should be properly calibrated for the specific construction and cross section of belts.

It is also possible to measure belt tension in V-belt drives using a method termed the "elongation method." This uses a measure of belt stretch to set belt tension. Oftentimes this value is in the ½–1½% range. The belt supplier must

Figure 55 Tensioning using force/deflection method.

Figure 56 Typical "direct reading" tension gauge.

must be contacted for the proper values for his specific belt since the values are quite sensitive (i.e., the cord modulus has a significant impact on the actual elongation percentage).

E. Drive Alignment

While proper design and tensioning of a drive has the major impact on the service life delivered, drive misalignment is only slightly less important.

This is particularly true in synchronous belts because of the very high cord modulus. The effects of misalignment are

Reduced belt life due to the higher edge cord stresses
Accelerated pulley and belt wear due to the increased nonuniform pressures
Premature failures due to instability of the belt in the drive (i.e., belt turn-over)

There are two types of misalignment:

1. Angular (see Fig. 57).
2. Parallel (see Fig. 58).

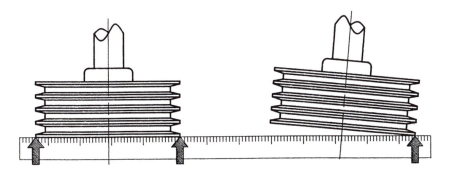

Figure 57 Angular misalignment—V-belts.

While both forms of misalignment can be detrimental to belts, the angular misalignment is especially severe on synchronous belts because of the forced edge cord elongation.

The best way to measure alignment is with the use of a straight edge or a piece of string. Figure 59 shows the correct method of using both tools.

The amount of permissible misalignment varies significantly depending on the application. For example, some lawn and garden equipment is misaligned as much as 6° to 8°. However, it should be remembered that this is a specially designed drive using extraordinary belt construction and special hardware to maintain belt stability. For the typical industrial drive, V-belts should no be misaligned more than 1/2° to obtain maximum belt life. This equates to

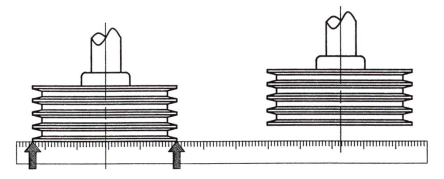

Figure 58 Parallel misalignment—V-belts.

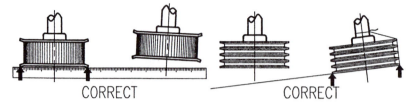

Figure 59   Use of a straight edge and string to measure misalignment.

approximately 1/8 in. offset per foot of span length. For synchronous and V-ribbed belts, the recommended misalignment is 1/4° or approximately 1/16 in. offset per foot of span length.

When checking the misalignment with either a straightedge or a string, it is best to check the alignment first from the driveR to the driveN and then repeat the same procedure from the driveN to the driveR. This cross-checks for both parallel and angular misalignment as well as combined parallel/angular misalignment.

10

FLUID POWER*

Vickers Training Center, Vickers, Incorporated

Rochester Hills, Michigan

ADAPTED BY DAVID W. SOUTH

I. INTRODUCTION

Like many branches of engineering, hydraulics is both ancient and modern. The use of the water wheel, for example, is so ancient that its invention precedes written history. On the other hand, the use of fluid under pressure to transmit power and to control intricate motions is relatively modern and has had its greatest development in the past two or three decades.

Power generation, the branch of hydraulics represented by the water wheel, does not concern us here. The steam engine, the internal combustion engine, the electric motor, and the water turbine all have performed an admirable job in supplying motive power. However, each lacks the mechanics to direct this power to useful work. The purpose of the chapter is to study the use of fluids under pressure in the transmission of power or motion under precise control.

"Why is industrial hydraulics necessary when we have at our disposal many well known mechanical, pneumatic and electric devices?" It is because a

*The material in this chapter was adapted by David W. South from the *Industrial Hydraulics Manual*, 3rd ed., prepared by the Vickers Training Center. Copyright ©1993 by Vickers®, Incorporated, Rochester Hills, MI. Vickers® is a Trinova Company. The editors wish to express their sincere thanks to Vickers, Incorporated, for granting permission for its use in this book. Additionally, the editors suggest that for a more comprehensive study on the fundamentals of fluid power, and related components, a copy of the Vickers manual be purchased.

confined fluid is one of the most versatile means of modifying motion and transmitting power known today. It is as unyielding as steel, and yet infinitely flexible. It will change its shape to fit the body that resists its thrust, and it can be divided into parts, each part doing work according to its size, and can be reunited to work again as a whole.

It can move rapidly in one part of its length and slowly in another. No other medium combines the same degree of positiveness, accuracy, and flexibility, maintaining the ability to transmit a maximum of power in a minimum of bulk and weight.

II. INTRODUCTORY OVERVIEW OF HYDRAULICS

The study of hydraulics deals with the use and characteristics of liquids. Since the beginning of time, man has used fluids to ease his burden. It is not hard to imagine a caveman floating down a river, astride a log with his wife—and towing his children and other belongings aboard a second log with a rope made of twisted vines.

Earliest recorded history shows that devices such as pumps and water wheels were known in very ancient times. It was not, however, until the seventeenth century that the branch of hydraulics with which we are to be concerned first came into use. Based upon a principle discovered by the French scientist Pascal, it relates to the use of confined fluids in transmitting power, multiplying force, and modifying motions. Pascal's law, simply stated, says this:

Pressure applied on a confined fluid is transmitted undiminished in all directions, and acts with equal force on equal areas, and at right angles to them.

This precept explains why a full glass bottle will break if a stopper is forced into the already full chamber. The liquid is practically noncompressible and transmits the force applied at the stopper throughout the container (Fig. 1). The result is an exceedingly higher force on a larger area than the stopper. Thus it is possible to break out the bottom by pushing on the stopper with a moderate force.

A. Pressure Defined

In order to determine the *total* force exerted on a surface, it is necessary to know the *pressure* or *force* on a *unit* of area. We usually express this pressure in *pounds per square inch,* abbreviated psi. Knowing the pressure and the number of square inches of area on which it is being exerted, one can readily determine the total force

(Force in pounds = pressure in psi × area in sq. in.)

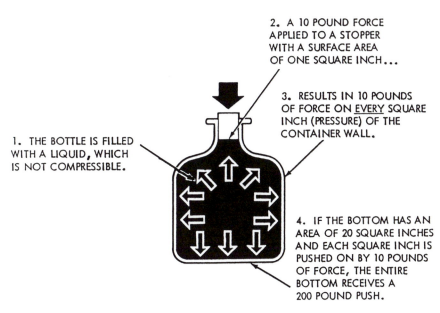

2. A 10 POUND FORCE APPLIED TO A STOPPER WITH A SURFACE AREA OF ONE SQUARE INCH...

3. RESULTS IN 10 POUNDS OF FORCE ON <u>EVERY</u> SQUARE INCH (PRESSURE) OF THE CONTAINER WALL.

1. THE BOTTLE IS FILLED WITH A LIQUID, WHICH IS NOT COMPRESSIBLE.

4. IF THE BOTTOM HAS AN AREA OF 20 SQUARE INCHES AND EACH SQUARE INCH IS PUSHED ON BY 10 POUNDS OF FORCE, THE ENTIRE BOTTOM RECEIVES A 200 POUND PUSH.

Figure 1 Pressure (force per unit area) is transmitted throughout a confined fluid. (Courtesy of Vickers, Incorporated, Rochester Hills, MI.)

B. Hydraulic Power Transmission

Hydraulics now could be defined as a means of transmitting power by pushing on a confined liquid. The input component of the system is called a *pump*; the output is called an *actuator*. The hydraulic system is not a *source* of power. The power source is a *prime mover* such as an electric motor or an engine that drives the pump.

C. Advantages of Hydraulics

Variable Speed. Most electric motors run at a constant speed. It is also desirable to operate an engine at a constant speed. The actuator (linear or rotary) of a hydraulic system, however, can be driven at infinitely variable speeds by varying the pump delivery or using a flow control valve.

Reversibility. Few prime movers are reversible. Those that are reversible usually must be slowed to a complete stop before they are reversed. A hydraulic actuator can be reversed instantly while in full motion without damage. A four-way directional valve or a reversible pump provides the reversing control, while a pressure relief valve protects the system components from excess pressure.

Overload Protection. The pressure relief valve in a hydraulic system protects it from overload damage. When the load exceeds the valve setting, pump delivery is directed to the tank with definite limits to torque or force output. The pressure relief valve also provides a means of setting a machine for a specified amount of torque or force, as in a chucking or clamping operation.

Small Packages. Hydraulic components, because of their high speed and pressure capabilities, can provide high power output with very small weight and size.

Can Be Stalled Without Damage. Stalling an electric motor will cause damage or blow a fuse. Likewise, engines cannot be stalled without the necessity for restarting. A hydraulic actuator, though, *can* be stalled without damage when overloaded and will start up immediately when the load is reduced. During stall, the relief valve simply diverts delivery from the pump to the tank. The only loss encountered is in wasted horsepower.

D. Hydraulic Oil

Any liquid is essentially noncompressible and therefore will transmit power in a hydraulic system. The name hydraulics, in fact, comes from the Greek, *hydor,* meaning "water" and, *aulos,* meaning "pipe." However, the most common liquid used in hydraulic systems is petroleum oil. Oil transmits power readily because it is only very slightly compressible. It will compress about one-half of 1 percent at 1000 psi pressure, a negligible amount in most systems. The most desirable property of oil is its lubricating ability. The hydraulic fluid must lubricate most of the moving parts of the components.

E. Pressure in a Column of Fluid

The weight of a volume of oil varies to a degree as the viscosity (thickness) changes. However, most hydraulic oils weigh from 55 to 58 pounds per cubic foot in normal operating ranges.

One important consideration of the oil's weight is its effect on the pump inlet. The weight of the oil will cause a pressure of about 0.4 psi at the bottom of a 1-foot column of oil (Fig. 2). For each additional foot of height, it will be 0.4 psi higher. Thus, to estimate the pressure at the bottom of any column of oil, simply multiply the height by 0.4 psi.

To apply this principle, consider the conditions where the oil reservoir is located above or below the pump inlet (Fig. 3). When the reservoir oil level is above the pump inlet, a positive pressure is available to force the oil into the pump. However, if the pump is located above the oil level, a vacuum equivalent to 0.4 psi per foot is needed to "lift" the oil to the pump inlet. Actually the oil is not "lifted" by the vacuum; it is forced by atmospheric pressure

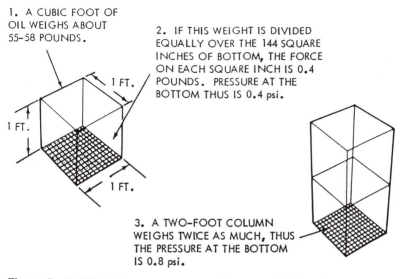

1. A CUBIC FOOT OF OIL WEIGHS ABOUT 55-58 POUNDS.

1 FT.

1 FT.

1 FT.

2. IF THIS WEIGHT IS DIVIDED EQUALLY OVER THE 144 SQUARE INCHES OF BOTTOM, THE FORCE ON EACH SQUARE INCH IS 0.4 POUNDS. PRESSURE AT THE BOTTOM THUS IS 0.4 psi.

3. A TWO-FOOT COLUMN WEIGHS TWICE AS MUCH, THUS THE PRESSURE AT THE BOTTOM IS 0.8 psi.

Figure 2 Weight of oil creates pressure. (Courtesy of Vickers, Incorporated, Rochester Hills, MI.)

into the void created by the pump inlet when the pump is in operation. Water and various fire-resistant hydraulic fluids are heavier than oil, and therefore require more vacuum per foot of lift.

F. Atmospheric Pressure Charges the Pump

The inlet of a pump normally is charged with oil by a difference in pressure between the reservoir and the pump inlet. Usually the pressure in the reservoir is atmospheric pressure, which is 14.7 psi on an absolute gauge. It then is necessary to have a partial vacuum or reduced pressure at the pump inlet to create flow.

G. Positive Displacement Pumps Create Flow

Most pumps used in hydraulic systems are classed as *positive displacement.* This means that, except for changes in efficiency, the pump output is constant regardless of pressure. The outlet is positively sealed from the inlet, so that whatever gets in is forced out the outlet port.

The sole purpose of a pump is to create flow; pressure is caused by a resistance to flow. Although there is a common tendency to blame the pump for loss of pressure, with few exceptions pressure can be lost only when there is a leakage path that will divert *all* the flow from the pump.

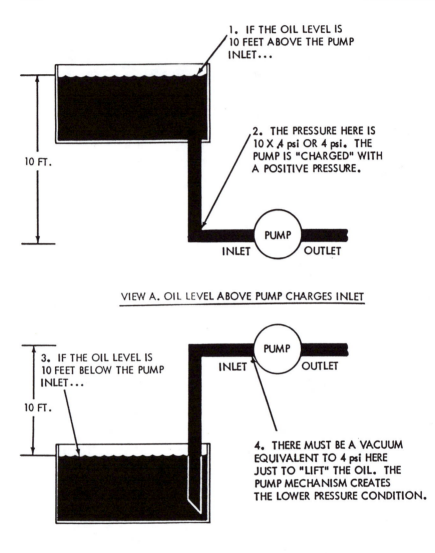

VIEW A. OIL LEVEL ABOVE PUMP CHARGES INLET

VIEW B. OIL LEVEL BELOW REQUIRES VACUUM TO "LIFT" OIL

Figure 3   Pump inlet locations. (Courtesy of Vickers, Incorporated, Rochester Hills, MI.)

H. How Pressure Is Created

Pressure results whenever the flow of a fluid is resisted. The resistance may come from (1) a load on an actuator or (2) a restriction (or orifice) in the piping.

I. Parallel Flow Paths

An inherent characteristic of liquids is that they will always take the path of least resistance. Thus, when two parallel flow paths offer different resistances, the pressure will increase only to the amount required to take the easier path.

J. Pressure Drop Through an Orifice

An orifice is a restricted passage in a hydraulic line or component, used to control flow or create a pressure difference (pressure drop).

In order for oil to flow through an orifice, there must be a pressure difference or *pressure drop* through the orifice. (The term *drop* comes from the fact that the lower pressure is always downstream). Conversely, if there is no flow, there is no difference in pressure across the orifice.

K. Pressure Indicates Work Load

Pressure is generated by resistance of a load. Pressure equals the force of the load divided by the piston area. We can express this relationship by the general formula:

$$P = \frac{F}{A}$$

In this relationship:

P is pressure in psi (pounds per square inch)
F is force in pounds
A is area in square inches

From this it can be seen that an increase or decrease in the load will result in a like increase or decrease in the operating pressure. In other words, *pressure is proportional to the load,* and a pressure gauge reading indicates the work load (in psi) at any given moment.

L. Force Is Proportional to Pressure and Area

When a hydraulic cylinder is used to clamp or press, its output force can be computed as follows:

$$F = P \times A$$

Again:

P is pressure
F is force in pounds

As an example, suppose that a hydraulic press has its pressure regulated at 2000 psi (Fig. 4) and this pressure is applied to a ram area of 20 square inches. The output force will then be 40,000 pounds or 20 tons.

M. Computing Piston Area

The area of a piston or ram can be computed by this formula

$$A = 0.7854 \times d^2$$

where
    A is area in square inches
    d is diameter of the piston in inches

The foregoing relationships are sometimes illustrated as shown to indicate the three relationships:

$F = P \times A$

$P = \dfrac{F}{A}$

$A = \dfrac{F}{P}$

N. Speed of an Actuator

How fast a cylinder travels or a motor rotates depends on its size and the rate of oil flow into it. To relate flow rate speed, consider the volume that must be filled in the actuator to effect a given amount of travel.

The relationship may be expressed as follows:

Speed = volume/time over area
Volume/time = speed × area
Area = volume/time over speed
v/t = cubic inches/minute
a = square inches
s = inches/minute

From this we can conclude: (1) that the force or torque of an actuator is directly proportional to the pressure and independent of the flow, and (2) that its speed or rate of travel will depend upon the amount of fluid flow without regard to pressure.

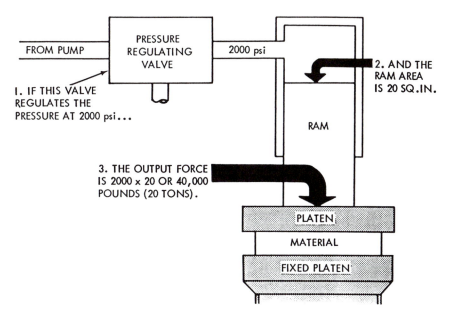

Figure 4 Force equals pressure multiplied by area. (Courtesy of Vickers, Incorporated, Rochester Hills, MI.)

O. Velocity in Pipes

The velocity at which the hydraulic fluid flows through the lines is an important design consideration because of the effect of velocity on friction.

Generally, the recommended velocity ranges are

Pump inlet line, 2 to 4 feet per second
Working lines, 7 to 20 feet per second

In this regard, it should be noted that

1. The velocity of the fluid varies *inversely* as the *square* of the inside diameter.
2. Usually, the friction of a liquid flowing through a line is proportional to the *velocity*. However, should the flow become turbulent, friction varies as the *square* of the velocity.

Friction creates turbulence in the oil stream and of course resists flow, resulting in an increased pressure drop through the line. Very low velocity is recommended for the pump inlet line because very little pressure drop can be tolerated there.

P. Determining Pipe Size Requirements

Two formulas are available for sizing hydraulic lines.

If the gpm (gallons per minute) and desired velocity are known, use this relationship to find the inside cross-sectional area:

$$\text{Area} = \frac{\text{gpm} \times 0.3208}{\text{velocity}} \quad \text{(in feet per second)}$$

When the gpm and size of pipe are given, use this formula to find what the velocity will be:

$$\text{Velocity} \quad (\text{feet per second}) = \frac{\text{gpm}}{3.117 \times \text{area}}$$

Q. Size Ratings of Lines

The nominal ratings in inches for pipes, tubes, etc., are not accurate indicators of the inside diameter. In standard pipes, the actual inside diameter is larger than the nominal size quoted. To select pipe, you'll need a standard chart which shows actual inside diameters. For steel and copper tubing, the quoted size is the outside diameter. To find the inside diameter, subtract twice the wall thickness (Fig. 5).

R. Work and Power

Whenever a force or push is exerted through a distance, *work* is done.

$$\text{Work} = \text{force} \times \text{distance}$$

Work is usually expressed in foot pounds. For example, if a 10-pound weight is lifted 10 feet, the work is 10 pounds × 10 feet or 100 foot pounds.

The preceding formula for work does not take into consideration how fast the work is done. The *rate* of doing work is called *power*. To visualize power, think of climbing a flight of stairs. The work done is the body's weight multiplied by the height of the stairs. But it is more difficult to run up the stairs than to walk. When you run, you do the same work at a faster rate.

$$\text{Power} = \frac{\text{force} \times \text{distance}}{\text{time}} \quad \text{or} \quad \frac{\text{work}}{\text{time}}$$

The standard unit of power is *horsepower,* abbreviated hp. It is equivalent to 33,000 pounds lifted 1 foot in 1 minute. It also has equivalents in electrical power and heat.

1 hp = 33,000 foot pounds/minute or 550 foot pounds/second

1 hp = 746 watts (electrical power)

1 hp = 42.4 btu/minute (heat power)

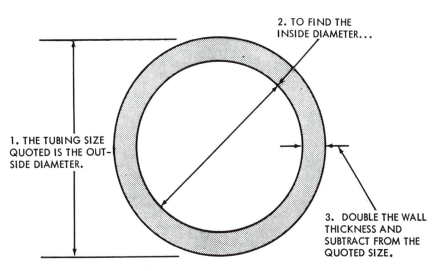

2. TO FIND THE
INSIDE DIAMETER...

1. THE TUBING SIZE
QUOTED IS THE OUT-
SIDE DIAMETER.

3. DOUBLE THE WALL
THICKNESS AND
SUBTRACT FROM THE
QUOTED SIZE.

Figure 5 Tubing inside diameter. (Courtesy of Vickers, Incorporated, Rochester Hills, MI.)

Obviously, it is desirable to be able to convert hydraulic power to horsepower so that the mechanical, electrical, and heat power equivalents will be known.

S. Horsepower in a Hydraulic System

In the hydraulic system, speed and distance are indicated by the gpm flow and force by pressure. Thus, we might express hydraulic power this way:

$$\text{Power} = \frac{\text{gallons}}{\text{minute}} \times \frac{\text{pounds}}{\text{square inches}}$$

To change the relationship to mechanical units, we can use these equivalents:

1 gallon = 231 cubic inches (in.^3)
12 inches = 1 foot

Thus:

$$\text{Power} = \frac{\text{gallon}}{\text{min.}} \times \left[\frac{231\ \text{in.}^3}{\text{gallon}}\right] \times \frac{\text{pounds}}{\text{in.}^2} \times \left[\frac{1\ \text{foot}}{12\ \text{in.}}\right]$$

$$= \frac{231\ \text{foot pounds}}{12\ \text{minutes}}$$

This gives us the equivalent mechanical power of 1 gallon per minute flow at 1 psi of pressure. To express it as horsepower, divide by 33,000 pounds/minute:

231 foot pounds/12 minutes over 33,000 foot pounds/minute = .000583

Thus, one gallon per minute flow at one psi equals 0.000583 hp. The total horsepower for any flow condition is:

$$hp = gpm \times psi \times 0.000583$$

or

$$hp = \frac{gpm \times psi}{1000} \times 0.583$$

or

$$hp = \frac{gpm \times psi}{1714}$$

The third formula is derived by dividing 0.583 into 1000.

T. Horsepower and Torque

It also is often desirable to convert back and forth from horsepower to torque without computing pressure and flow.

These are general torque—power formulas for any rotating equipment:

$$Torque = \frac{63025 \times hp}{rpm} \qquad hp = \frac{torque \times rpm}{63025}$$

III. PRINCIPLES OF POWER HYDRAULICS

A. Principles of Pressure

1. A Precise Definition

It has been noted that the term hydraulic is derived from the Greek word for water. Therefore, it might be assumed correctly that the science of hydraulics encompasses *any* device operated by water. A water wheel or turbine (Fig. 6), for instance, is a hydraulic device.

However, a distinction must be made between devices which utilize the impact or momentum of a moving liquid and those which are operated by pushing on a confined fluid, that is, by pressure.

A hydraulic device which uses the impact or kinetic energy in the liquid to transmit power is called a *hydrodynamic* device.

When the device is operated by a force applied to a confined liquid, it is called a *hydrostatic* device; pressure being the force applied distributed over the area exposed and being expressed as force per unit area (pounds per square inch or psi).

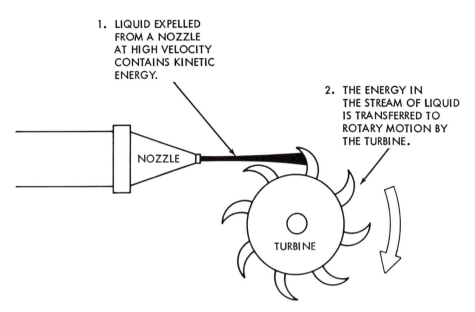

1. LIQUID EXPELLED
FROM A NOZZLE
AT HIGH VELOCITY
CONTAINS KINETIC
ENERGY.

2. THE ENERGY IN
THE STREAM OF LIQUID
IS TRANSFERRED TO
ROTARY MOTION BY
THE TURBINE.

NOZZLE

TURBINE

Figure 6 Hydrodynamic device uses kinetic energy rather than pressure. (Courtesy of Vickers, Incorporated, Rochester Hills, MI.)

2. How Pressure Is Created

Pressure results whenever there is a resistance to fluid flow or to a force which attempts to make the fluid flow. The tendency to cause flow (or the push) may be supplied by a mechanical pump or may be caused simply by the weight of the fluid.

It is well known that in a body of water, pressure increases with depth. The pressure is always equal at any particular depth due to the weight of the water above it. Around Pascal's time, an Italian scientist named Torricelli proved that if a hole is made in the bottom of a tank of water, the water runs out fastest when the tank is full and the flow rate decreases as the water level lowers. In other words, as the "head" of water above the opening lessens, so does the pressure.

Torricelli could express the pressure at the bottom of the tank only as "feet of head," or the height in feet of the column of water. Today, with the pound per square inch (psi) as a unit pressure, we can express pressure anywhere in any liquid or gas in more convenient terms. All that is required is knowing how much a cubic foot of the fluid weighs.

As shown in Fig. 7, a "head" or 1 foot of water is equivalent to 0.434 psi; a 5-foot head of water equals 2.17 psi, and so on. And as shown earlier, a head of oil is equivalent to about 0.4 psi per foot.

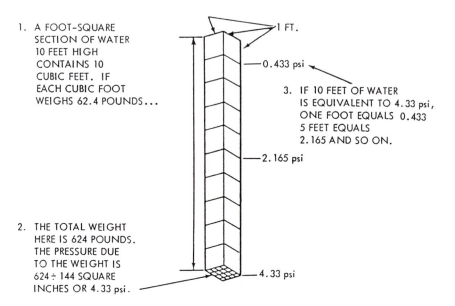

1. A FOOT-SQUARE
   SECTION OF WATER
   10 FEET HIGH
   CONTAINS 10
   CUBIC FEET. IF
   EACH CUBIC FOOT
   WEIGHS 62.4 POUNDS...

1 FT.

—0.433 psi

3. IF 10 FEET OF WATER
   IS EQUIVALENT TO 4.33 psi,
   ONE FOOT EQUALS 0.433
   5 FEET EQUALS
   2.165 AND SO ON.

—2.165 psi

2. THE TOTAL WEIGHT
   HERE IS 624 POUNDS.
   THE PRESSURE DUE
   TO THE WEIGHT IS
   624 ÷ 144 SQUARE
   INCHES OR 4.33 psi.

—4.33 psi

Figure 7   Pressure increases with depth in the fluid. (Courtesy of Vickers, Incorporated, Rochester Hills, MI.)

3.   Atmospheric Pressure

Atmospheric pressure is nothing more than pressure of the air in our atmosphere due to its weight. At seal level, a column of air 1 square inch in cross section and the full height of the atmosphere weighs 14.7 pounds (Fig. 8). Thus, the pressure is 14.7 psia. At higher altitudes, of course, there is less weight in the column, so the pressure becomes less. Below sea level, atmospheric pressure is more than 14.7 psi.

Note: Any condition where pressure is less than atmospheric pressure is called a vacuum or partial vacuum. A *perfect vacuum* is the complete absence of pressure or zero psia.

4.   The Mercury Barometer

Atmospheric pressure also is measured in inches of mercury (in. Hg) on a device known as a barometer. The mercury barometer (Fig. 9), a device invented by Torricelli, is usually credited as the inspiration for Pascal's studies of pressure. Torricelli discovered that when a tube of mercury is inverted in a pan of the liquid, the column in the tube will fall only a certain distance. He reasoned that atmospheric pressure on the surface of the liquid was supporting the weight of the column of mercury with a perfect vacuum at the top of the tube.

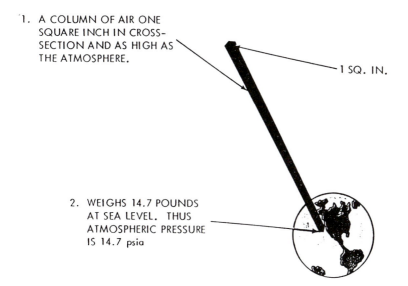

1. A COLUMN OF AIR ONE
   SQUARE INCH IN CROSS-
   SECTION AND AS HIGH AS
   THE ATMOSPHERE.

1 SQ. IN.

2. WEIGHS 14.7 POUNDS
   AT SEA LEVEL. THUS
   ATMOSPHERIC PRESSURE
   IS 14.7 psia

Figure 8 Atmospheric pressure is a "head" of air. (Courtesy of Vickers, Incorporated, Rochester Hills, MI.)

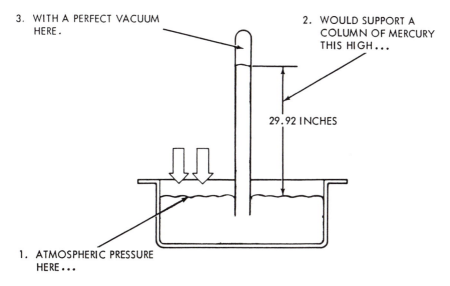

3. WITH A PERFECT VACUUM
   HERE.

2. WOULD SUPPORT A
   COLUMN OF MERCURY
   THIS HIGH...

29.92 INCHES

1. ATMOSPHERIC PRESSURE
   HERE...

Figure 9 The mercury barometer measures atmospheric pressure. (Courtesy of Vickers, Incorporated, Rochester Hills, MI.)

B. Principles of Flow

Flow is the action in the hydraulic system that gives the actuator its motion. Force can be transmitted by pressure alone, but flow is essential to cause movement. Flow in the hydraulic system is created by the pump.

How Flow Is Measured. There are two ways to measure the flow of a fluid: *Velocity* is the average speed of the fluid's particles past a given point or the average distance the particles travel per unit of time. It is measured in feet per second (fps), feet per minute (fpm), or inches per second (ips).

Flow rate is a measure of the *volume* of fluid passing a point in a given time. Large volumes are measured in gallons per minute (gpm). Small volumes may be expressed in cubic inches per minute.

Figure 10 illustrates the distinction between velocity and flow rate. A constant flow of 1 gallon per minute either increases or decreases in velocity when the cross section of the pipe changes size.

Flow Rate and Speed. The speed of a hydraulic actuator always depends on the actuator's size and the rate of flow into it. Since the size of the actuator will be expressed in cubic inches, use these conversion factors:

1 gpm = 231 cubic inches per minute
gpm = cubic inches per minute divided by 231
cubic inches per minute = gpm × 231

Flow and Pressure Drop. Whenever a liquid is flowing, there must be a condition of unbalanced force to cause motion. Therefore, when a fluid flows through a constant-diameter pipe, the pressure will always be slightly lower downstream with reference to any point upstream. The difference in pressure or *pressure drop* is required to overcome friction in the line.

Figure 11 illustrates pressure drop due to friction. The succeeding pressure drops (from maximum pressure to zero pressure) are shown as differences in head in succeeding vertical pipes.

Fluid Seeks a Level. Conversely, when there is no pressure difference on a liquid, it simply "seeks a level."

Laminar and Turbulent Flow. Ideally, when the particles of a fluid move through a pipe, they will move in straight, parallel flow paths. This condition is called *laminar* flow and occurs at low velocity in straight piping. With laminar flow, friction is minimized.

Turbulence is the condition where the particles do not move smoothly parallel to the flow direction. Turbulent flow is caused by abrupt changes in direction or cross section or by too high velocity. The result is greatly increased friction, which generates heat, increases operating pressure, and wastes power.

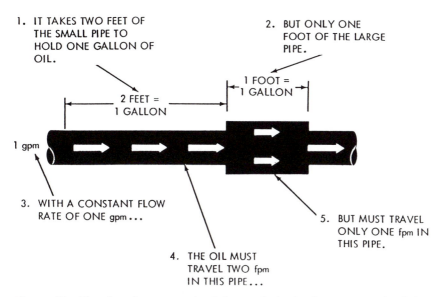

1. IT TAKES TWO FEET OF
THE SMALL PIPE TO
HOLD ONE GALLON OF
OIL.

2. BUT ONLY ONE
FOOT OF THE LARGE
PIPE.

1 FOOT =
1 GALLON

2 FEET =
1 GALLON

1 gpm

3. WITH A CONSTANT FLOW
RATE OF ONE gpm...

4. THE OIL MUST
TRAVEL TWO fpm
IN THIS PIPE...

5. BUT MUST TRAVEL
ONLY ONE fpm IN
THIS PIPE.

Figure 10 Flow is volume per unit of time; velocity is distance per unit of time. (Courtesy of Vickers, Incorporated, Rochester Hills, MI.)

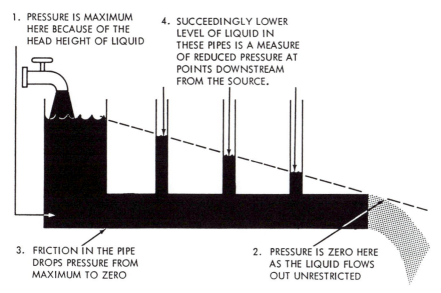

1. PRESSURE IS MAXIMUM
HERE BECAUSE OF THE
HEAD HEIGHT OF LIQUID

4. SUCCEEDINGLY LOWER
LEVEL OF LIQUID IN
THESE PIPES IS A MEASURE
OF REDUCED PRESSURE AT
POINTS DOWNSTREAM
FROM THE SOURCE.

3. FRICTION IN THE PIPE
DROPS PRESSURE FROM
MAXIMUM TO ZERO

2. PRESSURE IS ZERO HERE
AS THE LIQUID FLOWS
OUT UNRESTRICTED

Figure 11 Friction in pipes results in a pressure drop. (Courtesy of Vickers, Incorporated, Rochester Hills, MI.)

IV. HYDRAULIC FLUIDS

Selection and care of the hydraulic fluid for a machine will have an important effect on how it performs and on the life of the hydraulic components. The formulation and application of hydraulic fluids is a science in itself, far beyond the scope of this chapter.

A. Purposes of the Fluid

The hydraulic fluid has four primary purposes: to transmit power, to lubricate moving parts, to seal clearances between parts, and to cool or dissipate heat.

Power Transmission. As a power transmitting medium, the fluid must flow easily through lines and component passages. Too much resistance to flow creates considerable power loss. The fluid also must be as incompressible as possible so that action is instantaneous when the pump is started or a valve shifts.

Lubrication. In most hydraulic components, internal lubrication is provided by the fluid. Pump elements and other wearing parts slide against each other on a film of fluid. For long component life the oil must contain the necessary additives to ensure high antiwear characteristics. Not all hydraulic oils contain these additives.

Sealing. In many instances, the fluid is the only seal against pressure inside a hydraulic component.

Cooling. Circulation of the oil through lines and around the walls of the reservoir (Fig. 12) gives up heat that is generated in the system.

B. Quality Requirements

In addition to these primary functions, the hydraulic fluid may have a number of other quality requirements. Some of these are to

    Prevent rust
    Prevent formation of sludge, gum, and varnish
    Depress foaming
    Maintain its own stability and thereby reduce fluid replacement cost
    Maintain relatively stable body over a wide temperature range
    Prevent corrosion and pitting
    Separate out water
    Be compatible with seals and gaskets

These quality requirements often are the result of special compounding and may not be present in every fluid.

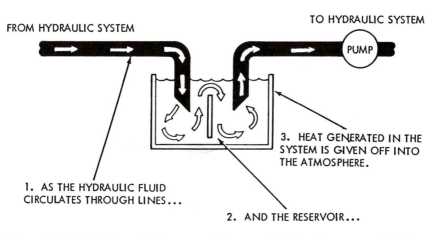

FROM HYDRAULIC SYSTEM

TO HYDRAULIC SYSTEM

PUMP

3. HEAT GENERATED IN THE SYSTEM IS GIVEN OFF INTO THE ATMOSPHERE.

1. AS THE HYDRAULIC FLUID CIRCULATES THROUGH LINES...

2. AND THE RESERVOIR...

Figure 12 Circulation cools the system. (Courtesy of Vickers, Incorporated, Rochester Hills, MI.)

V. VISCOSITY

Viscosity is the measure of the fluid's resistance to flow; or an inverse measure of fluidity. If a fluid flows easily, its viscosity is low. You also can say that the fluid is thin or has a low body. A fluid that flows with difficulty has a high viscosity. It is thick or high in body.

A. Viscosity a Compromise

For any hydraulic machine, the actual fluid viscosity must be a compromise. A high viscosity is desirable for maintaining sealing between mating surfaces. However, too high a viscosity increases friction, resulting in

High resistance to flow
Increased power consumption due to frictional loss
High temperature caused by friction
Increased pressure drop because of the resistance
Possibility of sluggish or slow operation
Difficulty in separating air from oil in reservoir

And should the viscosity be too low,

Internal leakage increases
Excessive wear or even seizure under heavy load may occur due to breakdown of the oil film between moving parts.
Pump efficiency may decrease, causing slower operation of the actuator
Increased temperatures result from leakage losses

B. SUS Viscosity

For most practical purposes, it will serve to know the relative viscosity of the fluid. Relative viscosity is determined by timing the flow of a given quantity of the fluid through a standard orifice at a given temperature. There are several methods in use. The most accepted method in this country is the Saybolt viscosity (Fig. 13).

The time it takes for the measured quantity of liquid to flow through the orifice is measured with a stopwatch. The viscosity in Saybolt universal seconds (SUS) equals the elapsed time.

Obviously, a thick liquid will flow slowly, and the SUS velocity will be higher than for a thin liquid, which flows faster. Since oil becomes thicker at low temperature and thins when warmed, the viscosity must be expressed as so many SUSs at a given temperature. The tests are usually made at 100°F or 210°F.

For industrial applications, hydraulic oil viscosities usually are in the vicinity of 150 SUS at 100°F. It is a general rule that the viscosity should never go below 45 SUS or above 4000 SUS, regardless of temperature. Where temperature extremes are encountered, the fluid should have a high viscosity index.

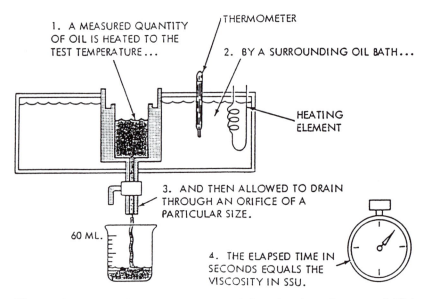

Figure 13 Saybolt viscometer measures relative viscosity. (Courtesy of Vickers, Incorporated, Rochester Hills, MI.)

C. SAE Numbers

SAE numbers have been established by the Society of Automotive Engineers to specify ranges of SUS viscosities of oils at SAE test temperatures. Winter numbers (5W, 10W, 20W) are determined by tests at 0°F. Summer oil numbers (20, 30, 40, 50, etc.) designate the SUS range at 210°F. Table 1 is a chart of the temperature ranges.

Table 1   SAE Viscosity Numbers for Crankcase Oils

SAE VISCOSITY NUMBER	Viscosity Units[a]	Viscosity Range[b]			
		At 0°F		At 210°F	
		Minimum	Maximum	Minimum	Maximum
5W	Centipoises	-	Less than 1,200	-	-
	Centistokes	-	1,300	-	-
	SUS	-	6,000	-	-
10W	Centipoises	1,200[c]	Less than 2,400	-	-
	Centistokes	1,300	2,600	-	-
	SUS	6,000	12,000	-	-
20W	Centipoises	2,400[d]	Less than 9,600	-	-
	Centistokes	2,600	10,500	-	-
	SUS	12,000	48,000	-	-
20	Centistokes	-	-	5.7	Less than 9.6
	SUS	-	-	45	58
30	Centistokes	-	-	9.6	Less than 12.9
	SUS	-	-	58	70
40	Centistokes	-	-	12.9	Less than 16.8
	SUS	-	-	70	85
50	Centistokes	-	-	16.8	Less than 22.7
	SUS	-	-	85	110

[a] The official values in this classification are based upon 210°F viscosity in centistokes (ASTM D 445) and 0°F viscosities in centipoises (ASTM D 260-2). Approximate values in other units of viscosity are given for information only. The approximate values at 0°F were calculated using an assumed oil density of 0.9 gm/cc at that temperature

[b] The viscosity of all oils included in this classification shall not be less than 3.9 cs at 210°F (39 SUS).

[c] Minimum viscosity at 0°F may be waived provided viscosity at 210°F is not below 4.2 cs (40 SUS).

[d] Minimum viscosity at 0°F may be waived provided viscosity at 210°F is not below 5.7 cs (45 SUS).

Courtesy of Vickers, Incorporated, Rochester Hills, MI.

D. Viscosity Index

Viscosity index is an arbitrary measure of a fluid's resistance to viscosity change with temperature changes. A fluid that has a relatively stable viscosity at temperature extremes has a high viscosity index (VI). A fluid that is very thick when cold and very thin when hot has a low VI.

VI. HYDRAULIC PLUMBING

A. Piping

Piping is a general term which embraces the various kinds of conducting lines that carry hydraulic fluid between components; plus the fittings or connectors used between the conductors. Hydraulic systems today use principally three types of conducting lines: steel pipe, steel tubing and flexible hose. At present, pipe is the least expensive of the three while tubing and hose offer more convenience in making connections and in servicing the "plumbing."

Pipes. Iron and steel pipes were the first conductors used in industrial hydraulic systems and are still used widely because of their low cost. Seamless steel pipe is recommended for hydraulic systems with the pipe interior free of rust, scale, and dirt.

Sizing Pipes. Pipe and pipe fittings are classified by nominal size and wall thickness. Originally, a given size pipe had only one wall thickness and the stated size was the actual *inside* diameter.

Later, pipes were manufactured with varying wall thicknesses: standard, extra heavy, and double extra heavy (Fig. 14). However, the outside diameter did not change. To increase wall thickness, the inside diameter was changed. Thus, the nominal pipe size alone indicates only the thread size for connections.

Pipe Schedule. Currently, wall thickness is being expressed as a *schedule* number. Schedule numbers are specified by the American National Standards Institute (ANSI) from 10 to 160 (Fig. 15). The numbers cover 10 sets of wall thickness.

For comparison, schedule 40 corresponds closely to standard. Schedule 80 essentially is extra heavy. Schedule 160 covers pipes with the greatest wall thickness under this system.

Tubing. Seamless steel tubing offers significant advantages over pipe for hydraulic plumbing. Tubing can be bent into any shape, is easier to work with, and can be used over and over without any sealing problems. Usually the number of joints is reduced.

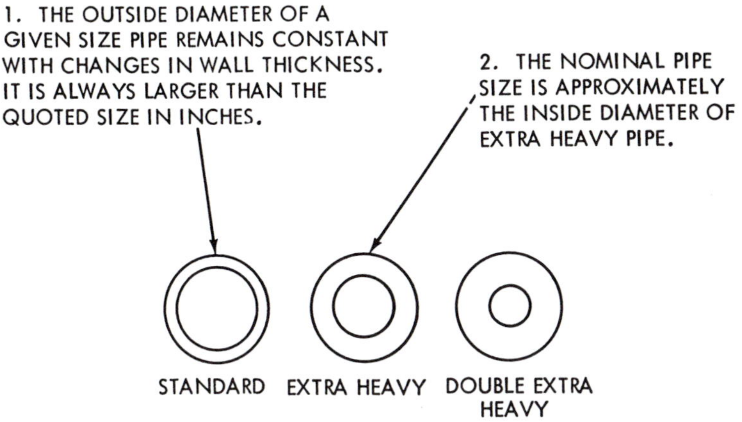

1. THE OUTSIDE DIAMETER OF A
GIVEN SIZE PIPE REMAINS CONSTANT
WITH CHANGES IN WALL THICKNESS.
IT IS ALWAYS LARGER THAN THE
QUOTED SIZE IN INCHES.

2. THE NOMINAL PIPE
SIZE IS APPROXIMATELY
THE INSIDE DIAMETER OF
EXTRA HEAVY PIPE.

STANDARD   EXTRA HEAVY   DOUBLE EXTRA
HEAVY

NOMINAL SIZE	PIPE O.D.	INSIDE DIAMETER		
		STANDARD	EXTRA HEAVY	DOUBLE EXTRA HEAVY
1/8	.405	.269	.215	
1/4	.540	.364	.302	
3/8	.675	.493	.423	
1/2	.840	.622	.546	.252
3/4	1.050	.824	.742	.434
1	1.315	1.049	.957	.599
1-1/4	1.660	1.380	1.278	.896
1-1/2	1.900	1.610	1.500	1.100
2	2.375	2.067	1.939	1.503
2-1/2	2.875	2.469	2.323	1.771
3	3.500	3.068	2.900	
3-1/2	4.000	3.548	3.364	
4	4.500	4.026	3.826	
5	5.563	5.047	4.813	4.063
6	6.625	6.065	5.761	
8	8.625	8.071	7.625	
10	10.750	10.192	9.750	
12	12.750	12.080	11.750	

Figure 14 Early classification of pipe well thickness. (Courtesy of Vickers, Incorporated, Rochester Hills, MI.)

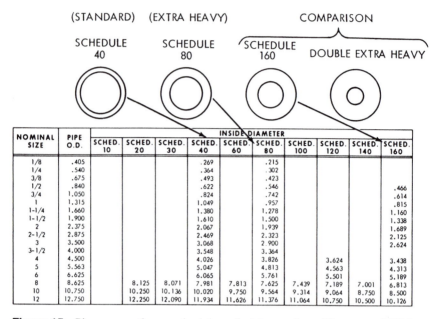

(STANDARD)   (EXTRA HEAVY)   COMPARISON

SCHEDULE   SCHEDULE   SCHEDULE
40   80   160   DOUBLE EXTRA HEAVY

NOMINAL SIZE	PIPE O.D.	SCHED. 10	SCHED. 20	SCHED. 30	SCHED. 40	SCHED. 60	SCHED. 80	SCHED. 100	SCHED. 120	SCHED. 140	SCHED. 160
1/8	.405				.269		.215				
1/4	.540				.364		.302				
3/8	.675				.493		.423				
1/2	.840				.622		.546				.466
3/4	1.050				.824		.742				.614
1	1.315				1.049		.957				.815
1-1/4	1.660				1.380		1.278				1.160
1-1/2	1.900				1.610		1.500				1.338
2	2.375				2.067		1.939				1.689
2-1/2	2.875				2.469		2.323				2.125
3	3.500				3.068		2.900				2.624
3-1/2	4.000				3.548		3.364				
4	4.500				4.026		3.826		3.624		3.438
5	5.563				5.047		4.813		4.563		4.313
6	6.625				6.065		5.761		5.501		5.189
8	8.625	8.125	8.071	7.981	7.813	7.625	7.439	7.189	7.001	6.813	
10	10.750	10.250	10.136	10.020	9.750	9.564	9.314	9.064	8.750	8.500	
12	12.750	12.250	12.090	11.934	11.626	11.376	11.064	10.750	10.500	10.126	

(INSIDE DIAMETER)

Figure 15 Pipes currently are sized by schedule number. (Courtesy of Vickers, Incorporated, Rochester Hills, MI.)

In low-volume systems, tubing will handle higher pressure and flow with less bulk and weight. However, it is more expensive, as are the fittings required to make tube connections.

Tubing Sizes. A tubing size specification refers to the *outside* diameter. Tubing is available in 1/16 in. increments from 1/8 in. to 1 in. O.D. and in 1/4 in. increments beyond 1 inch. Various wall thicknesses are available for each size. The inside diameter, as previously noted, equals the outside diameter less twice the wall thickness.

B. Flexible Hose

Flexible hose is used when the hydraulic lines are subjected to movement, for example, the lines to a drill head motor. Hose is fabricated in layers of synthetic rubber and braided fabric or wire (Fig. 16). Wire braids, of course, permit higher pressure.

The inner layer of the hose must be compatible with the fluid being used. The outer layer is usually rubber to protect the braid layer. The hose may have as few as three layers, one being braid, or may have multiple layers depending on the operating pressure. When there are multiple wire layers, they may alter-

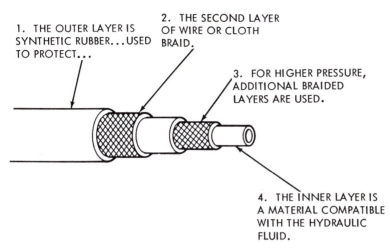

1. THE OUTER LAYER IS SYNTHETIC RUBBER...USED TO PROTECT...

2. THE SECOND LAYER OF WIRE OR CLOTH BRAID.

3. FOR HIGHER PRESSURE, ADDITIONAL BRAIDED LAYERS ARE USED.

4. THE INNER LAYER IS A MATERIAL COMPATIBLE WITH THE HYDRAULIC FLUID.

Figure 16 Flexible hose is constructed in layers. (Courtesy of Vickers, Incorporated, Rochester Hills, MI.)

nate with rubber layers, or the wire layers may be placed directly over one another.

VII. CONNECTIONS

A. Pipe Sealing

Pipe threads are tapered (Fig. 17) as opposed to tube and some hose fittings which have straight threads. Joints are sealed by an interference fit between the male and female threads as the pipe is tightened.

This creates one of the major disadvantages of pipe. When a joint is broken, the pipe must be tightened further to reseal. Often this necessitates replacing some of the pipe with slightly longer sections. However, the difficulty has been overcome somewhat by using teflon tape or other compounds to reseal pipe joints.

Special taps and dies are required for threading hydraulic system pipes and fittings. The threads are the "dryseal" type. They differ from standard pipe threads by engaging the roots and crests before the flanks, thus avoiding spiral clearance (Fig. 17).

B. Pipe Fittings

Since pipe can have only male threads, and it does not bend, various types of fittings are used to make connections and change direction (Fig. 18). Most

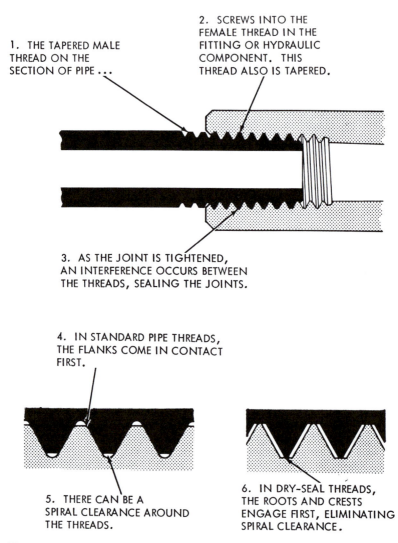

2. SCREWS INTO THE
FEMALE THREAD IN THE
1. THE TAPERED MALE   FITTING OR HYDRAULIC
THREAD ON THE   COMPONENT. THIS
SECTION OF PIPE ...   THREAD ALSO IS TAPERED.

3. AS THE JOINT IS TIGHTENED,
AN INTERFERENCE OCCURS BETWEEN
THE THREADS, SEALING THE JOINTS.

4. IN STANDARD PIPE THREADS,
THE FLANKS COME IN CONTACT
FIRST.

5. THERE CAN BE A
SPIRAL CLEARANCE AROUND
THE THREADS.

6. IN DRY-SEAL THREADS,
THE ROOTS AND CRESTS
ENGAGE FIRST, ELIMINATING
SPIRAL CLEARANCE.

Figure 17 Hydraulic pipe threads are of the dry-seal tapered type. (Courtesy of Vickers, Incorporated, Rochester Hills, MI.)

fittings are female-threaded to mate with pipe although some have male threads to mate with other fittings or with the ports in hydraulic components.

The many fittings necessary in a pipe circuit present multiple opportunities for leakage, particularly as pressure increases. Threaded connections are used up to 1¼ in. Where larger pipes are needed, flanges are welded to the pipe (Fig. 19). Flat gaskets or O-rings are used to seal flanged fittings.

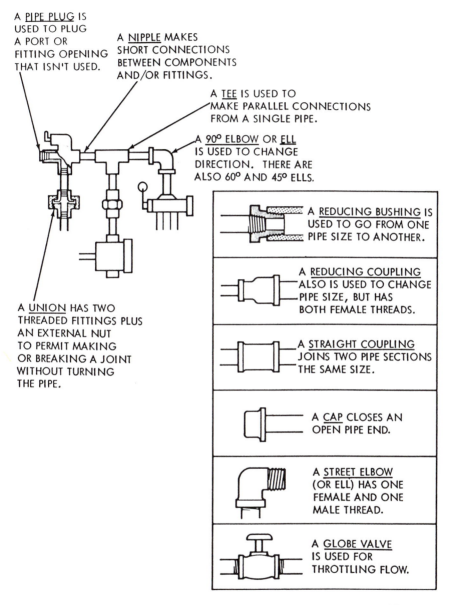

A PIPE PLUG IS USED TO PLUG A PORT OR FITTING OPENING THAT ISN'T USED.

A NIPPLE MAKES SHORT CONNECTIONS BETWEEN COMPONENTS AND/OR FITTINGS.

A TEE IS USED TO MAKE PARALLEL CONNECTIONS FROM A SINGLE PIPE.

A 90° ELBOW OR ELL IS USED TO CHANGE DIRECTION. THERE ARE ALSO 60° AND 45° ELLS.

A UNION HAS TWO THREADED FITTINGS PLUS AN EXTERNAL NUT TO PERMIT MAKING OR BREAKING A JOINT WITHOUT TURNING THE PIPE.

A REDUCING BUSHING IS USED TO GO FROM ONE PIPE SIZE TO ANOTHER.

A REDUCING COUPLING ALSO IS USED TO CHANGE PIPE SIZE, BUT HAS BOTH FEMALE THREADS.

A STRAIGHT COUPLING JOINS TWO PIPE SECTIONS THE SAME SIZE.

A CAP CLOSES AN OPEN PIPE END.

A STREET ELBOW (OR ELL) HAS ONE FEMALE AND ONE MALE THREAD.

A GLOBE VALVE IS USED FOR THROTTLING FLOW.

Figure 18 Fittings make the connections between pipes and components. (Courtesy of Vickers, Incorporated, Rochester Hills, MI.)

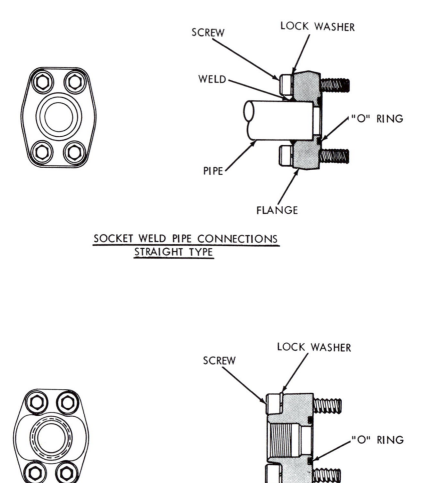

SOCKET WELD PIPE CONNECTIONS
STRAIGHT TYPE

THREADED PIPE CONNECTIONS
STRAIGHT TYPE

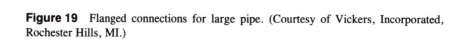

Figure 19 Flanged connections for large pipe. (Courtesy of Vickers, Incorporated, Rochester Hills, MI.)

C. Tube Fittings

Tubing is never sealed by threads, but by various kinds of fittings (Fig. 20). Some of these fittings seal by metal-to-metal contact. They are known as *compression* fittings and may be either the flared or flareless type. Others use O-rings or comparable seals. In addition to threaded fittings, flanged fittings also are available to be welded to larger sized tubing.

D. Hose Fittings

Fittings for hose are essentially the same as for tubing. Couplings are fabricated on the ends of most hose, though there are reusable screw-on or clamp-on connectors. It is usually desirable to connect the hose ends with union-type fittings which have free-turning nuts. The union is usually in the mating connector but may be built into the hose coupling. A short hose may be screwed into a rigid connection at one end before the other end is connected. A hose must never be installed twisted.

VIII. THE OIL RESERVOIR

This section and the following section (Filters and Strainers) deal with conditioning the fluid—that is, providing storage space for all the fluid required in the system plus a reserve, and keeping the fluid clean.

The storage space for the fluid, of course, is the oil reservoir. The fluid is kept clean by using strainers, filters, and magnetic plugs to the degree required by the conditions.

A. Reservoirs

With almost no location or sizing problems, the reservoir for a piece of shop equipment can usually be designed to perform a number of functions. It is first a storehouse for the fluid until called for by the system. The reservoir also should provide a place for air to separate out of the fluid and should permit contaminants to settle out as well. In addition, a well-designed reservoir will help dissipate any heat that is generated in the system.

Reservoir Construction. A typical industrial reservoir, conforming to industry standards, is shown in Fig. 21. The tank is constructed of welded steel plate with extensions of the end plates supporting the unit on the floor. The entire inside of the tank is painted with a sealer to reduce rust which can result from condensed moisture. This sealer must be chosen for compatibility with the fluid being used.

Breather. A vented breather cap is used on most reservoirs and should also contain an air-filtering system. In dirty atmospheres, an oil bath air filter may

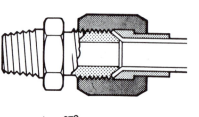

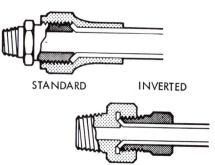

A. 37° FLARE FITTING

STANDARD   INVERTED

B. 45° FLARE FITTING

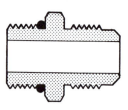

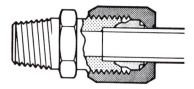

C. STRAIGHT THREAD "O"
    RING CONNECTOR

D. FERRULE COMPRESSION
    FITTING

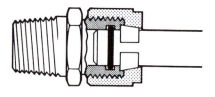

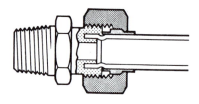

E. "O" RING COMPRESSION
    FITTING

F. SLEEVE COMPRESSION
    FITTING

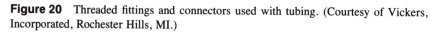

Figure 20 Threaded fittings and connectors used with tubing. (Courtesy of Vickers, Incorporated, Rochester Hills, MI.)

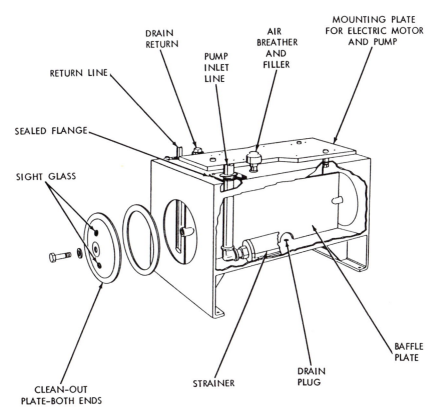

Figure 21 The reservoir is designed for easy maintenance. (Courtesy of Vickers, Incorporated, Rochester Hills, MI.)

be better. The filter or breather must be large enough to handle the air flow required to maintain atmospheric pressure whether the tank is empty or filled.

Baffle Plate. A baffle plate (Fig. 22) extends lengthwide through the center of the tank; it is usually about two-thirds the height of the oil level and is used to separate the pump inlet line from the return line so that the same fluid cannot recirculate continuously, but must take a circuitous route through the tank.

Thus, the baffle (1) prevents local turbulence in the tank, (2) allows foreign material to settle to the bottom, (3) gives the fluid an opportunity to get rid of entrapped air, and (4) helps increase heat dissipation through the tank walls.

IX. FILTERS AND STRAINERS

Hydraulic fluids are kept clean in the system principally by devices such as filters and strainers. Magnetic plugs also are used in some tanks to trap iron

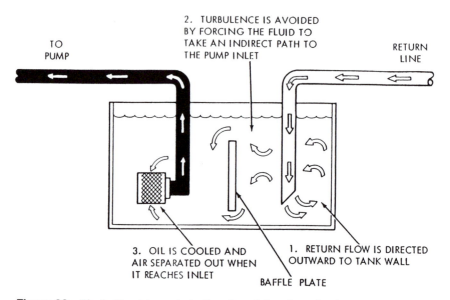

Figure 22 The baffle plate controls direction of flow in tank. (Courtesy of Vickers, Incorporated, Rochester Hills, MI.)

and steel particles carried by fluid. Recent studies have indicated that even particles as small as 1 to 5 microns have a degrading effect, causing failures in servo-systems and hastening oil deterioration in many cases.

A. Filter or Strainer

There will probably always be controversy in the industry over the exact definitions of filters and strainers. In the past, many such devices were named filters but technically classed as strainers. To minimize the controversy, the National Fluid Power Association (NFPA) gives us these definitions:

   Filter: A device whose primary function is the retention, by some porous medium, of insoluble contaminants from a fluid.
   Strainer: A coarse filter.

B. "Mesh" and Micron Ratings

A simple screen or a wire strainer is rated for filtering "fineness" by a *mesh* number or its near equivalent, *standard sieve* number. The higher the mesh or sieve number, the finer the screen.

   Filters, which may be made of many materials other than wire screen, are rated by *micron* size. A micron is one-millionth of a meter or 39-millionths of an inch. For comparison, a grain of salt is about 70 microns across. The smallest particle a sharp eye can see is about 40 microns. Figure 23 compares various micron sizes with mesh and standard sieve sizes.

C. Filtering Materials

Filtering materials are classified as mechanical, absorbent, and adsorbent.

Mechanical filters operate by trapping particles between closely woven metal screens or discs. Most mechanical filters are relatively coarse.

Absorbent filters are used for most minute-particle filtration in hydraulic systems. They are made of a wide range of porous materials, including paper, wood pulp, cotton, yarn, and cellulose. Paper filters are usually resin-impregnated for strength.

Adsorbent or *active* filters such as charcoal and Fuller's earth should be avoided in hydraulic systems, since they may remove essential additives from the hydraulic fluid.

X. CYLINDERS

Cylinders are *linear* actuators. By linear, we mean simply that the output of a cylinder is straight-line motion and/or force.

A. Types of Cylinders

Cylinders are classified as single- or double-acting and as differential or nondifferential. Variations include ram or piston and rod design; and solid or telescoping rods.

Ram Type Cylinder. Perhaps the simplest actuator is the ram type. It has only one fluid chamber and exerts force in only one direction. Most are mounted vertically and retract by the force of gravity on the load. Practical for long strokes, ram-type cylinders are used in elevators, jacks, and automobile hoists.

Telescoping Cylinder. A telescoping cylinder is used where the collapsed length must be shorter than could be obtained with a standard cylinder. Up to four and five sleeves can be used; while most are single-acting, double-acting units are available.

Standard Double-Acting. The double-acting cylinder is so named because it is operated by hydraulic fluid in both directions. This means it is capable of a power stroke either way. The standard double-acting cylinder is classed as a *differential* cylinder because there are unequal areas exposed to pressure during the forward and return movements.

RELATIVE SIZE OF MICRONIC PARTICLES

MAGNIFICATION 500 TIMES

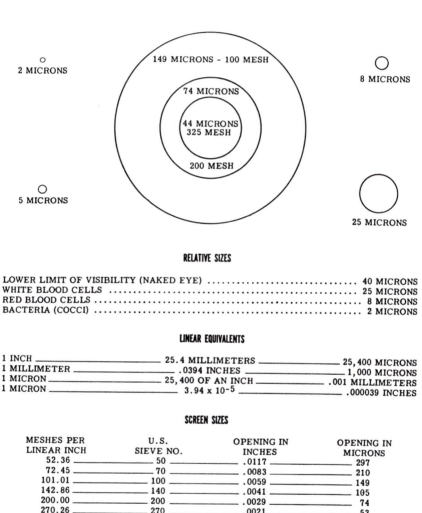

RELATIVE SIZES

LOWER LIMIT OF VISIBILITY (NAKED EYE) .... 40 MICRONS
WHITE BLOOD CELLS .... 25 MICRONS
RED BLOOD CELLS .... 8 MICRONS
BACTERIA (COCCI) .... 2 MICRONS

LINEAR EQUIVALENTS

1 INCH	25.4 MILLIMETERS	25,400 MICRONS
1 MILLIMETER	.0394 INCHES	1,000 MICRONS
1 MICRON	25,400 OF AN INCH	.001 MILLIMETERS
1 MICRON	3.94×10^{-5}	.000039 INCHES

SCREEN SIZES

MESHES PER LINEAR INCH	U.S. SIEVE NO.	OPENING IN INCHES	OPENING IN MICRONS
52.36	50	.0117	297
72.45	70	.0083	210
101.01	100	.0059	149
142.86	140	.0041	105
200.00	200	.0029	74
270.26	270	.0021	53
323.00	325	.0017	44
		.00039	10
		.000019	.5

Figure 23 A micron is 39 millionths of an inch. (Courtesy of Vickers, Incorporated, Rochester Hills, MI.)

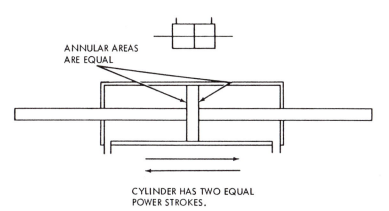

ANNULAR AREAS
ARE EQUAL

CYLINDER HAS TWO EQUAL
POWER STROKES.

Figure 24 The double-rod cylinder is double-acting but nondifferential. (Courtesy of Vickers, Incorporated, Rochester Hills, MI.)

Double-Rod Cylinder. Double-rod cylinders (Fig. 24) are used where it is advantageous to couple a load to each end, or where equal displacement is needed on each end. They too are double-acting cylinders but are classified as *nondifferential.*

B. Cylinder Construction

The essential parts of a cylinder (Fig. 25) are a barrel; a piston and rod; end caps; and suitable seals. Barrels usually are seamless steel tubing, honed to a

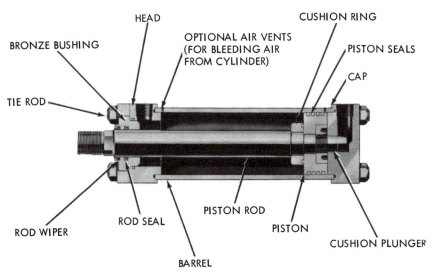

HEAD

CUSHION RING

BRONZE BUSHING

OPTIONAL AIR VENTS
(FOR BLEEDING AIR
FROM CYLINDER)

PISTON SEALS

CAP

TIE ROD

ROD WIPER

ROD SEAL

PISTON ROD

PISTON

CUSHION PLUNGER

BARREL

Figure 25 Cylinder construction. (Courtesy of Vickers, Incorporated, Rochester Hills, MI.)

fine finish on the inside. The piston, usually cast iron or steel, incorporates seals to reduce leakage between it and the cylinder barrel.

C. Cylinder Mountings

Various cylinder mountings (Fig. 26) provide flexibility in anchoring the cylinder. Rod ends are usually threaded for attachment directly to the load or to accept a clevis, yoke or similar coupling device.

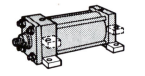 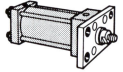 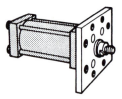

FOOT AND CENTERLINE LUG MOUNTS	RECTANGULAR FLANGE MOUNT	SQUARE FLANGE MOUNT

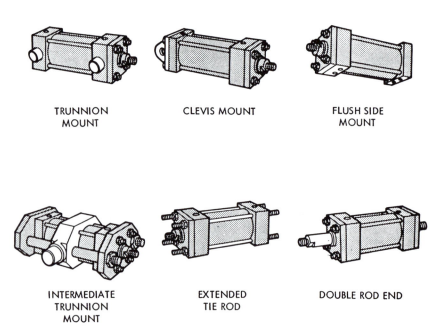

TRUNNION   CLEVIS MOUNT   FLUSH SIDE
MOUNT   MOUNT

INTERMEDIATE   EXTENDED   DOUBLE ROD END
TRUNNION   TIE ROD
MOUNT

Figure 26   Cylinder mountings. (Courtesy of Vickers, Incorporated, Rochester Hills, MI.)

XI. DIRECTIONAL CONTROLS

Directional valves, as the name implies, are used to control the direction of flow. Though sharing this common function, directional valves vary considerably in construction and operation. They are classified according to their principal characteristics, such as:

Type of internal valving element: poppet (piston or ball), rotary spool, and sliding spool.

Methods of actuation: cams, plungers, manual lever, mechanical, electric solenoid, hydraulic pressure (pilot operated), and others, including combinations of these.

Number of flow paths: two-way, three-way, four-way, etc.

Size: nominal size of pipe connections to valve or its mounting plate, or rated gpm flow.

Connections: pipe thread, straight thread, flanged, and back-mounted (sometimes called gasket or subplate-mounted).

A. Check Valves

A check value can function as either a directional control or a pressure control. In its simplest form, however, a check valve is nothing more than a one-way directional valve (Fig. 27). It permits free flow in one direction and blocks flow in the other.

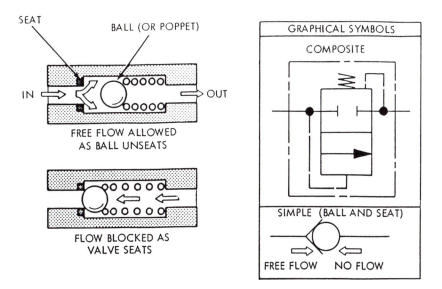

Figure 27 A check valve is a one-way valve. (Courtesy of Vickers, Incorporated, Rochester Hills, MI.)

B. Inline Check Valves

Inline check values are so named because they are connected into the line and the oil flows straight through. The valve body is threaded for pipe or a tubing connector and is machined inside to form a seat for the poppet or ball. A light spring holds the poppet seated in the normal closed position permitting the valve to be mounted in any attitude.

C. Right-Angle Check Valves

A heavier-duty unit, the right-angle valve, has a steel poppet and a hardened seat pressed into the iron body. It gets its name from the angle between the flow passage to the poppet and the passage away from the poppet.

D. Restriction Check Valves

A restriction check valve is a modification of a simple check valve. An orifice plug is placed in the poppet to permit a restricted flow in the normally closed position.

E. Pilot-Operated Check Valves

Pilot-operated check valves are designed to permit free flow in one direction and to block return flow, until opened by a pressure (pilot) signal.

F. Two-Way and Four-Way Valves

The basic function of two-way and four-way valves is to direct inlet flow to either of two outlet ports. In the four-way valve the alternate port is open to the tank port permitting return flow to the reservoir. In two-way valves the alternate port is blocked and the tank port serves only to drain leakage from within the valve.

Appendix A

Metric Systems of Measurement*

A metric system of measurement was first established in France in the years following the French Revolution, and various systems of metric units have been developed since that time. All metric unit systems are based, at least in part, on the International Metric Standards which are the meter and kilogram, or decimal multiples or sub-multiples of these standards.

In 1795, a metric system called the centimeter-gram-second (cgs) system was proposed, and was adopted in France in 1799. In 1873, the British Association for the Advancement of Science recommended the use of the cgs system, and since then it has been widely used in all branches of science throughout the world. From the base units in the cgs system are derived:

Unit of velocity = 1 centimeter per second.
Acceleration due to gravity (at Paris) = 981 centimeters per sec. per sec.
Unit of force = 1 dyne = 1/981 gram.
Unit of work = 1 erg = 1 dyne-centimeter.
Unit of power = 1 watt = 10,000,000 ergs per second.

*Adapted from the book, *Machinery's Handbook,* 24th ed., by Erik Oberg, Franklin D. Jones, Holbrook L. Horton, and Henry H. Ryffel, edited by Robert Green. Copyright 1992 by Industrial Press, Inc., New York. The editors would like to express their sincere thanks to Mr. Robert Green and Industrial Press for their gracious permission in letting us reproduce their material.

Another metric system called the MKS (meter-kilogram-second) system of units was proposed by Professor G. Giorgi in 1902. In 1935, the International Electrotechnical Commission (IEC) accepted his recommendation that this system of units of mechanics should be linked with the electro-magnetic units by the adoption of a fourth base unit. In 1950, the IEC adopted the ampere, the unit of electric current, as the fourth unit, and the MKSA system thus came into being.

A gravitational system of metric units, known as the technical system, is based on the meter, the kilogram as a force, and the second. It has been widely used in engineering. Because the standard of force is defined as the weight of the mass of the standard kilogram, the fundamental unit of force varies due to the difference in gravitational pull at different locations around the earth. By international agreement, a standard value for acceleration due to gravity was chosen (9.81 meters per second squared) which for all practical measurements is approximately the same as the local value at the point of measurement.

The International System of Units (SI). — The Conference Generale des Poids et Mesures (CGPM) which is the body responsible for all international matters concerning the metric system, adopted in 1954, a rationalized and coherent system of units, based on the four MKSA units (see above), and including the *kelvin* as the unit of temperature and the *candela* as the unit of luminous intensity. In 1960, the CGPM formerly named this system the Systeme International d'Unites, for which the abbreviation is SI in all languages. In 1971, the 14th CGPM adopted a seventh base unit, the *mole* which is the unit of quantity ("amount of substance")

In the period since the first metric system was established in France towards the end of the 18th century, most of the countries of the world have adopted a metric system. At the present time most of the industrially advanced metric-using countries are changing from their traditional metric system to SI. Those countries which are currently changing or considering change from the English system of measurement to metric, have the advantage that they can convert directly to the modernized system. The United Kingdom, which can be said to have led the now worldwide move to change from the English system, went straight to SI.

The use of SI units instead of the traditional metric units has little effect on everyday life or trade. The units of linear measurement, mass, volume, and time remain the same, viz. meter, kilogram, liter. and second.

The SI, like the traditional metric system, is based on decimal arithmetic. For each physical quantity, units of different sizes are formed by multiplying or dividing a single base value by powers of 10. Thus, changes can be made very simply by adding zeros or shifting decimal points. For example, the meter is the basic unit of length; the kilometer is a multiple (1000 meters); and the millimeter is a sub-multiple (one-thousandth of a meter).

In the older metric systems, the simplicity of a series of units linked by powers of ten is an advantage for plain quantities such as length, but this simplicity is lost as soon as more complex units are encountered. For example, in different branches of science and engineering, energy may appear as the erg, the calorie, the kilogram meter, the liter atmosphere or the horsepower hour. In contrast, the SI provides only one basic unit for each physical quantity, and universality is thus achieved.

As mentioned above, there are seven base-units, which are for the basic quantities of length, mass, time, electric current, thermodynamic temperature, amount of substance and luminous intensity, expressed as the meter (m), the kilogram (kg), the second (s), the ampere (A), the kelvin (K), the mole (mol) and the candela (cd). The units are defined in the accompanying table.

The SI is a coherent system. A system is said to be coherent if the product or quotient of any two unit quantities in the system is the unit of the resultant quantity. For example, in a coherent system in which the foot is the unit of length, the square foot is the unit of area, whereas the acre is not.

Other physical quantities are derived from the base units. For example, the unit of velocity is the meter per second (m/s), which is a combination of the base units of length and time. The unit of acceleration is the meter per second squared (m/s²). By applying Newton's second law of motion — force is proportional to mass multiplied by acceleration — the unit of force is obtained which is the kilogram meter per second squared (kgm/s²). This unit is known as the newton or N. Work, or force times distance is the kilogram meter squared per second squared (kgm²/s²), which is the joule, (1 joule = 1 newton-meter), and energy is also expressed in these terms. The abbreviation for joule is J. Power or work per unit time is the kilogram meter squared per second cubed (kgm²/s³), which is the watt (1 watt = 1 joule per second = 1 newton-meter per second.) The abbreviation for watt is W. The term horsepower is not used in the SI and is replaced by the watt which together with multiples and sub-multiples — kilowatt and milliwatt, for example — is the same unit as that used in electrical work.

The use of the newton as the unit of force is of particular interest to engineers. In practical work using the English or traditional metric systems of measurements, it is a common practice to apply weight units as force units. Thus, the unit of force in those systems is that force which when applied to unit mass produces an acceleration *g*, rather than unit acceleration. The value of gravitational acceleration *g* varies around the earth, and thus the weight of a given mass also varies. In an effort to account for this minor error, the kilogram-force and pound-force were introduced, which are defined as the forces due to "standard gravity" acting on bodies of one kilogram or one pound mass respectively. The standard gravitational acceleration is taken as 9.80665 meters per second squared or 32.174 feet per second squared. The newton is defined as "that force which when applied to a body having a mass of one kilogram, gives it an acceleration of one meter per second squared." It is independent of *g*. As a result, the factor *g* disappears from a wide range of formulas in dynamics. However, in some formulas in statics, where the weight of a body is important rather than its mass, *g* does appear where it was formerly absent (the weight of a mass of *W* kilograms is equal to a force of *Wg* newtons, where *g* = approximately 9.81 meter per second squared). Details concerning the use of SI units in mechanics calculations are given on page 95, and throughout the Mechanics section in this Handbook. The use of SI units in strength of materials calculations is covered in the section on that subject.

Decimal multiples and sub-multiples of the SI units are formed by means of the prefixes given in the following table, which represent the numerical factors shown.

Factors and Prefixes for Forming Decimal Multiples and Sub-multiples of the SI Units

Factor by which the unit is multiplied	Prefix	Symbol	Factor by which the unit is multiplied	Prefix	Symbol
10^{12}	tera	T	10^{-2}	centi	c
10^{9}	giga	G	10^{-3}	milli	m
10^{6}	mega	M	10^{-6}	micro	μ
10^{3}	kilo	k	10^{-9}	nano	n
10^{2}	hecto	h	10^{-12}	pico	p
10	deka	da	10^{-15}	femto	f
10^{-1}	deci	d	10^{-18}	atto	a

Standard of Length. — In 1866 the United States, by act of Congress, passed a law making legal the meter, the only measure of length that has been legalized by the United States Government. The United States yard is defined by the relation: 1 yard = $3600/3937$ meter. The legal equivalent of the meter for commercial purposes was fixed as 39.37 inches, by law, in July, 1866, and experience having shown that this value was exact within the error of observation, the United States Office of Standard Weights and Measures was, in 1893, authorized to derive the yard from the meter by the use of this relation. The United States prototype meters Nos. 27 and 21 were received from the International Bureau of Weights and Measures in 1889. Meter No. 27, sealed in its metal case, is preserved in a fireproof vault at the Bureau of Standards.

Comparisons made prior to 1893 indicated that the relation of the yard to the meter, fixed by the Act of 1866, was by chance the exact relation between the international meter and the British imperial yard, within the error of observation. A subsequent comparison made between the standards just mentioned indicates that the legal relation adopted by Congress is in error 0.0001 inch; but, in view of the fact that certain comparisons made by the English Standards Office between the imperial yard and its authentic copies show variations as great if not greater than this, it cannot be said with certainty that there is a difference between the imperial yard of Great Britain and the United States yard derived from the meter. The bronze yard No. 11, which was an exact copy of the British imperial yard both in form and material, had shown changes when compared with the imperial yard in 1876 and 1888, which could not reasonably be said to be entirely due to changes in Bronze No. 11. On the other hand, the new meters represented the most advanced ideas of standards, and it therefore seemed that greater stability as well as higher accuracy would be secured by accepting the international meter as a fundamental standard of length.

For more information on SI practice, the reader is referred to the following publications:

Metric Practice Guide, published by the American Society for Testing and Materials, 1916 Race St., Philadelphia, PA 19103.

ISO International Standard 1000. This publication covers the rules for use of SI units, their multiples and submultiples. It can be obtained from the American National Standards Institute, 11 West 42nd Street, New York, NY 10036.

The International System of Units, Special Publication 330 of the National Bureau of Standards — available from the Superintendent of Documents, U.S. Government Printing Office, Washington, DC 20402.

Traditional Metric Measures

Measures of Length

10 millimeters (mm)	= 1 centimeter (cm).
10 centimeters	= 1 decimeter (dm).
10 decimeters	= 1 meter (m).
1000 meters	= 1 kilometer (km.)

Square Measure

100 square millimeters (mm^2)	= 1 square centimeter (cm^2).
100 square centimeters	= 1 square decimeter (dm^2).
100 square decimeters	= 1 square meter (m^2).

Surveyor's Square Measure

100 square meters (m^2)	= 1 are (a).
100 ares	= 1 hectare (ha).
100 hectares	= 1 square kilometer (km^2).

Cubic Measure

1000 cubic millimeters (mm^3)	= 1 cubic centimeter (cm^3).
1000 cubic centimeters	= 1 cubic decimeter (dm^3).
1000 cubic decimeters	= 1 cubic meter (m^3).

Dry and Liquid Measure

10 milliliters (ml)	= 1 centiliter (cl).
10 centiliters	= 1 deciliter (dl).
10 deciliters	= 1 liter (l).
100 liters	= 1 hectoliter (hl).

1 liter = 1 cubic decimeter = the volume of 1 kilogram of pure water at a temperature of 39.2 degrees F.

Measures of Weight

10 milligrams (mg)	= 1 centigram (cg).
10 centigrams	= 1 decigram (dg).
10 decigrams	= 1 gram (g).
10 grams	= 1 dekagram (dag).
10 dekagrams	= 1 hectogram (hg).
10 hectograms	= 1 kilogram (kg).
1000 kilograms	= 1 (metric) ton (t).

Greek Letters

The Greek letters are frequently used in mathematical expressions and formulas. The Greek alphabet is given below.

A	α	Alpha	H	η	Eta	N	ν	Nu	T	τ	Tau
B	β	Beta	Θ	ϑ θ	Theta	Ξ	ξ	Xi	Y	υ	Upsilon
Γ	γ	Gamma	I	ι	Iota	O	o	Omicron	Φ	φ	Phi
Δ	δ	Delta	K	κ	Kappa	Π	π	Pi	X	χ	Chi
E	ε	Epsilon	Λ	λ	Lambda	P	ρ	Rho	Ψ	ψ	Psi
Z	ζ	Zeta	M	μ	Mu	Σ	σ ς	Sigma	Ω	ω	Omega

Table 1. International System (SI) Units

PHYSICAL QUANTITY	NAME OF UNIT	UNIT SYMBOL	DEFINITION
Basic SI Units			
Length	metre	m	Distance traveled by light in vacuo during 1/299,792,458 of a second.
Mass	kilogram	kg	Mass of the international prototype which is in the custody of the Bureau International des Poids et Mesures (BIPM) at Sèvres, near Paris.
Time	second	s	The duration of 9,192,631,770 periods of the radiation corresponding to the transition between the two hyperfine levels of the ground state of the cesium-133 atom.
Electric Current	ampere	A	The constant current which, if maintained in two parallel rectilinear conductors of infinite length, of negligible circular cross section, and placed at a distance of one metre apart in a vacuum, would produce between these conductors a force equal to 2×10^{-7} N/m length.
Thermodynamic Temperature	degree kelvin	K	The fraction 1/273.16 of the thermodynamic temperature of the triple point of water.
Amount of Substance	mole	mol	The amount of substance of a system which contains as many elementary entities as there are atoms in 0.012 kilogram of carbon 12.
Luminous Intensity	candela	cd	Luminous intensity, in the perpendicular direction, of a surface of 1/600,000 square metre of a black body at the temperature of freezing platinum under a pressure of 101,325 newtons per square metre.
SI Units Having Special Names			
Force	newton	$N = kg \cdot m/s^2$	That force which, when applied to a body having a mass of one kilogram, gives it an acceleration of one metre per second squared.
Work, Energy, Quantity of Heat	joule	$J = N \cdot m$	The work done when the point of application of a force of one newton is displaced through a distance of one metre in the direction of the force.
Electric Charge	coulomb	$C = A \cdot s$	The quantity of electricity transported in one second by a current of one ampere.
Electric Potential	volt	$V = W/A$	The difference of potential between two points of a conducting wire carrying a constant current of one ampere, when the power dissipated between these points is equal to one watt.
Electric Capacitance	farad	$F = C/V$	The capacitance of a capacitor between the plates of which there appears a difference of potential of one volt when it is charged by a quantity of electricity equal to one coulomb.

Table 1 (*Continued*). International System (SI) Units

PHYSICAL QUANTITY	NAME OF UNIT	UNIT SYMBOL	DEFINITION
SI Units Having Special Names			
Electric Resistance	ohm	$\Omega = V/A$	The resistance between two points of a conductor when a constant difference of potential of one volt, applied between these two points, produces in this conductor a current of one ampere, this conductor not being the source of any electromotive force.
Magnetic Flux	weber	$Wb = V\ s$	The flux which, linking a circuit of one turn produces in it an electromotive force of one volt as it is reduced to zero at a uniform rate in one second.
Inductance	henry	$H = V\ s/A$	The inductance of a closed circuit in which an electromotive force of one volt is produced when the electric current in the circuit varies uniformly at the rate of one ampere per second.
Luminous Flux	lumen	$lm = cd\ sr$	The flux emitted within a unit solid angle of one steradian by a point source having a uniform intensity of one candela.
Illumination	lux	$lx = lm/m^2$	An illumination of one lumen per square metre.

Table 2. International System (SI) Units with Complex Names

PHYSICAL QUANTITY	SI UNIT	UNIT SYMBOL
SI Units Having Complex Names		
Area	square metre	m^2
Volume	cubic metre	m^3
Frequency	hertz *	Hz
Density (Mass Density)	kilogram per cubic metre	kg/m^3
Velocity	metre per second	m/s
Angular Velocity	radian per second	rad/s
Acceleration	metre per second squared	m/s^2
Angular Acceleration	radian per second squared	rad/s^2
Pressure	pascal ‡	Pa
Surface Tension	newton per metre	N/m
Dynamic Viscosity	newton second per metre squared	$N\ s/m^2$
Kinematic Viscosity } Diffusion Coefficient }	metre squared per second	m^2/s
Thermal Conductivity	watt per metre degree Kelvin	W/(m °K)
Electric Field Strength	volt per metre	V/m
Magnetic Flux Density	tesla †	T
Magnetic Field Strength	ampere per metre	A/m
Luminance	candela per square metre	cd/m^2

* Hz = cycle/second.
† T = weber/metre^2.
‡ Pa = newton/metre^2.

Metric Conversion Factors
(Symbols of SI units, multiples and submultiples are
given in parentheses in the right-hand column)

Multiply	By	To Obtain
LENGTH		
centimetre	0.03280840	foot
centimetre	0.3937008	inch
fathom	1.8288*	metre (m)
foot	0.3048*	metre (m)
foot	30.48*	centimetre (cm)
foot	304.8*	millimetre (mm)
inch	0.0254*	metre (m)
inch	2.54*	centimetre (cm)
inch	25.4*	millimetre (mm)
kilometre	0.6213712	mile [U. S. statute]
metre	39.37008	inch
metre	0.5468066	fathom
metre	3.280840	foot
metre	0.1988388	rod
metre	1.093613	yard
metre	0.0006213712	mile [U. S. statute]
microinch	0.0254*	micrometre [micron] (μm)
micrometre [micron]	39.37008	microinch
mile [U. S. statute]	1609.344*	metre (m)
mile [U. S. statute]	1.609344*	kilometre (km)
millimetre	0.003280840	foot
millimetre	0.03937008	inch
rod	5.0292*	metre (m)
yard	0.9144*	metre (m)
AREA		
acre	4046.856	metre^2 (m^2)
acre	0.4046856	hectare
centimetre^2	0.1550003	inch^2
centimetre^2	0.001076391	foot^2
foot^2	0.09290304*	metre^2 (m^2)
foot^2	929.0304*	centimetre^2 (cm^2)
foot^2	92,903.04*	millimetre^2 (mm^2)
hectare	2.471054	acre
inch^2	645.16*	millimetre^2 (mm^2)
inch^2	6.4516*	centimetre^2 (cm^2)
inch^2	0.00064516*	metre^2 (m^2)
metre^2	1550.003	inch^2
metre^2	10.763910	foot^2
metre^2	1.195990	yard^2
metre^2	0.0002471054	acre
millimetre^2	0.00001076391	foot^2
millimetre^2	0.001550003	inch^2
yard^2	0.8361274	metre^2 (m^2)

* Where an asterisk is shown, the figure is exact.

Multiply	By	To Obtain
VOLUME (including CAPACITY)		
centimetre³	0.06102376	inch³
foot³	0.02831685	metre³ (m³)
foot³	28.31685	litre
gallon [U. K. liquid]	0.004546092	metre³ (m³)
gallon [U. K. liquid]	4.546092	litre
gallon [U. S. liquid]	0.003785412	metre³ (m³)
gallon [U. S. liquid]	3.785412	litre
inch³	16,387.06	millimetre³ (mm³)
inch³	16.38706	centimetre³ (cm³)
inch³	0.00001638706	metre³ (m³)
litre	0.001*	metre³ (m³)
litre	0.2199692	gallon [U. K. liquid]
litre	0.2641720	gallon [U. S. liquid]
litre	0.03531466	foot³
metre³	219.9692	gallon [U. K. liquid]
metre³	264.1720	gallon [U. S. liquid]
metre³	35.31466	foot³
metre³	1.307951	yard³
metre³	1000.*	litre
metre³	61,023.76	inch³
millimetre³	0.00006102376	inch³
yard³	0.7645549	metre³ (m³)
VELOCITY, ACCELERATION, and FLOW		
centimetre/second	1.968504	foot/minute
centimetre/second	0.03280840	foot/second
centimetre/minute	0.3937008	inch/minute
foot/hour	0.00008466667	metre/second (m/s)
foot/hour	0.00508*	metre/minute
foot/hour	0.3048*	metre/hour
foot/minute	0.508*	centimetre/second
foot/minute	18.288*	metre/hour
foot/minute	0.3048*	metre/minute
foot/minute	0.00508*	metre/second (m/s)
foot/second	30.48*	centimetre/second
foot/second	18.288*	metre/minute
foot/second	0.3048*	metre/second (m/s)
foot/second²	0.3048*	metre/second² (m/s²)
foot³/minute	28.31685	litre/minute
foot³/minute	0.0004719474	metre³/second (m³/s)
gallon [U. S. liquid]/min.	0.003785412	metre³/minute
gallon [U. S. liquid]/min.	0.00006309020	metre³/second (m³/s)
gallon [U. S. liquid]/min.	0.06309020	litre/second
gallon [U. S. liquid]/min.	3.785412	litre/minute
gallon [U. K. liquid]/min.	0.004546092	metre³/minute
gallon [U. K. liquid]/min.	0.00007576820	metre³/second (m³/s)
inch/minute	25.4*	millimetre/minute
inch/minute	2.54*	centimetre/minute
inch/minute	0.0254*	metre/minute
inch/second²	0.0254*	metre/second² (m/s²)

Multiply	By	To Obtain
VELOCITY, ACCELERATION, and FLOW (*Continued*)		
kilometre/hour	0.6213712	mile/hour [U. S. statute]
litre/minute	0.03531466	foot³/minute
litre/minute	0.2641720	gallon [U. S. liquid]/minute
litre/second	15.85032	gallon [U. S. liquid]/minute
mile/hour	1.609344*	kilometre/hour
millimetre/minute	0.03937008	inch/minute
metre/second	11,811.02	foot/hour
metre/second	196.8504	foot/minute
metre/second	3.280840	foot/second
metre/second²	3.280840	foot/second²
metre/second²	39.37008	inch/second²
metre/minute	3.280840	foot/minute
metre/minute	0.05468067	foot/second
metre/minute	39.37008	inch/minute
metre/hour	3.280840	foot/hour
metre/hour	0.05468067	foot/minute
metre³/second	2118.880	foot³/minute
metre³/second	13,198.15	gallon [U. K. liquid]/minute
metre³/second	15,850.32	gallon [U. S. liquid]/minute
metre³/minute	219.9692	gallon [U. K. liquid]/minute
metre³/minute	264.1720	gallon [U. S. liquid]/minute
MASS and DENSITY		
grain [1/7000 lb avoirdupois]	0.06479891	gram (g)
gram	15.43236	grain
gram	0.001*	kilogram (kg)
gram	0.03527397	ounce [avoirdupois]
gram	0.03215074	ounce [troy]
gram/centimetre³	0.03612730	pound/inch³
hundredweight [long]	50.80235	kilogram (kg)
hundredweight [short]	45.35924	kilogram (kg)
kilogram	1000.*	gram (g)
kilogram	35.27397	ounce [avoirdupois]
kilogram	32.15074	ounce [troy]
kilogram	2.204622	pound [avoirdupois]
kilogram	0.06852178	slug
kilogram	0.0009842064	ton [long]
kilogram	0.001102311	ton [short]
kilogram	0.001*	ton [metric]
kilogram	0.001*	tonne
kilogram	0.01968413	hundredweight [long]
kilogram	0.02204622	hundredweight [short]
kilogram/metre³	0.06242797	pound/foot³
kilogram/metre³	0.01002242	pound/gallon [U. K. liquid]
kilogram/metre³	0.008345406	pound/gallon [U. S. liquid]
ounce [avoirdupois]	28.34952	gram (g)
ounce [avoirdupois]	0.02834952	kilogram (kg)

* Where an asterisk is shown, the figure is exact.

Multiply	By	To Obtain
MASS and DENSITY (*Continued*)		
ounce [troy]	31.10348	gram (g)
ounce [troy]	0.03110348	kilogram (kg)
pound [avoirdupois]	0.4535924	kilogram (kg)
pound/foot^3	16.01846	kilogram/metre^3 (kg/m^3)
pound/inch^3	27.67990	gram/centimetre^3 (g/cm^3)
pound/gal [U. S. liquid]	119.8264	kilogram/metre^3 (kg/m^3)
pound/gal [U. K. liquid]	99.77633	kilogram/metre^3 (kg/m^3)
slug	14.59390	kilogram (kg)
ton [long 2240 lb]	1016.047	kilogram (kg)
ton [short 2000 lb]	907.1847	kilogram (kg)
ton [metric]	1000.*	kilogram (kg)
tonne	1000.*	kilogram (kg)
FORCE and FORCE/LENGTH		
dyne	0.00001*	newton (N)
kilogram-force	9.806650*	newton (N)
kilopond	9.806650*	newton (N)
newton	0.1019716	kilogram-force
newton	0.1019716	kilopond
newton	0.2248089	pound-force
newton	100,000.*	dyne
newton	7.23301	poundal
newton	3.596942	ounce-force
newton/metre	0.005710148	pound/inch
newton/metre	0.06852178	pound/foot
ounce-force	0.2780139	newton (N)
pound-force	4.448222	newton (N)
poundal	0.1382550	newton (N)
pound/inch	175.1268	newton/metre (N/m)
pound/foot	14.59390	newton/metre (N/m)
BENDING MOMENT or TORQUE		
dyne-centimetre	0.0000001*	newton-metre (N · m)
kilogram-metre	9.806650*	newton-metre (N · m)
ounce-inch	7.061552	newton-millimetre
ounce-inch	0.007061552	newton-metre (N · m)
newton-metre	0.7375621	pound-foot
newton-metre	10,000,000.*	dyne-centimetre
newton-metre	0.1019716	kilogram-metre
newton-metre	141.6119	ounce-inch
newton-millimetre	0.1416119	ounce-inch
pound-foot	1.355818	newton-metre (N · m)

* Where an asterisk is shown, the figure is exact.

875

Multiply	By	To Obtain
MOMENT OF INERTIA and SECTION MODULUS		
moment of inertia [kg · m²]	23.73036	pound-foot²
moment of inertia [kg · m²]	3417.171	pound-inch²
moment of inertia [lb · ft²]	0.04214011	kilogram-metre² (kg · m²)
moment of inertia [lb · inch²]	0.0002926397	kilogram-metre² (kg · m²)
moment of section [foot⁴]	0.008630975	metre⁴ (m⁴)
moment of section [inch⁴]	41.62314	centimetre⁴
moment of section [metre⁴]	115.8618	foot⁴
moment of section [centimetre⁴]	0.02402510	inch⁴
section modulus [foot³]	0.02831685	metre³ (m³)
section modulus [inch³]	0.00001638706	metre³ (m³)
section modulus [metre³]	35.31466	foot³
section modulus [metre³]	61,023.76	inch³
MOMENTUM		
kilogram-metre/second	7.233011	pound-foot/second
kilogram-metre/second	86.79614	pound-inch/second
pound-foot/second	0.1382550	kilogram-metre/second (kg · m/s)
pound-inch/second	0.01152125	kilogram-metre/second (kg · m/s)
PRESSURE and STRESS		
atmosphere [14.6959 lb/inch²]	101,325.	pascal (Pa)
bar	100,000.*	pascal (Pa)
bar	14.50377	pound/inch²
bar	100,000.*	newton/metre² (N/m²)
hectobar	0.6474898	ton [long]/inch²
kilogram/centimetre²	14.22334	pound/inch²
kilogram/metre²	9.806650*	newton/metre² (N/m²)
kilogram/metre²	9.806650*	pascal (Pa)
kilogram/metre²	0.2048161	pound/foot²
kilonewton/metre²	0.1450377	pound/inch²
newton/centimetre²	1.450377	pound/inch²
newton/metre²	0.00001*	bar
newton/metre²	1.0*	pascal (Pa)
newton/metre²	0.0001450377	pound/inch²
newton/metre²	0.1019716	kilogram/metre²
newton/millimetre²	145.0377	pound/inch²
pascal	0.00000986923	atmosphere
pascal	0.00001*	bar
pascal	0.1019716	kilogram/metre²
pascal	1.0*	newton/metre² (N/m²)
pascal	0.02088543	pound/foot²
pascal	0.0001450377	pound/inch²

* Where an asterisk is shown, the figure is exact.

Metric Conversion Factors (*Continued*)

Multiply	By	To Obtain
PRESSURE and STRESS (*Continued*)		
pound/foot²	4.882429	kilogram/metre²
pound/foot²	47.88026	pascal (Pa)
pound/inch²	0.06894757	bar
pound/inch²	0.07030697	kilogram/centimetre²
pound/inch²	0.6894757	newton/centimetre²
pound/inch²	6.894757	kilonewton/metre²
pound/inch²	6894.757	newton/metre² (N/m²)
pound/inch²	0.006894757	newton/millimetre² (N/mm²)
pound/inch²	6894.757	pascal (Pa)
ton [long]/inch²	1.544426	hectobar
ENERGY and WORK		
Btu [International Table]	1055.056	joule (J)
Btu [mean]	1055.87	joule (J)
calorie [mean]	4.19002	joule (J)
foot-pound	1.355818	joule (J)
foot-poundal	0.04214011	joule (J)
joule	0.0009478170	Btu [International Table]
joule	0.0009470863	Btu [mean]
joule	0.2386623	calorie [mean]
joule	0.7375621	foot-pound
joule	23.73036	foot-poundal
joule	0.9998180	joule [International U. S.]
joule	0.9999830	joule [U. S. legal, 1948]
joule [International U. S.]	1.000182	joule (J)
joule [U. S. legal, 1948]	1.000017	joule (J)
joule	.0002777778	watt-hour
watt-hour	3600.*	joule (J)
POWER		
Btu [International Table]/hour	0.2930711	watt (W)
foot-pound/hour	0.0003766161	watt (W)
foot-pound/minute	0.02259697	watt (W)
horsepower [550 ft-lb/s]	0.7456999	kilowatt (kW)
horsepower [550 ft-lb/s]	745.6999	watt (W)
horsepower [electric]	746.*	watt (W)
horsepower [metric]	735.499	watt (W)
horsepower [U. K.]	745.70	watt (W)
kilowatt	1.341022	horsepower [550 ft-lb/s]
watt	2655.224	foot-pound/hour
watt	44.25372	foot-pound/minute
watt	0.001341022	horsepower [550 ft-lb/s]
watt	0.001340483	horsepower [electric]
watt	0.001359621	horsepower [metric]
watt	0.001341022	horsepower [U. K.]
watt	3.412141	Btu [International Table]/hour

* Where an asterisk is shown, the figure is exact.

Metric Conversion Factors (Concluded)

Multiply	By	To Obtain
	VISCOSITY	
centipoise	0.001*	pascal-second (Pa · s)
centistoke	0.000001*	metre²/second (m²/s)
metre²/second	1,000,000.*	centistoke
metre²/second	10,000.*	stoke
pascal-second	1000.*	centipoise
pascal-second	10.*	poise
poise	0.1*	pascal-second (Pa · s)
stoke	0.0001*	metre²/second (m²/s)

TEMPERATURE

To Convert From	To	Use Formula
temperature Celsius, t_C	temperature Kelvin, t_K	$t_K = t_C + 273.15$
temperature Fahrenheit, t_F	temperature Kelvin, t_K	$t_K = (t_F + 459.67)/1.8$
temperature Celsius, t_C	temperature Fahrenheit, t_F	$t_F = 1.8\, t_C + 32$
temperature Fahrenheit, t_F	temperature Celsius, t_C	$t_C = (t_F - 32)/1.8$
temperature Kelvin, t_K	temperature Celsius, t_C	$t_C = t_K - 273.15$
temperature Kelvin, t_K	temperature Fahrenheit, t_F	$t_F = 1.8\, t_K - 459.67$
temperature Kelvin, t_K	temperature Rankine, t_R	$t_R = 9/5\, t_K$
temperature Rankine, t_R	temperature Kelvin, t_K	$t_K = 5/9\, t_R$

* Where an asterisk is shown, the figure is exact.

Miscellaneous Conversion Factors
(English Units)

Multiply	By	To Obtain
atmospheres	29.92	inches of mercury (32 deg. F.)
atmospheres	14.70	pounds/inch²
British thermal units/hour	12.96	foot-pounds/minute
circular mils	0.7854	square mils
feet of water (60 deg. F.)	0.8843	inches of mercury (60 deg. F.)
feet of water (60 deg. F.)	0.4331	pounds/inch²
feet/minute	0.01136	miles/hour
foot-pounds/second	0.07716	British thermal units/minute
gallons (U.S.) of water (60 deg. F.)	8.337	pounds of water (60 deg. F.)
gallons (U.S.)/second	8.021	feet³/minute
inches of mercury (32 deg. F.)	0.03342	atmospheres
inches of mercury (60 deg. F.)	1.131	feet of water (60 deg. F.)
inches of mercury (60 deg. F.)	0.4898	pounds/inch²
inches of water (60 deg. F.)	0.03609	pounds/inch²
knots (International)	1.151	miles (statute)/hour
miles/hour	88	feet/minute
miles (statute)/hour	0.8690	knots (International)
ounces (avoirdupois)	0.9115	ounces (troy)
ounces (troy)	1.097	ounces (avoirdupois)
ounces (troy)	0.06857	pounds (avoirdupois)
pounds (avoirdupois)	14.58	ounces (troy)
pounds of water (60 deg. F.)	0.01603	feet³
pounds of water (60 deg. F.)	0.1199	gallons (U.S.)
pounds/inch²	0.06805	atmospheres
pounds/inch²	2.309	feet of water (60 deg. F.)
pounds/inch²	2.042	inches of mercury (60 deg. F.)
pounds/inch²	27.71	inches of water (60 deg. F.)
square mils	1.273	circular mils

Metric and English Equivalents

Linear Measure

1 kilometer = 0.6214 mile.

1 meter = $\begin{cases} 39.37 \text{ inches.} \\ 3.2808 \text{ feet.} \\ 1.0936 \text{ yards.} \end{cases}$

1 centimeter = 0.3937 inch.
1 millimeter = 0.03937 inch.

1 mile = 1.609 kilometers.
1 yard = 0.9144 meter.
1 foot = 0.3048 meter.
1 foot = 304.8 millimeters.
1 inch = 2.54 centimeters.
1 inch = 25.4 millimeters.

Square Measure

1 square kilometer = 0.3861 square mile = 247.1 acres.
1 hectare = 2.471 acres = 107,639 square feet.
1 are = 0.0247 acre = 1076.4 square feet.
1 square meter = 10.764 square feet = 1.196 square yards.
1 square centimeter = 0.155 square inch.
1 square millimeter = 0.00155 square inch.

1 square mile = 2.5899 square kilometers.
1 acre = 0.4047 hectare = 40.47 ares.
1 square yard = 0.836 square meter.
1 square foot = 0.0929 square meter = 929 square centimeters.
1 square inch = 6.452 square centimeters = 645.2 square millimeters.

Cubic Measure

1 cubic meter = 35.315 cubic feet = 1.308 cubic yards.
1 cubic meter = 264.2 U.S. gallons.
1 cubic centimeter = 0.061 cubic inch.
1 liter (cubic decimeter) = 0.0353 cubic foot = 61.023 cubic inches.
1 liter = 0.2642 U.S. gallon = 1.0567 U.S. quarts.

1 cubic yard = 0.7646 cubic meter.
1 cubic foot = 0.02832 cubic meter = 28.317 liters.
1 cubic inch = 16.38706 cubic centimeters.
1 U.S. gallon = 3.785 liters.
1 U.S. quart = 0.946 liter.

Weight

1 metric ton = 0.9842 ton (of 2240 pounds) = 2204.6 pounds.
1 kilogram = 2.2046 pounds = 35.274 ounces avoirdupois.
1 gram = 0.03215 ounce troy = 0.03527 ounce avoirdupois.
1 gram = 15.432 grains.

1 ton (of 2240 pounds) = 1.016 metric ton = 1016 kilograms.
1 pound = 0.4536 kilogram = 453.6 grams.
1 ounce avoirdupois = 28.35 grams.
1 ounce troy = 31.103 grams.
1 grain = 0.0648 gram.

1 kilogram per square millimeter = 1422.32 pounds per square inch.
1 kilogram per square centimeter = 14.223 pounds per square inch.
1 kilogram-meter = 7.233 foot-pounds.
1 pound per square inch = 0.0703 kilogram per square centimeter.
1 calorie (kilogram calorie) = 3.968 Btu (British thermal unit).

Appendix B

Scientific, Technical, and Engineering Abbreviations

COMPILED BY DAVID W. SOUTH

£	Laplace operational symbol
Δ	Mass defect
μ	Micro $= 10^{-6}$
μA	Microampere
μC	Microcurie
μF	Microfarad
μm	Micrometer
μp	Microprocessor
μs	Microsecond
μW	Microwatt
σ	Boltzmann constant
Ω	Ohm
A (ang)	Angström units $= 10^{-10}$ m
A	Ampere
A	Mass number $= N + Z$
a	Atto $= 10^{-18}$
a.c.	Aerodynamic center
a-c	Alternating current
a-f	Audio frequency
a-m	Amplitude modulation
aa (AA)	Arithmetical average
abs	Absolute
ac	Alternating current

af	Audio frequency
AH	Ampere-hour
ahp	Air horsepower
am	Amplitude modulation
amp hr	Ampere-hour
amp	Ampere
amu	Atomic mass unit
antilog	Antilogarithm
approx	Approximately
at wt	Atomic weight
atm	Atmosphere
avdp	Avoirdupois
ave	Average
avg	Average
AWG	American Wire Gage
b	Barns
B	British thermal unit
B.G.	Birmingham gage
B.M.	Board measure; bench mark
B&S	Brown & Sharp (gage)
bar	Barometer
baro	Barometer
bbl	Barrels
Bé	Baumé (degrees)
bgd	Billions of gallons per day
Bhn	Brinell hardness number
BHN	Brinell hardness number
bhp	Brake horsepower
BLC	Boundary layer control
bmep	Brake mean effective pressure
BOD	Biochemical oxygen demand
bopress	Boiler pressure
bp	Boiling point
Bq	Becquerel
bsfc	Brake specific fuel consumption
Btu/h	Btu per hour
Btu	British thermal unit
Btuh	Btu per hour
bu	Bushel
BWG	Birmingham wire gage
c to c	Center-to-center
C	Coulomb
c	Velocity of light
c.f.	Centrifugal force
C.I.	Cast iron
C.N.	Cetane number

c.p.	Circular pitch; center of pressure
cal	Calorie
cc	Cubic centimer
CCR	Critical compression ratio
cd	Candle
cd	Cord
cd	Current density
cemf	Counter electromotive force
cf	Confer (or compare)
cfm	Cubic feet per minute
cfs	Cubic feet per second
cg	Center of gravity
cgs	Centimeter-gram-second
chem	Chemical
cfh	Cubic feet per hour
chu	Centigrade heat unit
cif	Cost, insurance, and freight
cir mil	Circular mils
cir	Circular
circ	Circular
cm^2	Square centimeter
cm	Centimeter
cmil	Circular mil
cm^3	Cubic centimeter
cndct	Conductivity
coef	Coefficient
col	Column
colog	Cologarithm
conc	Concentrate
const	Constant
cos	Cosine of
cos^{-1}	Angle whose cosine is, inverse cosine of
cosh	Hyperbolic cosine
$cosh^{-1}$	Inverse hyperbolic cosine of
cot	Cotangent of
coth	Hyperbolic cotangent of
$coth^{-1}$	Inverse hyperbolic cotangent of
covers	Coversed sine of
cp (CP)	Chemically pure
cp	Coef of performance
CPH	Closed-packed hexagonal
cpm	Cycles per minute
csc	Cosecant of
csch	Hyperbolic cosecant of
ctn	Cotangent of
cu m	Cubic meter

cu ft	Cubic foot
cu in.	Cubic inch
cu yd	Cubic yard
cu mm	Cubic millimeter
cu	Cubic
cycles/min	Cycles per minute
cyl	Cylinder
d^2tons	Breaking strength, d = chain wire diam, in.
d-c	Direct current
db	Decibel
Db	Decibel
dc	Direct current
def	Definition
deg	Degree
dia	Diameter
diam	Diameter
DO	Dissolved oxygen
doz	Dozen
DP	Diametral pitch
DPH	Diamond pyramid hardness
dr	Dram
DST	Daylight savings time
dwt	Pennyweight
DX	Direct expansion
e.g.	*Exempli gratia* (for example)
EAP	Equivalent air pressure
EDR	Equivalent direct radiation
eff	Efficiency
ehp	Effective horsepower
EHV	Extra high voltage
el	Elevation
elec	Electric
elong	Elongation
emf	Electromotive force
eng	Engine
engr	Engineer
engrg	Engineering
ENT	Emergency negative thrust
EP	Extreme pressure (lubricant)
eq	Equation
Eq	Equation
est	Estimated
etc.	Et cetera (and so forth)
eV	Electron volts
evap	Evaporation
exp	Expontential function of

exsec	Exterior secant of
ext	External
F	Farad
F.C.	Fixed carbon, %
F.C.C.	Face-centered cubic (alloys)
F.I.T.	Federal income tax
F.O.B.	Free on board
F.S.	Federal specifications
f-m	Frequency modulation
fbm	Board feet (feet board measure)
fc	Foot-candle
ff	Following (pages)
fhp	Friction horsepower
Fig.	Figure
fl	Fluid
fl	Foot-Lambert
fL	Foot-Lambert
fm	Frequency modulation
fnpt	Fusion point
fob	Free on board
FP	Fore perpendicular
fp	Freezing point
fpm	Foot per minute
fps	Foot per second
fps	Foot-pound-second (system)
freq	Frequency
FSB	Federal Specifications Board
fsp	Fiber saturation point
ft^2	Square foot
ft-lb	Foot-pound
ft/min	Foot per minute
ft/s	Food per second
ft	Foot
ft^3/s	Cubic feet per second
ft^3/min	Cubic feet per minute
ft^3/h	Cubic feet per hour
ft^3	Cubic foot
g	Acceleration due to gravity
g	Gram
gal/min	Gallon per minute
gal/s	Gallon per second
gal	Gallon
gc	Gigacycles per second
gcd	Greatest common divisor
gd	Gudermannian of
GEM	Gound effect machine

GFI	Gullet feed index
GMT	Greenwich Mean Time
gpcd	Gallon per capita day
gpd	Gallon per day; grams per denier
gpm	Gallon per minute
gps	Gallon per second
gpt	Grams per tex
gr	Grain
g · cal	Gram-calories
H	Henry
h	Hour
h	Planck's constant $= 6.624 \times 10^{-27}$ erg-s
H.T.	Heat treated
h-f	High frequency
h-p	High pressure
HEPA	High-efficiency particulate matter
hf	High frequency
hhv	High heat value
horiz	Horizontal
hp hr	Horsepower-hour
hp	High pressure
hp	Horsepower
hp · hr	Horsepower-hour
hr	Hour
HSS	High-speed steel
HTHW	High-temperature hot water
Hz	Hertz $= 1$ cycle/s (cps)
I.D.	Inside diameter
i.e.	*Id est* (that is)
i-f	Intermediate frequency
i-p	Intermediate pressure
ibid.	*Ibidem* (in the same place)
ID	Inside diameter
if	Intermediate frequency
ihp	Indicated horsepower
ihph	Indicated horsepower-hour
imep	Indicated mean effective pressure
Imp	Imperial
in.^{2}	Square inch
in./min	Inch per minute
in./s	Inch per second
in.-lb	Inch-pound
in.	Inch
int	Internal
intl	Internal
in.^{3}	Cubic inch

in. · lb	Inch-pound
ip	Intermediate pressure
ipm	Inch per minute
ipr	Inch per revolution
ips	Inch per second
IPS	Iron pipe size
isoth	Isothermal
J	Joule
J&P	Joists & planks
JP	Jet propulsion fuel
K	Degree Kelvin
k	Isentropic exponent; conductivity
K	Knudsen number
kB	Kilo Btu (1000 Btu)
kc	Kilocycles
kcps	Kilocycles per second
kg	Kilogram
kg · cal	Kilogram-calories
kg · m	Kilogram-meter
kip	1000 lb or 1 kilo-pound
kips	Thousands of pounds
km^2	Square kilometer
km	Kilometer
kmc	Kilomegacycles per second
kmcps	Kilomegacycles per second
kpsi	Thousands of pounds per square inch
ksi	One kip per square inch, 1000 psi ($lb/in.^2$)
kts	Knots
kV	Kilovolt
kVA	Kilovolt-ampere
KVA-h	Kilovolt-ampere/hour
kVah	Kilovolt-ampere/hour
kW	Kilowatt
kWh	Kilowatt hour
kwhm	Kilowatt-hour meter
L	Lambert, liter
l	Liter
L.B.P.	Length between perpendiculars
L.W.L.	Load water line
l-p	Low pressure
lat	Latitude
$lb/in.^2$	Pound per square inch
lb/hp	Pound per horsepower
lb ft, lb_f · ft	Pound-force foot
lb in, lb_f · in	Pound-force inch
lb	Pound

lb_f/ft^2	Pound-force per square foot
lcm	Least common multiple
lhv	Low heat value
lim	Limit
lin	Linear
liq	Liquid
lm/W	Lumen per watt
lm	Lumen
ln	Logarithm (natural)
loc. cit.	*Loco citato* (place already cited)
log	Logarithm (common)
LOX	·Liquid oxygen explosive
lp	Low pressure
LPG	Liquified petroleum gas
lpw	Lumen per watt
lpw	Lumens per watt
lx	Lux
m	Meter
M	Thousand; Mach number; moisture, %
m.c.	Moisture content
$M\Omega$	Megohm
mA	Milliampere
mA	Milliampere
math	Mathematics
max	Maximum
MBh	Thousands of Btu per hour
mc	Megacycles per second
Mcf	Thousand cubic feet
mcps	Megacycles per second
MeV	Million electron volts
mep	Mean effective pressure
MF	Maintenance factor
mF	Millifarad
mH	Millihenry
mhc	Mean horizontal candles
mi/h	Mile per hour
mi	Mile
MIL-STD	U.S. Military Standard
min	Minimum; minutes
mip	Mean indicated pressure
mks (MKS)	Meter-kilogram-second
MKSA	Meter-kilogram-second-ampere system
mL	Millilamberts
ml	Millilitre $= 1.000027 \text{ cm}^3$
mL	Millilitre $= 1.000027 \text{ cm}^3$
mlhc	Mean lower hemispherical candles

mm-free	Mineral matter free
mmf	Magnetomotive force
mm^3	Cubic millimeter
mo	Molecule
mol wt	Molecular weight
mol	Mole
mp	Melting point
MPC	Maximum permissible concentration
mph	Mile per hour
MRT	Mean radiant temperature
ms	Manuscript; milliseconds
msc	Mean spherical candles
Mu (μ)	Micron; micro
mv	Millivolt
MW day	Megawatt day
MW	Megawatt
MWT	Mean water temperature
m^3	Cubic meter
N	Number (in mathematical tables)
N	Number of neutrons; newton
n	Polytropic exponent
NA	Not available
nat	Natural
NEC	National Electrical Code
nm	Nautical miles
No.	Number
Nos.	Numbers
NPSH	Net positive suction head
N_S	Specific speed
NTP	Normal temperature and pressure
O.D.	Outside diameter (pipes)
O.H.	Open hearth (steel)
O.N.	Octane number
OD	Outside diameter (pipes)
op. cit.	*Opere citato* (work already cited)
oz-in.	Ounce-pinch
oz.	Ounce
p.	Page
P.C.	Propulsive coefficient
P.E.L.	Proportional elastic limit
p.m.	Post meridiem (after noon)
P.N.	Performance number
Pa	Pascal
pd	Potential difference
PE	Polyethylene
PEG	Polyethylene glycol

PETN	An explosive
pf	Power factor
php	Pound per horsepower
PIV	Peak inverse voltage
PM	Preventive maintenance
pot	Potential
pp	Pages
ppb	Parts per billion
PPI	Plan position indicator
ppm	Parts per million
press	Pressure
Proc.	Proceedings
PSD	Power spectral density, g^3/cps
psf	Pound-force per square foot
psi	Pound-force per square inch
psig	Pound per square inch gage
pt	Point; pint
PVC	Polyvinyl chloride
Q	10^{18}Btu
q.v.	*Quod vide* (which see)
qt.	Quart
r/min	Revolution per minute
r-f	Radiofrequency
r/min	Revolution per minute
r/s	Revolution per second
R-C	Resistor-capacitor
R	Degree Rankine; Reynolds number
R	Gas constant
r	Roentgens
R&D	Research & development
rad	Radius; radiation absorbed dose; radian
RBE	Obsolete; see rem
RDX	Cyclonite, a military explosive
rem	Roentgen equivalent man (formerly RBE)
rev	Revolutions
rf	Radiofrequency
rms	Root mean square
rms	Square root of mean square
rnd	Round
rpm	Revolution per minute
rps	Revolution per second
RSHF	Room sensible heat factor
rva	Reactive volt-ampere
ry	Railway
s	Boltzmann constant
s	Entropy

s	Second
S	Sulfur, %; siemens
sat	Saturated
scfm	Standard cubic foot per minute
SCR	Silicon-controlled rectifier
SE No.	Steam emulsion number
sec	Secant
sec	Second
Sec	Section
sech	Hyperbolic secant of
sech^{-1}	Inverse hyperbolic secant of
segm	Segment
sfc	Specific fuel consumption, lb per hphr
sfm	Surface feet per minute
shp	Shaft horsepower
SI	International System of Units
sin	Sine
sinh	Hyperbolic sine of
sinh^{-1}	Inverse hyperbolic sine of
sp ht	Specific heat
sp gr	Specific gravity
sp	Specific
SP	Static pressure
specif	Specification
spp	Species unspecified (botanical)
SPS	Standard pipe size
sq cm	Square centimeter
sq km	Square kilometer
sq in.	Square inch
sq ft	Square foot
sq	Square
sr	Steradian
SSF	Seconds Saybolt Furol
SSU	Seconds Saybolt Universal (same as SUS)
std	Standard
SUS	Saybolt Universal Seconds (same as SSU)
SWG	Standard Wire Gage (British)
T	Tesla
T.S.	Tensile strength; tensile stress
tan	Tangent
tanh	Hyperbolic tangent
tanh^{-1}	Inverse hyperbolic tangent of
TDH	Total dynamic head
TEL	Tetraethyl lead
temp	Temperature
THI	Temperature-humidity index (discomfort)

thp	Thrust horsepower
TNT	Trinitrotoluol (explosive)
torr	= 1 mm Hg = 1.332 millibars (1/760) atm
TP	Total pressure
tph	Tons per hour
tpi	Turns per inch
TR	Transmitter-receiver
Trans.	Transactions
ts	Tensile strength
tsi	Tons per square inch
ttd	Terminal temperature difference
UHF	Ultrahigh frequency
UL	Underwriters' Laboratory
ult	Ultimate
UMS	Universal maintenance standards
USS	United States Standard
USSG	U.S. Standard Gage
UTC	Coordinated Universal Time
V	Volt
V.M.	Volatile matter, %
VA	Volt-ampere
VC	Volt-coulomb
VCF	Visual comfort factor
VCI	Visual comfort index
vel	Velocity
vers	Versed sine
vert	Vertical
VHF	Very high frequency
VI	Viscosity index
viz.	Videlicet (namely)
vol	Volume
VP	Velocity pressure
vs (VS)	Versus
W	Watt
w.g.	Water gage
W.I.	Wrought iron
W&M	Washburn & Moen wire gage
Wb	Weber
Wh	Watthour
wk	Week
wt	Weight
Y.P.	Yield point
Y.S.	Yield strength; yield stress
yd	Yard
yd^3	Cubic yard
yr	Year

z	Atomic number; figure of merit
°	Degree
°C	Degree Centigrade
°F	Degree Fahrenheit

INDEX